2023 24th European Microelectronics and Packaging Conference & Exhibition (EMPC 2023)

Cambridge, United Kingdom
11-14 September 2023

IEEE Catalog Number: CFP2354H-POD
ISBN: 978-1-6654-8736-8

Copyright © 2023, IMAPS-Europe
All Rights Reserved

*** This is a print representation of what appears in the IEEE Digital Library. Some format issues inherent in the e-media version may also appear in this print version.

IEEE Catalog Number: CFP2354H-POD
ISBN (Print-On-Demand): 978-1-6654-8736-8
ISBN (Online):   978-0-9568086-9-1

Additional Copies of This Publication Are Available From:

Curran Associates, Inc
57 Morehouse Lane
Red Hook, NY 12571 USA
Phone: (845) 758-0400
Fax: (845) 758-2633
E-mail: curran@proceedings.com
Web: www.proceedings.com

1A: SUBSTRATES: LTCC AND HTCC

HIGH FREQUENCY BANDWIDTH TRANSITION FOR HTCC HERMETIC PACKAGES..........1
Emad Elrifai
EGIDE, France

APPLICATION OF REACTIVE BONDING METHODS ON LTCC SUBSTRATES..........4
Erik Wiss[1], Alexander Schulz[2], Adam Yuile[1], Jens Müller[2], Steffen Wiese[1]

[1]Saarland University, Chair of Microintegration and Reliability,
Germany; [2]TU Ilmenau, Electronics Technology Group, Germany

**LTCC-BASED CERAMIC SUBSTRATES FOR IDENTIFICATION
OF TRUSTWORTHY ELECTRONICS..........10**
Uwe Krieger[1], Annett Schroeter[1], Franz Bechtold[1], Gunter Hagen[2], Adrian Goldberg[3]

[1]VIA electronic GmbH, Germany; [2]KMS Technology Center GmbH, Germany; [3]Fraunhofer
Institute for Ceramic Technologies and Systems IKTS, Germany

1B: EMBEDDING

CONCEPTS FOR REALIZING HIGH-VOLTAGE POWER MODULES BY EMBEDDING OF SIC SEMICONDUCTORS..........13
Lars Böttcher, Andreas Ostmann, Thomas Löher, Manuel Seckel
Fraunhofer IZM, Germany

CERAMIC EMBEDDING OF SIC-SEMICONDUCTORS USING COFIRING TECHNOLOGY..........19
Steffen Ziesche[1], Jobin Varghese[1], Kathrin Reinhardt[1],
Birgit Manhica[1], Andreas Schletz[2]

[1]Fraunhofer IKTS, Germany; [2]Fraunhofer IISB, Germany

CHARACTERIZATION OF EMBEDDED AND THINNED RF CHIPS..........24
Ran Yin[1,2], Helmuth P. E. Morath[1,3], Christian Hoyer[3], Krzysztof Nieweglowski[1,2],
Karsten Meier[2], Karlheinz Bock[1,2]

[1]Centre for Tactile Internet with Human-in-the-Loop (CeTI), Germany;
[2]Institute of Electronic Packaging Technology, TU Dresden, Germany;
[3]Chair of Circuit Design and Network Theory, TU Dresden, Germany

2A: INTERCONNECT MATERIALS I

DEPOSITION OF FINE-PITCH INDIUM BUMPS ON SINGLE DIE..........30
Andreas Schneider[1], Navid Ghorbanian[1], David Burt[2], James Hollingham[1], Paul Booker[1], Toby G. Brookes[1], John D. Lipp[1], Marcus J. French[1]

[1]STFC-RAL, United Kingdom; [2]Kelvin Nanotechnology Ltd., United Kingdom

SILVER BONDING WIRE – AN ALTERNATIVE FOR MECHANICAL SENSITIVE CHIP CONFIGURATIONS IN AUTOMOTIVE ELECTRONICS PACKAGING..........35
Robert Klengel[1], Sandy Klengel[1], Noritoshi Araki[2], Motoki Eto[2], Teruo Haibara[2], Takashi Yamada[2], Jochen Feldmann[3], Ralph Binner[3], Henk Peters[3], Achim Scheer[3], Vincent Chee[3]

[1]Fraunhofer IMWS, Germany; [2]Nippon Micrometal Corporation, Japan; [3]ELMOS Semiconductor SE, Germany

COPPER PUMPING ANALYSIS FOR CU/SIO2 HYBRID BONDING USING IN-SITU SPM IMAGING..........41
Ali Roshanghias[1], Jaroslaw Kaczynski[1], Ude Hangen[2]

[1]Silicon Austria Labs GmbH, Austria; [2]Bruker Nano GmbH, Germany

UNDERSTANDING THE CONTACT RESISTANCE IN AN ACF BONDING..........45
Helge Kristiansen[1], Knut Eilif Aasmundtveit[2], Giang Nghiem[2], Molly Bazilchuk[3]

[1]Conpart AS, Norway; [2]University of South-Eastern Norway, Norway; [3]Ducky AS, Norway

2B: MEMS/SENSORS

MEMS MIRROR IN HERMETIC PACKAGE FOR ENHANCED PERFORMANCES..........50
Luca Maggi[1], Marco Del Sarto[1], Amedeo Maierna[1], Mark Shaw[1], Roberto Carminati[1], Gianluca Mendicino[1], Davide Rotta[2], Marco Chiesa[2], Aina Serrano[2], Antonella Bogoni[3]

[1]STMicroelectronics. Italy; [2]CamGraPhIC. Italy;
[3]Sant'Anna School of Adanced Studies, Italy

INNOVATIVE SILICON-CERAMIC (SICER) TECHNOLOGY FOR HIGH-STRENG-TH PRESSURE SENSOR APPLICATION USING DIFFERENT MANUFACTURING METHODS..........55
Cathleen Kleinholz[1], Michael Fischer[1], Nam Gutzeit[1], Andrea Cyriax[2], Michael Hintz[2], Thomas Ortlepp[2], Jens Müller[1]

[1]Technische Universität Ilmenau, Germany;
[2]CiS Forschungsinstitut für Mikrosensorik GmbH, Germany

FROM MEMS STRIP TO MEMS UNIT: A COMPREHENSIVE SIMULATION APPROACH TO WARPAGE..........61
Andrea Ratti[1], Daniele Simoncini[1], Annbel Adolfo[2], Marco Del Sarto[1], Patrick Fedeli[1], Alex Gritti[1], Luca Maggi[1], Teresa Napolitano[1], Mark Andrew Shaw[1], Jefferson Talledo[2]

[1]STMicroelectronics. Italy; [2]STMicroelectronics, Philippines

GAS PERMEABLE PROTECTION CAPS FOR WAFER LEVEL CHIP SCALE PACKAGING (WLCSP) OF MEMS ENVIRONMENTAL SENSORS..........67
Ole Behrmann, Thomas Lisec, Björn Gojdka
Fraunhofer ISIT, Germany

3A: SUBSTRATES – THICK FILM AND CU INTERCONNECTS

ADDITIVE METALLIZATION OF ALUMINA WITH COPPER-TITANIUM POWDER BLENDS FOR POWER ELECTRONIC APPLICATIONS..........71
Christoph Hecht[1],
Eric Schadow[1], Mario Sprenger[1], Felix Häußler[1], Thomas Stoll[2], Jörg Franke[1]

[1]Friedrich-Alexander-Universtität Erlangen-Nürnberg, Nuremberg, Germany;
[2]TUM School of Engineering and Design, Munich, Germany

NEGATIVE-TONE PHOTO-DEFINABLE POLYIMIDE WITH HIGH THERMAL STABILITY AND THICK FILM PROCESSABILITY..........77
Hitoshi Araki, Takayuki Kaneki, Yu Shoji, Chika Hibino
Toray Industries Inc., Japan

3B: ADHESIVES AND ENCAPSULANTS

EPOXY MOLDING COMPOUND BLEEDING REDUCTION ON SURFACE MOUNT SEMICONDUCTOR DEVICE..........81
Federico Leone[1], Fulvio Viviani[1], Hidetoshi Seki[2], Masami Ishii[2]

[1]STMicroelectronics, Italy; [2]Sumitomo Bakelite Singapore Pte. Ltd, Singapore

INFLUENCE OF THERMALLY AGED UNDERFILL ON FLIP-CHIP PACKAGES..........85
Kevin Cox, Ghassan Abu-Hamdeh, Matt Borden

Tektronix Component Solutions, United States of America

ADHESIVE SOLUTIONS FOR CLOSED CAVITY PACKAGING..........91
Patrick Schirmer, Severin Ringelstetter

DELO Industrie Klebstoffe GmbH & Co. KGaA, Germany

4A: INTERCONNECT MATERIALS II

DEVELOPMENT OF A STRETCHABLE AND REMOVABLE ELECTRICAL INTERCONNECTION SOLUTION FOR ULTRA-THIN ELECTRONIC COMPONENTS..........94
Auriane Despax-Ferreres[1], Pascal Tiquet[1], Jean-Charles Souriau[2], Vincent Jousseaume[2], Julia De Girolamo[1]

[1]Univ. Grenoble Alpes, CEA, Liten, Grenoble, France;
[2]Univ. Grenoble Alpes, CEA, Leti, Grenoble, France

UV LASER COPPER PAD SURFACE EXPOSURE FOR LASER DIRECT STRUCTURING (LDS) OF INTERCONNECTION..........100
Guendalina Catalano, Alessandro Mellina Gottardo, Riccardo Villa
ST microelectronics, Italy

FINE PITCH MICRO INDIUM BUMP INTERCONNECT FLIP CHIP BONDING..........106
Travis Scott
Finetech GmbH & Co.KG, Germany

4B: MEDICAL

INITIAL LIFE TEST OF SILICONE ENCAPSULATED FR4 PRINTED CIRCUIT BOARDS FOR PRE-CLINICAL ACTIVE IMPLANTS..........111

Ishpa Ali[1], Fei Xue[1], Carlos Perez Henriquez[1], Thomas Niederhoffer[1], Ahmad Shah Idil[1], Dai Jiang[2], Henry Thomas Lancashire[1]

[1]Department of Medical Physics and Biomedical Engineering, University College London, London, UK.; [2]Department of Electronic and Electrical Engineering, University College London, London, UK.

VOIDING IN PARYLENE-C ENCAPSULATION OF SURFACE MOUNT LEDS FOR AN OPTOGENETIC EPILEPSY NEUROPROSTHESIS..........115

Ahmad Shah Idil[1], Richard Bailey[2], Johannes Gausden[2], Antony O'Neill[2], Nick Donaldson[1]

[1]University College London, UK; [2]Newcastle University, UK

ASSEMBLY OF PRINTED INTERCONNECTS FOR IMMOBILIZED PROTEIN MICROFLUIDIC ASSAYS..........120

Qianwen Xu[1], Jeffery C. C. Lo[1], Yusong Guo[1], S. W. Ricky Lee[1,2,3]

[1]The Hong Kong University of Science and Technology, Hong Kong, PRC; [2]HKUST Shenzhen-Hong Kong Collaborative Innovation Research Institute, PRC; [3]HKUST LED-FPD Technology R&D Center at Foshan, PRC

FLEXIBLE HYBRID ELECTRONICS ON WEARABLE HEALTHCARE APPLICATION..........124

Ming-Hung Chen

ASE, Taiwan

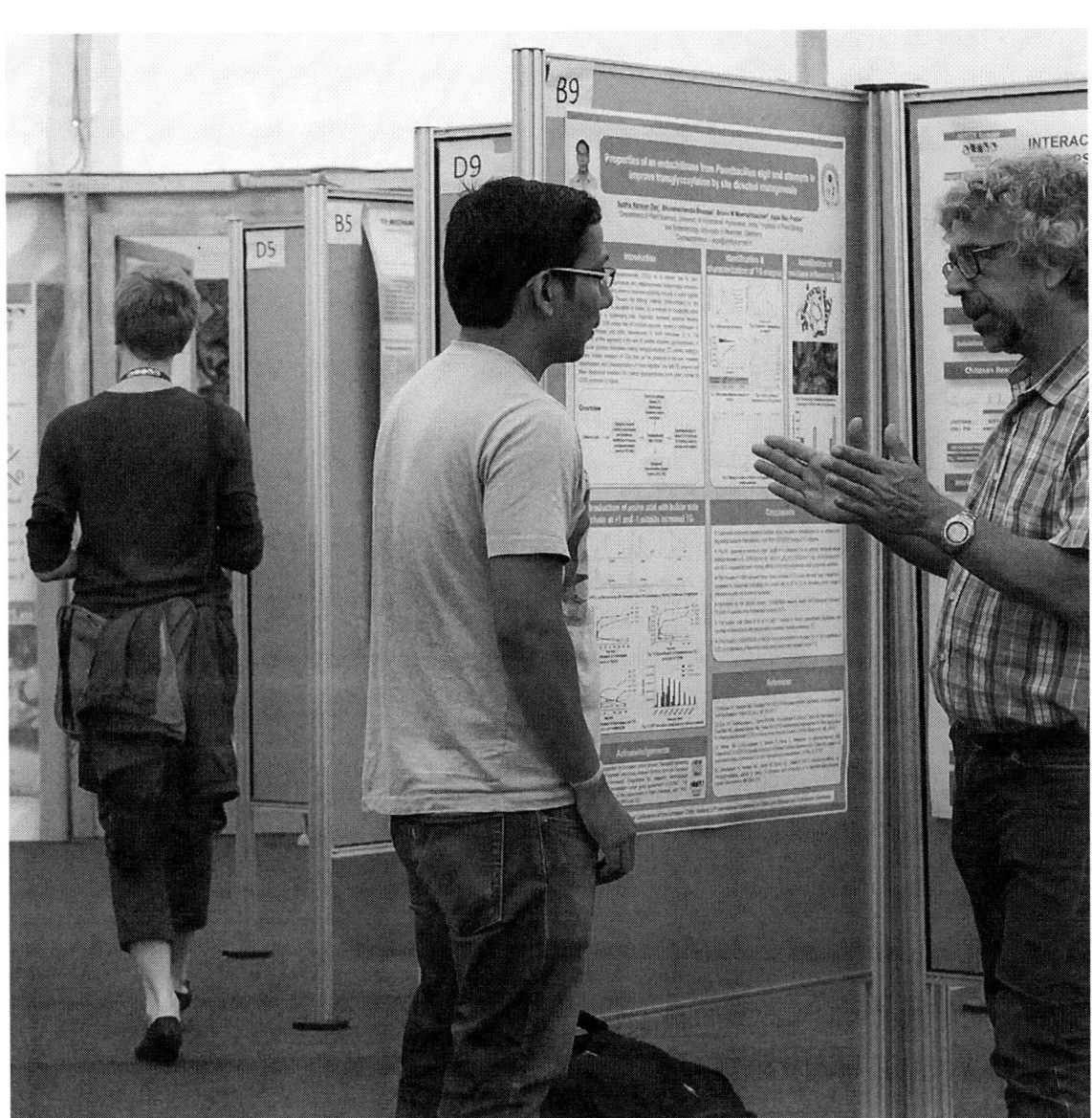

POSTER SESSION

01 TESTING OF ELECTROMIGRATION RESISTANCE OF COPPER AND SILVER THICK FILMS 129
Jiri Hlina[1], Jan Reboun[1], Marek Simonovsky[2], Ales Hamacek[1]
[1]University of West Bohemia, Faculty of Electrical Engineering, Czech Republic; [2]Elceram a.s., Hradec Kralove, Czech Republic

02 RELIABILITY TESTING OF RECYCLED SMD COMPONENTS REUSED IN E-TEXTILES AFTER AGEING BY WASHING CYCLES 134
Martin Hirman, Jiří Navrátil, Andrea Benešová, František Steiner
University of West Bohemia, Czech Republic

03 PROCESS WINDOW OF MINI-LED DISPLAY PANEL PACKAGING USING LASER ASSISTED BONDING TECHNOLOGY 139
Yong Sung Eom, Gwang-Mun Choi, Ki-Seok Jang, Ji-Ho Joo, Chan-Mi Lee, Jin-Heuk Oh, Seok-Hwan Moon, Kwang-Seong Choi
ETRI, Republic of South Korea

04 EFFECT OF SURFACE MICROSTRUCTURE ON JOINTS USING NANOPOROUS CU SHEET FOR POWER DEVICES 143
Hiroshi Nishikawa[1], Byungho Park[2], Mikiko Saito[3], Jun Mizuno[3]
[1]Joining and Welding Research Institute, Osaka Universit, Japan; [2]Graduate School of Engineering, Osaka University, Japan; [3]Research Organization for Nano & Life Innovation, Waseda University, Japan

05 INVESTIGATION OF ALUMINUM AND GOLD FLIP-CHIP BONDING FOR QUANTUM DEVICE INTEGRATION 147
Imants Cirulis[1], Uwe Zschenderlein[2], Silvia Braun[1], Moritz Radestock[1], Remi Pantou[1], Klaus Vogel[1], Franz Selbmann[1], Steffen Kurth[1], Bernhard Wunderle[1,2], Harald Kuhn[1,2]
[1]Fraunhofer Institute for Electronic Nano Systems, Germany; [2]Technical University Chemnitz, Germany

06 ADHESION COPPER/MOLDING COMPOUND: MODELING AND
CHARACTERIZATION 152
Marco Rovitto[1], Samuele Zalaffi[1], Carlo Passagrilli[1], Claudio Maria Villa[1],
Luca Andena[2], Stefano Mariani[2]
[1]STMicroelectronics, Italy; [2]Polytechnic University of Milan, Italy

07 A HIGH-DENSITY ORGANIC PACKAGE SOLUTION TO W-BAND SIGE
FLIP-CHIP APPLICATIONS 158
Fırat Altuntaş, Nihan Öznazlı, Olcay Kalkan, Emrah Koç
Aselsan A.Ş, Turkey

08 THERMAL DESIGN OF STACKED POWER MODULES FOR ELECTRIC
DRIVE APPLICATIONS..........162
Jianfeng Li, Yuekang Du, Xingzhi Wang, Liangjie Liu, Yong Pang, Feixiang Liu
Zhuzhou CRRC Times Electric UK Innovation Centre, United Kingdom

09 LAMINATION OF CAPACITIVE MICROMACHINED ULTRASONIC
TRANSDUCER ON A PIEZOELECTRIC ARRAY: PROCESS AND EVALUATION 168
Duy Hoang Le, Tung Manh, Lars Hoff
University of South-Eastern Norway (USN), Norway

10 ASSEMBLY OF ULTRA-THIN MEMS DEVICE ON DRIVER CHIP USING
ANISOTROPIC CONDUCTIVE FILM 172
Hoang-Vu Nguyen, Knut Eilif Aasmundtveit
University of South-Eastern Norway, Norway

11 AN INNOVATIVE CONFORMAL ELECTRONICALLY SCANNED ARRAY
ANTENNA FOR FULL 360° STEERABILITY IN THE KA-BAND 177
Peter Uhlig, Aline Friedrich, Markus Krengel, Winfried Simon, Oliver Litschke
IMST GmbH, Germany

12 THERMALLY CONDUCTIVE POLYMER COMPOSITES WITH HEXAGONAL
BORON NITRIDE FOR MEDICAL DEVICE THERMAL MANAGEMENT 181
Nu Bich Duyen Do[1], Kristin Imenes[1], Knut E. Aasmundveit[1], Hoang-Vu
Nguyen[1], Erik Andreassen[1,2]
[1]University of South-Eastern Norway, Norway; [2]SINTEF Industry, Norway

13   MICROSTRUCTURAL BASED RELIABILITY INVESTIGATION OF WATER-
AND SUSPENSION FREE PREPARED INTEGRATED ELECTRONIC PACKAGES..........189
Sandy Klengel, Robert Klengel, Tino Stephan
Fraunhofer IMWS, Germany

14   NUMERICAL STUDY ON THE INFLUENCE OF POLYIMIDE THICKNESS AND
CURING TEMPERATURE ON WAFER BOW IN WAFER LEVEL PACKAGING..........194
Prashant Kumar Singh[1,2], Patrick Rohlfs[1], Gunther Sandmann[1], Kashi
Vishwanath Machani[1], Dirk Breuer[1], Karsten Meier[2], Frank Kuechenmeister[1],
Marcel Wieland[1], Karlheinz Bock[2]
[1]GlobalFoundries Dresden Module One LLC & Co. KG, Germany; [2]Technische
Universität Dresden, Institute of Electronic Packaging Technology, Germany

15   IMPROVEMENT OF BONDING STRENGTH AND THERMAL SHOCK
RELIABILITY FOR AG SINTER JOINING DIRECT ON AL SUBSTRATE..........200
Chuantong Chen[1], Ran Liu[1], Koji Kobayashi[2], Hideyo Osanai[2], Zheng Zhang[1],
Katsuaki Suganuma[1]
[1]Osaka University; [2]DOWA POWER DEVICE Co., Ltd

16   INTERCONNECT STRESS TESTING AS A TOOL FOR ASSESSMENT OF
RELIABILITY OF MODERN PCB'S..........204
Marek Koscielski, Krzysztof Glinski, Dariusz Ostaszewski, Tomasz Klej,
Jan Oklej, Aneta Cholaj, Wojciech Steplewski, Stefan Galinski
Łukasiewicz Research Network - Tele and Radio Research Institute, Poland

17   HIGH SPEED TRANSMISSION CHARACTERISTICS ON GLASS BASED
INTERPOSERS..........209
Satoru Kuramochi, Masaya Tanaka, Takahiro Tai
Dai Nippon Printing, Japan

18   DESIGN, FABRICATION, AND CHARACTERIZATION OF A 4H-SIC CMOS
READOUT CIRCUIT FOR MONOLITHIC INTEGRATION WITH SIC SENSORS..........215
Romina Sattari, Henk van Zeijl van Zeijl, Guoqi Zhang
Department of Microelectronics, Delft University of Technology,
The Netherlands

5A: SINTERING I

RELIABILITY OF COPPER SINTERED INTERCONNECTS UNDER EXTREME THERMAL SHOCK CONDITIONS..........218
Sri Krishna Bhogaraju[1], Francesco Ugolini[2], Alessio Greci[2], Gordon Elger[1]
[1]Technische Hochschule Ingolstadt, Germany; [2]AMX Automatrix srl, Italy

RAPID SINTERING OF INKJET PRINTED CU COMPLEX INKS USING LASER UNDER AIR 223
Nihesh Mohan[1], Sri Krishna Bhogaraju[1], Juan Ignacio Ahuir-Torres[2], Hiren Kotadia[2,3], Gordon Elger[1]
[1]Institute of Innovative Mobility, Technische Hochschule Ingolstadt, Germany;
[2]School of Engineering, Liverpool John Moores University, UK;
[3]WMG, University of Warwick, UK

IMPROVING SEMICONDUCTOR RELIABILITY OF SILVER SINTER DIE ATTACH MATERIALS FOR LARGE DIE ON LEAD FRAME APPLICATIONS..........230
Ruud De Wit[1], Edsger Smits[2]
[1]Henkel Nederland BV, The Netherlands; [2]Chip Integration Technology Center (CITC), The Netherlands

5B: EQUIPMENT / INSPECTION

PICK AND PLACE OF SENSITIVE CHIPS WITH VACUUM-FREE GECOMER® TOOLS..........238

Lukas Lorenz[1], Thomas Ludewig[1], Kai Swiecinski[1], Henrik Ollmann[2], Amirabbas Razkordanisharahi[2], Volker Bock[1]

[1]Fraunhofer IPMS, Germany; [2]INNOCISE GmbH, Germany

RESEARCH OF CHIP PLACEMENT ACCURACY FOR FAN-OUT WLP USING A NOVEL SELF-ASSEMBLY STAGE..........244

Tadatomo Yamada, Ken Takano, Toshiaki Menjo, Shinya Takyu

LINTEC Corporation, Japan

ONE-PROBE NANOPROBING OF POWER DEVICES AND ELECTRONIC PACKAGES..........250

Chengliang Huang[1], Vignesh Viswanathan[1], Greg M. Johnson[2], Andreas Rummel[3], Heiko Stegmann[4], Elliott Andrew[5]

[1]Carl Zeiss Pte Ltd. Singapore; [2]Zeiss Microscopy, United States of America; [3]Kleindiek Nanotechnik, Germany; [4]Zeiss Microscopy, Germany; [5]Zeiss Microscopy, UK

6A: PHOTONICS AND OPTICS

STUDY OF SPATIAL DISTORTION IN INP NANOPHOTONIC MEMBRANES ON DIFFERENT CARRIER SUBSTRATES 255

Salim Abdi[1], Aleksandr Zozulia[1], Jeroen Bolk[2], Erik Jan Geluk[2], Yuqing Jiao[1], Kevin Williams[1]

[1]Eindhoven Hendrik Casimir Institute (EHCI), Eindhoven University of Technology; [2]Nanolab@tu/e, Eindhoven University of Technology

LASER-ASSISTED BONDING APPROACH FOR PHOTONIC INTEGRATION PROCESSES 260

Aleksandr Vlasov, Topi Uusitalo, Evgenii Lepukhov, Heikki Virtanen, Samu-Pekka Ojanen, Jukka Viheriälä, Mircea Guina

Tampere University, Finland

INTEGRATION OF MULTI-LITHOGRAPHY TECHNOLOGIES FOR THE FABRICATION OF FLEXIBLE OPTICAL LINK 264

Akash Sunilkumar Mistry, Krzysztof Kamil Nieweglowski, Karlheinz Bock

Technical University of Dresden, Germany

HYBRID LITHOGRAPHY FABRICATION OF SINGLE MODE OPTICS FOR SIGNAL REDISTRIBUTION AND COUPLING 269

David Weyers, Krzysztof Nieweglowski, Karlheinz Bock

Technological University Dresden, Germany

6B: SOLDER / SOLDERING

COMPARING THE SOLDERABILITY OF DIFFERENT SAC0307 COMPOSITE SOLDER PASTES..........277

Balázs Illés[1,2], Halim Choi[1], Agata Skwarek[2]

[1]Budapest University of Technology and Economics, Hungary;
[2]Łukasiewicz Research Network - IMiF, LTCC Technology and Printed Electronics Research Group, Poland

ANISOTROPIC SOLDER PASTE (ASP) MATERIAL SOLUTION FOR LASER ASSISTED BONDING (LAB) PROCESS..........282

Ki-Seok Jang, Yong-Sung Eom, Gwang-Mun Choi, Ji-Ho Joo, Jin-Hyuk Oh, Chan-Mi Lee, Yoon-Hwan Moon, Seok-Hwan Moon, Kwang-Seong Choi
Electronics and Telecommunication Research Institute, Republic of South Korea

DURABILITY OF LEAD-FREE SOLDER INTERCONNECTIONS FOR PRINTED CIRCUIT BOARD APPLICATIONS: COMPARING ENERGY-BASED THERMO-MECHANICAL FATIGUE MODELS..........286

Chien-Ming Huang, Jeffrey W. Herrmann
University of Maryland, United States of America

7A: SINTERING II

AN EXPERIMENTAL INVESTIGATION OF A FLEXIBLE SINTERED SILVER JOINT
FOR MICRO-JOINING BASED ON A DESIGN OF EXPERIMENTS..........292
Laurent Vivet[1], Lahouari Benabou[2], Olivier Simon[2]
[1]VALEO, THS Material Laboratory, France;
[2]UVSQ, University of Paris-Saclay, France

UNDERSTANDING CU SINTERING AND ITS ROLE ON CORROSION BEHAVIOUR
FOR HIGH-TEMPERATURE MICROELECTRONIC APPLICATION..........297
Juan Ignacio Ahuir-Torres[1], Sri Krishna Bhogaraju[2], Geoff West[3], Gordon Elger[2],
Hiren Kotadia[1,3]
[1]Liverpool John Moores University, UK;
[2]Technische Hochschule Ingolstadt, Germany; [3]The University of Warwick, UK

INSPECTION TECHNIQUES USING SCANNING ACOUSTIC MICROSCOPY FOR SIL-
VER SINTERING APPLICATIONS IN POWER ELECTRONIC MODULES..........302
Heaklig Ayala[1,2], Jose Ortiz-Gonzalez[1], Mohamed-Amer Karout[1], James Cotty[2],
Tim Rumney[2], Philip Mawby[1]
[1]University of Warwick, United Kingdom, [2]Custom Interconnect Limited, United
Kingdom

7B: HERMETIC/CONFORMAL COATINGS/TIMS

CHARACTERIZATION OF A NOVEL COST-EFFICIENT AND ENVIRONMENTALLY FRIENDLY GRAPHENE-ENHANCED THERMAL INTERFACE MATERIAL 309

Sihua Guo[1], Kristoffer Harr (Martinsen)[2], Amos Nkansah[2], Jiajia Chen[1], Zhiyang Shen[1], Murali Murugesan[2], Hongfeng Zhang[2], Lars Almhem[2], Arto Ahtonen[2], Jin Chen[3], Johan Liu[1,4,5]

[1]SMIT Center, Shanghai University, PRC; [2]SHT Smart High-Tech AB, Sweden; [3]Shanghai Ruixi New Materials High Tech Co. Ltd., PRC; [4]Electronics Materials and Systems Laboratory, Chalmers University of Technology, Sweden; [5]School of Energy and Materials Science, Shanghai Poly-Tech University, PRC

EVOLUTION OF GETTER TECHNOLOGY IN ELECTRONIC HERMETIC PACKAGING..........313

Luca Mauri, Giovanni Zafarana, Enea Rizzi, Alessio Corazza
SAES Getters, Italy

ADVANCES IN PARYLENE ADHESIVE BONDING FOR THE REALIZATION OF BIOCOMPATIBLE MICROSYSTEMS..........319

Franz Selbmann[1,2], Frank Roscher[1], Maik Wiemer[1], Harald Kuhn[1,3], Yvonne Joseph[2]

[1]Fraunhofer Institute for Electronic Nano Systems ENAS, Germany; [2]TU Bergakademie Freiberg, Institute for Electronic and Sensor Materials, Germany; [3]TU Chemnitz, Center for Microtechnologies, Germany

8A: FLIP CHIP

FLIP-CHIP INTERCONNECTS BASED ON SINGLE METAL-COATED POLYMER SPHERES..........325

Van Long Huynh, Knut Eilif Aasmundtveit, Hoang-Vu Nguyen

University of South-Eastern Norway, Norway

20µM COPPER MICRO-BUMP BONDING THROUGH A SILVER METALLIZATION FOR ADVANCED PACKAGING UNDER A LOW-PRESSURE CONDITION..........330

Zheng Zhang[1], M.-C. Hsieh[1], A. Suetake[1], H. Yoshida[1], R. Okumuara[1], N. Kagami[1], Kazamasa Okamoto[1], Chuantong Chen[1], Kei Hashizume[2], N. Hasegawa[2], R. Yoshida[2], H. Homma[2], K. Suganuma[1]

[1]SANKEN, Osaka University, Japan; [2]Okuno Chemical Industries Co. Ltd,

PRACTICAL RESULTS TO DEMONSTRATE AN INCREASE IN THE RELIABILITY OF FLIP CHIP CONNECTIONS BY ADDING NANOPARTICLES TO SOLDER..........333

David Harvey[1], Teresa Manzanera[2], Kangkana Baishya[3], Guangming Zhang[1], Mohd Arif Anuar[4], Y. C. Chan[1], Nduka Ekere[1], Derek Braden[5]

[1]Liverpool John Moores University, Liverpool, UK; [2]University of Liverpool, Liverpool, UK; [3]Assam Engineering College, Guwahati, India, [4]UniMAP, Perlis, Malaysia; [5]Aptiv, Coventry, UK

DEVELOPMENT AND CHARACTERIZATIONS OF FINE PITCH FLIP-CHIP INTERCONNECTION USING SILVER SINTERING..........337

Julie Gougeon[1,2], Céline Feautrier[1], Laurent Mendizabal[1], Jean-Charles Souriau[1], Mona Tréguer-Delapierre[2]

[1]CEA Leti, France; [2]ICMCB, France

8B: RELIABILITY AND SUSTAINABILITY

ANALYSIS OF THE IMPACT OF ENVIRONMENTAL CONDITIONS ON THE RELIABILITY IN 5G PCB ASSEMBLIES..........341

Hans Walter[1], Marius van Dijk[1], Julia-Marie Köszegi[1], Saskia Huber[1], Olaf Wittler[1], Michael Kaiser[1], Martin Schneider-Ramelow[1,2]

[1]Fraunhofer IZM, Germany; [2]Technische Universität Berlin, Germany

EVALUATION OF THE ENVIRONMENTAL IMPACT WITHIN SEMICONDUCTOR PACKAGING MATERIALS..........348

Andrew Bainbridge, Lewis Clark, Kathleen Grant, Jeff Kettle

University of Glasgow, United Kingdom

MEASUREMENT AND SIMULATION OF MECHANICAL STRENGTH OF BACK-END-OF-LINE LAYER IN ADVANCED CMOS DIES..........352

Bart Vandevelde[1], Kevin Cox[2], Reza Moloudi[1], Riet Labie[1], Jason Krantz[2], Matt Borden[2], Kris Vanstreels[1], Mario Gonzalez[1]

[1]imec, Belgium; [2]Tektronix, United States of America

9A: MACHINE LEARNING

APPLICATION OF MACHINE LEARNING METHODS FOR PROCESS OPTIMIZATION IN ELECTRONIC PACKAGING PROCESSES 357

Corinna Niegisch[1], Sabine Haag[1], Tanja Braun[2], Ole Hölck[2],
Martin Schneider-Ramelow[3]

[1]Robert Bosch GmbH, Germany; [2]Fraunhofer IZM Berlin, Germany;
[3]Technical University Berlin, Germany

PARTIAL DISCHARGE CHARACTERIZATION OF CERAMIC POWER ELECTRONICS CIRCUIT CARRIERS ASSISTED BY MACHINE LEARNING 363

Johannes Drechsel, Lars Rebenklau, Henry Barth

Fraunhofer IKTS, Germany

9B: SOLAR / SENSORS

IMPACT OF PAD LAYOUTS AND SOLDER VOLUME ON SELF-ALIGNMENT OF MICRO SOLAR CELLS..........373

Elisa Kaiser[1], Maike Wiesenfarth[1], Victor Vareilles[2], Henning Helmers[1]

[1]Fraunhofer ISE, Germany; [2]Université Grenoble Alpes, CEA LITEN, France

NOVEL LOW TEMPERATURE AND LOW PRESSURE SINTERING OF ADAS RADAR SENSOR ANTENNA STACK..........378

Sri Krishna Bhogaraju[1], Dirk Busse[2], Alexander Dahlbüdding[2], Philipp Hadrava[3], Hüseyin Erdogan[3], Gordon Elger[1]

[1]Technische Hochschule Ingolstadt, Germany; [2]Budatec GmbH, Germany; [3]Conti Temic microelectronics GmbH, Germany

10A: HIGH FREQUENCY

CHARACTERIZATIONS FOR EWLB (EMBEDDED WAFER LEVEL BALL GRID ARRAY)
ANTENNA IN MOLDED PACKAGE INTEGRATIONS IN 77GHZ AUTOMOTIVE
APPLICATIONS..........384
M.-C. Hsieh[1], F. Zhang[2], F. Zhu[2], K. Liu[2], L. Chua[3], Y. Lin[3], J. Damalerio[3], Hin Hwa Goh[3],
Kai Chong Chan[3], Zihao Chen[4]
[1]JCET Group Co. Ltd., Singapore; [2]Andar Technologies Co., Ltd.; [3]JCET Group Co.
Ltd.,, Singapore; [4]Harbin Institute of Technology, Shenzhen, PRC

A DUAL-BAND DUAL-POLARIZED 2X2 ANTENNA ARRAY WITH BEAMFORMING
FOR 5G AIP AND MMWAVE APPLICATIONS..........389
Sheng-Chi Hsieh, Wen-Chun Hsiao, Hong-Sheng Hong-Sheng, Cheng-Yu Ho,
Chen-Chao Wang
ASE group /Advanced Semiconductor Engineering, Taiwan

ANALYSIS AND CHARACTERIZATION OF CASTELLATED HOLES AS
RF INTERCONNECTS FOR MODULAR MILLIMETER-WAVE DEVICES..........394
Paul Perlwitz[1,2], Christian Tschoban[1], Ivan Ndip[1], Harald Pötter[1],
Martin Schneider-Ramelow[1,2]
[1]TU Berlin; [2]Fraunhofer IZM, Germany

HIGH-Q KU-BAND MICROSTRIP SPIRAL RESONATOR IN FAN-OUT WAFER-LEVEL
PACKAGING TECHNOLOGY FOR VCO APPLICATIONS..........401
M. Chernobryvko[1], M. P. Kaiser[1], K. S. Murugesan[1], D. Kuylenstierna[2], J.-M. Köszegi[1],
R. Gernhardt[1], T. Braun[1], I. Ndip[1], M. Schneider-Ramelow[1]
[1]Fraunhofer IZM Berlin, Germany; [2]Chalmers University of Technology, Sweden

10B: POWER MODULES

THERMAL-MECHANICAL ANALYSIS OF A POWER MODULE WITH PARAMETRIC MODEL ORDER REDUCTION..........405

Sheikh Rokibul Hassan[1], Pushparajah Rajaguru[1], Stoyan Stoyanov[1], Christopher Bailey[2]

[1]University of Greenwich, United Kingdom; [2]School of Electrical, Computer and Energy Engineering, Arizona State University, USA

THICKNESS EFFECT OF COPPER CLIPS ON POWER MODULE PACKAGING DESIGN..........411

H. Wan[1], N. Iosifidis[1], X. Zhang[1], R. Rong[2], M. Antoniou[1], P. Mawby[1]

[1]University of Warwick, United Kingdom; [2]MacMic Science & Technology Co., Ltd. China

HIGH FREQUENCY THIN FILM MAGNETICS-ON-SILICON WITH IMPROVED INDUCTANCE AND RESISTANCE..........416

Martin Sittner

Würth Elektronik eiSos Group, Germany

ENHANCED RELIABILITY FOR POWER MODULES VIA A NEW AG/SI SINTER JOINING STRATEGY..........421

Y. Liu[1], C. Chen[1], Koji S. Nakayama[1], Minoru Ueshima[2], Takeshi Sakamoto[2], Takuya Naoe[3], Hiroshi Nishikawa[3], Katsuaki Suganuma1

[1]Flexible 3D System Integration Laboratory, Osaka University, Japan; [2]Daicel Corporation, Japan, [3]Joining and Welding Research Institute, Osaka University, Japan

EMPC²³

24ᵗʰ European Microelectronics & Packaging Conference (EMPC 2023)

11 – 14 September 2023 | Cambridge, UK

iMAPS EUROPE

Licensed to IEEE Xplore for Publication

Copyright IMAPS-Europe

High frequency bandwidth transition for HTCC hermetic packages

Emad Elrifai
EGIDE SA
Bollène, France
eelrifai@fr.egide-group.com

Abstract—**This paper presents simulation and measurement of large frequency bandwidth transitions up 100GHz, we introduce two different transitions using multi-layer high temperature cofired ceramic technology for optical hermetic package applications. Measurements results for the first transition show an overall bandwidth up to 50GHz, and simulation results for the second presents 100GHz bandwidth.**

Keywords—HTCC, package, HB-CDM package, high bandwidth transition

I. INTRODUCTION (*HEADING 1*)

During the last years, as consequence of the outbreak of but not limited to mobile internet-based applications, 4k/8K streaming platforms, and virtual reality, the data traffic of communication networks has risen sharply, hence the need for long distance high-bandwidth communication networks has become a major issue for telecom market. Coherent optical networks can provide larger transmission capacity and much distance up to 100 times mor than non-coherent optical networks (thousands Km for coherent versus tens Km for non-coherent). The main difference between coherent and non-coherent optical networks is the coherent modulation on transmitter side with heterodyne detection technology on receiver side, coherent modulators are therefore vital components for such high-speed coherent communication networks. Since these coherent modulators need to be hermetically packaged, the optical internetworking forum (OIF) has introduced the specifications of high-bandwidth coherent driver modulator (HB-CDM) package [1], radio frequency (RF) performance for such telecom packages, play a key role in the packed component performance.

Most kind of hermetic packages are surface-mount device's (SMD's), use surface-mount technology (SMT) to be mounted directly on the surface of printed circuit board (PCB). SMT packages are either lead frame packages (use lead frame), or leadless packages use flex PCB (F-PCB), ball grid array (BGA) instead of lead-frame as SMT. As HB-CDM package has been classified into three classes 40, 60, and 80[1], Aside from RF bandwidth, the difference between the packaging of HB-CDM classes was the PCB mounted method. Various designs and technologies for modulators with different bandwidths up to 63GHz have been reported, for example a HB-CDM has been improved by optimize the RF ground led-frame that reduce the crosstalk and achieve the 54GHz [2], in [3] they present a 56GHz of overall electro-optic (EO) bandwidth.

Hereafter, we present high temperature co-fired ceramic HTCC package with two different mounted method, and two different radio frequency (RF) transition.

II. PACKAGE AND TRANSITION

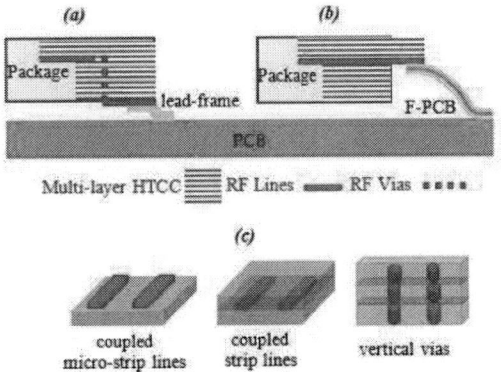

Fig.1 HTCC package. (a) lead-frame SMT (vertical transition)
(b) F-PCB SMT (straight transition)
(c) RF differential lines

Fig.1 presents standard RF HTCC package for optical telecommunication applications with two different SMT's Fig.1(a) lead frame, Fig.1(b) F-PCB, for lead frame case we need to drive the signal from the lead frame (top of package) into the package, so we use vertical vias, this transition (with vertical vias) can reach no more than 55GHz -3dB frequency bandwidth due to fabrication limits for vias diameter, ceramic sheet thickness, and lead-frame that produce impedance mismatch with HTCC package and create more losses.

To go further than 55GHz we need to overcome these parameters (lead frame, vias diameter and ceramic thickness), by reduce vias diameter below 80µm, consequently reduce ceramic thickness to be able to drill and fill these extra small vias, which is beyond actual fabrication limits.

As telecommunication network development require transmission bandwidth larger than 55GHz, package design engineers have turned to F-PCB Fig.1(b), which offer better impedance matching with HTCC package as direct result of absence of lead frame, and it doesn't need vertical vias to drive the signal into the package, where it can be brazed at the same ceramic sheet of interior RF traces, which means less changes in impedance and RF propagation modes than vertical transition for lead frame, so less losses and better transmission.

III. RESULS AND DISCUSSION

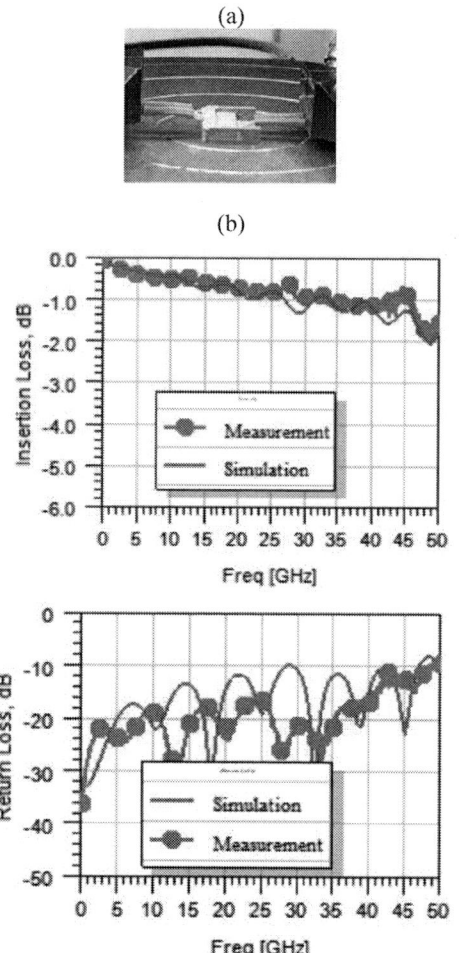

Fig.2 RF S-parameters results simulation Vs measurements for HTCC package with vertical transition befor lead frame brazing (a)measured structure (b)S-parameters results

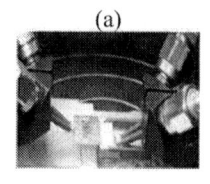

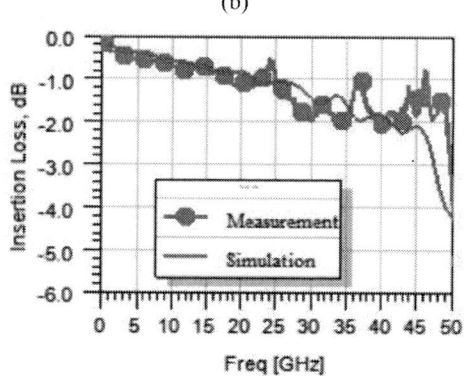

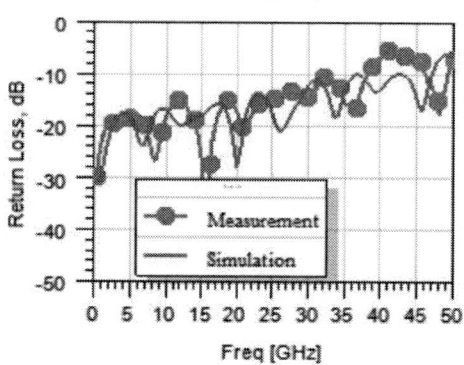

Fig.3 RF S-parameters results simulation Vs measurements for HTCC package (with vertical transition) +lead frame+PCB (a)measured structure (b)S-parameters results

Two packages for lead frame and F-PCB SMT's have been studied, Fig.2 presents s-parameters results of simulation and measurements for package with vertical transition (consist of coupled micro strip lines, coupled strip lines, and vertical vias) before brazing the lead frame, RF probes have been landed on the ceramic Fig.2(a), Fig.2(b) shows insertion and return loss results with -2dB insertion loss up to 50GHz and -10dB return loss up to 50GHz, also we can observe the accord the between simulation and measurement that confirm the simulation method.

In Fig.3 we present s-parameters results of simulation and measurements for the package with vertical transition after brazing lead frame and PCB, so RF probes have been landed on the ceramic of interior package and on the PCB that has been soldered with lead frame on the other side Fig.3(a), for this configuration we notice 1 dB degradation for insertion loss with -3dB up to 50GHzs Fig.3(b), we observe the degradation also in return loss results, as return loss > -10dB from 38GHz, this degradation explained by the mismatch impedance due to lead frame air gap with PCB and brazing.

As explained earlier, to extend the bandwidth further than 50GHz we replaced the lead frame SMT by F-PCB SMT, for F-PCB SMT the package transition consist of coupled micro strip line and coupled strip line without vertical vias, Fig.4 presents the simulated structure with s-parameters simulation results for HTCC package with F-PCB.

The HTCC package for F-PCB SMT has kept the transition principal dimensions as pitch between channels 2.4mm, but we had more flexibility on terms of RF traces dimensions of brazing zone, as we were committed to width minimum limits for package with lead frame SMT to be able to solder the lead frame. 7.6mm F-PCB length has been added with solder material to the HTCC package for simulation Fig.4(a). Fig.4(b) shows insertion loss and return loss simulation results, smooth insertion loss curve with -3dB up to 80GHz and -4dB at 100GHz, matched impedance between package and F-PCB can be clearly seen with return loss results where we are adapted around -15dB up to 100GHz.

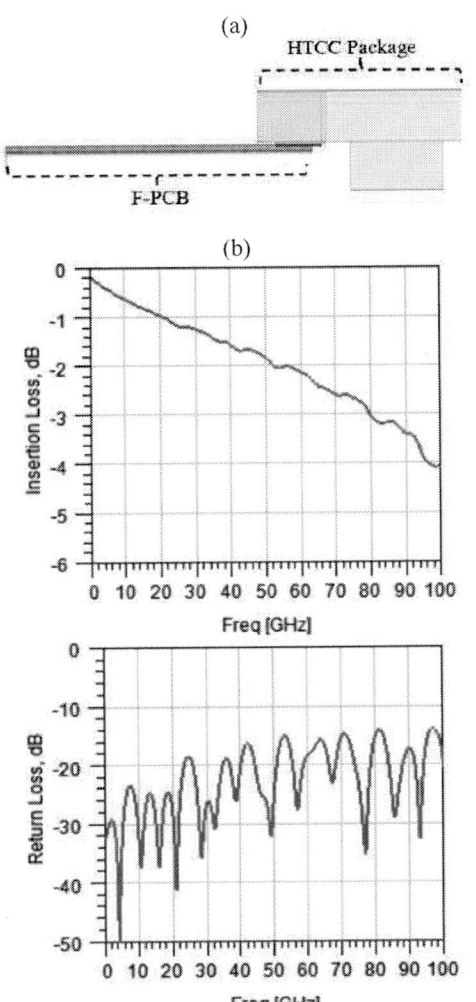

Fig.4 RF simulation results for package with F-PCB SMT. (a) simulated structure. (b) insertion and return loss results

IV. CONCLUSION

We presented and compared between two SMT methods for HTCC package (lead frame and F-PCB), 50GHZ bandwidth for lead frame, and an extended bandwidth up to 100GHz for F-PCB, these transition designs can be used for HTCC packages for optical telecom applications including the three classes of HB-CDM.

ACKNOWLEDGMENT

The author would like to thank EGIDE SA design and development teams for all support in HTCC packaging technology

REFERENCES

[1] OIF: 'Implementation Agreement for the High Bandwidth Coherent Driver Modulator (HB-CDM)'. https://www.oiforum.com/technical-work/hot-topics/high-bandwidth-coherent-driver-modulator-hb-cdm-2-0/.

[2] OZAKI, J. Crosstalk reduction between RF input channels of coherent-driver-modulator package by introducing enhanced ground lead structure. Electronics Letters, 2020, vol. 56, no 17, p. 893-895.

[3] OZAKI, J. High-speed modulator for next-generation large-capacity coherent optical networks. NTT Tech. Rev, 2018, vol. 16, no 4, p. 1-8.

RF bandwidth enlargement is clear between the two different mounted methods (lead frame and F-PCB), in fact using F-PCB as SMT gave us more flexibility in terms of design, the transition has no more vertical vias, and we can simulate and adapt the RF transition of package with F-PCB as one structure, using time domain reflectometry (TDR) simulation for such complicated structure with several changes in RF line modes was a good optimisation technique, as it enable us to identify the design problem by detecting mismatching impedance points along RF traces.

We are confident of our simulation methods and results for package with F-PCB mounted technology, also the accord between measurements and simulation for package with lead frame Fig.2(b) make us more confident, in the meantime we are working on developing the bandwidth beyond 100GHz, then fabrication and measurement.

Application of Reactive Bonding Methods on LTCC Substrates

Erik Wiss
Chair of Microintegration and Reliability
Saarland University
Saarbruecken, Germany
erik.wiss@uni-saarland.de

Alexander Schulz
Electronics Technology Group
TU Ilmenau
Ilmenau, Germany
alexander.schulz@tu-ilmenau.de

Adam Yuile
Chair of Microintegration and Reliability
Saarland University
Saarbruecken, Germany
adam.yuile@uni-saarland.de

Jens Müller
Electronics Technology Group
TU Ilmenau
Ilmenau, Germany
jens.mueller@tu-ilmenau.de

Steffen Wiese
Chair of Microintegration and Reliability
Saarland University
Saarbruecken, Germany
s.wiese@mx.uni-saarland.de

Abstract ⁻ This paper discusses the application of reactive bonding for the area of Low Temperature Cofired Ceramics (LTCC) assemblies. The goal is to reduce the thermal-mechanical stresses during soldering by transferring heat only locally to the solder joints without heating the entire component. Such a reactive multilayer system (RMS) consists of alternating nanolayers (10 - 300 nm) of at least two metal components which produce an exothermal reaction after ignition. Although the deposition of an RMS is established on silicon substrates for the use in micro-electromechanical systems (MEMS), it is very challenging to create them on LTCC substrates. One of the main obstacles is to overcome all issues connected with the significant roughness, because it is not an optimum territory to deposit nanolayers. In this paper, different methods like chemical mechanical polishing (CMP) and laser ablation, to modify the surface morphology, are presented. A direct relation between the morphology and the exothermal reaction can be observed. In addition, 3-D Computational Fluid Dynamics (CFD) simulations were conducted to analyze the process in more detail. These simulations make use of a shoebox model with different layers and an adjustable user-defined function for the heat release of the RMS to adapt the reaction front velocity and the combustion temperature to the experimental values.

Keywords ⁻ LTCC, RMS, reactive multilayer systems, Al-Ni, joining.

I. INTRODUCTION

The chemical reaction of nitrogen monoxide (NO) with ozone leads not only to the formation of nitrogen dioxide and oxygen, but also to the emission of radiation in the wavelength range between 600 to 2600 nm with a maximum in the infrared area at 1200 nm [1]. Measuring this radiation can be used to determine low concentrations of NO in gas flows. Therefore, a measurement device was developed and presented in [2]-[4] using LTCC (Low Temperature Cofired Ceramics) technology (see Fig. 1). It consists of three different modules: first, there is an ozone generator with a high voltage source, that is applied to a gas flow using embedded electrodes. Then, the ozone is transported via microfluidic channels to a chemiluminescence detection reaction chamber where it is mixed with the gas that contains an unknown amount of NO. An infrared transparent window is bonded onto the LTCC substrate, and a photo diode that measures the emitted radiation is placed on it. The last module is an exhaust gas treatment to decompose the ozone into nonhazardous oxide.

Therefore, a heating structure was embedded in the LTCC substrate and a platinum paste, that can be applied to the substrate, was used as a catalysator.

Fig. 1. Measurement device for low concentrations of nitrogen monoxide in gas flows, realized using LTCC technology.

The LTCC technology is an excellent candidate for such devices because it enables the possibility of three-dimensional structuring [5]-[7] and shows very good hermetic sealing properties [8], [9]. Additionally, it provides not only excellent robustness under harsh environmental and mechanical conditions [10], [11], but also chemical resistance [9]-[13]. Due to this properties, LTCC technology can even be used in automotive and aerospace. It is also possible to add passive elements using conventional surface mounting technology like reflow soldering. Therefore, both the component and the substrate need to be heated up to a temperature above the melting point of the used solder, which can be critical for heat sensitive components. Adding heat locally to the solder joints can be an option to avoid damages during this process e.g., through using a reactive multilayer system (RMS). While the use of an RMS is well established on silicon substrates [14], it is very challenging to deposit such a system on LTCC substrates due to their high roughness in the range of 0.4 to 1 ı m [15]-[17].

An RMS consists of at least two different materials that are applied on a substrate in an alternating way e.g., through magnetron sputtering [10]. The thickness of every single layer is in the range of 10 to 300 nm [19] and the total stack thickness varies in the range of 0.1 to 300 ı m [20]. After the system is ignited (e.g., by providing sufficient heat, by using an electrical spark or by using a laser pulse) the materials begin to intermix on atomic level building a new intermetallic phase. This leads to a self-propagating reaction [21] that is profoundly exothermal [22]. The released heat of this process can be used for joining processes by melting an additional solder layer, among other things. The propagation speed will depend on the materials used, the bilayer thicknesses, the ratio of the materials, the total stack thickness and the amount of heat released. However, the mechanical pre-processing of the substrates also has an influence on the speed of the reaction, what should be discussed here.

II. SAMPLE PREPARATION

A. Reference sample and laser ablation

As described in [23] the samples were prepared in different ways. For the reference samples, and the samples that were modified by laser ablation, a standard LTCC technology was processed using 6-layer DuPont 951 P2 resulting in a thickness of 817 nm ě 2.5 nm. Without further processing a roughness of 390 nm ě 16.6 nm was reached for the reference samples.

To achieve other roughnesses the surface of the LTCC substrate was laser ablated in a grid-like way using a 355 nm picosecond UV laser system. Different cutting distances between 5 and 30 ı m and a laser power between 0.5 W and 1 W were used, which resulted in a roughness between 572 nm ě 14.4 nm and 874 nm ě 20.5 nm. Finally, the samples were cut into pieces of 15 mm x 7 mm. After that a 20 nm thick layer of titanium was sputtered onto the samples as an adhesion layer, followed by the subsequentially deposition of the RMS. With a total thickness of 10 ı m, 100 bilayers were chosen with a thickness of 60 nm for the aluminum and 40 nm for the nickel. After processing the samples, they were ignited by an electrical spark using two electrodes and a power supply limited to 25 V and 4 A.

B. Chemical/mechanical lapping and polishing (CMP)

The other possibility for the sample preparation persists in using a CMP machine to achieve lower roughness. An 8-layer DuPont 951 P2 was processed in a standard LTCC technology resulting in a thickness of 1080 ı m. The substrates were lapped down chemically/mechanical on both sides to achieve a plane surface with a thickness of 810 ı m ě 5,5 ı m. Further polishing led to a roughness in the range between 105 nm ě 19.5 nm and 258 nm ě 16.5 nm. As before the samples were cut into smaller pieces and both the adhesion layer and the RMS were deposited.

C. Laser ablation with additional metallization

Some samples obtained an additional AgPd metallization between the LTCC substrate and the RMS [24]. Therefore, a standard LTCC technology was processed using 4-layer DuPont 951 PX resulting in a thickness of 839 ı m ě 2.6 ı m. The surface morphology was modified using the laser system in the same way as before. The reference sample (without further processing) reached a roughness of 525 nm ě 30 nm, whereas the other ones reached a roughness of 537 nm ě 12 nm (LS 3) and 773 nm ě 29 nm (LS 5). Finally, the samples were cut into pieces of 15 mm x 7 mm and the RMS was deposited.

D. Roughness characterization

The roughness of the surface was characterized by using a laser scanning microscope [23], [24]. An area of 250 ı m x 250 ı m was chosen for the measurements. The lowest roughness of about 100 nm was found for the CMP samples without metallization whereas the laser ablated samples show the highest roughness with a maximum of about 870 nm. Without processing the roughness is about 400 nm (see Fig. 2). For the samples with metallization, a roughness of 525 nm (reference) respectively 537 nm and 773 nm were reached (laser ablated samples).

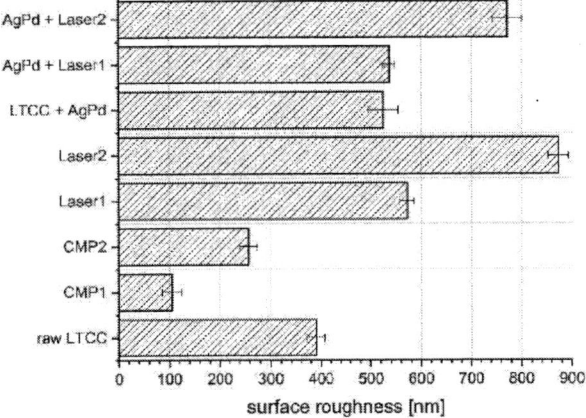

Fig. 2. Resulting surface roughness for some samples with different preparation methods, measured by LSM [23], [24].

III. CFD MODEL

A. CFD model description

To achieve more information about the reactive joining process a suitable CFD model was already developed and presented in [25]-[27] (see Fig. 3). It consists of the aluminum-nickel bilayer (blue), the LTCC substrate (green) and the isolation layer (green) between, some simulations also contain a solder layer. The base of the model has a primary surface of 4 mm x 4 mm with a surrounding air environment of the dimensions 10 mm x 10 mm x 5 mm to avoid interactions between the region of interest and the boundary conditions. Due to stability and convergence reasons no titanium layer was considered in the simulation model.

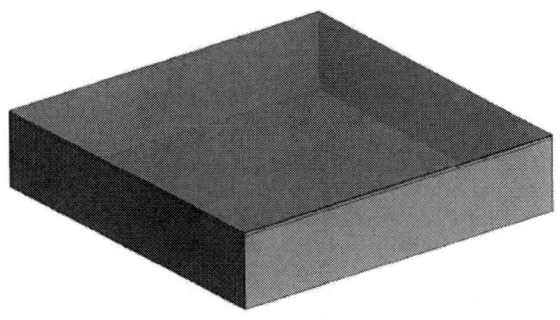

Fig. 3. The CFD simulation model mainly consists of the reactive multilayer system (blue) and the LTCC substrate with an isolation layer (green) [25].

The material properties and the dimensions of the individual layers are shown in Table I. The properties of the LTCC substrate and the isolation layer were assumed to be identical (GreenTape™ DuPont DP951), which is why the effective thickness of the LTCC substrate is 860 μm. The properties of the RMS match with an unreacted multilayer [28], and the properties for the solder and the silicon chip were taken from the ANSYS Fluent material databases.

Table I. Dimensions and properties of the layers used in the CFD simulation model [26], [28].

layer	thickness [μm]	ρ [kg/m³]	λ [W/m·K]	c_p [J/kg·K]
LTCC	825	3100	3.3	600
Isolation	35	3100	3.3	600
RMS	30	5500	152	830
Solder	200	7000	63.2	230
Chip	400	2500	100	710

A 3x3 matrix of temperature probes was used to measure the temperatures during the movement of the reaction front. The probes in the same vertical position (e.g., P1, P2, P3) give information about the reaction velocity (through the time delay in the recorded values), and the probes in the same horizontal direction (e.g., P1, P4, P7) can be used to determine the peak temperature. For additional information the temperatures were also measured in the middle of the RMS, at the interface between the RMS and the solder, and at distances of 10 and 20 μm in the solder starting from the interface.

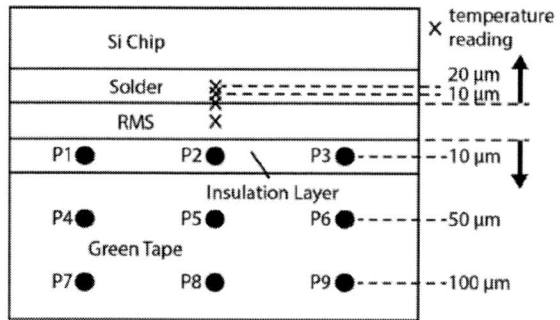

Fig. 4. Locations of the platinum temperature probes (P1 to P9) and the measurement points in the solder and the RMS.

All the structures were meshed in ANSYS Workbench 2023R1 with a length of 100 μm for the mesh edges in x- and y-direction, and with a length of 10 μm in the z-direction leading to a total number of 5,001,000 cells. The time-step size was fixed to 10⁻⁶ s for 10,000 time-steps which results in a simulation time of 10 ms. For a more detailed view see [25], [26]. To simulate the propagation of the reaction front, a probability density function in the form of (1) was used.

$$\text{(1)}$$

The four parameters A, B, C and D are used to control the heat release function, where A and B approximately correspond to the amplitude and width, C is an offset for the alignment with the leading edge of the reactive multilayer system and D corresponds to the velocity.

IV. RESULTS

A. Experimental analysis

As described in [23], all the samples showed a good adhesion behavior before ignition (see Fig. 5). After ignition, a peel-off effect could be observed at both the unmachined reference sample and the CMP prepared samples with low roughness (105 – 390 nm). In contrast the samples that were laser ablated with a roughness between 572 nm and 874 nm still show a good adhesion even after ignition, but SEM images show the occurrence of micro cracks within the RMS. Both the micro cracks and the peel-off effect are mainly caused by the different coefficients of thermal expansion of the materials. As an additional factor, there is a shrinkage in volume of approximately 12 % in the RMS due to lattice-spacing reduction [29] which may also affect the adhesion.

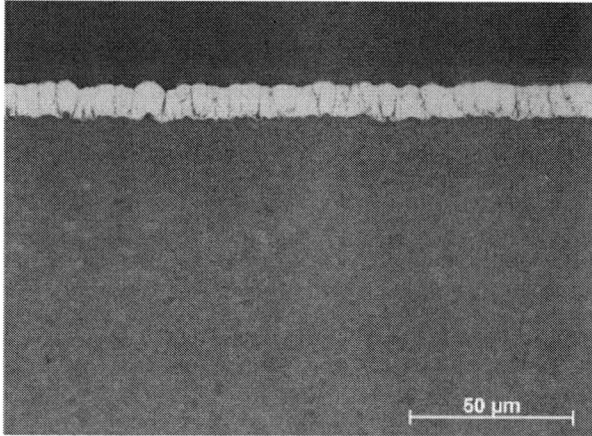

Fig. 5. Reflected light microscopy image reveals a good adhesion of the reactive multilayer on the LTCC substrate (blue) before ignition.

To determine the peak temperatures and the reaction speed, high velocity pyrometer measurements and high-speed camera measurements were done simultaneously [23], [24]. The pyrometer measurements show that the unmachined LTCC substrate reaches the highest peak temperature (1092 °C), followed by the samples that were prepared by CMP with nearly the same temperature (1067 and 1062 °C). The samples that were prepared by laser ablation reach the lowest temperatures (811 and 803 °C). It is remarkable that there is nearly no difference within the laser ablated samples although they differ in their roughness (572 nm to 873 nm). It can be assumed that the higher temperatures occur due to the peel-off of the RMS: the heat cannot be dissipated through the air as fast as if the RMS was still in contact with the substrate.

In case of the laser ablated samples with AgPd metallization, the lift-off effect was not as pronounced as in the other samples without the metallization. There is a punctual adhesion of the reactive multilayer, whereas some areas are separated from the metallization (see Fig. 6). The reference sample reaches a peak temperature of 854 °C whereas the laser ablated samples reach 847 and 739 °C. This temperature drop (compared to the first samples) may be caused by the additional metallization layer that absorbs some of the released heat.

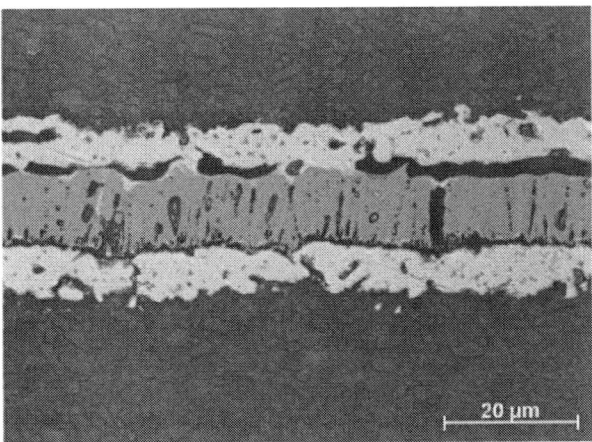

Fig. 6. Reflected light microscopy image of a reactive multilayer system between two LTCC substrates (blue) with AgPd metallization (bright areas) after ignition. There is only punctual adhesion of the RMS, whereas most areas of the RMS separated from the metallization.

The high-speed measurements show a velocity of around 5 m/s for the reaction front of both the reference sample and the CMP prepared samples. In contrast, the laser ablated samples show a reduction in velocity down to 3.7 m/s (roughness 572 nm) and 2.9 m/s (roughness 873 nm). The same effect can be observed on the samples with AgPd metallization: the velocity reduces from 3.7 m/s (roughness 525 nm) to 2.8 m/s (roughness 537 nm) and 2.2 m/s (roughness 773 nm). The measured values are clearly represented in Table II.

Table II. Overview of the reached peak temperatures and reaction front velocities depending on different substrate preparation methods and resulting surface roughness [23], [24].

	substrate preparation	surface roughness [nm]	peak temp. [éC]	reaction front velocity [m/s]
LTCC without metallization	CMP 1	105	1067	5.1
	CMP 2	257	1062	4.7
	Ref.	390	1092	5.1
	LS 1	572	811	3.7
	LS 2	873	803	2.9
LTCC with metallization	AgPd + Ref.	525	854	3.7
	AgPd + LS 3	537	847	2.8
	AgPd + LS 5	773	739	2.2

B. CFD comparisons

Simulations were performed for a model with solder (Fig. 7) and for a model without solder (Fig. 8). A velocity for the reaction front of 1 m/s and a time-step size of 10^{-6} s was chosen for the temperature recordings.

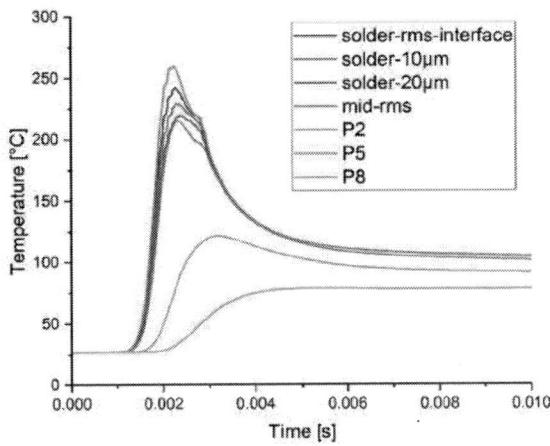

Fig. 7. Temperature contour plot of the interface between the solder and RMS, within the solder and the RMS, and at the temperature probes P2, P5 and P8 for a velocity of 1 m/s.

As depicted in Fig. 7 the maximum temperature of about 260 éC was reached within the RMS which is only slightly above the required temperature of 240 éC [26]. At the interface between the RMS and the solder it reaches around 240 éC and the lowest value with around 220 éC is reached at a distance of 20 ı m within the solder.

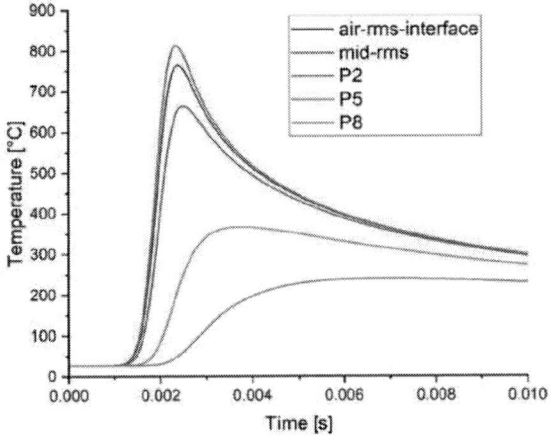

Fig. 8. Temperature contour plot for the model without solder of the interface between the air and RMS, within the RMS and at the temperature probes P2, P5 and P8 for a velocity of 1 m/s.

Fig. 8 shows the temperature plot for the model without a solder layer. The reached temperature within the RMS is raised by more than 200 % up to 820 éC while the temperature at the interface between the surrounding air and the RMS is above 750 éC. By comparison to the previous model, it can be assumed that the melting and solidification of the solder is responsible for this significant drop of the RMS peak temperature, so the amount of solder deposited on the RMS should be a good parameter to control the peak temperature during reactive bonding. A similar relation was already considered by Wang [30].

V. CONCLUSIONS

This paper discusses the application of reactive bonding methods on LTCC substrates. Therefore, a reactive multilayer system (RMS) consisting of alternating nanolayers of aluminum and nickel with a total thickness of 100 µm was deposited after two different surface modifications were applied. Then, the samples were electrically ignited which was recorded by a high velocity pyrometer and a high-speed camera for temperature and velocity measurements. The first preparation method is chemical mechanical polishing (CMP) of a LTCC substrate, without metallization, which lowers the surface roughness down to around 100 nm. Compared to the reference sample (roughness 390 nm), the CMP does not seem to influence the peak temperature or the reaction front velocity. In addition, there is a peel-off of the RMS structure. The second method uses a picosecond laser to modify the LTCC substrate morphology, whereby the roughness can be increased up to 873 nm. This leads to a decrease in the peak temperature by nearly 300 °C and a decrease in the reaction front velocity by around 27 - 43 %. In case of the laser ablated surfaces no peel-off effects could be observed but some micro cracks in the RMS could be found. Both the peel-off effects and the micro cracks occur due to different coefficients of thermal expansion and the volume shrinkage during the exothermal reaction. An additional reason for the problems in adhesion is the lower thermal conductance of the LTCC substrate compared to silicon substrates.

Laser ablation was also done for samples with an additional AgPd metallization between the LTCC substrate and the deposited RMS. After ignition, less peel-off effects could be observed. The drop in the RMS peak temperature is not as high as in the other version, but the velocity of reaction front was slowed down by around 24 - 41 %. The experimental data show that the adhesion still has to be improved.

CFD simulations were conducted to analyze the process in more detail. A shoebox model with the different layers (LTCC, RMS and solder) was presented and compared to another model without a solder layer. An adjustable user-defined function for the heat release of the RMS during the reaction was used to adopt the reaction front velocity and the combustion temperature to the experimental values. The temperature during the exothermal reaction was recorded at different positions with a time-step size of 10^{-6} s. The measurements suggest that the solder layer can be used to control the peak temperature during the reactive bonding process.

ACKNOWLEDGMENT

This research work was supported by funding from the Deutsche Forschungsgemeinschaft (DFG, German Research Foundation) – Project ID 426204742.

The authors would like to offer many thanks to Andreas Ruh for preparing and performing the cross-sections.

REFERENCES

[1] P. N. Clough and B. A. Thrush, `Mechanism of Chemiluminescent Reaction between Nitric Oxide and Ozone,_ Transactions of the Faraday Society, 1967, vol. 63, pp. 915-925.

[2] T. Geiling, T. Welker, H. Bartsch and J. Müller, `Design and Fabrication of a Nitrogen Monoxide Measurement Device Based on Low Temperature Co-Fired Ceramics,_ International Journal of Applied Ceramic Technology, 2012, vol. 9, no. 1, pp. 37-44.

[3] T. Welker, T. Geiling, H. Bartsch and J. Müller, `Design and Fabrication of Transparent and Gas-Tight Optical Windows in Low-Temperature Co-Fired Ceramics,_ International Journal of Applied Ceramic Technology, 2013, vol. 10, no. 3, pp. 405-412.

[4] T. Geiling, T. Welker, C. Ehrling and J. Müller, `Design, Fabrication, and Operation of a Nitrogen Monoxide Measurement Device Based on

LTCC,_ Journal of microelectronics and electronic packaging, 2012, vol. 9, no. 4, pp. 171-177.

[5] K. Malecha and L. J. Golonka, `Microchannel Fabrication Process in LTCC Ceramics,_ Microelectronics Reliability, 2008, vol. 48, no. 6, pp. 866-871.

[6] H. Bartsch de Torres, C. Rensch, M. Fischer, A. Schober and M. Hoffmann, `Thick Film Flow Sensor for Biological Microsystems,_ Sensors and Actuators A: Physical, 2010, vol. 160, no. 1-2, pp. 109-115.

[7] P. Ciosek et al., `Monitoring of Cell Cultures with LTCC Microelectrode Array,_ Analytical and bioanalytical chemistry, 2009, vol. 393, pp. 2029-2038.

[8] J. Wilde, `B4. 1-Trends in Assembly and Packaging of Sensors,_ Proceedings SENSOR 2009, vol. 1, 2009, pp. 205-210.

[9] Y. Fournier, `3D Structuration Techniques of LTCC for Microsystems Applications,_ EPFL, 2010.

[10] P. Uhlig et al., `LTCC Short Range Radar Sensor for Automotive Applications at 24 GHz,_ Proceedings of the IMAPS, 2004.

[11] L. Golonka, P. Bembnowicz, D. Jurków, K. Malecha, H. Roguszczak and R. Tadaszak, `Low temperature co-fired ceramics (LTCC) microsystems,_ Optica Applicata, 2001, vol. 41, no. 2, pp. 383-388.

[12] A. Bittner and U. Schmid, `The Porosification of Fired LTCC Substrates by Applying a Wet Chemical Etching Procedure,_ Journal of the European Ceramic Society, 2009, vol. 29, no. 1, pp. 99-104.

[13] T. Thelemann, M. Fischer, A. Groß and J. Müller, `LTCC-based Fluidic Components for Chemical Applications,_ Journal of microelectronics and electronic packaging, 2007, vol. 4, no. 4, pp. 167-172.

[14] J. Braeuer, J. Besser, M. Wiemer and T. Gessner, `A novel technique for MEMS packaging: Reactive bonding with integrated material systems,_ Sensors and Actuators A: Physical, 2012, vol. 188, pp. 212-219.

[15] H. Jantunen, T. Kangasvieri, J. Vähäkangas and S. Leppävuori, `Design aspects of microwave components with LTCC technique,_ Journal of the European Ceramic Society, 2003, vol. 23, no. 14, pp. 2541-2548.

[16] M. Matters-Kammerer et al., `Material properties and RF applications of high k and ferrite LTCC ceramics,_ Microelectronics Reliability, 2006, vol. 46, no. 1, pp. 134-143.

[17] A. Bittner, A. Ababneh, H. Seidel and U. Schmid, `Influence of the crystal orientation on the electrical properties of AlN thin films on LTCC substrates,_ Applied surface science, 2010, vol. 257, no. 3, pp. 1088-1091.

[18] A. S. Rogachev and A. S. Mukasyan, `Combustion of Heterogeneous Nanostructural Systems (Review),_ Combustion, Explosion and Shock Waves, 2010, vol. 46, pp. 243-266.

[19] S. Sen, M. Lake, N. Kroppen, P. Farber, J. Wilden and P. Schaaf, `Self-propagating exothermic reaction analysis in Ti/Al reactive films using experiments and computational fluid dynamics simulation,_ Applied Surface Science, 2017, vol. 396, pp. 1490-1498.

[20] D. P. Adams, `Reactive Multilayers Fabricated by Vapor Deposition (Review),_ Thin Solid Films, 2015, vol. 576, pp. 98-128.

[21] A. J. Gavens, D. Van Heerden, A. B. Mann, M. E. Reiss and T. P. Weihs, `Effect of intermixing on self-propagating exothermic reactions in Al/Ni nanolaminate foils,_ Journal of Applied Physics, 2000, vol. 87, no. 3, pp. 1255-1263.

[22] K. T. Rai P et al., `CFD analysis of exothermic reactions in Al-Au nano multi-layers foils,_ Materiali in tehnologije, 2011, vol. 45, no. 4, pp. 335-338.

[23] A. Schulz et al., `Characterization of Reactive Multilayer Systems deposited on LTCC featuring different surface morphologies,_ MikroSystemTechnik Congress 2021; Congress. VDE, 2021, pp. 1-5.

[24] A. Schulz, A. Ruh, S. Wiese and J. Müller, `Characterization of Reactive Multilayer Systems deposited on LTCC screen printing pastes featuring different surface morphologies suitable for reactive joining applications,_ Ceramic Interconnect and Ceramic Microsystems Technologies (CICMT) 2022; Congress, 2022.

[25] E. Wiss, A. Yuile, A. Schulz, J. Müller and S. Wiese, `Reactive Die Bonding on LTCC Substrates – Analysis by CFD Simulation,_ 24th International Conference on Thermal, Mechanical and Multi-Physics Simulation and Experiments in Microelectronics and Microsystems (EuroSimE). IEEE, 2023, pp. 1-5.

[26] A. Yuile, A. Schulz, E. Wiss, J. Maßler and S. Wiese, `The Simulated Effect of Adding Solder Layers on Reactive Multilayer Films Used for Joining Processes,` Applied Sciences, 2022, vol. 12, no. 5, p. 2397.

[27] A. Yuile, A. Schulz, J. Maßler and S. Wiese, `Analysis of selective bonding processes using reactive multi-layers for system integration on LTCC based SiPs,` Microsystem Technologies, 2022, vol. 28, no. 9, pp. 1995-2009.

[28] S. Liang, Y. Zhong, S. Robertson, A. Lio, Z. Zhou and C. Liu, `Investigation of Thermo-mechanical and Phase-change Behavior in the Sn/Cu Interconnects during Self-Propagating Exothermic Reaction Bonding,` 2020 IEEE 70th Electronic Components and Technology Conference (ECTC). IEEE, 2020, pp. 269-275.

[29] T. Namazu, K. Ohtani, K. Yoshiki, S. Inoue, `Crack propagation direction control for crack-less solder bonding using Ni/Al flash heating technique,` 2011 16th International Solid-State Sensors, Actuators and Microsystems Conference. IEEE, 2011, pp. 1368-1371.

[30] J. Wang et al., `Joining of stainless-steel specimens with nanostructured Al/Ni foils,` Journal of applied physics, 2004, vol. 95, no. 1, pp. 248-256.

LTCC-based Ceramic Substrates for Identification of Trustworthy Electronics

Uwe Krieger
RnD Department
VIA electronic GmbH
Hermsdorf, Germany
u.krieger@via-electronic.de

Annett Schroeter
RnD Department
VIA electronic GmbH
Hermsdorf, Germany
a.schroeter@via-electronic.de

Franz Bechtold
RnD Department
VIA electronic GmbH
Hermsdorf, Germany
bechtold@via-electronic.de

Gunter Hagen
KMS Technology Center GmbH (KMS)
Dresden, Germany
g.hagen@kms-automation.de

Adrian Goldberg
Microsystems. LTCC and HTCC
Fraunhofer Institute for Ceramic
Technologies and Systems (IKTS)
Dresden, Germany
adrian.goldberg@ikts.fraunhofer.de

Abstract — **The technology of low temperature cofired ceramics (LTCC) is already established in the production of reliable and robust electronic assemblies (e.g., multi-chip modules, system-in-package) as well as in safety-critical areas. With the incorporation of additional features based on surface profiles modular protection functions and barriers will be developed as a contribution to trustworthy electronics. Both obvious and hidden structures are investigated in a modular way, which in combination ensures a one-to-one tamper-proof identification. Keywords — trustworthy electronics, obvious and hidden structures, embossing, LTCC**

I. BACKGROUND

In an increasingly digitalised world, electronic counterfeiting and manipulation pose a great threat to the sovereignty of companies and business locations and the security of society, possibly accompanied by significant economic losses [1, 2]. In the research project VE-CeraTrust identification and modular protection functions and barriers will be developed and investigated as a contribution to trustworthy electronics. Low Temperature Cofired Ceramic (LTCC) multilayer technology is applied for special applications, e.g., which requires long term stability, harsh environment stability or active cooling. The embossing of LTCC-tape has been demonstrated already in several investigations [3,4]. Nevertheless, additional equipment or processing steps did not lead to commercial application.

II. MOTIVATION AND EXPERIMENTAL

A. LTCC Manufacturing

The scope within the project is to incorporate additional security and functionalities during the running production, without time or cost intensive processing. The tape casted unfired ceramic material is used to build up multilayer boards with integrated current lines, filled vias and passive components. In Fig. 1 the process flow of low temperature cofire ceramic (LTCC) manufacturing is shown. Punching equipment is widely used for the creation of vias which were finally filled by screen printing method. The so called green tape consists of alumina and glass powder mixed with organic components acting as binder and plasticiser agent [5].

Several paste systems on the basis of gold or silver for current lines are available. Furthermore milling (not shown in Fig. 1) can be applied to build up cavities after lamination. The

ceramic substrates are sintered typically within box furnaces at approx. 850°C peak temperature. In addition, layers on top may be screen printed and treated in postfiring processing followed by optical inspection, electrical testing and singulation of the modules.

In the last 25 years many designs and different applications for LTCC were realized by VIA electronic GmbH. The adaption of design, adjustment of processing parameters or the selection of material system are considered to realize the best performance and high quality in the final product and application.

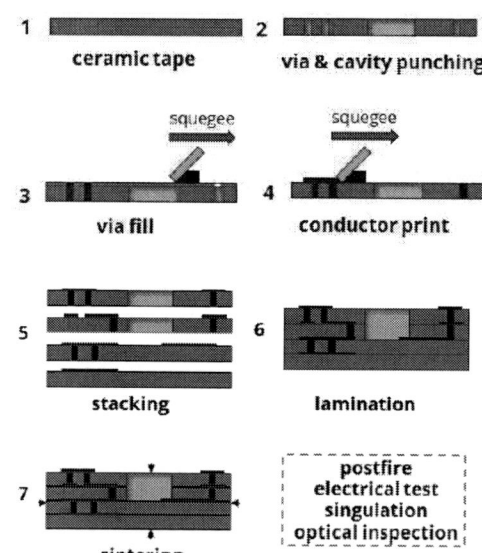

Fig. 1. Process flow of LTCC manufacturing: before or after step 3 or before step 7 embossing was applied (using punching equipment).

B. Embossing of Tape

In LTCC technology the green tapes with typical thicknesses from 50µm up to 250µm are punched to realize vias or cavities, this means through holes are created with pins typically in a range from 50µm up to several millimeters in

---

This work is supported by the German Federal Ministry of Education and Research (BMBF / 16ME0330K), www.via-electronic.de

diameter or opening. In our investigations we used a standard punching equipment from KMS to create a defined surface pattern by embossing the pin into the surface of laminated layer stack (two or four layers of laminated green tape). The LTCC material were used, namely Micromax™ GreenTape™ 951PX from company Celanese ("951") and green tape from KOA Group Japan ("KGS") with nominal thickness of 250μm and 160μm. The tapes were laminated with standard processing parameters to achieve a typical layered stack.

The embossing with the pins created a plastic deformation of the green tape in the shape of the pin itself (cylindrical bore with dead end). Pattern were realized by multiple embossings, e.g., lines by pitches lower then the pin diameter. Different pitches were applied to investigate the readability and improve the topology after sintering (e.g., 90% and 110% of the pin diameter). The pattern were related to data matrix codes with 16 x 16 „cells" and the name of the project „VE CeraTrust" (see Fig. 2). The size of the data matrix codes differs with the pin diameter (100μm, 80μm and 50μm) and by the applied pitch.

Fig. 2. Data matrix code used as reference for the embossing, "VE VeraTrust" (vertically flipped)

In addition, three different target values of embossing were investigated to evaluate the robustness of the green tape. Due to the fact that the depth of embossed cells differs within the data matrix code according to close proximity of cells (embossed / not embossed cells) a range of achieved depth will be reported. The target value of embossed depth were: 10μm / 40μm / >50μm.

The embossing of the tape was realized at three different processing steps in the LTCC process flow (see also Fig. 1):

1. Embossing after lamination → sintering

2. Screen printing of paste → embossing → sintering

3. Embossing → screen printing of paste → sintering

III. INSPECTION METHODS

Optical inspections were applied with standard equipments, which are available for production such as optical and digital microscopy (e. g. Keyence VHX-7000). The depth of the cells of embossed data matrix codes were measured with laser scanning microscopy at IKTS to verify the measurements achieved by optical inspections.

In addition, X-ray microscopy (Cheetah System from the Company Comet Yxlon GmbH) was applied by IKTS, Dresden. The background of this measurement was to evaluate and detect optically „unvisible" data matrix code pattern, which were covered with metal paste after embossing (embossing → screen printing of paste → sintering). The embossed cells of the data matrix code were filled with metal paste after the embossing.

The readability of the embossed data was checked with Keyence SR-X100 code reader, adapted with optical lens. The analysis was supported by AI software algorithms. However, lighting plays the important role and best performance was achieved with dark field illumination approach.

IV. RESULTS AND DISCUSSION

The typical shrinkage in lateral direction and related thickness of the substrates were detected after sintering.

The results of three different embossings within the process flow, as described in section II, are reported in the following.

A. Embossing of Tape followed by Sintering

In the first experiments embossings were performed with 100μm and 80μm pin diameter in GreenTape™ 951PX, see Fig. 3.

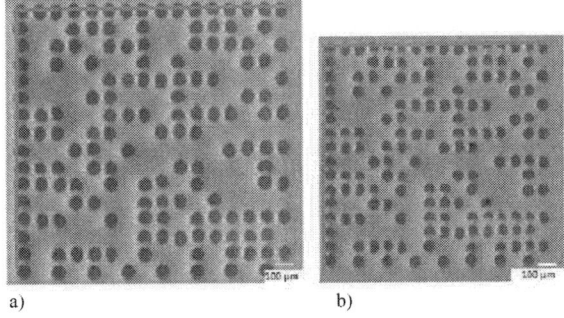

a)   b)

Fig. 3. Embossed data matrix codes in Micromax™ GreenTape™ 951PX after sintering, 110% pitch between the cells: a) 100μm pin diameter (dimensions≈1.4mm ˣ 1.4mm), b) 80μm pin diameter, (≈1.2mm ˣ 1.2mm)

In the finder pattern at the top and left side of the code (highlighted by green line) the reduced pitch results in slight deformation of the former cylindrical cells. Nevertheless the code was successfull readable with the described equipment and software.

The embossed depth measured with laser scanning microscopy and optical microscopy were in the range of 30μm to 45μm (see Fig. 4) after sintering.

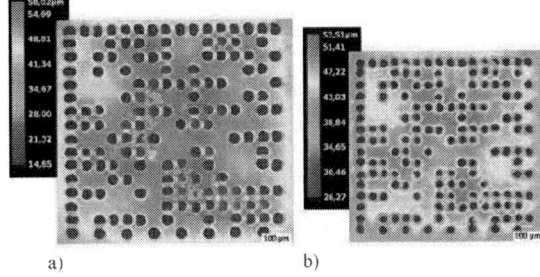

a)   b)

Fig. 4. 3D images of data matrix codes after sintering, a) 100μm pin diameter, b) 80μm pin diameter, samples corresponding to Fig. 3

With the variation of embossing depth the plastic deformation of the tape increases. In case of target values of embossing depth >>50μm the deformation and risk increase to destroy the green tape. This leads to particle generation and for some cells tape material is pressed into the neighbouring cells. Thus, the control and monitoring of the embossing depth will be of high interest in the upcoming investigations.

B. Screen printing of Paste followed by Embossing and Sintering

To increase the flexibility and readability of the data matrix code we investigated the embossing of metal paste on top of the green tape. The metal paste was screen printed and dried with standard conditions followed by the embossing of the data matrix codes. The achieved layer thicknesses of the metal paste were printed layer: ≈26μm, dried layer: ≈13μm, sintered layer: ≈8μm. Fig. 5 shows optical images without and with applied metal paste.

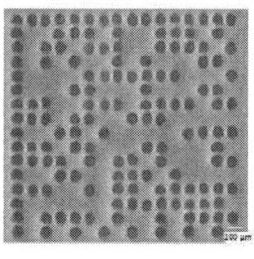 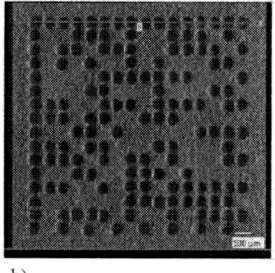

a) b)

Fig. 5.   Embossing of data matrix codes a) without and b) with metal paste as top layer, 100μm pin diameter, optical image after sintering

The embossing was successfully achieved on the layer stack as a combination of plastic deformation of paste and tape material in one processing step. The embossing depth was not significantly changed or reduced due to the metal paste. The readability was successfully approved for this type of embossing as well. As a conclusion the embossing of dried metal paste is feasable within the described parameters.

C. Embossing followed by Screen Printing of Metal Paste

The starting point of the investigations were embossed data matrix codes in green tape with lowest embossing depth of approx. 10μm. This type of data matrix codes were covered by screen printing with a homogenous metal paste layer and corresponding thickness of printed layer: ≈26μm, dried layer: ≈13μm, sintered layer: ≈8μm. The scope was to fully cover the matrix code and prevent any detection via optical inspection. After sintering the samples were investigated by X-ray microcopy by IKTS, Dresden - Fig. 6 b) shows typical images.

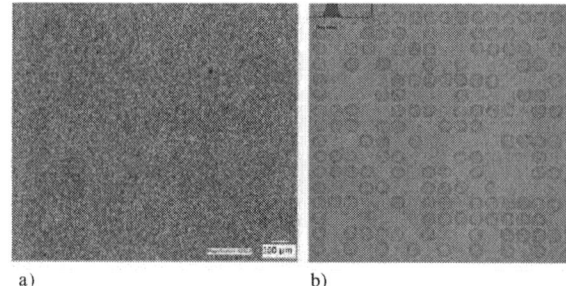

a) b)

Fig. 6.   a) optical image of data matrix code covered with metal paste after embossing, b) X-ray inspection image, 100μm pin diameter

Comparing the two images in Fig. 6 we conclude that the coverage was successful and hidden structures can be easily incorporated into the LTCC technology to increase truthworthy of electronics. These structures might be detected only by high efforts or intensive reverse engineering and may work as individual security features to track products.

V.   CONCLUSION

The embossing of green tape and the creation of readable data matrix codes within the standard process flow of LTCC technology was demonstrated. Standard punching equipment was used and lowest pin diameter of 50μm was applied. The embossing depth below 50μm results in stable processing. To increase the contrast and readability also embossing of dried metal paste layers on top was investigated.

The coverage of embossed data matrix codes with metal paste was successfully shown and hidden structures can be easily incorporated into the ceramic substrates, which might be detected only by high efforts or intensive reverse engineering and may work as individual security features for products and minimize the risk of counterfeiting.

ACKNOWLEDGMENT

The presented work was financially supported through the Federal Ministry of Education and Research under project VE-CeraTrust / 16ME0330K. The authors would like to thank all collaboration partners, especially Fraunhofer Institute for Ceramic Technologies and Systems (IKTS, Dresden) providing the measurements and KMS Technology Center GmbH for supporting the embossing. www.VE-CeraTrust.de

REFERENCES

[1]   VDMA press release, 01.06.2022, VDMA: Measures against product piracy in mechanical engineering pay off https://www.thenewsmarket.com/news/vdma--measures-against-product-piracy-in-mechanical-engineering-pay-off/s/c4f490e9-c2df-4403-8494-e011b9482550, access date 14.07.2023

[2]   Yasin, M.; Rajendran, J.; Sinanoglu, O.; Ismail, M. & Sawan, M.: Trustworthy Hardware Design: Combinational Logic Locking Techniques, Cham: Springer International Publishing, 2019

[3]   Bartsch, H.; Albrecht, A.; Hoffmann, M.; Müller, J.: "Microforming process for embossing of LTCC tapes", J. Micromech. Microeng., 2011, 22, 015004

[4]   Shan, X.; Soh, Y. C.; Shi, C.W.P.; Tay, C. K.; Chua, K. M.; Lu, C.W.: "Large-area patterning of multilayered green ceramic substrates using micro roller embossing", J. Micromech. Microeng., 2008, 18 065007

[5]   Imanaka, Y.: Multilayered Low Temperature Cofired Ceramics (LTCC) Technology, Springer, 2004

Concepts for realizing High-Voltage Power Modules by Embedding of SiC Semiconductors

Lars Böttcher
System Integration and Interconnect
Technologies
Fraunhofer IZM
Berlin, Germany
Lars.Boettcher@izm.fraumhofer.de

Andreas Ostmann
System Integration and Interconnect
Technologies
Fraunhofer IZM
Berlin, Germany
Andreas.Ostmannn@izm.fraumhofer.de

Thomas Löher
System Integration and Interconnect
Technologies
Fraunhofer IZM
Berlin, Germany
Thomas.Loeher@izm.fraumhofer.de

Manuel Seckel
System Integration and Interconnect
Technologies
Fraunhofer IZM
Berlin, Germany
Manuel.Seckel@izm.fraumhofer.de

Abstract— Today's power electronics modules typically consist of a ceramics substrate (DBC – Direct Bond Copper), carrying IGBTs, diodes or MOSFETs. These semiconductors are soldered or sintered to the ceramics and their top sides are interconnected by thick Al wires. An integration of further components or functions on the DBC substrate is difficult or even not possible. Therefore, driver circuits and controllers have to be mounted to a separate substrate, typically an organic Printed Circuit board (PCB). The PCB has to be connected to the DBC by wires or pins. The mechanical integration of the whole system requires a bulky housing.

Embedding of power semiconductors like SiC MOSFET allows a significant size reduction, improved electrical performance and a high degree of reliability of such modules.

In the last years, the capability of PCB embedding technology for the realization of low and high voltage power modules was demonstrated. Fraunhofer IZM together with partners from the industry demonstrated the feasibility for different applications, from single die SiC packages to high voltage automotive traction inverters.

In this paper, a general overview about innovative power module technologies will be given and the embedding of power semiconductors will be introduced in detail. It will address all most relevant point, like the demands on the semiconductor, the thermal and electrical considerations as well as the demands on the used materials

To address the integration of the required driver circuits and controllers, the idea of modularization such electronics systems will also be presented. Here already packaged components will be used and embedded into PCB layers too. As a result, a modular approach to form a complete system will be developed. Different functional layers, e.g. power switches, logic modules, will be formed and finally stacked und connected to form the system. The concept and first realized demonstrators will be discussed.

Keywords—Embedding, PCB technology, SiC MOSFET, Power module

I. INTRODUCTION TO INNOVATIVE POWER MODULE TECHNOLOGIES

The PCB based embedding of power semiconductors like SiC MOSFET or GaN provides several advantages. Beside the potential of miniaturization and the possibility of three-dimensional approaches, the improvement of the electrical performance is a mayor benefit. The replacement of bond wires to contact the semiconductor with the use of direct copper connections, result in a very short connection length and with this in a significant reduction of parasitic inductance. As a result, the switching behavior can be improved and switching losses are reduced. The addressed applications can be manifold: (1) Single die packages, what we call "Prepackages", only containing one power semiconductor to form a very robust and improved package solution, that could be handled like an SMD component, (2) power modules, half bridge or complete inverter structures, for electrically improved systems, or (3) 3D modular power electronics, which provides the possibility to combine a power module with needed logic and driver functionality in a compact and robust monolithic block.

Looking on this technology, several companies are offering embedding approaches for power modules, most of them out of the PCB world. Schweizer Electronic (DE) and AT&S (AT) do have a long experience with their p² Pack [1] or ECP technology [2] respectively. In addition, other PCB manufactures gain more interest in this technology topic, like Würth electronic [3] or Unimicron Germany [4]. The basic principle of all the different ways to realize embedded component packages or modules is quite similar for all, and will be described in the following in detail.

II. EMBEDDING TECHNOLOGY

A. Variations of Power Embedding

In order to address high-voltage applications, a suitable electric isolation needs to be added to the embedded structure. On the other hand, this isolation layer also needs to ensure a sufficient thermal management, since e.g. a single SiC can generate more than 100W losses.

The method that will fit best into a PCB production environment would be an organic isolation layer [Fig. 1]. Such a film could be easily implemented into the PCB process flow. The organic material needs to provide sufficient breakdown voltage for e.g. 1.200V applications and a good thermal conductivity. However, these organic layers are limited in their thermal conductivity. Best in class provide 8 to 10 W/mK according to the datasheets. With this limitation, maximum losses and power density of modules are limited.

Fig. 1. Schemactic of embedded structure with organic isolation layer

To address higher power densities the integration of ceramic substrates needs to be considered. These could be silicon nitride (Si3N4) AMB (active metal brazing), DBC (direct bond copper) or DPC (direct plated copper) substrates. These substrates are providing superior electrical isolation and thermal conductivity up to 90W/mK. Here the challenge is to combine the ceramic material with the PCB process sequence. In order to overcome this challenge, the ceramic substrate should be kept as small as possible and will be embedded into the PCB structure similar like the power component itself.

Fig. 2. Schemactic of embedded structure with ceramic isolation layer

In Fig. 2 a schematic example for the use of an embedded ceramic is shown. In this case, the ceramic substrate is embedded into a base PCB substrate and the SiC die is assembled to this substrate and is embedded using several layers of epoxy prepreg material. Warpage control of the resulting substrate after embedding is a critical point and needs to be balanced, and is mainly determined by the size and density of the ceramic in the substrate.

For both variations, the by the power component generated heat, must be removed effectively from the bottom side of the chip. That is why a sufficient heat spreading, e.g. by thick copper layer, is important to realize an effective transfer at the heat sink or cooler.

B. Semiconductor

Typical values of SiC MOSFETs used in recent projects are an area of around 25mm², thickness between 100µm and 400µm and maximum blocking voltage of 1.200V with currents up to 100A. In order to electrically and thermally design correct, the conductive losses needs to be known. These are determined by the on-state resistance (R_{DSon}). With modern generation SiC MOSFETs, R_{DSon} can be as low as around 12mΩ. More importantly is the temperature behavouir of R_{DSon}, since SiC will be operated with high junction temperatures of e.g. 175°C. At this operation state R_{DSon} can easily double or triple, which need to be considered for a proper thermal layout.

In order to be compatible with the embedding process, the components need to have a specified contact metallization. SiC power semiconductors have back and front contacts, typically the drain contact is the back-side, whereas gate and source are on the top side of the chip. The back contacts of most power semiconductors have by default a silver finish, which provides an optimal match with the Ag-sintering process, used for die attach. The top contacts (gate, source) of power semiconductors are typically delivered with a 1 to 3 µm

thick aluminum metallization. This metallization is not compatible with the micro via drilling process using UV or IR laser: The micro vias are cut through 5 µm copper (on top) and the 50 to 80 µm cured prepreg (glass fiber epoxy compound). During the laser drilling around 1 to 2 µm of the contact metallization can be removed by the laser ablation process. Additionally, cleaning processes, required for the copper metallization, can remove up to 1 µm of copper. Therefore the contact pad of the embedded component needs to provide a compatible metallization. Hence, a 10-12 µm thick electroplated copper metallization is applied, ideally in a wafer level process. This 10 to 12µm copper pad metallization provides a sufficiently thick pad for a reliable micro via drilling and metallization process.

C. Embedding materials

For the embedding process for high voltage applications, two classes of materials need to be considered: (1) The embedding material itself and (2) thermally higher conductive isolators enabling the thermal management.

As the embedding material epoxy prepreg or copper clad (CCL) materials are used. Prepreg is the PCB initial material. It consists of a glass cloth of different types, different thickness of the glass cloth and the glass fiber itself, which is impregnated with a certain resin system. This resin is still not fully cured, but in a partially cured stage (B-stage). In this stage, the resin is still able to melt and flow while applying temperature and pressure and with that, filling any gaps within the circuit board and around the power component, which has to be embedded. Certain properties are mandatory, which are the following:

- **High thermal stability:** Since SiC MOSFETs will be operated at high junction temperatures, the embedding material needs to be capable of operating at these temperatures. Standard FR4 materials do provide glass transition temperatures (T_g) between 120 and 170°C. At this temperature, the polymer system changes from a hard (glassy) state to a soft state. That is why this temperature should not be exceeded in operation.

- **Sufficient Breakdown voltage:** Currently 850V inverter modules are of high interest, typically 1.200V SiC MOSFET will be used for this application. With these given numbers, the used embedding material should provide breakdown voltages of at least 2 to 3 times of these values.

- **Coefficient of thermal expansion (CTE):** The CTE should be well balanced to the different materials, which will be present in the embedded structure. Thermo-mechanical modelling will help to determine a well-fitting material selection.

Typically used materials for the embedding of power semiconductors do provide T_g up to 280°C and higher, as well as CTE between 5 and 9ppm/K. A careful testing of breakdown voltages is of high importance, since data sheet values may indicate only limited numbers, which will not give sufficient information for the material in operation. Finally yet importantly, all selected materials will need to be suitable to be processed in standard PCB equipment.

As already described thermal management and electrical isolation can be realized by ceramic substrates or organic highly thermal conductive layers. Here again, the breakdown voltage and processability needs to be considered for their selection.

D. Process flow

The complete process flow for the embedding of power electronic devices like IGBT, MOSFET, SiC transistors or others is based on printed circuit board technology in combination with a suitable die attach method. Also, the material properties of the used FR4 materials are of paramount importance. Like already described, beside suitable thermo-mechanical characteristics, they need to provide the ability to withstand higher operational temperatures, because of the losses and resulting temperatures of the embedded power semiconductors.

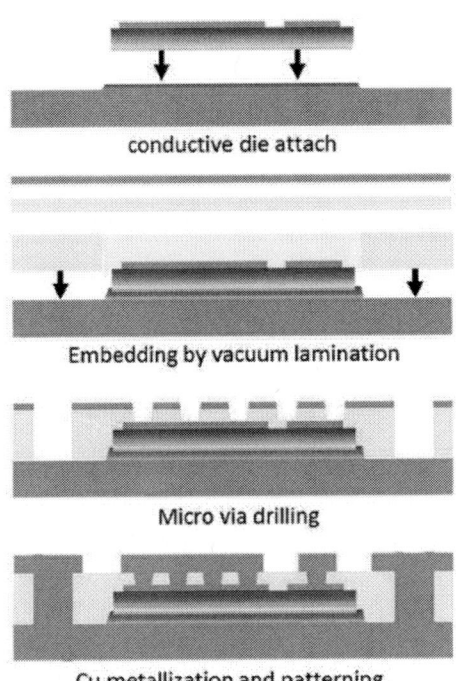

conductive die attach

Embedding by vacuum lamination

Micro via drilling

Cu metallization and patterning

Fig. 3. PCB based embedding of power devices

The general process flow is shown in Fig. 3. On a suitable substrate the power components are assembled using a conductive die attach. This substrate could be a simple copper foil or leadframe, a high current PCB or even a base substrate with integrated electrical isolation like a ceramic substrate. The preferred method for mounting the semiconductors would be a low pressure and low temperature Ag sintering process. This method provides some advantages: First, it is capable to assemble the dies on a large panel format, enabling a highly parallel processing. In addition, the die bond accuracy, which is important for the following process steps, mainly depends on the accuracy of the die bond process, since, other than for a soldering process, the dies stay most accurate at their position during the sintering process. Additionally the Ag sinter layer provides an excellent thermal ($\geq$100W/mK) and electrical contact ($\leq$0,008mΩcm) to the substrate. If applicable, an Ag sinter glue could be used also. This would be possible if electrical and thermal demands are fitting with the characteristics of these glues. The main benefit of these glue, which consist of a metal-organic Ag system and a resin system for mechanical reinforcement, is, that they do not need

pressure to be processed, and sinter/curing temperatures are not exceeding 200°C.

The embedding itself is done by vacuum lamination of FR4 prepreg (glass fiber with B-stage epoxy resin) layers. Structured prepreg layers are used to compensate the height of die attach and die, and full layers provide the required isolation to the electrical layer above the die. Additionally a copper foil is applied which is needed for the metallization process. The embedding itself takes place by lamination of the stack in a standard multi-layer lamination press.

Laser via drilling is used to create contacts to the embedded die, which needs to be controlled and parameterized carefully, in order not to damage the chip. Blind via to the thick copper substrate or thicker copper layer are typically made by mechanical drilling. Subsequently the micro via are filled in a copper plating process, followed by the structuring of the electrical layer by lithography and etching of the copper.

The appearance of the printed circuit board with embedded components is the same as a conventional circuit board. Hence, in subsequent processes either additional signal layers can be processed, solder mask and surface finish can be applied, in order to mount further components on top of the embedded module, or additional heat spreaders or embedded modules of different types can be added onto the module [5].

III. APPLICATIONS

A. Single Die Packages – Prepackages

The most simple embedded power semiconductor structure is a package, which contains only one die. Here, the target is to create a robust housing for further production steps to form e.g. a half bridge structure, which uses the advantages of the embedding technology, nearly adding no thermal impedance and no inductance. Such packages are already on the market for some years, and gained more and more interest within project for module applications over the last time. One widely known example is GaNSystems GaN prepackage, Fig. 4. These packages are available in 100V and 650V classes and are manufactured in PCB embedding technology.

Fig. 4. GanSystems embedded HEMT

At Fraunhofer IZM different types of SiC prepackages were developed (Fig. 5). One example for modularization of these packages is given in [6]. These prepackages are manufactured using the described process flow.

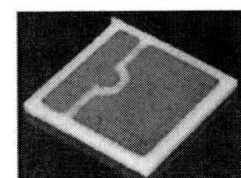

Fig. 5. SiC prepackages

15

On a thin copper foil the SiC MOSFET was assembled using a silver sinter die attach. A simple and robust package was formed by means of embedding technology, carrying gate and source contact on top and drain contact on the bottom side.

For the development of new power electronics concepts, prepackages with their own safety isolation become of interest. Therefore, a ceramic substrate will be integrated into the embedded like schematically shown in Fig. 6.

Fig. 6. Schematic of embedded (SiC) prepackage with integarted safety isolation

The process sequence will remain the same, with the difference, that in addition to the semiconductor, the ceramic substrate will be assembled to the base copper. As the ceramic substrate, a DPC with a 320μm Si₃N₄ ceramic and 100μm copper on each side is used in this example. In Fig. 7, a cross section of the resulting package is shown. These prepackages are designed for 850V applications using 1.200V SiC MOSFET. In the following modularization these prepackages can be handled like a SMD component.

Fig. 7. Prepackage with embedded SiC MOSFET and Ceramic isolation

B. SiC Inverter with organic isolation

An early project for the realization of a SiC inverter with up to 100kW switching power targeted the use of an organic isolation layer. Instead of a ceramic substrate, a polymeric material with a thermal conductivity of around 3W/mK was used. The schematic of the module is shown in Fig. 8

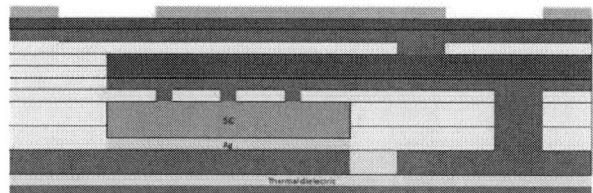

Fig. 8. Schematic SiC inverter with organic isolation

Within this module two times four SiC MOSFET (HS/LS) per phase were embedded. The switching current for each semiconductor is 50A. The resulting module combines the improved electrical characteristics, short electrical connections resulting in low inductance and lower switching losses, with the possibility to mount additional components, primary DC link capacitors, shunts, gate driver and connectors, directly and close to the switching cell. The final half bridge module is shown in Fig. 9.

Fig. 9. SiC halfbridge module with assembled components

C. SiC Inverter module with driver and logic integration

In the German funded project „SiCModul", a new concept for power module fabrication was developed. The idea was to integrate safety insolation by a ceramic substrate and to manufacture two separate layers, power core and logic board, and join them together to finalize the module.

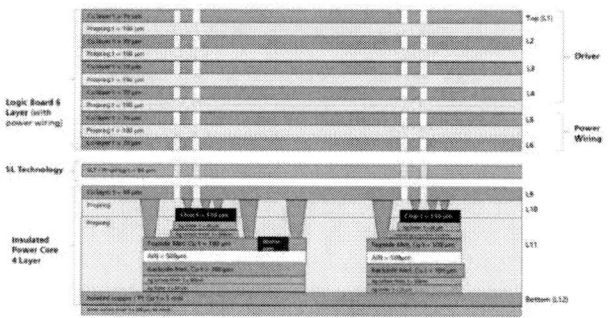

Fig. 10. Concept of SiC inverter with driver and logic integration

These power modules are fabricated based on the layer structure shown in Fig. 10. The fabrication was divided into two parts: the logic PCB and the power PCB with the embedded SiC MOSFET. The logic PCB consists of a 6-layer structure, with through holes, micro via and thick copper of 70μm in each layer. On the bottom side, partial silver plating was performed in the areas where the joining by sinter lamination technology (SLT) between logic and power PCB is made later.

The idea of sinter lamination technology (SLT) is to simultaneously produce an electrical (or thermal) contact, and to fill the cavities between the contacts with resin of the prepreg (PCB starting material) in one process step. In this process, a silver sintering paste is applied to one of the respective joining partners, usually by stencil or screen printing. To fill the cavities, prepreg layers are produced which contain openings at the positions where Ag is to be sintered. The entire stack of joining partners with Ag sinter paste and prepregs is then placed oriented to each other. The joining process takes place in a multilayer laminating press. In the first phase, Ag sintering takes place, as well as flowing and thus filling of the cavities. In the second phase, the epoxy resin of the prepregs is completely thermally cross-linked. After this lamination process, the joining of the individual layers is completed.

For the power PCB, 1mm thick copper is used as substrate. A DPC ceramic and the SiC MOSFET are mounted on this by means of Ag sintering and subsequently embedded and electrically wired using PCB embedding technology. The assembly is carried out in two steps. The DPC ceramic is mounted on the prepared 1mm copper substrate. This is done by stencil printing of the Ag sintering paste, placement of the DPC using automatic placement machines and subsequent sintering process (Fig. 11a). In the second step, the Ag sintering paste is applied to the sintered DPC using stencil printing, and the SiC MOSFET is assembled and then sintered (Fig. 11b, c).

a)   b)   c)

Fig. 11. Ag Sinter sequence DPC and SiC on 1mm copper

After that, the assembled substrate is ready for the embedding process. DPC and SiC MOSFET embedding is done by placing patterned prepreg layers containing the outline of the DPC and chip, and a full area prepreg layer for insulation and copper foil. This is followed by lamination in a PCB multilayer lamination press. Laser micro vias are fabricated to make the electrical contacts to the embedded SiC MOSFET, which are subsequently copper metallized. The copper layer is patterned subtractive by the fabrication of a photoresist mask and subsequent wet chemical etching. Finally, partial deposition of immersion silver (imAg) is performed on the areas where the Ag sintered connections to the logic PCB are made. After this, the power PCB is also available for joining with the logic PCB.

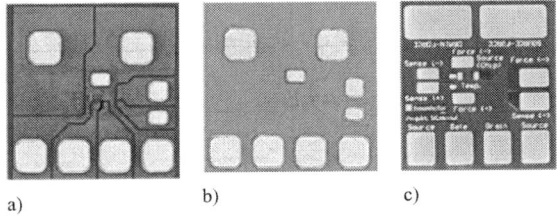

a)   b)   c)

Fig. 12. Joining of Power coer and logic board by SLT

The joining of the two individual layers is done as shown pictorially in Fig. 12. First, the power PCB is placed on a press plate with pins at defined positions, which serve to align the individual layers to each other (Fig. 12a). Then a filler prepreg with cutouts in the areas to be sintered is applied (Fig. 12b). This serves to fill the cavities and to bond the two PCBs. Finally, the logic PCB (Fig. 12c), additional separation, and pressing materials are applied. The entire press stack is then loaded into the multilayer lamination press and pressed under temperature and pressure control.

The resulting module is shown in the cross sectional view in Fig. 13

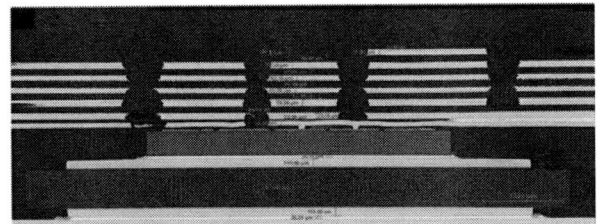

Fig. 13. Cross section of power module

D. GaN HEMT Modular Driver and Logic Integration

A new concept for the driver and logic integration into a compact power module was addressed in the German project "3D Leistungselektronik"(3D power electronics). The basic idea is to divide the complete module into different functional layers and finally join them to create a kind of monolithic block. The functional layers should be (1) the motherboard, (2) the driver module and (3) the power module, all together mounted to a heatsink. The schematic of the concept is illustrated in Fig. 14.

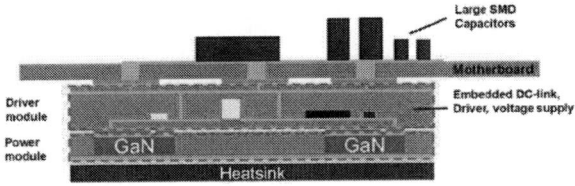

Fig. 14. Schematic of the 3D power system

For the driver module, also an embedding technology is used. Instead of bare dies already packaged SMD components, active and passive, are used. On a multi-layer core, SMD components are assembled by soldering (Fig. 15), and in the following embedded using epoxy prepreg layers. For the power modules, GaN prepackages are used. The electrical isolation is realized by a DPC ceramic. Prepackage and DPC are assembled using Ag silver sintering (Fig. 16), and are embedded by the use of prepregs in PCB embedding technology.

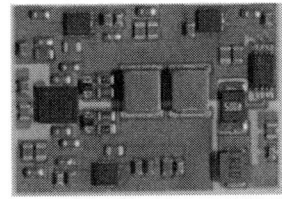

Fig. 15. SMD assembly for logic module

Fig. 16. GaN prepackages assembled to DPC and substrate

In Fig. 17 a cross section of the logic module is shown, with the different SMD parts visible.

Fig. 17. Cross section of embedded logic module

To join the driver and logic module, sinter-lamination technology (SLT) is used once again. In a similar way as already described, the different layers will be stacked and laminated in a multi-layer lamination press. Fig. 18 illustrates a cross section through the complete stack, showing the GaN prepackages and the highest SMD component in the driver module.

Fig. 18. Cross section of the Logic-Power module

The resulting Logic-Power module, with x/y dimension of only 3.5cm by 2.5cm and 4.5mm thickness, provides a monolithic and highly miniaturized component (Fig. 19).

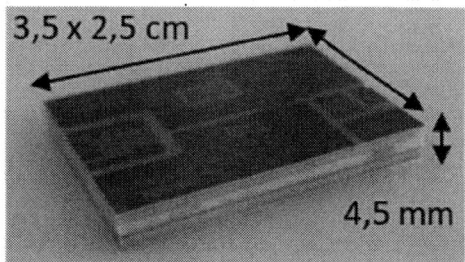

Fig. 19. Final Logic-Power module

As the final step, the complete system is assembled by soldering the Logic-Power module face down on the backside of the motherboard (Fig. 20). The backside of the power module provides the interface for the heatsink assembly.

Fig. 20. Logic-Power modules soldered to bottom side of motherboard

On the top side of the motherboard, the larger SMD components, which cannot be embedded, are assembled as well as control components and connectors. The resulting complete three-phase power module is shown in Fig. 21.

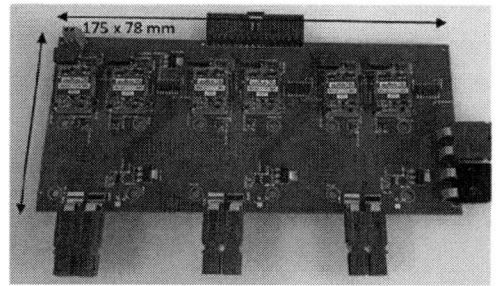

Fig. 21. Complete module with all top side motherboard assemblies

With the intended concept for a new kind of power modules, it was possible to demonstrate a way to realize a compact and highly miniaturized module in a robust monolithic block, carried on a motherboard. With the integration of the electrical isolation within this module, a best possible thermal path could be realized. In addition, a very robust and highly reliable construction was addressed. To join the different functional layers the novel sinter-lamination technology was applied. It is intended to develop this concept further to provide a platform for scalable configurations. Although GaN HEMTs were used in this project, the concept can equally be applied to SiC MOSFET applications.

IV. SUMMERY

Embedding of power electronic components provides a large benefit in terms of reliability, volume reduction and electrical performance. Direct copper connection and the resulting short connection length lead to a significant reduction of parasitic inductance, resulting in improved switching behavior and switching losses. Proper thermal and electrical design is a key enabler for such high performance power modules.

Meanwhile, there are many different ways to realize innovative power semiconductor packages and modules, based on PCB embedding technologies, developed. Which technology fits best often is defined by the demands of the application.

A new topic is 3D modular power electronics. This technology approach can provide compact embedded power electronics systems, and can be realized by use of established processes. It is also intended to provide a scalable platform for different configurations.

V. ACKNOWLEDGMENT

The authors would like to thank for the financial support within:

Project SiCModul, FKZ 16EMO0258
Project 3D Leistungselektronik, IGF-20183 BG - FE 1

VI. REFERENCES

[1] https://schweizer.ag/en/technologies-solutions/pcb-technologies/semiconductor-embedding-systems/p2-pack

[2] https://ats.net/en/technologies/ecp-technology/

[3] https://www.we-online.com/en/products/printed-circuit-boards/embedding#i4339

[4] https://unimicron-germany.de/news-leaser-en/embedding-technology.html

[5] L. Böttcher, et al.: 3D Modular Power Electronic Systems, based on Embedded Components, Device Packaging Conference, March 5-8, 2018, Fountain Hills, AZ, USA.

[6] C. Marczok, et al.: Low inductive SiC Mold Module with direct cooling, PCIM Europe 2019, Nuremberg, Germany

Ceramic embedding of SiC-semiconductor using cofiring technology

Steffen Ziesche
Microsystems, LTCC and HTCC
Fraunhofer IKTS
Dresden, Germany
steffen.ziesche@ikts.fraunhofer.de

Kathrin Reinhardt
Thick-Film Technology and Functional
Printing
Fraunhofer IKTS
Dresden, Germany
kathrin.reinhard@ikts.fraunhofer.de

Jobin Varghese
Microsystems, LTCC and HTCC
Fraunhofer IKTS
Dresden, Germany
jobin.varghese@ikts.fraunhofer.de

Birgit Manhica
Microsystems, LTCC and HTCC
Fraunhofer IKTS
Dresden, Germany
birgit.manhica@ikts.fraunhofer.de

Andreas Schletz
Hybrid integration
Fraunhofer IISB
Erlangen, Germany
andreas.schletz@iisb.fraunhofer.de

Abstract - The presented work demonstrates our current state in embedding of SiC power electronic dies using Ceramic Multilayer Technology. The novelty of the approach presented includes the cofiring of a semiconductor element with ceramic base material, which demands for a reduced sintering temperature of ceramics (using LTCC or ULTCC based ceramics) in combination with an increased temperature stability of the semiconductor.

Keywords—power electronics, embedding, SiC, LTCC, ULTCC, Multilayer Ceramics

I. INTRODUCTION

SiC-based semiconductors are currently used in the form of diodes and transistors. Within the application power electronics, the devices are used as field effect transistors (MOSFET) or Schottky-Diodes. Conventional assembly and connection technology involves soldering or metal sintering the semiconductor to a metalized ceramic substrate, with electrical contacting on the top using wire bonding technology. For encapsulation, the semiconductor can be injected with a plastic. More innovative approaches integrate the semiconductor into a polymer-based circuit board. Regardless of the type of semiconductor, the current assembly and connection technology (housing, potting compound and connection contacts) can only be operated up to temperatures of 200 °C for a short time and up to 150 °C for a long time. In addition, the housing materials currently used have poor corrosion resistance and mismatched thermal linear expansion. The proposed solution involves developing and establishing a high temperature-stable assembly without principal temperature limit and connection technology for SiC. For this purpose, the components are embedded in a multilayer ceramic package under ceramic-specific firing temperatures, sintered and at the same time electrically and thermally contacted.

II. STATE OF THE ART

A. Ceramic Multilayer Technology

Ceramic Multilayer Technology is an established method for the manufacturing of ceramic microcircuits [1]. The process (Fig. 1) uses ceramic green tapes, manufactured by tape casting, and functional pastes as semi-finished products. During a blanking step the ceramic tapes are cut into defined sizes (up to 8x8 sq inch) and geometrically structured by punching (creation of via holes) or lasing. The via holes will afterwards be filled by stencil printing of a metallization paste, followed by screen printing of a predefined conductor layout. The individually structured and functionalized tapes will be stacked, laminated and sintered. During the sintering process the laminated stack transforms into a highly robust monolithic ceramic body, while the functional structures can be buried into the ceramic. The process allows not only for the integration of conducting structures, but also for the embedding of other passive functionalities like resistors, capacitors or inductors.

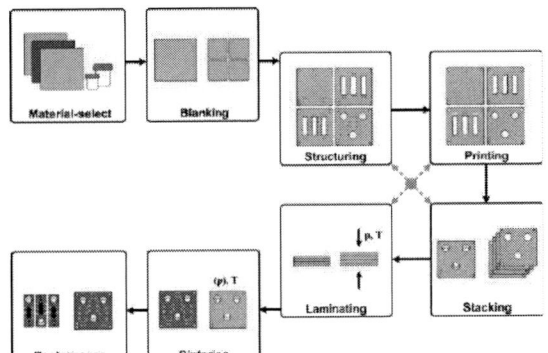

Fig. 1. Multilayer Ceramic Manufacturing Process

B. Semiconductor design and selection

SiC semiconductors are used in power electronics mainly for switching high electrical currents (up to several 100 A) and high voltages (up to several kV) in drives and energy converters. Because of its larger band gap, SiC has advantages compared to state of the art Si: higher junction temperatures ($Tj \geqslant 250$ ° C, no practical temperature limit by intrinsic carrier density), higher power density and higher switching speed and switching frequency with lower switching losses. Due to the high thermal stability of the SiC, the power and power loss density can be increased during operation, resulting in significantly higher operating temperatures (T > 400 °C) of the chip. The SiC material can endure higher operation temperatures, the first limiting factor is the surface metallization of the SiC chip, which has to withstand these high temperatures. As part of the semiconductor development various metals or metal combinations were examined as suitable high-temperature process-capable surface platings. SiC chips with selected layer stacks for the power metallization were chosen for the embedding investigations.

III. LTCC EMBEDDING TECHNOLOGY

The initial approach of our work was the integration of a SiC semiconductor device in a pre-metallized and pre-structured LTCC green tape stack using commercially available LTCC tapes. After cofiring at the sintering temperature of the ceramic, a monolithic ceramic pre-package with embedded SiC device is generated (Fig. 2).

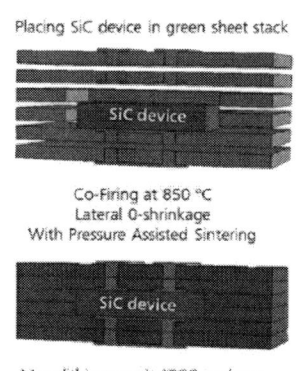

Fig. 2. Illustration of the approach: Embedding a SiC semiconductor device in a pre-metallized LTCC multilayer during cofiring using pressure assisted sintering

That approach requires a precise control of the shrinkage during cofiring. Especially, a lateral shrinkage of ≤ 1 % is important for a selective interconnection between the ceramic package and SiC device [2, 3]. Thus a pressure assisted sintering process [4,5] is applied to constrain the shrinkage in the x/y-direction.

The investigation [6] was performed with three commercial LTCC materials: GT 951 (DuPont©), L8 (Ferro©) and 9k7 (DuPont©). These materials have a recommended firing temperature of 850 °C. Further we selected two commercially available Ag via-fill pastes and two commercially available Ag conductor pastes for screen-printing. The embedding experiments were executed with SiC dummy dies with 2 and 5 mm edge length showing ohmic behavior. They were fabricated from a n-doped SiC wafer with a dopant concentration of $\geq 10^{18}$ cm^3. Top and bottom surface of the SiC dies were covered with a processed ohmic contact Ni$_2$Si layer and a power metallization layer. For the investigation of the embedding process we developed a 2 inch LTCC layout which enables the integration of four equal SiC chips in one process run. After laminating the green substrate with four integrated SiC chips they were co-fired in a box furnace (ATV PEO 603 - Sinterpresse, company ATV Technologie GmbH) under applied sinter pressure p_s. After the firing process the four segments of each substrate were separated by wafer dicing and in addition two segments were cut in the middle of the embedded chip within the same process to

enable the documentation of the cross-section. A cross-section of SiC dummy chips with 2 mm edge length embedded in a 2 inch LTCC substrate at specified parameter combinations is illustrated and rated in Table I.

TABLE I. Embedding results for SiC dummy chips with 2 mm edge length at specified parameter combinations

material	GT 951
SiC chip size	2 mm square
parameter: HR; T$_{max}$; t$_d$; p$_s$	2 K/min; 850 °C; 60 min; 40 kPa
cross-section	
rating	feasible

For the mechanical/electrical embedding investigations we expanded the layout and the technology of the 2-inch LTCC sample preparation by the mechanical embedding of inner and outer contact pads as well as via structures to interconnect four SiC dies with 5 mm edge length inside the LTCC pre-package during sintering. For the embedding process we chose the material 9K7 and we reduced the firing temperature to 750°C at a heating rate of 5 K/min, a sintering pressure of 40 kPa and a dwell time of 2h to treat the SiC device with care. Adapted conductor paths connected the vias with the inner contact pads of the SiC device. We fabricated pre-packages with electrical interconnects between top and bottom side (Fig. 3) without any obvious defects.

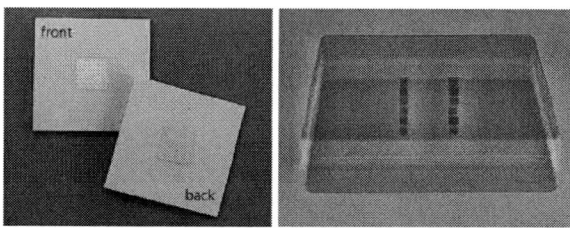

Fig. 3. Single pre-packages fabricated in LTCC 9K7 with embedded SiC dummy chips having 5mm edge length (component left, X-Ray image right)

Additionally, we evaluated the electrical interconnections of LTCC pre-packages with integrated SiC chips of different power metallizations (Table II) by measuring the resistance between top and bottom side of the package [7], [8]. For best package performance the resistance should be similar to the resistance of the ohmic dummy SiC dies (< 1 Ω).

TABLE II. Characteristics of the fabricated LTCC Pre-packages

	Fabricated pre-packages		
	Batch 1	Batch 2	Batch 3
Die generation	Power metallization A)	Power metallization B)	Power metallization B)
Resistance R (Ω), Package + SiC die	3.3 to 8.3	1.0 to 1.8	1.7 to 2.6

In-situ measurements of the resistance during heating exhibit the onset of a degradation at 600 °C for the investigated power metallizations. At all, these effects are less distinctive at power metallization (B) than at (A) due to the implementation of an extra diffusion barrier.

Besides further optimization of the power metallization regarding higher temperature stabilities new materials with lower sintering temperatures had to be developed und qualified for the embedding process.

IV. ULTCC EMBEDDING TECHNOLOGY

ULTCC (Ultra Low Temperature Cofired Ceramics) are a new class of multilayer ceramic materials made of glass-ceramic composites. They can be sintered at very low temperatures (400 °C to 700 °C), making their manufacturing process very energy efficient. These materials seem to be an elegant choice to reduce the temperature load of the semiconductor chips, while maintaining the design freedom of the Multilayer Ceramic manufacturing process.

For ULTCC development, appropriate materials and mixing approaches were derived. The approach concentrated on glass-ceramic composites. These are characterized by a mixture of inert ceramic powders (for example Al_2O_3 or SiO_2) and powder of a low-melting glass. The glass-ceramic composites are densified by non-reactive liquid phase sintering, whereby the glass melts, partially wets the ceramic particles and causes particle rearrangement and pore closure. This is majorly attributed to capillary forces and/or viscous flow. Corresponding material compositions were in a first step analyzed in the form of compacts with focus on density, sintering temperature and microstructure. Table III lists the relevant ULTCC compositions studied and compares the calculated density and the calculated coefficient of thermal expansion (TEC) corresponding to the blend ratios set.

TABLE III. ULTCC compositions under investigation

Description	Volume ratio filler [%]	Volume ratio glass [%]	Calc. Density [g/cm³]	Calc. TEC [ppm/K]
ULTCC65	65 %	35 %	3,7	3,1
ULTCC60	60 %	40 %	3,9	3,4
ULTCC55	55 %	45 %	4,0	3,7
ULTCC40	40 %	60 %	4,6	4,5

For the analysis of the sintering temperatures and the microstructure, compacts (Ø10 mm x approx. 2 mm) were produced from these powder mixtures with a pressure of 25 MPa and fired at 20 K/min at temperatures from 500 °C to 900 °C (holding time 30 min each). Subsequently, the density of the sintered bodies and the microstructure was analyzed. The plot of the relative density as a function of the firing temperature (Fig. 4) shows, that the temperature of maximum densification increases with increasing ceramic filler content. With a relative density of 92 % at 550 °C firing temperature, the composition ULTCC40 showed high potential for low temperature sintering while adjusting a microstructure without open porosity.

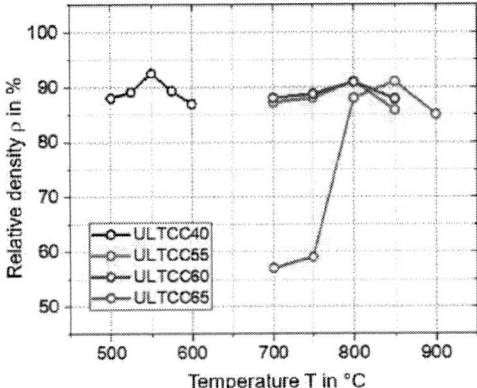

Fig. 4. Determination of the sintering temperature of ULTCC compositions with varying glass fractions

Based on an adapted composition named mULTCC40, the tape casting process for the ULTCC material was developed. For this purpose, a castable slurry was prepared by adding solvents, binders and plasticizers. In addition to setting the viscosity for the slurry, the challenge in this step was to select binder materials which can already be completely thermally extracted (debinding) below the sintering start of the ULTCC material. Subsequently, the slurry was deposited on a carrier film using a film casting system, dried and qualified. In a total of five iterative development stages, the slurry composition and film casting parameters were varied for realizing satisfactory film qualities. Fig. 5. visualizes important process steps during this development.

Fig. 5. Process steps in the development of tape casting for ULTCC materials

For testing the substrate manufacturing the green tapes were cut, punched, laminated and subsequently fired. Fig. 6 shows a laminate consisting of 4 layers with integrated 0,2 mm punched alignment marks for shrinkage evaluation.

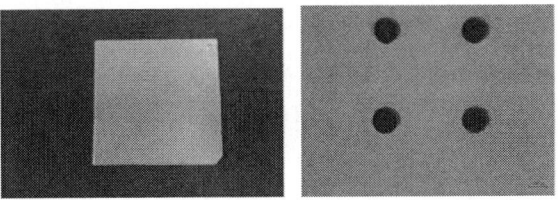

Fig. 6. ULTCC-based laminate (2"x2" size) with alignment marks for shrinkage measurement (left); punched 0,2 mm via holes (right)

Subsequently the material was sintered in an unconstrained sintering process (without pressure assistance). Table IV shows the determined density, calculated porosity and the measured water absorption of the material sintered at 480 °C/30min.

TABLE IV. Characterization of mULTCC40 after unconstrained sintering

Sample	Sintering profile	Density in g/cm³	Porosity in %	Water absorption in %
mULTCC40	480°C/ 30 min	5,18	< 1%	0,02

In Fig. 7 the microstructure of the sintered material is presented. The low magnification represents the entire substrate with the magnified pictures are focusing on the center and the top region of the substrate. The results show that the sample is well compacted in agreement with the density and water absorption measurements. The residual glass phase in combination with the filler materials can be clearly identified. Besides the initial filler materials (Al_2O_3 and SiO_2) additional secondary phases have formed. They improve the mechanical stability of the ceramic and provide stability during further firing steps.

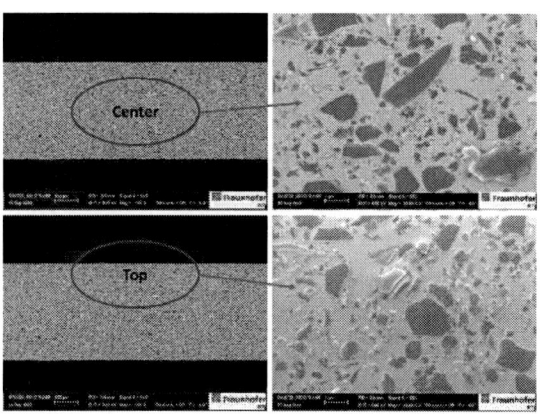

Fig. 7. Microstructure of mULTCC40 sintered at 480°C/30 min

Due to the need for a precise shrinkage control in the chip embedding step (lateral shrinkage of ≤ 1 %) the ULTCC material must be fired in a pressure sintering process (compare to section III). Thus, an additional shrinkage investigation of mULTCC40 material was performed by a thermal mechanical analysis using the vertical dilatometer TMA 402 F1/F3 Hyperion (Netzsch-Gerätebau GmbH). Cylindrical samples (Ø5 mm x approx. 2 mm) were laser cut out of a ULTCC laminate. During the experiments a uniaxial pressure of 40 kPa was applied to the samples, while the axial shrinkage (ε_z), which corresponds directly to the measured strain, was determined. Fig. 8 shows the results of the TMA measurement.

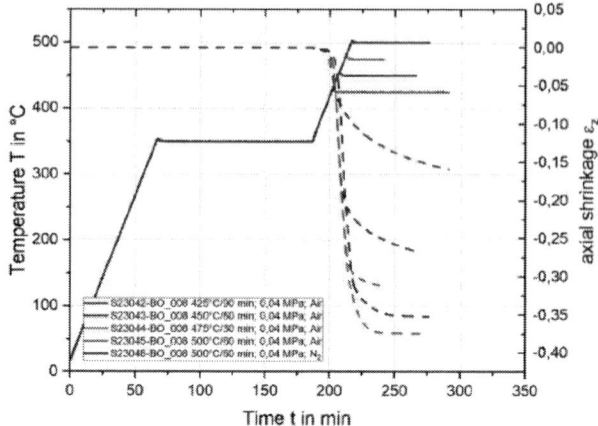

Fig. 8. TMA analysis of mULTCC40 with variation of sintering profile and atmosphere

After a debinding phase at 350°C the ceramic starts to shrink at 400°C until the sintering ends at approx. 500°C. For a dense sintered state an axial shrinkage of about 35% was determined, which must be considered in the design of the future package. The shrinkage behavior under nitrogen atmosphere is comparable to the measurement in air, which is favorable for the stability of the SiC chip power metallization.

Based on these results first mechanical integration experiments for embedding SiC dummy chips in ULTCC material were conducted. SiC dies were embedded in 5 Layer ULTCC laminates, isostatically laminated at 5 MPa and sintered in air (Fig. 9).

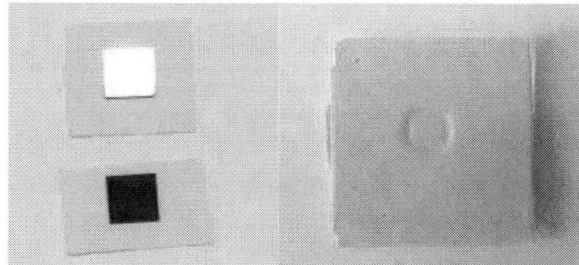

Fig. 9. ULTCC packaging of SiC dummy chips (left: before lamination step, right: after cofiring)

The first results show, that crack and delamination free embedding of SiC dummy chips is possible within ULTCC base material. Nevertheless, further studies related to appropriate shrinkage control and implementation of metallization structures must be done.

V. CONCLUSION AND OUTLOOK

In this contribution we introduced a new approach to embed SiC semiconductor devices in full ceramic pre-packages to provide higher operation temperatures and better switching properties for power electronic applications. Initial investigations showed that the embedding of SiC power chips can be successfully realized in LTCC manufacturing technology using a pressure assisted sintering process. A major drawback of LTCC materials are the high sintering temperatures (> 700°C) leading to a significant temperature load on the power metallization of the SiC chips. This favors degradation and limits their electrical performance. Besides optimizing the temperature stability of the metallization, the reduction of the ceramic sintering temperature is a decisive step to solve these problems. Thus, a new type of low temperature sintering ceramics (ULTCC) was investigated. ULTCC materials were identified, synthesized, transferred into ceramic green tapes and characterized regarding processability, sintering behavior and microstructural properties. The investigation shows that a sintering of these materials below 500°C is possible, combined with the achievement of an attractive pore free microstructures. First mechanical embedding tests were successfully completed. However, the future work on this topic has to deal with additional challenges as follows:

- Further characterization of the physical properties of selected ULTCC materials,
- Development of new metallizations pastes with suitable constrained shrinkage behavior adapted to ULTCC materials,
- Optimization of sintering parameters to reduce free spaces on the sides walls of the embedded SiC dies using sinter simulations and
- Verification of the approach with real SiC semiconductor devices (transistor) regarding interconnection, isolation and switching behavior at temperatures > 300 °C.

ACKNOWLEDGMENT

This work was supported by the market-oriented strategic pre-competitive research (MAVO) from the Fraunhofer-Gesellschaft through grant "MESiC".

REFERENCES

[1] Y. Imanaka, "Multilayered Low Temperature Cofired Ceramics (LTCC) Technology, 2005 Springer Science + Business Media, Inc. ISBN: 0-387-23130-7

[2] M. Sobocinski, L. Khajavizadeh, M. Andersson, A. L. Spetz, J. Juuti, H. Jantunen, "Performance of LTCC embedded SiC gas sensors," in Procedia Engineering, Vol. 20, 2015, pp. 253-256.

[3] S. Gebhardt, D. Ernst, B. Bramlage, M. Flössel, A. Schönecker, "Integrated Piezoelectrics for Smart Microsystems - a Teamwork of Substrate and Piezo," in Advances in Science and Technology, Vol. 77, 2013, pp. 1-10.

[4] A. Mohanram, S.-H. Lee, G. L. Messing, D. J. Green, "Constrained Sintering of Low-Temperature Co-Fired Ceramics," in J. of the American Ceramic Society, Bd. 89, Nr. 6, Juni 2006, pp. 1923-1929.

[5] B. Brandt, H. Naghib-Zadeh, T. Rabe, "Improved Co-Firing and Dielectric Tape Based on Master Sintering Curve Predictions and Shrinkage Mismatch Calculations," in J. of the American Ceramic Society, Vol. 96(3), March 2013, pp. 726-730.

[6] C. Lenz, S. Ziesche, A. Schletz, H. L. Bach, T. Erlbacher "Real embedding process of SiC devices in a monolithic ceramic package using LTCC technology" in Electronics System-Integration Technology Conference (ESTC), 2020.

[7] C. Lenz, S. Ziesche, K. Reinhardt, S. Körner, A. Schletz, H.L. Bach, I. Schmidt, M. Simon-Najasek „Development and characterization of a monolithic ceramic pre-package for SiC-semiconductor devices based on LTCC technology" in Electronics System-Integration Technology Conference (ESTC), 2022

[8] S. Ziesche, S. Körner, K. Reinhardt, A. Schletz, B. Bayer, I. Schmidt, M. Simon-Najasek „Embedding of SiC-Semiconductor Devices in Full Ceramic LTCC Pre-packages for PE-applications" ECPE Workshop, Graz/Austria, 23.11.2022

Characterization of Embedded and Thinned RF Chips

1st Ran Yin
Centre for Tactile Internet with Human-in-the-Loop (CeTI)
Institute of Electronic Packaging Technology
Technische Universität Dresden, Germany
ran.yin@tu-dresden.de

2nd Helmuth P. E. Morath
Centre for Tactile Internet with Human-in-the-Loop (CeTI)
Chair of Circuit Design and Network Theory
Technische Universität Dresden, Germany
helmuth.morath@tu-dresden.de

3rd Christian Hoyer
Chair of Circuit Design and Network Theory
Department of Electrical and Computer Engineering
Technische Universität Dresden, Germany
christian.hoyer1@tu-dresden.de

4th Krzysztof Nieweglowski
Centre for Tactile Internet with Human-in-the-Loop (CeTI)
Institute of Electronic Packaging Technology
Technische Universität Dresden, Germany
krzysztof.nieweglowski@tu-dresden.de

5th Karsten Meier
Institute of Electronic Packaging Technology
Department of Electrical and Computer Engineering
Technische Universität Dresden, Germany
karsten.meier@tu-dresden.de

6th Jens Wagner
Centre for Tactile Internet with Human-in-the-Loop (CeTI)
Chair of Circuit Design and Network Theory
Technische Universität Dresden, Germany
jens.wagner@tu-dresden.de

7th Frank Ellinger
Centre for Tactile Internet with Human-in-the-Loop (CeTI)
Chair of Circuit Design and Network Theory
Technische Universität Dresden, Germany
frank.ellinger@tu-dresden.de

8th Karlheinz Bock
Centre for Tactile Internet with Human-in-the-Loop (CeTI)
Institute of Electronic Packaging Technology
Technische Universität Dresden, Germany
karlheinz.bock@tu-dresden.de

Abstract—**This work studies the effect of thinning down chips with transmission line structures for flexible embedding technology. Test chips with $90\,\Omega$ grounded coplanar waveguide (G-CPW) transmission lines have been designed and manufactured to characterize the high-frequency performance before and after embedding in sheet molding compound (SMC), and thinning down from $300\,\mu m$ to $20\,\mu m$. Molding first technology enables single-die or multiple-die thinning with the embedding material while being held in the exact position, where the subsequent fan-out copper (Cu) redistribution layer (RDL) structuring realizes the direct contacts to the embedded chip, followed by semi-additive manufacturing of traces. The embedded and thinned chips are characterized by S-parameter measurements in the frequency range from $20\,MHz$ to $65\,GHz$ using a vector network analyzer (VNA). The results show the feasibility of the examined embedding and thinning processes which are suitable for millimeter wave (mmWave) components as shown by the example of a G-CPW transmission line without degrading its RF performance in terms of characteristic impedance and attenuation.**

Index Terms—**chip embedding, chip thinning, flexible packaging, millimeter wave, tactile internet, transmission line**

I. INTRODUCTION

In the context of the Tactile Internet, a digital twin (DT) acts as an interface between a human and a machine [1]–[3]. It enables the real-time transmission of tactile sensations via a network, where the DT serves as a virtual image of physical objects or processes that are continuously updated. To allow the human to control the DT, a wireless body area network (WBAN) is used consisting of multiple tactile nodes (N) as shown in Fig. 1(a). Each tactile node has sensors that capture various physiological data such as heart rate, body temperature, movements, contacts, and contact force. This information is pre-processed and wirelessly transmitted to a central unit (C) where it is further processed and analyzed. In addition to sensors a tactile node also consists of an antenna and an application-specific integrated circuit (ASIC) including a radio frequency (RF) transceiver and a data processing unit. The packaging concept of such a node is illustrated in Fig. 1(b).

There are three main requirements for the tactile node: First, for a truly tactile experience, the nodes must be as thin and flexible as possible, forming an electronic skin. Second, they must be as small as possible. Third, depending on the application scenario and the number of sensors and actuators, and the connection of other wearables such as smart glasses to the WBAN, high-bandwidth data streams must be sent and received. The first requirement can be met using thinned electronics connected by a flexible redistribution layer

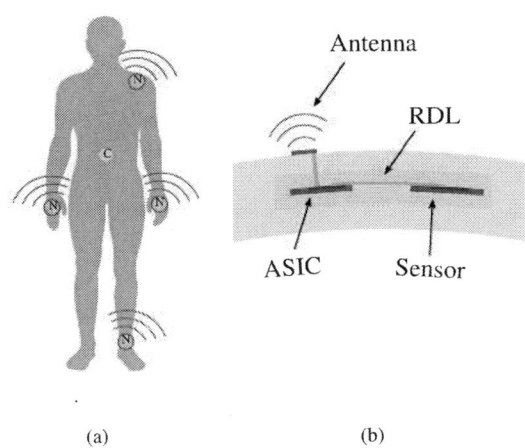

(a)   (b)

Fig. 1. Human with wireless body area network consisting of tactile nodes (a) and packaging concept of a flexible tactile node (b).

(RDL). The last two points can be addressed by operating the WBAN at mmWave frequencies (30 GHz-300 GHz), e.g. around 60 GHz, as suggested in [4]. In addition to a large amount of available bandwidth, the size of an antenna and other passives used in the transceiver scale inversely with frequency [5]–[11].

Flexible ASICs can be realized either by inherently flexible semiconductors consisting of non-silicon organic and inorganic semiconductors [12]–[15] or by thinning fully processed conventional semiconductors. The latter approach has the advantage of high carrier mobilities in established high-speed silicon technologies needed for operation at mmWave frequencies [16]. Thinned down to 20 µm, silicon chips get flexible and even more stable during bending compared to larger thicknesses between 50 µm to 100 µm [17]–[19]. At mmWave frequencies, thinning down these structures impacts not only the characteristics of active devices such as transistors but also passives such as transmission lines [20], [21].

In this work, the process of embedding and thinning of silicon chips with transmission line structures is studied, and the effect of thinning down is investigated by characterization of grounded coplanar waveguide (G-CPW) structures representing passive structures of an ASIC. The process has been intentionally designed to be compatible with conventional semiconductor manufacturing technology.

II. METHODOLOGY

To study the impact of thinning on the RF performance of passive structures on silicon chips, test chips with 90 Ω G-CPW transmission lines have been designed and analyzed. The analysis has been carried out before and after embedding and thinning down. In this work, fully processed ASICs are used which are thinned from approximately 300 µm to a total thickness of 20 µm and encapsulated using the molding first technique. Section II-A discusses the process of chip embedding and thinning. In Section II-B the test vehicle is presented.

(a) Side view of original (left) and thinned bare die (right).

(b) Edge chipping failure, to be prevented by embedding first technology

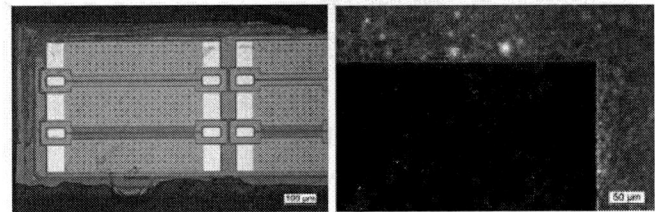

(c) Top view (left) and back view (right) of the thinned embedded die.

Fig. 2. The microscope images of the transmission line test-structures with 300 µm and 20 µm thickness. In (a) the side view of the original thickness and thinned bare die is shown, and in (b) the edge chipping of the bare die after thinning is shown. in (c) the front side (left) and backside (right) of the embedded die after thinning is shown. The chip size is 0.55 mm × 1.45 mm.

A. Chip Embedding and Thinning Process

One of the most critical steps in flexible packaging is the thinning of the ASIC. A comparison of the bare test chips before and after thinning is shown in Fig. 2(a). One major issue when thinning silicon die under 50 µm is the damage on the edge due to the mechanical stress on the bare die, as the edge chipping failure shown in Fig. 2(b). In contrast to the back-thinning technology of the entire wafer, embedding first technology is developed and evaluated here for a small amount and application-specific process. Embedding of the die in a sheet mold compound (SMC) before thinning effectively protects the die from edge chipping and warpage. SMC is the molding material in pre-preg shape. It has good fluidity and can realize the molding of complex structures, with low thermal conductivity, and good corrosion resistance [22]. The microscope image of the embedded die is shown in Fig. 2(c).

In the following, the suggested process flow for thinning of an embedded die is shown in Fig. 3 and explained in detail. The main idea is to embed both the ASIC and the sensor in SMC, then thin them down simultaneously. As a lab-based process that was developed on purpose, this embedding first technology is not designed for large-scale manufacturing.

In the first step of the process Fig. 3(a), the ASIC and sensor are placed on a temporary adhesive tape and embedded in SMC on a hot plate with a defined temperature profile [23].

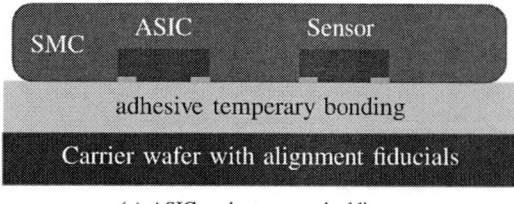

(a) ASIC and sensor embedding.

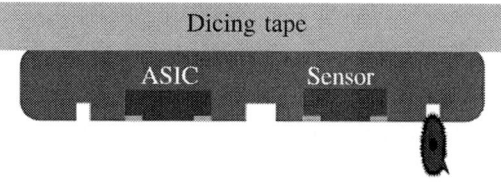

(b) Pre-cut before thinning.

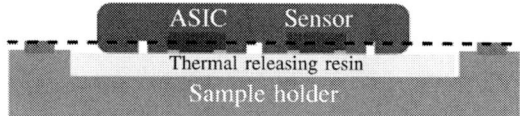

(c) Thinning process with a sample holder.

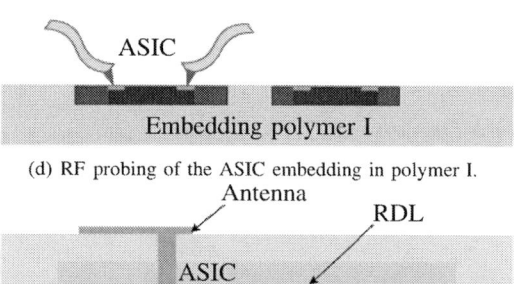

(d) RF probing of the ASIC embedding in polymer I.

(e) Structuring and embedding in polymer II.

Fig. 3. Proposed process flow of chip embedding and thinning: (a) Chip embedding in SMC with accurate die placement on temporary adhesive tape, (b) pre-dicing before grinding, (c) thinning of the embedded ASIC and sensor with pre-cut marks, (d) characterization of thinned and embedded ASIC, (e) embedding in polymer II and dielectric structuring and metallization.

Alignment marks were used during the placement as the front of both the ASIC and the sensor must be aligned by means of distance and angle.

After the hard cure of SMC, a pre-cut of 20 μm trench depth on these embedded devices at the front side was made by a dicing saw to indicate the final thickness of the chips. In the next step Fig. 3(c), the backside thinning process is done on a mechanical grinder. To do this, the embedded node is detached from the temporary dicing tape and attached to a sample holder using thermal-releasing resin as a temporary bonding layer. In addition, a ceramic spacer (shown in green) is placed on the outer edges of the sample holder to prevent tilting and allow for more homogeneous thinning to the desired thickness. The grinding papers with abrasive particles size from 20 μm to 1 μm were used to make gradual thinning without cracking

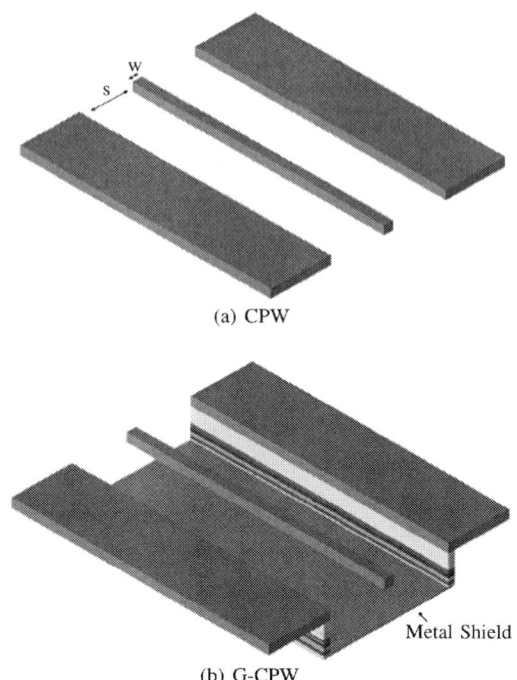

Fig. 4. Sketch of (a) coplanar waveguide (CPW) and (b) grounded coplanar waveguide (G-CPW) transmission line with signal conductor width w and side wall spacing s.

the chip. After each step, the complete removal of the spiral grinding track is inspected under a microscope. The grinding process continues until the pre-cut trenches are visible from the backside. The final polishing step is then performed with the VibroMet vibratory polisher to minimize any remaining micro damage. This produces a stress-reduced surface without the need for the hazardous electrolytes required by electropolishers. Then ASIC and sensor are released from the sample holder and cleaned with acetone.

To validate the effect of thinning and embedding, the RF characteristics of these structures are analyzed by on-wafer S-parameter measurements as indicated in Fig. 3(d), before further processing. The measurement procedure and results are discussed in Section III.

Next, the sample is ready for further flexible embedding and metallization processes. In [24] the DC characterization for the proposed RDL structures is presented. Recent simulation results demonstrate that embedding in the neutral bending plane of a flexible system further reduces the mechanical stress on the chip effectively [25]. To embed these devices in flexible embedding polymer I and II, the elements are precisely placed on an adhesive tape with the top side down, and the liquid state polymer is poured from the back side, then cured with UV exposure. A subsequent semi-additive fan-out Cu-RDL structuring realizes the direct contacts to the embedded chips and connects to through mold via (TMV) to the antenna as the complete system shown in Fig. 3(e).

The embedding materials provide step stiffness. After hard cure, SMC has Young's modulus of 24.2 GPa, which protects

26

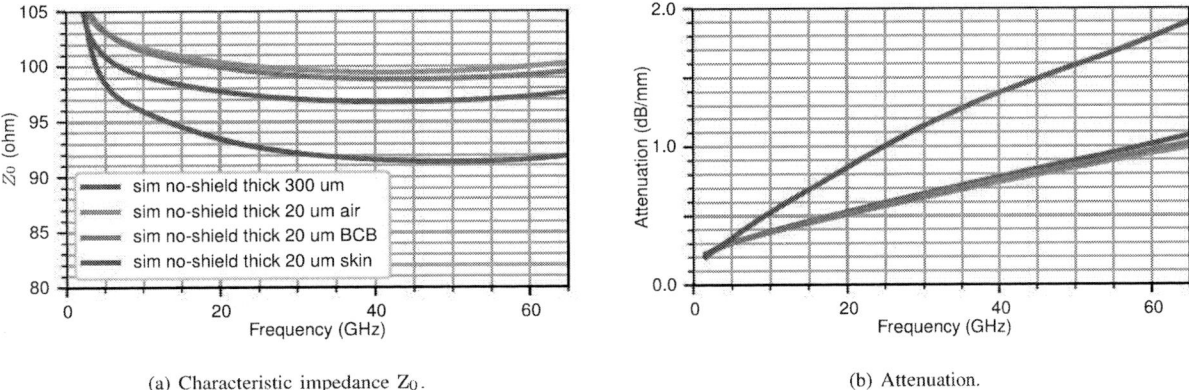

(a) Characteristic impedance Z_0.

(b) Attenuation.

Fig. 5. Simulated characteristic impedance Z_0 and attenuation of a CPW transmission line without ground (GND)-shield for 300 μm and 20 μm substrate thickness with different materials underneath the thinned transmission line.

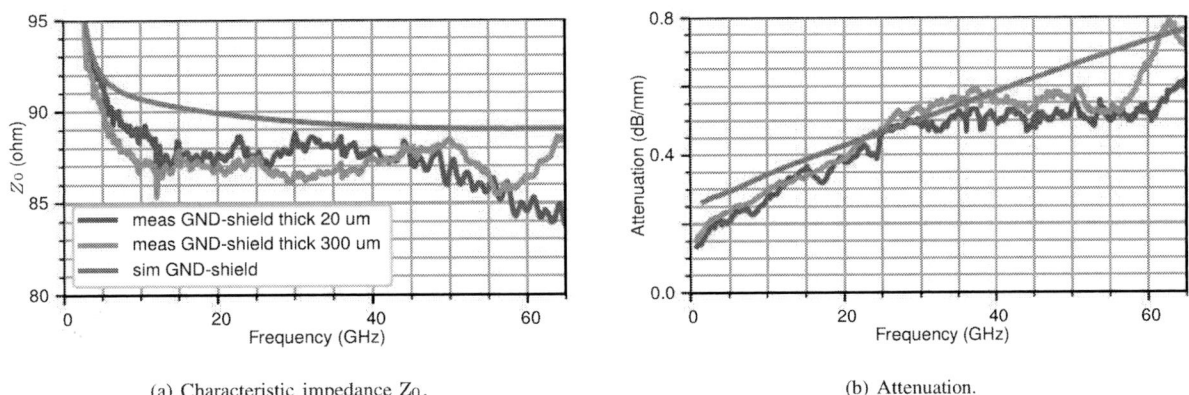

(a) Characteristic impedance Z_0.

(b) Attenuation.

Fig. 6. Measured and simulated characteristic impedance Z_0 and attenuation of a transmission line with the ground (GND)-shield for 300 μm and 20 μm substrate thickness.

the chips with higher stiffness, and also reduces mechanical stress in the interconnect structures between the RDL and the device, resulting in increased reliability. As a unit, the tactile node is encapsulated in two different types of polymer. Polymer I is polyimide, it is an excellent dielectric material with Young's modulus of 8.5 GPa. While polymer II is a flexible resin with Young's modulus of 13 MPa. It is believed, the different stiffness properties create a gradient so that the bending stress is released gradually. The Young's modulus of bulky silicon with the orientation of [110] is 168 GPa, while the thinned silicon has the mechanical strength of 3 GPa [26]. With the active thinned chip sitting in the neutral bending plane, recent simulation results demonstrate that it further reduces the mechanical stress on the chip in a flexible system effectively [25]. Therefore, SMC provides sufficient stiffness within the tactile node just surrounding the ASIC, and the metallization structures of the RDL see less stress while in a bent state.

B. Test vehicle

A typical ASIC consists of passive and active components. For a better understanding of any performance change after chip embedding and thinning, it is important to investigate active and passive components separately. At mmWave frequencies transmission lines are typically used for impedance matching and connections between circuits and pads. Thus, for the characterization of any active component transmission lines are typically involved and their effects can be de-embedded later on. Therefore, this work focuses on the characterization of thinned transmission lines while thinned active components will be subject to future research. A cross-section of a coplanar waveguide (CPW) is shown in Fig. 4(a) with signal conductor width w and side wall spacing s. Compared to microstrip lines the ground metal of CPWs is alongside the signal conductor instead of underneath it [27]. When thinning a CPW the characteristic impedance Z_0 and attenuation of the line change depending on the material underneath the line. Thus, these two parameters are chosen for representing the RF performance of the ASIC.

Fig. 5(a) shows the simulated Z_0 for a thick and thinned CPW while Fig. 5(b) depicts the attenuation of the lines, respectively. In the case of the 20 μm thick CPW different materials were placed underneath it. With air underneath, the highest Z_0 and lowest attenuation can be observed. A 300 μm

27

thick layer of BCB as a high-quality dielectric material with a permittivity ϵ_r of 2.65 [28] underneath the transmission line results in similar characteristics compared to air. When simulating the transmission line on top of a human body model according to [29], Z_0 deviates by less than 5 % compared to the 300 μm thick version. However, the attenuation of the thinned line on the human body at 60 GHz is 0.8 dB/mm higher than the attenuation of the thick line without a ground shield. This deviation increases with frequency and might degrade the performance of ASICs operating at mmWave frequencies significantly.

As a result, we suggest having a metal ground layer under the signal conductor for thinned transmission lines as shown in Fig. 4(b) shielding the line from the substrate. Such a line is called a grounded coplanar waveguide (G-CPW). As a proof of concept, the BEOL chip with several G-CPW lines of different lengths and a characteristic impedance Z_0 of 90 Ω is used to characterize the RF performance. Measurements are made before and after embedding and thinning, allowing a comparison of both performances. The length of the G-CPW lines ranges from 500 μm to 100 μm, as shown in the photo of the chip in Fig. 2(a).

III. CHARACTERIZATION AND DISCUSSION

Thinning of the ASIC for flexible packaging is critical not only because of the fragile mechanical properties of the ultra-thin silicon chip but also because the performance of mmWave components is sensitive to the material properties influenced by the thinning process. Embedding the ASIC in the SMC before thinning reduces the mechanical stress on the die and effectively protects the die from edge chipping and warpage. However, it introduces thermal stress to the components. On the other hand, the molding material SMC has good under-filling properties, and there is a risk of covering the contacting pads during the molding process. Thus, the topography of the die is designed in a way that a line with an oxidation layer is at the highest plane at the edge of the die. During the molding process, it has direct contact with the adhesive layer and prevents the SMC from under-filling the test structures successfully.

To characterize the RF performance Z_0 and attenuation are calculated from on-wafer S-parameter measurements according to [30]. The measurements were performed by a vector network analyzer from 20 MHz up to 65 GHz. Characterizing flexible circuits at mmWave frequencies in the bent state is challenging because conventional methods such as on-wafer probing are not feasible without dedicated measurement equipment and are therefore content for some future research. The pads were de-embedded by measuring two identical lines with different lengths, 500 μm and 100 μm, using the method proposed in [31]. The resulting characteristics correspond to a line with an effective length of 400 μm without pads.

Fig. 6(a) shows measured and simulated Z_0 for a thick and thinned G-CPW while Fig. 6(b) depicts the attenuation of the lines, respectively. It can be observed that despite the thermal stress and mechanical stress introduced by the embedding and thinning processes, it has no effect on both characteristics. This ensures well-defined characteristics independent of the underlying material and shows the feasibility of applying the embedding first technology for mmWave components.

IV. CONCLUSION

As an important part of the flexible packaging of tactile nodes, thinning processes of chips representing ASIC and sensors as tactile nodes were carried out successfully with embedding first technology to prevent edge chipping. The transmission characteristics of the G-CPW transmission lines were measured before and after thinning. The results show the assessed RF performance characteristics did not degrade through the process flow, it shows the feasibility of the examined embedding and thinning processes is suitable also for mmWave components. As for future work, the RF performance of thinned active components with semi-additive Cu-RDL at the deformed state is yet to be investigated. Furthermore, the mechanical and thermo-mechanical reliability, e.g. through bending cycles.

ACKNOWLEDGMENT

Funded by the German Research Foundation (DFG, Deutsche Forschungsgemeinschaft) as part of Germany's Excellence Strategy – EXC 2050/1 – Project ID 390696704 – Cluster of Excellence "Centre for Tactile Internet with Human-in-the-Loop" (CeTI) of Technische Universität Dresden.

REFERENCES

[1] C. von Lengerke, A. Hefele, J. A. Cabrera, and F. H. P. Fitzek, "Stopping the data flood: Post-shannon traffic reduction in digital-twins applications," in *NOMS 2022-2022 IEEE/IFIP Network Operations and Management Symposium*. IEEE, apr 2022.

[2] F. H. Fitzek, S.-C. Li, S. Speidel, T. Strufe, and P. Seeling, "Frontiers of transdisciplinary research in tactile internet with human-in-the-loop," in *2021 17th International Symposium on Wireless Communication Systems (ISWCS)*. IEEE, sep 2021.

[3] I. A. Tsokalo, D. Kuss, I. Kharabet, F. H. P. Fitzek, and M. Reisslein, "Remote robot control with human-in-the-loop over long distances using digital twins," in *2019 IEEE Global Communications Conference (GLOBECOM)*. IEEE, dec 2019.

[4] H. Morath, X. An, P. V. Testa, F. Ellinger, J. Wagner, and K. Bock, "Designing a 60 GHz Sub-Milliwatt transceiver for wireless Body-Area-Networks," in *European Wireless 2021 (EW 2021)*, Verona, Italy, Nov. 2021.

[5] H. P. E. Morath, P. V. Testa, J. Wagner, and F. Ellinger, "A 3.6 mW 60 GHz low-noise amplifier with 0.6 ns settling time for duty-cycled receivers," *IEEE Microwave and Wireless Components Letters*, vol. 31, no. 8, pp. 977–980, aug 2021.

[6] L. Steinweg, P. V. Testa, C. Carta, and F. Ellinger, "A 213 GHz 2 dBm Output-Power Frequency Quadrupler with 45 dB Harmonic Suppression in 130 nm SiGe BiCMOS," in *ESSCIRC 2021 - IEEE 47th European Solid State Circuits Conference (ESSCIRC)*, Sep. 2021.

[7] V. Rieß, P. Stärke, C. Carta, and F. Ellinger, "An Integrated mm-Wave Quadrature Up-Conversion Mixer Based on a Six-Port Modulator," in *2019 14th European Microwave Integrated Circuits Conference (EuMIC)*, Oct. 2019, pp. 176–179.

[8] P. Stärke, V. Rieß, C. Carta, and F. Ellinger, "Continuous 360° Vector Modulator with Passive Phase Generation for 140 GHz to 200 GHz G-Band," in *2019 12th German Microwave Conference (GeMiC)*. IEEE, Mar. 2019.

[9] C. Hoyer, J. Wagner, and F. Ellinger, "A 60 GHz VCO with 654 MHz direct Frequency Modulation Bandwidth in 0.13-μm SiGe BiCMOS," in *2021 International Conference on Electrical, Computer, Communications and Mechatronics Engineering (ICECCME)*, Oct. 2021.

[10] C. Hoyer, F. Protze, J. Wagner, and F. Ellinger, "A Reconfigurable 60-GHz VCO with -103.2 dBc/Hz Phase Noise in a 0.13-μm SiGe BiCMOS Technology," in *2022 Asia-Pacific Microwave Conference (APMC)*. IEEE, Nov. 2022.

[11] D. Fritsche, S. Li, N. Joram, C. Carta, and F. Ellinger, "Design and characterization of a 190-GHz voltage-controlled oscillator," in *2016 46th European Microwave Conference (EuMC)*. IEEE, Oct. 2016.

[12] H. Zhang, J. Li, D. Liu, S. Min, T.-H. Chang, K. Xiong, S. H. Park, J. Kim, Y. H. Jung, J. Park, J. Lee, J. Han, L. Katehi, Z. Cai, S. Gong, and Z. Ma, "Heterogeneously integrated flexible microwave amplifiers on a cellulose nanofibril substrate," *Nature Communications*, vol. 11, no. 1, p. 3118, Jun 2020.

[13] U. Zschieschang, R. Hofmockel, R. Rödel, U. Kraft, M. J. Kang, K. Takimiya, T. Zaki, F. Letzkus, J. Butschke, H. Richter, J. N. Burghartz, and H. Klauk, "Megahertz operation of flexible low-voltage organic thin-film transistors," *Organic Electronics*, vol. 14, no. 6, pp. 1516–1520, Jun. 2013.

[14] T. Meister, K. Ishida, C. Carta, N. Munzenrieder, and F. Ellinger, "Flexible electronics for wireless communication: A technology and circuit design review with an application example," *IEEE Microw. Mag.*, vol. 23, no. 4, pp. 24–44, Apr. 2022.

[15] T. Meister, F. Ellinger, J. W. Bartha *et al.*, "Program FFlexCom — High frequency flexible bendable electronics for wireless communication systems," in *2017 IEEE International Conference on Microwaves, Antennas, Communications and Electronic Systems (COMCAS)*, Nov. 2017, pp. 1–6.

[16] H. Sharifi, K. Shinohara, W. Ha, J. Kang, M. Montes, C. McGuire, J. May, and H. Kazemi, "Microwave and millimeter-wave flexible electronics," in *2014 IEEE MTT-S International Microwave Symposium (IMS2014)*, 2014, pp. 1–3.

[17] R. S. Dahiya and S. Gennaro, "Bendable Ultra-Thin Chips on Flexible Foils," *IEEE Sensors Journal*, vol. 13, no. 10, pp. 4030–4037, 2013.

[18] Z. Cao, A. Goritz, M. Stocchi, M. Wietstruck, C. Hoyer, L. D. Steinweg, C. Carta, F. Ellinger, B. Tillack, and M. Kaynak, "An advanced finite element model for BiCMOS process oriented ultra-thin wafer deformation," *IEEE Transactions on Semiconductor Manufacturing*, vol. 35, no. 1, pp. 2–10, Feb. 2022.

[19] C. Hoyer, L. Steinweg, Z. Cao, V. Rieß, L. Li, F. Protze, C. Carta, J. Wagner, M. Kaynak, B. Tillack, and F. Ellinger, "Bendable 190-GHz Transmitter on 20-m Ultra-Thin SiGe BiCMOS," *IEEE Journal on Flexible Electronics*, vol. 1, no. 2, pp. 122–133, 2022.

[20] L. Wang, "Design and Technology of Ultra Thin Chip Packages for High-Frequency Applications up to 60 GHz," Ph.D. dissertation, Ghent University, 2013.

[21] Y. S. Kim, N. Maeda, H. Kitada, K. Fujimoto, S. Kodama, A. Kawai, K. Arai, K. Suzuki, T. Nakamura, and T. Ohba, "Advanced wafer thinning technology and feasibility test for 3d integration," *Microelectronic Engineering*, vol. 107, pp. 65–71, jul 2013.

[22] R. Yin, K. Nieweglowski, K. Meier, and K. Bock, "Characterization of material adhesion in redistribution multilayer for embedded high-frequency packages," in *2022 IEEE 9th Electronics System-Integration Technology Conference (ESTC)*. IEEE, sep 2022, pp. 321–325.

[23] R. Yin, P. Clara, K. Nieweglowski, K. Meier, and K. Bock, "Process developments on sheet molding and redistribution deposition for cu-pillar chips," in *2022 IEEE 45th International Spring Seminar on Electronics Technology (ISSE)*. IEEE, may 2022.

[24] R. Yin, K. Nieweglowski, K. Meier, and K. Bock, "Embedding of thinned RF chips and electrical redistribution layer characterization," in *2022 IEEE 24th Electronics Packaging Technology Conference (EPTC)*. IEEE, dec 2022.

[25] S. Li, Y. Su, and R. Li, "Splitting of the neutral mechanical plane depends on the length of the multi-layer structure of flexible electronics," *Proceedings of the Royal Society A: Mathematical, Physical and Engineering Sciences*, vol. 472, no. 2190, p. 20160087, jun 2016.

[26] S. Endler, T. Hoang, E. Angelopoulos, H. Rempp, C. Harendt, and J. Burghart, "Mechanical characterisation of ultra-thin chips," in *2011 Semiconductor Conference Dresden*. IEEE, 2011, pp. 1–4.

[27] F. Ellinger, *Radio Frequency Integrated Circuits and Technologies*. Springer, 2008.

[28] S. Costanzo, I. Venneri, G. Di Massa, and A. Borgia, "New technologies and antenna design concepts at millimeter-wave bands," in *2009 3rd European Conference on Antennas and Propagation*, 2009, pp. 3801–3805.

[29] K. Islam, T. Hossain, M. M. Khan, M. Masud, and R. Alroobaea, "Comparative design and study of a 60 GHz antenna for body-centric wireless communications," *Computer Systems Science and Engineering*, vol. 37, no. 1, pp. 19–32, 2021.

[30] U. Stehr, L. F. Centeno, Y. Ni, H. O. Jacobs, and M. A. Hein, "Rf properties of stretchable transmission line structures," in *2020 German Microwave Conference (GeMiC)*, 2020, pp. 272–275.

[31] A. Mangan, S. Voinigescu, M.-T. Yang, and M. Tazlauanu, "De-embedding transmission line measurements for accurate modeling of IC designs," *IEEE Transactions on Electron Devices*, vol. 53, no. 2, pp. 235–241, feb 2006.

Deposition of Fine-Pitch Indium Bumps on Single Die

Andreas Schneider
Technology
UKRI-STFC Rutherford Appleton Laboratory (RAL)
Didcot OX11 0QX, U.K.
andreas.schneider@stfc.ac.uk

Navid Ghorbanian
Technology
UKRI-STFC Rutherford Appleton Laboratory (RAL)
Didcot, U.K.
navid.ghorbanian@stfc.ac.uk

David Burt
Kelvin Nanotechnology Ltd.
70 Oakfield Av.

Glasgow G12 8LS, U.K.
dave@kntnano.com

James Hollingham
Technology UKRI-STFC RAL
Didcot, U.K.

Paul Booker
Technology UKRI-STFC RAL
Didcot, U.K.

Toby G. Brooks
Technology UKRI-STFC RAL
Didcot, U.K.

John D. Lipp
Technology UKRI-STFC RAL
Didcot, U.K.

Marcus J. French
Technology UKRI-STFC RAL
Didcot, U.K.

Abstract— Photolithographic lift-off at wafer level is a well-established method for depositing indium bumps onto pixels of electronic components such as sensors and application-specific integrated circuits (ASIC). These indium bumps form interconnects between pixels of the components during flip-chip bonding to make devices such as radiation detectors. Such components are not always available on wafer-scale, and single dies need to be processed in small-scale production, R&D, or for Cd(Zn)Te detector material. Photolithography on single dies is ineffective due to the edge-bead of photoresist that compromises perfect indium deposition onto pixels near the periphery of the die. To utilize the maximum surface area of a die, a rigid mask with apertures (shadow mask) is an adequate substitute for the photoresist. Different mask designs and production techniques for fine-pitch indium bump arrays are investigated by testing electroformed stencils or micromachined silicon (Si) membranes as shadow masks. For the indium deposition, the sensor and ASIC dies are clamped into custom-made jigs that include shadow masks. Alternatively, dies can be adhered to shadow masks with soluble adhesive. The apertures of a mask are aligned to the pixel array of the detector component. Bump arrays with 250μm- and 100μm-pitch were demonstrated with electroformed stencils. For smaller pitch (55μm) and bump diameter of ~20μm, specialized masks from Si membranes are tested. Components with successfully deposited indium bump arrays were subsequently flip-chip bonded at room temperature without reflowing the indium. A bond yield of at least 99.9% pixels for such flip-chip bonded radiation detector was shown using the read-out signal from exposure with a flat field X-ray radiation of an Am-241 sealed source.

Keywords—shadow mask, indium bumps, flip-chip bonding, fine-pitch, radiation detectors

I. Introduction

X-ray and gamma-ray detectors for imaging and spectroscopy in scientific experiments at synchrotron light sources and X-ray Free Electron Lasers (XFEL) are often pixel detectors that are hybridized sensors with application-specific integrated circuits (ASIC). For advanced experiments at modern and next-generation synchrotrons and XFELs, these detectors need to be suitable for high radiation energy and

UKRI STFC's Centre for Instrumentation has funded this research.

extreme photon flux [1,2]. For quite some years, high-quality radiation grade $CdZn_xTe_{(1-x)}$ ($0 \le x \le 0.1$) has been the choice of material [1,3,4,5] for those sensors due to its electrical and radiation absorption properties. For desirable high spatial resolution, small pixels on a 55μm pitch are required. Such detectors are largely utilized with Si sensors hybridized to Medipix ASICs [6]. First detectors with CdTe and Cd(Zn)Te hybridized to 55μm-pitch ASICs were presented as early as 2004 [7]. Those were fabricated on wafer-scale and using PbSn for interconnects. In addition, processes are carried out on single die but no details of the hybridization/flip-chip bonding were given [7]. Currently commercially available small pixel detectors are fabricated by Advacam with CdTe sensors and Timepix3 ASIC [8] for specialized medical applications (positron emission tomography). High-quality detector-grade Cd(Zn)Te is provided by Redlen in the form of single dies that are fabricated by selecting monocrystalline defect-free areas from ingots [9]. Processing such single die with photolithography using negative photoresist (PR) in lift-off processes of subsequently deposited indium for interconnects is not trivial [10]. The edge-bead of PR compromises perfect indium deposition onto pixels near the periphery of the die. Processes using a rigid mask with apertures (shadow mask) – as shown by other research groups for deposition of thin layers [11] – are a suitable alternative. The shadow-mask technique presented here enables the deposition of μm-high indium bumps on single die without photolithography and with fewer, simpler process steps. The paper presents different shadow masks that will be capable for small-pitch size of 55μm. This technique is important when single dies are processed in small scale production or for R&D projects that require high yield.

II. Process Preparation

The hybridization of sensors and ASICs for radiation detectors is conventionally carried out with single sensor and ASIC dies in a flip-chip bonding process. This requires the deposition of some form of bond material for each pixel between sensor and ASIC. For large pixel pitch (~250μm) this can be achieved by printing electrically conductive adhesive onto the sensor pixel array and subsequently the sensor is flip-chip bonded to the matching array of gold ball studded pixels of an ASIC [12]. However, those processes have limitations due to the size of bumps that can be formed. For smaller pitch, indium bump bonding is preferred. Also solder bumps such as eutectic Sn-Pb [13] may be suitable for this process. Conventionally this is carried out on wafer scale and involves a large number of process steps [14]. Wafer processing allows the spinning coating of negative PR and subsequent high-precision and fine-structure (pitch) patterning using photolithography. On such wafers indium is deposited and after stripping the PR indium bumps remain on the desired pixel position (lift-off). RAL conventionally achieves 20μm-∅ indium bumps with 55μm-pitch on 8-inch wafer using this lift-off method. However, in a following fabrication step individual sensor/ASIC dies have to be singulated from these wafers using mechanical dicing or similar processes which is delicate because the existing indium bumps can be easily damaged. For the flip-chip bonding, the indium deposition/bumping is carried out on both sensors and ASICs -- the malleable indium bumps can form the interconnects between sensor and ASIC pixels in the bonding process. Often an additional indium reflow process is executed to improve alignment between sensor and ASIC pixel arrays as well as better bond formation [14]. However, STFC-RAL demonstrate that such reflow process is often not necessary, and a cold indium welding process [15] can be used instead.

A process that enables the indium bump deposition onto single die without photolithography will reduce several process steps that conventionally used as described in [14]. In addition, indium bumping of CZT dies that are frequently used for high-quality radiation-grad sensor material will be much simpler without the disadvantages of photolithography on single dies as explained in the introduction of this paper. Instead using PR, these single sensor/ASIC dies need to be masked using a solid stencil with apertures (shadow mask) which will be aligned with the pixels of the component. Consequently, indium bumps will match and reveal the same aperture array pattern on the component after the evaporation and the removal of the shadow mask..

A. Masking of Sensors and ASIC Dies

For masking dies with shadow masks, two different types of mask material were studied. Initially for aperture pitches ≥100μm, electroformed nickel stencils fabricated by ASMPT to RAL's pixel design were chosen. ASMPT's "DEK Electroform process" [16] stencils which are usually utilized for printing solder paste (surface mount technology for miniLED dies) is specified for minimum aperture size of 50μm and a minimum gap of 50μm between apertures that is equivalent to a 100μm-pitch. The smallest thickness of such stencil is 23μm. These dimensions are compatible with most radiation detector pixel arrays and hence suitable for shadow masks of large indium bump pattern (≥100μm-pitch). Electroformed masks used for this study here were 40μm to 50μm thick. According to ASMPT [16], laser cut stencils are not able to meet the same requirements as those electroformed stencils. RAL designed customized jigs for holding single dies with a clamping mechanism that mounts the shadow mask to the die. The alignment of the aperture array of the shadow mask to the pixel array on sensor/ASIC dies is carried out optically before the mask is fixed to the jig. Further details are given in [17,18].

For detector applications with pitch <100μm currently micro-machine Si membranes are tested for indium bumping with shadow masks. Si membranes with thickness 50μm and 100μm, respectively, were fabricated on ~380μm-thick 4-inch Si wafers by Kelvin Nanotechnology. Ltd. Rrecesses for 9 membranes with arrays of 256 × 257 apertures (55μm-pitch) were etched on each wafer. For initial test purposes, the diameter of the apertures ranges from 25μm to 45μm. Due to the fragility of the membranes, those membranes were not used with the custom-made jigs yet, instead single individual dies are temporarily attached across the membranes of the Si wafer with soluble adhesive. The adhesive will be applied in dedicated areas of the wafer in close vicinity of the membranes. Three different adhesives were examined (IPA-diluted adhesive, acetone-diluted adhesive, and double-sided tape comprising similar solid adhesive). The IPA-diluted adhesive generates a viscous fluid that was dispensed with an ASYMTEK X1010 dispenser using an Auger screw. The acetone-diluted adhesive is less viscous than the IPA-diluted adhesive mix and, therefore, was not used with the ASYMTEK dispenser. Instead, droplets of the mixture were dispensed manually from a needle syringe. Alternatively, a thin film was spread with a doctor blade and it was investigated to transfer material from this film to surrounding areas of membrane of the shadow masks using a stamp or similar applicator. The adhesive on tape is based on the same material as the diluted adhesives and is a solid adhesive that is commercially available as a single-sided tape. In order to tape die and mask together, two of those adhesive tapes were joint together with a 3M double-sided tape creating an approx. 500μm thick double-sided tape. The adhesive of such tape is soluble in acetone.

B. Indium Bump Deposition

For the indium deposition, a MPS800 thermal evaporator with integrated in-situ Argon (Ar) plasma system was used. The plasma system allows stripping oxide layers and other residues from the substrates prior to the metal deposition (plasma cleaning). After plasma cleaning, a ~10nm thin chromium barrier/adhesion layer is deposit followed by the indium deposition of a thick (~6μm) layer. The thickness of the indium layer is monitored by a quartz crystal microbalance. Up to 8-inch wafers as well as the custom-made jigs for single die can be mounted on a rotating table of the evaporator. With a special adaptor (and depending on the size of the dies) about 16 jigs with shadow masks can be mounted on this table. The distance between crucibles with indium to the table is approx. 1m for achieving a good uniformity of an indium layer. Hence the MPS800 evaporator can be utilized for indium deposition with standard wafer-scale photolithographic lift-off processes as well as for deposition on single die using shadow masks.

C. Flip-Chip Bonding

The flip-chip bonding of sensor to ASIC dies is carried out on either a SET FC150 or FC300 bonder at RAL. Even though the FC300 has the capability of an in-situ reflow process with formic acid that reduces any surface of indium oxide before bonding, an indium cold welding bonding process at room temperature without reflow and formic acid was carried out in the FC150 for any hybrid detector. The bond force is typically 0.5gf/pixel so that the insignificant and negligible oxide layer of the indium bump surface is ruptured and a sufficient bond between pixels is formed. Usually after hybridization of sensor and ASIC, the ASIC is wire bonded to a PCB and connected to a data-acquisition system.

III. RESULTS

Utilizing the shadow mask technique with electroformed stencils, a prototype indium bumped hybrid radiation detector with 250μm pixel pitch was successfully demonstrated [18]. Applying an exposure of the detector with a flat field X-ray radiation of an Am-241 sealed source, a significant signal was detected in 99.9% of pixels of the detector that was fabricated

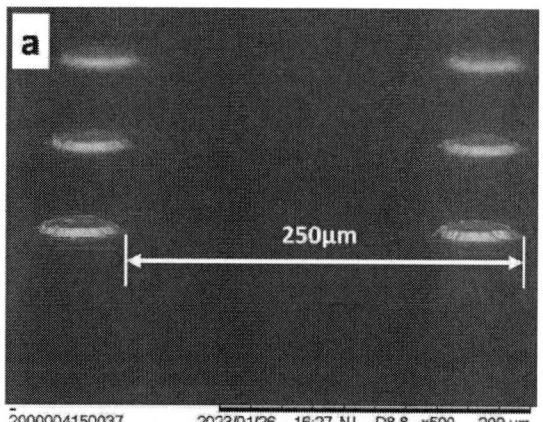

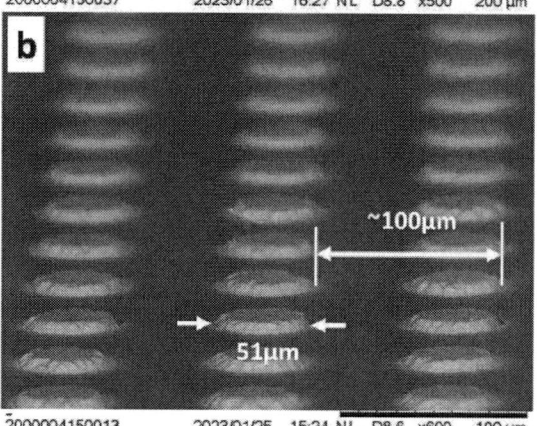

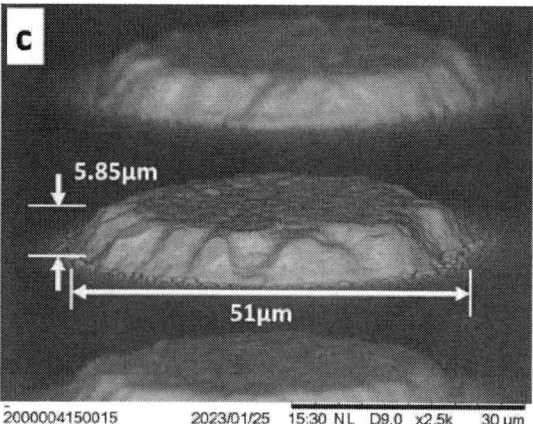

Fig. 1. a) 250μm-pitch ASIC with indium bumps; b) indium bumps on 100μm-pitch; c) detail of a bump.

with the shadow mask method [18]. The indium interconnect bond yield is potentially higher than 99.9% because diminishing signal in detector pixels can also be contributed by defects in sensor and ASIC that cannot be distinguished from missing bonds with the test used here. A detailed SEM image of such bumps with 250μm-pitch on an 80×80 pixel ASIC is shown in Fig.1a. Electroformed shadow masks are limited to an aperture pitch of 100μm. Initial trials on test substrates achieved indium bump arrays with ~100μm-pitch (Fig.1b+c) with those electroformed stencils and mechanical clamping of shadow masks to dies in dedicated jigs. Each indium bump (~50μm diameter, ~6μm high) has a small, insignificant amount of thin indium (width of ~3μm) around the bumps. This additional halo of thin indium material is potentially caused by indium penetrating under the edges of the mask apertures because the shadow mask is not in full contact to the die surface. In contrast, such halos were not observed for indium bumps on wafers that were prepared with a standard photolithography lift-off process. A spin-coated PR covers a wafer surface perfectly with only a minute intended undercuts of the openings in the PR that facilitate the lift-off of PR and indium. It is important to take these additional indium halos into account for the pixel array design on sensors and ASIC (pixel size > 56μm) when using electroformed shadow masks. Nevertheless, this is still sufficient for any 100μm-pitch device.

For trials with 55μm-pitch shadow masks, two 4-inch Si wafers with membranes were fabricated. Recesses were initially etched to create the membranes. On the opposite unetched side of the wafer apertures for membranes and additional pattern (dimples and trenches) adjacent to the membrane areas were photolithographically formed and etched until aperture openings in the membrane occurred. The dimples around the membranes are intended to confine adhesive in these areas for die attachment and trenches are intended to protect adhesive from flowing onto the membranes. Nine different membrane configuration per wafer were constructed (Tab. 1). In addition to the membrane wafers, a further dummy wafer with the same pattern for apertures, dimples, and trenches but without the membrane recesses was produced for testing the different adhesives for die attachment. All three wafers are shown in Fig. 2 with an example of apertures on a 50μm-thick membrane in the inset of Fig.2.

TABLE I.   MEMBRANE APERTURES AND DIMPLES/TRENCHES

#	Membrane apertures	Adhesion area		
	Diameter Ø	Trench width	Dimple Ø	Dimple pitch
1	25μm	20μm	20μm	80μm
2	35μm	5μm	20μm	40μm
3	45μm	10μm	20μm	60μm
4	35μm	20μm	5μm	80μm
5	35μm	5μm	5μm	20μm
6	35μm	10μm	5μm	40μm
7	40μm	20μm	10μm	80μm
8	35μm	5μm	10μm	20μm
9	30μm	10μm	10μm	40μm

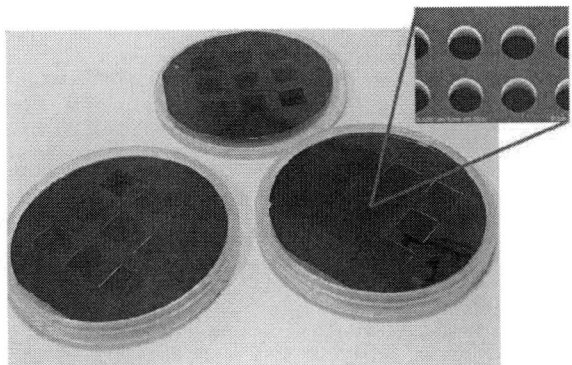

Fig. 2. Three 4-inch wafers for tests with 55μm-pitch shadow masks - inset shows an SEM image of apertures (35μm-diameter, 55μm-pitch on ~50μm-thick membrane).

Using these Si wafers with membranes, a die attachment with dissolvable adhesives to the wafers seems to be a better solution rather than handling and mounting the fragile membranes dies into the customized jig described above. Three different adhesives (IPA-diluted, acetone-diluted, double-sided tape) which were tested compares as follows:

IPA-diluted adhesive (2/3 mix in weight): Small droplets of viscous solution that were dispensed through an Auger screw of the ASYMTEK dispenser was not easy to control and was inconsistent in shape, height, and amount that resulted on the Si wafers. There is some tendency to achieve smaller droplets with increased gap (0.6mm) between dispensing needle and Si surface as well as shorter dispensing time (0.5sec). These adhesive dots vary in height from 100μm to 300μm and in diameter from 0.8mm to 1.6mm. Further optimization achieved smaller dots with a height between 25μm and 55μm, but these dots did not always form uniform circular dots. When just dispensed and left drying, the dimple diameter and pitch on the patterned surface area where adhesive was applied did not have any significant effect on the dot height. Remelting of these dried dots (initial height approx. 38μm) changes size and shape to approx. 3μm to 5.3μm in height on a flat surface after heating on a hot plate at 150°C for ~15min. Due to surface tension, these remelted dots are often taller at the edges than in the middle of the dot, sometimes resulting in a diminishing/vanishing amount of adhesive in the center of the dot. Dots on surfaces patterned with dimples are usually thinner compared to the flat surface because the dimples take up some of the adhesive. Therefore, the remaining dot height has a slight tendency to be smaller with increasing dimple diameter but is insignificantly influenced by the dimple-pitch.

Acetone-diluted adhesive (1/1 mix in weight): This type of mixture was too fluid to be used in the ASEMTEK dispenser. Dispensing such mixture manually from a syringe in small quantities allows to form small droplets up to a height of 400μm; especially with mixtures that contain less acetone. Keeping droplets of this mixture at room temperature for several hours or heating at 90°C for at least 10min on a hot plate does not evaporate all acetone, and droplets remained soft. This may create difficulties to correctly attach a die to the mask due to outgassing. A longer drying process at higher temperature (150°C for ~15min) forms more solid adhesive dots with a height between 5μm to 15μm which are more likely to perform a good adhesion of dies to shadow masks.

To further improve control over the dot height and size of this adhesive (1/1 mix), a thin film was created with a doctor blade on a rotating plate and then small portions of the thin adhesive film were transferred with a stamp to the shadow mask. This creates features of adhesive which are confined to the area of the stamp impression. The typical height is between 5μm to 16μm but can also occasionally be as high as 45μm or no adhesive at all in parts of the stamped area. The height is independent of the area where the adhesive was deposit with the stamp (flat Si or patterned with dimples). After reheating at ~150°C for several minutes, these adhesive spots decrease slightly in height by a few μm when those were stamped on flat Si or area with dimples of 5μm diameter. Only in areas with dimples of diameter ≥10μm the adhesive almost completely retracts into the dimples leaving only a minute amount of adhesive on the surface in this area. The dimple pitch has no significant influence on height or shape. This mixture stamped on areas with dimples of diameter ≥10μm seems to be well suited to attach dies to shadow masks with a minimal gap.

Double-sided adhesive tape: The constructed tape had a far better thickness consistency compared with the droplets but is approx. 500μm thick. This increases the distance between shadow mask and die considerably, potentially increasing halos of indium around the bumps as seen for indium bumps created with electroformed masks (Fig. 1c). The tape is easy to use and experiments with glass slides attached to each other with this tape, were separated again by simply leaving the assembly in acetone for approx. 13 hours.

Due to current unavailability of the indium evaporator and expected upgrade of the equipment, no indium deposition could be carried out with the 55μm-pitch Si wafer masks so far. We expect further results by the time of the EMPC'23 conference.

IV. CONCLUSION

Applying electroformed stencils as shadow masks demonstrated that indium bumps of 250μm- and 100μm-pitch arrays can be produced on individual dies. The shadow mask aperture array is aligned to the pixel array of the die and clamped together in a custom-made jig that can be mounted into an indium evaporator. To date this method was employed for flip-chip bonding a prototype radiation detector with 250μm pixel pitch indicating an interconnect bond yield better than 99.9%. To further decrease the pixel pitch suitable for 55μm-pitch radiation detectors, shadow masks of Si membranes with apertures were produced on 4-inch wafers. Three different soluble adhesives (IPA-diluted, acetone-diluted, double-sided tape) are currently tested for adhering single dies temporarily to such Si shadow masks instead of mounting Si membrane and die in a custom-made jig. A process with double-sided tape or stamping acetone-diluted adhesive from a thin film onto the shadow mask seems to be promising for the alternative attachment. This is a considerable step towards the fabrication of radiation detectors with 55μm-pitch made from single sensor and ASIC dies which will be further tested in future.

ACKNOWLEDGMENT

The authors would like to acknowledge STFC-RAL's Detector Development Group for testing the prototype detector.

References

[1] T. Hatsuia, H. Graafsma, "X-ray imaging detectors for synchrotron and XFEL sources," IUCrJ (International Union of Crystallography Journal), vol. 2, pp. 371–383, Jan. 2015.

[2] N. Tartoni, "X-ray detectors at Diamond Light Source, evolution and future challenges," JINST, vol. 17, C11003, Nov. 2022.

[3] S. Del Sordo, L. Abbene, E. Caroli, A. M. Mancini, A. Zappettini, P. Ubertini, "Progress in the Development of CdTe and CdZnTe Semiconductor Radiation Detectors for Astrophysical and Medical Applications," Sensors, vol. 9, pp. 3491-3526, May 2009.

[4] L. Davydov, P. Fochuk, A. Zakharchenko, V. Kutny, A. Rybka, "Improving and Characterizing (Cd,Zn)Te Crystals for Detecting Gamma-Ray Radiation," IEEE Trans. Nuc. Sci., vol. 62, pp. 1779-1784, Aug. 2015.

[5] R. O. Pak, K. V. Nguyen, C. Oner, T. Chowdhury, K. C. Mandal, "Characterization of Cd0.9Zn0.1Te single crystals for radiation detectors," 2015 IEEE Nuc. Sci. Symp. and Med. Imag. Conf., San Diego, CA, USA, pp. 1-7, Oct. 2015, doi: 10.1109/NSSMIC.2015.7582273.

[6] E. N. Gimenez, R. Ballabriga, M. Campbell, I. Dolbnya, I. Horswell, "Evaluation of the Radiation Hardness and Charge Summing Mode of a Medipix3-based detector with Synchrotron Radiation," IEEE Nuc. Sci. Symp. and Med. Imag. Conf., Knoxville, TN, USA, 1976-1980, Oct. 2010, doi: 10.1109/NSSMIC.2010.5874120.

[7] M. Fiederle, H. Braml, A. Fauler, J. Giersch, J. Ludwig, K. Jakobs, „Development of Flip-Chip Bonding Technology for (Cd,Zn)Te," IEEE Trans. Nuc. Sci., vol. 51 pp. 1799-1802, Aug. 2004.

[8] D. Turecek, J. Jakubek, E. Trojanova, L. Sefc, V. Kolarova, "Application of Timepix3 based CdTe spectral sensitive photon counting detector for PET imaging," Nuc. Inst. and Meth. in Phys. Res. A, vol. A 895, pp. 84-89, April 2018 & https://advacam.com/1624.html.

[9] H. Chen, S. A. Awadalla, K. Iniewski, P. H. Lu, F. Harris, "Characterization of large cadmium zinc telluride crystals grown by traveling heater method," J. Appl. Phys., vol. 103, pp. 014903 1-5, Jan. 2008.

[10] C. Koch, T. J. Rinke, Photolithography – Basics of Microstructuring, 2nd ed. Siegl Druck & Medien GmbH & Co. KG, Friedrichshafen, 2020, pp. 64.

[11] W. Decker, R. Belan, V. D. Heydemann, S. Armstrong, T. Fisher, "Novel Low Pressure Sputtering Source and Improved Vacuum Deposition of Small Patterned Features Using Precision Shadow Masks," Society of Vacuum Coaters 59th Annual Technical Conference, Indianapolis, IN, U.S.A., p. 95-100, May 2016.

[12] A. Schneider, M.C. Veale, D.D. Duarte, S.J. Bell, M.D. Wilson, "Interconnect and bonding techniques for pixelated X-ray and gamma-ray detectors," JINST vol. 10, C02010, Feb. 2015.

[13] J. Salmi, J. Salonen, "Solder bump flip chip bonding for pixel detector Hybridization," presented at Workshop on Bonding and Die Attach Technologies, CERN, Geneva, Switzerland, June 2003, https://ssd-rd.web.cern.ch/bond/talks/1-04_Salmi.pdf (VTT).

[14] C. Broennimann et al., "Development of an indium bump bond process for silicon pixel detectors at PSI," Nucl. Instrum. Meth. A, vol. A 565, pp. 303-308, May 2006.

[15] J. D. Lipp, "STFC Existing Bump-bonding Programme," presented at UK Bump Bonding Workshop, Daresbury, May 2012, https://indico.cern.ch/event/191355/ .

[16] Datasheet in https://www.smt.asmpt.com/en/products/process-support-products/stencil-technologies/ .

[17] A. Schneider, D. Beckett, N. Ghorbanian, S. P. Cross, M. C. Veale, "A Study on Fine-Pitch Convertors for Radiation Detectors with Interposers as an Alternative to Through Silicon Via Technology," IEEE 9th Electronics System-Integration Technology Conference (ESTC), Sibiu, Romania, Sept. 2022, https://ieeexplore.ieee.org/document/9939380 .

[18] A. Schneider, N. Ghorbanian, J. Osborn, S. P. Cross, J. D. Lipp, M. J. French "Bump deposition techniques for hybrid X-ray detectors," JINST, vol. 18, C06009, June 2023.

Silver bonding wire – an alternative for mechanical sensitive chip configurations in automotive electronics packaging

Robert Klengel*, Sandy Klengel*, Sebastian Tismer*, Noritoshi Araki**, Motoki Eto**, Teruo Haibara**, Takashi Yamada**,
Jochen Feldmann***, Achim Scheer***, Ralph Binner***, Henk Peters***
*Fraunhofer-Institute for Microstructure of Materials and Systems IMWS,
Electronic Materials and Components
Walter-Huelse-Strasse 1, 06120 Halle, Germany
Email: robert.klengel@imws.fraunhofer.de
**Nippon Micrometal Corporation
158-1 Sayamagahara, Iruma-City, Saitama 358-0032, Japan
Email: ntaraki@nmc-net.co.jp
***ELMOS Semiconductor SE
Heinrich-Hertz-Strasse 1, 44227 Dortmund, Germany
Email: jochen.feldmann@elmos.com

Abstract— Silver (Ag) bond wires are mainly used in consumer products as a reliable low-cost alternative to gold (Au) bond wires. In previous investigations, we have shown that silver wires also have potential for use in the automotive sector [1, 2, 3, 4], where the materials are exposed to particularly harsh environmental conditions. In order to understand the influence of these conditions in the electronic package on the silver-aluminum material system, we carried out extensive studies on test packages built on laboratory level. For the future development of challenging automotive devices, it will be important to have a cooperation between semiconductor package developer and material supplier with an assist of characterization specialists. In this project, we formed a three-party project in which characterization specialists, package developer, and wire manufacturer collectively conduct a feasibility study of a new Ag wire material for automotive application. Two wire types (standard Ag wire and new type of Ag wire GX2s) were chosen for the investigation in direct comparison regarding long-term reliability. Samples were built under mass production conditions and then subjected to various reliability tests following AEC-Q006 requirements. Subsequently, mechanical characterizations by means of ball shear and pull tests as well as high resolution SEM microstructure analyses on cross-section samples are applied.

In the paper, results of the different tests will be summarized, but a more in-depth analysis will be given to the temperature cycle test (TCT), where the most significant differences between the sample variations occurred. Overall, the results show the importance of the knowledge for the material reliability upfront a device qualification process (robustness validation) and give further guidance for material selection.

Keywords— *silver wire bonding, TCT, temperature cycle test, long term reliability, automotive electronics packaging*

I. INTRODUCTION & MOTIVATION

In the past decade, gold has been widely replaced by copper as a bonding wire material for ultrasonic ball-wedge bonding. The main driver here was the cost factor due to the enormous difference in the price of the raw materials. Copper is ideal as a bonding wire contact material because it has very good electrical and thermal conductivity as well as high stiffness. However, the copper wire material also has disadvantageous properties. Due to the higher hardness and

stiffness, there is an increased risk of chip cratering effects when bonding on particularly thin semiconductor metallization or for chip-to-chip bonding. Silver wire materials can bridge the gap between gold and copper bond wires here. Like copper, it has very good electrical and thermal properties, but has more favorable mechanical characteristics for wire bonding applications. Table 1 compares relevant properties of the three bonding wire raw materials [1, 2, 3].

Typically, silver wire materials are used in LED applications, where copper wires cannot be applied due to the necessary reflectance and in consumer electronics for challenging chip configurations, where the mechanical properties of the copper material can become a problem. In our previous publication [4], we have already investigated the reliability of silver wire bond contacts under harsh conditions, such as automotive applications, and were able to show that specially enhanced silver wire materials can also perform reliable in such application scenarios. The study now available is a follow-up to the results of the last publication. Based on the positive findings of the investigations on dummy samples manufactured at the laboratory level, an extensive study was now carried out on functional standard packages manufactured under mass production conditions. These were subjected to a variety of reliability tests based on the requirements of the AEC-Q006 standard. The aim is to demonstrate the reliability of a silver wire bonded mass production component for automotive applications.

II. SILVER WIRE BONDING

Silver wire bonding in the ball-wedge process is carried out in the common process steps as also known from gold or copper wire bonding. Due to its tendency to oxidize, the Free

TABLE I. Comparison of bond wire raw materials [1, 2, 3]

Property / Material	Gold	Copper	Silver
Electrical resistivity [μΩm]	0.024	0.017	0.016
Thermal conductivity [W/(m*K)]	318	401	428
Elastic modulus [10^{11}Pa]	0.88	1.36	1.01
Ultimate strength [MPa]	100	210	170

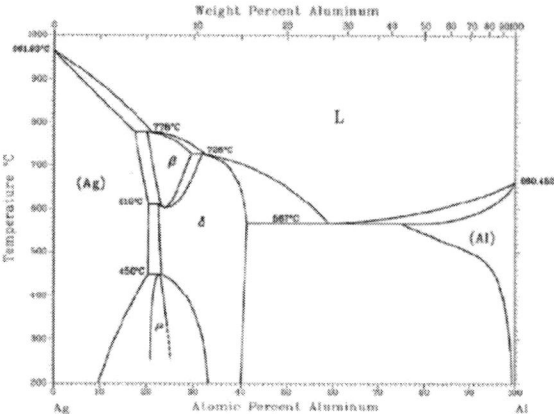

Fig.1   Silver-Aluminum binary phase diagram [5]

Air Ball (FAB) is formed under inert gas (forming gas N_2 + 5% H_2 or pure N_2). All common chip-side metallization and substrate-side bond metallization used for the gold and copper wires can be considered as bond surfaces.

According to the phase diagram (see Fig.1), in the silver-aluminum wire bond system, there are two types of intermetallic compound (IMC), Ag_2Al (δ) and Ag_3Al (μ). Each IMC exhibits different property and thus different sensitivity to certain degradation mechanisms. In our previous publication [4], we have identified and investigated these IMCs intensively by means of transmission electron microscopy (TEM) and have discussed relevant corrosion inducing/suppressing mechanisms.

III. RELIABILITY STUDY

In automotive applications, the materials are exposed to particularly harsh environmental conditions. In order to understand the influence of these conditions in the electronic package on the silver-aluminum material system, we carried out this extensive study. The focus is on the correlation beween the mechanical stability of the chip-side wire bond contacts and the material interaction processes in the interface depending on the silver wire material and the epoxy mold compound (EMC).

A. Sample Preparation

For the present study, two silver wire materials with 25µm in diameter were contacted to an Al-0.5wt%Cu chip metallization (1.0 µm thickness) by thermo-sonic ball/wedge

TABLE II. Reliability test conditions

Reliability test	Test conditions
MSL-1	125°C for 24 hours → 85°C + 85%rh for 168 hours → 3x 270°C reflow
AC	121°C + 100%rh + 29psig for 96/192 hours
TCT	-55/150°C (30 min/cycle) for 1,000/2,000/3,000 cycles
Biased HAST	0-3.3V bias Voltage at 130°C + 85%rh for 96/192 hours
HTSL 1	150°C for 1,000/2,000/3,000/4,000 hours
HTSL 2	175°C for 1,000/2,000 hours

bonding using forming gas. The first wire material is a commercially available standard silver wire, denoted as stdAg. The second material is the GX2s wire from Nippon Micrometal Corporation (NMC) which has optimized corrosion resistance by additive elements and a low resistivity (2.4 μ Ω cm). The used component is a QFN32L5 package provided by ELMOS with two silicon-based dies (see Fig. 2). Two different EMC were used – a low halide containing standard EMC (<<5ppm) and a sulfur-free variant. The complete assembly (die placement and attach, wire bonding, molding, singulation etc.) was processed in a standard ELMOS mass production assembly line. Fig. 2 shows the package configuration and the wire bond layout.

B. Reliability Testing

The components were subjected to various reliability tests following AEC-Q006 requirements. Table I lists the tests performed with the individual conditions. Samples of all variations (both wire types and both EMC) were investigated in initial state by pull and shear testing as well as by means of microstructural cross section analyses to ensure sufficient and comparable starting condition of interface quality. Moisture sensitivity level (MSL-1) was performed as pre-conditioning before autoclave testing (AC) testing, temperature cycle testing (TCT) and biased highly accelerated stress test (bHAST). High temperature storage (HTSL) was applied at 150°C and 175°C.

C. Analysis Methods

Ball shear and pull tests were conducted on samples in initial condition (as bonded) as well as in intermediate steps and at the end of each reliability test. Two parts with 35 wire bond contacts each were prepared for both pull and shear tests. For this, the mold compound was removed with laser IC opener and mixed acid followed by acetone wash to gain access to the wire bond contacts. Fig. 3 shows the sample condition after mold compound removal exemplarily. For microstructural analysis the test specimens were cross sectioned using broad ion beam preparation. The method uses an argon ion beam to mill cross sections with a high surface

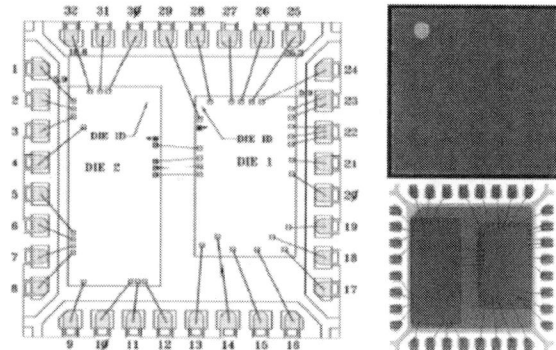

Fig 2   Scheme of QFN component used for the study (left), optical image and 2D x-ray image with wire bonding layout visible (right)

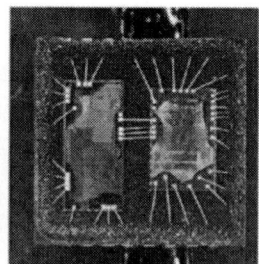

Fig 3   Decapsulation prior pull and shear testing: condition after laser opening (left); condition after chemical removal of EMC (right)

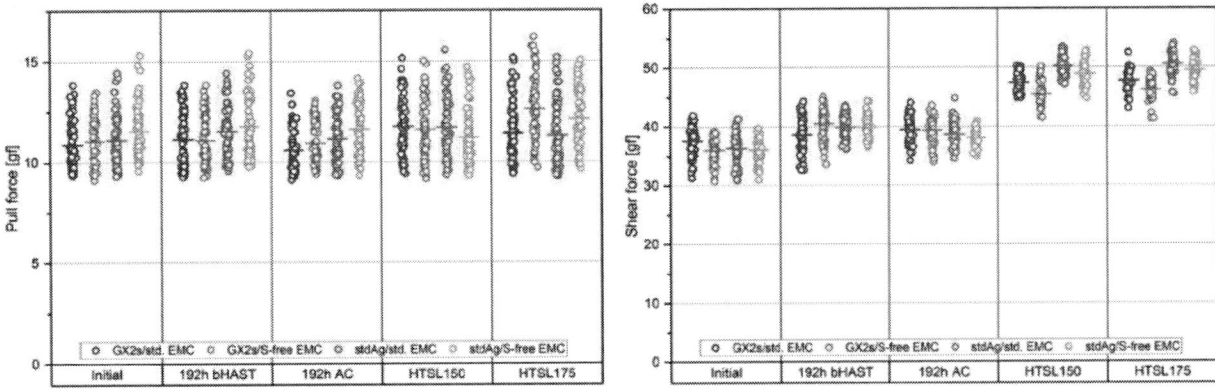

Fig. 4   Results of wire pull testing (left) and ball shear testing (right) in initial state and after different reliability tests.

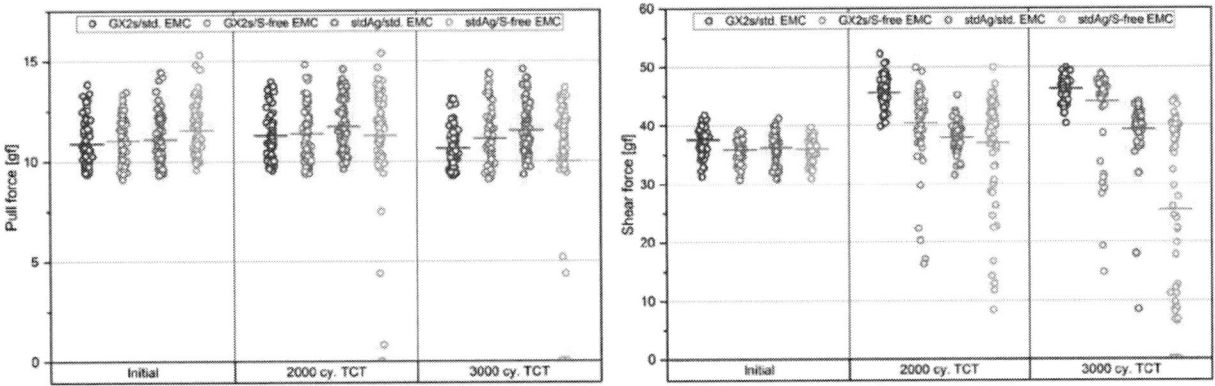

Fig. 5   Results of wire pull testing (left) and ball shear testing (right) in initial state and after 2000 resp. 3000 cycles of TCT.

quality and to obtain cleanly polished surfaces of the interface between the wire bond materials and substrate metallization. Following the preparation, high resolution microstructural analyses were performed by SEM.

IV. RESULTS AFTER RELIABILITY TESTING

A. Results After AC Testing, Biased HAST and HTSL

Both in the initial condition after bonding and molding, and after reliability testing, ball pull and shear testing was performed to determine the contact strength. The diagrams in Figure 4 show the results after AC testing, bHAST and HTSL at 150°C/ 175°C. For all these test conditions a very similar behavior could be determined for the different material variations. No significant differences in the stability of the wire bond contacts were found, neither depending on the wire types nor with regard to the epoxy mold compounds. For both the pull test and the shear test, the mean values are at a very comparable level and also with regard to the scatter there are only minor differences. All variants showed very stable and reliable interface connection. The microstructure analyses after ion beam cross-section preparation confirmed these results. All contacts showed good formation of intermetallic phases. In no case significant degradation phenomena, such as void formation or corrosion damage, were detected.

B. Results After TCT

The condition of the contact interfaces after thermal cycling differs significantly from the results of the other reliability tests (see Fig. 5). With sulfur-free EMC, the scatter of shear strengths for the contacts of both wire materials

increases noticeably already after 2000 cycles. There are a few outliers with significantly reduced values, whereby this trend is more pronounced for the std.Ag wire. With the standard EMC, both wire types show no changes in the shear test after 2000 cycles compared to the initial values. For pull testing, both EMC variants with GX2s wire behave completely inconspicuously, and the std.Ag wire with standard EMC also shows no decrease in force compared to the initial condition.

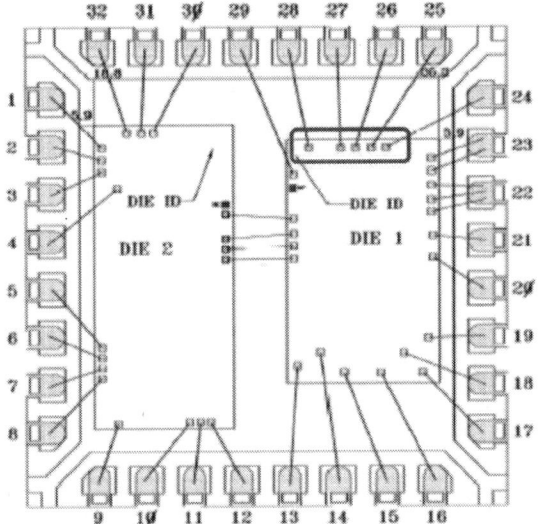

Fig. 6   Package layout with marked area of increased damage potential

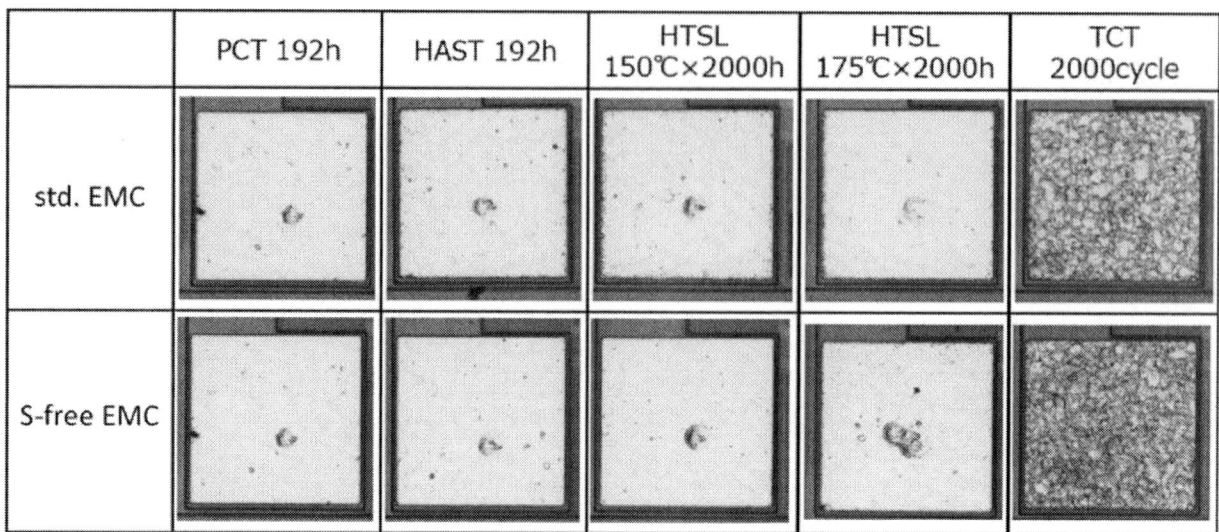

	PCT 192h	HAST 192h	HTSL 150℃×2000h	HTSL 175℃×2000h	TCT 2000cycle
std. EMC					
S-free EMC					

Fig. 7   *Differences in surface appearance of non-bonded Aluminum pad metallization depending on reliability test condition.*

In contrast, the std.Ag wire with S-free EMC exhibits sporadic outliers with low pull forces.

After 3000 cycles of thermal cycling, the interface stability continued to degrade significantly. In the pull test, the GX2s wire with both EMC variants and the std.Ag wire with the standard EMC again show no changes from the initial state. In the combination of std.Ag wire with S-free EMC, as in the condition after 2000 cycles, a few outliers with significantly lower values occur. The development of the shear strength shows the clearest differences. The GX2s wire with the standard EMC is the only combination where the bond contacts show stable test values. The std.Ag wire with

standard EMC shows some outliers with significantly reduced values. With the S-free EMC, both wire types show a partly drastic reduction in shear strength, although the behavior is significantly more pronounced with the std.Ag wire and also results in sporadic lift-off.

Two aspects were particularly striking in the development of the contact strengths under load with temperature cycling. Firstly, the differences as a function of the EMC were unexpected. Secondly, the question arose as to why some bond contacts show a significant reduction in strength, while the majority remain in a very stable condition. The correlation of the reduced strength values with the position in the package

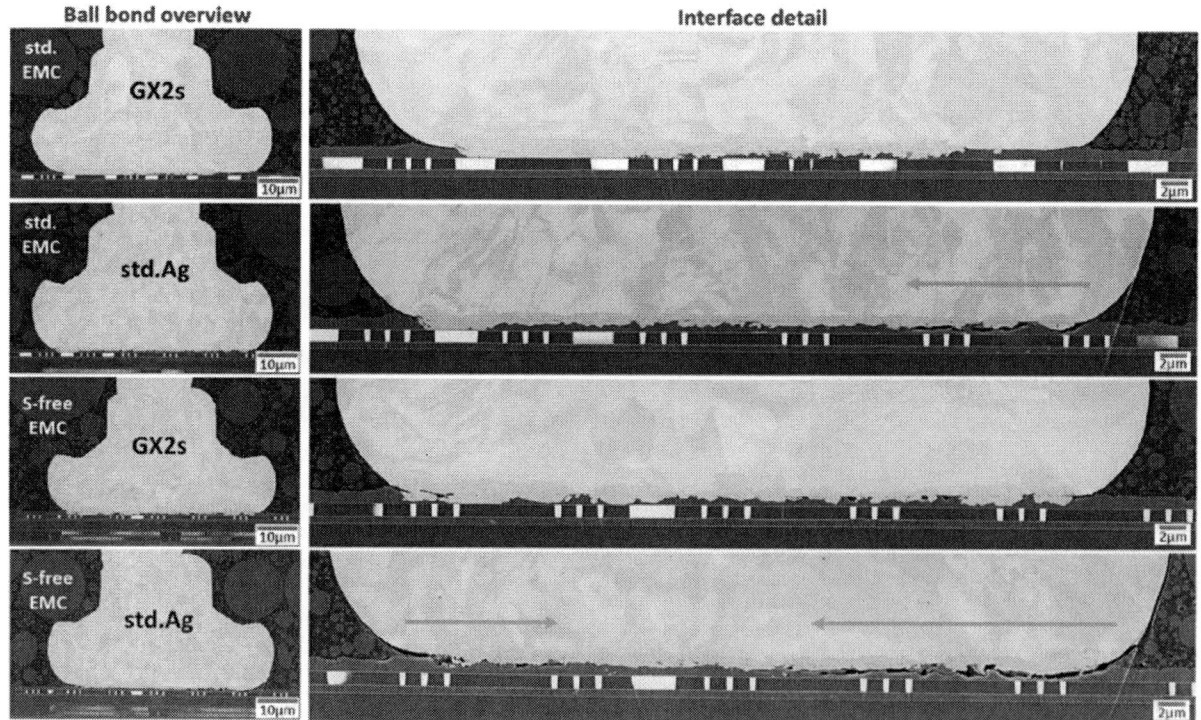

Fig. 8   *Results of SEM cross section analyses on wire bond contacts after 2,000 cycles of TCT (-55°C/ 150°C).*

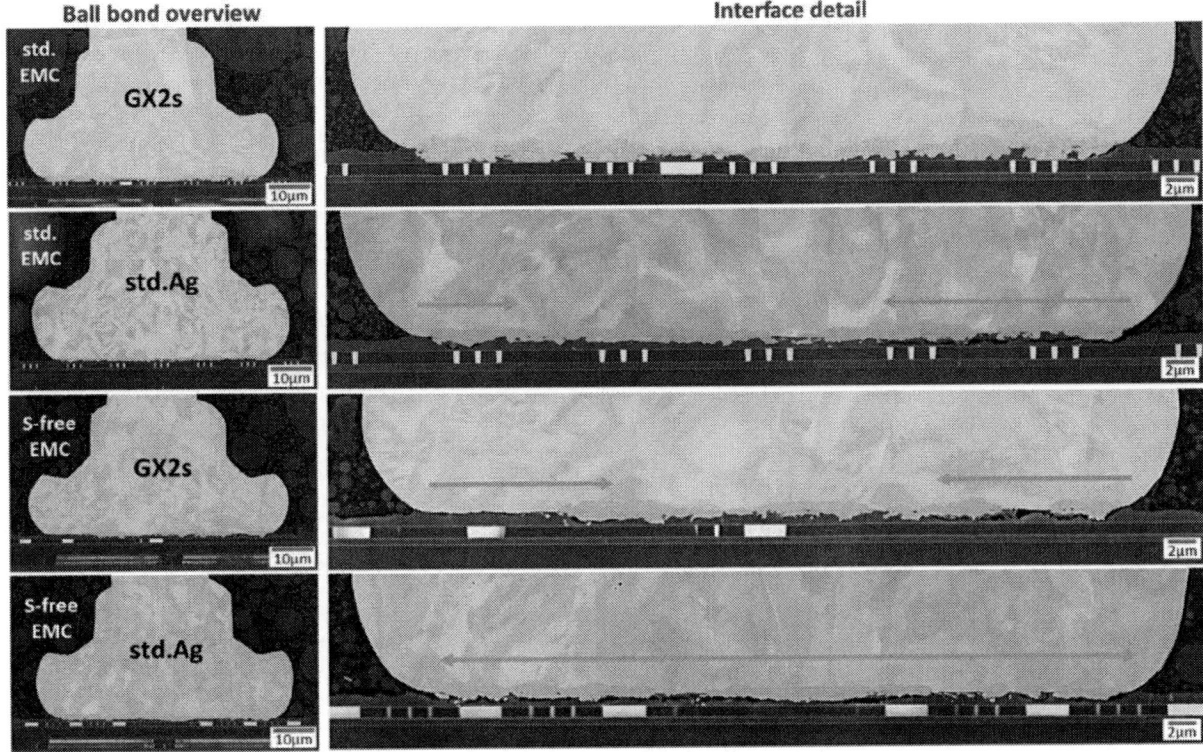

Fig. 9   Results of SEM cross section analyses on wire bond contacts after 3,000 cycles of TCT (-55°C/ 150°C).

revealed that there is an accumulation of lower values in the area of pins 25 to 28 (see Fig. 6). Furthermore, it was noticed during the investigation that an neighboring bond pad, which is not contacted, shows a significantly higher roughness after loading with temperature changes than after all other reliability tests. Figure 7 shows these differences as a function of the loading condition. Based on these contact findings, additional cross-sections were prepared on this contact sequence using ion beam technology and subsequently the damage pattern was analyzed microstructurally using SEM.

Figures 8 and 9 show the interface degradation on one exemplary wire bond contact per material variation from the

relevant chip area after 2000 and after 3000 cycles of TCT, respectively. After 2000 cycles, no degradation can be seen for the contacts of the GX2s wire with either EMC variant. In contrast, the std.Ag wire already shows clear damage at this stage, especially in combination with the S-free EMC. The blue arrows in the figures mark the progress of degradation. The condition after 3000 cycles shows a significant increase in damage. The std.Ag wire shows a complete lift-off in combination with the S-free EMC. The GX2s wire also shows significant interface weakening with this EMC, but the contact is not completely separated. With the standard EMC, The contacts of the std.Ag-wire show progressive damage, while

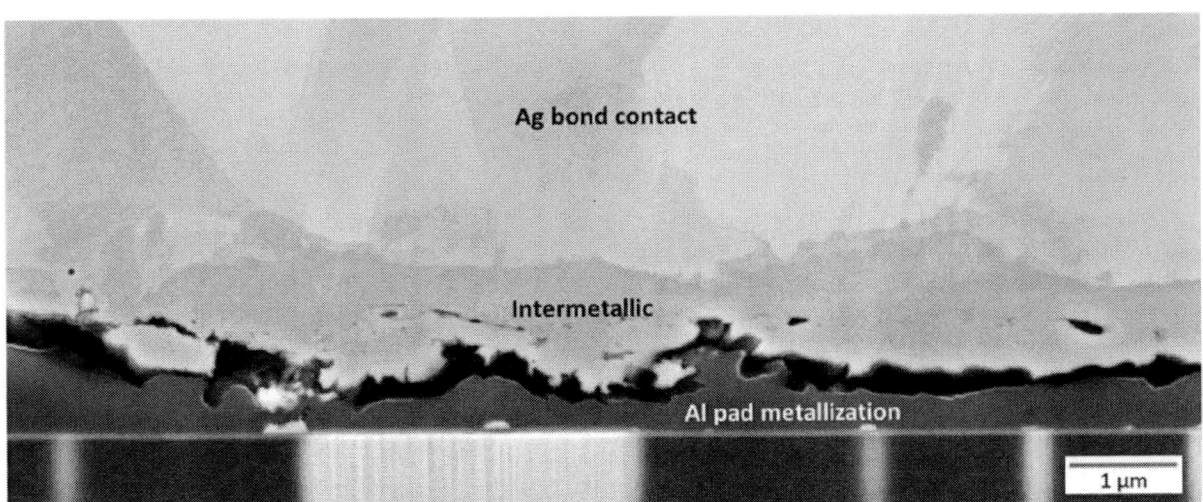

Fig. 10   Degraded interface of wire bond contact after 3000 cycles TCT; std.Ag wire and S-free EMC

39

no degradation is evident for the GX2s. These results correlate excellently with the results of the pull and shear tests.

However, the concrete failure interface is unexpected. Already in the overviews, but even better in the detailed picture in Figure 10, it can be seen that the crack runs between the intermetallic compound region (IMC) and the remaining aluminum pad metallization. In contrast to halide-induced corrosion damage of wire bond contacts, it is not the IMC that is weakened here, but the remaining bond metallization appears to be degraded. A gap is formed between this and the IMC, which is not filled with reaction products. Since the phenomenon occurs only at temperature cycling, thermo-mechanical deformation of the aluminum metallization is likely. However, at this stage, the damage mechanism is not fully understood. Further analyses to clarify this will follow.

V. SUMMARY AND DISCUSSION

An extensive study was conducted to investigate the reliability of silver bond wires in harsh environments (e.g. automotive applications). Two different silver wire materials (std.Ag and GX2s) and molding compounds (standard EMC and S-free EMC) were selected for this purpose. Standard automotive components were assembled under mass production conditions and subsequently subjected to various reliability tests based on the AEC-Q006 standard. In the initial state without aging as well as in all test levels, mechanical characterizations of the contact strength were performed by means of pull and shear tests (after decapsulation of the package) as well as microstructural analyses of the interconnect interfaces by means of SEM (after ion beam cross-sectioning).

Except for cyclic temperature loading in the TCT, the four wire-EMC variations showed very stable and comparable contact behavior for all reliability tests. No significant reductions occurred with respect to mechanical strength. This correlates very well with the microstructure analyses, in which no relevant degradation could be detected. From 2000 cycles TCT on, a reduction of the pull forces and shear strengths can be detected for the components with the S-free EMC and the std.Ag wire. For the parts with GX2s wire, only a slight decrease in shear values can be detected, but no change can be observed in the pull test. Even after 3000 cycles, no reduction in the contact stability of the GX2s wire is evident in the pull test, and the shear strengths have only decreased marginally. The picture is different for the std.Ag wire. For this, a considerably higher number of contacts with low shear values is recorded after 3000 cycles of TCT. As already after 2000 cycles, complete contact lift-offs without detectable shear force also occur in part. The components with the standard EMC are significantly more robust. Only the std.Ag wire showed reduced shear strength after 3000 cycles.

The matching of the low contact strengths detected in the mechanical tests with the corresponding positions in the component revealed an accumulation of weakened bond contacts in a specific area. In this area, microstructure analyses were carried out on cross-sectioned wire bond interfaces in order to investigate the damage pattern and the degradation progress as a function of the number of cycles. The results regarding the degradation level correlate very well with the findings from the mechanical tests. With increasing test duration, the damage pattern shows a progressive separation between the IMC and the aluminum pad metallization remaining under the contact. Probably, starting from the contact edge, a roughening and associated pore formation occurs in the interface due to the thermomechanical stress arising from a large CTE (coefficient of thermal expansion) difference between the Al metallization and underneath Si die. With further loading, a gap forms. No signs of corrosive material degradation have been observed so far. The influence of the position, which obviously exists, has not yet been investigated further.

In summary, all material variations showed very good reliability for most stress tests and showed no relevant contact degradation. Only the cyclic temperature changes in the TCT showed a significant difference in the test variations. A particularly severe bond damage was observed for the std.Ag wire in combination with the S-free EMC. Degradation was also found for the GX2s wire, but to a much lesser extent, which was also confirmed by the microstructure analyses. Thus, the GX2s wire shows the more robust reliability behavior. The damage pattern observed still leaves some questions open with regard to the degradation mechanism. Further investigations must be carried out in this regard. The influences of the EMC material, the wire type and the position in the component also need to be examined in more detail in order to fully understand the mechanism of the damage process and to derive improvement actions.

REFERENCES

[1] ASM Metals Reference Book, 3rd Edition, ASM International 1993.

[2] Metal Data Book, 4th Edition, Japan Institute of Metals and Materials, 2004.

[3] A. M. Howatson, P. G. Lund, J. D. Todd, Engineering Tables and Data, William Clowes & Sons Ltd, London, 1972.

[4] R. Klengel, S. Klengel, J. Schischka, T. Stephan, M. Eto, N. Araki, T. Yamada, "Corrosion effects and reliability improvement of silver wire bonded contacts in automotive application", Electronic Components and Packaging Conference (EMPC), 2021, pp. 1-7.

[5] T.B. Massalski, Binary Alloy Phase Diagrams, 2nd Edition, ASM International 1990.

Cu Pumping Analysis during Cu/SiO₂ Hybrid Bonding using In-situ SPM Imaging

Ali Roshanghias
Silicon Austria Labs GmbH
Villach, Austria
ali.roshanghias@silicon-austria.com

Jaroslaw Kaczynski
Silicon Austria Labs GmbH
Villach, Austria
Jaroslaw.Kaczynski@silicon-austria.com

Ude Hangen
Bruker Nano GmbH
Aachen, Germany
Ude.Hangen@bruker.com

Abstract— The assessment of Cu pumping (also known as Cu extrusion) during thermal annealing is vital information for the successful execution of hybrid bonding as well as defect-free processing of the through silicon vias (TSV). Unpredicted Cu pumping can pose major reliability issues. Correspondingly, in this study, high-temperature analysis of Cu pumping was conducted utilizing in-situ scanning probe microscopic (SPM) imaging. Cu / SiO₂ surfaces with recessed and protruded Cu topographies were produced by chemical mechanical polishing (CMP) and used for Cu pumping investigations. The amount of Cu pumping upon thermal annealing up to 400 °C and cooling down to room temperature was precisely quantified. The SPM results were compared with FEM simulation results, and a numerical equation for Cu pumping was proposed, accordingly. It was shown that by using in-Situ SMP imaging, valuable information on the behavior of hybrid Cu / dielectric surfaces can be generated.

Keywords—Cu pumping, SPM, Nanoindentation, Hybrid bonding, Cu Extrusion, TSV, BEOL, CMP

I. INTRODUCTION

Hybrid bonding and Through Silicon Via (TSV) technologies are the two key processes for 3D integration enabling "more than Moore" scaling. Hybrid bonding has recently emerged as the preferred direct wafer-level bonding approach for creating high-density interconnects. TSVs are also widely utilized for both 2.5 D interposer-type integration as well as 3D IC stacking. Hybrid bonding consists of dielectric surface activated bonding (SAB) at low temperatures followed by thermal annealing up to 400 °C to facilitate copper (Cu) bonding. For a successful hybrid bonding, the chemical-mechanical polishing (CMP) processing of the Cu/ dielectric plays a critical role where the surface roughness, uniformity and dishing depth can be influenced. Moreover, the annealing process parameters can highly impact the Cu-Cu bonding yield [1-5].

Hybrid bonding is usually executed using two wafers with recessed Cu surfaces (with dishing heights ranging from 5 to 10 nm) surrounded by a dielectric layer. The dielectric layer can be SiO₂, SiN, or SiCN. Recently, the employment of recessed Cu on one wafer and protruded Cu on the other wafer has received attention. This is owing to the creation of huge tensile stress after cooling in recessed-recessed bonds degrading the yield and long-term reliability of the bonds. The protruded-recessed bond can facilitate an optimized bondline shape for hybrid bonding [3]. The advantage of protruded-recessed Cu bonds for further downscaling of hybrid bonds to sub-μm dimensions was also reported before [4]. The relatively limited Cu expansion at sub-μm dimensions might be a concern for typical recessed-recessed hybrid bonding. The smaller gap size of protruded-recessed Cu bonds might be advantageous to avoid excessive annealing temperatures [4].

The assessment and control of Cu pumping during annealing is necessary for the successful execution of hybrid bonding as well as the processing of the through silicon vias (TSV) [6]. Cu pumping, which stems from Cu expansion due to differential CTE between Cu and the surrounding dielectric, is a stress release mechanism at high temperatures. The dominant mechanism for Cu pumping is diffusional creep [6, 7]. In fact, the coefficient of thermal expansion (CTE) of Cu is much higher (16.7 ppm/C) than the one of Si (2.3 ppm/C) and typical dielectric materials such as SiO₂/ SiN and SiCN. Consequently, Cu, which is confined within the dielectric or Si, will undergo compressive stress during any high-temperature processing. The weakest boundary where this stress can be released is the open surface on top of the via in hybrid bonding or the thin back end of line (BEOL) pad in TSV. Consequently, Cu extrusion in the vertical direction can be observed during any pre- or post-CMP annealing of TSVs, as well as heating associated with dielectric-deposition during the fabrication of BEOL or redistribution layer (RDL). Cu pumping increases with increasing temperature leading to the deformation and damage to the BEOL layers above in TSVs. Due to the plastic deformation of Cu, this effect cannot be completely reversed upon cooling to ambient temperature and only a very small elastic portion can be retracted [6].

Cu pumping had previously been studied using room-temperature metrology tools before and after thermal annealing using scanning white light interferometry (WLI) and atomic force microscopy (AFM). However, due to Cu's rapid oxidation at elevated temperatures, in-situ analysis of Cu pumping was not comprehensively investigated [4-8]. Usually, the amount of Cu pumping at high temperatures was estimated from simulation data. Additionally, both AFM and WLI have limitations when used as an in-line wafer-level process-controlling tool. For example, AFM is a relatively slow technique with a restricted scanning area, which does not easily allow for obtaining statistical information [6]. WLI, on the other hand, is a quick, contact-free, wafer-level technique. However, WLI is sensitive to surface optical properties, which is an issue for hybrid (Cu/dielectric) surfaces with different amounts of light reflection. In particular, for transparent layers such as SiO₂, WLI analysis can be challenging, as the optical properties and thickness of the SiO₂ layer should be known. Another approach is the deposition of a thin metal layer (e. g. 30 nm PVD Ta) on top of the sample [6].

In this study, an alternative methodology for in-situ analysis of Cu pumping based on scanning probe microscopy using nanoindentation imaging is introduced and investigated.

II. EXPERIMENTAL

In this study, 300 mm silicon test wafers with a Cu pad diameter of 4 μm and a pitch size of 10 μm surrounded by SiO_2 were used, as shown in Fig. 1. The area fraction of Cu at the surface of the chips was 12,5%. By using two different CMP recipes, wafers with Cu protrusion of + 13 nm and wafers with Cu recession of -9 nm relative to SiO_2 surface were generated. The effects of CMP processing parameters (e.g. Cu/ barrier polishing slurries, pads, polishing pressure, duration, and polishing rate selectivity) on the topography of Cu were described in [9].

In-situ scanning probe microscopy (SPM) was carried out using a Hysitron TI Premier nanoindenter (Bruker Corporation, USA). SPM is a contact-mode profilometry technique that employs a piezo scanner to sweep the indenter tip across the sample surface, monitoring any changes caused by topography or surface roughness [10]. A Berkovich indenter was used to perform in-situ SPM imaging. An extremely low contact force of 2 μN and a scan rate of 1 Hz was applied during raster scanning of the samples, while an area of 25 × 25 μm was analyzed in different temperatures. Both recessed and protruded Cu wafers were examined.

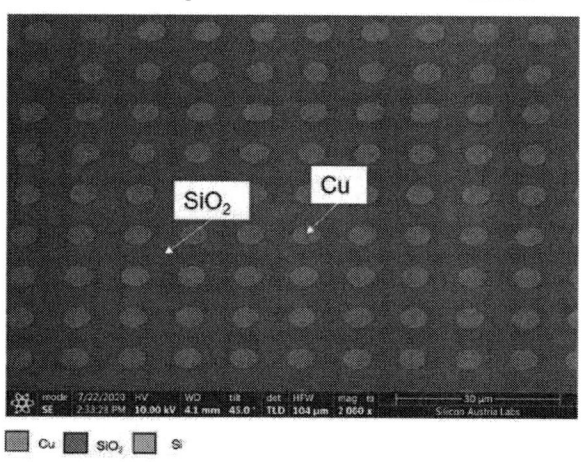

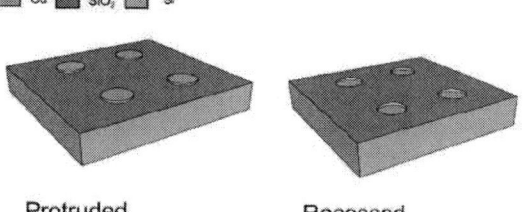

Fig. 1. An SEM image of the sample used in this study with a Cu pad diameter of 4 μm (top) and schematic demonstrations of the generated protruded and recessed topographies after CMP processing (bottom).

The samples were placed in a heating stage with resistive heating elements and kept in an N_2 environment during the measurements. SPM was performed up to 400 °C, while the temperature of the sample was controlled using a precisely calibrated PID feedback loop. Each SPM mapping took less than a minute. Also, the heating time to 400°C was less than a minute. The high-temperature SPM imaging setup and the

temperature analysis of the setup indicating the uniform temperature distribution over the sample are shown in Fig. 2.

In addition to the experiments, Cu pumping was also simulated using the finite element method (FEM) in ANSYS 21.2. Both simulation results and SPM measurements were compared and used to define a formula for Cu pumping which will be discussed later.

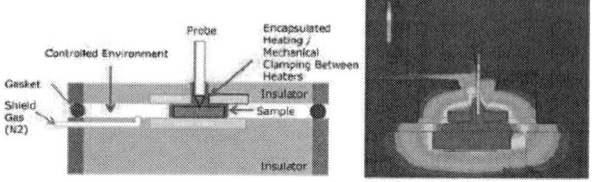

Fig. 2. High-temperature SPM imaging setup (left) and the temperature analysis of the setup indicating the uniform temperature distribution over the sample (right)

III. RESULTS

Fig. 3 depicts the 3D topographical maps of the recessed and protruded samples before, during, and after 400 °C annealing. An area of 25 × 25 μm consisting of 8 Cu pads was selected for these investigations.

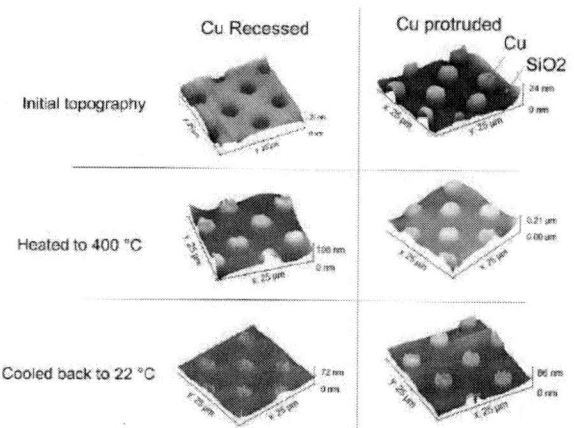

Fig. 3. SPM mapping results of two types of Cu/SiO_2 topographies (recessed and protruded) before annealing, heated to 400 °C and after cooled back to room temperature

Correspondingly, the section profiles of the samples are plotted in Fig. 4. As shown here, Cu in the recessed samples expands to 40 nm at 400 °C and retains 9 nm above the SiO_2 surface after cooling. Similarly, in protruded samples, the Cu pumping reaches 65 nm at 400 °C and remains 28 nm above the surface after cooling to ambient temperature. It can be inferred from these results that Cu pumping is an irreversible extrusion of Cu due to the occurrence of plastic deformation upon annealing which relaxed the high compressive stress in Cu [11].

The FEM simulation results of Cu pumping are also shown in Fig. 5. All materials are considered linear isotropic. Temperature variations in Young's modulus of Cu, as well as the coefficients of thermal expansion of SiO_2, Cu, Si and Ti, were also taken into consideration. The nonlinear stress-strain relationship of the Cu was implemented in the simulation by using the Ramberg-Osgood model [12, 13].

42

pads It can be estimated that the maximum amount of Cu pumping ($h_{max@\ 400\ °C}$) was around 51 ± 3 nm for both recessed and protruded Cu.

IV. DISCUSSION

As demonstrated in the previous section, utilizing in-situ metrology can be a valuable asset for providing reliable simulation input parameters and validating simulation output. Employing the empirical data obtained from SPM profilometric results and a feedback loop to the FEM model can provide a deeper understanding of the underlying mechanical behavior and expansion mechanisms of Cu. Given that, an attempt has been made to derive numerical equations for the maximal Cu pumping (h_{max}) at the annealing temperature as a function of Young's moduli and coefficient of thermal expansion (CTE) of Cu and dielectric materials. For this purpose, other dielectric materials such as SiN and SiCN were also fabricated and examined. For instance, the maximum amount of Cu pumping at 300 °C ($h_{max@300\ °C}$) can be expressed as a function of Young's moduli of Cu (E_{Cu}) and the dielectric material (E_d) as well as the CTE of the dielectric (α_d) as follows:

$$h_{max@\ 300°C} = A(\alpha_d) + B(\alpha_d) * ln\ (E_d/E_{Cu}),\qquad(1)$$

where

$$A(\alpha_d) = 40.415 + 1.3771 * 10^6 * \alpha_d \qquad(2)$$

and

$$B(\alpha_d) = 8.3373 + 1.3177 * 10^5 * \alpha_d\ .\qquad(3)$$

As inferred from Eq. 1, the amount of Cu pumping has a logarithmic relationship to E_d/E_{Cu}. This means that a hybrid surface consisting of Cu with the highest stiffness and dielectric with the lowest stiffness can yield the minimum amount of Cu pumping. In other words, the greater the stiffness of the dielectric, the greater the Cu pumping; since the expansion of Cu in the radial direction is more constrained.

The proposed formula holds great potential in optimizing both TSV processing and hybrid bonding in terms of Cu pumping prediction, ensuring reliable performance in various applications. Future work is currently underway to extend this equation including additional parameters such as temperature dependency, Poisson's ratio, and Cu microstructure, etc. In fact, Cu microstructure such as crystal orientation, and grain size can influence the amount of Cu pumping. In a previous study, a correlation between Cu pumping and Cu microstructure based on the grain size at the top of TSV was reported, where the TSVs with a higher number of grains showed a higher amount of Cu pumping. This was attributed to the fact that high-angle grain boundaries enhance both diffusion creep rate and dislocation creep rate, rendering higher Cu pumping [11].

V. CONCLUSION

In this study, Cu pumping in hybrid Cu/ SiO$_2$ surfaces was analyzed and quantified using in-situ SPM imaging at different temperatures. The maximum amount of Cu pumping was found to be 51 ± 3 nm for both recessed and protruded Cu topographies. The irreversible deformation of Cu after cooling back to room temperature was also verified. The experimental results showed good agreement with the FEM simulation results. By implementing both empirical data from SPM and

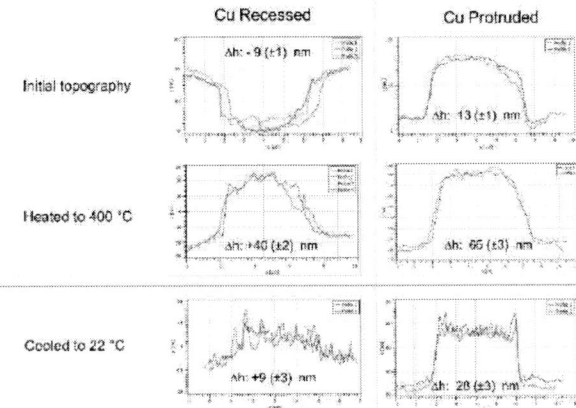

Fig. 4. SPM profilometric analysis of the samples upon thermal annealing

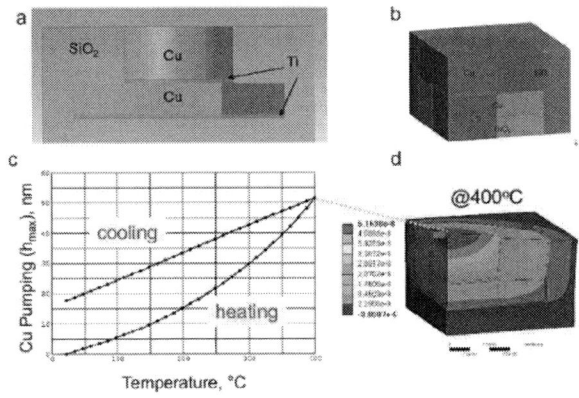

Fig. 5. FEM simulation results of Cu pumping upon heating to 400 °C. (a) and (b) illustrate the model geometry, (c) shows a gradual increase of Cu deformation upon heating to 400 °C and cooling back to room temperature. (d) presents the region with the maximum Cu deflection (h_{max})

TABLE I THE SUMMARY OF SPM AND SIMULATION REUSLTS

Recessed Cu	Condition	Simulation Δh [nm]	Experiment Δh [nm]
	@ 22 °C (initial)	-9.00	-9 ±1
	@ 400 °C	42.3	40 ±3
	@ 22 °C (after cooling)	8.6	+9 ±3
Protruded Cu	Temperature [°C]	Simulation Δh [nm]	Experiment Δh [nm]
	@ 22 °C (initial)	13	13 ±1
	@ 400 °C	64.5	65 ±3
	@ 22 °C (after cooling)	30.1	28 ±4

Table 1 summarizes the SPM experimental and FEM simulation results, implying a good agreement between them. Here Δh was defined as the average height difference of Cu in the middle of the pad to the SiO$_2$. As mentioned before, the reported measurement values are the average of those for 8 Cu

the results of a parametric simulation study, Cu pumping at high temperatures was consequently presented as a numerical function of Young's moduli and coefficient of thermal expansion (CTE). The proposed equation (Eq.1) provides a deeper understanding of Cu pumping and the underlying material parameters, laying a solid foundation for future investigations.

ACKNOWLEDGMENT

This work has been supported by Silicon Austria Labs (SAL), owned by the Republic of Austria, the Styrian Business Promotion Agency (SFG), the federal state of Carinthia, the Upper Austrian Research (UAR), and the Austrian Association for the Electric and Electronics Industry (FEEI).

REFERENCES

[1] Ohba, T., 2013, September. Wafer level three-dimensional integration (3DI) using bumpless TSV interconnects for tera-scale generation. In 2013 International Semiconductor Conference Dresden-Grenoble (ISCDG) (pp. 1-4). IEEE.

[2] Rudolph, C., Hanisch, A., Voigtländer, M., Gansauer, P., Wachsmuth, H., Kuttler, S., Wittler, O., Werner, T., Panchenko, I. and Wolf, M.J., 2021, June. Enabling D2W/D2D hybrid bonding on manufacturing equipment based on simulated process parameters. In 2021 IEEE 71st Electronic Components and Technology Conference (ECTC) (pp. 40-44). IEEE.

[3] Inoue, F., Iacovo, S., El-Mekki, Z., Kim, S.W., Struyf, H. and Beyne, E., 2021. Area-Selective Electroless Deposition of Cu for Hybrid Bonding. IEEE Electron Device Letters, 42(12), pp.1826-1829.

[4] Beyne, E., Kim, S.W., Peng, L., Heylen, N., De Messemaeker, J., Okudur, O.O., Phommahaxay, A., Kim, T.G., Stucchi, M., Velenis, D. and Miller, A., 2017, December. Scalable, sub 2μm pitch, Cu/SiCN to Cu/SiCN hybrid wafer-to-wafer bonding technology. In 2017 IEEE International Electron Devices Meeting (IEDM) (pp. 32-4). IEEE.

[5] Li, Y. and Goyal, D. eds., 2020. 3D Microelectronic Packaging: From Architectures to Applications (Vol. 64). Springer Nature.

[6] De Wolf, Ingrid, Kris Croes, O. Varela Pedreira, Riet Labie, Augusto Redolfi, Myriam Van De Peer, Kris Vanstreels, Chukwudi Okoro, Bart Vandevelde, and Eric Beyne. "Cu pumping in TSVs: Effect of pre-CMP thermal budget." Microelectronics Reliability 51, no. 9-11 (2011): 1856-1859.

[7] Su, Fei, and Bowen Zhang. "Copper pumping behaviors of TSV and experimental investigations of the mechanism." In 2020 21st International Conference on Electronic Packaging Technology (ICEPT), pp. 1-4. IEEE, 2020.

[8] Su, Fei, Xiaoxu Pan, Pengfei Huang, Yong Guan, Jing Chen, and Shenglin Ma. "Influence of copper pumping on integrity and stress of through-silicon vias." IEEE Transactions on Components, Packaging and Manufacturing Technology 6, no. 8 (2016): 1221-1225.

[9] Wang, S., Zhang, H., Tian, Z., Liu, T., Sun, Y., Zhang, Y., Dong, F. and Liu, S., 2022. Optimization of Cu protrusion of wafer-to-wafer hybrid bonding for HBM packages application. Materials Science in Semiconductor Processing, 152, p.107063.

[10] Dickinson, E., and Jeffrey P. "Probing more than the surface." Materials Today 12.7-8 (2009): 46-50.

[11] De Messemaeker, J., Pedreira, O.V., Philipsen, H., Beyne, E., De Wolf, I., Van der Donck, T. and Croes, K., 2014, May. Correlation between Cu microstructure and TSV Cu pumping. In 2014 IEEE 64th Electronic Components and Technology Conference (ECTC) (pp. 613-619). IEEE.

[12] Ramberg, W. and Osgood, W.R., 1943. Description of stress-strain curves by three parameters (No. NACA-TN-902).

[13] Uddin, M.M., Mondal, D. and Herrington, P.D., 2018, November. Finite Element Simulation of Backward Micro Extrusion for Annealed Copper. In ASME International Mechanical Engineering Congress and Exposition (Vol. 52019, p. V002T02A043). American Society of Mechanical Engineers.

Understanding the contact resistance in an ACF bonding

Helge Kristiansen
Conpart AS
Dragonveien 54
2013 Skjetten
helge@conpart.no

Giang M..Nghiem
Dept. of Microsystems
University of South-Eastern Norway
Vestfold, Norway
Giang.Nghiem@usn.no

Molly Bazilchuk
Ducky AS
Kjøpmannsgata 51, 7011 Trondheim
molly@ducky.eco

Knut E. Aasmundtveit
Dept. of Microsystems
University of South-Eastern Norway
Vestfold, Norway
kaa@usn.no,
ORCID: 0000-0001-9003-1916

Abstract— Anisotropic Conductive Film (ACF) bonding, the technique of choice for display interconnects, allows very fine pitch interconnect at a moderate process temperature, typically at the expense of a non-negligible interconnect resistance. Modern ACFs have conducting particles in a thin layer of high-viscosity adhesive. This allows even finer pitch but comes at the risk of high interconnect resistance due to trapped adhesive between pad and conductive particle. Conductive particles with spikes are designed to penetrate such a trapped adhesive layer.

In this paper, we compare spiky conductive particles with the traditional, spherical, ones. We measure interconnect resistance of individual particles in a nano-indentation test setup, as well as the interconnect resistance of ACF bonding using the two different particles, in daisy-chain measurements.

Interconnect resistance of individual spiky particles is more than an order of magnitude higher than the one for spherical particles. However, when mixed in an adhesive and used for ACF bonding, the interconnect resistance of ACF with spiky particles is somewhat lower than that of the ACF with spherical particles. The interconnect resistance of ACF with spherical particles is two orders of magnitude higher than that of the individual particles, whereas the interconnect resistance of ACF with spiky particles is comparable to the individual-particle resistance.

We conclude that trapped adhesive is a major concern for fine-pitch ACF bonding, and that the use of spiky conductive particles has the potential to overcome these challenges.

Keywords—anisotropic conductive adhesive, ACF bonding, metal-coated polymer spheres (MPS), fine-pitch bonding, trapped adhesive

I. INTRODUCTION

Anisotropic Conductive Film (ACF) is today the standard interconnect technology for display applications. The technology allows a very fine pitch interconnect and a moderate temperature during the assembly process. To improve the fine pitch capabilities of the technology, double layer ACF has been introduced, where the particles are confined to a thin layer of adhesive with extra high viscosity. The aim is to increase the area density of conducting particles

and at the same time avoid uncontrolled flow of the conductive particles during bonding.

The Achilles heel of the technology, is that the contact resistance is a concern since advanced display technologies require higher current densities than the traditional ones. Contact resistance has been addressed experimentally and by modelling. ACF interconnect resistance is typically higher than what is predicted by a simplified model of a metal-coated polymer particle compressed between two contact pads [1][2][3], implying that additional factors play a role. The contact resistance depends on several factors, the main are:

- Number of particles and their state of compression
- Effective conductivity of particle metallisation
- Effective contact area (given by the alignment accuracy)
- Particle deformation and contact topography
- Presence of particle insulation
- Trapped adhesive between particle and the respective contacts

Experimental measurements of contact resistance typically include a number of uncertainties:

- Bump and pad design makes the effective contact area (and hence number of particles) strongly dependent on bonding alignment
- Unknown or highly variable particle deformation due to the roughness of the bump surface and lack of bond planarity

Previous studies [2][3][4] have indicated that trapped adhesive film at the contact area is a major cause for the interconnect resistance being higher than expected, but a thorough experimental investigation is lacking. The resistance of single particles (without adhesive present) has been measured by electromechanical testing as a function of deformation, adding new insight [5]. In this paper, we compared such single-particle resistance (as measured in electromechanical testing) with ACF interconnect resistance (as measured for daisy chains in ACF bonded test vehicles) for two different metal-coated polymer particles.

This work is funded by the Norwegian Research Council through the project "Novel Particle for Display Interconnect", project number 228453 and through the Norwegian Micro- and Nano-Fabrication Facility (NorFab, project number: 295864).

II. Experimental

A. Design and manufacturing of test circuits

The test circuits were based on a silicon die measuring 10 mm x 2 mm, mounted onto a glass substrate. The silicon die has Al wiring, Si_3N_4 passivation and 10 μm high electroplated gold bumps. The test dies were fabricated by Taiwan Semiconductor Manufacturing Company (TSMC). The test vehicle included 5 daisy chains all containing 406 connections, each chain with different bump size. The daisy chains were divided into 12 segments of different lengths and was measured using a probe card. Total resistance in the measurement system including probes and their contact to the probe pads was below 10Ω. This is less than 5% of the resistance measured in the segments and is therefore neglected. The size of the gold bumps in the different daisy chains are 30x10, 35x10, 40x10, 40x15 and 60x15μm² respectively. To minimize the variation of the bump gap, one of the test chip wafers were planarized using a DAS 8920 surface planar at Disco in Taiwan. The surface topography of the test chip bumps was examined by a Wyko NT9100 interferometer.

The glass substrate was made by Industrial Technology Research Institute (ITRI) using Ti/Al/Ti metallization stack covered with ITO using a Ti/Al/Ti metallisation stack covered with an ITO layer. The substantial over-sized contact pads (80x90μm²) on the glass substrate allows some degree of misalignment without changing on the effective contact area. For more information on the test circuits, we refer to [6].

B. Conductive particles and ACF

For this experiment two different ACFs based on an epoxy system, were prepared by Hitachi Chemicals, specified in Table 1. The total adhesive thickness was 15 μm, with all the particles confined in a 4 μm thick layer of a higher viscosity resin.

Two different types of conductive particles were used in this experiment, both based on the same acrylic polymer core of 3.0 μm. One was plated with "Spiky Ni" by Hitachi Chemicals, the other with a traditional Ni-Au coating plated by Conpart. The surface roughness of the nickel coating of the spiky particle is of the order of hundred nanometres. Additionally, a fraction of the area of small (≈200 nm) polymer particles is added to the surface to improve the dispersion in the adhesive and avoid particle to particle contact, see Figure 1.

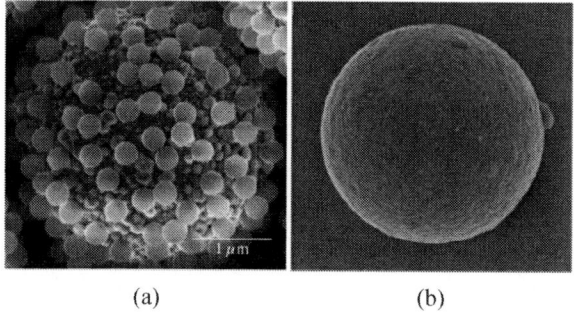

(a)   (b)

Fig. 1. a) Polymer core particle with spiky Ni plating. The surface is covered with small (≈ 200 nm) polymer particles. b) Polymer core particles with Ni-Au plating.

TABLE I.   ACF SYSTEM SPECIFICATION

ACF system	Type of conductive particle
ACF 1	Smooth Ni-Au
ACF 2	Spiky Ni
ACF Components	
Adhesive	Acrylic
Ni-Au	3.0 μm Acrylic core, coating 100nm Ni/20nm Au
Spiky Ni	3.0 μm Acrylic core, coating 150nm Ni
Particle density	16 -18,000 particles per mm²

C. Nanoidentation

The electrical properties of the two types of conductive particles have been measured using electrical nanoindentation, see Figure 2. The particle under test is placed between an indenter tip made of platinum-iridium and a gold coated silicon substrate, surrounded by normal air. During the test, the particle is compressed by a load increasing with 1.2 mN/s for 5 s, a hold stage of 2 s followed by unloading with the same rate. The total resistance from tip to substrate as well as the deformation is measured every 50 ms, as the indenter-tip is compressing the particle. A fixed resistance range of 100 Ω was chosen to minimize the time of measurement. In this way, the electrical performance of individual conductive particles was tested as function of mechanical deformation in a way mimicking the ACF bonding process, but crucially without the presence of an adhesive resin. This method is described in detail in [5].

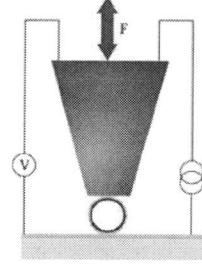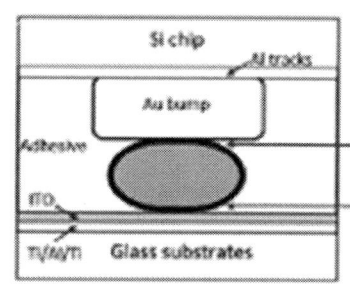

Fig. 2. Comparing Nano-indentation and ACF bonding

D. Assembly and testing of test circuits

The test circuits were assembled at Hitachi Chemicals research center. ACF was laminated to the glass / ITO substrate at 90°C for 3 seconds. After removing the carrier film, the test chip was aligned and pre-bonded to the glass substrate. The test vehicles were then moved to the bonding unit where the actual bonding took place at 150°C for 5 seconds. A thin PTFE film was inserted between the component and the bonding head.

Each component bonded to the patterned substrate was inspected using differential interference contrast (DIC) microscopy which makes it possible to observe the imprint from the particle on the pad from below, through the pad metallisation. The number of conductive particles were counted for several specified pads. A few components were also bonded to a non-patterned area of the glass that made it possible to inspect with traditional microscopy.

TABLE II. MEASURED BUMP AREA [μm²] FOR THE DIFFERENT DAISY CHAINS

Daisy chain	900	600	400	350	300
	Avg.	Avg.	Avg.	Avg.	Avg.
Planarized bump	920	570	330	290	250
Un-planarized bump	810	510	315	270	225

TABLE III. AVERAGE NUMBER OF CAPTURED PARTICLES AT DIFFERENT BUMP SIZES FOR DIFFERENT ACFs

	Number of particles									
	900μm²		600μm²		400μm²		350μm²		300μm²	
Type of ACF	Avg.	Std	Avg.	Std	Avg.	Std	Avg.	Std	Avg.	Std
ACF 1	17	3	11	3	7	2	6	2	6	2
ACF 2	15	3	9	2	6	2	5	2	4	1

Note- The statistics for ACF 1 and ACF 2 was measured on planarized IC samples.

The electrical resistance of the daisy chains was measured using 4-point measurement using a probe card. A Keithley 3706 System Switch / Multimeter was used for the measurements.

After electrical testing, some of the components were moulded in epoxy and cross-sectioned by mechanical grinding and polishing. The polishing process was consistently conducted in the direction from ITO toward Au bump and at a very slow speed to minimize any metal smearing. This also allowed for the measurement of the bond gap between bump and pad for the different bump sizes and bonding pressures.

III. RESULTS AND DISCUSSION

A. Planarization

Figure 3 shows the typical topography of the small (300 μm²) and large (900 μm²) bumps. Table 2 shows test chip bump area before and after planarization process. The variation in bump height between different bump sizes is less than 0.1μm, both for the planarized and un-planarized test chip. The bump sidewalls are slightly inclined, and hence the planarization of the bumps slightly increases the bump area, see Table 2. The planarization process also seems to create a slight smearing of the soft gold bump, see Figure 4. The bump area for the planarized bumps is 5 to 15% larger than for the un-planarized bumps.

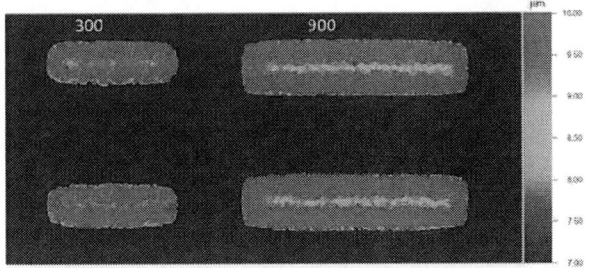

Fig. 3. Interferometry measurement of un-planarized bump surface at sizes 300 and 900 μm²

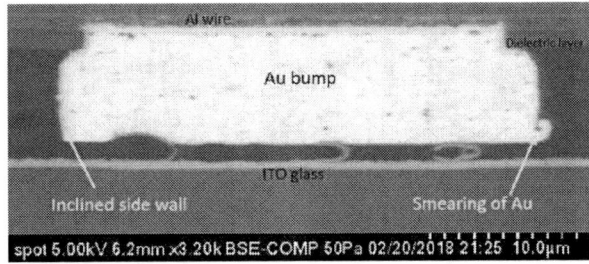

Fig. 4. Cross-section of planarized bump for a nominal bump size of 350μm²

B. Particle distributions

Table 3 shows the number of conducting particles captured between the bump and ITO electrodes is slightly higher (~10-20%) for ACF 1 than for ACF 2. All measurements in Table 3 gives a particle density within the range specified for the ACFs (Table 1), showing that there is no significant flow of particles during ACF bonding.

C. Mechanical testing of conductive particles

Ten particles of each type were tested in the nano-indenter. Typical mechanical results are shown in Figure 5. There is a significant difference in the compression curve, "shifting" the Spiky Ni particles towards a larger "deformation" in the very early stage of the test. Within 50 ms the position of the indenter tip suddenly moves more than 300 nm, that is reflected in the rest of the measurement curve. This sudden displacement is typically described as a "pop-in". This is probably caused by a displacement of the tiny polymer particles on the nickel surface.

Further increasing the deformation of the particles, the nickel layer will fracture, typically around 15-20% particle deformation [7]. This is observed in the measurements causing a sudden "pop-in". On the Ni-Au and Spiky Ni coated particles this is observed in the region of 400 to 550nm and 1000-1200nm respectively.

In this test, we observed fracture of the polymer core in 3 of the 10 NiAu particles and 6 of the 10 Spiky Ni particles.

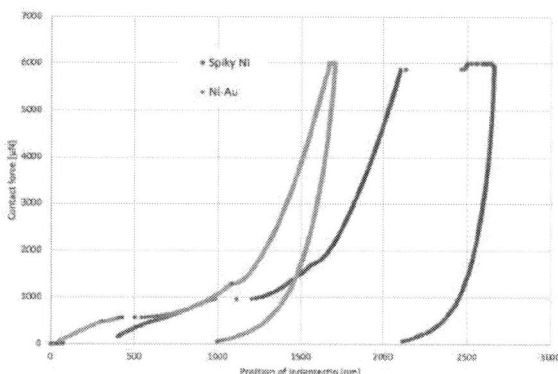

Fig. 5. Typical mechanical behaviour of the Spiky Ni particles (blue) and Ni-Au coated particles (green).

D. Electrical testing

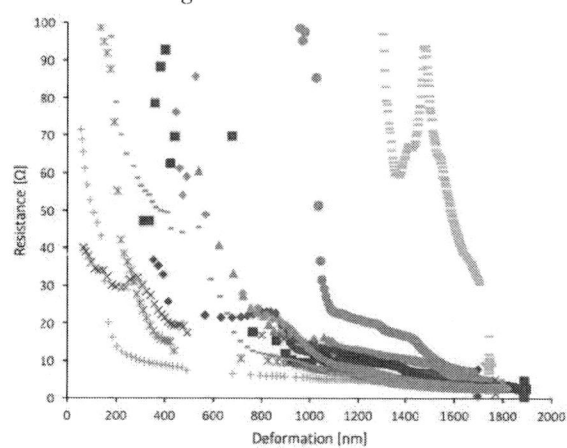

a)

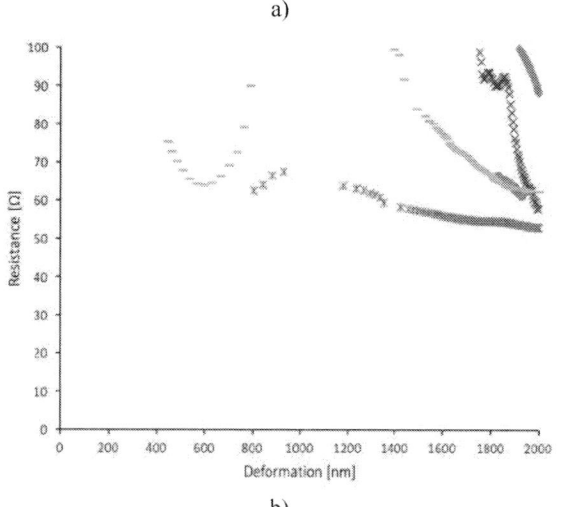

b)

Fig. 6. The resistance-deformation characteristics of individual Ni-Au particles on an Au-coated silicon substrate (a) and spiky Ni particles on same substrate (b). Each curve represents a single particle.

The electrical results from the nano-indentation testing shows a significant difference in the behaviour of the two different particles, see figure 6. The NiAu particles shows a large variation in resistance at small deformations but converging towards a value slightly below 3Ω as the deformation reaches 50%. This value is an order of magnitude higher than what would be expected from a first order theoretical model. The contact resistance of the Spiky Ni particles is even much higher, in fact, only a few of these reach a contact resistance below 100Ω before the polymer core fractured.

E. Deformation and planarization

Due to the test chip design, we observed a significantly thicker bond-line at the side with the large bond pads compared to the side of the small bond pads. This appeared consistently throughout all the components, irrespectively of the type of ACF and bonding pressure. This significantly complicated the interpretation of the results.

F. ACF bonding

The resistance of all the daisy chain segments was measured and the average contact resistance was calculated by dividing with the number of contacts in the segment. A typical contact resistance for the individual particle is then estimated by multiplying the contact resistance by the average number of particles for the given bump size.

In Figure 7 we present results from four different components, including the two different types of particles as well as planarized (P) and non-planarized (NP) bumps, all bonded with a force of 40N. All the results are displayed as the contact resistance per particle. It is clear from the results that the smaller pads provide lower contact resistance per particle. This is due to the asymmetry of the component which causes a tilt towards the smaller pads, and hence increased particle deformation.

Despite a much higher resistance in the nano-indenter test, the spiky Ni particles obtained a lower contact resistance in the ACF test than the NiAu particles.

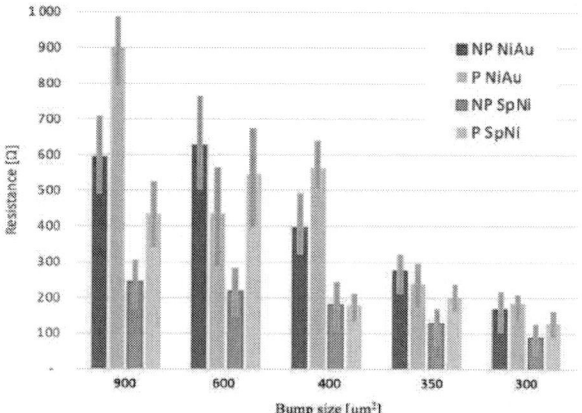

Fig. 7. Resistance per conductive particle, comparing NiAu and Spiky Ni particles as well as planarized and non-planarized bumps. All results are for a bonding force of 40N.

In Figure 8 we have compared the contact resistance for the two different particles in the nano indenter and in ACF bonding as a function of bond gap. We see that in the ACF test the NiAu particles the contact resistance was approximately

48

two orders of magnitudes higher than in the nano-indenter test. This large difference in contact resistance for the Ni-Au particles can be explained by the fact that during the bonding, the adhesive film between the conducting particles, and the bump and the pad respectively, needs to be squeezed out to provide a good electrical contact. With the smooth surface of the NiAu particles there seems to remain a thin layer of the adhesive in one or both of these contact areas. This problem has been further increased by the introduction of the doubler layer adhesive where the viscosity of the particle layer has been increased to minimize the flow of the particles.

On the other hand, for the spiky Ni particles there is no big difference in the contact resistance measured by the two different test methods. This is likely because the spiky surface more easily penetrates the adhesive layer. Additionally, the thicker nickel layer increased the stiffness of the spiky particles.

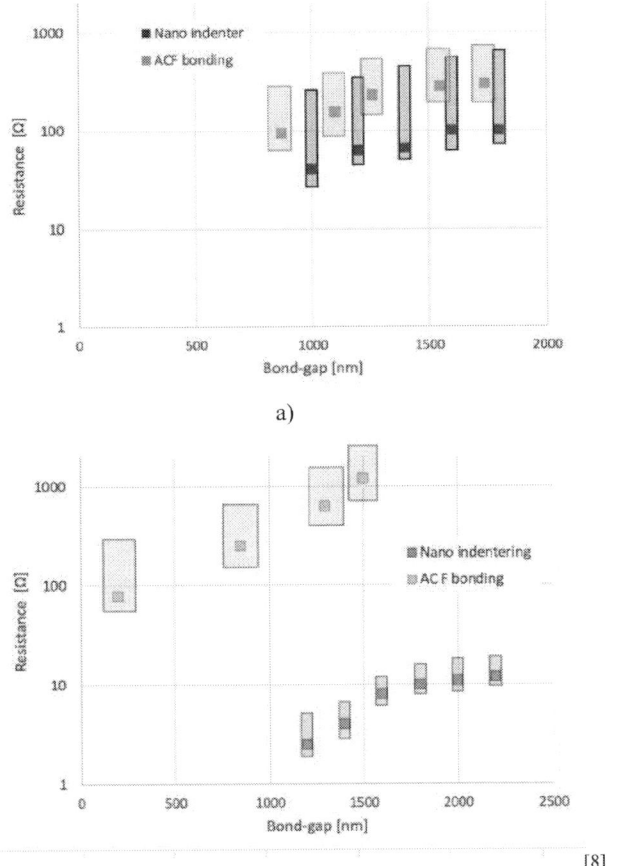

a)

b)

[8]

Fig. 8. : Comparing the contact resistance measured in nano-indentation and in an ACF bonding as a function of bond gap for, a) spiky Ni particles and b) Ni-Au coated particles. The bumps on both components have been planarized.

IV. CONCLUSION

ACF bonding allows very fine pitch interconnects, but comes at the expense of higher interconnect resistance. The use of double layer adhesive with high viscosity of the layer containing particles risks increasing the interconnect resistance due to adhesive trapped at the interfaces.

We demonstrate in this paper that traditional, spherical conductive particles will indeed face this challenge, as the interconnect resistance can increase two orders of magnitude compared to the individual particles. Using spiky particles, a lower ACF bonding contact resistance is achieved, due to puncturing of the adhesive layer.

ACKNOWLEDGMENT *(Heading 5)*

Takashi Nakazawa, Kazuya Matsuda and Masaru Tanaka, at Hitachi Chemical for the manufacturing of the ACF and providing the bonding of the test samples. Grand Cheng and Josh Chen at Disco Taiwan for providing the planarization of the gold bumps. The Norwegian Research Council for funding of the project "Novel Particle for Display Interconnect", project number 228453 and for and the Norwegian Micro- and Nano-Fabrication Facility (NorFab, project number: 295864).

REFERENCES

[1] M. Chin, K. A. Iyer, and S. J. Hu, "Prediction of electrical contact resistance for anisotropic conductive adhesive assemblies," *IEEE Trans. Components Packag. Technol.*, vol. 27, no. 2, pp. 317–326, 2004.

[2] J. H. Constable, "Analysis of the Constriction Resistance in an ACF Bond," *IEEE Trans. Components Packag. Technol.*, vol. 29, no. 3, pp. 494–501, 2006.

[3] C. N. Oguibe, S. H. Mannan, D. C. Whalley, and D. J. Williams, "Conduction mechanisms in anisotropic conducting adhesive assembly," in *Proceedings. The First IEEE International Symposium on Polymeric Electronics Packaging, PEP '97 (Cat. No.97TH8268)*, 1997, pp. 249–258.

[4] R. L. Jackson and L. Kogut, "Electrical Contact Resistance Theory for Anisotropic Conductive Films Considering Electron Tunneling and Particle Flattening," *IEEE Transactions on Components and Packaging Technologies*, vol. 30, no. 1, pp. 59–66, 2007.

[5] M. Bazilchuk, S. R. Pettersen, H. Kristiansen, Z. Zhang, and J. He, "Electromechanical characterization of individual micron-sized metal coated polymer particles", Journal of Applied Physics 119, 245102 (2016); doi: 10.1063/1.4954218

[6] G. M. Nghiem, K. E. Aasmundtveit, H. Kristiansen and M. Bazilchuk, "Anisotropic Conductive Film (ACF) bonding: effect of interfaces on contact resistance," 2018 7th Electronic System-Integration Technology Conference (ESTC), 2018, pp. 1-5, doi: 10.1109/ESTC.2018.8546414.

[7] Fracture of micrometre-sized Ni/Au coated polymer particles, He J Y, Helland T, Zhang Z L, Kristiansen H., Journal of Physics D: Applied Physics 2009; 42(8): 085405 (5pp)

MEMS Mirror in Hermetic Package for Enhanced Performances

Marco Del Sarto
STMicroelectronics
Cornaredo, Italy
marco.delsarto@st.com

Luca Maggi
STMicroelectronics
Agrate, Italy
luca.maggi@st.com

Amedeo Maierna
STMicroelectronics
Cornaredo, Italy
amedeo.maierna@st.com

Mark Shaw
STMicroelectronics
Agrate, Italy
mark.shaw@st.com

Roberto Carminati
STMicroelectronics
Cornaredo, Italy
roberto.carminati@st.com

Gianluca Mendicino
STMicroelectronics
Cornaredo, Italy
gianluca.mendicino@st.com

Davide Rotta
CamGraPhIC srl
Pisa, Italy
Davide.Rotta@camgraphic-technology.com

Marco Chiesa
CamGraPhIC srl
Pisa, Italy
Marco.Chiesa@camgraphic-technology.com

Aina Serrano Rodrigo
CamGraPhIC srl
Pisa, Italy
Aina.Serrano.Rodrigo@camgraphic-technology.com

Antonella Bogoni
Sant'Anna School of Adavanced Studies
Pisa, Italy
antonella.bogoni@santannapisa.it

Abstract—**Ceramic substrate coupled to metal lid has been selected for realizing hermetic package for MEMS mirror. The component passed the fine and gross leak tests and show good performance with improved Q-factor**

Keywords—MEMS mirror, hermetic package

I. INTRODUCTION (*HEADING 1*)

New emerging applications in consumer electronics, Augmented Reality (AR), Virtual Reality (VR) and in automotive field for autonomous driving are asking for new optical components with enhanced features [1]. Laser Beam Scanning (LBS) using MEMS mirror is suitable for addressing these requirements. By the way there is the need to overcome the current performances, mainly limited by the fact that MEMS mirror is operating in open air environment. MEMS mirror enclosed in hermetic package could enhance their performance, especially the Q-factor and the voltage required for maximum angle of deflection

In the present paper the design of hermetic package and method of realization are described. The experimental results are shown and the enhancement in performances, especially about Q-factor are presented.

Package design and structure are presented in section II, the method of realization is explained in section III, section IV shows the characterization results of the new component, section V summarizes the conclusion.

II. PACKAGE DESIGN AND STRUCTURE

A. Package Design

It's well known [2] that an issue of spurious reflections occurs in every optical package. The known solution is to tilt the cap so that the spurious reflections are no more parallel to the light reflected by the mirror. the effect is schematically presented in Fig. 1 and Fig. 2

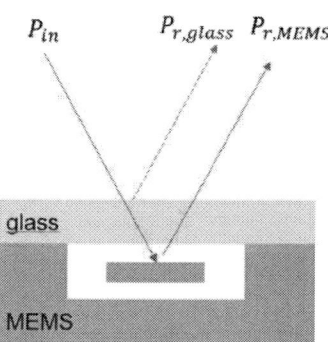

Fig. 1. Spurious image

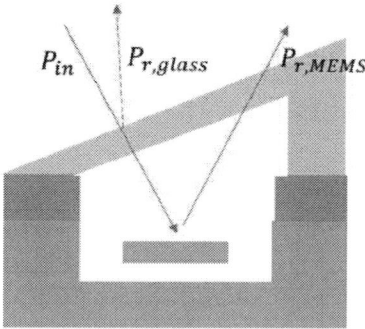

Fig. 2. Ghost image avoided by tilt cap

We simulated the effectiveness of this solution using a optical software [3]. Simulation results are presented in Fig. 3 where is clearly visible that the intensity of ghost image, created by spurious reflection, is negligible compared to light reflected by mirror.

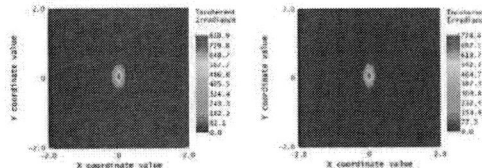

Fig. 3. Optical simulation of flat cap (on the left) and of tilted cap (on the right). The ghost image is clearly vsible on the left, while it completely disapperas on the right

For the investigation about the performance in hermetic package, two different mirrors have been selected- Both are activated by piezoelectric actuation [4], but they are characterized by different size, targeting possible different applications. The MEMS die 1 is 4 x 2.4 x 0.67 mm^3, with mirror diameter = 1.1 mm, the MEMS die 2 is 5.5 x 5 x 0.9 mm^3, with mirror diameter = 3 mm. Representation of the two dice are presented in Fig. 4 and Fig. 5.

Fig. 4. MEMS mirror 1

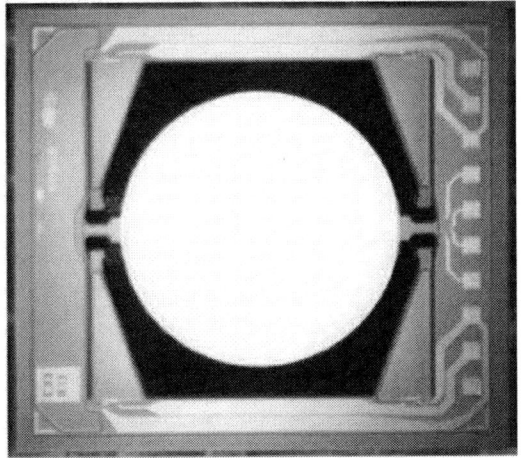

Fig. 5. MEMS mirror 2

A unique package, suitable for accommodating both the dice, has been developed.

B. Package Structure

Package is based on ceramic carrier- coupled to metal lid, suitable for soldering operations. Ceramic base is 2 layer substrate with size 8 x 7.24 x 0.5 mm^3. On the top surface of ceramic base, a solder ring is realized. The lid is made of kovar, plated with NiAu. The optical glass is sintered to the metallic case. Lid size is 7.30 x 6.58 x 3.10 (the highest side) mm^3. Soldering between lid and ceramic is performed using a

ring shaped preform. The selected alloy is SAC305 (Sn 96.5% Ag 3% Cu 0.5%). For this alloy, liquidus and solid temperature are respectively 220 °C and 217 °C.

Complete package structure is presented in Fig. 6

Fig. 6. Strucutre of the pacakge

TABLE I.   PACKAGE ELEMENTS

lement ID	Description
1	Optical window
2	Metal lid
3	Material for sintering the window
4	Die
5	Solder preform
6	Wire bonding
7	Ceramic substrate

III. PACKAGING PROCESS

The dice have been attached to the ceramic substrate using a conductive epoxy resin which shows excellent stability up to 300 °C and is compatible with NASA outgassing limits [5]. Dedicated pick-up tools with a gimbal have been designed to avoid damaging the MEMS and improve the attachment parallelism. Particular care has been dedicated to avoiding the contamination of the metals by resin bleeding, where necessary the attached devices have been cleaned with Ar plasma. Both dice withstand a non-destructive shear test with a maximum applied force of 10 kgf performed according to MIL-STD-883-2 standards, Method 2019.10 [6].

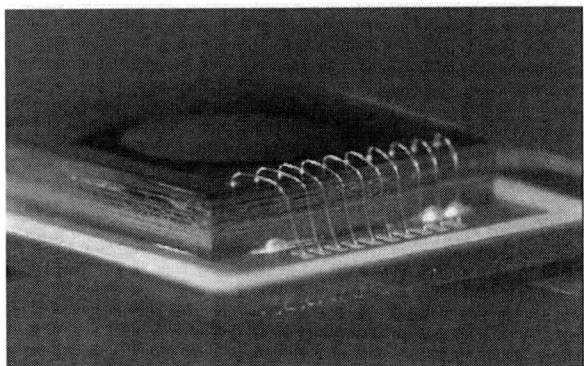

Figure 7 - Die attach and wire bonding of a MEMS mirror

The electrical connections are realized by automatic ball bonding using 20 μm gold wire. An automated routine to process batches of 9 packages has been developed. In order to ensure a safe clearance from the lid, the loop shape has been controlled with a height kept in the 200-250 μm range. Pull tests have been performed according to MIL-STD-883-2 standards, Method 2011.10 [7] on devices annealed with the same thermal profile used to solder the lid. The measured breaking force is 4.04 gf or higher, i.e. well above the threshold limit. Fig. 7 shows a MEMS mirror 2, attached and bonded to the ceramic substrate.

Figure 8 – Vision assisted alignment of the lid to the package + preform

The solder preform and the lid have been aligned to the package ring by vision-assisted pattern recognition (Fig. 8), using specially designed tools. The lid is then pressed on top of the preform (Fig. 9), with the package heated, to promote some tacking and avoid displacements of the lid before the soldering process.

Figure 9 - Placement of the lid on top of the preform

The assemblies have been transferred to a reflow oven in N_2 atmosphere using a silicon wafer as a substrate. The temperature profile is shown in Fig. 10: First the temperature is raised to 150 °C and the system is vacuum pumped below 5 Pa for 30 minutes so that any volatile residue can outgas from the epoxy. After refilling the chamber with nitrogen, the temperature is raised to 200 °C and formic acid is fluxed into the chamber to reduce any oxides. The peak temperature of the reflow process is 250 °C. During the solidification phase it is important to avoid pressure changes in the chamber, which may cause displacements of the lid and adversely affect the solder joint. For this reason, the cooling is performed with a limited N_2 flux although this can result in a longer time above liquidus (TAL).

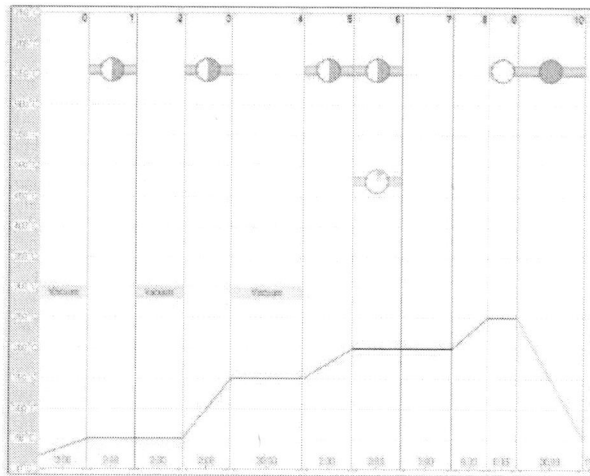

Figure 10 - Temperature profile of the solder reflow process

IV. CHARACTERIZATION

Parts passed the fine and gross leak test[8], exhibiting a leak rate < 1.5 E-8 atm cc/s. Then the package has been mounted on a board suitable for characterization, shown in Fig. 7.

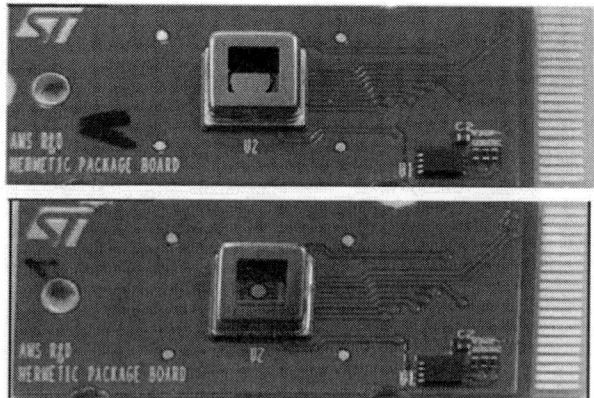

Fig. 7. Characterization board

A measurement station has been used for collecting mirror opening angle in different operating conditions. The setup is composed by a laser, a camera, a projection screen, a function generator and an oscilloscope, as reported in Figure 1. Laser light hits the mirror surface through a dedicated aperture in the projection screen. The projected line is acquired using a camera. By processing the collected picture, it is possible to extract the line length and to obtain the opening angle of the device.

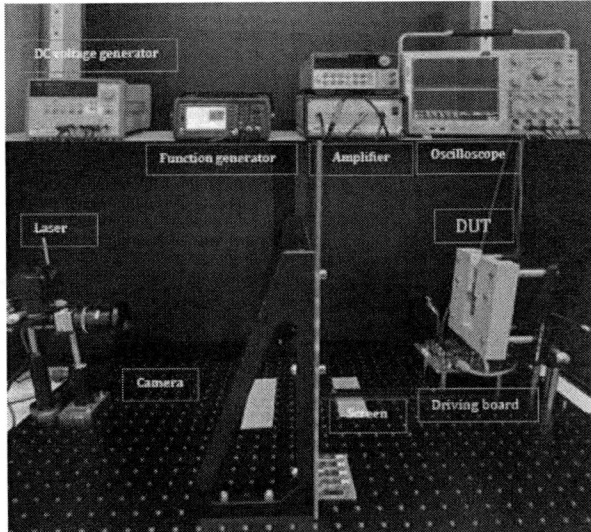

Figure 11 Measurement station picture

The performance of those devices has been evaluated by analyzing the frequency response function (FRF) of the functional mode. The FRF has been obtained by collecting the opening angle at different frequency values around its resonance and at fixed voltage. The measurements have been performed with an automated setup developed for this purpose. Hermetic and not hermetic packages have been compared for MEMS mirror. Effectiveness of the package solution with soldered tilted cap has been proved with two MEMS mirror layouts.

The performance enhancement has been tested on MEMS mirror 1. A performance comparison has been done considering two different samples: one in a hermetic package and one in open air. As visible in Fig. 8, the hermetic package can provide an improvement in quality factor and consequently decrease in the voltage required to achieve the maximum deflection angle. In particular, the device actuated in the hermetic package shows a performance improvement up to 40%.

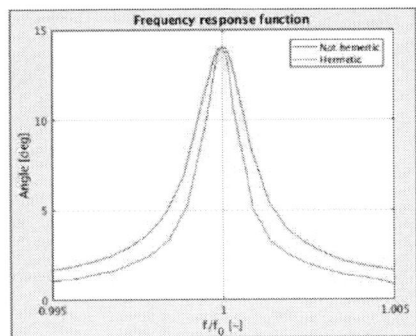

Fig. 8.   Frequency response function comparison of MEMS mirror 1

Several mirror 1 samples have been tested. As visible from graph in Figure 9, still some the spread is visible in the performances,

The packages are sealed at 1 bar in nitrogen atmosphere at 250 °C, that means that at room temperature the *internal pressure is 0.5 – 0.6 bar. The root cause of the reported spread is linked to air ingression in sealed package, that follows an exponential decay law [9]. So, even if the threshold of leak test is passed (< 1.5 E-8 atm cc/s), there is air ingression in the package in few weeks. For guaranteeing that pressure inside the package remains low we have to select parts with leak rate around E-10 atm cc/s. this value assures that air ingression takes more than 10 years for filling the volume.

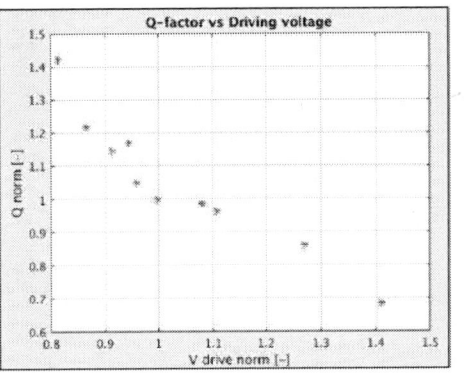

Fig. 9.   Driving voltage versus quality factor for different Mirror 1 samples

Mirror design 2 has been tested in hermetic package with pressure level equivalent to the ambient one. The frequency response functions collected display a very similar trend, As visible from the graph in Fig. 9. Therefore, the performances in the hermetic and not hermetic one are comparable.

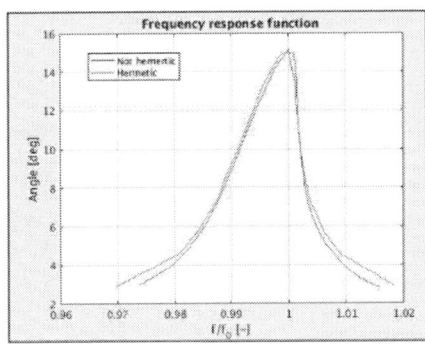

Fig. 10. Frequency response function comparison of MEMS mirror 2

This test confirms that the MEMS mirror 2 can be actuated in a protected environment also at pressure level comparable to the ambient one. In this condition, an effective barrier can be used to protect the device from environmental elements.

V. CONCLUSION

Package approach has been proved to be suitable for obtaining hermetic packaging of MEMS mirrors. Experimental results show an improvement with respect to the case of MEMS mirrors working at open air. The enhancement in performances is in line with expectations and pave the way to the use of LBS for new emerging applications.

VI. AKNOWLEDGMENT

We thank A. Disha, V. Di Bari, S. Talamo, and G. Cucinella at IMT s.r.l., Valenzano, Italy for providing the leak tests.

REFERENCES

[1] Urey H., Wine D. W., Osborn T. D.,"Optical performance requirements for MEMS-scanner based microdisplays",Proc. SPIE Vol.4178,pp 176-185 (2000)

[2] U. Hofmann, C. Eisermann, H.-J. Quenzer, J. Janes, C. Schroeder, O. Schwarzelbach, B. Jensen, L. Ratzmann, T. Giese, F. Senger, J. Hagge, M. Weiss, B. Wagner, and W. Benecke "MEMS scanning laser projection based on high-Q vacuum packaged 2D-resonators", Proc. SPIE 7930, MOEMS and Miniaturized Systems X, 79300R (14 February 2011)

[3] ZEMAX Optic Studio, www.zemax.com

[4] N. Boni, R. Carminati, G. Mendicino and M. Merli, "Quasi-static PZT actuated MEMS mirror with 4x3mm2 reflective area and high robustness," in Proc. SPIE 11697, MOEMS and Miniaturized Systems XX, 1169708, 2021.

[5] https://outgassing.nasa.gov/

[6] MIL-STD-883-2 method 2019.10 (2018).

[7] MIL-STD-883-2 method 2011.10 (2018).

[8] MIL-STD-883-1 method 1014.17 (2018).

[9] L. Fang and L. A. Menk, "On Gas Ingression of Hermetic Packages," in IEEE Transactions on Components, Packaging and Manufacturing Technology, vol. 9, no. 6, pp. 1038-1044, June 2019, doi: 10.1109/TCPMT.2019.2915164.

Innovative Silicon-Ceramic (SiCer) technology for high-strength pressure sensor applications using different manufacturing methods

Cathleen Kleinholz [1], Michael Fischer [1], Nam Gutzeit [1], Andrea Cyriax [2], Michael Hintz [2], Thomas Ortlepp [2], Jens Müller [1]

[1] Electronics Technology Group, Technische Universität Ilmenau, 98693 Ilmenau, Germany
[2] CiS Forschungsinstitut für Mikrosensorik GmbH, 99099 Erfurt, Germany
cathleen.kleinholz@tu-ilmenau.de

Abstract—A variety of manufacturing methods can be used to assemble silicon-based piezoresistive pressure sensors, such as anodic bonding or silicon direct bonding. To achieve high-strength pressure sensors, a high quality interface connection is required. The combination of silicon and ceramic creates a new innovative composite system known as Silicon-on-Ceramics, short SiCer. By using a special developed bondable Low Temperature Co-fired Ceramic tape, which corresponds to the thermal expansion coefficient of Si, the combination of both materials is possible through a sintering process. Advantages of the SiCer technology are high temperature stability, manufacturing at wafer-level and a high bond strength at the interface. Different manufactured SiCer-based pressure sensors were fabricated, electrically and mechanically characterized and compared with anodic bonded and silicon direct bonded pressure sensors.

Keywords—piezoresistive pressure sensor, SiCer, LTCC, ceramic, silicon, low pressure assisted sintering, burst pressure

I. INTRODUCTION

The production of piezoresistive pressure sensors divides into different steps, e.g. wafer processing including realization of measuring bridge and membrane as well as creation of an air channel through a back plate. Induced pressure through the channel results in a deformation of the membrane and the corresponding bending can be electrically evaluated via the measuring bridge. The connection of the wafer with the back plate can be achieved with various technologies, such as anodic bonding [1], [2] or silicon direct bonding [2], [3]. These methods are well established and easily accessible. However, with those manufacturing methods the achieved burst pressure measurements are in range of 100 – 200 bar.

The Silicon-Ceramic (SiCer) technology opens new possibilities for the fabrication of pressure sensors by combining the advantages of Micro-Electro-Mechanical-Systems (MEMS) and Low-Temperature Co-fired Ceramic (LTCC) technology. Characteristics of the SiCer composite substrate are high-strength bond connection at the interface depending on the used sinter method [4], temperature stability, individual pre-processability and reduced form factor. A comparison of different manufactured pressure sensors and the influence on the sensor thickness resulting from the respective back plate is shown in Fig. 1. Special manufactured LTCC tapes (bondable ceramic tapes - BCT, source: Fraunhofer IKTS), which are explicitly adapted to the expansion coefficient of Si, are used for the material composite of silicon and ceramics. The compound is achieved

The authors thank the Federal Ministry of Education and Research (BMBF) for the financial support of this research project

in a sintering process; whereby a distinction is made between high pressure assisted sintering (HPAS), low pressure assisted sintering (LPAS) and other sintering processes such as a modified procedure based on tape-on substrate process.

Figure 1 Pressure sensor thickness in correlation to the manufacturing method (from left to right: anodic bonding, silicon direct bonding and SiCer technology)

II. FABRICATION METHODS FOR PRESSURE SENSORS

A. Previous manufacturing methods

Piezoresistive differential pressure sensors can be manufactured using various technologies, such as anodic bonding, silicon direct bonding or glass frit bonding.

Anodic bonding (AB) works without an intermediate layer by applying pressure, temperature and an electric field to the substrate to create a stable bond connection. In this process, the silicon wafer is bonded to a glass wafer containing alkali ions. Bonding takes place at temperatures between 300°C and 400°C and an additionally applied potential of 200 V to 1000 V. The connection is formed due to a thin SiO_2 layer at the bond interface, which is created by the reaction of oxygen on the Si-surface. [2]

With silicon direct bonding (SDB), two wafers are aligned with one another and initially pressed against each another. Adhesion is generated by hydrogen bonding and van der Waals forces. The resulting connection point continues to develop from the center over the rest of the area. A plasma-treatment of the Si-surface [5] enables low-temperature SDB at room temperature. The final bond strength is achieved through a tempering step at 200°C to 400°C, in which ionic bonds are formed between both Si-wafers [2]. This method does not require an intermediate layer to create a stable connection. [2], [3]

For glass frit (GF) bonding, a low melting glass paste is applied as an intermediate layer to a wafer via screen printing and bonded to a second wafer at temperatures of up to 400 – 600°C. During the bonding process, the glass paste is heated until it is liquid and the wafers are pressed together. When it cools down, the glass solidifies and a mechanically stable, hermetic connection is created. [2]

B. Manufacturing process of SiCer substrates

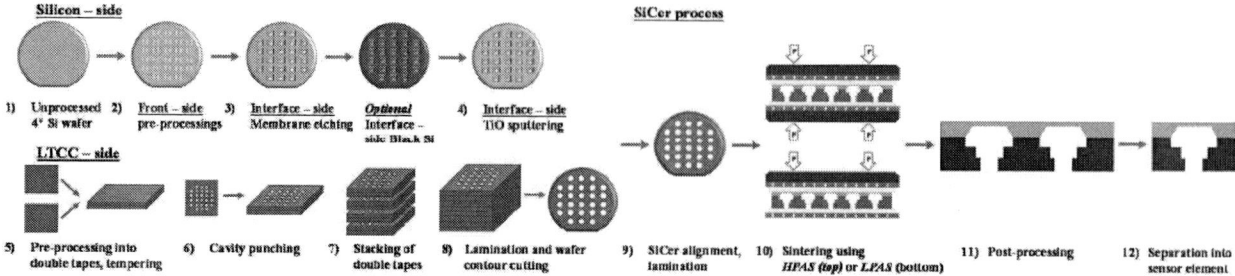

Figure 2 Schematically manufacturing process for SiCer-based pressure sensors

The SiCer process is divided into 3 main parts (Fig. 2), the respective/individual pre-processing of the Si side and the LTCC side as well as the SiCer combining/joining process.

A 4-inch Si-wafer (350 µm thickness, <100> crystal orientation, p doped) is uses as the basis for the fabrication of the pressure sensor (1). Various processes such as lithography, etching and implantation are performed on the front side to achieve the functionality of the sensor (2). The later interface side is structured by etching with Potassium hydroxide solution (KOH) to generate 800 µm x 800 µm cavities with a membrane thickness of around 50 µm (3). Black silicon can be applied additionally onto the interface-side to enhance the bond strength [6]. To improve wetting [7] of the bond-interface a ~25 nm titanium is applied on the Si-wafer and thermally oxidized to ~50 nm thick titanium oxide. [4], [6]

The LTCC process begins with cutting the tape in the required processable size (115 mm x 115 mm) with consideration of the casting direction. This is necessary to compensate the casting direction related shrinkage deviation. Therefore, two BCT are stacked with a 90-degree rotation of the casting direction and isostatically pre-laminated with lamination parameter set (abbr. LPS): 1 (82°C, 10 MPa, 10 min process time including 4 min pre-heating) into a double tape layer stack (5). The created double tapes are tempered at 80°C for 10 min to drive out easily volatile solvents which stabilizes the stack for subsequent processing steps in the green state (no further shrinkage). The number of required tape-layers varies depending on the application. For the pressure sensor, 4 ceramic double tapes are necessary to create the channel structure. The cavities in the double tapes are punched in accordance with the respective layout (6). If required further pre-processing steps like screen-printing, via filling or laser structuring can be carried out. After pre-processing each double tape, the layers are stacked on a metal plate with stacking pins using alignment vias (7). A planar plate is placed on top of the ceramic to protect the cavities from deformation, followed by a silicone mat to smoothen sharp edges and thus allows an equal pressure distribution as well as prevents a tearing of the vacuum-sealed bag during lamination. The substrate is isostatically laminated with LPS: 2 (82 °C, 21 MPa, 15 minutes process time, including 4 min preheating) and a 100 mm wafer contour is cut out of the BCT laminate stack (8) with a picosecond laser (microSTRUCT C v2.0, 3D-Micromac AG). [4], [6]

In the next steps the pre-processed Si-wafer and Cer-laminate are joined together. Both materials must be precisely aligned to each other with a low lateral positioning tolerance (9). For this purpose, an optical-based adjustment and stacking machine is applied. Special structures on both materials are used as alignment marks. After stacking of the Si-wafer and Cer-laminate a sacrificial High Temperature Co-fired Ceramic-tape (HTCC) is placed on top of the ceramic and isostatic lamination (LPS: 2) is used to create a temporary SiCer compound substrate. The HTCC sacrificial tape prevents the lateral shrinkage of the outer surface of the LTCC during the sintering process and is used as release layer to avoid sticking to the pressure plates later on. The final SiCer composite substrate is created by sintering (10), either HPAS (pressure range: 200 – 800 kPa, ATV PEO 603 sintering press) or with LPAS (pressure range: 1.8 – 15 kPa, ATV PEO 603). The Si-side is facing upwards in the sintering furnace to prevent the flow of the softening ceramic into the Si-cavities during the sintering phase. After sintering post-processing including cleaning steps, ultrasonic microscopy, X-ray examination and wafer-bow measurement at wafer-level (11) are carried out. Finally, the wafer is separated into individual 2 mm x 2 mm sensor elements using wafer sawing (12). [4]

III. SENSOR DESIGN, FABRICATION AND EVALUATION

A. Pressure sensor design and SiCer realization concept

A schematic structure of the pressure sensor with a straight cavity in the back plate is shown in Fig 3 a-c. For the presented investigations, a subdivision into different wafers is made: a) electrically functional wafers with a ~50 µm thick membrane, b) non-electrically functional wafers with a ~200 µm thick membrane and c) unstructured wafers without membranes. According to the respective experiments, a distinction between the 3 Si-wafer types is made. Unstructured wafers are used to verify the sintering profiles and to evaluate the bond-interface, non-electrically functional Si-wafers for application related burst pressure tests and electrically functional Si-wafers for electrical characterization of the sensor. Both Si-wafers with an etched membrane contain 966 individual sensor elements per wafer, each 2 mm x 2 mm in size with a bonding frame of ~3.5 µm.

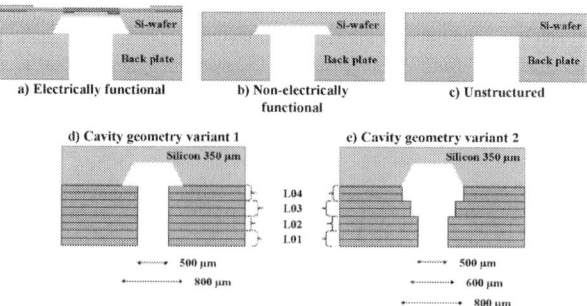

Figure 3 Pressure sensor structures (a-c) and approach to determine optimal cavity geometry in the ceramic back plate (d-e)

Previously only a straight cavity in the back plate could be examined due to the manufacturing method. LTCC multilayer technology and individual pre-processing of the ceramic tapes open up new possibilities for the back plate cavity. In addition to the straight ceramic cavity (Fig. 3 d, variant 1), a gradually widening cavity towards the Si-wafer (Fig. 3 e, variant 2) is examined and will be investigated in chapter III. B. The straight cavity consists of punched holes with a diameter of 500 µm, whereby for the gradually widening cavity 3 different cavity diameters 500 µm, 600 µm and 800 µm (from outer ceramic layer to silicon interface) were used. The cavity size in the layer L01 is predetermined at 500 µm, as this is the appropriate size for the burst measurement test station. The ceramic laminate for variant 1 was laminated in one process. For variant 2 a sequential lamination was initiated to ensure the stability of the varying cavity geometries. The L01 and L02 will first be laminated together using LPS: 2 (L012), afterwards L03 is stacked on top of the newly created L012 laminate stack and laminated using LPS: 1 (L0123). Separately the lamination of the Si-wafer and L04 is carried out with LPS: 1 (SiL04) and finally both individual created laminated parts (SiL04 and L0123) are combined using LPS: 2. Because of the different applied lamination pressures a stronger connection between the individual ceramic layers is achieved and the channel structure is strengthened for the following sintering process.

B. Manufacturing and evaulation of SiCer-based pressure sensors using HPAS process

The SiCer process was developed at Technische Universität Ilmenau and has been continuously advanced for more than 15 years [6], [8]. Due to the REACH-RoHS regulations, organic components of the SiCer-compatible ceramic tapes had to be replaced. This material change influenced the bonding properties between LTCC and Si. Therefore, the entire SiCer process had to be re-evaluated and optimized. The previously unsatisfactory bond-interface should noticeably improve by using HPAS. A first SiCer test with HPAS (Fig. 4) and the pressure sensor layout was performed. The optical inspection (a) as well as the bond-interface analysis (b) were satisfactory.

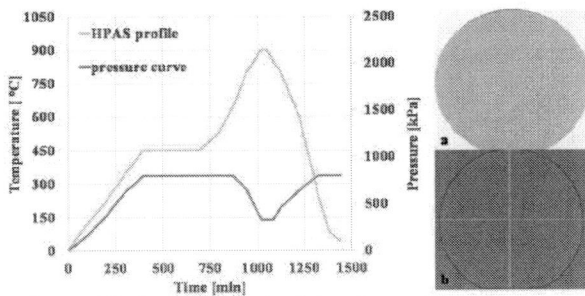

Figure 4 High pressure assisted sintering profile and resulting SiCer-substrate shown in a) optical inspection of the Cer-side and b) ultrasonic microscopy analysis of the SiCer bond-interface

Special non-electrically functional Si-wafers with a ~200 µm thick membrane were used for this examination to determine the influence of the cavity geometry using burst pressure test. A machine-specific wedge error led to an uneven pressure distribution, which resulted in damaging ~½ of the sensor elements and thus reduced the yield of usable sensor elements. Nevertheless, SiCer sensor elements without visible impairments of each cavity variant were used for burst tests to evaluate this manufacturing method. During the burst pressure

test three fractures/events occurred: destruction of the sample (Si membrane or SiCer interfaces), testing up to machine limit of 400 bar without destroying the sample and other sources of error (e.g. detachment from the holder of the testing machine, not evaluated). With layout variant 1, bursting values of 335 – 400 bar (average: 378.5 bar and standard deviation: 18.7 bar) were achieved, whereas with variant 2 the results were in the range of 353.5 – 400 bar (average: 390.3 bar and standard deviation: 11.8 bar). The burst pressure test showed that a gradually widening ceramic cavity (variant 2) leads to slightly higher burst values as well as smaller scattering and will be exclusively used for all subsequent SiCer experiments.

Various artefacts were detected (Fig. 5) by optical inspection. The interaction of high temperatures and very high pressures creates dislocation lines in the Si (Fig. 5 – 1), deformations of the Si-membrane (2a) compared to an ideal Si-membrane (2b) and squashed/blocked ceramic channels (3 and 4). The occurrence of dislocation lines in the Si can lead to premature functional failures in the pressure sensor.

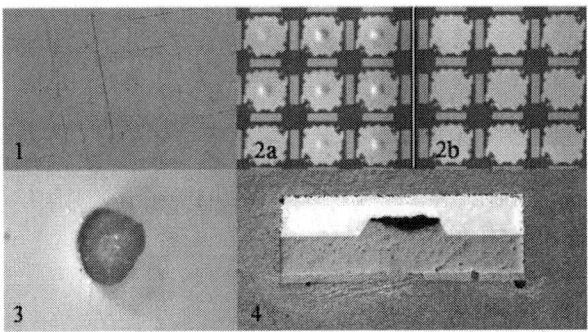

Figure 5 Imperfections found during evaluation of HPAS manufactured SiCer-based pressure sensors

C. Process adjustments and optimizations for LPAS method

Due to the mentioned imperfections of the manufactured SiCer-based pressure sensors, the sintering process was changed from HPAS to LPAS. For test purposes and because of cost reasons, unstructured Si-wafers without etched membranes were used for the sintering tests. The respective sintering results are shown in Fig. 6. A direct application of the HPAS temperature profile with a lower applied pressure resulted in noticeable cracking in the ceramic and a little to no bonded interface (Fig. 6 a).

A vast comprehensive process optimization of sintering temperature profiles was conducted during the project period [9]. Various factors that influence the sintering result were investigated, such as sintering temperature, heating- and cooling-rates, temperature plateaus and layout dependency. During the optimization, special attention was paid to: generating a fully bonded interface, crack-free ceramic layers depending on various layout designs as well as compatibility with different metallization pastes for soldering and wire bond applications. The detailed optimization will be presented in a separate work (in progress [10]). The developed sintering profiles (increased and optimized profile) were also tested with the pressure sensor layout.

By increasing the burn-out and sintering temperature by constant 40 K, cracking in the ceramic could be eliminated and the interface improved noticeably (Fig. 6 b). However, the very high density of cavities obviously still leads to delamination due to induced mechanical stress at the edges. For following investigations, the number of sensor elements was reduced to 340 (Fig. 6 c), the number of BCT double tape layer were reduced from 4 to 3 and the temperature profile was changed to a more constant temperature gradient (Fig. 6, optimized LPAS profile – green curve). After successful manufacture, later increased to 460 and 586 (Fig. 6 d).

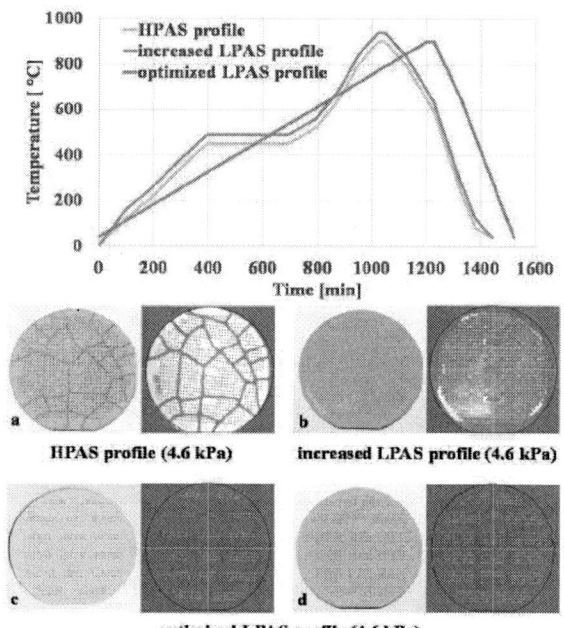

HPAS profile (4.6 kPa)   **increased LPAS profile (4.6 kPa)**

optimized LPAS profile (4.6 kPa)
with cavity variation to increase bond-interface and determine ideal number and spacing of cavities

Figure 6 Process adjustment and layout variation of the ceramic cavity number to achieve a homogeneously bonded interface

D. Electrical characterization to examine the influence of the sintering temperature

The sintering tests have shown that increasing the temperature by 40 K leads to an improvement of the interface. Therefore, examinations with electrically functional Si-wafers (membrane thickness: ~ 50 µm) were carried out to determine the temperature influence on the pressure sensor's measuring bridge. For this purpose, standard pressure sensor samples (Silicon-Silicon, abbr. SiSi – SDB ref.) were manufactured using silicon direct bonding and electrically characterized to enable a realistic comparison. The Si part on which the measuring bridge is located was examined in 3 variants (blue-untempered/standard, green: tempered with 900°C and yellow: tempered with 940°C; Fig. 7).

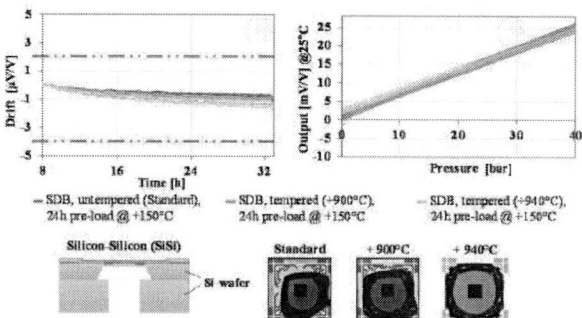

Figure 7 Electrical measurements of the zero point stability/drift with different temperatures using standard SiSi pressure sensor samples and respective burst fracture pattern of the Si-membrane

Based on these samples, electrical measurements at wafer level, testing of the bond interface using ultrasonic microscopy, pressure characteristics, sensitivity and hysteresis measurements, and burst pressure tests were carried out. It was found that the increase in the sintering temperature does not cause any electrical impairment of the sensor.

Finally, burst tests were carried out on the 3 pressure sensor samples to determine the influence of the temperature preload on the Si wafer. Values of 158 – 204 bar for the standard SDB, 157 – 203 bar for Si with a pre-load at 900°C and 145 – 160 bar for Si with a pre-load at 940°C could be verified. In all three variants, the membrane ruptured during the burst test, which indicates a stable connection at the bond interface. An influence of the sintering temperature at 900°C on the performance of the sensor cannot be detected either in the electrical measurements or in the bursting values. Using the increased sintering temperature shows no influence on the electrical functionality and a lower standard deviation was found in the bursting tests carried out, but the bursting values achieved are below the permissible limit of 160 bar.

E. Modified SiCer approach: Substrate-on-Tape-on-Substrate (SoToS)

In order to maintain the yield of the sensor elements on the wafer, a modified process based on the tape-on-substrate method [11] was developed parallel to the sinter optimization. For the modified SiCer approach, two separately fabricated substrates were bonded with an additional unsintered ceramic layer by laminating and sintered a second time to form a SiCer substrate (Substrate-on-Tape-on-Substrate, SoToS). Two approaches were examined for the SoToS procedure (Fig. 8).

In variant 1, a BCT double tape is laminated (LPS: 2) to the Si wafer and then sintered, and the ceramic stack consisting of layers 1-3 is also sintered separately from the SiCer substrate. The final pressure sensor is then produced by laminating (LPS: 2) a green double tape between the two already sintered substrates and a third sintering process.

In variant 2, the ceramic part is sintered first, then the unsintered ceramic double tape layer is laminated (LPS: 2) directly to the interface of the two substrates and the final sensor is produced in a second sintering step. Both SoToS variants were first built with Si-wafers without cavities and tested for their suitability with the burst pressure test.

The results of the bursting tests were in average 188 bar (min.: 136 bar, max.: 202 bar) for variant 1 and 151 bar (min.: 54 bar, max.: 201 bar) for variant 2. A smaller deviation was noticed with variant 1 in comparison to variant 2. Based on the resulted measurements variant 1 was chosen for manufacture of a SiCer-based pressure sensor using a full functional Si-wafer.

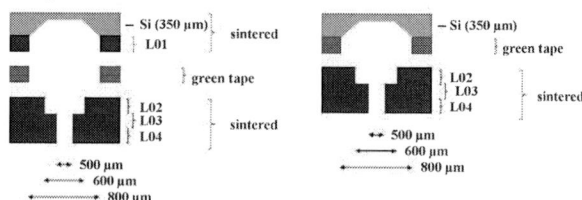

Variant 1: sinter steps in total = 3   **Variant 2: sinter steps in total = 2**

Figure 8 Two approaches with the SoToS method to manufacture SiCer-based pressure sensors

58

IV. RESULTS AND DISCUSSION

After successfully examined implementation variants (chapter III.) electrically functional SiCer-based pressure sensors with a widening cavity in the ceramic back plate were manufactured:

– SiCer #1: HPAS, 800 kPa/ 260 kPa
– SiCer #2: HPAS, 800 kPa/ 260 kPa, pre-sawn Si
– SiCer #3: increased LPAS, 14.7 kPa, pre-sawn Si
– SiCer #4: optimized LPAS, 4.6 kPa and
– SiCer #5: SoToS variant 1, optimized LPAS, 4.6 kPa.

All pressure sensors (wafer-level) were first cleaned, the bond-interface optically inspected using ultrasonic microscopy, and precise positioning of both materials determined using X-ray analysis. The X-ray image (Fig. 9) shows the varying cavity diameters in the ceramic and the KOH-etched square cavity in the Si-wafer. The alignment was successfully performed and the ceramic cavities are centered under the Si cavity with a slight stacking misalignment.

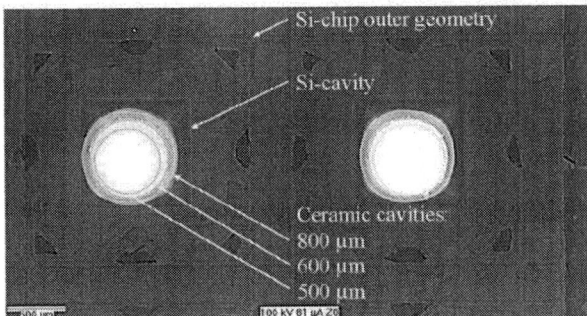

Figure 9 X-ray image of a SiCer pressure sensor after sintering process to evaluate the alignment accuracy

An important topic in MEMS sensor technology are the effects of aging, which directly affect the signal stability of the sensor systems. By measuring the zero-point stability (drift) of the piezoresistive test structures, the relaxation of stresses in the assembly process, e.g. due to creep or micro-crack growth, as well as the drifting of critical contact resistances or of the measuring resistors themselves, could be monitored with the highest application relevance. A quantitative reference level was achieved through the combination of thermal pre-aging. For the functional verification of the SiCer pressure sensor, the wafers were separated with a correspondingly adapted combined sawing process with a low chipping. The assembly was carried out onto special test sockets or evaluation boards, avoiding assembly stresses. The basic physical principles of the sensor and the processes as well as the properties of the SiCer material system and the possible mutual interactions were taken into account in the investigations. The electrical characterizations were carried out with a constant measuring current of 300 µA. Fig. 10 shows the measuring curves of bridge resistance [kOhm], sensitivity [mV/V], pressure characteristic [mV/V] and characteristic deviation [%FS]. All manufactured pressure sensor variants were electrically characterized. The illustration is limited to three selected pressure sensor variants; two SiCer pressure sensors (green: HPAS – SiCer #2; blue: LPAS – SiCer #3) and the standard SDB reference sensor (black – SDB ref.) without a temperature preload for comparison. The different measurement runs showed that the signal curves of the SiCer-based pressure sensors were comparable to those of the standard SDB sensor and are within the tolerance range. The adaptation of the

sintering temperature profile depending on the reduced sintering pressure resulted in a significant improvement in the sensor performance. The deviation of the non-linearity of both SiCer (#2 and #3) is negligible for this limit range.

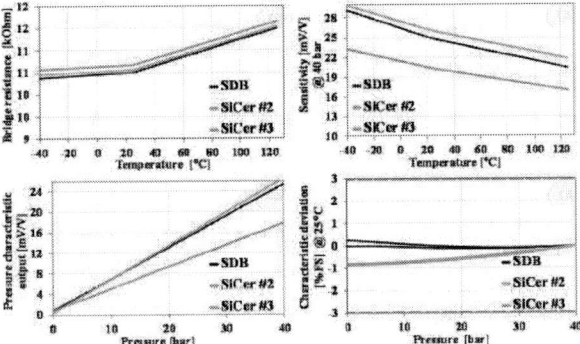

Figure 10 Measuring curves of the electrical characterization of the individual pressure sensors

The adhesion of the interface and the resulting burst strength of the assembled pressure sensors are tested with burst pressure method. All sensors were mounted onto special test specimens and subjected to pressures of up to 400 bar, which is the maximal capacity of the burst measuring station. For the pressure sensor a minimum burst value of 160 bar should be achieved in order to ensure the correct functioning of the pressure sensor. As reference, an anodic bonded pressure sensor, a standard SDB pressure sensor and glass frit bonded pressure sensors are included in Fig. 11 to allow a comparison and classification with the SiCer burst measurements. Predominantly a fracture in the membrane of the pressure sensor occurred and it can be assumed that the bond interface has higher strength than the membrane. A possible explanation for a membrane fracture could be previous damage to the membrane due to thermal pre-aging or process-related handling of the Si wafer, which induces tension. This would also explain why parts of the Si surface bursted as well. The GF #2 reference samples, which were single chip bonded achieved burst values over the higher burst values than GF #1 (wafer-level bonded), nevertheless both GF pressure sensors fractured at the bond interface and are not ideal for this specific pressure sensor manufacturing.

A significant influence of the sintering process in correlation of the resulting burst pressure measurements is visible. Both HPAS SiCer #1 & #2 samples showed very low burst values and the fracture occurred in the membrane. Because of high temperatures and high sintering weight during the sinter process various defects occurs and lead to premature failure of the Si membrane of the SiCer pressure sensors. SiCer #3 shows a rather large range with around 80 bar of the examined burst pressure and fractures occurred in the membrane again. SiCer #4 with the optimized sintering profile and a sintering pressure of 4.6 kPa reached burst values of 153 bar to 193 bar and the fracture occurred at the adhesive connection of the sample to the test specimen, which indicates that the actual burst strength of the composite pressure sensor could not be determined because of premature failure. Compared to the other pressure sensor samples where the fracture mainly occurred at the membrane, the SoToS (SiCer #5) sensors fractured within the ceramic. This indicates that the green tape layer, which is used as a bond layer between the two sintered parts, is the weakest part in the sample and it can be considered that the membrane is apparently stronger than the ceramic and withstand burst pressures up to 193 bar. An influence of sintering temperatures and resulting stress on the membrane could not be proven with the burst pressure tests.

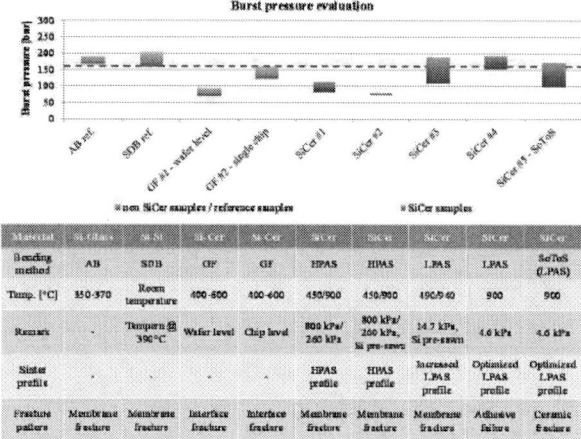

Material	St. Glass	St. Si	Si Cer	Si Cer	SiCer	SiCer	SiCer	SiCer	SiCer
Bonding method	AB	SDB	GF	GF	HPAS	HPAS	LPAS	LPAS	SoToS (LPAS)
Temp. [°C]	350-370	Room temperature	400-600	400-600	450/900	450/900	490/940	900	900
Remark	-	Tempern @ 390°C	Wafer level	Chip level	800 kPa/ 260 kPa	800 kPa/ 260 kPa, Si pre-etch	14.7 kPa, Si pre-etch	4.6 kPa	4.6 kPa
Sinter profile					HPAS profile	HPAS profile	Increased LPAS profile	Optimized LPAS profile	Optimized LPAS profile
Fracture pattern	Membrane fracture	Membrane fracture	Interface fracture	Interface fracture	Membrane fracture	Membrane fracture	Membrane fracture	Adhesive failure	Ceramic fracture

Figure 11 Comparison of different manufactured pressure sensors

V. CONCLUSION AND OUTLOOK

The measured burst values in this examination with differently produced SiCer-based pressure sensors showed a strong influence by the choice of the sintering process. HPAS (pressure range: 200 – 800 kPa) leads to imperfections in the sensor such as dislocation lines in the silicon and a deformation of the air channel as well as lower burst values because of the mentioned pre-damages of the membrane. The most promising burst values were achieved with low pressure assisted sintering (pressure range: 1.8 – 15 kPa) and the Substrate-on-Tape-on-Substrate (SoToS) process. LPAS manufactured SiCer pressure sensors achieved burst values around 110 – 193 bar. The SiCer SoToS method resulted in values of 99 – 174 bar, but the fracture took place inside the ceramic and not at the bond interface or the membrane as with the other sensors. SiCer #4 and #5 are comparable to the AB and SDB reference samples, which proves that the innovative SiCer technology can be used to manufacture high-strength pressure sensors. The advantage of the SiCer technology for pressure sensor applications is the possibility of individual designing of the ceramic cavity and therefore optimize the pressure distribution inside the sensor and increase the burst pressure resistance. Additionally, the overall thickness of the pressure sensor has been significantly reduced with the SiCer composite. Furthermore, depending on the application, an insulating capability of the pressure sensor may be preferred. SiCer-based pressure sensors possesses insulating capability due to the properties of the ceramic, whereas this can only be achieved with SDB sensors through additional steps.

Since the burst values of the SiCer SoToS approach showed very promising results, further examinations should continue to increase the bonding strength inside the ceramic layer, as the bond interface as well as the Si membrane withstood burst pressures up to 193 bar. An explanation for the partial disconnection within the ceramic layers can be found in the warpage of the ceramic part during the first sintering. To rule this theory out, tests with freely sintered ceramic parts are planned, whereby the shrinkage of the ceramic tape must first be determined in order to be able to adjust the layout accordingly. Furthermore, the process sequence with regard to the silicon side must be examined

more closely in order to avoid possible pre-damage of the Si membrane and thus further improve the bursting values of the Si membrane that have been achieved so far.

ACKNOWLEDGMENT

The authors gratefully thank the Federal Ministry of Education and Research (BMBF) for the financial support of the project (03WKDG03A, 03WKDG03E and 03XP0276D). Furthermore, we thank I. Koch and B. Müller (Electronics Technology Group, Technische Universität Ilmenau) for the technical support with the ceramic manufacturing and analysis of the substrates as well as D. Flock (Group of Materials for Electrical Engineering and Electronics, Technische Universität Ilmenau) for cross-section preparations. We also thank R. Grellmann and D. Hanig for the support with separation of the sensor elements, K. Pischek for SAM examination and analysis as well as S. Jagomast for carrying out the burst pressure measurements (CiS Forschungsinstitut für Mikrosensorik GmbH).

REFERENCES

[1] X. Zheng, W. Chen and X. Chen, "Stress in Si-glass anodic bonding and its effect on silicon piezoresistive pressure sensor," 2010 IEEE 5th International Conference on Nano/Micro Engineered and Molecular Systems, 2010, pp. 524-527, doi: 10.1109/NEMS.2010.5592452

[2] M. Hönle, "Direct wafer bonding with transparent conductive oxides /Direktes Waferbonden mit transparenten leitfähigen Oxiden", Dissertation, Johannes Kepler Universität Linz, 2021

[3] R. Täschner, "Process and sensor development for production-ready silicon-based pressure sensor systems suitable for high temperatures /Prozess- und Sensorentwicklung für fertigungsgerechte hochtemperaturtaugliche Drucksensorsysteme auf Siliziumbasis", Dissertation, Technische Universität Ilmenau, 2018

[4] C. Kleinholz, A. Cyriax, M. Hintz, J. Müller and T. Ortlepp, "Manufacture of high-strength differential pressure sensor using SiCer technology," 2022 International Conference on Electronics Packaging (ICEP), Sapporo, Japan, 2022, pp. 137-138, doi: 10.23919/ICEP55381.2022.9795558

[5] CiS Forschungsinstitut für Mikrosensorik GmbH, "Low temperature silicon direct bonding", https://www.cismst.de/news-2012-05-15/, last access: 07/19/2023

[6] M. Fischer et al., "Silicon-Ceramic Composite Substrate: A Promising RF Platform for Heterogeneous Integration," in IEEE Microwave Magazine, vol. 20, no. 10, pp. 28-43, Oct. 2019, doi: 10.1109/MMM.2019.2928675

[7] S. Gropp, M. Fischer, J. Mueller and M. Hoffmann, "Wetting behaviour of LTCC and glasses on nanostructured silicon surfaces during sintering," Micro-Nano-Integration; 6. GMM-Workshop, Duisburg, Germany, 2016, pp. 1-5.

[8] M. Fischer et al., "Investigations of Metal Systems in a Silicon Ceramic Composite Substrate for Electrical and Thermal Contacts as well as Associated Mounting Aspects", 12th IMAPS/ACerS International Conference and Exhibition on Ceramic Interconnect and Ceramic Microsystems Technologies (CICMT), April 19-21. Denver, USA 2016, pp 107-110

[9] Bundesministerium für Bildung und Forschung, "HIPS – High Performance Sensorsysteme durch Verbindung von Siliziumtechnologie und keramischer Mehrlagentechnik", https://www.innovation-strukturwandel.de/strukturwandel/shareddocs/entries/de/Unternehmen Region/Wachstumskerne/wachstumskerne-2019-2022/hips_942.html, last access: 07/26/2023

[10] C. Kleinholz et al., "Sinter optimization on metallization pastes for assembly technologies using Silicon-Ceramic-based composite substrates", 2023, unpublished

[11] F. K. Patterson et al., "Tape on substrate, a new systems approach for manufacturing multilayer hybrid circuits," Proceedings. Japan IEMT Symposium, Sixth IEEE/CHMT International Electronic Manufacturing Technology Symposium, Nara, Japan, 1989, pp. 147-151, doi: 10.1109/IEMTS.1989.76126.

From MEMS Strip to MEMS Unit: a Comprehensive Simulation Approach to Warpage

Andrea Ratti
STMicroelectronics
Agrate, Italy
andrea.ratti@st.com

Daniele Simoncini
STMicroelectronics
Agrate, Italy
daniele.simoncini@st.com

Annabel Adolfo
STMicroelectronics
Calamba, Philippines
annabel.balahadia@st.com

Marco Del Sarto
STMicroelectronics
Castelletto, Italy
marco.delsarto@st.com

Alex Gritti
STMicroelectronics
Agrate, Italy
alex.gritti@st.com

Patrick Fedeli
STMicroelectronics
Castelletto, Italy
patrick.fedeli@st.com

Luca Maggi
STMicroelectronics
Agrate, Italy
luca.maggi@st.com

Teresa Napolitano
STMicroelectronics
Arzano, Italy
teresa.napolitano@st.com

Mark Andrew Shaw
STMicroelectronics
Agrate, Italy
mark.shaw@st.com

Jefferson Talledo
STMicroelectronics
Calamba, Philippines
jefferson.talledo@st.com

Abstract - The paper suggests a simulation flow methodology suitable for MEMS package designers that can help R&D teams to develop industrially feasible products by achieving consistent warpage prediction results, both at unit and strip levels. Strip-level-wise, one of the most important findings of this activity is the direct impact of raw material selection. Unit-level-wise, it allows checking a priori and in different conditions (inside the socket or mounted on a PCB for example) stress distribution on different elements of the assembled MEMS unit.

Keywords — Strip Warpage; MEMS; package warpage; thermo-mechanical FEM simulations; MEMS packaging.

I. Introduction

MEMS (Micro Electro-Mechanical Systems) package design is the challenge to couple sensing or actuating principles expressed by a silicon structure with the external environment [1]. It must accommodate both the emerging push of the miniaturization trends and the rise of more complex MEMS dies. Consequently, silicon content inside the package is continuously increasing, leading to the growing relevance of the warpage phenomenon [2]. Warpage consists of bending of the structures once undergone to temperature stress path caused by material properties mismatch and is one of the major concerns which accompany semiconductor products throughout their manufacturing life [3]. High warpage values may affect MEMS performances in many contexts:

- Assembly (strip level warpage) → workability issues

- Testing (unit level warpage) → socket to MEMS interaction - calibration issues

- Soldering (unit level warpage) → MEMS to board compatibility issues

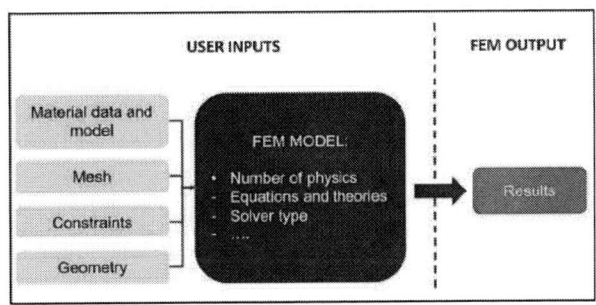

Fig. 1. Schematic overview of FEM focusing on user inputs and FEM outputs.

It starts in the Front-End, when the first layers are deposited on bulk crystalline wafers; it changes concavity, shape, order of magnitude, it is carried out and addressed in different hi-tech machines, but it never disappears, even during its final application on customer PCBs (Printed Circuit Board).

This paper intends to describe the warpage evolution and the measurement techniques exploited for warpage assessment in Back-End. MEMS products are assembled in panels and are typically moulded. These structures suffer from warpages as well as single assembled units. This structural bending affects the package interaction with the testing machine and PCBs.

As reported in Fig. 1, this study can be approached with varying parameters, study setups and assumptions, that may lead to consistently different results. Consequently, one of the hardest parts of the task is to select the ones that allow the best fit with experimental data, starting from scratch.

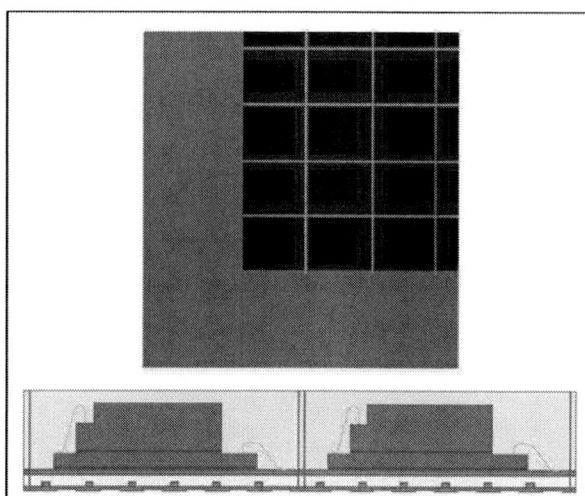

Fig. 2. Above, schematic top view of a strip portion. Blue part is the strip rail, green the sawing street, black the units. Below, schematic view of a strip section: green is the solder mask, beige the core material, orange the metals, black the DAF, dark grey the silicon, gold the wires and light grey the moulding compound.

II. STRIP WARPAGE

Most of the motion MEMS devices are assembled in panels called "strips" to increase productivity. As shown in Fig. 2, strips are made of single units (their overall number depends on the size of the device itself) juxtaposed one to the other, sawing streets (parts of the strip interposed among units, where the singulation is performed using a rotating saw blade to separate the individual devices when the assembly is completed) [4] and the strip rail (the outer part of the strip, used as guides for handling within the assembly equipment). The strip stack-up is usually made of a couple of elements (Fig. 2) - solder mask, metal, core material, die attach film, silicon, and cured moulding compound - but the relative amount of these elements can vary consistently depending on the device.

Stacking up many different materials may lead to warpage issues during assembly. The reason behind this is mostly a Coefficient of Thermal Expansion (CTE) mismatch among the different elements, meaning that parts already stacked together, thus constrained, tend to expand and contract differently when the strip undergoes the several thermal cycles needed during the assembly (or to solder the device onto the customer board) [4]. Internal stress (any dimensional change of the MEMS - being MEMS mechanical sensors and actuators - due to stress can have implications on the device performance) and physical bending/tilting (this often implies workability problems, leading to machine downtimes and production yield losses) of the strip itself are generated. Excessive warpage may cause not only lower productivity but even the inability to produce: before moulding it can generate problems in strip handling or clamping during die attach and wire bonding; after moulding it is not possible to singulate within package tolerances.

Fig. 3. Strip warpage classification.

Marketing trends consisting of continuous device height reduction foster the vulnerability toward this phenomenon: in particular, a substrate thickness reduction amplifies the bending possibility. Being the strips in a rectangular shape, warpage can be properly classified into some categories: coil set warpage, crossbar warpage and twisted warpage. Each of them can show positive or negative curvature (Fig. 3).

Related to different needs in production line, the assembly flow is divided in two main phases: Front of Line (FOL) and End of Line (EOL). FOL includes all the assembly steps before moulding, whereas EOL covers all the remaining ones. This distinction is important since warpage magnitudes depend not only on the applied temperature, but also on the applied constraints typical of each assembly step (internal, such as the die attach, or external such as the shape of the strip carrier). Until the strip is considered part of FOL, layout design modifications can be applied and are effective to correct the warpage phenomenon, whereas once in EOL a change of materials is required as an effective way to compensate for warpage.

It is now clear that being able to predict and manage the warpage has thus become a key factor in ensuring productivity and reducing costs and time-to-market. Warpage has become a criterion for the choice of assembly materials. To speed up this selection process, the possibility of forecasting the impact of certain materials on the overall stack-up should be investigated. Moreover, a clear picture of the stress distribution inside the strip ensures better performances of the final device and the possibility to forecast some kind of failures.

A. FEM model setup

The Finite Element Method (FEM) is the tool we selected to perform this predictive analysis, even though MEMS structures are not always easy to mesh. They are full of high aspect ratio features, that cannot be easily managed: lots of (more than 90%) good small elements – element aspect ratio < 5 – are needed to accurately mesh thin layers, consequently requiring high computational power. To overcome this problem, simplifications should be performed on the model geometry, the numerical models exploited to fit the material properties and the kind of simulation carried out.

Geometric simplifications concern different parts of the units, and consequently of the strip:

- Substrate: based on the I/O quantity and routing complexity, the strip substrate can have two or more metal layers. For motion moulded MEMS, these are the typical stack-up structures: solder mask bottom, metal bottom, core, metal top, solder mask top. It is evident then that the strip itself is a composite material, obtained by lamination. This is probably the most difficult to mesh part of the die/strip, and consequently, different approaches have been tested. It is possible to group them into two main categories: Hard Geometrical Simplification (HGS) and Soft Geometrical Simplification (SGS):

  o HGS: it consists of the substitution of the overall substrate with a single bulky parallelepiped of the same height as the real substrate (Fig. 4A). The material applied to this bulky element is fictitious, and it features properties that result from the application of composite materials mixing rules to the original stack up. The most important ones to be calculated are the Young Modulus ($E_{composite}$) and the Thermal Expansion Coefficient ($\alpha_{composite}$) [5]:

$$\alpha_{composite} = \frac{\alpha_1 E_1 V_1 + \alpha_2 E_2 V_2 + \cdots}{E_1 V_1 + E_2 V_2 + \cdots} \quad (1)$$

$$E_{composite} = E_1 V_1 + E_2 V_2 + \cdots \quad (2)$$

This method doesn't consider the influence of the geometry of the different parts the substrate is made of on the overall warpage behaviour. Moreover, the exploited formulas were developed for composite materials, where fibre and matrix are clearly divided and oriented. The results obtained using this procedure don't fit the experimental ones as expected.

- ○ SGS: it consists of the substitution of the overall substrate with multiple stacks of bulky parallelepipeds of the same height as core material and prepreg (Fig. 4B1). Stock material is considered, both for prepreg and core materials, where the volume occupied by metallic layers is assigned to the core. Strip warpage wise, the contribution of metallic layers can be considered negligible [6]. The results obtained exploiting these simplifications are the ones that better fit the experimental ones.

- Dice: here the level of detail can be increased more and more. Strip level wise, to get good results it is enough to represent them as bulky blocks of monocrystalline silicon. Moreover, there is no need of representing wire bodings, being low in the amount of metallic material and not shaped in a bulky way. Die Attach Film (DAF) and Film on Wire (FOW) should be introduced as bulky parallelepipeds (Fig. 4B2).

- Mould: it can be shaped as the remaining part of the stack up, up to the nominal thickness of the package (Fig. 4B3). Together with silicon and core material, it is one of the main responsible for the final shape of the overall strip. Laser marking is not considered.

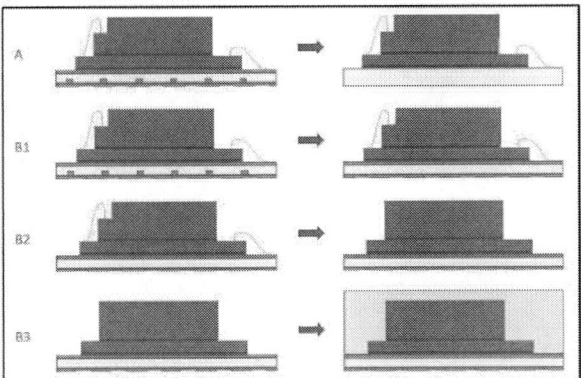

Fig. 4. A: substrate HGS simplification. B1: substrate SGS simplification. B2: dice simplification. B3: mould compound simplification.

Fig. 5. Automatic Optical Measurement machine.

Simplifications of the kind of analysis carried out consist of the performance of a static study instead of a dynamic one. It is assumed that the behaviour of the system being simulated does not depend on time, and the applied loads are constant. Also in this case, choice was made due to the lack of experimental data that described the dynamic behaviour of materials and to reduce the computational load of the analysis.

Simplifications on the numerical models exploited to fit the material properties affect all the materials considered in the study: they have been all considered linear elastic isotropic materials. This choice was made due to the lack of experimental data that described the real behaviour of materials.

To check for the accuracy of the results and to validate the FEM model, several measurements were done on the warped strips. The system exploited for this analysis is an Automatic Optical Measurement machine, that can scan the whole strip and provide a 3D map of the curved strip (Fig. 5). This kind of investigation is standard only after post mould curing. Consequently, it was possible to tune the FEM model at this step of the process only (temperature profile spans from 298K up to 523K). This is just one of the two moments of the process of interest, where high warpage values are highlighted: critical warpage values can also be seen after the first die attach (still part of the FOL).

B. Results

Results of the simulations performed with the simplifications above are reported in Fig. 6 and correlated with the experimental results assessed via automatic optical measurements.

The accuracy of the results varies depending on the device: being the outer dimensions of the strip constant as well as the sawing street dimensions and the strip rail dimensions, the number of devices hosted by strips varies significantly depending on their footprint. Selected simplifications imply a small mismatch from the real overall strip behaviour: replicating this mismatch a higher number of times implies a greater deviation from the targeted value. In this case, the meaning of the result relies on the relative comparison of the different tested solutions at design phase. In the plot (Fig. 6), the ideal case (perfect match) is represented by the red dotted line.

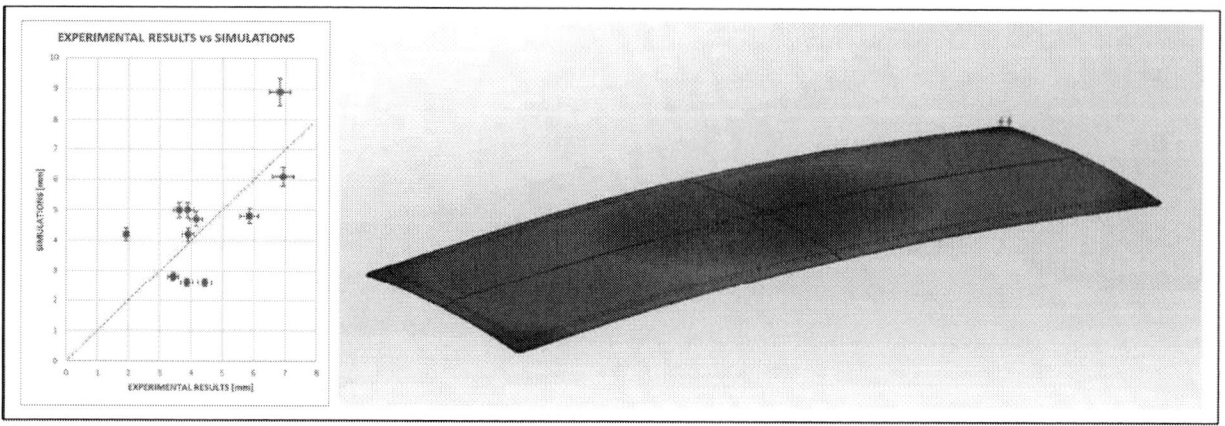

Fig. 6. Above, experimental results compared to simulated ones. The footprint of the simulated devices ranges from 2mmx2mm to 5mmx5mm. Dots placement variation is set to 5%. Below, an example of FEM simulation strip warpage output.

III. UNIT SIMULATION AND EXPERIMENTAL DIC MEASUREMENTS CORRELATION

Assumptions for the analysis reported in this paragraph are:

- This flow doesn't provide a complete connection from strip to unit simulation yet: they are currently still independent. It will be evaluated if it's worth connecting them, also considering the following statements. Anyhow, this flow is suitable for all the packages, not only for the ones assembled in panels.

- The stresses of each unit depend on the manufacturing processes. The boundary conditions for sure impact the final unit stress state and this can lead to a not negligible simplification. Therefore, each additional assembly step should be simulated in a sequential way which feasibility has been proven but the flow is not mature yet.

- Since each unit undergoes a sawing process which might introduce or release some stresses itself, µDIC (Micro Digital Image Correlation) measurements have been performed on a few units. By comparing unit simulations results with these experimental data, it is possible to achieve a reasonable level of confidence on unit level stresses and deformations.

Fig. 7 represents the generic schematic workflow developed in the unit level simulations. It can be grouped in three main steps: package and design simulation, µDIC Measurements and correlation and package deformation transfer to MEMS.

A. Pakage design and simulation

Simulations at the unit level have been performed avoiding the 3D CAD model simplifications required to study warpage at the strip level.

Local deformations are not negligible assessing the stresses induced on the MEMS. Fig. 8 reports an example of the package structures analysed: fully detailed metals, solder mask, core material and die geometry have been considered.

Warpage oriented package simulations have been set up constraining three out of four package corners (free condition) and ranging the temperature according to the investigated process steps [7].

B. µDIC Measurements and correlation

µDIC is a widely used and well-known optical measuring technique to evaluate strain and displacement on a specimen [8, 9]. In our application, a stereo microscope has been exploited. Data have been extracted from the samples following the 4 lines path highlighted in Fig.9 on the left.

The output of the analysis has been found slightly noisy. Consequently, a post-processing Fourier analysis-based fitting curve has been applied to the data.

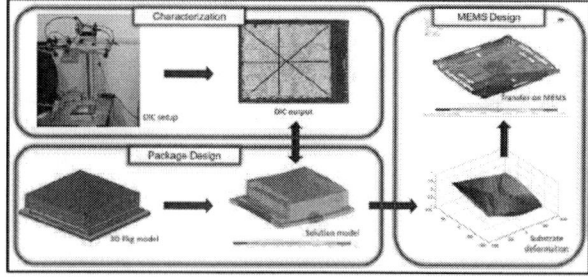

Fig. 7. MEMS unit level simulation workflow.

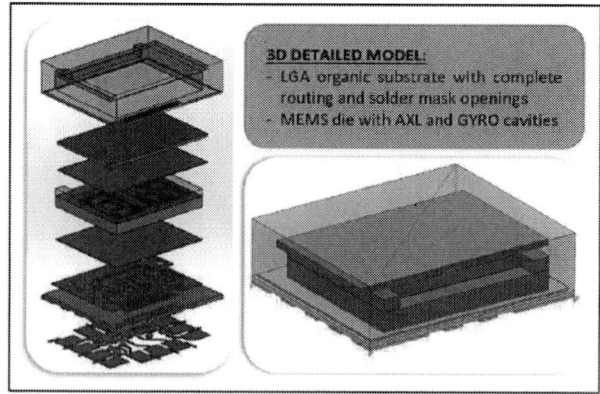

Fig. 8. 3D detailed and exploded model example (on the left) and model representation during simulation with constrains applied (on the right).

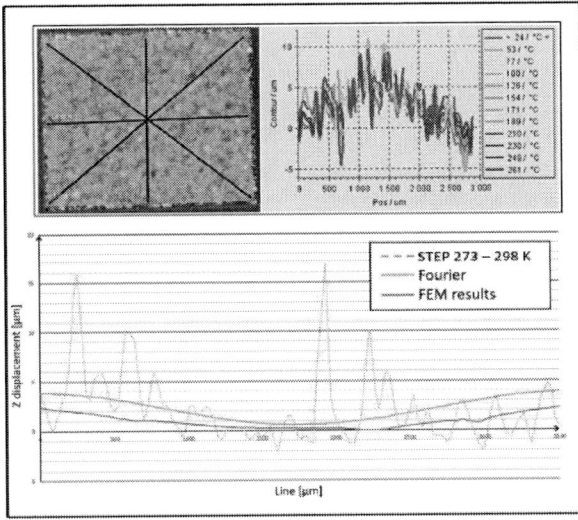

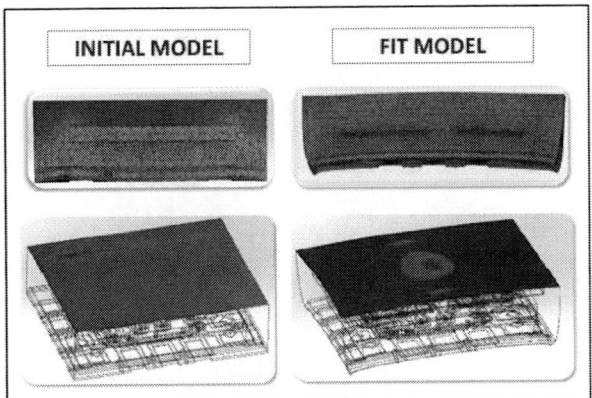

Fig. 11. Initial mode (left) vs fit model (right) with corresponding contact pressure difference.

Fig. 9. Upper Box: lines exported from the the μDIC measurements (left) and z displacement example of a specimen. Lower Box: comparison curves between experimental data (light blue line), Fourier analysis (orange line) and FEM results (blue line).

This methodology was found to be effective and consistent with the results coming from the previously described unit-level simulations (Fig. 9 reports the acquisitions made on one of the paths under evaluation). The simulated curves and the experimental ones have been investigated for a few devices showing consistency between the expected results and the FEM model as shown in Fig. 9.

C. Package deformation transfer to MEMS

Once the model has been fitted and the FEM results were comparable with the experimental data, the following step has been to understand the impact of this deformation on the device's performance.

Belonging MEMS and packages to different orders of magnitude, it was not trivial to bridge the reciprocal impact. The solution has been found in modelling the MEMS cavities inside the silicon performing package level simulation and then transferring the obtained displacement to the MEMS itself as reported in Fig. 10. The MEMS substrate deformation can cause a change in the capacitance between the stator and rotor electrodes, affecting the sensitivity of the gyroscope.

IV. UNIT SIMULATION IN SOCKET

During testing, the socket pushes the MEMS during its insertion, inducing a deformation of the package and of the substrate. The aim of this simulation is to establish a flow able to simulate the interaction between sockets and MEMS and to predict the level of stress and deformation of the device during testing. Socket design and optimization are essential for MEMS performance, being gyroscopes calibration and testing opportunely calibrated through their exploitation. It consists in contacting the I/O of the device with some needles, the so-called "pogo pin", and applying a defined stimulus.

The simulation setup consisted not only of the application of the pogo pin force on the substrate and of the top surface constraining. Since the μDIC analysis showed a bending in the package, it was implemented in the simulation. This step was essential to align the results with the expected ones (Fig. 11).

It was discovered that the yaw axis is not influenced by testing conditions whereas the roll worsens as the applied pogo force increases (Fig. 12).

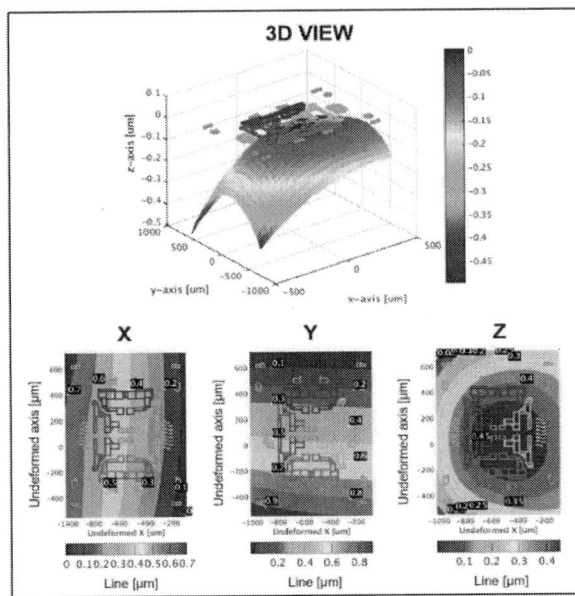

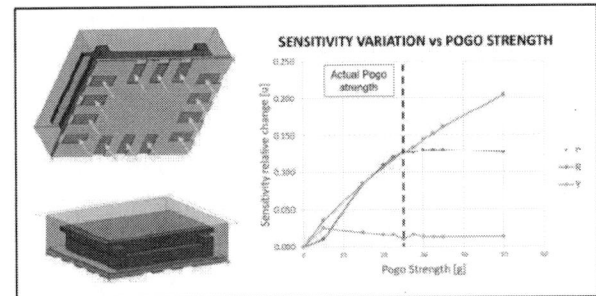

Fig. 12. Package to socket interaction model (left) and outputs (right).

Fig. 10. Example of gyroscope substrate deformation extrapolated by the performed package simulation.

Fig. 13. Results of thermo-mechanical soldering simulations performed on a pressure sensor. 3D CAD model, stress map of the section of the overall assembly and stress map of the MEMS present inside the package.

V. UNIT SIMULATION ON PCB

Before device usage, even though it is not part of the assembly process, the soldering on the PCB board may represent a final critical step for the MEMS sensor or actuator. The temperature needed for this purpose is relatively high)(about 573K). Once again, the whole stack up of materials may undergo a constrained expansion (or restriction), this time even more bounded due to the presence of the PCB (usually much thicker – and consequently rigid – than the device itself).

Simulations are performed to assess the stress distributions onto the most sensible parts of the MEMS, to avoid any damages and performance corruption.

It is challenging to perform a stress distribution data experimental extraction on an assembled and installed device. Consequently, instead of targeting the output of a certain amount of stress, simulations often try to assess the consistency between the amount of stress generated by a consolidated design and the new generation of the device.

The 3D CAD model exploited for these simulations consisted of a complete device (same degree of detail exploited for unit simulations) mounted onto a PCB through solder paste (Fig. 13).

CONCLUSIONS

Comparing the experimentally measured values during different MEMS assembly steps and predicted ones obtained via FEM, an in-depth study of the warpage phenomenon has been performed. A way to consistently simulate warpage at die and strip levels has been found and validated. Predictive analysis integrated into the package R&D design flow can help to prevent or mitigate warpage happening during production and related issues.

ACKNOWLEDGMENT

This work has been possible thanks to the efforts and the support of several colleagues from STMicroelectronics.

A special thanks to STMicroelectronics CAE group and Roseanne Duca coordinating it.

REFERENCES

[1] Zhang, Xueren, Tong Yan Tee, and Jing-en Luan, "Comprehensive warpage analysis of stacked die MEMS package in accelerometer application." 2005 6th International Conference on Electronic Packaging Technology. IEEE, 2005

[2] W. Chuang and W. Chen, "Study on the Strip Warpage Issues Encountered in the Flip-Chip Process", Materials, Vol. 15, 2023, DOI: 10.3390/ma15010323

[3] Xueren Zhang and Tong Yan Tee, "Advanced Warpage Prediction Methodology for Matrix Stacked Die BGA during Assembly Processes." 2004 Electronic Components and Technology Conference, Las Vegas, NV, USA, 2004, DOI: 10.1109/ECTC.2004.1319399

[4] R. Duca and M. O. Ghidoni, "Design for Package Miniaturization for a MEMS Pressure Sensor," 2019 20th International Conference on Thermal, Mechanical and Multi-Physics Simulation and Experiments in Microelectronics and Microsystems (EuroSimE), Hannover, Germany, 2019, pp. 1-5, doi: 10.1109/EuroSimE.2019.8724506

[5] M. Chae and E. Ouyang, "Strip warpage analysis of a flip chip package considering the mold compound processing parameters," 2013 IEEE 63rd Electronic Components and Technology Conference, Las Vegas, NV, USA, 2013, pp. 441-448, doi: 10.1109/ECTC.2013.6575609

[6] M. Rovitto, C. M. Villa, "Novel methodology for real time thermal expansion characterization on ball grid array substrate stack-up materials", Microelectronics Reliability Volumes 100–101, September 2019, DOI:10.1016/j.microrel.2019.113478

[7] M. Del Sarto, R. Duca, A. Maierna, N. Manca, M. O. Ghidoni, Teresa Napolitano, "Multiphysic Simulations for MEMS Sensor Package", 2019 21st Electronics Packaging Technology Conference (EPTC).

[8] M. Del Sarto, R. Duca, A. Maierna, N. Manca, M. O. Ghidoni and T. Napolitano, "Multiphysic Simulations for MEMS Sensor Package," 2019 IEEE 21st Electronics Packaging Technology Conference (EPTC), Singapore, 2019, pp. 688-695, doi: 10.1109/EPTC47984.2019.9026616

[9] M. Del Sarto, T.Napolitano and N.Manca,"Characterization of MEMS Inertial Measurement Unit Package with Digital Image Correlation" , AISEM 2021, LNEE918, pp.188-194, 2023

Gas permeable protection caps for wafer level chip scale packaging (WLCSP) of MEMS environmental sensors

Ole Behrmann
MEMS Applications
Fraunhofer-Institute for Silicon
Technology ISIT
Itzehoe, Germany
ole.behrmann@isit.fraunhofer.de

Thomas Lisec
MEMS Applications
Fraunhofer-Institute for Silicon
Technology ISIT
Itzehoe, Germany
thomas.lisec@isit.fraunhofer.de

Finn Klingbeil
MEMS Applications
Fraunhofer-Institute for Silicon
Technology ISIT
Itzehoe, Germany
finn.klingbeil@isit.fraunhofer.de

Niklas Kyoushi
MEMS Applications
Fraunhofer-Institute for Silicon
Technology ISIT
Itzehoe, Germany
niklas.kyoushi@isit.fraunhofer.de

Björn Gojdka
MEMS Applications
Fraunhofer-Institute for Silicon
Technology ISIT
Itzehoe, Germany
bjoern.gojdka@isit.fraunhofer.de

This work presents new porous environmental protection caps designed for wafer level chip scale packaging (WLCSP) of MEMS environmental sensors. The caps consist of gas-permeable microstructures formed from loose aluminum oxide powder solidified by a ceramic thin film, grown using atomic layer deposition (ALD). For the first time, the full process flow of the proposed approach is demonstrated by manufacture of the cap wafer followed by substrate bonding using glass-frit technology. By analyzing the influence of the caps on the response time of MEMS humidity sensors, this study proves the viability of the proposed packaging technology. In addition, the ability to further functionalize the porous caps is demonstrated by the deposition of a superhydrophobic polymer thin film by chemical vapor deposition.

Keywords - Sensor, MEMS, Gas Sensor, Pressure Sensor, Cap, Package, Wafer Level Package, Humidity Sensor, Chip Scale Package

I. INTRODUCTION

This work builds upon the previously reported "PowderMEMS" fabrication process, used for creating three-dimensional porous microstructures in MEMS applications [1–3]. MEMS environmental sensors require protection from dust, condensing moisture, and mechanical damage during operation while maintaining gas exchange with the environment [4–6]. Typically, this protection involves mounting the sensor die on a lead-frame and then over molding it to create an open cavity above the sensing area. For humidity sensors, a gas-permeable hydrophobic polymer membrane seals this cavity, while pressure sensors use a filter element or gel-like substance.

In this study, we introduce a novel concept for simultaneously capping multiple MEMS environmental sensors through a single wafer-bonding step. Leveraging the unique PowderMEMS technology, we create gas-permeable porous microstructures within a silicon wafer (Figure 1(a)). This cap wafer is then bonded to the MEMS device wafer, and the capped sensors are released using standard wafer dicing techniques (Figure 1(b)). To demonstrate the efficacy of this approach, we compare the output signals of capped and uncapped MEMS humidity sensors at the chip level.

PowderMEMS structures can possess pore sizes ranging from tens of nanometers to several micrometers. Their significant internal surface areas offer the advantage of functionalizing with thin films to achieve specific surface properties. By employing particles with a mean size in the micrometer range, we can create micron-sized pores that facilitate rapid gas exchange while still allowing the deposition of functionalization layers without the risk of

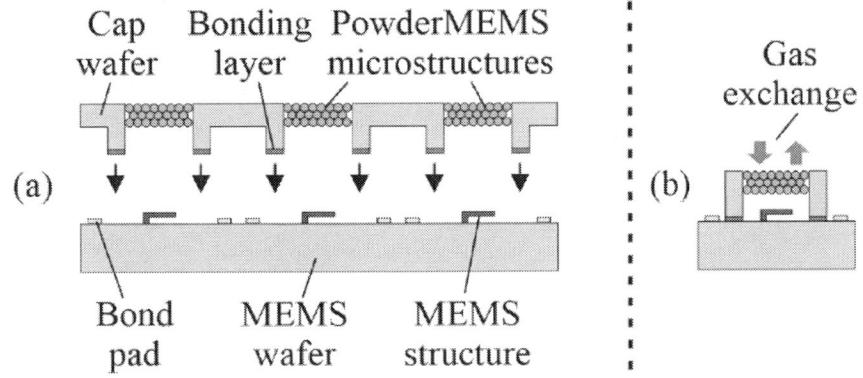

Figure 1: (a) PowderMEMS Cap- and MEMS-wafers before bonding. (b) Chip scale package (CSP) after bonding and dicing.

clogging.

Nevertheless, the presence of pores in PowderMEMS structures also results in high hydrophilicity, which is undesirable for environmental protection caps meant to shield the encapsulated MEMS from condensing moisture. To address this limitation, we investigate the hydrophobation of PowderMEMS cap structures through the deposition of Parylene C polymer thin films.

II. MATERIALS AND METHODS

A. PowderMEMS

In Figure 2, an overview of the PowderMEMS process is presented, demonstrating the creation of porous micro-structures within 200 mm silicon wafers with a thickness of 725 µm. The process begins with standard lithography to pattern a photoresist, serving as a mask for the subsequent deep reactive ion etching (DRIE) of 400 µm cavities using SPTS Pegasus equipment (Figure 2(a)).

Subsequently, loose Al_2O_3 particles with a mean diameter of 7 µm (D_{50}) are carefully deposited into these etched cavities (Figure 2(b)). The next step involves subjecting the wafers to 750 cycles of binary low-temperature (75 °C) atomic layer deposition (ALD). This ALD process utilizes trimethyl-aluminum (TMA) and water as precursors, resulting in the growth of a 75 nm thick Al_2O_3 layer on both the powder and other exposed surfaces (Figure 2(c)).

This ALD layer facilitates the agglomeration of the previously loose powder, transforming it into mechanically rigid porous microstructures. Notably, the porous micro-structures also become mechanically connected to the interior surfaces of the microcavities due to the ALD layer.

Following this, the wafers are inverted, and standard lithography is employed to pattern a second soft mask on the backside. In a subsequent step, another DRIE process is

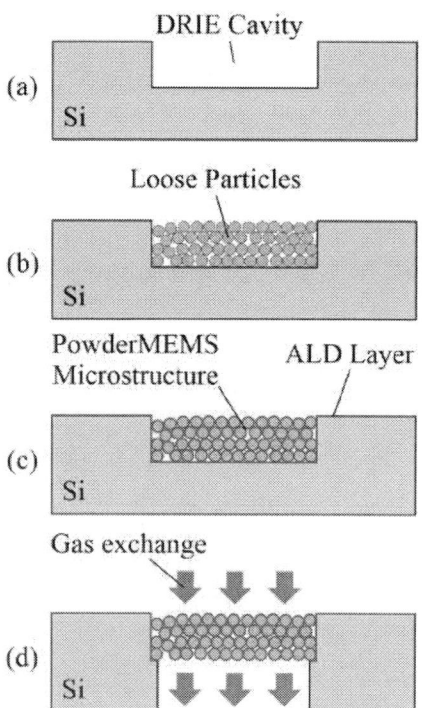

Figure 2: Major steps of the PowderMEMS Process.

performed, removing the remaining silicon from the backside. The ALD layer at the bottom of the cavities acts as an etch-stop during this process (Figure 2(d)).

Finally, a final Reactive Ion Etching (RIE) step using Applied Materials P5000 equipment is carried out to eliminate the bottom ALD layer and enable gas flow through the porous structures.

B. Substrate bonding

The PowderMEMS cap wafer is bonded to a second 200 mm substrate using a standard glass-frit process. To facilitate visual inspection of the bond seams, a Borofloat 33 glass wafer is chosen as the second substrate. In brief, glass-frit paste is screen printed onto the glass wafer. The glass wafer is then heated in an oven to remove the polymeric binder from the glass-frit paste. After alignment of the two wafers on a bond aligner (Süss BA8), glass-frit substrate bonding is performed in a substrate bond tool at 425 °C (Süss SB8).

C. CVD

Parylene C was applied onto the embedded porous micro-structures using a conventional chemical vapor deposition (CVD) procedure. 1.08 g of the Parylene C precursor di-para-xylylene were evaporated under vacuum conditions and allowed to polymerize on the chips resulting in a film thickness of about 200 nm. The substrates were kept at room-temperature.

D. Measurement setup

The cap arrays were diced into individual caps, which were subsequently hand-mounted onto the inlet port of a commercial MEMS humidity sensor (SHT-35, Sensirion, Switzerland) using silicone glue. To conduct the measurements, a sensor enhanced with a PowderMEMS cap, and an unmodified reference sensor were both connected in parallel to a USB-interface (SEK-SensorBridge, Sensirion, Switzerland). Throughout the measurements, both sensors remained in close proximity to each other.

III. RESULTS AND DISCUSSION

A. PowderMEMS processing

The proposed caps were successfully manufactured, as showcased in Figure 3. The 200 mm wafer features embedded PowderMEMS membranes of various geometries.

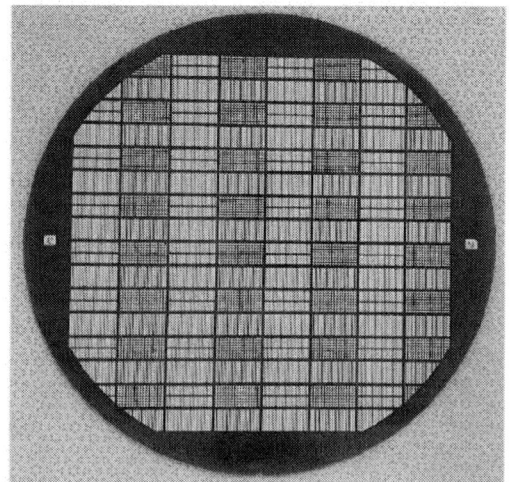

Figure 3: Top view of a 200 mm PowderMEMS cap wafer

Figure 4: Top (left) and bottom (right) view of an 8x16 array of PowderMEMS environmental protection caps.

Figure 6: Wetting behavior of the porous microstructures before (left) and after (right) hydrophobation.

For further investigation, circular structures with a diameter of 1 mm were selected. For a more detailed view, Figure 4 exhibits an 8x16 cap array from both the top and bottom angles, providing insight into the precise positioning of PowderMEMS structures within vertical silicon channels.

B. Substrate bonding

Substrate bonding was successfully performed, and no defects were observed on the PowderMEMS cap wafer. Figure 5 gives an overview of the backside of the bonded wafer stack. The glass-frit bond seams are visible through the transparent Borofloat 33 wafer. Visual inspection confirmed the success of the substrate bonding process.

C. Hydrophobation

In Figure 6, the wetting characteristics of the Powder-MEMS microstructures are depicted both before and after the process of hydrophobation. In the initial state (Figure 6, left), the structures exhibit easy wetting, causing water droplets to be drawn into them. However, following the hydrophobation by Parylene C deposition (Figure 6, right), the structures become impermeable to water. Instead of penetrating the surface, water forms a droplet with a contact angle exceeding 120°, providing clear evidence of the successful hydro-phobation process. Finally, the caps were immersed in water at a depth of 1 meter for a duration of 30 minutes and no water ingress was observed.

D. Humidity sensor performance

In this experiment, both a modified humidity sensor (Figure 7) and an unmodified sensor were subjected to repeated exposure to humid air (Figure 8). Notably, the performance of the modified sensor closely mirrored that of the unmodified sensor. This outcome indicates that the PowderMEMS structure's μm-sized pores facilitate rapid gas exchange, and the proposed approach has minimal impact on the sensor's response time constant.

IV. SUMMARY AND OUTLOOK

This study showcases the fundamental production process and capabilities of PowderMEMS environmental protection caps designed for MEMS gas and pressure sensors. The successful fabrication of these caps on 200 mm Si-wafers was demonstrated, followed by full wafer substrate bonding.

In addition, hydrophobation by deposition of Parylene C thin films was performed on single caps. These caps were then mounted on commercial MEMS humidity sensors and their influence on the sensor response time constant was evaluated. The change in sensor time constants was found to be only very minor. These results demonstrate the validity of the proposed wafer-level packaging approach. Finally, the hydrophobic caps were immersed in water at a depth of 1 meter for a duration of 30 minutes after which no water ingress was observed. In further work, environmental testing towards long term stability of the caps according to industrial standards such as IPxx and AEC-Q100 is planned. Additionally novel concepts for the integration of catalytic materials for the removal of interfering gases will be explored.

Figure 7: Micrograph of a single cap assembled on a commercial MEMS humidity sensor (SHT-35, Sensirion).

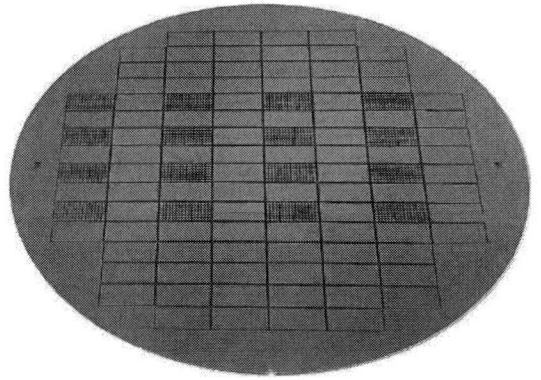

Figure 5: Bottom view through the glass wafer. The glass-frit bond seams are visible.

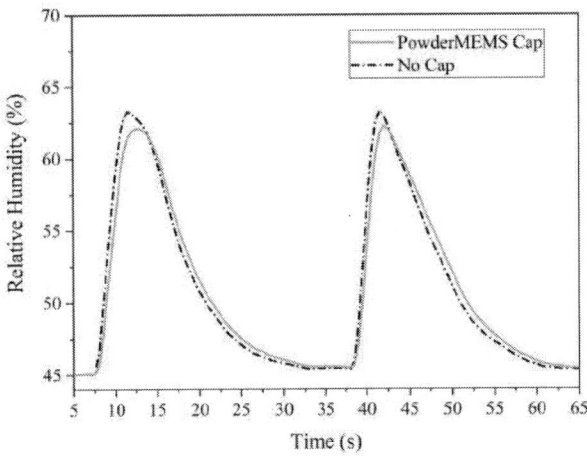

Figure 8: Performance of a commercial MEMS humidity sensor modified with a 400 μm thick gas permeable PowderMEMS environmental protection cap (solid line) compared to an unmodified sensor (dotted line).

ACKNOWLEDGMENT

This work was funded in part by the European Regional Development Fund (LPW-E/1.2.2/1305).

REFERENCES

[1] T. Lisec, O. Behrmann, and B. Gojdka, "PowderMEMS—A Generic Microfabrication Technology for Integrated Three-Dimensional Functional Microstructures," *Micromachines*, vol. 13, no. 3, p. 398, 2022, doi: 10.3390/mi13030398.

[2] O. Behrmann, T. Lisec, and B. Gojdka, "Towards Robust Thermal MEMS: Demonstration of a Novel Approach for Solid Thermal Isolation by Substrate-Level Integrated Porous Microstructures," *Micromachines*, vol. 13, no. 8, 2022, doi: 10.3390/mi13081178.

[3] C. Kostmann, T. Lisec, M. T. Bodduluri, and O. Andersen, "Automated Filling of Dry Micron-Sized Particles into Micro Mold Pattern within Planar Substrates for the Fabrication of Powder-Based 3D Microstructures," *Micromachines*, vol. 12, no. 10, p. 1176, 2021, doi: 10.3390/mi12101176.

[4] Y. Ma, J. Kaczynski, C. Ranacher, A. Roshanghias, M. Zauner, and B. Abasahl, "Nano-porous aluminum oxide membrane as filtration interface for optical gas sensor packaging," *Microelectronic Engineering*, vol. 198, pp. 29–34, 2018, doi: 10.1016/j.mee.2018.06.013.

[5] R. N. Dean *et al.*, "Porous Ceramic Packaging for a MEMS Humidity Sensor Requiring Environmental Access," *IEEE Trans. Compon., Packag. Manufact. Technol.*, vol. 1, no. 3, pp. 428–435, 2011, doi: 10.1109/TCPMT.2010.2101230.

[6] S. Raible, D. Briand, J. Kappler, and N. F. de Rooij, "Wafer Level Packaging of Micromachined Gas Sensors," *IEEE Sensors J.*, vol. 6, no. 5, pp. 1232–1235, 2006, doi: 10.1109/JSEN.2006.881355.

Additive metallization of alumina with copper-titanium powder blends for power electronic applications

Christoph Hecht
Institute for Factory Automation and
Production Systems
Friedrich-Alexander-Universität
Erlangen-Nürnberg
Nuremberg, Germany
christoph.hecht@faps.fau.de

Eric Schadow
Institute for Factory Automation and
Production Systems
Friedrich-Alexander-Universität
Erlangen-Nürnberg
Nuremberg, Germany
eric.schadow@gmx.de

Mario Sprenger
Institute for Factory Automation and
Production Systems
Friedrich-Alexander-Universität
Erlangen-Nürnberg
Nuremberg, Germany
mario.sprenger@faps.fau.de

Felix Häußler
Institute for Factory Automation and
Production Systems
Friedrich-Alexander-Universität
Erlangen-Nürnberg
Nuremberg, Germany
felix.haeussler@faps.fau.de

Thomas Stoll
Professorship of Laser-based Additive
Manufacturing
TUM School of Engineering
and Design
Munich, Germany
thomas.stoll@tum.de

Jörg Franke
Institute for Factory Automation and
Production Systems
Friedrich-Alexander-Universität
Erlangen-Nürnberg
Erlangen, Germany
joerg.franke@faps.fau.de

Abstract— **Additive manufacturing shows great potential to further increase the performance of power electronic modules through novel packaging concepts. Such an approach is the integrated manufacturing of metal-ceramic substrates by means of laser powder bed fusion of metals (PBF-LB/M). With this layered additive manufacturing process, planar ceramic substrates can be metallized and electrically functionalized for power electronic applications. In this paper Al_2O_3 ceramic substrates are metallized via PBF-LB/M by selectively melting applied powder layers with software defined geometries. The investigated powders are mixtures of copper and titanium powders with 1 wt.%, 5 wt.% and 10 wt.% titanium in order to enable bonding by creating a titanium-oxide reaction layer at the interface to the ceramic. Shear tests and microstructural investigations show that a subsequent heat treatment increases the adhesion and density of the metallization. By energy dispersive X-ray microscopy (EDX) the partial formation of reaction layers is detected.**

Keywords—Active Metal Brazing, Laser powder bed fusion, Power electronics, Substrates

I. INTRODUCTION

The increasing electrification of the entire energy and mobility sector, due to the existing climate change significantly caused by CO_2 emissions from the combustion of fossil fuels, is being driven forward with high priority in most industrialized countries. Only recently, it was decided within the EU that only CO_2 emission free engines will be permitted for the public motor vehicle sector from 2035 onwards [1] paving the way for electrically driven cars. For the industrial sector, greenhouse gas emissions can be reduced by up to 95 %, if a 60 % electrification rate is achieved [2]. In addition, the expansion of renewable energies such as solar and wind power also increasingly requires power electronic modules for the conversion of the electrical energy to match the requirements of electrical consumers like motors or heaters.

The presented work received funding from the Deutsche Forschungsgemeinschaft (DFG, German Research Foundation) – 434962551; 442921285.

A. Substrate technologies in power electronics packaging

Most commonly used as substrate materials in power electronic modules are metal-ceramic-interconnects due to an improved performance with regard to ampacity, insulation, thermal management and reliability characteristics compared to printed circuit boards based on organic materials. Such high power electronic substrates are commonly manufactured via two different processes:

- Direct Bonded Copper (DBC)
- Active Metal Brazing (AMB)

whereas AMB-substrates are more used for high performance applications with high requirements on thermo-mechanical stability and thermal conductivity. [3] Both technologies are described in the following:

The DBC-technology is based on an eutectic bonding of copper foils on alumina (Al_2O_3), zirconia toughened alumina (ZTA) or aluminum nitride (AlN) in a burning process in the range of 1065 °C and 1083 °C. At this temperature range, the eutectic of Cu_2O at the Cu-ceramic-interface forms a liquid wetting on the substrate [4]. The Cu-foils are either pre-oxidized or oxidized during the burning process in the furnance to form Cu_2O at the surface. With the wetting of the Cu_2O-eutectic on the ceramic, phases like $CuAl_2O_4$ (spinell) or $CuAlO_2$ (delafossit) are emerging that ensure a strong bonding between the two materials with a high bonding strength. [5] [6] [7]

The technology of Active Metal Brazing utilizes an active filler material to ensure a bonding between laminated copper foils and ceramic substrates. The filler material nowadays mostly contains a silver-based alloy with a share of 1 to 10 wt.% of reactive elements like titanium, hafnium or zirconium, that enable a wetting of the active solder (filler material) on the ceramic substrate [8]. Industrially used ceramic materials are mostly non-oxide ceramics like AlN or silicon nitride (Si_3N_4) due to the superior material properties with respect to mechanical strength and/or thermal conductivity. Nevertheless, both substrates are very expensive compared to oxide-based ceramics, which makes substrate materials like Al_2O_3 interesting for future research and development concerning novel metallization techniques. The bonding mechanism is based on the emergence of a reaction zone between the ceramic substrate and the used active solder [9] whereas on the metal side (copper) an adequate wetting is given anyway, due to the lower surface energy of the metal, determined by the persistence of free electrons in the material. In the reaction zone between the active solder and the oxide-based ceramic different Magneli-phases, like Ti_3O_5, Ti_4O_7, TiO_2 [8] as well as $Ti_3Al[O]$-phases [10] are emerging that resist mechanical shear stresses of up to 180 MPa [11].

Nevertheless, both technologies have in common that they are exclusively used for planar metallization in the power electronic sector. Moreover, both methods require long process chains with different manufacturing steps like laminating, application of the active solder at AMB-substrates, burning, etching and cleaning with the usage of stencils and/or lithographic technologies. [3] Furthermore, according to the goal to apply sustainable technologies in the future, the usage of etching technologies shall be replaced with respect to ecological aspects. Additionally, just a horizontal integration of active and passive electronic elements for the power electronic device is possible, as the substrate itself is planar. As the power density of future power electronic devices needs to further increase to maintain or even reduce the spatial dimensions of the devices, a 3D- and additional vertical integration of electronic components will be necessary to keep up with the rising need for electric energy.

B. Additive Manufacturing: Laser Powder Bed Fusion

Generally, additive manufacturing processes are divided into seven categories, with powder bed fusion being the most important category for the material class of metals [12]. Layer by layer applied metal powder is melted with a beam source such as a laser or an electron beam. The schematic layout of a system for powder bed fusion of metals with a laser-based system (PBF-LB/M) is shown in Fig. 1.

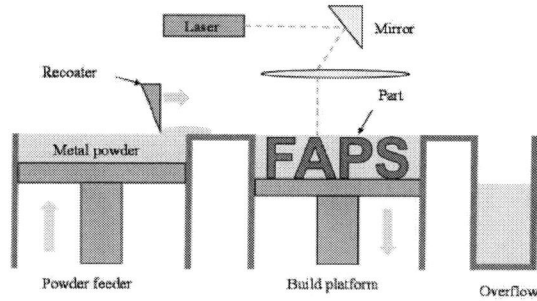

Fig. 1. Powder bed fusion of metals with a laser-based system (PBF-LB/M) [13].

Thin powder layers with a thickness between 20 and 100 micrometers are drawn from the powder reservoir onto the build platform with a recoater system. The layer is then irradiated with a finely focused laser according to the part cross-section and the build platform is lowered by one layer thickness. By building up a large number of layers, the component geometry is thus created, and all the unmelted

powder can be reused after a sieving process. In the process, the parts are built up on a metallic build platform, so that subsequent separation is necessary. By replacing the metallic build platform with other substrates, metallizations can also be produced [13] as applied in this study to ceramic substrates. This way the ceramic can be functionalized for use in power electronic modules.

II. MACHINE AND MATERIALS

The investigations were performed on the PBF-LB/M-Machine Mlab Cusing R of the manufacturer Concept Laser, which is equipped with a fiber laser emitting at a wavelength of 1070 nm up to 500 W. The diameter of the laser spot on the build platform was set to 35 µm. In order to decrease the thermal shock during solidification and to enhance diffusion processes at the metal-ceramic interface, the building chamber was equipped with a heating module, which allowed preheating of the ceramic substrate to 300 °C. [14] The ceramic substrate was positioned in a milled pocket of an adapter plate, which was mounted onto the build platform. The residual oxygen in the argon flooded process chamber was held below 50 ppm, which was monitored with an Orbitalum ORBMax.

The processed powder was copper ETP of manufacturer Ecka Granules mixed with 1 wt.%, 5 wt.% and 10 wt.% titanium powder of manufacturer TLS Technik GmbH & Co. Spezialpulver KG. The copper powder is labeled with AK < 0.045 mm and has a particle size distribution of $d_{10} = 11.60$ µm and $d_{90} = 42.85$ µm. [15] The titanium powder is characterized with a d_{10} of 10 µm and a d_{90} of 45 µm according to the data sheet. Both powders have a spherical shape as shown in Fig. 2. The blending process was performed manually and with a vibratory plate for at least 30 minutes. The metallized substrates were $40\times40\times1$ mm³ Rubalit 708S plates of manufacturer CeramTec. Shear testing was conducted based on DIN EN 15340 with a XYZTEC Condor 150-3, which can apply a maximal shear force of 400 N. For subsequent thermal treatments, the tube furnance Gero Carbolite GHA 12/300 was flooded with argon 5.0 to reach a residual oxygen content of less than 10 ppm in the tube to avoid oxidation of the metallizations.

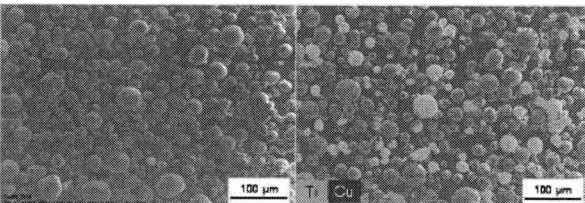

Fig. 2: SEM image and elemental mapping of powder with 10 wt.% titanium.

III. METALLIZATIONS WITH VARYING TITANIUM CONTENT

Due to the high affinity of titanium to react with oxygen, an influence of the titanium content in the metal powder mixture on the adhesive strength of the metallization on the Al_2O_3 ceramic is expected. This aspect will be investigated within this study.

A. Experimental design

Metallization studies were conducted according to Table I. For each parameter combination one cuboid metallization with a quadratic base of 5 mm edge length and a height of

360 µm was printed, which results from 12 layers of 30 µm thickness. In total 450 cubes on 51 ceramics were printed as one ceramic was metallized with up to 9 cubes. The samples were built up with an island exposure strategy, dividing the 5x5 mm² cross-section of the cuboids in the exposure plane into 25 1x1 mm² squares. In order to improve the powder deposition of the first layer, thin shims were placed between the build platform and the adapter plate.

TABLE I.   Full factorial experimental design.

Parameter	Min.	Max.	Increment
Power in Watt	30	40	2
Scan speed in mm/s	550	750	50
Hatch in µm	10.5	24.5	3.5
Ti share in wt.%	1, 5, 10		

B. Adhesion

All samples within the experimental design according to Table I were subjected to shear testing. With a maximum load cell force of 400 N and a specimen cross-section of 25 mm², a maximal adhesive strength of 16 MPa can be obtained. The boxplot in Fig. 3 shows the results of the shear tests.

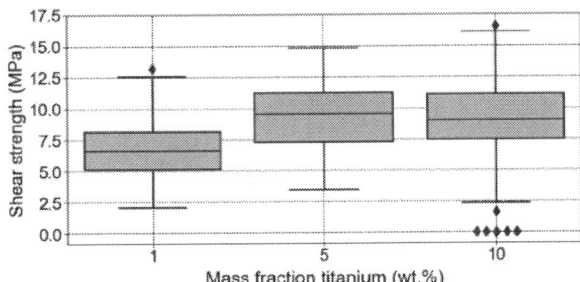

Fig. 3. Adhesion box plots of powder mixtures obtained by shear testing.

With an one-way analysis of variances (ANOVA) the obtained differences in mean adhesive strength between the different titanium shares are checked for statistical significance at a level of significance of 5 %. The mean value of the samples with 1 wt.% titanium differs significantly from the samples with 5 wt.% and 10 wt.% titanium. Between 5 wt.% and 10 wt.% no significant difference between the mean values of adhesion are obtained. So, the increase of titanium within the powder system improves the adhesion only within the range of 1 wt.% to 5 wt.% The mean values for all powder mixtures are depicted in Fig. 4.

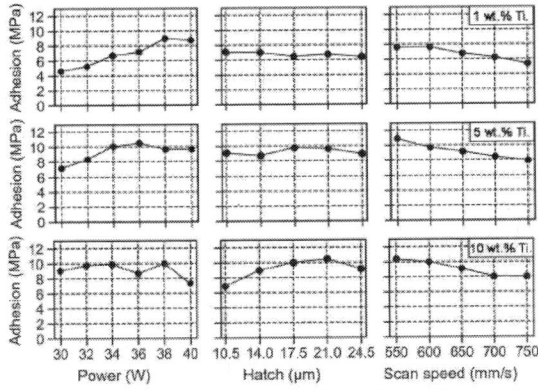

Fig. 4. Mean adhesion at factor levels according to the experimental design: 1 wt.% titanium (top), 5 wt.% titanium (middle), 10 wt.% titanium (bottom).

At 1 wt.% the laser power and scan speed have a strong impact on the adhesion in such a way, that higher laser powers and lower scan speeds result in stronger adhesion. At varying hatches, the adhesion alters only slightly within less than 1 MPa. Therefore, within the considered experimental design the adhesion improves with increasing energy input. For each parameter, the highest mean value is obtained for parameter range limiting values and therefore no optimal parameters are obtained. For 5 wt.% titanium, a laser power of 34 W and 36 W results in samples with a mean adhesion of more than 10 MPa. Similar to 1 wt.% the hatch has less impact on the adhesion. The strongest bonding is obtained at 550 mm/s. This applies also to powders with 10 wt.% titanium. In contrast to 1 wt.% and 5 wt.%, the hatch has a strong impact and the laser power less impact at 10 wt.% titanium. Adhesion maximizing parameters within the investigated parameter range are calculated based on a regression model. The result can be obtained from Table II. The increasing laser power with decreasing share of titanium can be subjected to the high reflectivity of copper as more power is required for fusion. With regard to the scan speed, strongest adhesion is obtained at 550 mm/s. At lower scan speed, the material maintains longer in the molten state and diffusion processes are enhanced. Due to the high adhesive strength and low titanium content, metallizations with 5 wt.% titanium will be further investigated in heat treatment studies. To check the reproducibility of the identified process parameters for 5 wt.% titanium given in Table IV, a control group of 27 pads on three different ceramics is manufactured and sheared off. Due to machine limitation, the laser power was rounded to an integer value of 38 W. The obtained mean adhesion of 15.2 MPa is even higher than the model prediction of 13.2 MPa. The obtained standard deviation is 2.6 MPa.

TABLE II.   Adhesion maximizing parameters.

Parameter	Share of titanium in wt.%			Unit
	1	5	10	
Power	40	38.7	31.5	Watt
Scan speed	550	550	550	mm/s
Hatch	10.5	24.5	16.7	µm
Predicted adhesion of regression model	10.1	13.2	13.2	MPa

IV. HEAT TREATMENT

Subsequent heat treatments according to Table III are performed on samples with 5 wt.%. The edge length of the quadratic base area was reduced from 5 mm to 3 mm in order to allow measurements up to 44.4 MPa with the load cell of the shear testing equipment, which is limited to a shear force of 400 N.

TABLE III.   Experimental design for heat treatment.

Parameter	Heat treatment		
	1	2	3
Peak temperatur in °C	850	850	950
Dwell time at peak temperature in minutes	5	60	60

A. Cross-section analysis

The effect of the heat treatment on the metallization can be observed in cross-section analysis shown in Fig. 5. The relative density increases according to Table IV. The values are obtained by evaluating the share of white and black pixels after conversion to a binary image.

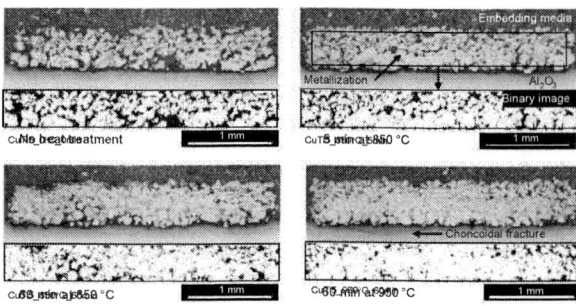

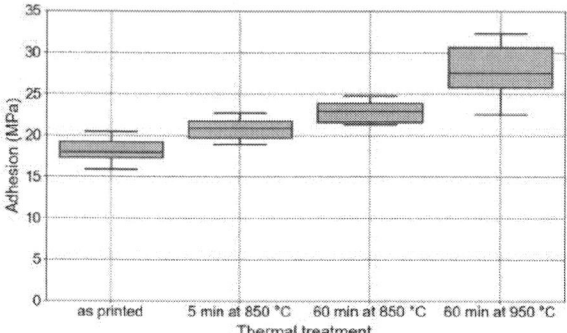

Fig. 5. Cross-section incident light micrographs of heat-treated samples and associated binary images for relative density measurements.

This increase in relative density can be subjected to time and temperature dependent solid-state diffusion processes, which intensify at high temperatures. [16] At 950 °C conchoidal fractures in the ceramic occur, which are a consequence of thermo-mechanical stresses in brittle materials like Al_2O_3. [17]

TABLE IV.   Relativ densities obtained by optical measurement.

Treatment	Relative density in %
No treatment	72.2
1	82.4
2	87.1
3	93.8

B. Shear tests

The reduction of the base area of the metallization to 9 mm² allows to measure adhesion up to 44.4 MPa. For each heat treatment 12 samples on one ceramic are manufactured and shear tested. The resulting box plots are shown in Fig. 6. By decreasing the base area of the metallization, stronger adhesion is obtained as the mean adhesion increases by 3 MPa to 18.2 MPa in the as printed state in comparison to the control group in section 3B with a quadric base of 5 mm edge length. With increasing dwell times and temperatures, the mean adhesion improves to 20.8 MPa, 22.8 MPa and 27.8 MPa. At 950 °C the adhesion fluctuates more strongly as shown in the box plot in Fig. 6. This can be subjected to conchoidal fractures in the ceramic interface, which are observed in cross-section analysis in Fig. 5.

Fig. 6. Adhesion box plots of heat-treated samples with 5 wt.% titanium.

C. Microstructure

The impact of the heat treatment on the microstructure was further analyzed via EDX analysis. The elemental mapping of O, Cu and Ti at the cross-section is shown in Fig. 7. The element distribution of the sample without heat treat shows that no homogeneous alloying by the laser process is obtained, because undissolved titanium particles are embedded in the copper matrix.

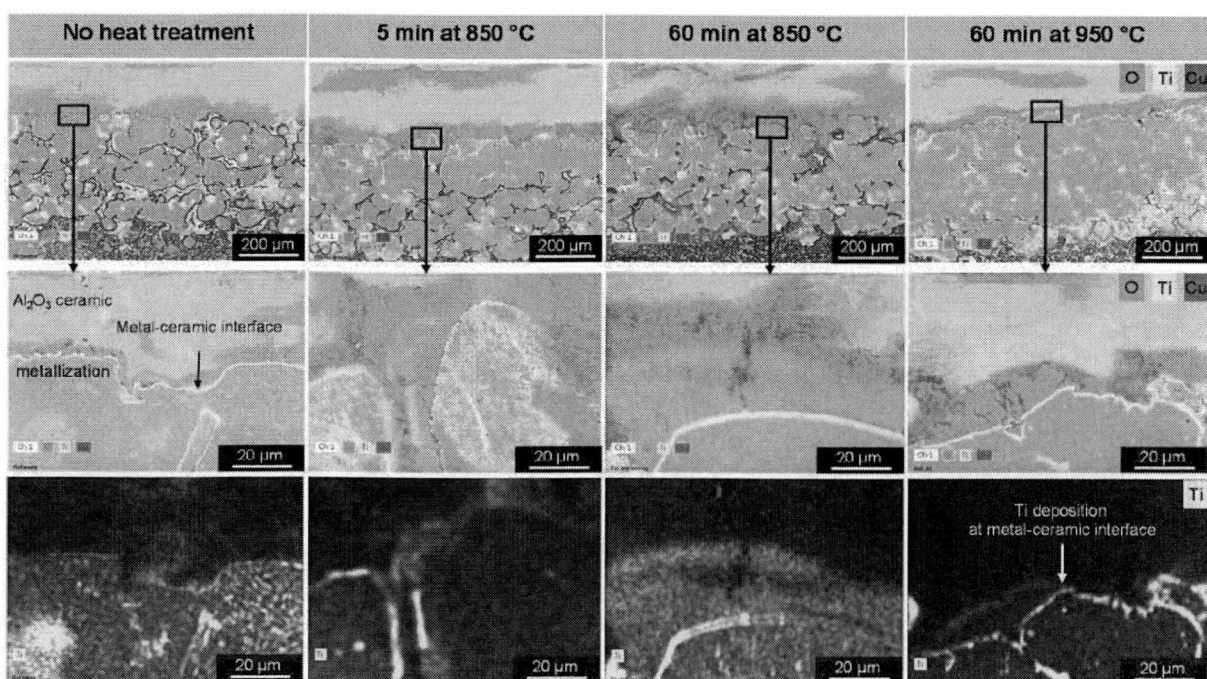

Fig. 7. Elemental mapping of O, Cu and Ti at cross-section after heat treatment of specimen with 5 wt.% titanium.

TABLE V. Elemental composition at positions according to Fig. 8.

| Element | Elemental composition in at.% | | | | | | | | |
| | Position | | | | | | | Mean | Std. dev. |
	1	2	3	4	5	6	7		
Al	4.11	3.69	11.13	2.54	2.53	2.31	2.7	4.14	3.15
C	9.49	12.07	9.74	13.93	13.7	13.39	13.41	12.25	1.89
Cu	39.99	29.29	14.6	29.71	29.9	29.85	30	29.05	7.43
O	14.73	21.94	44.61	18.11	19.47	18.25	18.37	22.21	10.10
Si	0	0.85	0	0	0	0	0	0.12	0.32
Ti	31.68	32.16	19.91	35.71	34.4	36.19	35.52	32.22	5.71

Nevertheless, the detailed element map at the interface shown on the bottom left in Fig 7 shows, that alloying is achieved to some extent across the cross-section by laser powder bed fusion. The distribution at the metal-ceramic interface changes after the samples are subjected to thermal treatments. Titanium deposits at the interface and locally a thin layer is formed. This can be observed in more detail in Fig. 8. The formation of such reaction layers at the interface is well known from active metal brazing processes [18]. An elemental analysis at the marked positions within this interface layer is shown in Table V.

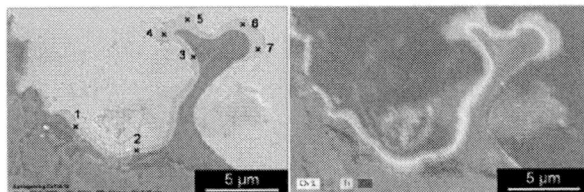

Fig. 8. Cross-section SEM images and elemental mapping at metal-ceramic interface after heat treatment for 60 minutes at 950 °C.

The ratio of detected Cu, Ti and O is an indicator for the phase at the interface, which is likely to be a material system based on Cu, Ti and O with dissolved Al like Cu_2Ti_2O [18] or Cu_3Ti_3O, which is mentioned in [19]. The formation of a reaction layer serves to explain the improved adhesion after thermal treatment.

V. SUMMARY

In this paper a novel approach on manufacturing metal-ceramic-substrates by means of laser powder bed fusion of a powder mixture with titanium as an active element is presented. The adhesion and microstructure of locally fused copper powders with titanium shares of 1 wt.%, 5 wt.% and 10 wt.% was analyzed. Based on a regression model, adhesion optimized parameter values for all powder mixtures are calculated. For 1 wt.% titanium, the adhesion maximizing parameters are parameter range limiting factors, which indicates, that with higher energy input during the laser powder bed fusion process stronger adhesion can be obtained. Further investigations are required to optimize the adhesion for 1 wt.% titanium. For 5 wt.% and 10 wt.% titanium an optimal laser power to maximize the adhesion was identified. Samples with 5 wt.% titanium were then subjected to different heat treatments at varying temperatures and dwell times. The mean adhesion of 18 MPa without heat treatment was improved to 27.8 MPa after a heat treatment at 950 °C for 60 minutes. At 950 °C conchoidal fractures due to thermo-mechanical stresses in the brittle ceramic were observed in cross-section analysis, leading to a more pronounced variation in adhesive strength. Elemental mapping showed locally distributed reaction layers at the metal-ceramic interface after heat treatment at 950 °C for 60 minutes. So these layers seem to increase the adhesion of the metallization.

ACKNOWLEDGMENT

Funded by the Deutsche Forschungsgemeinschaft (DFG, German Research Foundation) – 434962551; 442921285.

REFERENCES

[1] GERMAN GOVERNMENT. EU-Umweltrat: Nur noch CO2-frei fahren [online]. 2023. https://www.bundesregierung.de/breg-de/schwerpunkte/europa/verbrennermotoren-2058450. Accessed: 6th July 2023.

[2] BERGER, C. Das Erreichen der Klimaziele durch Elektrifizierung [online]. springerprofessional.de. 3rd April 2019, https://www.springerprofessional.de/klimawandel/energie/das-erreichen-der-klimaziele-durch-elektrifizierung/16592690. Accessed: 6th July 2023

[3] JILLEK, W. and G. KELLER. Handbuch der Leiterplattentechnik. Bad Saulgau: Eugen G. Leuze Verlag; Carl Hanser Verlag, 2003. Hanser eLibrary. ISBN 9783874803304

[4] BURGESS, J.F., C.A. NEUGEBAUER and G. FLANAGAN. The Direct Bonding of Metals to Ceramics by the Gas - Metal Eutectic Method. Journal of The Electrochemical Society, 1975, 122(5), S. 688-690. ISSN 1945-7111. doi:10.1149/1.2134293

[5] YOSHINO, Y. and H. OHTSU. Interface Structure and Bond Strength of Copper-Bonded Alumina Substrates. Journal of the American Ceramic Society, 1991, 74(9), S. 2184-2188. ISSN 0002-7820. doi:10.1111/j.1151-2916.1991.tb08281.x

[6] KIM, S.T. and C.H. KIM. Interfacial reaction product and its effect on the strength of copper to alumina eutectic bonding. Journal of Materials Science, 1992, 27(8), S. 2061-2066. ISSN 1573-4803. doi:10.1007/BF01117918

[7] HE, H., R. FU, D. WANG, X. SONG and M. JING. A new method for preparation of direct bonding copper substrate on Al2O3 . Materials Letters, 2007, 61(19-20), S. 4131-4133. ISSN 0167-577X. doi:10.1016/j.matlet.2007.01.036

[8] NICHOLAS, M.G., Hg. Joining of ceramics. London: Chapman and Hall, 1990. Advanced ceramic reviews. ISBN 0412367505

[9] EUSTATHOPOULOS, N. Wettability at High Temperatures. London: Elsevier Science & Technology, 1999. Pergamon Materials Ser. v.Volume 3. ISBN 9780080543789

[10] KELKAR, G.P. and A.H. CARIM. Phase Equilibria in the Ti-Al-O System at 945oC and Analysis of Ti/Al2O3 Reactions. Journal of the American Ceramic Society, 1995, 78(3), S. 572-576. ISSN 1551-2916. doi:10.1111/j.1151-2916.1995.tb08216.x

[11] HONGQI, H., J. ZHIHAO and W. XIAOTIAN. The influence of brazing conditions on joint strength in Al2O3/Al2O3 bonding. Journal of Materials Science, 1994, 29(19), S. 5041-5046. ISSN 1573-4803. doi:10.1007/BF01151094

[12] DIN EN ISO/ASTM 52900. 2022. Additive Fertigung – Grundlagen – Terminologie. Berlin: Beuth Verlag GmbH

[13] SYED-KHAJA, A., D. SCHWARZ and J. FRANKE. Advanced substrate and packaging concepts for compact system integration with additive manufacturing technologies for high temperature applications. In: 2015 IEEE CPMT Symposium Japan (ICSJ).

[14] STOLL, T., KIRSTEIN, M. and FRANKE, J.. A novel approach of copper-ceramic-joints manufactured by selective laser melting. In

Material Technologies and Applications to Optics, Structures, Components, and Sub-Systems IV (Vol. 11101, pp. 53-63). SPIE.

[15] STOLL, T. Laser Powder Bed Fusion von Kupfer auf Aluminiumoxid-Keramiken. 2023.FAU Studien aus dem Maschinenbau Band 420, Erlangen: FAU University Press. ISBN 978-3-96147-632-9

[16] KANG, S.-J.L. Sintering. Densification, Grain Growth, and Microstructure. Oxford: Elsevier Science & Technology, 2005. ISBN 9780080493077.

[17] VALDEZ-NAVA, Z., D. KENFAUI, M.-L. LOCATELLI, L. LAUDEBAT and S. GUILLEMET. Ceramic substrates for high voltage power electronics: past, present and future. In: 2019 IEEE International Workshop on Integrated Power Packaging (IWIPP). Toulouse, France, April 24-26, 2019. Piscataway, NJ: IEEE, 2019, S. 91-96. ISBN 978-1-5386-7610-3

[18] KRITSALIS, P., L. COUDURIER and N. EUSTATHOPOULOS. Contribution to the study of reactive wetting in the CuTi/Al2O3 system. Journal of Materials Science, 1991, 26(12), S. 3400-3408. ISSN 1573-4803. doi:10.1007/BF01124693

[19] HANSEN, M., K. ANDERKO and H.W. SALZBERG. Constitution of Binary Alloys. Journal of The Electrochemical Society, 1958, 105(12), S. 260C. doi: 10.1149/1.2428700

Negative-tone Photo-definable Polyimide with High Thermal Stability and Thick Film Processability

Takayuki Kaneki
Electronic & Imaging Materials
Research Laboratories
Toray Industries, Inc.
Otsu, Shiga, Japan
Takayuki.Kaneki@mail.toray

Yu Shoji
Electronic & Imaging Materials
Research Laboratories
Toray Industries, Inc.
Otsu, Shiga, Japan
yu.shoji.r7@mail.toray
ORCID 0000-0001-6290-701X

Chika Hibino
Electronic & Imaging Materials
Research Laboratories
Toray Industries, Inc.
Otsu, Shiga, Japan
chika.taniguchi.z9@mail.toray

Hitoshi Araki
Electronic & Imaging Materials
Research Laboratories
Toray Industries, Inc.
Otsu, Shiga, Japan
hitoshi.araki.u8@mail.toray
ORCID 0000-0002-9632-7811

Abstract´ High voltage power semiconductor requires higher polyimide insulation properties. We have designed a polyimide with high thermal stability, high light transparency and alkali dissolvable polyimide that is capable of passivating thick layer of polyimide with large topographys. A novel negative-tone photo-definable polyimide was formulated with cross linking reagent that can easily decompose during cure and a photo-initiator that initiates well during lithography. Our novel Negative photo sensitive polyimide, PSPI-1, has a patternable capability up to 16 mm thickness on silicon with 350 mJ/cm². PSPI-1 showed higher patternability than of the conventional polyimide. The cured film showed excellent properties as follows: 310 ℃ glass transition temperature (Tg), 500 ℃ 5% weight loss temperature, 40% elongation, no short observation during bias-HAST (130 ℃. 96%, 100hr, 100V 15/30 mm line&space) and good adhesion to Si and Cu substrates. These results indicated a possibility for this novel PSPI-1 to be applicable for passivating power semiconductor.

Keywords´ Photo-definable polyimide, Thick film, High thermal stability, Passivation layer, Power semiconductor

I. INTRODUCTION

Recently, low loss, high temperature and high voltage power semiconductors are required for high efficient energy use, especially, for electric vehicle application. [1] Polyimide is widely used as a passivation material of semiconductor because of its high thermal and electrical stability; and good adhesion [2]; however, these requirements are becoming more strict in recent years. High voltage power semiconductor requires higher polyimide insulation properties; for in another word, thicker polyimide passivation layer is required because thick insulator shows high voltage resistivity in general. Generally, positive tone photosensitive polyimide (PSPI) is utilized used for the semiconductor processes due to its nature of photolithography function, which reduces the process steps -to reduce the cost of manufacturing. However, since PSPI lack in thermal and electrical properties when compared to non-photosensitive polyimide due to additives for photolithography, choosing PSPI was not favorable. Thick pattern formation is also one of the challenges for PSPI. Due to itslight absorption nature of imide group, PSPI tends to need more light energy during lithography processes.

In this paper, we will introduce and demonstrate the high thermal stability and thick patternability of PSPI-1. This novel PSPI-1 showed thick film patternability up to 16 mm, high thermal stability (Td5=500 ℃, Tg=310 ℃), excellent adhesion to silicone and copper substrate, and high reliability with high voltage.

II. EXPERIMENT

A. Polyimide precursor synthesis

Polyimide precursor resins were synthesized by polyaddition of tetracarboxylic dianhydrides with diamines. The prescribed amount of diamines was poured into a 4 neck flask with a mechanical stirrer, a thermometer and an inlet pipe for nitrogen flow. N-methyl-2-pyrridone (NMP, Mitsubishi Chem.) was added into the 4 neck flask. Under nitrogen flow, the flask was heated to 40 ℃. The prescribed amount of tetracarboxylic dianhydrides was added into the flask with NMP. After the mixture in the flask was stirred for 1 hr at 40 ℃, the prescribed amount of N,N-dimethylformamide dimethyl acetal was added into the flask and stirred for 2 hr.

After cooling the polyimide precursor solution to room temperature, the solution was poured into the water to precipitate the polyimide precursor. The obtained polyimide precursor precipitate was collected by filtration. The polyimide precursor precipitate was washed by water 3 times. The obtained polymer was dried at 50 ℃ for 72 hours in a convection oven.

B. Preparation of polyimide precursor solution

A Polyimide precursor solution was obtained by following procedure. 10g of polyimide precursor was dissolved by g-butylolactone (GBL) to prepare polyimide precursor solution at concentration of 30 ¨ 40 wt%. Then, 10~40 wt% crosslinker and 1~10 wt% photoinitiator were added to those solution.

C. Patterning process

The novel polyimide material (see Section II-B) was coated on Si wafer and Cu/Si wafer. After prebake

(120 ℃/3min on hot plate), the specimen was exposed by i-line stepper with patterned mask and baked 120 ℃/1min as post exposure bake. After development by alkali aqueous developer, the coated material was cured at 380 ℃/ 1hr under nitrogen atmosphere. The cured wafer was cut by diamond cutter to observe pattern profile by scanning electron microscope (SEM). Polyimide precursor was transformed to polyimide by the thermal cure process. The cured film was evaluated for various properties (see Section II-D).

D. Measurement of mechanical and thermal properties

Polyimide film of thickness 16 μm was obtained as in Section II-C without patterning. The coefficient of thermal expansion (CTE) and glass translation temperature (Tg) were measured by thermal mechanical analysis (TMA) equipment (SEIKO, TMA/SS6000) at heating rate 5 ℃/min under N_2 flow. Young's modulus, elongation and strength of the polyimide film were measured by Universal testing machine (Orientech, TENSILON RTM-100) at rate of 5 mm/min. Mechanical properties measurement were conducted according to JIS K7127 standards. 5% Thermal decomposition temperature was measured by thermogravimetric analysis (TGA) (Shimazu, DTG-60)

E. Thermogravimetric analysis of polymerized cross linker

2g of cross linker and 0.05g of thermal initiator were dissolved in 3g of propylene glycol monomethyl ether acetate in a plastic vial. The solution was poured in aluminum cup and heated by hot plate for curing. The cure condition was 130℃ for 10 min and 200 ℃ for 30 min. Measurement of thermal stability of the resin was performed by TGA under nitrogen gas.

F. UV spectra measurement of polyimide precursors

40 wt% polyimide precursor solutions dissolved in GBL were coated on glass substrate by spin coating, then prebaked (120 ℃/3min on hot plate). UV spectra of the films on glasses were measured by UV spectroscopy (Hitachi, U2910).

G. Adhesion strength test

The cured polyimide on wafer was obtained as in Section II-C without patterning. Tape adhesion test of polyimide on Si and Cu was conducted according to ASTM D 3359-87 Method B standards. The criterial is as follows:

5: no peel

4: peel area is less than 5%

3: peel area is over 5%, less than 15%

2: peel area is over 15%, less than 35%

1: peel area is over 35%, less than 65%

0: peel area is over 65%

This test was conducted before and after Pressure Cooker Test (PCT) (ESPEC, EHS-221MD). The condition of PCT was 121 ℃, 2atm, 100% RH.

H. Reliability tests-1 (bias-HAST)

Patterned Cu pads and comb electrode were fabricated on 8 inches Si wafer with SiNx surface by sputtering, electroplating and photolithography process. The test element group (TEG) was singulated by blade dicer.

The polyimide precursor solution was coated at 10 μm thickness on the TEG substrate as shown in Fig. 1. The dimensions of line, space and height of copper were 15 μm, 30 μm and 4 μm, respectively. Then biased HAST (High Accelerated Stress Test) was carried out at 130 ℃ for 96 hr with 100 V in 85 %RH. The TEG substrate was fabricated as in Fig. 2.

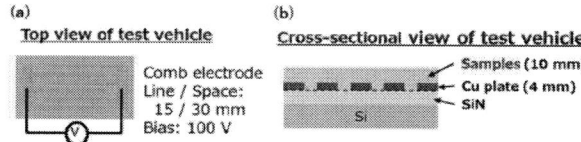

Fig. 1. (a) Top of view of TEG. (b) Cross section of TEG.

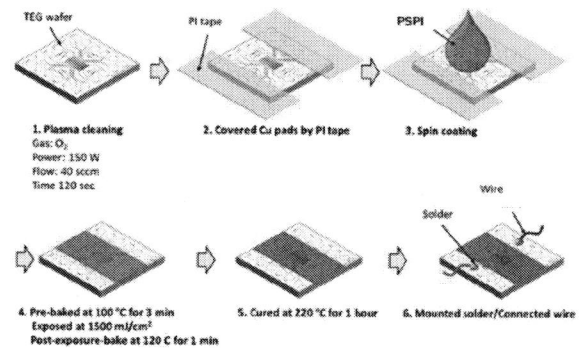

Fig. 2. Procedure of test vehicle fabrication for biased HAST. (a) Top of view of test vehicle. (b) Cross section of test vehicle.

III. RESULT AND DISSUSSION

A. Design concept of polyimide structure and additive

At first, we focused on thick film patternability with photolithography, which is the biggest challenge for PSPI. In general, polyimide and polyimide precursor absorb lots of i-line (365nm) light derived from imide and amide group of the polymers. We selected polyimide precursor for thick film patterning because of weaker light absorption than polyimide. One of the methods to decrease the light absorption was the aliphatic group introduction to polyimide main chain. However, this method decrease thermal stability. To overcome the trade-off problem, we designed a novel aromatic diamine which had electron withdraw group (EWG) as monomer of polyimide. Mechanisms of light absorption of imide and amide groups are intramolecular charge transfer from nitrogen atom to carbonyl group [3]. By reducing the electron density on nitrogen atom by EWG, intramolecular charge transfer of polyimide became weaker. We synthesized two different polyimide precursor and derived it from diamine with and without EWG (Fig. 3). Fig.4 shows UV spectra of PI-A and PI-B, transmittance of PI-A was around 60% and PI-B was less than 1%. PI-A is higher transparency at 365nm than PI-B.

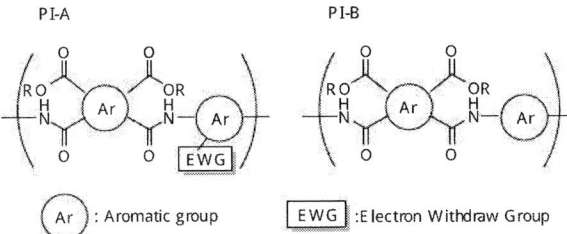

Fig. 3. Modified chemical structure of PI-A and PI-B. PI-A was aromatic polyimide precursor with EWG. PI-B was without EWG.

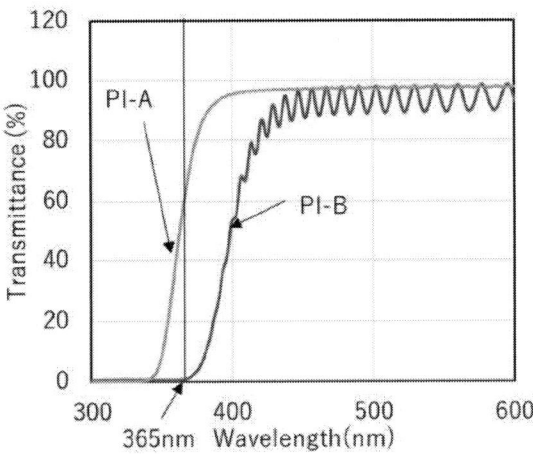

Fig. 4. UV spectra of PI-A and PI-B film (Film thickness=10 mm).

Another challenge of PSPI for power semiconductor was a high thermal stability after cure. Because PSPI includes cross linker and photo initiator for photo-definable property, thermal stability of PSPI is lower than that of polyimide itself. Although the cross linker is essential for photolithography, residue and ash of cross linker in cured film causes lower thermal stability. Our material design strategy is to fully thermal decompose the cross linker during cure. We measured the thermal stability of polymerized cross linkers by TGA, then the samples, which consisted of dipentaerythritol hexaacrylate (DPHA), (well-known cross linker for negative-tone photo definable material) and our new cross linker (CL-A). Fig. 5 shows the result of TGA. Poly(DPHA), that was polymerized resin of DPHA, which had remained 60 wt% of its form even after 380éC (60min) thermal treatment. On the other hand, poly(CL-A) had decomposed after 380éC. This result indicates that CL-A decomposes after cure and doesn't have negative effect to cured film for thermal stability.

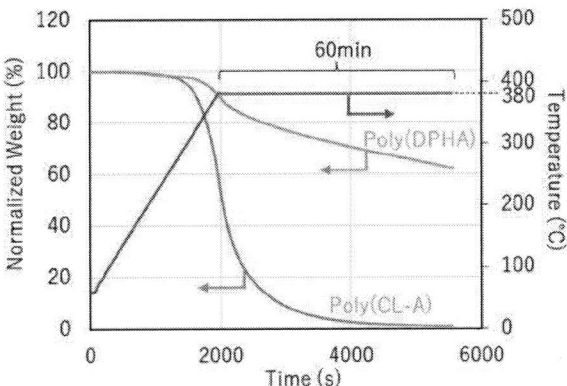

Fig. 5. TGA diagram of poly(DPHA) nad poly(CL-A). Thermal degradation during cure (380éC@60min in N_2) was measured.

B. Patterning, mechanical and thermal properties of new PSPI

We developed novel negative-tone photo-definable polyimide (PSPI-1) made of PI-A with high thermal stability, high light transparency, alkali dissolvable and decomposable CL-A during thermal curing process.

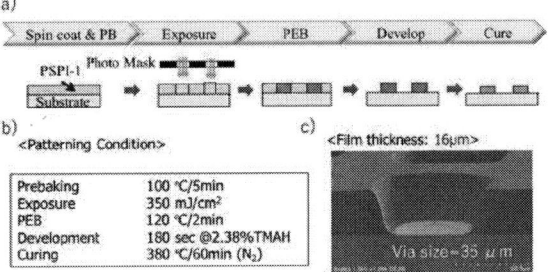

Fig. 6. (a) Process flow of photolithography process using PSPI-1. (b) Process conditions of patterning. (c)) SEM image of 35 ㎛ via from cross section view..

PSPI-1 is patternable up to 16 mm thickness on silicon wafer with 350 mJ/cm² (Fig. 6) by alkali aqueous developer. Upper limit thickness of conventional PSPI for power semiconductor is usualyy under 10 mm in case of i-line photolithography. PSPI-1 showed thicker patternability than conventional PSPI.

TABLE 1 summarizes thermal, mechanical, and electrical properties of PSPI-1's cured film (cured condition was 380 éC/60min). The cured film showed excellent properties as follows: 310 éC glass transition temperature (Tg), 500 éC 5% weight loss temperature, 40% elongation and good adhesion on Si and Cu substrate even after PCT. Dielectric strength was over 450 kV/mm and volume resistivity was over 5 E+16 ohm x cm before and after PCT. Because these properties of PSPI-1 were close to those of PI-A itself, it is assumed that CL-A was decomposed during the cure as expected. Properties of PSPI-1 is expected to be applicable for passivation layer of power semiconductors.

79

TABLE I.   PROPERTIES OF PSPI-1

			PSPI-1
Photo-definable type			Nega. Type
Cure Temp.			380 éC
Tg			310 éC
Td5			500 éC
CTE			47 ppm
Young´s Modules			3.2 GPa
Tensile strength			150 Mpa
Elongation			40 %
Adhesion	Si	PCT 0h	5
		PCT 100h	5
	Cu	PCT 0h	5
		PCT 100h	5
Dielectric Constant			2.91 (1kHz)
Dielectric Loss			0.0022 (1kHz)
Dielectric Strength			475 KV/mm
Volume resistivity			5.68 E+16 W·m

C. Reliability test results of PSPI-1

We confirmed insulation reliability of PSPI-1 by biased HAST. Usually, voltage of biased HAST is less than 20V and 2~20 mm line and space (Electric field strength is under 1.5 V/mm). However, biased HAST of PSPI-1 was performed at 100V of 15 mm line & 30 mm space TEG (Electric field strength was 3.3 V/mm). As shown in Fig.7, no short and open failure occurred after 100 hours storage at 130 éC with 100 V in 85 %RH. This result indicated that PSPI-1 had enough insulating performance even after a reliability test with high voltage condition.

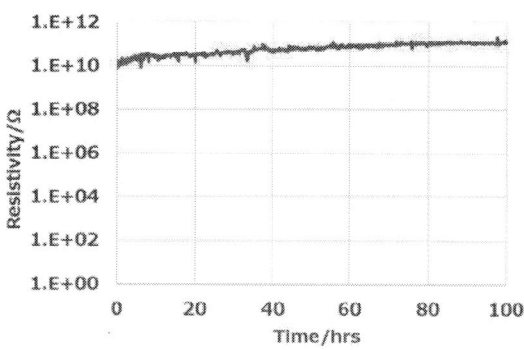

Fig. 7.   Resistivity of PSPI-1 during the treatment at a condition of 130 éC, 85 RH% (HAST condtion), and 100 V bias for 100 hrs.

IV. CONCLUSION

Because high voltage power semiconductor requires higher polyimide insulation property,,thicker polyimide passivation layer is required. We have designed a polyimide with high thermal stability, high light transparency and alkali dissolvable polyimide to achieve thick patternability. We also examined high thermal degradable cross linker to be decomposed during cure. Furthermore, this novel negative-tone photo-definable polyimide was formulated by the a new set of novel polyimide, thermal degradable cross linking reagent and photoinitiator. Our novel negative-tone photo-definable polyimide (PSPI-1) has a patternability up to 16 mm thickness on silicon with 350 mJ/cm^2 and the cured film showed excellent properties as follows: 310 éC Tg, 500 éC 5% weight loss temperature, 40% elongation, no short observation of bias-HAST (130 éC. 96%, 100hr, 100V 15/30 mm line&space) and good adhesion to Si and Cu substrates. These results indicate possibilities for this novel (PSPI-1) to be an applicable passivation material for power semiconductor.

REFERENCES

[1]   T, Kimoto, Proc. Jpn. Acad., Ser. B vo.98, No.4, pp 161-189, 2022. DOI: 10.2183/pjab.98.011

[2]   S. Zelmat, M.-L. Locatelli, T. Lebey, and S. Diaham, Microelectronic Engineering vol. 83, pp. 51-54, 2006. DOI: 10.1016/j.mee.2005.10.050

[3]   C.L.Tsai, H.Ju. Yen, and G.S. Liou, Reactive and Functional Polymers, vol. 108, pp. 2-30, 2016 . DOI:/10.1016/j.reactfunctpolym.2016.04.021

Epoxy Molding Compound bleeding reduction on surface mount semiconductor device

Fulvio Viviani
Backend Manufactoring and Technology R&D
STMicroelectronics
Agrate Brianza, Italy
fulvio.viviani@st.com

Federico Leone
Backend Manufactoring and Technology R&D
STMicroelectronics
Agrate Brianza, Italy
federico.leone@st.com

Hidetoshi Seki
Information & Telecommunication Materials Research Laboratory
Sumitomo Bakelite
Singapore
hseki@sumibe.co.jp

Masami Ishii
Information & Telecommunication
Materials Research Laboratory
Sumitomo Bakelite
Singapore
ishii-masami@sumibe.co.jp

Abstract – **Epoxy Molding Compound (EMC) bleed consists of a transparent layer of resin which mainly could occur during molding injection at material packing stage. Resin bleeds on exposed pad of a Surface Mount Device (SMD) can significantly impact the solderability performances of the package, causing failure of the product. De-flashing process after molding is generally performed to eliminate the resin bleed. However, de-flashing can be performed on post-plated leadframes without critical drawback, while it is not recommended on pre-plated leadframes due to the high risk of damaging the finishing surface. In this study, an alternative approach aimed to reduce or eliminate molding compound bleed by optimizing the inorganic part of the molding compound formulation is presented. Besides the quality improvement, the presented approach provides an economical advantage since, by tuning the properties of the filler inside the EMC, it is possible to eliminate the de-flashing process from the assembly flow of a package and consequently reduce the manufacturing cost.**

Keywords - epoxy molding compound, molding compound bleed, solderability, de-flashing.

I. INTRODUCTION

Nowadays, in the semiconductor market, two main Surface Mount Device (SMD) package families are used: full plastic and exposed pad/slug (Fig. 1). On full plastic type the die pad on which the die is bonded is completely embedded in the molding compound. Whereas in the exposed pad type, the die pad enhances the thermal dissipation of the product, representing a surface of contact between the die and the environment. For both types of SMD package families, two types of leadframe finishing can be used: post-plated and pre-plated. For the post-plated family (i.e., Bare Cu/Ag spot) the plating process is mandatory to grant the solderability of the package on Printed Circuit Board (PCB). For the pre-plated family, the plating process can be skipped thanks to the multiple layers finishing structure (for instance, NiPdAu) that preserves the solderability of the package on PCB, enhancing the total production cycle time. Focusing on exposed pad package type, one of the requirements is the cleanness of the die pad. One concern is the Epoxy Molding Compound (EMC) bleeding, a residue that can form nearby the end of EMC injection step, at the material packing stage. Molding

compound bleeds on exposed pad of Surface Mount Device can significantly impact the solderability performances of the package causing failure of the product. On the exposed pad with post-plated leadframes, epoxy molding compound bleed can be easily removed/reduced by de-flashing process but on pre-plated leadframes finishing, the de-flashing process can induce cosmetic or functional defect of the product. In this study an alternative solution to eliminate/reduce the epoxy molding compound bleeds without the usage of de-flashing process was evaluated. The selected test vehicle is a Thin Quad Flat Package (TQFP) exposed pad package, with pre-plated leadframe finishing.

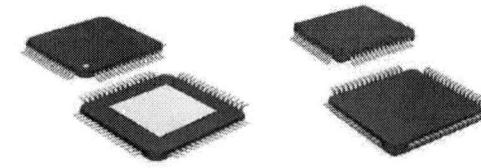

Fig. 1. Images of an exposed pad QFP (left) and a full plastic QFP (right).

II. MATERIALS AND METHOD

Typically epoxy molding compound formulation includes organic and inorganic contents (Ellis, 1993). The organic content is made of polymeric molecules that react during molding and curing processes, forming mutual cross-linking. The inorganic content (generally a silica-based filler) takes the predominant part in the molding compound formulation (around 80% in weight or more). The fillers are loaded to modify several EMC features including the flowability (i.e., the rheology), thermomechanical and electrical properties, and to add bulk (mass). Depending on the application of the package, the EMC manufacturer controls the filler content, the size and distribution, which can be represented by Gaussian curve, and the maximum particle size. In this study the effect of filler on molding compound flowability, viscosity and so bleeding was investigated by adjusting fillers sphericity and sizes and in some cases by intentionally adding a given content of submicron filler (size < 1μm), that normally is not present in the formulation, without changing the total filler content in the formulation.

TABLE I. FORMULATION COMPARISON BETWEEN DIFFERENT EMC GRADES

Leg #	#1	#2	#3	#4	#5	#6	#7
Main filler (average size > 10um)	STD	STD	STD	STD	STD	STD	STD
Fine filler (average size < 5um)	STD	UP ↑	UP ↑	UP ↑↑	UP ↑↑	STD	STD
Sub-micron filler (average size < 1um)	No	No	No	No	No	Yes	Yes
Filler sphericity	STD x 1.15	STD x 1.15	STD	STD x 1.15	STD	STD x 1.15	STD

In order to verify the effect of filler sphericity and sizes on exposed pad molding compound bleeding, seven different epoxy mold compound formulations (Legs) were prepared by the EMC manufacturer as shown in Table I. The relative contents are not specified to preserve the manufacturer confidentiality.

A. DEFINITION OF THE FILLER SPHERICITY

The sphericity of the filler is defined according to the equation appearing in Fig. 2. The projected contour (L) and projected area (A) are measured by a particle analyzer, while the circle's contour (L') is the perimeters of a perfect circumference. The sphericity is then defined as the squared ratio between the real perimeter of the filler and the perimeter of an ideal circle.

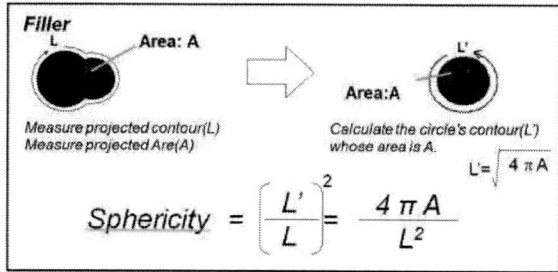

Fig. 2. Definition of Filler Sphericity.

B. MEASUREMENT METHOD OF EMC BLEEDING

The molding compound bleeding of the seven legs was measured by using a dedicated mold chase, as represented in Fig. 3. This tool is made of one resin pot, two runners and seven cavities for each runner. Every cavity has a vent located at the opposite side with respect to the gate to allow the evacuation of the air during the injection of the molding compound and the bleeding during the compaction phase. The process parameters are kept fixed for every trial.

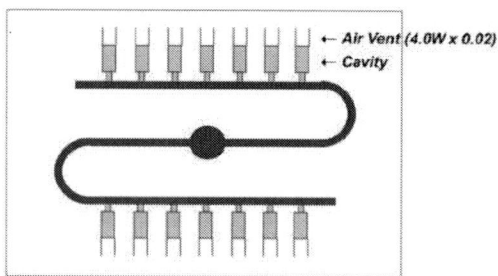

Fig. 3. Mold die tool with 14 cavities.

III. RESULTS AND DISCUSSION

The evaluation was split into two parts: the first step was the assessment of the bleeding and a test on package at the EMC manufacturer; the second test was the evaluation of the workability properties on a real device at STMicroelectronics (STM).

A. EVALUATION AT EMC MANUFACTURER

Fig. 4 shows the comparison of the seven EMC in terms of bleeding length, collected using Fig. 3 tool.

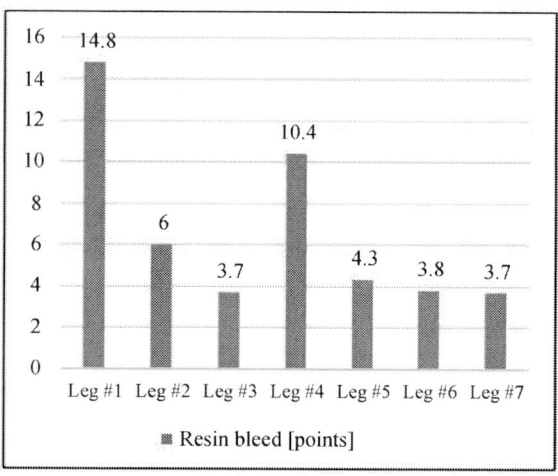

Fig.4. Molding compound bleed measurements with dedicated mold tool.

From the couples LEG2/LEG3 and LEG4/LEG5, which have the same content of main filler, fine filler and no sub-micron filler, it is evident that the reduction of the filler sphericity (moving from "STD x 1.5" to "STD") cause a reduction of the resin bleeding. From the couple LEG1/LEG6 we can observe the significant reduction of the bleeding length simply by introducing the sub-micron filler inside the resin formulation. Finally, LEG7 confirms that the implementation of a lower sphericity together with a sub-micron filler allows a further reduction of the bleeding. This test provided indication on how to modify the resin formulation theoretically. However, in order to confirm that these changes do not impact significantly the rheology of the resin, the Flow Pressure was measured and confirmed to be within the specification limits of the reference formulation (LEG1). In addition, an assembly was performed on a Quad-Flat-Package (QFP) exposed pad package to correlate the bleeding measurement obtained by the dedicated tool with the EMC bleeding on a real package.

On LEG1 (reference grade), LEG3, LEG6, LEG7 (having the lowest bleeding) the workability assessment was completed with the visual inspection of the plastic body (incomplete fill check) and wire sweep measurement. The results presented in Table II and Fig. 5 confirm the previous findings, assuring also that a change in the formulation does not imply any drawback in quality.

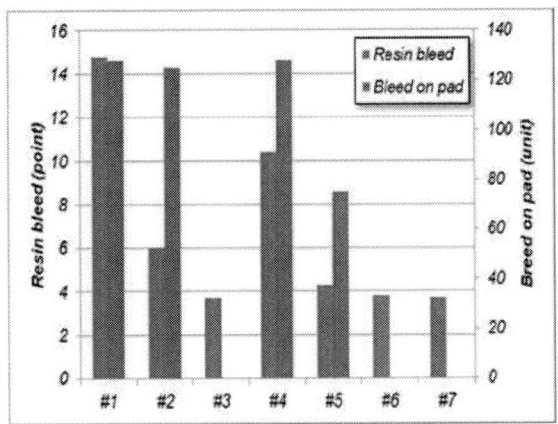

Fig. 5. Relationship between molding compound bleed and bleed on pad.

B. STM EVALUATION

After the evaluation at the resin manufacturer, the seven grades were tested by STM in a production environment. A new workability assessment was performed, adding external voids and delamination checks after post mold curing (PMC). A first screening to select the best candidates was performed by molding all the seven legs in the same conditions (i.e., molding parameters, material utilization, machine, mold chase type) and assembly flow. Table III shows the first screening results. The trials confirmed the preliminary evaluation results conducted by the supplier. LEG1, LEG2, LEG4 and LEG5 showed severe bleeding on the exposed pad. LEG6 had only 1 failure, while all the units of LEG3 and LEG7 passed the bleeding check (0/128 pcs). Having passed all the quality checks and being the best EMC candidate according to both STM and resin supplier assessments, LEG7 was finally selected. Fig. 6 shows a visual comparison of the bleeding phenomena before (LEG1) and after (LEG7) the resin formulation improvement.

Fig. 6. LEG1 (left) and LEG7 (right) molding compound bleeding comparison.

Subsequently, LEG7 samples were submitted to standard reliability assessment to verify the absence of any package quality issue, passing all the requirements. To complete the full qualification, LEG7 has been further evaluated reproducing the typical corner conditions of mass production scenarios, both for material properties, utilization and equipment conditions. Results confirmed LEG7 to eradicate bleeding without any negative drawback on the quality of the final product.

C. RESULTS EXPLANATION

Summarizing, a bleeding optimization can be achieved by modifying the formulation of the EMC. LEG7 is a combination of two factors: the reduction of the filler sphericity together with the addition of a sub-micron filler content in the formulation. A lower sphericity triggers phenomena of mechanical interlocking and indentation, producing higher friction of the molecules within the resin flow. This is confirmed by the increase of the Koka viscosity, moving from the reference LEG to LEG6 (Table II). The submicron filler produces a similar effect, but according to a different mechanism: it easily fills the gap between the exposed pad and mold chase due to its small particles size, resulting in a local increase of the EMC viscosity during material packing stage (Fig. 7). Opposing more resistance to the resin flow, the bleeding can be drastically reduced.

TABLE II.   SUPPLIER EVALUATION AND RESULTS

LEG	#1	#2	#3	#4	#5	#6	#7
Flow Pressure [MPa]	2.1	2.3	2.5	2.6	2.4	2.2	2.6
Koka viscosity [Poise]	116	151	170	154	175	140	163
Resin bleed [points]	14.8	6	3.7	10.4	4.3	3.8	3.7
Bleed on pad [units]	128/128	125/128	0/128	128/128	75/128	0/128	0/128
Wire Sweep (%)	<2		<2			<2	<2
Incomplete fill [units]	1/128		0/128			0/128	0/128

TABLE III.   STM EVALUATION AND RESULTS

LEG	#1	#2	#3	#4	#5	#6	#7
Bleed length [points]	14.8	6	3.7	10.4	4.3	3.8	3.7
MAX wire sweep (%)	2		4			2	2
Bleed on pad [units]	128/128	125/128	0/128	128/128	75/128	1/128	0/128
Incomplete fill [units]	0/128		0/128			0/128	0/128
External voids [units]	0/128		0/128			0/128	0/128
Delamination after PMC [units]	1/128		0/128			1/128	0/128

IV. CONCLUSION

The present paper shows a feasible and advantageous solution to reduce/prevent epoxy molding compound bleeding on exposed pad for pre-plated leadframe. It was proven that a lower filler sphericity and the addition of a submicron filler inside the EMC formulation have significant impact on this phenomenon. The study was developed in two main phases.

- An assessment at the EMC manufacturer: thanks to a dedicated tool it was possible to measure the effect of the resin formulation changes on the length of the bleeding: the candidates were later molded on a QFP exposed pad package.
- An assessment at STM: seven grades prepared by the supplier were tested in a production environment, confirming the improvement, and assuring the absence of quality issues for the final product.

At the end of the evaluation, LEG7 was chosen as the best formulation to eradicate the molding compound bleeding on exposed pad without affecting other key material properties and performances.

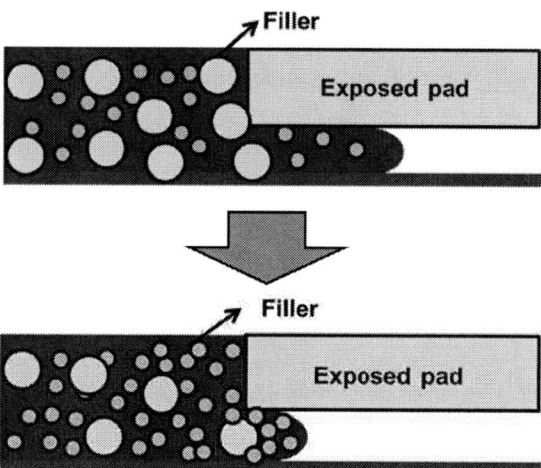

Fig.7. Effect of sub-micron filler on molding compound bleed.

The proposed approach grants the possibility to remove the de-flashing operation that could damage pre-plated leadframes. There are two immediate consequences: the improvement of the reliability of SMD on PCB and the reduction of the assembly costs.

REFERENCES

Chao, S., Liaw, Y., & Chou, J.-H. (2021). The Effect of Filler Shape, Type, and Size on the Properties of Encapsulation Molding Componenets. *Electronics*. doi:10.3390/electronics10020098

Ellis, B. (1993). *Chemistry and Technology of Epoxy Resins*. (B. Ellis, Ed.) Springer-Science+Business Media. doi:10.1007/978-94-011-2932-9

Haleh Ardebili, J. Z. (2019). *In Materials and Processes for Electronic Applications - Encapsulation Technologies for Electronic Applications* (2nd ed., Vols. 6 - Characterization of encapsulant properties). (J. Z. Haleh Ardebili, Ed.) William Andrew Publishing. doi:B978-0-12-811978-5.00006-7

Kaya, B. (2018). Concept Development and Implementation of Online Monitoring Methods in the Transfer Molding Process for Electronic Packages. Berlin: Technische Universität Berlin » Publications. Retrieved from https://depositonce.tu-berlin.de//handle/11303/8508

Influence of Thermally Aged Underfill on Flip-Chip Packages

Kevin Cox
Tektronix Component Solutions
Beaverton, USA
kevin.cox@tektronix.com

Ghassan Abu-Hamdeh
Tektronix Component Solutions
Beaverton, USA
ghassan.abu-hamdeh@tektronix.com

Matt Borden
Tektronix Component Solutions
Beaverton, USA
matt.borden@tektronix.com

Abstract— Underfill material plays a crucial role in flip-chip package reliability, thus it necessary to accurately understand the mechanical properties. The material itself is nonlinear and is dependent on degree of cure, temperature, strain rate and thermal aging, among others. In this work, two underfill materials were characterized for thermal expansion and viscoelastic properties over a wide temperature range. A second set of each material was thermally aged through flip-chip processing steps and 250 thermal cycles and then characterized. Temperature dependent thermal expansion and time-temperature dependent moduli curves were generated to represent each material type. These material models were then implemented into a nonlinear finite element model of a flip-chip package to evaluate package stress interaction through processing steps and thermal cycling. The influence that each underfill material had with and without thermal aging was analyzed with respect to typical package failure mechanisms.

Keywords—materials characterization, viscoelastic, nonlinear finite element analysis

I. INTRODUCTION

Over the last few years, Tektronix Component Solutions (CSO) has been working to advance package reliability through selecting optimal material sets, using best-in-class simulation methods, and refining processing techniques. The reliability of flip-chip packages depends heavily on the underfill (UF) materials used to bond the IC to the substrate. UF plays a crucial role in redistributing mechanical stresses and protecting the chip from environmental factors. Hence, understanding the impact of UF material properties on the reliability of flip-chip packages is paramount. Highly stiff UFs can create excessive stresses on the die and the bonded interface leading to die fracture or UF delamination/fracture, while low stiffness UF don't transfer enough stress and may lead to early C4 fatigue failures or die ULK fractures underneath C4 bumps. In addition to being highly temperature dependent, UF materials are sensitive to the degree of cure, thermal aging, moisture exposure and exhibit creep and viscoelasticity – the modulus is dependent on time (strain rate).

A. Background

The mechanical properties specified on UF technical datasheets can be misleading. For example, the moduli values are often specified for strain rates that are 1-2 orders of magnitude faster than what a typical flip-chip package experiences during thermal cycle qualification testing. This results in poor design and simulation models since the UF modulus and glass transition temperature (T_g) are highly strain rate dependent. In common qualification testing protocols, flip-chip packages are thermally cycled at a rate of 5 C/min (12 sec/C) [1]. A typical UF has a CTE of 30 ppm/C below T_g and 125 ppm/C above T_g. Thus, the thermal strain rate at 5 C/min would be between ~10^{-3} and 10^{-4} % per second. Therefore, it is most relevant to characterize and report UF properties at these strain rates when exploring stresses in flip-chip package processing and thermal cycling conditions. Many studies have been performed previously confirming the significance of UF viscoelasticity as well as thermal aging on package performance [2], [3], [4], [5], [6], [7]. In this work, the focus was on evaluating the UF before and after thermal aging produced from real processing conditions and real thermal cycling qualification loads.

CSO has recently investigated numerous UF material options for optimal performance in advanced-node flip-chip packages. The result of that study identified the two 'best materials" to which further work was required. The follow-on work to identify the best material is described herein. This paper addresses the effects thermal aging on UF coefficient of thermal expansion (CTE) and viscoelasticity and how it influences common flip-chip package failure mechanisms.

The remainder of this section provides an overview of the flip-chip package used for the evaluation. The subsequent sections describe the methods used to characterize the materials and evaluate the impact of thermal aging on key material properties. Thereafter, a nonlinear finite element model and results are discussed which compares how the different UF material properties affected key stress metrics associated with package failure mechanisms CSO has encountered: excessive package warpage, ULK fracture and UF delamination.

B. Description of Package

The device under consideration in this research was an in-house designed advanced node CMOS flip-chip package that goes into a Tektronix product. The package incorporated a 10x8 mm IC with a 180-micron bump pitch and 800-micron BGA pitch on a 25x25 mm substrate. The package geometry is shown in Fig. 1.

II. METHODS

This section first describes the materials characterization testing procedure and evaluation techniques. That is followed by a description of the simulation model of the flip-chip package that was created to evaluate key metrics related to critical failure mechanisms.

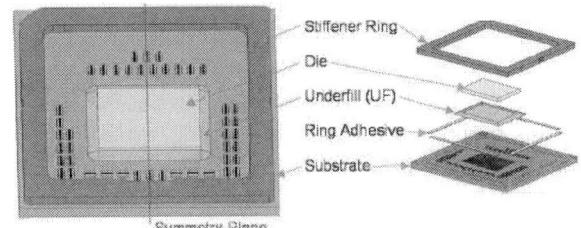

Fig. 1. Description of flip-chip package

Material	Thermal treatment	Group
UFA	Cure only	A1
UFA	Cure + thermal aging	A2
UFB	Cure only	B1
UFB	Cure + thermal aging	B2

TABLE I. MATERIAL GROUP NOMENCLATURE

A. Materials and preparation

Material coupons were fabricated for two different UF materials (UFA and UFB). The coupon size was approximately 50x10x4 mm. Each material was divided into two sets: Set 1 (cure only), and Set 2 (cure + thermal aging) to yield 4 unique material sets described in TABLE I.

Curing was performed as described on the UF supplier datasheets. The thermal aging included standard flip-chip processing steps followed by pre-conditioning to MSL 4, and then 250 thermal cycles. Thermal cycling was between 125 C and -40 C with 15-minute dwells at the high/low temperatures and 5 C/min ramp rates. These conditions matched typical processing and qualification test requirements for Tektronix products [1]. Coupon dimensions and masses were recorded before and after thermal cycling to evaluate material shrinkage. Thereafter, the specimens were sent for CTE and rate-dependent modulus (viscoelasticity) testing over a relevant temperature range.

B. Thermal Expansion Testing

The CTEs of cured resins have strong dependency on temperature, most notably near their respective glass transition temperatures. To characterize this for UFA and UFB, 10x10 mm square specimens were cut from the prepared coupons and placed in a thermal chamber. The test chamber was purged with nitrogen and was thermally ramped from room temperature to -60 and up to 260 C at a rate of 5 C/min. Specimen displacement was measured with a probe which applied a 50 mN force. Three specimens were tested for each Group. Resulting displacement over temperature curves were generated for each test group.

C. Viscoelasticity Testing

To characterize the time-temperature dependency of the material groups, the relaxation moduli were measured at different temperatures across the temperature range. The specimens were loaded in torsion (shear) to 0.01% strain. Force decay was recorded between 0.01 and 600 seconds. Fig. 2 illustrates the force decay method, where the relaxation modulus (a function of time) can be calculated by multiplying the stress (a function of time) and the applied strain (0.01%). The test was run for each specimen at 10 degree C intervals between -60 and 250 C. As previously mentioned, typical strain rates in the UF component (from thermal expansion alone) during thermal cycling are very low. With the 0.01% strain applied in the testing, the UF materials in a package would be best represented by the relaxation moduli values bounded roughly by the 10 to 100-second relaxation times (to yield a strain rate of 0.001% − 0.0001 %/sec). Representative modulus over temperature curves at varying strain rates are shown in Fig. 3 where the influences of relaxation time (strain rate) and temperature are clearly visible.

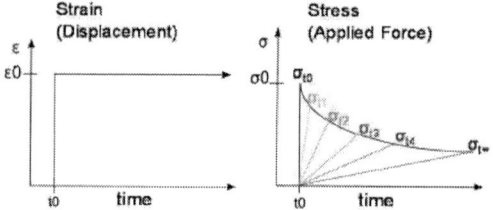

Fig. 2. Stress-relaxation test method

From the test data, time-temperature master curves were generated for each material Group. Each master curve was then fit with a 20-term Prony Series and a Williams-Landel-Ferry shift function. With viscoelasticity and CTE models generated across the full temperature range, each material Group could then be properly represented in the flip-chip simulation model described in the next section.

D. Simulation Model

The assembly shown in Fig. 1 was modelled in SolidWorks and imported to ANSYS Mechanical for structural simulations. Due to symmetry, the package was sectioned through the center vertically with respect to Fig. 1 to create a symmetry boundary condition. The structural simulation utilized quadratic elements and large deflections in a multi-step analysis to capture the thermal loading steps involved in flip-chip processing and thermal cycling. The "element birth and death" feature of ANSYS was utilized to "activate" the Stiffener Ring and Ring Adhesive at a specific point during the simulation to match the real-world assembly process. The thermal loading is displayed in Fig. 4. The simulation started with the cure of UF at 165C; the package was at zero stress. The package was cooled to room temperature, relaxed for 15 minutes, then ramped to 165C where the Stiffener Ring and Ring Adhesive became active and bonded to the rest of the package. The package then was cooled room temperature, followed by a thermal cycle from 125 to -40 and back to 125C.

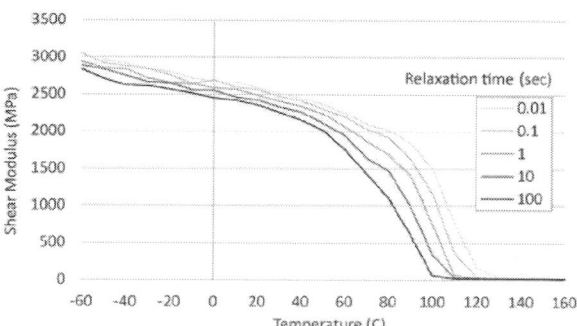

Fig. 3. Shear modulus versus temperature at different relaxation times

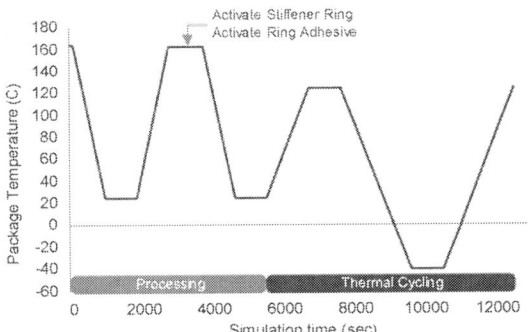

Fig. 4. Thermal Load applied in simulation

A sub-model of a single bump was also designed to provide more accurate results at the corner C4 bump region. Specifically, the sub-model allowed for the extraction of normal stresses in the die back-end-of-line (BEOL) layers and could be used in future studies for C4 bump fatigue predictions. C4 fatigue however was out of scope for this work since that has not been a frequently encountered failure mechanism at CSO. The design of the sub-model is shown beside the global model in Fig. 5 with each component assigned a number that is specified in the materials summary: TABLE II. As is standard for Ansys sub-models, the deformations and stresses came from boundary conditions applied to the circumference of the sub-model which were extracted at every step from the global model displacement results [8]. In addition, the same thermal load from Fig. 4 was applied to the sub-model.

A summary of the material type applied to each component in the global- and sub-model is provided in TABLE II. where temperature-dependent properties are designated with "(T)" and Poisson's ratio is designated with "v". The die, UF, substrate and ring attach adhesive incorporated temperature dependent CTE values. Material properties for the die BEOL were calculated based on averaging the actual USG, Low-K and ELK stack-up as described in [9]. The UF, ring attach adhesive and substrate moduli values were derived from the respective material's time-temperature master curves and were each represented in ANSYS by a multi-term Prony Series and a Williams-Landel-Ferry shift function.

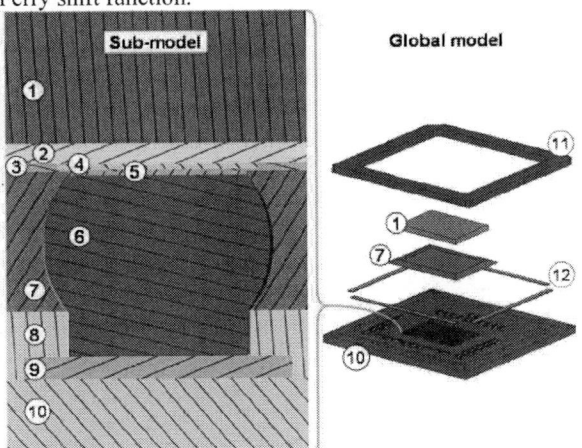

Fig. 5. Simulation sub-model and component identification #

TABLE II.   SUMMARY OF MATERIALS IN SIMULATION MODEL

#	Component	Material	Key Properties
1	Die (bulk)	Silicon	Orthotropic Modulus, CTE(T), v
2	Die BEOL	Composite	Orthotropic Modulus Orthotropic CTE, v
3	Die Passivation	Polyimide	Modulus, CTE, v
4	Die C4 Pad	Aluminum	Modulus, CTE, v
5	Die UBM	Copper	Bilinear Modulus, CTE, v
6	Solder Bumps	SnAg	Viscoplasticity(T), CTE, v
7	Underfill	Thermoset Resin	Viscoelasticity(T), CTE(T), v
8	Substrate Solder Mask	Thermoset	Modulus(T), CTE(T), v
9	Substrate C4 Pad	Copper	Bilinear Modulus, CTE, v
10	Substrate	Composite	Viscoelasticity(T), CTE, v
11	Stiffener Ring	Copper	Modulus, CTE, v
12	Ring Adhesive	Thermoset	Viscoelasticity(T), CTE(T), v

Like the UF materials, the substrate and ring attach adhesive master curves were developed by performing stress-relaxation testing over the relevant temperature range; however, the influence of aging on these materials was out of scope in this work. The substrate, due to its orthotropic nature, required performing testing in both in-plane directions. Finally, the solder bumps incorporated Anand viscoplastic properties [10]. Additional details about the simulation model are provided in [11].

III. RESULTS AND DISCUSSION

In this section, results from the material characterization tests are discussed first followed by the simulation results.

A. Material Characterization Tests

1) Physical Properties

The length, width, thickness, and mass of each test specimen was recorded before and after thermal aging. With the length of the coupon being the largest dimension, it represented the best metric for comparison since it minimized influence of measurement tool uncertainty. Four specimens were measured for each group. As shown in TABLE III. , thermal aging imparted no significant shrinkage or change in mass after the initial cure. The values in the table show percent change with respect to A1 (row 2) and B1 (row 3).

TABLE III.   CHANGES TO SPECIMEN DIMENSIONS AND MASS AFTER THERMAL AGING

Group	Delta Length	Delta Mass
A2	+0.08%	+0.05%
B2	-0.05%	-0.17%

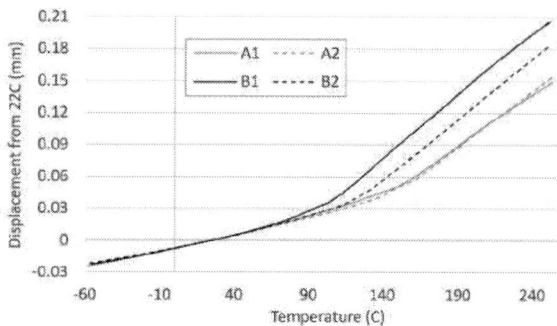

Fig. 6. Thermal expansion versus temperature for all groups

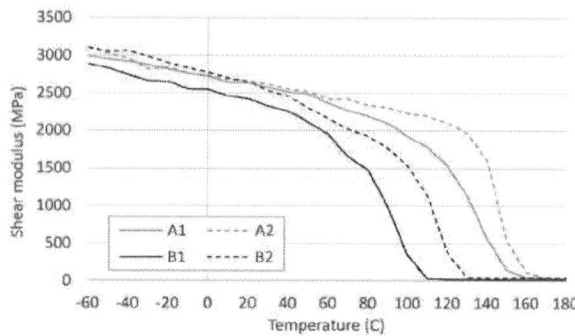

Fig. 7. Shear modulus over temperature for all groups at 10 seconds relaxtion time

2) Thermal Expansion

Specimen displacement for each material group is plotted in Fig. 6 across the relevant temperature range with displacements taken relative to Room Temperature. Each curve represents the average response of the 3 coupons tested in each group. The slope of the curve at any specific temperature is equal to the material's CTE. All curves exhibited typical responses for cured resins: bilinear behavior with a lower CTE (flatter slope) below T_g and a higher CTE (steeper slope) above T_g. TABLE IV. summarizes the changes to the average CTE between -40 and 125C after thermal aging. In the table 4th column, B1 is calculated with respect to A1 and B2 is with respect to A2.

Material A exhibited a stable response with respect to thermal aging; both CTE and T_g remained within close proximity of the initial, as-cured properties. Material B exhibited a strong dependency on thermal aging; the T_g increased by 16% and because of this, the average CTE between -40 and 125 C decreased by 19%. This is a rather common occurrence for thermosetting resins: additional heat and time may further increase a resin's cross-linking which yields a stiffer, more brittle material with a higher T_g and a lower CTE.

From a package reliability perspective, low CTE and high T_g are typically good traits for UF materials. In this regard, the results showed that while UFB improved after thermal aging, it never became as good as UFA. UFA was therefore preferrable for this aspect, and the fact that it was more stable makes it easier and more predictable to work with.

3) Viscoelasticity

To provide a simple visual comparison, Fig. 7 shows the relaxation moduli versus temperature plotted for the different material groups at a single relaxation time (10 seconds). TABLE V. shows the average shear modulus of the materials between -40 and +125 C.

Thermal aging increased T_g for UFA by ~15 C and decreased the span of the T_g transition region, while the modulus stayed rather constant outside of the transition region. For UFB, thermal aging increased the modulus consistently across all temperatures as well as shifted the Tg higher by ~20 C. When comparing UFA and UFB, again it was clear that UFA was less influenced by thermal aging. Before aging, UFB had lower T_g and moduli at all temperatures especially in the transition region. After thermal aging, the moduli were similar below 40C. Above 40C, UFA was stiffer, exhibited a higher T_g and a narrower transition zone.

B. Simulation

Results from both the global model and sub-model are described in this section. The global model provided die (and package) deflections and peak stresses on the BEOL layer near the die corners. Stress on the BEOL was taken on the surface in contact with the UF as a metric for delamination at this interface. The sub-model provided results at the corner C4 bump region for BEOL fracture (ULK fracture or delamination). The most significant stress in BEOL layers is in the normal (through-thickness) direction as described in [9], [12]. The results are summarized in the following sub-sections.

1) Package deflection

Characteristic package warpage (displacement in the through-thickness direction) is shown at -40 C (maximum deflection) in Fig. 8. Values of the die warpage (maximum – minimum) were recorded for each material group at -40 C and are presented in TABLE VI. The change in die deflection due to thermally aged UF properties was rather minor with UFA and UFB deflections both changing by less than 5%. A clearer result was that UFB produced less warpage, possibly because it was softer and thereby allowed for more shear strain between the die and substrate which reduced bending magnitude.

TABLE IV.   CHANGE TO AVERAGE CTE AFTER THERMAL AGING

Group	Avg CTE (ppm/C)	Change from cure only	Change from UFA
A1	34.9	---	---
A2	33.1	-5.2% -1.8 ppm/C	---
B1	46.6	---	+33.6%
B2	37.4	-19.6% -9.1 ppm/C	+13.3%

TABLE V.   CHANGE TO SHEAR MODULUS AFTER THERMAL AGING

Group	Avg Shear Modulus (MPa)	Change from cure only	Change from UFA
A1	2407	---	---
A2	2523	+4.8%	---
B1	1795	---	-25.4%
B2	2196	+22.3%	-13.0%

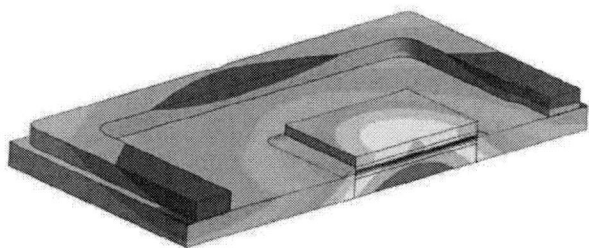

Fig. 8. Package warpage (deflection in thickness direction)

TABLE VI.   CHANGES TO DIE WARPAGE AFTER THERMAL AGING

Group	Die Warpage (mm)	Change from cure only	Change from UFA
A1	29.6	---	---
A2	28.8	-2.9%	---
B1	25.2	---	-14.8%
B2	26.3	+4.1%	-8.7%

2) Delamination Stress

Stress on the BEOL component was recorded near the corner of the die: the specific location is represented by the orange triangle region shown in Fig. 9. As shown in the figure, the region did not extend all the way to the die edges/corner to avoid influence from stress singularities at the (component's) corner nodes. The stress results are presented in TABLE VII. The maximum principal stress was selected as the stress metric since delamination at this surface would likely be due to (in-plane or through-thickness) tensile loads.

UFA was found to produce lower delamination stresses than UFB both before and after thermal aging. Thermal aging increased the UFA BEOL stress by ~6%. Considering UFB, the delamination stress remained nearly constant after thermal aging. This was a surprising result since major changes were found from material testing before and after thermal aging, with average CTE decreasing ~20% and average modulus increasing ~22%. It appeared that these changes may have offset each other. However it isn't as simple as comparing average values since T_g and ramp rates also played important roles. Furthermore, the substrate and C4 bump materials also had their own complex behaviors which, together with all the other components, contributed to highly complex package stress interactions.

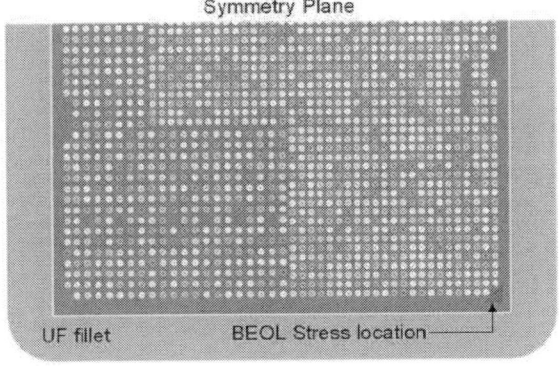

Fig. 9. Location of delamination stress metric on BEOL surface

TABLE VII.   CHANGES TO DELAMINATION MAX PRINCIPAL STRESS AFTER THERMAL AGING

Material	Delamination Max Principal Stress (MPa)	Change from cure only	Change from UFA
A1	71.1	---	---
A2	75.3	+5.9%	---
B1	80.3	---	+12.9%
B2	79.5	-1.0%	+5.6%

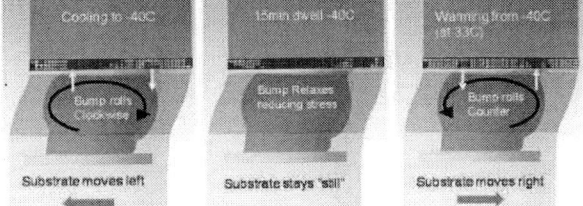

Fig. 10. Cross-section of normal stress in BEOL layers at corner C4 bump

3) BEOL fracture

A summary of the maximum through-thickness stress in the BEOL layers is presented in TABLE VIII. Through the simulation load steps, the location of maximum normal stress depended on whether the package was heating up or cooling down as depicted in Fig. 10. The normal stress was created from the moment applied by the C4 bump to the BEOL. The influence of dwell time can clearly be seen between the left and center images where the solder relaxed (creeped) under the high load, and thereby reduced the BEOL stress. Upon heating from -40C, the maximum normal stress moved to the opposite side of the bump and approached a similar but smaller magnitude as that found at -40 C.

Thermal aging increased BEOL stress for both UFA and UFB. Reductions in UF CTE after aging meant less vertical compression to the solder bumps when cooling the package and also less compression to the C4 pad and BEOL layer. This could be one reason why BEOL normal stresses increased. As was found for delamination stress, UFA with its lower average CTE and higher average modulus than UFB yielded lower BEOL normal stresses both before and after thermal aging.

IV. SUMMARY AND CONCLUSIONS

Two materials were considered for a flip-chip package UF. This work characterized the CTE and viscoelasticity of each UF before and after thermal aging to understand how the mechanical properties of the materials changed over time. Temperature-dependent CTE curves and Time-Temperature dependent moduli curves were created to represent the materials in nonlinear finite element simulations.

TABLE VIII.   CHANGES TO BEOL NORMAL STRESS AFTER AGING

Material	BEOL Normal Stress (MPa)	Change from cure only	change from UFA
A1	38.0	---	---
A2	40.8	+7.2%	---
B1	42.1	---	+10.7%
B2	44.4	+5.4%	+8.9%

The simulations incorporated the materials into the Underfill component of an advanced node CMOS flip-chip package. The model included relevant processing loads followed by a thermal cycle that the device would experience in qualification testing. The simulation output metrics associated with common critical failure mechanisms in flip-chip packages and provided an understanding of the influence thermal aging had on each material and the determination of which material was best. Thermally aging the UF materials influenced key mechanical properties by up to ~20%, while critical package stress results were influenced by ~5 – 10%. The following conclusions were made:

- Thermal aging did not impart a significant shrinkage or change in mass to either UF material.

- The CTE of UFA was lower and remained lower than UFB after thermal aging. The CTE and of UFB was significantly influenced by thermal aging, reducing almost 20% from its initial value.

- It was found from viscoelasticity characterizations that thermal aging increased T_g for UFA by ~15C. For UFB, aging increased the modulus consistently across all temperatures as well as shifted the T_g higher by ~20 C. UFA was the stiffer material.

- UFA was less influenced by thermal aging than UFB.

- Thermal aging did not have a major influence on die warpage.

- UFA performed better in both simulation stress metrics for both the pristine and thermally aged material states. However, UFB yielded a package with less warpage.

- Package performance with UFB generally improved after thermal aging with respect to simulated failure metrics, but never matched the performance provided by UFA.

REFERENCES

[1] JEDEC JESD22-A104F.01, "Temperature Cycling," 2023.

[2] Z. Qian, J. Wang, J. Yang and S. Liu, "Visco-elastic-Plastic Properties and Constitutive Modeling of Underfills," *IEEE Transactions on Components and Packaging Technologies*, vol. 22, no. 2, pp. 152-157, 1999.

[3] F. Feutsel, S. Wiese and E. Meusel, "Time-dependent material modeling for finite-element analyses of flip chips," *Proc. of 52nd Electrical Components Technologies Conference*, pp. 1548-1553, 2000.

[4] C. Lin, J. C. Suhling and P. Lall, "Evolution of the stress-strain and creep behavior of underfill encapsulants with aging," *Proceedings ASME 2009 InterPACK Conference*, 2009.

[5] F. X. Che, X. R. Zhang and L. Ji, "Thermal Aging Induced Underfill Degradation and Its Effect on Reliability of Advanced Packaging," *IEEE 70th Electronic Components and Technology Conference*, pp. 1525-1532, 2020.

[6] X. Q. Shi, Y. L. Zhang, W. Zhou and X. J. Fan, "Effect of Hygrothermal Aging on Interfactial Reliability of Silicon/Underfill/FR-4 Assembly," *IEEE Transactions on components and packaging technologies*, vol. 31, no. 00, pp. 94-103, 2008.

[7] P. Lall, M. Kasturi, Y. Zhang, H. Wu, J. Suhling and E. David, "Effect of Underfill Property Evolution on Solder Joint Reliability in Automotice Applications," *Proceedings of the IEEE Electronic Components and Technology Conference*, pp. 456-466, 2022.

[8] T. Zhang, S. Rahman, K. K. Choi, K. Cho, P. Baker, M. Shakil and D. Heitkamp, "A Global-Local Approach for Mechanical Deformation and Fatigue Durability of Microelectronic Packaging Systems," *Journal of Electronic Packaging*, vol. 127, pp. 179-189, 2007.

[9] M. Gonzales, B. Vandevelde, A. Ivankovic, V. Cherman, B. Debecker, M. Lofrano, I. De Wolf, G. Beyer, B. Swinnen, Z. Tokei and E. Beyne, "Chip Package Interaction (CPI): Thermo Mechanical challenges in 3D Technologies," *IEEE 14th Electronics Packaging Technology Conference*, pp. 547-551, 2012.

[10] Z. Cheng, L. Wang, J. Wilde and K. Becker, "Viscoplastic Anand Model for Solder alloys and its application," *Soldering & Surface Mount Technology*, vol. 12, pp. 31-36, 2000.

[11] K. Cox, J. Krantz, M. Borden and S. Tonthat, "Validating Flip Chip package models through experimental deflection measurements," in *IMAPS Device Packaging Conference*, Fountain Hills, 2023.

[12] M. Lane, "Interface Fracture," *Annual Review of Materials Research*, vol. 33, pp. 29-54, 2003.

Adhesive solutions for closed cavity packaging

Patrick Schirmer
Product Management
DELO Industrial Adhesives
Germany, Windach
Patrick.Schirmer@delo.de

Severin Ringelstetter
Engineering
DELO Industrial Adhesives
Germany, Windach

Abstract— The trend towards autonomous driving demands continuously increasing safety requirements and hence reliable components such as CMOS image sensors or communication devices. To provide such components, manufactures are facing ever greater challenges as the sensor packages need to be airtight sealed over their whole lifetime. Typical defects that can be observed when using conventional lid attach materials for closed cavity packages in tests at automotive level (e.g., according to AEC Q100) are pop-up effects and delamination. To avoid these defects and to meet the increasing reliability requirements of the semiconductor industry, DELO Industrial Adhesives has developed special adhesives that not only ensure reliable bonding while keeping narrow and high bondlines, but also improve process stability. Features such as light/heat B-stage or light fixation help to keep the attached lid in place during heat curing and other subsequent steps like temperature cycling, humidity storage and even reflow.

Keywords—CMOS, image sensor, autonomous driving, DMS, LIDAR, ADAS, automotive, adhesive, bonding, MSL, closed cavity, airtight, sealed, AEC Q100, dual cure, B-Stage, reflow

I. INTRODUCTION

DELO responds to the needs of the semiconductor industry with the strategic development of new adhesives for airtight packaging, mainly for end applications like automotive image sensors. Previously existing solutions cannot withstand the ever-increasing requirements towards the needs of trends like autonomous driving whereby the AEC Q100 standard plays an important role [1]. The main challenge is to compensate the changing pressures inside the cavity while keeping the cavity airtight. In addition, outgassing must be reduced to a minimum to avoid contamination of other components inside the package [2].

Figure 1: DELO Image Sensor Illustration

1) Defining the challenges

When working on "Lid Attach onto Housing applications", as shown in Figure 1, the same defect patterns often occur. A major challenge is the airtight design of the image sensors. Due to temperature changes, the air volume contracts at low temperatures, whereas it expands at high temperatures. Since the air volume is entrapped, the pressure decreases or increases. This leads to particular stress in areas where different materials are connected to each other, e.g., by adhesive processes. Typical defect patterns are cracks, which cause the joint to break, or tilt, in which the cover glass becomes misaligned. Both error patterns are judged as defects and must be avoided [1].

Furthermore, outgassing and bleeding of any kind are not acceptable [2]. All components that settle in the sensor in form of precipitation are an unacceptable problem. For example, this can cause corrosion on the wires or impair the image quality of the sensor.

Therefore, it is necessary to develop an adhesive solution that does not cause any of the described defects both during adhesive curing and in subsequent use.

2) Image Sensor assembly & reliability testing

The image sensors are typically manufactured at OSATs, which use equipment of the semiconductor industry. The singulated cover glasses are provided on a wafer and the packages are prefabricated on strips. In the first step, the image sensor dies are bonded into the cavities and electrically contacted with wire bonds. In the following process step, the lids are bonded. This is done on machinery such as the Besi Datacon Evo 2200, which is also used in this way in development at DELO. The machine is loaded with the housings as well as the lid-wafer, and the individual sensor cavities are then closed with the lids in a pick-and-place process. The dispensing process for the lid attach adhesive also takes place within the system. Typically, this is followed by the heat curing process in a convection oven.

Once the sensors are finished, they undergo a variety of reliability tests [3]. Of great importance from an adhesive point of view are those that can cause aging or degradation of the adhesive. These can be, among others, storage according to JEDEC MSL or also based on the AEC Q-100 standard. Both stress the packages with high temperature changes as well as the influence of humidity.

After the storage is completed, the condition of the respective package is finally examined. Special attention is paid to the sealing as well as to the general condition of the package. In addition to the visual inspection via microscope, the non-visible bottom interface of the adhesive line can also be checked for delamination using acoustic scanning microscope. Failure patterns can be identified under the microscope, for example. Here, among other things, delamination between glass and adhesive layer become visible. A good and a poor specimen is shown in Figure 2 a and b, respectively.

In order to also make delamination between the housing and the adhesive layer visible, an acoustic microscope is used. This allows the package to be analyzed non-destructively. Delamination becomes visible in the form of light/dark distinctions, as shown in Figure 3 b. For comparison reasons Figure 3 a shows a bondline without delamination. The inspection is completed with the so-called "Red Ink Test", which finally ensures the tightness of the package by means of ink penetration.

In addition to the tightness and condition of the adhesive layer, the cover glass and image sensor are also examined for outgassing to ensure the optical performance of the sensor.

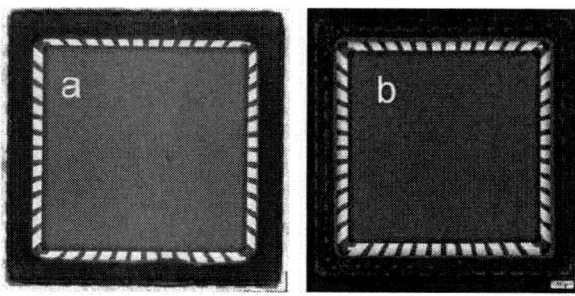

Figure 2: Bondline inspection by means of microscope on DELOs test vehicle

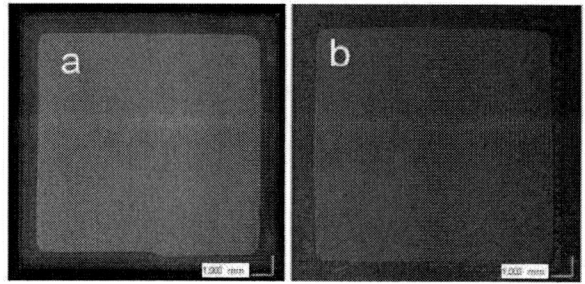

Figure 3: Bondline inspection by means of acoustic microscope on DELOs test vehicle

II. ADHESIVE DEVELOPMENT

For a targeted development of new adhesives, a precise set of requirements is essential. In addition, the definition of the curing process is also of great importance to identify or exclude suitable adhesive chemistries at an early stage. Furthermore, the tests of the adhesives for suitability in the application are defined and specified.

A. Curing process

Typically, heat curing adhesives are used in this application area [5]. The curing temperature range lies in a window between +100 °C and +150 °C with a curing time, depending on the chemistry, of approx. 1 hour to 3 hours. To counteract adhesive spreading, which can often be observed with heat curing adhesives due to viscosity reduction at elevated temperatures, dual curing is also considered in the investigations. The aim is to pre-fix the adhesive bead by fast UV curing within seconds to improve its dimensional stability before moving to final heat curing.

B. Definition of mechanical requirements

To define the target properties of new adhesives, the substrate materials and the conditions of use are first analyzed in depth. The two main bonding partners are the image sensor housing and the glass lid, which in most cases is a filter glass. The two main materials used for the housings are ceramics and epoxy mold [4]. To analyze the behavior of the materials at different temperatures, FEM simulation models, illustrated in Figure 4, are created which are fed with the respective material characteristics. Adhesives of different mechanical properties are tested on these test vehicles.

Since both, the filter glass and the housing materials, have very low expansion coefficients [6], the development of a low-modulus system with Young´s Modulus below 100 MPa and a low glass transition temperature of about -50 °C proved to be useful. The flexible adhesive is chosen to compensate pressure differences that arise during temperature changes, thus avoiding pop-up defects or delamination.

Figure 4: Filter glass deformation at elevated temperature in FEM simulation

C. Material properties check

The Young's modulus and the glass transition range are investigated by means of a DMA (Dynamic Mechanical Analysis) over the temperature range from -55 °C up to +125 °C (according to AEC Q 100 Grade 2). The thermal expansion of the adhesives is determined using TMA (thermomechanical analysis). The data obtained are fed into the simulation model and tested again for suitability in the closed cavity package application. In addition, the new developments are subjected to adhesion tests in form of die shear (silicon die, 2x2 mm², polished) on the relevant substrate materials such as glass or ceramics.

D. Built of dummy image sensors & failure analysis

To ensure that the new developments also properly work in the application, dummy packages, as shown below in Figure 5, very similar to the final product were set up and used as internal test vehicles. The manufacturing process mimics the production conditions of the CMOS image sensors and uses comparable equipment such as the Besi Datacon EVO 2200 for assembly. After testing the bondline shape such as width and height via 3D microscopy in the liquid state, testing is also carried out after final curing. As described above, microscopy, acoustic microscopy and red ink test are available for this purpose. If the packages are intact after finishing the curing process, they pass through the qualification process based on the AEC Q100 standard with a subsequent repeated check of the bondline. AEC Q100 reliability testing includes MSL storage plus additional temperature cycling, high temperature storage, and moisture storage of varying intensity depending on the quality requirements.

Figure 5: DELOs test vehicle for closed cavity packaging

III. CONCLUSION

After completion of all tests, DELO DUALBOND materials, based on a patented chemistry, seem to be very suitable for airtight bonding of image sensors and similar closed cavity packages. Its high flexibility, with a Young's modulus of < 10 MPa at room temperature in combination with its glass transition temperature of below -50 °C (which is outside the usual range of application), as shown in Figure 6, make it possible to compensate for the changing pressures inside the package. These required properties can be achieved thanks to DELO's patented chemistry and are therefore unique. Even after completion of the AEC Q100 tests, based on Grade 2, no defects could be detected in the bondline which proves the compatibility with typical Semicon standards. Contents of this investigation are MSL3, 1000 temperature cycles from -55 °C to +125 °C, 1000 hours of storage at +125 °C and a temperature / humidity storage for 1000 hours at +85 °C / 85 % r.h..

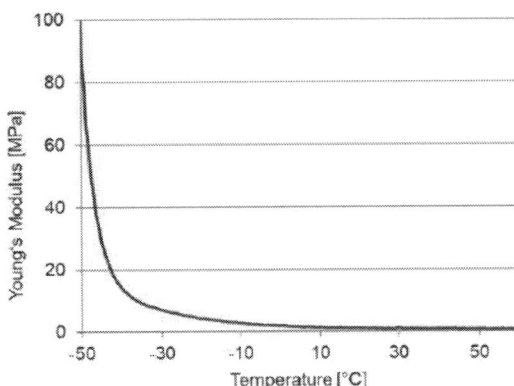

Figure 6: Young's Modulus over Temperature

ACKNOWLEDGMENT

Special thanks goes to Theresa M. Schmid from DELO, who supervised and drove the developments on the engineering side.

REFERENCES

[1] Teoh Eng Kang, Alastair Attard and Jonathan Abela, "CMOS IMAGE SENSOR PACKAGING TECHNOLOGY FOR AUTOMOTIVE APPLICATIONS", Minapad 2019, 2019

[2] Yeh Yee Kee, Kei Lin Sek, Lei Zhu, Younan Hua and Xiamomin Li, „A Novel CMOS Image Sensor Package Cover Glass White Stain Material Identification Metrology by TOF-SIMS", 2021 IEEE International Symposium on the Physical and Failure Analysis of Integrated Circuits (IPFA), 2021

[3] Tianshen Zhou, Shuying Ma, Daquan Yu, Ming Li and Tao Hang, "Development of Reliable, High Performance WLCSP for BSI CMOS Image Sensor for Automotive Application", Sensors 2020, 2020

[4] A. El Gamal and H. Eltoukhy, "CMOS image sensors", IEEE Circuits and Devices Magazine (Volume: 21), 2005

[5] Yukinari Abe and Kazuki Iwaya, „Pre-curable adhesive for Image Sensor Packages", Polytronic 2007 - 6th International Conference on Polymers and Adhesives in Microelectronics and Photonics, 2007

[6] Markus Schindler, "Novel Materials for MEMS Packaging: MEMS Die Attach and ASIC Die Coating and Encapsulation", 2018 IEEE 38th International Electronics Manufacturing Technology Conference (IEMT), 2018

Development of a stretchable and removable electrical interconnection solution for ultra-thin electronic components

Auriane Despax-Ferreres
Univ. Grenoble Alpes, CEA, Liten,
F-38000
Grenoble, France
0000-0003-3207-033X

Pascal Tiquet
Univ. Grenoble Alpes, CEA, Liten,
F-38000
Grenoble, France
pascal.tiquet@cea.fr

Jean-Charles Souriau
Univ. Grenoble Alpes, CEA, Leti,
F-38000
Grenoble, France
jean-charles.souriau@cea.fr

Vincent Jousseaume
Univ. Grenoble Alpes, CEA, Leti,
F-38000,
Grenoble, France
vincent.jousseaume@cea.fr

Julia De Girolamo
Univ. Grenoble Alpes, CEA, Liten,
F-38000
Grenoble, France
julia.degirolamo1@cea.fr

Abstract—

Currently, the solutions for interconnecting electronic components with their active face facing on substrates are based on metallic soldering. The mechanical contacts are therefore rigid. In order to enhance the reliability of the bonding, underfill is usually used to redistribute the thermo-mechanical stress created by the Coefficient of Thermal Expansion (CTE) mismatch between the silicon chip and substrate. Underfill is done with epoxy resins containing silica fillers (SiO2). As a result, the removal of components is no longer possible. Moreover, this solution is not suitable for devices integrating ultra-thin silicon components (<100 µm) hybridized on flexible substrates that may be subject to deformation. This is the case, for example, of medical "patches" worn on the person and continuously solicited. Indeed, the rigid contact points are likely to break. To address this issue, we are developing a thin anisotropic conductive and stretchable adhesive film inspired by the adhesion of the gecko. Thanks to the microstructuration of its toes involving about 1 million setae, the gecko can develop a large contact surface and thus a large force of attraction by the multiplication of van der Waals interactions. In this work, this "dry adhesion" based on the principle of "contact splitting", was implemented in order to improve the adhesion of a flexible interconnection. For this purpose, the surface of a polydimethylsiloxane (PDMS) film was structured with micrometric mushroom-shaped patterns known to be the most efficient form of contact. To this end, silicon molds with varying mushroom geometries were used to shape the PDMS (with different cap and pillar diameters) and the adhesion force microstructured films were assessed (shear and pull-off experiments). To make these films locally conductive through the thickness, a conductive composite was prepared and locally deposited in the mold. One approach we investigated, was using a screen-printing mask. This approach has been implemented and characterized using electrical tests (I-V measurements) in order to select the most suitable films to make a flexible interconnection.

Keywords— Interconection, flexible, gecko-like adhesion, ACA

I. INTRODUCTION

This work presents an innovative solution for assembling electronic components for applications where mechanical deformation such as bending and stretching must be tolerated. For example, thin flexible electronic devices can follow the shape of the human body in motion. In addition, this work shows a removable solution to facilitate the replacement of a defective component. During the two last decades, the packaging of electronic component has to satisfy high interconnection density, high data throughput, miniaturization, and reliability needs while keeping a low cost. Flip-Chip interconnection of chips, i.e. with the active face facing the substrate, offers the best solution to meet these requirements. The main techniques involve soldering micro-bump between the chip and the substrate to create an electrical and mechanical bond. In order to enhance the reliability, underfill is usually used to redistribute the thermo-mechanical stress created by the Coefficient of Thermal Expansion (CTE) mismatch between the silicon chip and substrate. Underfill materials are usually epoxy bases. Once cross-linked, the material is hard and not tolerate deformation. Moreover, the chip cannot be disassembled. Another interconnection solution is to use conductive adhesives. There are two types of conductive adhesives: anisotropic conductive adhesives (ACA) and isotropic conductive adhesives (ICA). Polymers used for these type of interconnections are thermoset with a high Young's modulus as epoxy [1] or acrylic adhesives [2] mixed with conductive fillers. In ICAs, the proportion of conductive fillers is very high: around 80 % weight or higher [3]. ACAs are electrically conductive in the z-axis due to an adding of conductive fillers in low proportion in the polymer matrix [2]. In case of ACAs, films (or pastes) require high pressure and temperature for processing. Pressure is necessary to connect conductive fillers and obtain an electrical continuity and temperature is necessary for the polymer curing (typically T> 100°C). Nevertheless, this approach is still difficult to bend and impossible to stretch. Moreover, there are some difficulties to remove easily the component.

Dry adhesion based on van der Waals interactions [9] is an alternative to process conductive adhesive at room temperature. In nature, dry adhesion was observed for insects or lizards and rely on the microstructuration of their feet [4]. Thanks to SEM image technology, the hierarchical microstructuration of the gecko lizard toes was observed as divided in three levels: lamellae, setae and spatulae [5] which enable to establish a very large contact area with surfaces [10]. Dry adhesives inspired from the gecko have been studied for around 20 years. Scientists groups tried to mimic by simplifying the gecko microstructuration because of its complexity for technical processes [6–10]. In 2007, Del Campo *et al.* determined that the mushroom shape was the ideal contact shape to obtain a maximum pull-off adhesion [11]. In the literature, most of the microstructured films are built with polymers [8,9], particularly elastomers such as poly(dimethylsiloxane) (PDMS) which have often been chosen for their flexibility, stretchability and ability to be molded into very small structures [14].

To obtain a conductive material with dry adhesive properties, conductive ICA-like gecko films were developed from the last 10 years. The targeted applications were mainly the fabrication of EEG-ECG electrodes applied directly to the skin [15]. To that aim, elastomer composites were prepared including silver powders, carbon powders or carbon nanotubes (CNT) mixed with graphene, and were molded in a mushroom shapes [15-17]. All these conductive microstructured films are fully electrically conductive. Short-circuits come out of a multiple of electrical pads are in contact, the electrical way is blocked. Yet in microelectronic there is a necessity to fabricate a non-continuous conductive films to avoid short-circuit in the system.

In this work, we aimed to create a flexible, stretchable, reversible, anisotropic gecko-like interconnection, adapted for the processing of flexible substrates at room temperature. The concept is presented in Figure 1 .

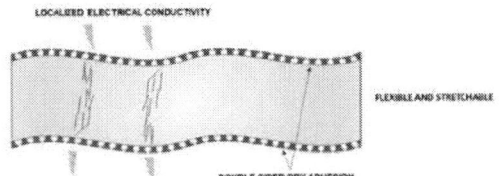

Figure 1 Scheme of the concept of the gecko anisotropic interconnection film

Here, the fabrication process of a conductive and mushroom-shaped microstructured PDMS is described. First, silicon master molds were manufactured. Then, composites based on carbon nanotubes (CNT) and PDMS were prepared and molded. In this work, the electrical conductivities of structured and unstructured composite films are compared. Then, the adhesion of a CNT-PDMS structured film is compared to a pristine PDMS structured film. Finally, a first attempt to localize the conductive composite in the film is presented.

II. EXPERIMENTAL METHODS

A. Conductive and Microstructured PDMS-mushroom

Fabrication of conductive dry adhesives mushroom-shaped molds were fabricated from Silicon On Insulator (SOI) (100 mm - orientation <100>) wafers with a 2 μm-thick BOX (Burried Oxide layer) and a 15 μm thick device layer. To define the location of mushrooms and pillars, a positive photosensitive resin AZ4562 was spin-coated on the wafer, cured and subsenquently exposed to UV (365 nm) through a mask with apertures diameters of 10 μm and spaces of 16 μm in a hexagonal arrangement. After the development of the photosensitive resin, the Si device layer was etched till the BOX in a ICP Oxford tool using SF_6 and CH_2F_2. The BOX was then etched by HF vapor to define the cap diameter.

Prior to PDMS replica molding, a fluorinated silane was deposited on the SOI master molds by a vapor phase process in a hermetically sealed chamber. Then, liquid precursors of PDMS (Sylgard 184 from Dow) were poured into the molds and spread out by spin-coating. The spin-coating parameters were adjusted to obtain films with a backing layer thickness of approximately 190 μm. The curing of PDMS was carried out at 100°C for 30 min and the microstructured PDMS film were peeled from the SOI mold. To prepare a conductive PDMS, multiwall carbon nanotubes with a mean aspect ratio of 830 (diameter between 6 to 13 nm) were purchased from Sigma-Aldrich. The dispersion protocol of CNTs in the PDMS matrix was inspired from Kim *et al.* [18]. A proportion of 1-wt% (relative to PDMS) of CNTs was added in a beaker with isopropyl(alcohol) (IPA) with a ratio of 1:100 in weight. The mixture was homogenized in an ultra-sonic bath for 10 min to separate the CNT bundles. Then 1-wt % (relative to the CNT and PDMS proportions) of a low viscosity Poly(dimethylsiloxane-co-methylphenylsiloxane) (purchased from Sigma Aldrich) , was added to the suspension of CNT/IPA and mixed using a planetary mixer, followed by sonication for 10 min. Finally, IPA was evaporated using a rotary evaporator and the crosslinking agent was added in the mixture. Lastly, the CNT-PDMS mixture was degassed under vacuum and poured into mushroom-shaped molds.

The localization of the conductive composite was performed by screen-printing. The uncured CNT-PDMS composite was deposited through a stencil directly on a SOI master mold patterned with pillars. The stencil was composed of round apertures of 300 μm diameter and a spacing of 500 μm. After the screen printing process, the CNT-PDMS areas were thermally cured and pristine PDMS was deposited by spin coating, cured and demolded following the procedure presented in the experimental section.

B. Measurement Test Methods

Electrical characterizations of the conductive films along the thickness direction were performed by placing the films (microstructured and unstructured) between two flat copper electrodes with a surface of 1 cm². To insure a good contact between the electrodes and the film, a weight (880 g) was added on the top of the box tester electrode.

Current-Voltage curves (I-V) were recorded using an electrical measurement bench developed internally. A forward and backward voltage from 0 to 10 V and from 10V to 0V was applied to the electrodes with a step of 100 mV. Several cycles were performed and the electrical conductivity was extracted from the 4th cycle using the following formula:

$$\sigma = \frac{t}{R \times S}$$

were σ is the electrical conductivity of the composite (S.m^{-1}), R is the resistance (Ω), S the surface (0.0001 m^2) and t is the film thickness (m).

To assess the adhesion performances of the microstructured films, pull-off test were achieved by contacting the mushroom structured PDMS with a flat glass substrate. Before the test, the glass substrates were cleaned with IPA and covered with a protective film to avoid particles contamination. The microstructured film (surface: 5x5 mm^2) was fixed on a piece of glass which was already pasted on aluminum sample holder. Pull-off tests were carried out on a tensile machine Syntax 12 from 3R (Recherches et Réalisations Rémy) equipped with a 100 N load cell. Once positioned between the jaws, the samples were first compressed to a preload of 2.3 N at a velocity of 0.1 mm/min. Then, the sample was pulled-off with a velocity of 0.1 mm/min. The adhesion strength of the microstructured films was determined at the maximum of the pull-off force.

III. RESULTS AND DISCUSSION

A. Non-conductive PDMS Microstructured Films

Scanning Electron Microscopy (SEM) pictures of a mushroom-shaped PDMS film are shown in Figure 2. A dense array of mushrooms arranged in a hexagonal pattern is clearly visible (Figure 2 a.). The PDMS mushrooms are well aligned and defined without collapsing which is favorable in terms of adhesion as all mushrooms can contribute to establish van der Waals interactions. The backing layer which is the part supporting the mushrooms, has a thickness of around 190 μm. From the Figure 2 b., the dimensions of individual mushroom can be measured precisely: the mushroom pillar height is 15 μm, the cap diameter is 18 μm, the pillar diameter is around 10 μm and the cap thickness is 2 μm.

Figure 2 SEM image of a micro-structured PDMS film a. Top view of the film b. Zoomed-in side view of the PDMS mushrooms

A second objective was to perform a molding of a conductive CNT-PDMS. Firstly, a conductive film without structured part was made. Then, once the viscosity parameter adapted for molding, the mushroom-shaped were molded with CNT-PDMS. The next part shows the electrical and mechanical characterization for CNT-PDMS films.

1) CNT-PDMS based film

In order to prepare a conductive microstructured film, a study was first performed involving the preparation of unstructured conductive PDMS films by introducing MWCNTs with different wt% in the mixture and determining the electrical

percolation threshold. To be able to mold the CNT/PDMS mixture, a minimum CNT content (low viscosity) must be aimed for and the quality of the dispersion of CNT in the elastomer matrix is therefore a critical aspect. The conductivity of the composites with different filler contents was assessed using I-V measurements. Figure 3 shows an example of I-V curves for an unstructured and a structured sample (composite PDMS – CNT (1 wt%)).

Whatever the CNT loading (in the range investigated), the I-V curves are nonlinear. The electrical resistivity varies with the voltage, the resistance decreases with the voltage increases. This behavior is typical of inhomogeneous materials containing conductive fillers in a polymer matrix and was already observed on composites of PDMS/CNT [19]. Wang et al. [25] have determined that the nonlinear behavior was mainly due to electron hopping effect and tunneling effect for relatively low voltages. In this study the resistance was determined by fitting the I-V curves to a straight line with an imposed start at 0.

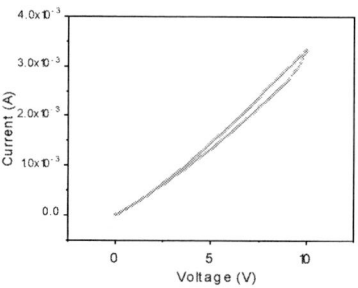

Figure 3 I/V curves measurement for a CNT-PDMS structured film

Figure 4 presents the variation of the electrical conductivity as a function of the filler content (mass fraction) for unstructured composites (with CNT mean aspect ratio of 830). For a MWCNTs mass fraction of less than 0.4 wt% the composite has a conductivity of around 10^{-12} S/m. At the opposite, for a higher mass fraction of MWCNTs of 1wt% and above, the conductivity rises sharply up to 0.1 S/m. The region in between corresponds to the percolation transition. In a conductive composite, the percolation threshold is the lower content of fillers to obtain an electrical path through the composite.

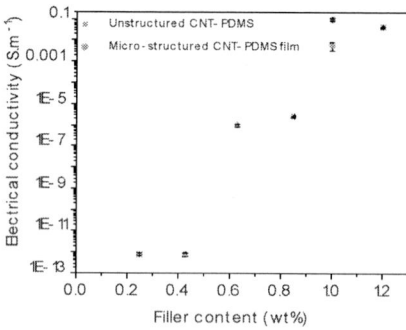

Figure 4 Variation of the electrical conductivity as a function of the filler content for CNT-PDMS composites

96

The electrical conductivity of a composite depends on volume or mass fraction of the conductive fillers and can be described by the following equation [20] :

$$\sigma \propto \sigma_0 (p - p_c)^t$$

with σ electrical conductivity of the composite, σ_0 electrical conductivity of the fillers, p volume fraction of CNT, p_c percolation threshold and t a critical exponent which depends on the connectivity and aspect ratio of the fillers

The percolation threshold can be estimated from Figure 4 and is between 0.4 and 0.6 wt% of MWCNTs in the PDMS. The percolation threshold depends strongly on the filler aspect ratio but also on their dispersion in the matrix. The presence of MWCNTs aggregates observed on SEM image (Figure 5) increase the percolation threshold. Indeed, the estimation of the percolation threshold as a function of the aspect ratio (833), is around 0.17 wt% of CNT in a matrix. Here, the aspect ratio corresponding to this percolation threshold is between 240 and 350. A similar result was observed for Ag nanowires (NWs with a mean aspect ratio of 220) in a polymer matrix were a discrepancy of a factor 2 was evaluated between the experimental results and the theoretical one [21]. To prepare a microstructred conductive film we have chosen a MWCNT mass fraction well above the percolation threshold i.e. 1 wt%.
For that proportion, the mixture of uncured PDMS and MWCNTs prepared with the protocol described in the experimental method section, is sufficiently fluid to be molded. Figure 5 shows a cross section of a CNT-PDMS microstructured film. The mushrooms can be distinguished on the upper side of the film and do not show any defects or collapse. In the backing layer, lighter zones appear corresponding to aggregates of MWCNTs suggesting that the mixing process can still be optimized. Some mushrooms on the film surface appear shinier confirming the presence of MWCNTs. Finally, one can notice an irregular thickness of the backing-layer resulting from the difficulty to spin-coat the mixture of PDMS/MWCNTs.
The electrical measurement appears dependent of the surface finish. Indeed, the electrical conductivity decreases from 10^{-1} S.m^{-1} for the unstructured flat conductive film to 10^{-2} S.m^{-1} for the microstructured rough CNT-PDMS film (Figure 5). It should be noticed this electrical conductivity is estimated using the real surface in contact with the electrode on the mushroom face (i.e. 0.13 cm²). The lower conductivity measured for microstructured CNT-PDMS composites could be due to an imperfect contact with the copper electrode, the backing layer flatness being not optimized. Kim *et al.* have obtained a slightly higher conductivity of 0.6 S.m^{-1} for similar composite (with 1 wt% CNT with AR of 1000 in PDMS) and even 1 S.m^{-1} at 1 wt% of CNT and graphene fillers [16]. It

Figure 5 SEM image of a cross-section of a structured CNT-PDMS composite

should be mentioned that higher loadings cannot be used to perform structured conductive composites. Indeed, PDMS mixed with a high CNT loading is too viscous to be used for the filling of micrometer structures.

To confirm the dry adhesion properties of these micro-structured films, adhesion tests in pull-off configuration were performed. Pull-off adhesion tests performed normal to a given surface, are a classical tests used to characterized gecko inspired dry adhesives [10], [22], [23] and usually imply a compression step followed by a retraction step. During the compression step the mushrooms are pressed against the glass substrate enabling them to maximize the contact area and as a result improve the adhesion of the film. In this study, the preload value of 2.3 N was is the limiting force value to prevent mushroom buckling. The Figure 6 shows the pull off force - displacement curve obtained for a microstructured CNT-PDMS film with compressive forces shown as positives and tensile forces shown as negatives.

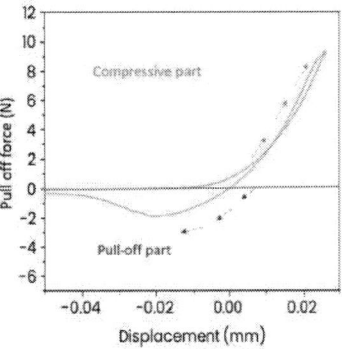

Figure 6 Graph of a pull-off test for CNT-PDMS (1wt%) micro-structured film

The adhesion strength of a microstructured film is determined as the maximum force of the pull-off part. On this example, the adhesion force is 1.9 N.cm^{-2}.

To assess the impact of the presence of fillers in the film on adhesion strength, the micro-structured CNT-PDMS film was compared to the same micro-structured film made of pure PDMS (Figure 7). Several successive measurements were performed on each sample under identical conditions to assess the repeatability of the measurement six times. To facilitate the visualization of the results, the six strength-displacement curves in the pull-off part are shown as positive.

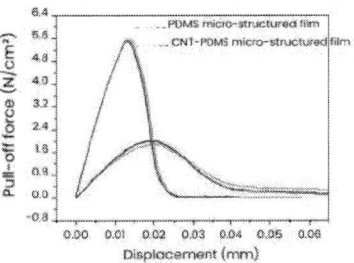

Figure 7 Graph of the comparative adhesion force in pull-off for PDMS micro-structured film and CNT-PDMS structured film

For each sample, the obtained curves overlap and the maximum pull-off forces are repeatable. The pristine microstructured PDMS shows a superior adhesion strength with a maximum adhesion force of 6 N.cm^{-2} compared to 2 N.cm^{-2} when 1% CNT fillers were added in the PDMS-matrix. In addition, CNT-PDMS curves appear more spread with a maximum pull-off shifted to higher values of displacement than for the microstructured PDMS film.

These results differ from those obtained by Seong *et al.* who demonstrated that the pull-off adhesion force was almost unchanged by the addition of low mass fraction of CNT in the PDMS matrix (0.5-2 wt%) for a given mushroom geometry [24]. We assume that in our case, this reduction of pull-off force could be the result of a misalignment of the sample with the substrate result of the thickness variation of the backing layer (shown in Figure 5) and the lack of flatness of the sample. Booth *et al.* observed that for misalignment of 0.1° the adhesion measurement type assigned is a peel-like test [25]. Thus, to maintain an adhesion force around 6 N/cm², a smooth backing layer is mandatory. Indeed, a film with a rough backing layer pasted on the glass is still irregular (composed with bumps and hollows). Hence, the mushroom contact is not homogeneous on all the sample surface. Finally, the adhesion is characterized only on the contact surface which is possible only on bumps areas.

2) Localization of Composite in Pillar Structured Film

The fabrication of a removable anisotropic interconnect requires to localize conductive areas at certain points in the microstructured film.

Figure 8 shows a picture of the obtained film with localized CNT-PDMS areas (filler loading of 1 wt%).

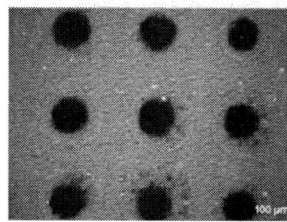

Figure 8 Optical microscope image of pillar-shaped microstructured film with conductive CNT-PDMS inserts.

In the picture, one can distinguish 9 large black zones corresponding to CNT-PDMS composite whereas lighter zones correspond to pristine PDMS. The small dots on the picture are the micrometrics pillars. A circle around the pillar of CNT-PDMS is observed, which could be a disturbance zone for the pillar PDMS.

For a better understanding of the microstructured film architecture, SEM observations were carried out on the cross section of the film (Figure 9).

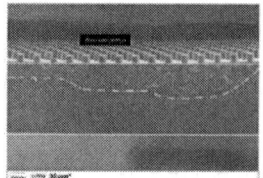

Figure 9 SEM image of a slice of a localized pillar structured film

In this picture, the conductive composite of CNT-PDMS is clearly visible in the upper part of the backing layer (just below the pillars). To guide the eye, a dashed yellow line was added and delimits this area. The lower part of the backing layer is a pure PDMS layer indicating that the conductive areas do not pass through the entire thickness of the film. As a result, the electrical conductivity of this microstructured film in z-axis is very high (10^{-10} S/m). These preliminary results are not surprising since the non-conductive PDMS was poured over the conductive areas and completely covers them. To avoid this situation, strategies are currently implemented to reduce the excess PDMS over the conductive areas and chemically or physically etch the remaining PDMS.

IV. CONCLUSIONS AND PERSPECTIVES

In this work we prepared and molded a fully conductive composite based on high aspect ratio MWCNTs and PDMS. To this end, composite formulation and mixing protocols were optimized to obtain a homogeneous and fluid mixture with a low mass fraction of MWCNTs (1 wt%). The electrical conductivity obtained for such composites is 10^{-1} S.m^{-1} but decreased by a decade for microstructured films (10^{-2} S.m^{-1}). The adhesion performances of the mushroom-shaped microstructured conductive films were characterized by pull-off tests and compared to a non-conductive equivalent. An adhesion force of 1.9 N.cm^{-2} was obtained for the microstructured conductive composite (CNT 1 wt% - PDMS) whereas for a similar mushroom geometry, the adhesion force on structured PDMS is 6 N.cm^{-2}. This deterioration in adhesion and electrical conductivity could be explained by the lack of planarity of the composite films inducing misalignment during the pull-off test and a decreased contact area with electrode during the electrical characterization. This limitation will be the subject of further works especially to improve the smoothness of the backing layer. Finally, firsts encouraging results for the localization of the MWCNT-PDMS composite in microstructured films were obtained. Inserts of conductive MWCNTs-PDMS were successfully screen printed on a silicon mold. Further work will be performed to obtain the z-axis conductivity in the microstructured films.

ACKNOWLEDGMENT

This work was supported by the CEA of Grenoble and the University of Grenoble Alpes with the help of the "Plateforme Technologique Amont" in Grenoble, with financial support from the CNRS Renatech network".

REFERENCES

[1] R. Aradhana, S. Mohanty, et S. K. Nayak, « A review on epoxy-based electrically conductive adhesives », *Int. J. Adhes. Adhes.*, vol. 99, p. 102596, juin 2020, doi: 10.1016/j.ijadhadh.2020.102596.

[2] P. J. Opdahl, « Anisotropic Conductive Adhesives », in *Handbook of Visual Display Technology*, J. Chen, W. Cranton, et M. Fihn, Éd., Cham: Springer International Publishing, 2016, p. 1533-1541. doi: 10.1007/978-3-319-14346-0_65.

[3] M. J. Yim, Y. Li, K. Moon, K. W. Paik, et C. P. Wong, « Review of Recent Advances in Electrically Conductive Adhesive Materials and Technologies in Electronic Packaging », *J. Adhes. Sci. Technol.*, vol. 22, n° 14, p. 1593-1630, janv. 2008, doi: 10.1163/156856108X320519.

[4] S. N. Gorb, M. Sinha, A. Peressadko, K. A. Daltorio, et R. D. Quinn, « Insects did it first: a micropatterned adhesive tape for robotic applications », *Bioinspir. Biomim.*, vol. 2, n° 4, p. S117-S125, déc. 2007, doi: 10.1088/1748-3182/2/4/S01.

[5] « Macroscale adhesion of gecko setae reflects nanoscale differences in subsurface composition ». https://royalsocietypublishing.org/doi/epdf/10.1098/rsif.2012.0587 (consulté le 21 juin 2023).

[6] B. Aksak, M. P. Murphy, et M. Sitti, « Adhesion of Biologically Inspired Vertical and Angled Polymer Microfiber Arrays », *Langmuir*, vol. 23, n° 6, p. 3322-3332, mars 2007, doi: 10.1021/la062697t.

[7] D. Santos, S. Kim, M. Spenko, A. Parness, et M. Cutkosky, « Directional Adhesive Structures for Controlled Climbing on Smooth Vertical Surfaces », in *Proceedings 2007 IEEE International Conference on Robotics and Automation*, avr. 2007, p. 1262-1267. doi: 10.1109/ROBOT.2007.363158.

[8] Z. Wang, P. Gu, et X. Wu, « Gecko-inspired bidirectional double-sided adhesives », *Soft Matter*, vol. 10, n° 18, p. 3301, 2014, doi: 10.1039/c3sm52921e.

[9] S. Li, H. Tian, J. Shao, H. Liu, D. Wang, et W. Zhang, « Switchable Adhesion for Nonflat Surfaces Mimicking Geckos' Adhesive Structures and Toe Muscles », *ACS Appl. Mater. Interfaces*, vol. 12, n° 35, p. 39745-39755, sept. 2020, doi: 10.1021/acsami.0c08686.

[10] H. Yi, M. Kang, M. K. Kwak, et H. E. Jeong, « Simple and Reliable Fabrication of Bioinspired Mushroom-Shaped Micropillars with Precisely Controlled Tip Geometries », *ACS Appl. Mater. Interfaces*, vol. 8, n° 34, p. 22671-22678, août 2016, doi: 10.1021/acsami.6b07337.

[11] A. del Campo, C. Greiner, et E. Arzt, « Contact Shape Controls Adhesion of Bioinspired Fibrillar Surfaces », *Langmuir*, vol. 23, n° 20, p. 10235-10243, sept. 2007, doi: 10.1021/la7010502.

[12] H. Im, J. Kim, S. Han, et T. Kim, « Process, Design and Materials for Unidirectionally Tilted Polymeric Micro/Nanohairs and Their Adhesion Characteristics », *Polymers*, vol. 8, n° 9, p. 326, sept. 2016, doi: 10.3390/polym8090326.

[13] R. Hensel, K. Moh, et E. Arzt, « Engineering Micropatterned Dry Adhesives: From Contact Theory to Handling Applications », *Adv. Funct. Mater.*, vol. 28, n° 28, p. 1800865, 2018, doi: 10.1002/adfm.201800865.

[14] M. Lamblet, E. Verneuil, T. Vilmin, A. Buguin, P. Silberzan, et L. Léger, « Adhesion Enhancement through Micropatterning at Polydimethylsiloxane−Acrylic Adhesive Interfaces », *Langmuir*, vol. 23, n° 13, p. 6966-6974, juin 2007, doi: 10.1021/la063104h.

[15] F. Stauffer *et al.*, « Skin Conformal Polymer Electrodes for Clinical ECG and EEG Recordings », *Adv. Healthc. Mater.*, vol. 7, n° 7, p. 1700994, avr. 2018, doi: 10.1002/adhm.201700994.

[16] T. Kim, J. Park, J. Sohn, D. Cho, et S. Jeon, « Bioinspired, Highly Stretchable, and Conductive Dry Adhesives Based on 1D-2D Hybrid Carbon Nanocomposites for All-in-One ECG Electrodes », *ACS Nano*, vol. 10, n° 4, p. 4770-4778, avr. 2016, doi: 10.1021/acsnano.6b01355.

[17] J. Krahn et C. Menon, « Electro-Dry-Adhesion », *Langmuir*, vol. 28, n° 12, p. 5438-5443, mars 2012, doi: 10.1021/la2048882.

[18] J. H. Kim *et al.*, « Simple and cost-effective method of highly conductive and elastic carbon nanotube/polydimethylsiloxane composite for wearable electronics », *Sci. Rep.*, vol. 8, n° 1, p. 1375, janv. 2018, doi: 10.1038/s41598-017-18209-w.

[19] C. H. Liu et S. S. Fan, « Nonlinear electrical conducting behavior of carbon nanotube networks in silicone elastomer », *Appl. Phys. Lett.*, vol. 90, n° 4, p. 041905, janv. 2007, doi: 10.1063/1.2432283.

[20] « 978-3-662-02403-4_5.pdf ».

[21] « V. Bedel - Mise en oeuvre et optimisation d'un revêtement conducteur poly(époxy) / films submicroniques d'argent pour la protection foudre de structures composites aéronautique (2018) ».

[22] L. Heepe et S. N. Gorb, « Biologically Inspired Mushroom-Shaped Adhesive Microstructures », *Annu. Rev. Mater. Res.*, vol. 44, n° 1, p. 173-203, 2014, doi: 10.1146/annurev-matsci-062910-100458.

[23] Y. Ma *et al.*, « Remote Control over Underwater Dynamic Attachment/Detachment and Locomotion », *Adv. Mater.*, vol. 30, n° 30, p. 1801595, 2018, doi: 10.1002/adma.201801595.

[24] M. Seong *et al.*, « Adhesion of bioinspired nanocomposite microstructure at high temperatures », *Appl. Surf. Sci.*, vol. 413, p. 275-283, août 2017, doi: 10.1016/j.apsusc.2017.04.036.

[25] J. A. Booth, M. Bacca, R. M. McMeeking, et K. L. Foster, « Benefit of Backing-Layer Compliance in Fibrillar Adhesive Patches—Resistance to Peel Propagation in the Presence of Interfacial Misalignment », *Adv. Mater. Interfaces*, vol. 5, n° 15, p. 1800272, 2018, doi: 10.1002/admi.201800272.

UV laser copper pad surface exposure for Laser Direct Structuring (LDS) of interconnection

Guendalina Catalano
Back-End Manufacturing &
Technology
ST Microelectronics
Agrate Brianza (MB), Italy
guendalina.catalano@st.com

Alessandro Mellina Gottardo
Back-End Manufacturing &
Technology
ST Microelectronics
Agrate Brianza (MB), Italy
alessandro.mellinagottardo@st.com

Riccardo Villa
Back-End Manufacturing &
Technology
ST Microelectronics
Agrate Brianza (MB), Italy
riccardo.villa@st.com

Abstract— Microelectronics market demands strong miniaturization and high-quality packages solutions; therefore, companies and researchers start thinking about alternative methods to create more flexible and performing interconnections which are able to reduce package dimensions. Laser direct structuring (LDS) technology creates a conductive pattern on MID (Moulded Interconnected Device) with the combination of laser layout structuring and electroplating bath. LDS can be applied to leadframe based IC packages to realize interconnections in alternative to standard wires connection. Feasibility study on copper (Cu) pad covered by moulding compound is conducted varying laser parameters as power, frequency and scanning speed to obtain a 80μm diameter via on die. Die via is connected through a trace with a 200μm diameter via on lead to realize the interconnection path, filled with copper after plating process. UV 355nm wavelength laser is used to obtain a complete exposition of Cu surface without silicon damage. Visual inspection with optical microscope and SEM analysis method are performed to evaluate the integrity of metal pads and interconnection layout.

Keywords—LDS (laser direct structuring), DCI (direct copper interconnection), laser, semiconductor packages

I. INTRODUCTION

LDS (Laser direct structuring) [1] technology is mostly used to create conductive patterns on MID (Moulded interconnected Device). MID technology is easily adaptable even for a variety of innovative electrical device designs. [2] Special fillers, such an organometallic complex or inorganic spinel compounds, are combined in the MID used in LDS. The filler is exposed by the laser, which causes a physicochemical reaction; the metal atoms that are exposed from this reaction serve as the catalytic nuclei for electroless plating. [3]

With the combination of laser layout structuring, electroless plating and electroplating plating baths it is possible to create interconnections on molding compound surface. Electroplating is an alternative method to realize copper interconnections proved to be better than methods such as stencil printing, dispensing and mechanical pressing using as filling material copper, silver pastes and copper rivets. In fact, these latter methods have shown problems of filling the structures made by the laser due to the size of the particles of the pastes and showed the best results with silver paste. [4]

LDS technology can be investigated as an alternative to wire bonding to realize interconnection for leadframe based packages. Using a laser machine to create vias and traces, which reflect CAD drawings, is crucial for the realization of copper interconnections. To achieve this goal, it is necessary to find laser parameters that allow to remove the amount of resin required to expose the device without damaging it, thus allowing to obtain dimensions that enable copper plating, which completely fills the cavities.

A recent study demonstrated the possibility of realizing vias, with a diameter of 90μm and 250μm deep with an approximate aspect ratio of 3, through resins that contained alumina or silica particles as fillers using a 355nm UV laser.[5]

Since 1995, when solid state UV laser was introduced in the laser drilling market, it has been considered an excellent solution because it couples well with a wide spectrum of materials, especially with most microelectronic materials including copper and dielectrics and also allowing to obtain via diameter lower than 100μm. [6]

Recently B. Tan et al. [7] demonstrate that is possible to obtain micro vias with a diameter of 20μm and 100μm deep always using a UV laser. However, the aspect ratio is too high to allow acceptable structures to be obtained by electrodeposition.

The purpose of this paper is to explore the feasibility of the application of LDS technology for the realization of copper interconnection starting from the developing of the laser process on top of molding compound coupled with copper pad on die. Laser feasibility process has been studied to realize good aperture of the via on molding compound with good exposure of Cu pad metal.

This paper present first the experimental section that deepens into the materials, characterization methods and process used and finally analyzes the results obtained.

II. EXPERIMENTRAL SECTION

A. Material preparation

Bare copper leadframe with LDS molding compound and die with Cu pads metals was used for this investigation. UV picosecond laser with 355nm wavelength is used to obtain laser vias with a target of 80μm diameter for die, 200μm for lead and 10μm depth for traces on molding compound top surface.

B. Characterization

Visual inspection on Cu surface is performed with high resolution optical microscope able to detect defectivity as mold residuals, die pad damaging and layout was also investigated.

Optical microscope was used to investigate the morphology of the samples and to measure vias diameter, instead with the usage of profilometer traces depth was measured.

Additional SEM inspection is conducted to double check on Cu surface integrity. Data analysis was carried out with JMP software.

The data of the three main structures created by the laser were considered for analysis: Die vias, lead vias and traces (Fig 1). Laser process quality is governed by a lot of parameters related to the laser source itself as wavelength, power, beam diameter, related to hardware, as for example scanner speed along axes, motion table travel velocity and software as programmable drilling strategy, mark delay and so on. Scanning speed, power and number of laser repetitions have been identified as laser parameters which most influence the geometries of the structures. Scanning speed is defined as the speed of scanner head for lasering the layout; laser power is the output power at the end of the optical chain: the combination of laser source radiation with related beam shaping and scanner head focal lens; number of laser repetition indicates the number of laser passes repeated on the same layout. A stable working point to realize the full LDS structure was identified and validated and starting from that condition, a DoE has been designed for process window identification using a classical response surface design with a central composite design model.

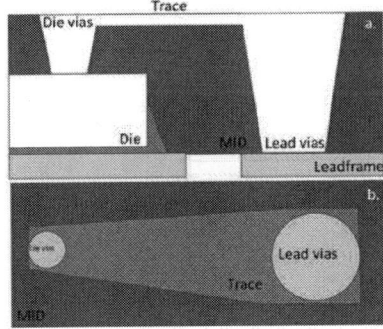

Fig. 1. Side (a) and top (b) view of the three main structure created by laser: Die Vias, Lead vias and traces.

III. RESULTS AND DISCUSSION

A. Die vias

Die drilling is the most critical laser process step due to the low thickness of the copper metal on the die so possibility of breakage and damage of the die itself is very high with consequent exposure of the silicon. (Fig 2)

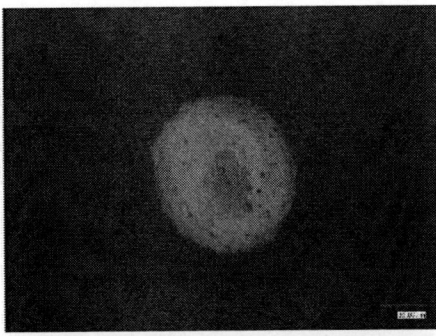

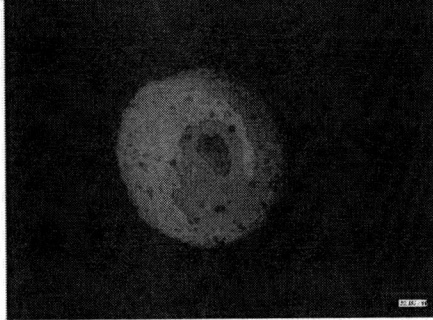

Fig. 2. Microscope images of die breakages detected on the bottom of die vias

Initially, a central laser working point for die drilling was identified using a power of 1.6W, scanning speed of 400mm/s and a number of repetitions equal to 2. To identify a more stable working point, it was decided to set the DoE by varying the Power of 0.6W, the repetitions of 1 and the scanning speed of 100mm/s.

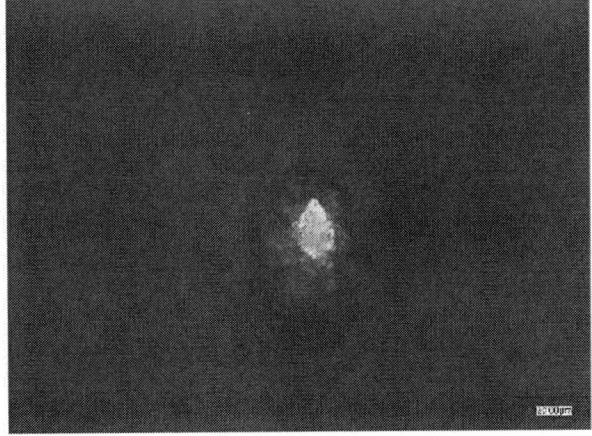

Fig. 3. Microscope image of die vias not fully open by laser due to low parameters

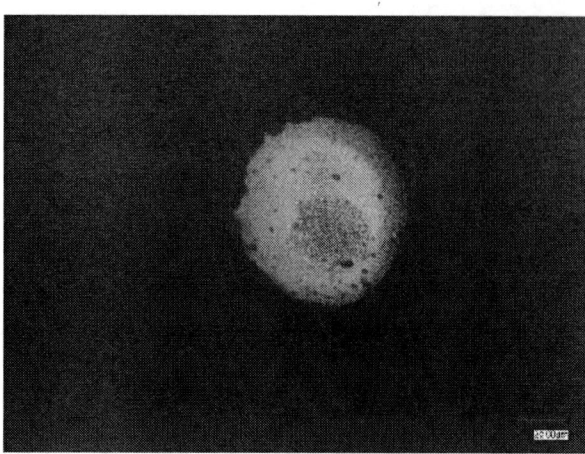

Fig. 4. Microscope image of die vias realized with the set of best laser parameters

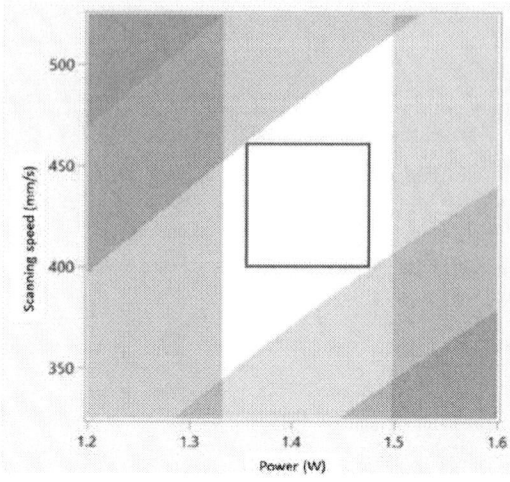

Fig. 6. JMP contour profiler for die vias. The blue square identifies the process window found by DoE

The best repetition parameter to obtain the desired die vias dimensions has been found to be 2, as a repetition is not able to guarantee a sufficient opening of the MID and in some cases not to expose the copper of the pads properly (Fig 3), while 3 repetitions increase the risk of damaging the die and exposing silica (Fig 2).

From the response of the DoE, it can be deduced that the variability is too high as regards the analysis of mold residuals and die breakages. On the contrary JMP analysis of bottom and top diameter showed a good prediction. (Fig 5)

Therefore, by fixing 2 repetitions, the DoE showed that the best parameters to use to meet the dimensional requirements of the die vias are 1.6W and 500mm/s. However, it was possible to identify a process window, the blue square on figure 6, with scanning speed varying from 408 to 450mm/s and power from 1.33 to 1.5W.

Powers and scanning speeds lower than those determined by the DoE, similarly to 1 repetition, have shown limits to meet the minimum size requirements of the die vias and in opening the MID, which in some cases was not completely removed or opened (Fig 3).

On the contrary, parameters greater than those found thanks to the DoE showed damage of the die (Fig 2) due to the overetched of the copper pads and geometric dimensions of the vias above the design limits, considering to maintain usual die pad dimensions and geometries combined with the new Cu metal finishing.

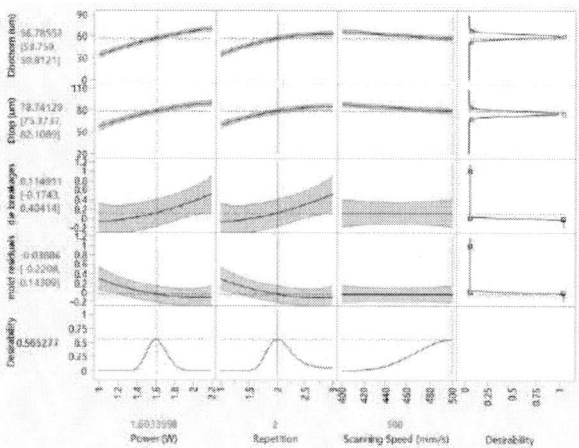

Fig. 5. JMP prediction profiler for die vias

B. Lead vias

With a copper thickness of over 200μm of the leadframe available, the realization of the Lead Vias, compared to the Die Vias, proved to be less critical.

After identifying a laser working point using a power of 5W, a scanning speed of 500mm/s and a number of repetitions equal to 5, to identify a feasibility windows, it was decided to set the DoE by varying the Power of 1W, the repetitions of 1 and the scanning speed of 50 mm/s.

In this case the number of repetitions turned out to be a non-significant parameter. The laser parameters that guarantee to match the target size of the diameters of the Lead Vias in the most accurate way are a Power of 4.8W and a Scanning speed of 550 mm/s.

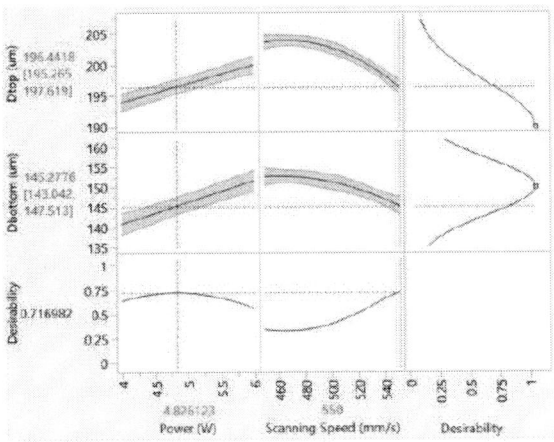

Fig. 7. JMP prediction profiler for lead vias

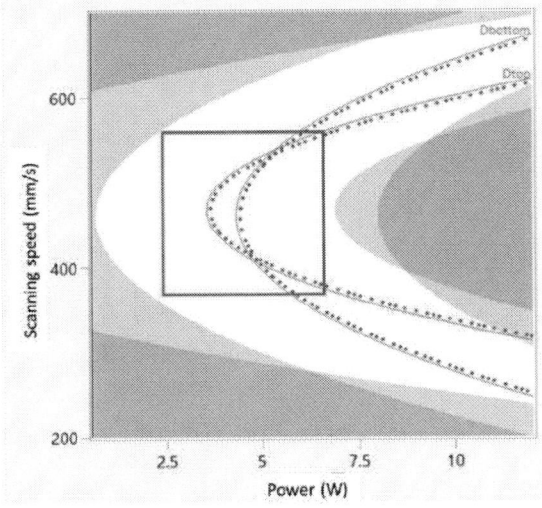

Fig. 8. JMP contour profiler for lead vias. The blue square identifies the process window found by DoE

Figure 8 shows a wide process window that could be identified thanks to the DoE and that allows to meet the dimensional requirements. Compared to the Die Vias in this case no copper breakages or mold residuals were detected, and the variability is quite different from the Die Vias prediction profiler. For the lead vias the variability of diameters, shown in the prediction profiler (Fig 7), is better. Also, the desirability is better compared to die vias due to the absence of mold residuals and die breakages. However, using a low power and a high scanning speed it's possible to observe a very small amount of mold residuals (Fig 9), which allow to obtain a good quality of vias and connections, made by plating.

In figure 10 it's possible to observe that using high power and low scanning speed in the middle of the vias an overetched copper zone is present.

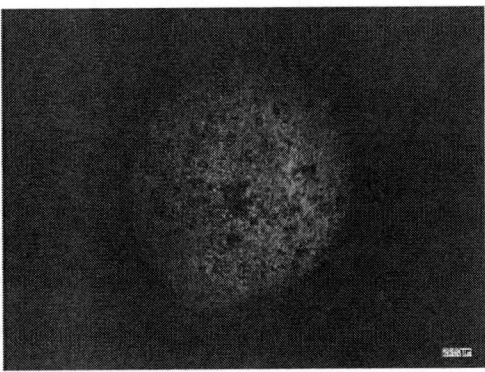

Fig. 9. Microscope image of mold residuals on the bottom of lead vias

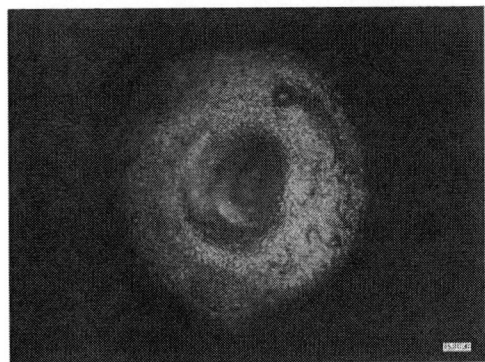

Fig. 10. Microscope image of lead vias realized with HH parameters

C. Traces

Traces are made by hatching MID and then applying plating to deposit copper directly on it. It differs from vias structures realization in which laser drills on a copper substrate to expose it.

A laser working point was initially identified using a power of 1.5W and a scanning speed of 1000 mm/s. In comparison to the die and lead vias a lower power and a much higher scanning speed were used because there is no copper but only MID.

The repetitions have been set at 1 to better investigate power and scanning speed as it has been verified that by setting 2 repetitions it is possible to obtain the same size, in terms of depth, that by setting double power and half scanning speed.

The depth of the traces made was investigated, setting a DoE by varying the power of 0.5W and the scanning speed of 500 mm/s.

The prediction profiler (Fig 11) showed the best desirability between JMP analysis, of the three structure realized by laser, performed and also a good depth variability. It also showed a linear power trend with low slope and a parabolic scanning speed trend with the best parameter, identified by the DoE to meet the target depth, at the least inclined point of the curve.

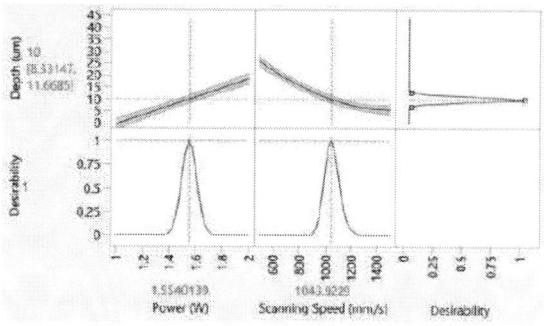

Fig. 11. JMP prediction profiler for lead vias

Fig. 13. Profilometer images of traces

It was also possible to identify a large working area on the contour profiler in which was possible to obtain the target depth of the traces. However, the best and wide process windows was identified by the blue square in figure 12.

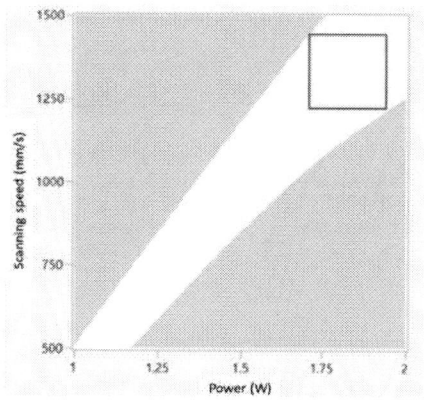

Fig. 12. JMP contour profiler for lead vias. The blue square identifies the process window found by DoE

A more accurate analysis of the depth of the traces was carried out using a profilometer.

Figure 13 shows an average depth of 10μm on top and a deeper average depth, around 40μm, on bottom.

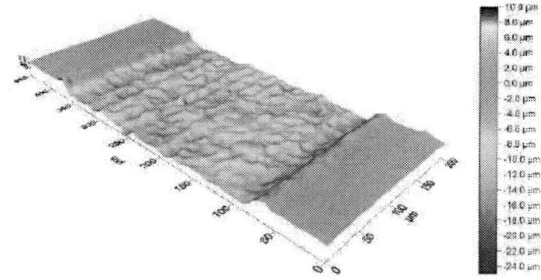

Feasibility investigation showed good results in terms of laser patterning consistency respect to CAD drawing and vias opening for some laser parameters combination. One set up of laser parameters has been identified for realization of the interconnection and plating process of laser structures shows good results. Figure 1 shows the result of via opened with laser and filled with Cu plating

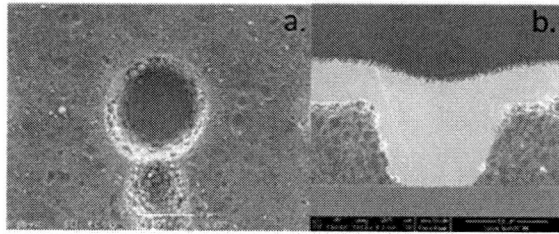

Fig. 14. Die laser via through moulding compound (a) and laser vias filled with plating bath (b)

IV. CONCLUSIONS

Application of laser direct structuring technology on leadframe based packages allows to obtain new type of interconnection without wires with this methodology,

Laser direct structuring technology can be applied in leadframes based packages solution to create customized interconnections which allow to exploit wider die surface and reduce package size. The application of this innovative technology for semiconductor interconnection is the results of a consistent combination of design rule from molding, laser and plating. Laser process parameters identification to realize traces, lead via and die via involved great effort, starting from the preliminary working point to the investigation of the process window, with the definition of quality acceptance criteria. Wide working area have been identified for traces and lead realization matching the target need to test plating process: 10μm depth of trace and 200μm diameter for lead. Die vias process parameters identification was the most critical because of the high risk of Cu metals breakages and mold residuals. In any case also for die vias a small stable working area has been identified to obtain 80μm diameter dimensions without defects. The identification of laser process parameters for each structure necessary to create the interconnection represents the first step for the creation of more efficient and flexible leadframe based packages.

ACKNOWLEDGMENT

We gratefully thanks to dr. Roberto Tiziani and dr. Michele Derai for the continuous and constant support on the activity described in this paper and for the full collaboration and patience showed during the experimental activity. It was a great privilege to work under their guidance.

REFERENCES

[1] Technical Brochure, "3-Dimensional Circuitry, Laser Direct Structuring Technology (LPKFLDSTM) for Moulded Interconnect Devices", LPKF Laser & Electronics AG (2009)

[2] K. Ratautas, A. Jagminienė, I. Stankevičienė, E. Norkus and G. Račiukaitis, "Laser Assisted Selective Metallization of Polymers," 2018 13th International Congress Molded Interconnect Devices (MID), Würzburg, Germany, 2018, pp. 1-3.

[3] Zhang, Y., Kontogeorgis, G. M., & Hansen, H. N. (2011). An explanation of the selective plating of laser machined surfaces using surface tension components. Journal of Adhesion Science and Technology, 25(16), 2101–2111.

[4] H. L. Bach et al., "Vias in DBC Substrates for Embedded Power Modules," CIPS 2018; 10th International Conference on Integrated Power Electronics Systems, Stuttgart, Germany, 2018, pp. 1-5.

[5] M. Murai et al., "Development of a laser processing technology for high thermal radiation multilayer module," 56th Electronic Components and Technology Conference 2006, San Diego, CA, USA, 2006, pp. 4 pp.

[6] Weisheng Lei, S. Raman and Curry Chow, "UV Laser Solutions for Electronic Interconnect and Packaging," 2005 6th International Conference on Electronic Packaging Technology, Shenzhen, China, 2005, pp. 1-7.

[7] B. Tan and K. Venkatakrishnan, "Interconnection microvia drilling using solid state UV laser," LEOS 2006 - 19th Annual Meeting of the IEEE Lasers and Electro-Optics Society, Montreal, QC, Canada, 2006, pp. 887-

Fine Pitch Micro Indium Bump Interconnect Flip Chip Bonding

Travis Scott
Product Management
Finetech GmbH & Co.KG
Berlin, Germany
travis.scott@finetech.de

Abstract—**Quantum computing processors, micro LED displays and Focal Plane Array (FPA) imaging and detector devices such as infrared (IR) thermal imaging sensors are seeing higher demand as more practical applications requiring these components are coming into research and development, industrial, military and consumer markets. This paired with higher pixel and Qubit count and interconnect density, on larger and larger chips is driving hybridization and monolithic integration in these technologies. This is showing a marked increase in demand for fine pitch micro bump interconnection flip chip die bonding. However, some critical challenges facing these technologies are; larger component sizes mean higher density interconnections over increasing surface area. Sub-micron accuracy required to align fine pitch micro interconnect arrays. This together with the challenges facing the materials that are becoming the industry standard for these applications, such as the requirement for the assembled components to remain stable in extreme conditions such as cryogenic application environments. Combined with low loss high strength mechanical / electrical interconnect requirements on components containing sensitive materials, structures and unmatched coefficient of thermal expansion (CTE) means that processing gases such as formic acid or high temperature reflow bonding can no longer be used to bond these devices. These challenges mean that the industry is fast approaching the limitations of even state-of-the-art die bonders and die bonding methods on the market today. This paper is going to highlight these challenges and the methods used to address them to produce large format, high density IR thermal imaging FPA devices, Quantum processors and micro LED displays using fine pitch micro Indium bump array interconnections that meet today's industry requirements.**

Keywords— Focal Plane Array / Quantum / Super – computing / Fine pitch micro Indium bump / flip chip die bonding / hybridization / Monolithic – integration

I. BACKGROUND

The application material of fine pitch micro Indium bump arrays, such as quantum computer processors, Infrared (IR) thermal imaging focal plane arrays (FPA) and micro LED displays covered by this paper usually consists of one or more components bonded to a substrate using a fine pitch array of vacuum deposited micro Indium bumps.

There are many challenges involved in bringing these components together to form a fully functional assembly, such as co-planarity, accuracy, interconnect yield, mechanical/electrical strength and interconnect quality while maintaining a competitive throughput in the industry.

To ensure a highly repeatable high yield flip-chip bonding process that meets these requirements, a number of solutions were developed and tested to address these challenges. As well as addressing tertiary issues such as handling and cleaning of the material and reducing oxides and organic materials on the Indium or UBM interconnect surfaces were trailed and or solved to reduce the time required to prepare the material for bonding and reduce the risk of damaging the material via manual handling.

In this paper the details and solutions to these challenges will be outlined.

II. FLIP CHIP DIE BONDING OF FINE PITCH MICRO INDIUM BUMP ARRAYS

Die Bonding process

Due to the high cost of the application materials, Infrared (IR) Focal Plane Arrays (FPA) and their Read Out Integrated Circuit (ROIC) components, and Quantum processors as well as the time and difficulty involved to produce high quality functional material, it was vital that the initial bonding trials can produce successful parts with minimal trial and error or failure.

The following processes and preparations as well as software and hardware solutions were used in the trial bonds to achieve the desired results.

The trials described here were all carried out on a FINEPLACER® femto 2 automated die bonder.

Die bonder requirements

Accuracy requirements for bonding 5 µm bump, 15 µm pitch, 640x512 resolution Indium bump array on 10x8mm components:

- Post bond accuracy < 0.5 µm @ 3 sigma over 10-25mm components

- Automatic systematic error correction of the die bonder is required to maintain post bond accuracy over multiple days and high temperature duty cycles within the die bonder

- Controlled handling and accurate fine force control during touching down to components as well as the initial touchdown and force control during the bonding process is critical to maintain post bond accuracy and mitigate any damage to the Indium bump array or sliding / shearing during the bonding.

Bond Process

Bond force

- Low bond forces are required to mitigate damage to the delicate indium bumps. Low forces used in component handling in the die bonder in the range or 0.05 N – 1.0 N showed no visible change to the bump surfaces.

- Fine control of force ramps in the bonding profile were desired to ensure no slipping or shearing occurred during force build up _ 0.1 – 0.5 N/s was used and showed stable results.

Bond profile

The temperature and force bond profile depends on the chosen bonding method.

- Cold compression Bonding = room temp - 90°C
- Thermal Compression Bonding = 100 - 164°C
- Reflow Bonding = 165°C +
- Formic acid reflow bonding = ~210°C
- FAC reduction at ~210°C and bonding at ~165 - 180°C

Material preparation and handling

Compression die bonding of fine pitch micro indium bump arrays usually consists of one or more components, and a substrate. In this paper we will use the example of an Infrared thermal imaging sensor (IR Sensor) and Substrate (ROIC) both containing an interconnect layer of fine pitch micro Indium bump array (Indium bump to bump bonding), as this is the example with the smallest bumps and bump array pitch over the largest surface area, making it one of the more challenging applications to address.

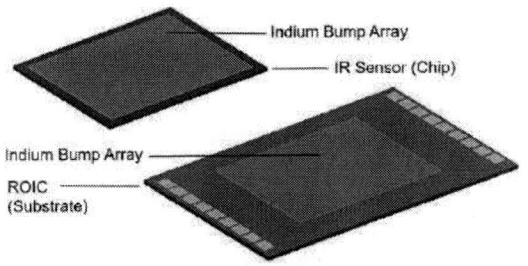

Fig. 1. IR FPA and ROIC components

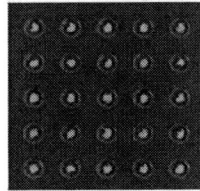

Fig. 2. 5 μm bump, 15 μm pitch Indium bump array

Material

The fine pitch micro Indium bump array is usually grown on the component and/or substrate material using a vacuum-evaporation deposition process supported by a photolithography exposure mask of photoresist. In some cases, after the bump array has been grown on the wafer, the deposition mask is removed and a fresh layer of photoresist is deposited to protect the delicate indium bumps from physical damage, prevent oxidization and prolong the shelf life of the material.

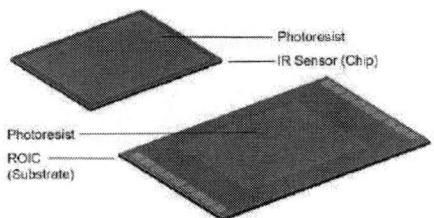

Fig. 3. Photoresist-coated components

Fig. 4. Photoresist-coated Indium bumps

Preparation

The first challenge most encounter when preparing to bond fine pitch micro indium bump arrays, is the removal of the protective photoresist layer and the cleaning and removal of all contamination and particles (Particles > 1.0 μm) from the Indium bump array area. To address this challenge a material cleaning and preparation station was developed to allow for a seamless tweezer free chemical cleaning and material kitting process, to prepare material and packs for automated bonding of multiple assemblies with minimal handling of the components to reduce risk of contamination or damage.

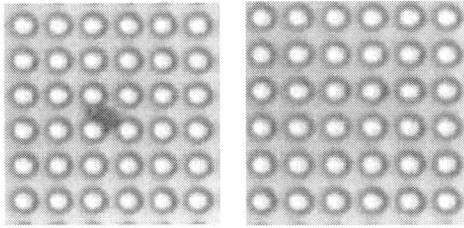

Fig. 5. ~10 μm contamination removed between Indium bumps

Handling

Handling the material during the cleaning, kitting and material to die bonder loading process poses the greatest risk of damaging or contamination of the components and more critically the Indium bump array. To address this challenge a tray handling and loading system that completely removes the requirement of tweezer handling was developed. Once the components have been loaded into the handling trays they can be easily cleaned, inspected, flipped to be presented "Indium up" or "Indium down" for the automated die bonding process and loaded into the die bonder, all without the use of tweezers, this method significantly reduced the occurrence of particle contamination, photoresist and acetone residues while reducing the chance of damaging the indium bump array or dropping the components. This also reduced the material preparation time by ~ 50% for a single assembly, but allows for enough component material to be cleaned and prepared for 4-6 assemblies, which is a significant improvement in process time and risk reduction.

All materials of the handling pack can aggressively be cleaned with acetone without risk of generating particles or contamination. The functional elements of the kit can also withstand high compression bond forces, high reflow bond temperatures and formic acid process gas and plasma cleaning meaning that it is suitable for both cold compression bonding or formic acid reflow bonding.

For standard components (640x512 resolution, 10x8mm) for example, the kit will allow for the simultaneous cleaning, kitting, flipping and loading of 4-6 components and 4-6 substrates to the bonding area.

All clamping and work holding was supplied by vacuum and the pockets were defined to either avoid all contact with the Indium bump array or sensitive areas on the components or limiting the "fall gap" of the component during flipping to <500 μm ensuring that the maximum force exerted on the Indium bump array would only be the weight of the component itself.

Tooling requirements

The tooling required to carry out the automated die bonding of fine pitch micro indium bump arrays will be broken down into tooling for the die handling and tooling for the substrate handling, but the mechanical requirements for both chip and substrate handling and their contact/bonding surfaces are very similar.

Material and Flatness

The materials and as well as the flatness of the tooling and bonding surfaces required to withstand high force and high temperature bonding environments with corrosive processing gases such as formic acid were a critical consideration during this project.

As the technology for fine pitch micro bump growth/deposition develops the tendency is for the interconnect size to decrease as the density increases, meaning that bump height and diameter is steadily heading into the ranges of single microns and potentially sub-micron bumps or interconnects in the near future.

In the more extreme case of IR FPAs with component sizes are getting larger, 20 - 50 mm, coated almost completely in Indium bump arrays of ~5-7 μm bumps, 15 μm pitch, with an ROIC component containing only metallised pads or vice versa, leaving only 3-4 μm bump height across the bond.

Due to this, one of the main requirements of the component handling and die bonding surfaces is flatness. In these extreme cases bonding surfaces need to meet flatness requirements below 0.5 – 1.0 μm over a surface area larger than the components themselves (< 0.5 μm over 20mm for example).

There are a range of solutions to solve the flatness requirement. These range from precision lapped and polished surfaces, compound tools made of two or more components/materials and speciality materials for the bonding interface surfaces depending on the chosen process. For the reflow and formic acid processes materials such as Copper Tungsten (WCu) were evaluated for higher coefficient of thermal expansion (CTE), and composite Tungsten (W) tools were tested for cold compression bonding to preserve flatness..

Vacuum structure

A bonding impression is when the resulting image of a sensor has a "ghost" pattern or regular pattern breaking the uniform image of the sensor. This is usually caused by non-uniform force distributed across the bonding surface.

A common example of this in sensor bonding is the shape of the vacuum structure, used to hold the component on the bond tool, being transferred and seen in the final output image

of the sensor as a result of less force being applied in the gap between tool and component left by the vacuum structure.

Larger vacuum channels can also deform and bend the chip itself into the vacuum channel if the chip is thin enough and the vacuum structure is too wide.

To mitigate this phenomenon, tooling and work holding were all defined with the narrowest and smallest vacuum channel depending on the tool material and component size. This thinner vacuum structure also showed improved surface flatness specifically around the vacuum structures.

Modules

The automated FINEPLACER® die bonder system used for the experiments was fitted with the following modules to address bonding specific challenges. Each module aims to address a specific challenge faced when carrying out cold, reflow and formic acid reflow flip chip bonding of fine pitch micro bumps.

Heated Modules

For thermal compression bonding, reflow bonding and formic acid reflow bonding processes the bonding surfaces of the bond head and substrate carrier may be required to reach the elevated temperatures required to melt and reflow the Indium bumps (165°C +) as well as reaching the activation temperature required to reduce the oxide layer found on the Indium using formic acid (~ 210°C).

Tool Tip Changer Module

To allow multiple tools to be used in a bonding process the tool tip module was used. Allowing the automated loading and unloading of operation specific tools, for example, and the automated systematic error calibration could be carried out using the calibration tool and then the bonder could immediately proceed to the automated bonding process.

This was useful specifically during the high temperature duty cycle processes. To maintain very high accuracy.

Tool Levelling Station for co-planarity

Due to the height of the micro bump array as well as the large component size, co-planarity is vital to the bonding process, to allow for even pressure and a parallel, even bond line across the entire bump array.

To compensate and correct the co-planarity of the tooling, a passive levelling method was used that was observed to correct tool co-planarity to within < 0.5 μm over a 25mm tool surface.

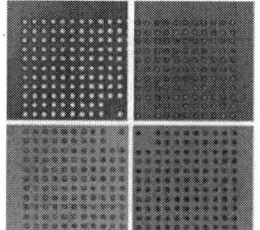
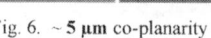
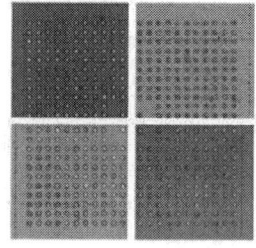

Fig. 6. ~ **5 μm** co-planarity   Fig. 7. < **1 μm** co-planarity

Inert Gas / Formic Acid

For reflow and formic acid reflow bonding processes the bonding area was supplied with inert gases to create an inert environment using Nitrogen (N2), or a processing gas like formic acid to reduce oxides on metal surfaces to expose fresh Indium bonding surfaces/materials.

To reduce oxide on the Indium bumps or to reform the pyramid shaped bumps into more uniform and less amorphous spheres the Indium can be reflowed in a formic acid environment with a laminar flow to carry away the oxides.

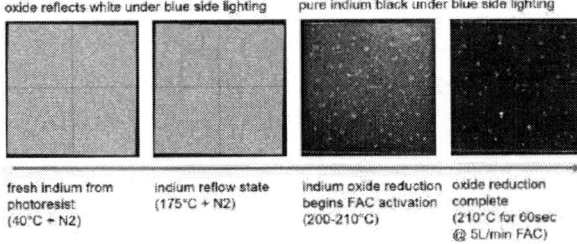

Fig. 8. Indium oxide reduction under FAC process gas

Once formic acid reflow is complete the components can be observed to contain far less oxide, with a majority of the surface being pure Indium or a very thin layer of oxide covering the Indium making for a higher strength and higher electrical quality bond surface.

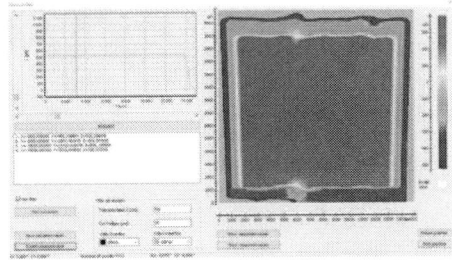

Fig. 9. Indium oxide reduction under FAC and cooled to solid state under N2

Laser Height Measurement

The automated Fineplacer system used for bonding is fitted with a Laser Height Module for various measurement, evaluation and qualification functions.

Fig. 10. Laser height measurement (Surface Height Measurement)

The module can be used to measure and evaluate tool surfaces, bonding surfaces and component surfaces for flatness and co-planarity or the height between the top surface of the substrate and component using 3 or more points in a point height measurement process, or a more detailed line scanning process (pictured above) to generate a topographical map of the surface being measured with a sub-micron resolution.

The results of these measurements were useful when evaluation the trial bond for flatness, co-planarity and estimating the bond line thickness of assemblies.

III. RESULTS

Accuracy

To develop the tooling requirements, bonding process and fine tune the bonding parameters a transparent chip with indium bump array was bonded to an ROIC with indium bump array. This trial was also used to asses post bond accuracy.

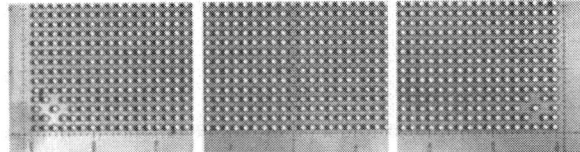

Fig. 11. ~ 4-5 μm offset error_ observed through transparent component 130N over 5 μm bump, 15 μm pitch 10x8mm Indium bump array. Substrate bumps (black) can be seen between chip bumps (white)

Fig. 12. < 1 μm offset error_ observed through transparent component 130N over 5 μm bump, 15 μm pitch 10x8mm Indium bump array. Chip and Substrate bumps perfectly aligned, only top chip bumps (white) can be seen.

By optimizing the tooling, cleaning, process preparation, and bonding parameters in combination with systematic error correction, it was possible to consistently bond the components within 1μm over many trials per day and maintain this quality over extended periods of time ($1 - 2$ weeks) without having to adjust the system.

Flatness and co-planarity

Optimizing the hardware and tooling design, in addition to the passive levelling of the tools enabled consistent, repeatable and controllable flatness and co-planarity in the bonds.

The flatness and co-planarity achieved on the 5 μm bump, 15 μm pitch 8x10mm IR FPA trials was consistently below 1 μm.

 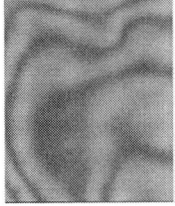

Fig. 13. ~ **5-6 μm** co-planarity error   Fig. 14. ~ **1 μm** co-planarity

(Newton ring effect, each 4 lines is estimated to be ~ 1 μm. Observed through transparent component, 130N over 5 μm bump, 15 μm pitch 10x8mm Indium bump array)

Interconnect Quality

Real IR FPAs were bonded using these methods and underwent functional testing and stress testing for operation in cryogenic environments. Unfortunately these results cannot be shared but the results showed and interconnection yield of above 90 - 95%+ of the components with good dark current characteristics while withstanding high stress duty and temperature cycles.

IV. Discussion

Fine Pitch Micro Indium Bump Arrays

Indium bump arrays are seeing larger adoption in quantum computing with very demanding requirements for co-planarity and control of bond line thickness as well as high quality electrical interconnections. IR FPAs are showing a constant drive for higher resolution meaning smaller bumps, higher density and larger surface areas, and finally the higher consumer market demand for μLEDs or Micro LEDs means that much higher through put will be required for fine pitch indium interconnects.

These requirements will become quite demanding for the die bonders on the market today and will drive further innovation to solve these challenges in novel ways.

V. Conclusion

Fine pitch micro Indium bump bonding is a challenging application, but using the methods described above it was possible to take challenging bonding process and streamline it into a more stable version of a low volume production process, removing risk of damage or failure while improving yield and bond quality in a more controllable and repeatable manner.

The FINEPLACER® femto 2 automated die bonder and the tooling and modules in combination with the cleaning and handling methods developed for this process is an Ideal solution for low level production or RnD development of any fine pitch micro Indium bump array application for IR FPAs, quantum computing processors or μLED displays.

VI. References

[1] International Data Corporation (IDC), (2020): "Worldwide Quantum Computing Spending Guide" https://www.idc.com/getdoc.jsp?containerId=prUS48414121

[2] Quantum Industry Report, (2020): "2020 Quantum Industry Survey" https://www.globenewswire.com/en/news-release/2020/11/10/2123434/28124/en/Global-Quantum-Technology-Market-Report-2020-Much-More-than-Computing-the-Market-will-Reach-21-6-Billion-by-2025.html

[3] Gartner, (2021): "Hype Cycle for Emerging Technologies" https://infocert.digital/analyst-reports/2021-gartner-hype-cycle-for-emerging-technologies/#menuid2

[4] Yole Développement, (2021): "MicroLED Displays - Market, Industry and Technology Trends" https://s3.i-micronews.com/uploads/2021/08/YINTR21162-MicroLED-Displays-Market-Industry-and-Technology-Trends-2021-Sample.pdf

[5] R. K. Bhan (2019): "Recent infrared detector technologies, applications, trends and development of HgCdTe based cooled infrared focal plane arrays and their characterization" http://dx.doi.org/10.1016/j.opelre.2019.04.004

Initial Life Test of Silicone Encapsulated FR4 Printed Circuit Boards for Pre-Clinical Active Implants

Ishpa Ali[è1], Fei Xue[è1], Carlos Perez Henriquez[è1], Thomas Niederhoffer[1], Ahmad Shah Idil[1], Dai Jiang[2], & Henry T. Lancashire[*1]

[1]Department of Medical Physics and Biomedical Engineering, University College London, WC1E 6BT, London, UK.
[2]Department of Electronic and Electrical Engineering, University College London, WC1E 7JE, London, UK.
email: * h.lancashire@ucl.ac.uk

Abstract´ Silicone encapsulated FR4 printed circuit boards may provide a rapid solution for protecting pre-clinical prototype implant electronics. Interdigitated electrodes (IDEs) with and without solder coating were manufactured on Cu-FR4 laminates and silicone encapsulated (N = 14). IDEs were aged in saline and change in impedance was measured. Solder coated IDEs had stable 1 kHz impedances throughout the aging period with promising lifetimes for pre-clinical prototypes. A single uncoated IDE failed with a fall in impedance and verdigris. Other uncoated IDEs showed increasing impedance and dark copper(II) oxide. Failures attributable to contaminants and moisture ingress are under investigation.

Keywords´ life test, encapsulation, printed circuit board, impedance, FR4, solder.

I. INTRODUCTION

Pre-clinical prototype active medical implants for in vitro or animal investigations require miniaturisation, low cost, and rapid solutions for connectors [1], cables, electrode arrays, and printed circuit boards (PCBs). For clinical use active implant electronics must be protected from moisture in the body which can cause corrosion and eventual electronic failure [2]. Hermetic encapsulation with airtight packages made from metals, ceramics, and glasses is commonly used to achieve reliability over long implantation times due to their low gas permeability [3,4]. Non-hermetic polymer encapsulation is an attractive alternative to hermetic packages for prototype preclinical implants. Silicone rubber encapsulation shows high longevity for packaging circuits formed of hermetically packaged components [2], dependent upon the silicone-substrate adhesion, cleanliness, and void-free encapsulation [5,6]. Good silicone adhesion has been achieved to alumina ceramic substrates. To accelerate active implant innovation, it would be valuable to use prototype PCBs in pre-clinical studies including traditional glass-fibre-reinforced epoxy laminate (FR4) materials [7]. Promising lifetimes for silicone encapsulated FR4 PCB implants have been demonstrated [7,8]. FR4-silicone adhesion appears promising under accelerated aging with mean time to failure of 25 days at 100éC in simulated gastric fluid, extrapolated to almost 6 years at 37éC, although FR4 underperforms compared with a ceramic substrate [9]. Despite promising adhesion results we have observed corrosion at metal on FR4 surfaces within aged active implants: Fig. 1 shows copper corrosion and verdigris under silicone observed on an implant immersed in saline and continually operated for 15 days at 100éC and then stored in saline at 21éC for 8 months, corrosion was observed following the 8-month period. We hypothesise that, despite void-free silicone encapsulation, moisture can condense at the Cu-FR4 interface leading to corrosion. This paper investigates the performance of silicone encapsulation of Cu-FR4 PCB prototypes under accelerated aging conditions.

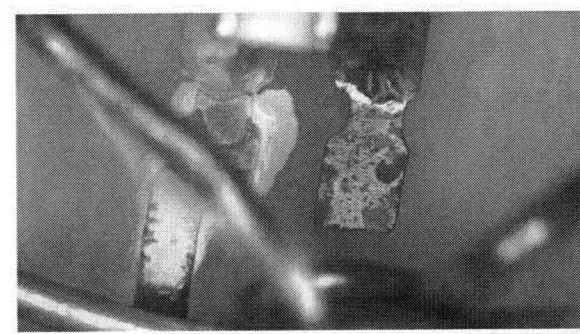

Fig. 1. Corrosion and verdigris observed on a silicone encapsulated Cu-FR4 laminate PCB in a prototype active device following accelerated aging in saline.

II. MATERIALS AND METHOD

A. Materials

Cu-FR4 laminates were acquired from Holders Technology Medical grade silicone (MED-6015) was acquired from Avantor-NuSil. MED-6015 was chosen due to its low viscosity and promising performance as an encapsulant for integrated circuits. Polishing paper up to P1200 was acquired from Agar Scientific. Solder with water soluble flux (HYDX 60EN, Multicore/Locktite), and additional water soluble flux were used (ORH1, CW8300, Chemtronics). Phosphate buffered saline was prepared from deionized water (ħ13 M'n.cm) and tablets (Sigma Aldrich) to give pH = 7.4.

B. Sample Preparation

Copper interdigitated electrodes (IDEs) were produced from 0.8 mm thickness Cu-FR4 laminates by photolithography and wet etching. IDEs are coplanar electrode pairs between which impedance or leakage current may be measured. IDE dimensions were: 40 mm B 0.3 mm interdigitated tracks; 9 interdigitations per electrode; 0.3 mm inter-track spacing; and 50 mm B 15 mm outer dimensions, Fig. 2. A surrounding shield conductor was used in addition to the IDE electrode pair. IDEs were not solder mask coated, leaving conductive copper tracks exposed, Fig. 2. IDEs were either solder coated, or uncoated copper (n=6 and n=8 respectively). Prior to solder coating copper samples were cleaned by burnishing with polishing paper. All samples were soldered to silicone coated stranded copper wire at test pads. Immediately following soldering samples were cleaned in warm deionized water to remove flux. All samples were cleaned by ultrasonication in a detergent (Teepol-L) and sodium phosphate solution [6,9,10], followed by further

---

è These authors contributed equally to this work.

Fig. 2. a) The interdigitated electrode design. b) As produced interdigitated electrodes with (left) and without (right) solder coating. c) Silicone encapsulated interdigitated electrode.

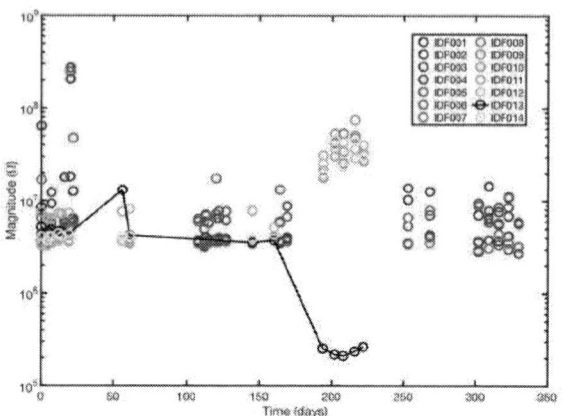

Fig. 3. 1 kHz Impedance change with time for solder coated (ID001 and 003 to 007) and uncoated (ID002 and 008 to 014) Cu-FR4 interdigitated electrodes with silicone encapsulation.

ultrasonication in deionized water, acetone, and isopropyl alcohol. Samples were rinsed with deionized water, and cleanliness was accepted at conductivities less than 120 nS.cm^{-1}. Connecting wires were surrounded with a silicone tube which was embedded into the silicone encapsulation at the sample end [11]. Immediately prior to encapsulation sample surfaces were cleaned with air plasma for 2 minutes at 0.5 mBar. Cleaned IDEs were encapsulated with medical grade silicone by centrifuge vacuum moulding to ensure conformal, void-free, coating of the sample surface [1]. Silicone was cured at 80éC for at least 4 hours to ensure complete curing. Connecting wires were soldered to PEEK bungs and exposed metal surfaces were protected with room-temperature-vulcanizing silicone as previously described [11].

C. Sample Aging

Silicone encapsulated IDEs were aged at 21éC and 67éC in phosphate buffered saline in a previously developed accelerated ageing and life-test apparatus [11]. Changes in aging temperature were due to equipment availability. Solder coated IDEs were observed for up to 330 days, and uncoated Cu IDEs were observed for up to 222 days. Solder coated IDEs (and uncoated sample 2) were initially aged at 67éC for 1 month, followed by: 86 days not in saline; 194 days aging at 21éC; and 1 month at 67éC. Uncoated Cu IDEs (except sample 2) were initially aged for 194 days at 21éC followed by 1 month at 67éC.

D. Electrochemical Impedance Spectroscopy

Electrical impedance spectroscopy (EIS) was repeated regularly during aging. EIS was carried out with either a Wayne-Kerr 6500B Impedance Analyser during 21éC aging or a Solartron Modulab XM potentiostat and frequency response analyser during 67éC aging. EIS was carried out from 20 Hz to 10 kHz with 50 mV $_{p-p}$ excitation amplitude with the WK 6500B, and from 10 mHz to 100 kHz with 50 mV $_{p-p}$ excitation with the Modulab XM.

E. Moisture Absorption

The moisture absorption of Cu-FR4 laminate IDEs and silicone encapsulated IDEs was measured by change in mass. Dried samples (n = 5) were placed in deionized water at 21éC, change in mass was measured at intervals, sample surfaces

were dried using a lint-free cloth to remove surface water prior to each measurement. Sample dry mass was measured following drying in a conventional oven followed by a vacuum oven for at least 2 hours.

F. Data Analysis

Impedance at 1 kHz was calculated for all samples throughout the measurement period for comparison. Where given data are mean ě standard deviation.

III. RESULTS

A. Impedance Changes

Change in 1kHz impedance with time is plotted in Fig. 3. Solder coated IDEs were observed to have a stable 1 kHz impedance throughout the aging period, whether aged at 21éC or 67éC. Initial 1 kHz impedance was 4.45 ě 1.34 M'n, after 330 days 1 kHz impedance was 4.94 ě 1.34 M'n.

For uncoated Cu IDEs 1 kHz impedances were relatively constant during aging at 21éC for 194 days. Initial 1 kHz impedance was 4.36 ě 1.52 M'n. Following aging at 67éC

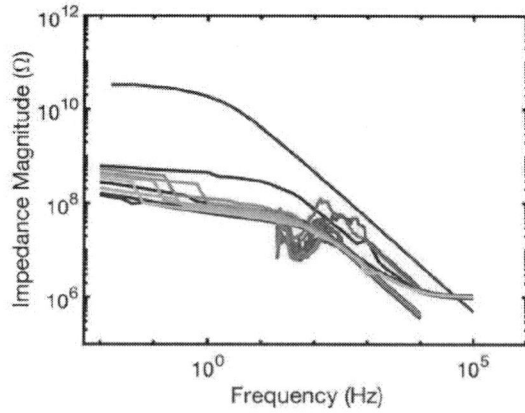

Fig. 4. Example impedance spectrum from a single solder coated Cu-FR4 interdigitated electrode (ID007). Time increase is shown as change from darker (purple) to lighter (orange). Measurements made with WK6500B have a narrower frequency range.

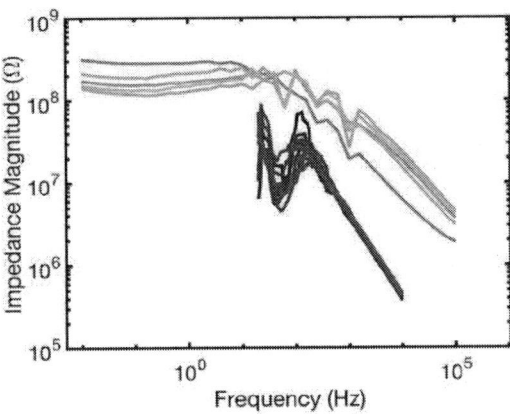

Fig. 5. Example impedance spectrum from a single uncoated Cu-FR4 interdigitated electrode (ID008). Time increase is shown as change from darker (purple) to lighter (orange). Measurements made with WK6500B have a narrower frequency range.

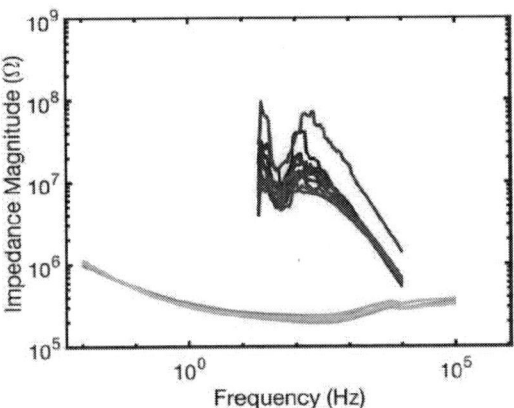

Fig. 6. Impedance spectrum from ID013 showing impedance decrease with increasing time (from darker purple to lighter orange). Measurements made with WK6500B have a narrower frequency range.

Fig. 7. Micrograph of bright surface on a solder coated Cu-FR4 interdigitated electrode with silicone encapsulation following aging. Imaged at 6.4Ⅹ magnification.

changes in uncoated Cu IDE impedance were observed: for most of the sample impedances were observed to increase to 31.2 ě 4.81 M'n after 222 days total aging; however, for a single sample impedance was observed to decrease to 264 k'n at 222 days.

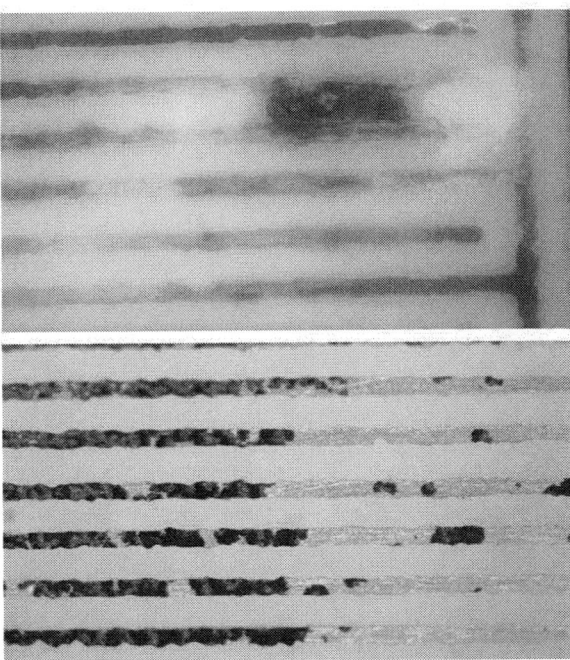

Fig. 8. Micrographs of uncoated Cu-FR4 interdigitated electrodes with silicone encapsulation. Above, ID013 with virdigris and precipitation at the FR4-silicone interface. Below, dark copper(II) oxide discoloration on the copper surface. Imaged at 16Ⅹ magnification.

Example impedance spectra are given for solder coated (Fig. 4) and uncoated (Fig. 5) IDEs which were observed to be representative of the respective groups. At high frequency impedance was observed to fall with increasing frequency, representative of the IDE capacitance. Low frequency impedance for both sample types was observed to plateau at >100 M|, this plateau was not observed with WK6500B measurements due to the instrument minimum frequency, and noisy spectra due to accuracy at high impedance / low current. The solder coated IDE spectrum showed a high frequency plateau when measured during 67éC aging, which was not observed in the limited frequency range during 21éC measurements. No clear difference in solder coated IDE impedance was observed between impedances at 1 kHz measured using different instruments. Uncoated sample 1 kHz impedance measured during 67éC aging (with Modulab XM) was approximately 1 order of magnitude greater than measured at 21éC (with WK6500B).

The impedance spectrum of the single uncoated IDE which was observed to decrease during aging is shown in Fig. 6. Impedance at low frequency was observed to fall with time during 21éC aging. All spectra recorded during 67éC aging were more representative of a resistive response, with constant phase element-like response at low frequencies and a possible inductive contribution at high frequencies.

B. Sample Observations

Solder coated IDEs were observed to have bright solder throughout the aging period, Fig. 7; however, surrounding regions of uncoated copper, not forming part of the IDE, were observed to have dark discolorations attributable to copper(II) oxide.

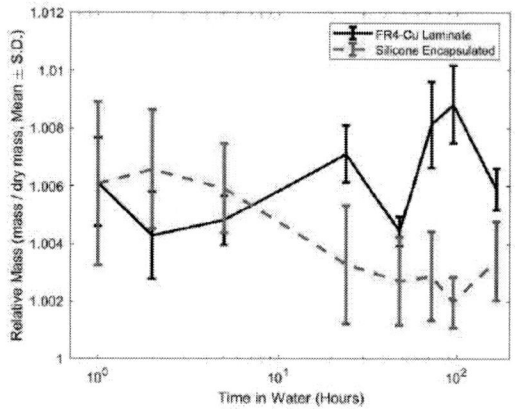

Fig. 9. Relative mass over time in deionized water for FR4-Cu laminate and silicone encapsulated IDEs.

Visual inspection showed that the single uncoated Cu IDE with reduced impedance was exhibiting verdigris, on the copper surface, with corrosion products precipitated between tracks at the FR4-silicone interface, Fig. 8. On inspection other uncoated Cu IDEs were observed to have dark copper(II) oxide at the copper surface, Fig. 8.

C. Moisture Absorption

Following soaking in deionized water for 168 hours the change in bare FR4-Cu laminate sample mass was 1.0059 ± 0.0007 relative to dry mass (unitless, calculated as $mass_{wet} / mass_{dry}$) and silicone encapsulated sample mass was 1.0034 ± 0.0014. FR4-Cu laminate absorbed $0.59\pm0.07\%$ moisture by mass, and silicone $0.34\pm0.14\%$ moisture by mass. Some variation in moisture absorption was observed over time, Fig. 9.

IV. Discussion

Silicone coated Cu-FR4 IDEs with silicone encapsulation were aged in saline with heat acceleration. Solder coated IDEs were observed to have stable 1 kHz impedance over 330 days including 2 months accelerated aging at 67éC, Fig. 3. Uncoated IDE 1 kHz impedance was observed to vary: most samples showed an impedance increase during 67éC aging which must be investigated further; and a single sample showed an impedance decrease associated with copper oxidation, Fig. 8. Copper oxidation in the single uncoated sample appears similar to the failure observed within a prototype implant continually operated for 15 days at 100éC, Fig. 1. We hypothesise that the observed verdigris and copper(II) oxide formation is due to moisture ingress into voids due to FR4 porosity or at the Cu-FR4-silicone interface at the edge of copper tracks, local corrosion and resulting osmotic pressure led to FR4-silicone delamination and leakage paths between the electrodes and a reduction in impedance [4]. We observed that both FR4-Cu laminates and silicones absorb water and this can be expected to change the dielectric properties and therefore IDE impedance. Sectioning and elemental analysis of failed and as produced samples will be essential to check for the presence of voids at the FR4-Cu interface and for contaminants and corrosion products. The observed verdigris is also attributable to insufficient surface cleaning [6], where contaminants act as a water nucleation

point, and indicates the presence of chloride, sulphate or carbonate ions. Solder coating achieved increased IDE lifetimes with no failures apparent and is therefore promising for pre-clinical prototype implants where short device lifetimes (months) are required, for example for basic neuroscience. The survival of solder coated IDEs does not support the hypothesis of failure due to FR4 porosity, which would still be present beneath the copper layer, instead the solder may cover and smooth pinholes or other irregularities in the copper surface which are not filled during silicone encapsulation. Silicone rubber is a promising material due to its observed adhesion to FR4 [9]. However, alternative encapsulation methods can provide sufficient lifetimes for pre-clinical prototype active implants including epoxies, parylene C, polyimide, and liquid crystal polymers [2,12].

Acknowledgment

The authors thank N. Donaldson for equipment access.

References

[1] H. T. Lancashire, M. Habibollahi, D. Jiang, and A. Demosthenous, `Evaluation of Commercial Connectors for Active Neural Implants,_ in 2021 10th International IEEE/EMBS Conference on Neural Engineering (NER), May 2021, pp. 973¯976. doi: 10.1109/NER49283.2021.9441072.

[2] T. Stieglitz, `Implantable Device Fabrication and Packaging,_ in Handbook of Neuroengineering, N. V. Thakor, Ed., Springer Nature, 2023, pp. 289¯337. doi: 10.1007/978-981-16-5540-1_102.

[3] R. Traeger, `Nonhermeticity of Polymeric Lid Sealants,_ IEEE Transactions on Parts, Hybrids, and Packaging, vol. 13, no. 2, pp. 147¯152, Jun. 1977, doi: 10.1109/TPHP.1977.1135193.

[4] A. Vanhoestenberghe and N. Donaldson, `Corrosion of silicon integrated circuits and lifetime predictions in implantable electronic devices,_ J. Neural Eng., vol. 10, no. 3, p. 031002, May 2013, doi: 10.1088/1741-2560/10/3/031002.

[5] P. E. K. Donaldson, `The essential role played by adhesion in the technology of neurological prostheses,_ International Journal of Adhesion and Adhesives, vol. 16, no. 2, pp. 105¯107, May 1996, doi: 10.1016/0143-7496(95)00031-3.

[6] P. Kiele, J. Hergesell, M. Bahler, T. Boretius, G. Suaning, and T. Stieglitz, `Reliability of Neural Implants´ Effective Method for Cleaning and Surface Preparation of Ceramics,_ Micromachines, vol. 12, no. 2, Art. no. 2, Feb. 2021, doi: 10.3390/mi12020209.

[7] D. Jiang et al., `A Versatile Hermetically Sealed Microelectronic Implant for Peripheral Nerve Stimulation Applications,_ Frontiers in Neuroscience, vol. 15, 2021, doi: 10.3389/fnins.2021.681021.

[8] L. Lonys et al., `Design and implementation of a less invasive gastrostimulator,_ European Journal of Translational Myology, vol. 26, no. 2, Art. no. 2, Jun. 2016, doi: 10.4081/ejtm.2016.6019.

[9] L. Lonys et al., `Silicone rubber encapsulation for an endoscopically implantable gastrostimulator,_ Med Biol Eng Comput, vol. 53, no. 4, pp. 319¯329, Apr. 2015, doi: 10.1007/s11517-014-1236-9.

[10] A. Carnicer-Lombarte, H. T. Lancashire, and A. Vanhoestenberghe, `In vitro biocompatibility and electrical stability of thick-film platinum/gold alloy electrodes printed on alumina,_ J Neural Eng, vol. 14, no. 3, p. 036012, Jun. 2017, doi: 10.1088/1741-2552/aa6557.

[11] N. Donaldson, C. Lamont, A. S. Idil, M. Mentink, and T. Perkins, `Apparatus to investigate the insulation impedance and accelerated life-testing of neural interfaces,_ J. Neural Eng., vol. 15, no. 6, p. 066034, Oct. 2018, doi: 10.1088/1741-2552/aadeac.

[12] S. L. C. Au, F.-Y. B. Chen, D. M. Budgett, S. C. Malpas, S.-J. Guild, and D. McCormick, :Injection Molded Liquid Crystal Polymer Package for Chronic Active Implantable Devices With Application to an Optogenetic Stimulator´, IEEE Transactions on Biomedical Engineering, vol. 67, no. 5, pp. 1357¯1365, May 2020, doi: 10.1109/TBME.2019.2936577.

Voiding in Parylene-C Encapsulation of Surface Mount LEDs for an Optogenetic Epilepsy Neuroprosthesis

1st Ahmad Shah Idil
Medical Physics & Bioengineering
University College London
London, United Kingdom
a.shahidil@ucl.ac.uk

2nd Richard Bailey
Electrical & Electronic Engineering
Newcastle University
Newcastle, United Kingdom
richard.bailey2@newcastle.ac.uk

3rd Enrique Escobedo-Cousin
Electrical & Electronic Engineering
Newcastle University
Newcastle, United Kingdom
enrique.escobedo-cousin@newcastle.ac.uk

4th Johannes Gausden
Electrical & Electronic Engineering
Newcastle University
Newcastle, United Kingdom
johannes.gausden@newcastle.ac.uk

5th Antony O'Neill
Electrical & Electronic Engineering
Newcastle University
Newcastle, United Kingdom
anthony.oneill@newcastle.ac.uk

6th Nick Donaldson
Medical Physics & Bioengineering
University College London
London, United Kingdom
n.donaldson@ucl.ac.uk

Abstract—It is often assumed that Parylene-C encapsulation films are void-free & conformal by nature of their vapour deposition. We present a case study in an optrode device where voiding occurs; we argue the voids originate from the geometry of the substrate and from the use of surface mount components. The segment under study consists of a commercial LED bonded with Au/Au thermosonic bonding onto a custom silicon substrate. We demonstrate via micro-sectioning that there is polymer voiding behaviour (1) in sub-LED surfaces; and (2) within through-hole vias in the silicon substrate (channels designed to allow polymer vapour ingress during deposition). By making a comparison to the solutions found in the IC industry around tungsten vapour filling of blind vias, we present geometric solutions for failure mitigation.

Keywords—optrodes, neuroprostheses, Parylene-C, voids, encapsulation, micropackaging

I. INTRODUCTION

The CANDO (Controlling Abnormal Neural Dynamics with Optogenetics) project aimed to treat focal epilepsy with an optogenetic brain implant and its associated gene therapy [1]. The brain implant consisted of an array of 16 optrodes which was to be implanted in the epileptic focus. Each optrode was made of silicon and was designed to penetrate cortical tissue. Discrete surface mount commercial LEDs were chosen as the light source due to their high efficiency (~40%), low cost, and low form factor (50μm thick). Prior investigations into custom "super thin" LEDs (3-5μm thick) with an outer seal ring cathode had inadequate efficiency (~3%) (Fig. 1). Monolithic LEDs built into the substrate such as those in [2] were outside the scope of our manufacturing capability.

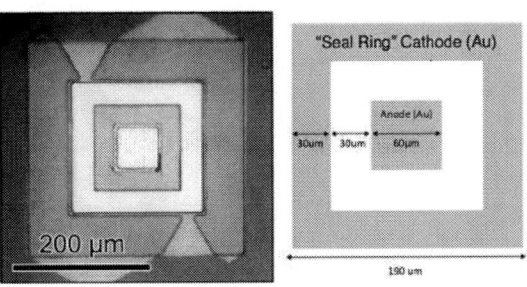

Fig. 1. Custom LEDs from our collaborator at the University of Strathclyde (not used in final implant due to low efficiency); note the "seal ring" cathode, intended to seal out moisture from diffusing in and shorting the anode and cathode.

The commercial LEDs chosen had only a spacing of 50μm between the two bond pads, and were any moisture present, device failure could occur. Thus, a method of encapsulating not only the surface mount LED, but also more importantly its anode and cathode would be required in the wet, corrosive environment of the human body. The device was protected by Parylene-C encapsulation, an established method of protecting electronics intended for use in implanted devices, such as neuroprostheses. The polymer is used due to its properties of chemical stability, non-toxicity & its ability to create low stress conformal films [3].

For polymer encapsulants applied in liquid phase, such as liquid silicone rubber, techniques such as vacuum centrifuging have been developed to ensure the encapsulant covers all surfaces and is void-free [4]. For encapsulants applied in vapour phase, such as Parylene-C, the conformal films are often assumed to be void-free; this paper examines this assumption and examines its consequences. A comparison is made to the voiding behaviour of tungsten films in blind silicon vias to form an understanding of the phenomenon [5].

II. THEORY

A. Water Permeability in Polymers & Diffusion in Voids

While gravimetrically measured water uptake (as a function of immersion time & temperature) is often used as a measure of a polymer's moisture sensitivity, a more useful and thorough measure is its water permeability. A polymer's water permeability is a function of its: (1) diffusion coefficient; and its (2) solubility (or 'partition') coefficient. Both parameters are temperature & pressure dependent. From [7] the formula for water permeability (P) is:

$$P = \frac{q\tau}{At(VP_1 - VP_2)} = Q_f\tau \qquad (1)$$

where q is the quantity of gas diffusing through a film of thickness τ, area A, in time t, and with a partial pressure difference between the diffusing gas on either side of the membrane as $(VP_1 - VP_2)$.

If we take the human cortex, or any implant environment generally, we may globally assume 100% humidity, constant temperature (37°C), and constant pressure (1 atm).

Locally, we may find minimal variations in temperature originating from (1) natural body fluctuations; and (2) from heat generated from a device. The latter is limited to 1°C increase in the cortex by regulation. Thus, we may assume the temperature-dependency to be constant and ignore.

On the other hand, a local pressure gradient across a polymer encapsulation, i.e. ($VP_1 - VP_2$) from Eq. 1, may be developed by the following phenomena: (1) osmotic pressure from ionic contaminants on the surface of the device, which would form a concentration gradient, drawing in water; (2) applied external pressure; and (3) a polymer void formed under vacuum creating a negative pressure gradient when the device is then exposed to atmosphere.

Regarding (1): the osmotic pressure of various surface contaminants has been explored in [6]; this is controlled for by rigorous surface cleansing. Regarding (2): this would be an unusual situation for a neuroprosthesis to experience. The third case (3) is of most significance to us – a void formed under vacuum will, due to the pressure difference across the encapsulation (the membrane), cause water vapour to diffuse through the membrane, which will then condense into liquid water, through which current leakage may occur.

Table 1 below taken from [7] lists the water permeability constants of various common polymers. Parylene-C is often referred to as a "moisture barrier" [8]; while it is an order of magnitude less permeable than silicone, it is still water permeable, as all polymers are, due to their network structure.

TABLE I.   WATER PERMEABILITY CONSTANTS[1] [7]

Polymer	Permeability Constant
Silicones	$2.4 - 4.3 \times 10^{-8}$
Poly(chloro-p-xylylene) (Parylene -C)	2.0×10^{-9}
Cellulose acetate	$1.2 - 3.1 \times 10^{-6}$
Polystyrene	1.2×10^{-7}
Polyurethanes	$1.3 - 4.7 \times 10^{-8}$
Low-density polyethylene	$1.1 - 1.8 \times 10^{-8}$
High-density polyethylene	$1.6 - 1.8 \times 10^{-9}$
Epoxies	$1.0 - 1.3 \times 10^{-8}$
Poly(ethylene terephthalate) (Mylar)	1.0×10^{-8}
Polypropylene	$3.8 - 4.0 \times 10^{-9}$
Poly(tetrafluoroethylene) (Teflon)	3.5×10^{-9}
Poly(vinylidene chloride) (Saran)	3.1×10^{-9}

The unit of permeability here is defined as $P = \frac{cm^3_{STP} \cdot cm}{cm^2 \cdot s \cdot cmHg}$ where cm^3_{STP} is the 'standard cubic centimeter' (a unit of amount of gas rather than a unit of volume). The equivalent SI unit would be the GPU (gas permeance unit) where $1\ GPU = 3.35 \frac{10^{-10} mol}{m^2 \cdot s \cdot Pa}$.

In dimensional analysis, permeability is an L^2/T unit, and thus theoretically, it would take 10 times longer for a void in Parylene-C to fill with water than one in silicone, all other factors equal.

B. Chemical Vapour Deposition of Parylene-C

Parylene-C, or poly(chloro-para-xylylene) is formed from the reaction of chloro-para-xylylene monomers into a polymer with a backbone chain of chloro-para-benzenediyl rings (–C_6ClH_3–) connected by 1,2-ethanediyl bridges (–CH_2–CH_2–). The process of deposition is called the Gorham process, a chemical vapour deposition (CVD) process involving 3 steps: sublimation, pyrolysis & deposition. Firstly, the precursor Parylene-C dimer is sublimed; the dimer is then pyrolised into 2 monomers at high temperature; finally, the monomers polymerise onto contact surfaces of the device at room temperature into linear polymer chains [9].

C. The Sub-LED Through-Hole Via

The creation of a through-hole via underneath the LED between the anode & cathode pads was introduced (1) to prevent shorts between the anode/cathode; and (2) as a technique for enhancing the Parylene-C's encapsulation by allowing a pathway for flow of the monomer vapour.

Because the LED sits on top of this through-hole via, it effectively turns into a blind via; thus we may make a comparison to the work in [5] which is on voiding in tungsten-filled blind vias for IC manufacture.

D. Step Coverage

Step coverage is the parameter that describes the conformality of a set of deposition conditions, where at 100% step coverage every surface exposed to the gas phase experiences identical deposition rates and is thus ideally conformal and voidless [5]. Voidless deposition requires sufficient mass transport of reactants into contact surfaces, and removal of reactants away from surfaces. For a blind via, step coverage can be defined in two ways:

$$step\ coverage = 1 - A(L_0/R_0) \qquad (2)$$

where A is a function of the deposition conditions (which for Parylene-C deposition are a function of temperature and pressure of Parylene-C monomer vapour); R_0 & L_0 are the initial radius and depth of the via respectively. Alternatively, step coverage can be experimentally determined by cross sections, via the formula:

$$step\ coverage = (x/y) * 100\% \qquad (3)$$

where x is the wall thickness of the void, and y is the maximum thickness of the deposited film inside the void (see Fig. 2 below).

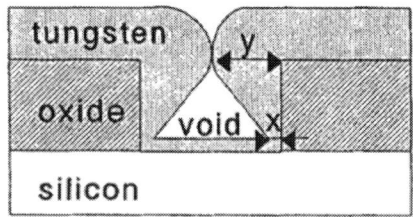

Fig. 2.   Definition of step coverage of a film in a blind via using the example of metallic tungsten films in silicon [5].

---

[1] Measured at 23°C on 3-8×10⁻³ cm thick films.

Parylene-C deposition produces no by-products, and thus in terms of mass transport considerations, the accumulation of depleted reaction products is not a consideration. In comparison, the vapour deposition of tungsten relies on the reduction of tungsten hexafluoride (WF_6) with hydrogen gas (H_2) yielding metallic tungsten with a by-product of hydrogen fluoride vapour (HF) which must be continuously exhausted in order for film deposition to proceed forward [10].

III. METHOD

A. Device Fabrication

Optrode devices were fabricated in-house in the Newcastle University cleanrooms on a silicon substrate (see Fig 3). The LEDs used were Cree® TR2227™ (CreeLED, USA). There were 2 LEDs & 2 recording electrodes on each optrode tine. Appropriate design of photoresist layer (SPR 220) allowed the growth of a 5-8um pillar of gold over the anode and cathode pads using electroplating. A thermosonic fine-placer tool was then used (150°C, 1.5N) to bond each LED to its pads (300mW, 1s). The sub-LED via was created using DRIE during the same step as device singulation. The gap between the anode and the cathode pads was 50µm. The via was ⌀50µm in diameter.

B. Parylene-C Encapsulation

Device encapsulation was carried out in a Labcoater Parylene deposition system (PDS Labcoater 2010, Speciality Coating Systems Inc, USA). 17.5g of Parylene-C precursor powder was used per deposition run; this was repeated twice for a goal film thickness of 20µm. Adhesion promoter (A-174 silane) was added to the chamber before each deposition. The dimer powder was sublimated (175°C, 1.0 torr) and passed through a furnace stage for monomerisation (690°C, 0.5 torr) before low-vacuum deposition (20 min, 25°C, 0.1 torr).

C. Micro-Sectioning

Device micro-sectioning was performed via a commercial service called Perfect Edge™ (MCS Ltd, UK). It is a non-mechanical technique that uses an inert gas plasma. Images were produced with SEM. Devices were micro-sectioned along the longitudinal axis of the optrode (see Fig 3).

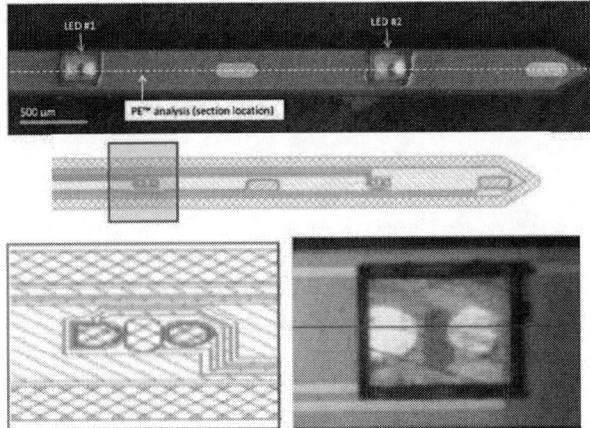

Fig. 3. TOP: photograph of one Parylene-C coated optrode tine with 2 LEDs & 2 oval recording electrodes visible; MIDDLE: KLayout view of same optrode tine with box around detail view; BOTTOM-LEFT: KLayout detail view of the anode & cathode bonding pads, and the sub-LED through-via; BOTTOM-RIGHT: microscope image of LED #1 on the device shaft, with the red line the direction of the micro-section cut.

IV. RESULTS

We present results for the micro-sections of two LEDs: LED #1 and LED #2 (see Fig. 5-10).

For LED #1, the Parylene-C encapsulation was observed to be conformal and coated all surfaces, though unequally. Two kinds of voids were observed: (1) a larger void in the via underneath the LED, approximately 12.5µm wide and 30µm deep (see Fig. 5); and (2) a thin void underneath the LED approximately 0.8µm thick and 10µm long (see Fig. 6). Additionally, around voids were visible around the gold bond pads (see Fig. 7).

For LED #2, the Parylene-C encapsulation was observed to be conformal on the top surfaces, and the through-hole via was void-free (see Fig. 8). However, on the underside of the LED, and around the gold bond pillars large voids were observed: voids 1.8µm thick and 60µm long (see Fig. 9, 10).

V. DISCUSSION

A. Void Formation

The failure of the Parylene-C vapour to continue polymerising in volumes narrower than ~1-2µm suggests that the reaction is preferentially occurring elsewhere. As there are no reaction products that need to be evacuated, it is likely that the monomer is simply depleted in those spaces where we can observe voids. We can deduce that the vapour pressure in the chamber is inadequate to supply new reactants. Thus, for our deposition parameters a geometric limit (~1-2µm) exists.

Furthermore, the void in the sub-LED through-hole via is similar to that reported in silicon vias filled using tungsten chemical vapour deposition as described in [5]. Figure 4 is intentionally reproduced from [5] upside-down; this is to draw a comparison to Figure 5, whereby in the former, a blind via in silicon is incompletely filled with metallic tungsten, and in the latter, a through-hole via becomes a blind via as a result of being blocked by an LED, and this pseudo-blind via is similarly incompletely filled with Parylene-C.

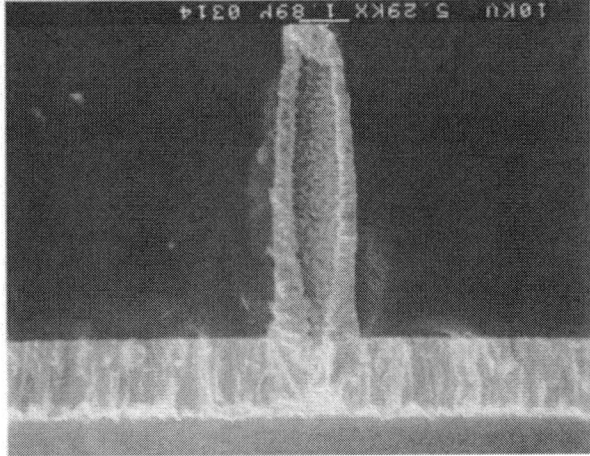

Fig. 4. A large void in metallic tungsten film for via filling of a blind via in silicon. (Intentionally presented upside-down by the author.) [5]

In LED #1, the aspect ratio of the via was 50/200µm = 0.25 (i.e., $R_0 = 25$µm & $L_0 = 200$µm). The void in the via had a maximum width of 12.5µm; thus, the step coverage = (50-12.5)/50 = 75% (using Eq. 3). Substituting this into Eq. 2 resulted in A = 0.03125.

117

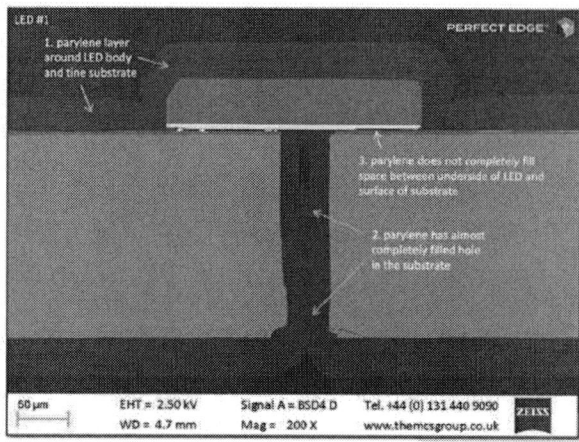

Fig. 5. LED #1 (overall view) – A void in the sub-LED through-hole via approximately 12.5μm wide and 30μm deep.

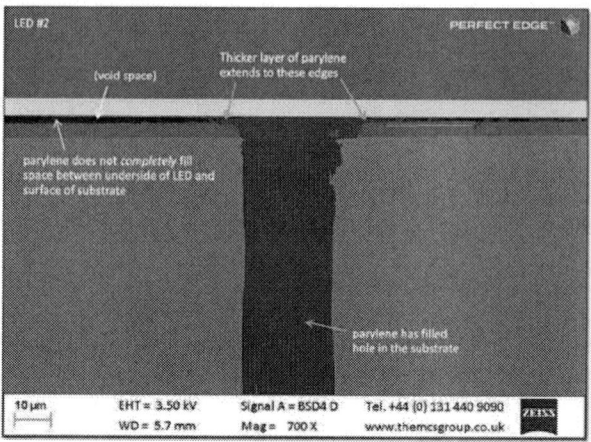

Fig. 8. LED #2 (overall view) – The sub-LED through-hole via is void-free; compare to Fig. 5 where there is a void.

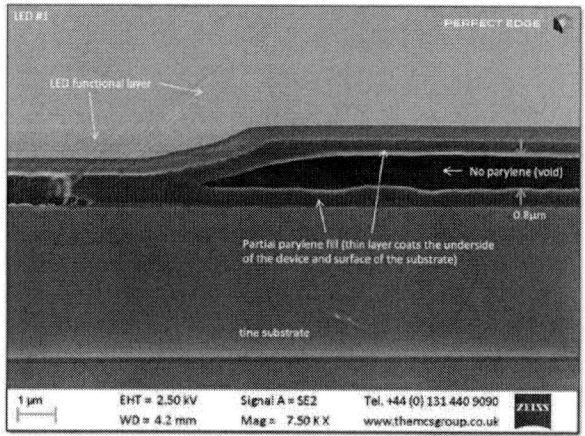

Fig. 6. LED #1 (underside view) – a thin void on the underside of the LED ("sub-LED") 0.8μm thick and approximate 10μm long.

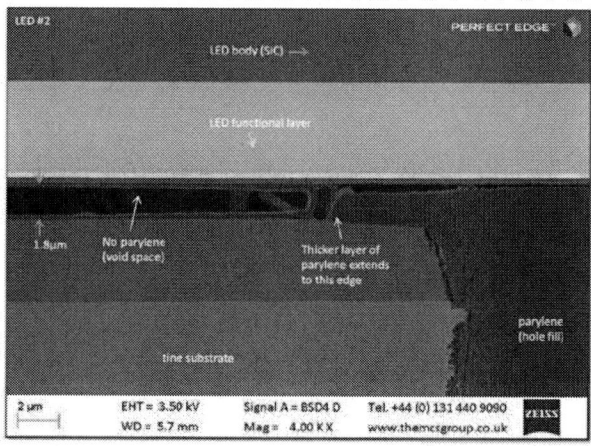

Fig. 9. LED #2 (underside view) – a thin void on the underside of the LED ("sub-LED") 1.8μm thick and approximate 60μm long.

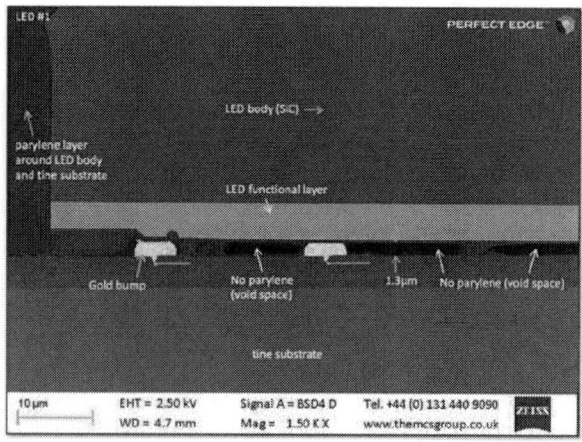

Fig. 7. LED #1 (gold pad view) – Around the critical areas of the gold pads, voids are present and the metal is not fully encapsulated.

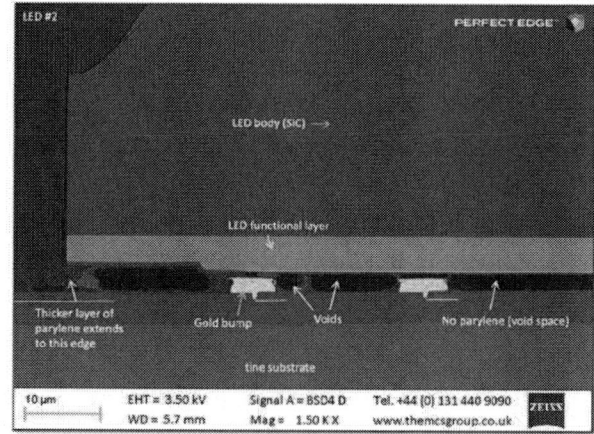

Fig. 10. LED #2 (gold pad view) – Around the critical areas of the gold pads, voids are present and the metal is not fully encapsulated.

By utilising the concept of step coverage introduced in [5] we can offer some geometric solutions to this problem.

B. Geometric Solutions

For applications that require surface mount components with adjacent bond pads, where designers would like to use sub-component through-hole vias between pads to aid in encapsulation, where deposition parameters are fixed, we present some solutions to the voiding problem in and Figure 11 & Table 2 below:

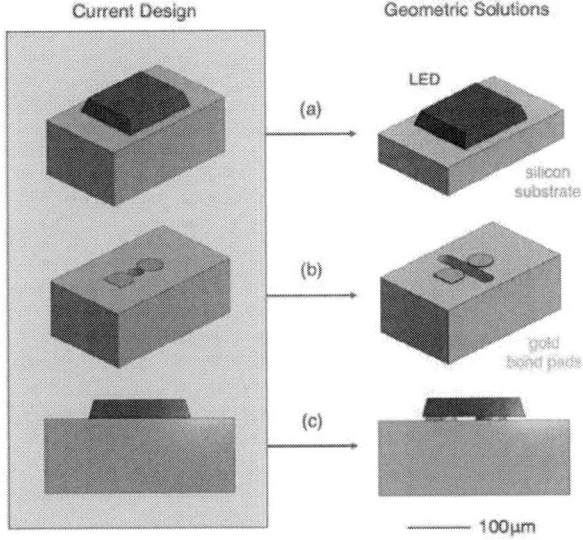

Fig. 11. Hypothetical 300µm long segments of the optrode; 3 geometric solutions to voids in Parylene-C film: (a) substrate thinning; (b) via-widening; and (c) gold bond pad heightening.

TABLE II. GEOMETRIC SOLUTIONS TO VOIDS

Solution	Advantages	Disadvantages
substrate thinning	reduces overall device profile; increases step coverage by decreasing L_0	additional wafer level process; weakens the device for insertion
via-widening	increases step coverage by increasing effective R_0; no longer blind-via, acts as through-hole for gas flow	large stress concentrations at thin edges of shank; much weaker for buckling during insertion
gold bond pad heightening	no longer blind via, larger path for vapour to flow; void-free sub-LED encapsulation	increases overall device profile; difficult to electroplate multiple microns of gold

With substrate thinning we can increase the aspect ratio of vias (e.g., thinning the substrate by half to 100µm would increase aspect ratio from 0.25 to 0.5); for the calculated value of A (0.03125) the step coverage would increase from 75% to 93.75% and thus the maximum void width would theoretically decrease from 12.5 to 1.6µm. The disadvantage of substrate thinning is the technical difficulty of the process, and the weakening of the device mechanically.

Via-widening beyond the edges of the surface mount component is intended to allow the via to behave as a true through-hole via, rather than as a pseudo-blind via.

Theoretically, the Parylene-C vapour would have more of a pathway to flow. The weakness of this is that it would induce large stress concentrates in the shank.

Lastly, gold pad heightening is intended to overcome the ~1-2µm geometric limit of the deposition process, by lifting the component higher off the substrate (10µm in Fig. 11). The main disadvantage is the increase in the device profile, which is not desirable for an implantable device.

A technical solution may require one or multiple of these modifications to an existing design.

VI. CONCLUSION

We cannot assume that Parylene-C vapour will form conformal, void-free encapsulations over all surfaces in surface mount components. Specific geometries and deposition parameters may lead to voids in vias and in the undersides of surface mount components. For devices used in harsh environments, such as implanted devices, these voids may lead to device failure in the event of moisture ingress. For polymer encapsulation to be successful in these applications, we must predict such voiding and mitigate with geometric solutions, such as those suggested in this paper.

ACKNOWLEDGMENT

The authors extend their thanks to Professor Anne Vanhoestenberghe, School of Biomedical Engineering & Imaging Sciences, King's College London, for her valuable discussion and encouragement in the development of this manuscript.

REFERENCES

[1] Zaaimi, B., Turnbull, M., Hazra, A., Wang, Y., Gandara, C., McLeod, F., ... & Jackson, A. (2022). Closed-loop optogenetic control of the dynamics of neural activity in non-human primates. Nature Biomedical Engineering, 1-17.

[2] Wu, F., Stark, E., Ku, P. C., Wise, K. D., Buzsáki, G., & Yoon, E. (2015). Monolithically integrated µLEDs on silicon neural probes for high-resolution optogenetic studies in behaving animals. Neuron, 88(6), 1136-1148.

[3] Von Metzen, R. P., & Stieglitz, T. (2013). The effects of annealing on mechanical, chemical, and physical properties and structural stability of Parylene C. Biomedical microdevices, 15, 727-735.

[4] Donaldson, P. E. K., & Sayer, E. (1975). A vacuum centrifuge for void-free potting of implantable hybrid microcircuits in silicone. Medical and biological engineering, 13, 595-596.

[5] Schmitz, J. E. J., & Hasper, A. (1993). On the mechanism of the step coverage of blanket tungsten chemical vapor deposition. Journal of the Electrochemical Society, 140(7), 2112.

[6] Donaldson, N., Baviskar, P., Cunningham, J., & Wilson, D. (2012). The permeability of silicone rubber to metal compounds: relevance to implanted devices. Journal of Biomedical Materials Research Part A, 100(3), 588-598.

[7] Spivack, M. A., & Ferrante, G. (1969). Determination of the water vapor permeability and continuity of ultrathin Parylene membranes. Journal of The Electrochemical Society, 116(11), 1592.

[8] Li, W., Rodger, D., Menon, P., & Tai, Y. C. (2008). Corrosion behavior of parylene-metal-parylene thin films in saline. ECS Transactions, 11(18), 1.

[9] Marszalek, T., Gazicki-Lipman, M., & Ulanski, J. (2017). Parylene C as a versatile dielectric material for organic field-effect transistors. Beilstein Journal of Nanotechnology, 8(1), 1532-1545.

[10] Korolev, Y. M. (2015). Deposition of tungsten by reduction of its hexafluoride with hydrogen under the stoichiometric component ratio: an environmentally pure production process. Russian Journal of Non-Ferrous Metals, 56, 149-154.

Assembly of Printed Interconnects for Immobilized Protein Microfluidic Assays

Qianwen Xu
Smart Manufacturing Thrust
Systems Hub
Hong Kong University of Science and
Technology
Hong Kong, China
qxuak@connect.ust.hk

S. W. Ricky Lee*
Smart Manufacturing Thrust
Systems Hub
Hong Kong University of Science and
Technology
Guangzhou, China
HKUST Shenzhen-Hong Kong
Collaborative Innovation Research
Institute
*rickylee@ust.hk

Yusong Guo
Division of Life Science
Hong Kong University of Science and
Technology
Hong Kong, China
guoyusong@ust.hk

Jeffery C. C. Lo
HKUST Foshan Research Institute for
Smart Manufacturing
Hong Kong University of Science and
Technology
Hong Kong, China
jefflo@ust.hk

Abstract—**Microfluidics has attracted significant attention for biological applications, particularly for high-throughput screening and multiplexing. However, conventional interconnect techniques are limited in their ability to fabricate the hundreds or thousands of microfluidic interconnects needed to process multiple samples for high-throughput microfluidics. To address this challenge, coaxial printing was invented to enable rapid, customizable, and scalable microfluidic interconnects. This paper focuses on the assembly process of coaxial printed interconnects for immobilized protein microfluidic assays. We demonstrate the immobilization of protein into a 1 x 1 microwell microfluidic unit with printed connectors to showcase the process.**

Keywords—*Microfluidic, Interconnects, High-throughput, 3D Printing, Protein Immobilization*

I. INTRODUCTION

The field of microfluidics has experienced rapid growth and has exciting applications in various areas of research and industry, including drug delivery [1], point-of-care diagnostics [2], lab-on-a-chip systems [3], and environmental monitoring [4]. Compared to traditional laboratory methods, microfluidics can reduce sample costs, enhance mass transfer and chemical/biological reactions, and provide highly controlled and precise fluid handling, which can improve the accuracy and reproducibility of analyses [5]. Additionally, microfluidics enables high-throughput screening [6] and multiplexing (see Fig. 1), allowing for the handling of large numbers of samples or compounds. This capability can accelerate the discovery of new drugs or biomarkers and reduce the time and resources required for experiments.

Microfluidics has the potential to revolutionize the way we study and understand biological systems. By enabling the isolation and analysis of individual cells, it provides insights into cellular heterogeneity and function, which is particularly important for applications such as cancer research and regenerative medicine [7]. Moreover, microfluidics is a valuable tool for studying the complex networks of protein interactions that underlie biological processes and for identifying new targets for drug discovery and development. Overall, microfluidics has the potential to greatly enhance our understanding of biological systems and accelerate the discovery of new treatments for diseases.

The fabrication of microfluidic devices relies on microfabrication techniques such as photolithography, replica molding, bonding, and functionalization [8]. These techniques are used to construct channels, chambers, and other features that manipulate fluids at the micro-scale. Surface coatings or other modifications are often necessary to enable specific biological or chemical interactions. For biosensors, biomolecules such as proteins or nucleic acids must be immobilized on the surface using methods such as covalent bonding, physical adsorption, affinity, or entrapment. Covalent bonding is the most efficient immobilization method with various chemistries. Microcontact printing (μCP) [9] is an effective approach to creating desired patterns by transferring the patterns of polydimethylsiloxane (PDMS) stamps to substrates with high resolution and precision.

Reliable interconnects play a vital role in communicating between microfluidic parts, including sample feeding, channel processing, and analysis. In the early days of microfluidics, tubes were manually inserted into chip holes using adhesives [10]. However, due to the lack of precise control, adhesives could spread to channels and cause clogging. Subsequently, adapters such as Luer-lock fittings [11] and clamp fixtures [12] [13] were employed to facilitate interconnection. However, these approaches often have cumbersome interfaces or large pitch sizes, which work for only several or tens of microfluidic connections but face challenges for high-throughput microfluidics with hundreds or thousands of connections. Scalability of microfluidic interconnects is an urgent need when processing multiple samples for high-throughput screening. This means that interconnects must be easily replicated or modified to accommodate different sample volumes or assay formats.

3D printing has the potential to enable rapid and customizable fabrication of microfluidic devices, including interconnects. In our previous works [14] [15], we introduced coaxial printing to print hollow connectors on device orifices, serving as microfluidic interconnects. This approach is fixture-free, scalable, robust, and reliable. In this paper, we focus on the assembly process of printed interconnects for immobilized protein microfluidic assays, as shown in Fig. 2. We immobilize proteins in a 1 x 1 microwell region via covalent bonds and employ microcontact printing to create a single-circle protein pattern for demonstration.

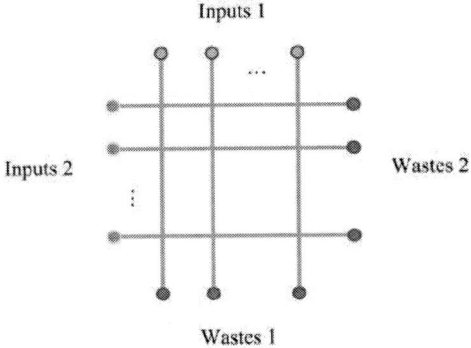

Fig. 1. Schematic of multiplexed microfluidic matrix for high-throughput screening.

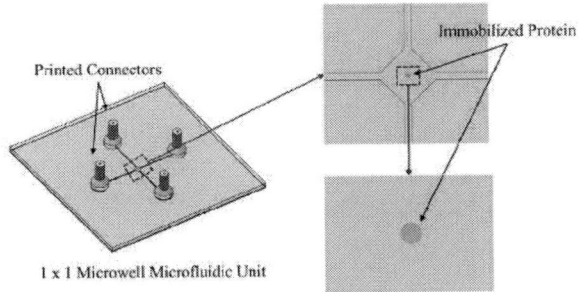

Fig. 2. Schematic of demonstrated 1 x 1 microwell immobilized protein assay with printed connectors in this study.

II. MATERIALS & METHODS

A. Materials

Photoresist (AZ4620, AZ Electronic Materials, Luxembourg) and developer (FHD-5, FUJIFILM Electronic Materials Co., Ltd, Japan) were used to fabricate the silicon mold. Trichloro (1H,1H,2H,2H-perfluorooctyl) silane 97% (Sigma-Aldrich, USA) was used to modify the silicon mold surface to make it hydrophobic. PDMS (SYLGARD 184) was purchased from Dow Corning Corporation, USA. For glass functionalization, (3-Aminopropyl)triethoxysilane (APTES) and tween-20 were purchased from Sigma-Aldrich, USA. 1-Ethyl-3-(3-dimethylaminopropyl)carbodiimide (EDC), N-Hydroxysuccinimide (NHS), mPEG-silane (MW = 1000), phosphate buffered saline (PBS), and 2-(N-morpholino)ethanesulfonic acid (MES) were purchased from Aladdin, China. The fluorescent-labeled protein, FITC-BSA, was purchased from Thermal Fisher Scientific, USA. The printing material, LOCTITE 3491, was purchased from Henkel AG & Company, Germany.

B. Preparation of PDMS Channels/Stamps

PDMS channels and stamps were prepared using a standard soft lithography process, which involved using patterned silicon molds to transfer desired layouts onto PDMS parts. Two types of Si molds were fabricated for this study: convex molds for microwell channels and concave molds for microcontact printing. Fig. 3 shows the process flow for fabricating Si molds and PDMS parts.

After sulfuric cleaning (10:1 H_2SO_4:H_2O_2), Si wafers were spin-coated with a positive photoresist (AZ 4620) at a speed of 4000 rpm for 30 seconds. A 6 μm thick PR layer was formed on the wafers after soft-baking at 90°C for 30 seconds.

The PR-coated wafers then underwent photolithography and develop steps to create patterns for etching. Following deep reaction ion etching (DRIE), the wafers were deposited with trichloro (1H,1H,2H,2H-perfluorooctyl) silane by vacuum evaporation for 1 hour.

For PDMS casting, a PDMS mixture (10:1) was poured onto the Si molds and degassed in a vacuum chamber. After 2 hours of curing at 80°C, the PDMS parts were diced into pieces (20 x 20 mm²). Fig. 4 shows the PDMS channel layout, where the channel width is 100 μm and the crossed microwell is a diamond shape with a width of 500 μm. The diameter of the inputs/wastes is 500 μm, and 2 mm half-circles are designed for aligning. Fig. 5 shows the stamp layout, involving a single circle with the diameter of 100 μm and 4 alignment marks.

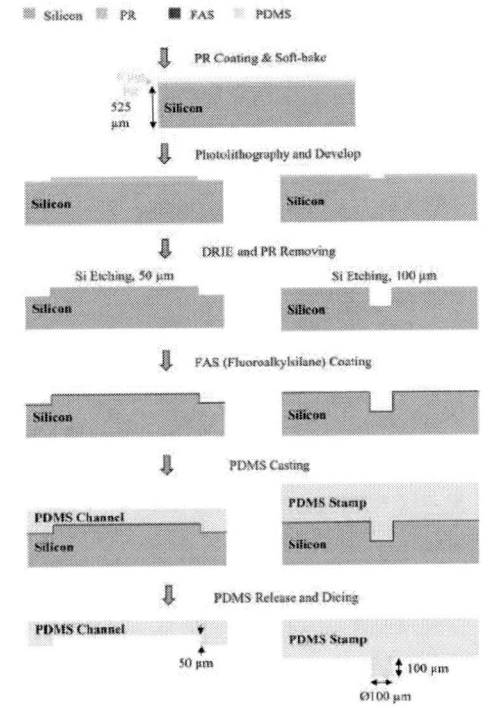

Fig. 3. Process flow for fabricating Si molds and PDMS parts.

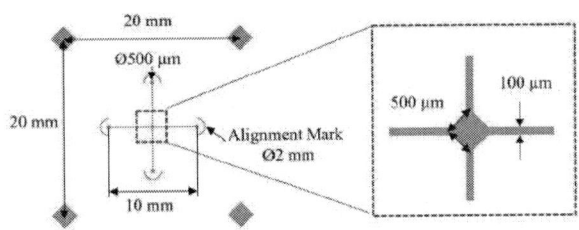

Fig. 4. Design layout of PDMS channels.

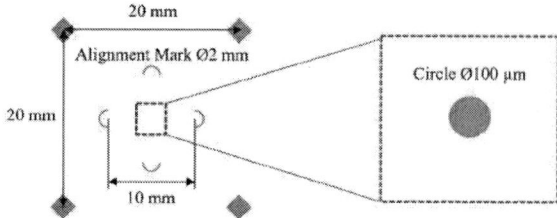

Fig. 5. Design layout of PDMS stamps.

C. Protein Immobilization on Glass

The procedure for protein immobilization on glass follows protocols described in published literature [16]. Glass substrates were cleaned with isopropanol (IPA) in an ultrasonic bath for 5 minutes, rinsed with deionized water (DI water), and then dried with N_2 gas. The glasses were then treated with O_2 plasma (100 W, 3 min) for surface hydroxylation.

Simultaneously, micropatterned stamps were inked with 20 µL of 2% APTES (dissolved in water) for 5 minutes. After drying with N_2 gas, the inked stamps were brought into contact with the plasma-treated glasses for 15 seconds. To minimize non-specific adsorption, the APTES-patterned glasses were subsequently treated with a 2% mPEG-silane solution (dissolved in water) for blocking.

Finally, the APTES-patterned glasses were incubated with an EDC/NHS activated protein solution for protein immobilization via covalent bonding for 30 minutes at room temperature (RT). For EDC/NHS activation, the protein solution was mixed with an EDC/NHS (1:1, 20 mg/mL) mixture for 30 minutes at RT. To demonstrate protein immobilization, we used a fluorescent-labeled protein, FITC-BSA (excitation/emission, 491/515 nm), at a concentration of 1 mg/mL in our experiments.

D. Assembly of Immobilized Protein Microfluidic Assay

The assembly process for immobilized protein assays is depicted in Fig. 6. Openings were drilled on the glass substrates for subsequent connector printing. A UV-assisted coaxial printing system was used for direct printing of microfluidic connectors. During printing, the coaxial nozzle deposited two types of materials, where the outer fluid was UV-curable adhesive, and the inner fluid was sacrificial water. The UV-curable adhesive was cured as a cylindrical hollow tube, while the water passed through the openings for drainage. Details of the printing process can be found in our previous papers [14].

After connector printing, APTES was patterned onto the other side of the glass substrates using the µCP method mentioned above. Then, after O_2 plasma treatment (100 W, 3 min), the APTES-patterned glass with printed connectors and the prepared PDMS channel substrates were bonded together to form a unit. Blocking was performed by flowing a mPEG-silane solution into the channel to block the un-patterned area for protein adsorption. Lastly, the protein solution was activated with an EDC/NHS mixture and flowed into the channel for protein immobilization.

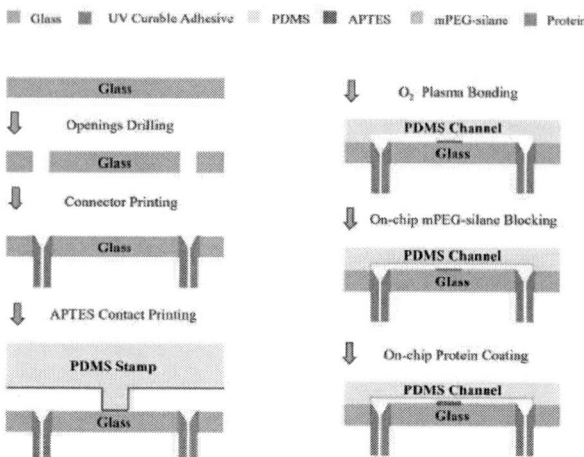

Fig. 6. Assembly process for 1 x 1 microwell immobilized protein assays with printed connectors.

III. RESULTS & DISCUSSIONS

To ensure the effectiveness of chemical grafting with relative treatments, we conducted water contact angle measurements and performed certain characterizations and experimental trials. As shown in Fig. 7, the water contact angle of untreated glasses was approximately 25°. We then treated the glasses with oxygen plasma for 3 minutes at 100 W, which decreased the water contact angle to approximately 9° due to the existence of hydroxyl groups on the surfaces. The plasma-treated glasses were subsequently soaked in a 2% APTES solution for 1 hour. The water contact angle increased to approximately 91°, representing the successful grafting of the APTES linker.

Then, we assembled the 1 x 1 microwell immobilized protein assays with 4 printed connectors, serving as interconnections at inputs or wastes. 1 x 1 microwell PDMS channels and single-circle PDMS stamps were prepared by soft lithography process described in Fig. 3. Glass substrates were drilled with 4 openings, and printed connectors were directly printed on the openings.

Four half-circle alignment marks were designed to fit the openings on glass. With their assistance, a single-circle APTES pattern was contact printed onto the microwell region of glass substrates with printed connectors. By O_2 plasma treatment, the patterned glass with printed connectors was bonded to the 1 x 1 microwell channel substrate to form a unit.

Subsequently, a 2% mPEG-solution and EDC/NHS activated FITC-BSA solution were flowed into the channel to finish blocking and protein immobilization. Fig. 8a shows an overview picture of the microfluidic unit filled with red ink, while Fig. 8b and 8c are the zoom-in views of a printed connector and the crossed microwell, respectively. The inner diameter of the connectors is approximately 240 µm, and the outer diameter is approximately 1 mm. Fig. 8d is the fluorescence image of the immobilized FITC-BSA circle (Ø100 µm) inside the microwell region (500 x 500 µm).

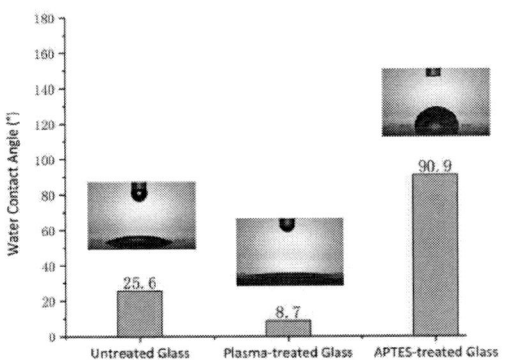

Fig. 7. Water contact angle measurements of glasses with relative treatments.

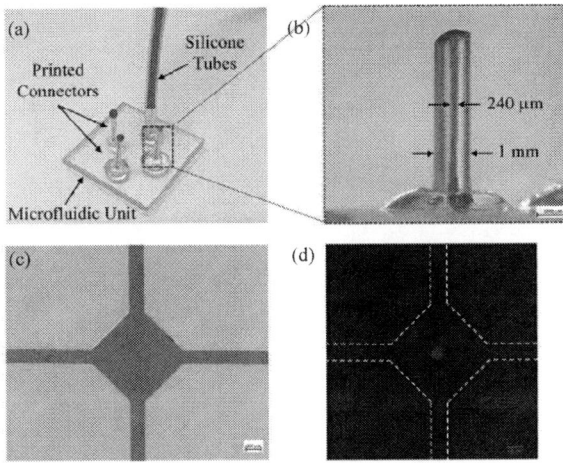

Fig. 8. (a) Overview picture of the 1 x 1 microwell immobilized protein assay; (b) zoom-in view of a printed connector; (c) zoom-in view of the crossed microwell; (d) the fluorescence image of the immobilized FITC-BSA circle (Ø100 μm) inside the microwell region (500 x 500 μm).

IV. CONCLUSIONS

The use of microfluidics for biological applications, particularly for high-throughput screening and multiplexing, has become increasingly popular. However, conventional interconnect techniques are limited in their ability to fabricate the hundreds or thousands of microfluidic interconnects needed for high-throughput microfluidics.

To address this challenge, coaxial printing as a rapid, customizable, and scalable technique was developed for microfluidic interconnects. The assembly process for immobilized protein microfluidic assays is detailed, including the use of printed connectors for interconnections at inputs or wastes, the protein immobilization methods via chemical grafting and the protein patterning with microcontact printing. The paper demonstrates the successful immobilization of protein into a 1 x 1 microwell microfluidic unit with printed connectors. For further studies, functional proteins such as antigens or antibodies can be immobilized into the device for detection. Additionally, future studies can extend the microfluidic unit to n x n microwells for high-throughput screening applications towards protein-to-protein and protein-to-cell interactions.

Overall, the paper provides valuable insights into the use of coaxial printing for microfluidic interconnects and the assembly process of them for immobilized protein microfluidic assays.

REFERENCES

[1] S. T. Sanjay *et al.*, "Recent advances of controlled drug delivery using microfluidic platforms," *Adv. Drug Deliv. Rev.*, vol. 128, pp. 3–28, Mar. 2018, doi: 10.1016/j.addr.2017.09.013.

[2] S.-M. Yang, S. Lv, W. Zhang, and Y. Cui, "Microfluidic Point-of-Care (POC) Devices in Early Diagnosis: A Review of Opportunities and Challenges," *Sensors*, vol. 22, no. 4, pp. 1-33, Feb. 2022, doi: 10.3390/s22041620.

[3] S. Haeberle and R. Zengerle, "Microfluidic platforms for lab-on-a-chip applications," *Lab Chip*, vol. 7, no. 9, pp. 1094–1110, 2007, doi: 10.1039/b706364b.

[4] M. Yew, Y. Ren, K. S. Koh, C. Sun, and C. Snape, "A Review of State-of-the-Art Microfluidic Technologies for Environmental Applications: Detection and Remediation," *Glob. Challenges*, vol. 3, no. 1, pp. 1-13, Jan. 2019, doi: 10.1002/gch2.201800060.

[5] S. Halldorsson, E. Lucumi, R. Gómez-Sjöberg, and R. M. T. Fleming, "Advantages and challenges of microfluidic cell culture in polydimethylsiloxane devices," *Biosens. Bioelectron.*, vol. 63, pp. 218–231, Jan. 2015, doi: 10.1016/j.bios.2014.07.029.

[6] G. Du, Q. Fang, and J. M. J. den Toonder, "Microfluidics for cell-based high throughput screening platforms—A review," *Anal. Chim. Acta*, vol. 903, pp. 36–50, Jan. 2016, doi: 10.1016/j.aca.2015.11.023.

[7] A. Sontheimer-Phelps, B. A. Hassell, and D. E. Ingber, "Modelling cancer in microfluidic human organs-on-chips," *Nat. Rev. Cancer*, vol. 19, no. 2, pp. 65–81, Feb. 2019, doi: 10.1038/s41568-018-0104-6.

[8] A.-G. Niculescu, C. Chircov, A. C. Bîrcă, and A. M. Grumezescu, "Fabrication and Applications of Microfluidic Devices: A Review," *Int. J. Mol. Sci.*, vol. 22, no. 4, pp. 1–26, Feb. 2021, doi: 10.3390/ijms22042011.

[9] S. Alom Ruiz and C. S. Chen, "Microcontact printing: A tool to pattern," *Soft Matter*, vol. 3, no. 2, pp. 168–177, 2007, doi: 10.1039/B613349E.

[10] A. C. Glavan *et al.*, "Rapid fabrication of pressure-driven open-channel microfluidic devices in omniphobic RF paper," *Lab Chip*, vol. 13, no. 15, pp. 2922–2930, 2013, doi: 10.1039/c3lc50371b.

[11] Eden Tech, "Connector Kit." https://eden-microfluidics.com/eden-materials/luer-lock/.

[12] A. Chen and T. Pan, "Fit-to-Flow (F2F) interconnects: Universal reversible adhesive-free microfluidic adaptors for lab-on-a-chip systems," *Lab Chip*, vol. 11, no. 4, pp. 727–732, 2011, doi: 10.1039/c0lc00384k.

[13] Z. Yang and R. Maeda, "Socket with built-in valves for the interconnection of microfluidic chips to macro constituents," *J. Chromatogr. A*, vol. 1013, no. 1–2, pp. 29–33, 2003, doi: 10.1016/S0021-9673(03)01125-7.

[14] Q. Xu, J. C. C. Lo, and S. W. R. Lee, "Directly printed hollow connectors for microfluidic interconnection with UV-assisted coaxial 3D printing," *Appl. Sci.*, vol. 10, no. 10, pp. 1–13, 2020, doi: 10.3390/APP10103384.

[15] Q. Xu, J. C. C. Lo, and S. W. R. Lee, "Characterization and evaluation of 3D-printed connectors for microfluidics," *Micromachines*, vol. 12, no. 8, 2021, doi: 10.3390/mi12080874.

[16] H. Li *et al.*, "Aminosilane micropatterns on hydroxyl-terminated substrates: Fabrication and applications," *Langmuir*, vol. 26, no. 8, pp. 5603–5609, 2010, doi: 10.1021/la9039144

Flexible Hybrid Electronics on Wearable Healthcare Application

Ming-Hung Chen*, Wei-Hao Chang, Tun-Ching Pi, Wei-Chun Lee, Jen-Chieh Kao and Yung-I Yeh
Corporate R&D, Advanced Semiconductor Engineering, Inc.
Kaohsiung, Taiwan
Oudi_Chen@aseglobal.com

Abstract—24/7 real-time body signal monitoring for decease check and life quality enhancement is becoming a megatrend these years. However, the precious yet limited clinical facility and medical professionals cannot fulfill the emerging needs out-of-hospital. A wearable device such like smart watch utilizing biosensor-integrated System-in-Package (SiP) module can provide a real time and personal-based bio data collection as well as better health management through AI assistance. For performing a more comfortable experience, flexible hybrid electronics with ECG SiP embedded was proposed with lower profile and higher structure flexibility. The patch durability by 100,000 times rolling test as simulating daily use condition was also examined.

Keywords — Flexible Hybrid Electronics (FHE), System-in-Package (SiP), Electrocardiography (ECG), Patch

I. INTRODUCTION

Wearable electronics empowered by System-in-Package (SiP) technology become wide spread for almost all time and immediate communication as well as bio-information could be analysis through AI for instant care, especially in post pandemic era of COVID-19. However, current wearable electronics, such like smart watches and smart earbuds, with rigid hard case in appearance achieved bio signals harvest through a proximity and unstable interface make a poor signal-to-noise ratio and limited stich time and located at terminal of human body. Main bio-signals from heart and lung, which are central of circulation and respiration system respectively, can still be monitoring through either hospital based complex system or tight fasten sensor box link to portable systems under special clinical care and comment by doctors. Unfortunately, most of sudden disease occurred and fatal accidence cannot be alarmed in the scheduled time of monitor system utilizing. A 24/7 wearable solution for bio signal monitoring through extendable sensors integration as well as wireless module obtains signal access to nearby mobile phone for providing instant care and health analysis becoming an emerging issue. Current wearable ECG product, as shown in Fig. 1(a), with thick & rigid appearance induce a foreign body sensation and bondage while use. New concept for sensors on skin draw a great attention and various approaches were proposed [1-6] which mostly stayed at research stage.

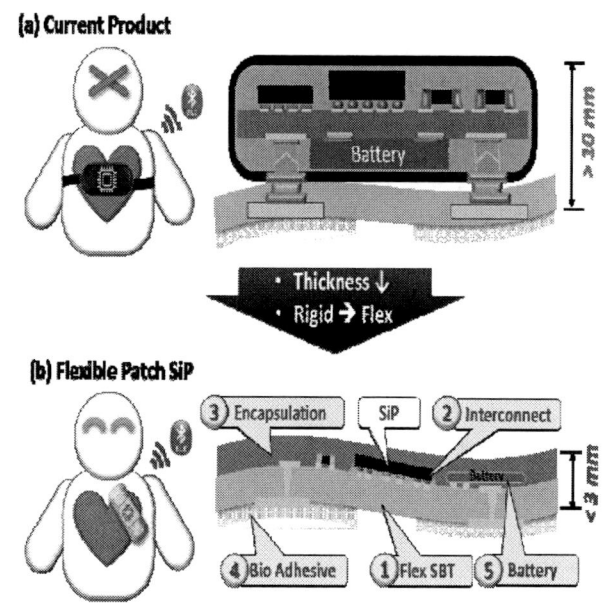

Fig. 1. ECG product as exeample. (a) current ECG product with rigid, thick and belt fasten for long time health monitoring ; (b) Flexible patch SiP provide a comfort experience for use.

In this study, a patch-type flex SiP with slim and bendable structure coat with flexible and bio compatible encapsulant, illustrated in Fig. 1(b), was considered as an achievable solution to market. This patch achieved through 5 key process technologies including a 2-layer flexible printed circuit (FPC), interconnect material for surface mount technology (SMT), flex encapsulation, bio-adhesive for patch immobilize and bio signal transportation as well as the feasibility of embedded battery in SiP. Moreover

The examination criteria of patch was setting as pass rolling test for 100,000 times in R ≤ 60 mm which adapt to required curvature of human chest without structure delamination and ECG monitoring showed clear signal character that comparable to result of conventional 1-lead system. In order to reduce foreign material sensation, the total thickness should be reduced to less than 3 mm.

II. EXPERIMENTAL

A. Flexible Printed Circuit

Flexible printed circuit composed by substrate polymer and conductive pattern with redistribution layer (RDL) and pad for component landing. Substrate polymer was mainly focus on polyimine (PI, thickness = 150 um) material due to bendable property and wide spread in electrical package application. Conductive material was electroplating copper (EP-Cu) as reference and new silver pastes with printing process as alternative solution for higher process flexibility on prototyping and integration feasibility study. 2 kinds of silver pastes utilized for stability test. Ag paste A (AT-XB80, Giga Solar Materials, Taiwan) composed with silver particle in micro size and Ag paste B (RD0117B, ACI Materials, US) composed with nano/micro particle mixture, respectively.

B. Interconnection

The interconnection was formed through surface mount technology in this study. Daisy-chain die (2 types: 4.2 x 4.2 mm^2 and 8.4 x 8.4 mm^2 with pad diameter = 150 um in both cases) and passive components (0603 in size of 1.6 x 0.8 mm^2; 0201 in size of 0.6 x 0.3 mm^2) were mount onto designed FPC. Due to reduce thermal damage and lower shrinkage under polymer substrate process, interconnect material curing temperature setting 150°C in maximum. The interconnection material in this study including low temperature solder paste (T6, Heraeus, Germany) and Ag paste C (TM1079, ACI Materials, US). Underfill (HiTech CU13-3150, MacDermid Alpha, US) in some cases was applied at the die bottom for increasing die strength during rolling test.

C. Encapsulation

Epoxy molding compound (EMC) play a key role for package protection and component isolation. Conventional EMC with high modulus in range of 5~30 GPa perform a rigid and high stiffness and not suitable for Flex SiP application. Silicone with tunable and extremely low modulus (1~100 Mpa) as well as high chemical resistance and bio-compatible was considered for encapsulation material. In this study, silicone with modulus < 30 Mpa (RTA4024, Howren, Taiwan) was focused for rolling durability test.

D. Bio-adhesive

The customized designed bio-adhesive tape composed with adhesive tape and conductive hydrogel in which electrical signal of ECG can pass through RDL and accessed to SiP module. System integration test by check character signal of ECG was obtained.

E. Battery

Lithium Polymer (Li-Po) with flexible and slim shape show a great potential for battery embedded ; Button cell battery (also called "Coin battery") is another candidate due to relative high power density and compact size. The battery with DC 3.7 V and capacity should larger than 200 mAh were defined for stable power supply in 1-week wearable use.

F. Roller test machine

Rolling test in this study was achieved in Yuasa's desktop model endurance system (Yuasa System Co., Japan). The test condition of rolling radius was in range of 1.5~60 mm with 50 rpm and loading in 500 g. Rolling criteria was 100,000 times.

III. RESULTS AND DISCUSSION

To realize patch-type flex SiP, material selection of substrate and interconnection would be most critical and should be clarify by step. Structure protection / strengthen material such as underfill and encapsulant were derived a synergy performance with substrate and interconnect material. Bio-adhesive and battery solution as peripheral infrastructure affect the system integration level as to application condition.

A. Bare FPC for Rolling Test

In order to meet rolling criteria, RDL design for robust electrical transmission structure was studied. 4 types printing wire with various line width and shapes were considered. As Fig 2, Type 1 was straight line with width 0.3 mm; Type 2 was also straight but larger width of 0.5 mm; Type 3 and Type 4 were the same width to Type 1 but different shape. These non-straight shape were for reducing bending stress and prevent line crack.

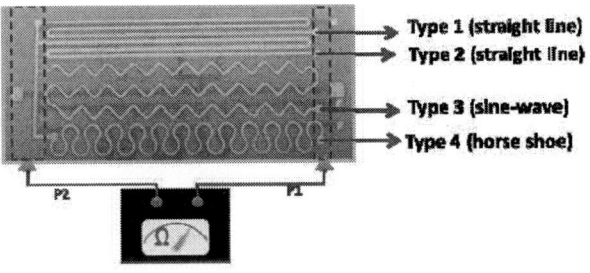

Fig. 2. RDL design test with various line type: Type 1 (Straight / width 0.3 mm) ; Type 2 (straight / width 0.5 mm) ; Type 3 (Sine-wave / width 0.3 mm) ; Type 4 (horseshoe / width 0.3 mm)

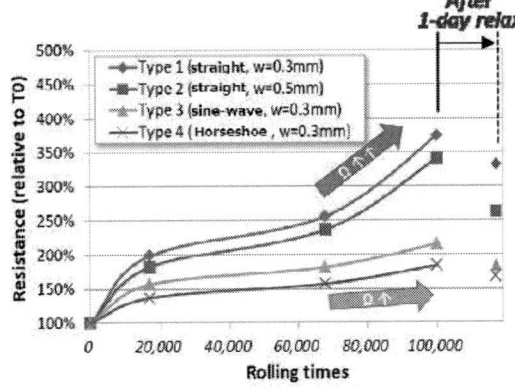

Fig. 3. RDL resistance increasment relative to T0. Type 3 and Type 4 perform a relative lower resistance change and betterstability after 15-hr relaxing.

125

The resistance change result as shown in Fig. 3. For line width comparison in Type 1 and Type 2, the resistance increasing trend is the same, means line width cannot provide a stable RDL. For various line shape in Type 1, Type 3 and Type 4, both sine-wave and horse shoe could significant inhibit the resistance increase in higher rolling times where were also less resistance change after 15-hr relaxing. As considering of design difficulty and high density request, sine-wave pattern should be a better choice than horse shoe.

An alternative RDL process of conductive paste printing was further studied since current EP-Cu RDL with longer process time and resource consumption in lithography and wet etching steps. Various Ag pastes was applied with sine-wave shape and compare to EP-Cu as reference. Paste A composed with micro size Ag particle where Paste B was mixture of nano/micro particle and obtain a lower contact resistance, the sample prepared as shown in Fig. 4. The resistance change was shown in Fig 5 and Paste A with single micro Ag particle obtain not only a high resistance at T0 but also resistance instability after rolling test. Paste B perform a comparable result to EP-Cu under the same rolling times and it would be more process application for using paste printing process on new flex patch.

B. Rainforced FPC for Rolling Test

Bare FPC could be exist in system while playing as a RDL layer. As talk to components mounting and interconnect formation, underfill and encapsulation were required in following process in real application. In this section, a high density SMT with daisy-chain dies and passive components was achieved on EP-Cu/PI substrate and 2 types of interconnect material including Ag paste C and low temperature solder were utilized. The test vehicle with SMT process as shown in Fig. 6(a) and the resistance monitoring for die and passive components was separated due to different challenge and possible underfill applied at die bottom. In case of underfill apply, shown in Fig. 6(b), the gap for underfill filling is 120 um. As patch encapsulated with silicone, the test piece would proceed rolling test as setup in Fig. 6(c).

In Fig. 7, low temperature as interconnect with silicone encapsulant, no matter underfill applied or not, perform a stable resistance status even up to 100,000 times rolling. However, in Ag paste C cases, insufficient structure strength without underfill induced a rapid fail. In material aspect, low temperature solder with better stiffness than Ag paste C and underfill support can enhance the system durability while Ag paste used.

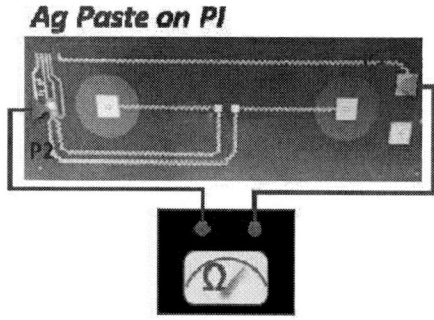

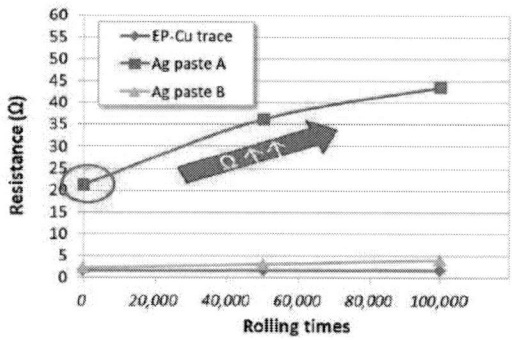

Fig. 4. RDL materials comparison in sine-wave shape of ECG patch: A sample build.

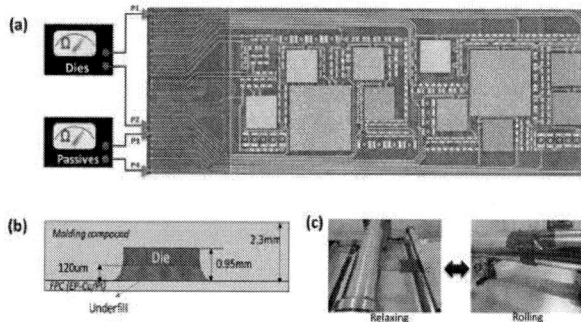

Fig. 6. Test vehile for high desity daisy-chain die and passive component with silicone encapsulated: (a) resistance monitoring setup; (b) test vehicle cross-section view; (c) rolling test setup.

Fig. 5. Resistance increasement of various RDL material. Paste B with more stable resistance than Paste A.

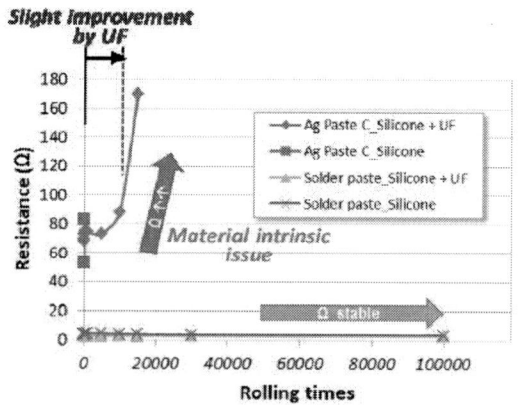

Fig. 7. Resistance chage under different structure.

C. Bio-adhesiive Integration

Bio-adhesive composed with adhesive and hydrogel was designed to fit the test vehicle in this study. Conventional ECG product was utilized for the new patch interface evaluation. Bio-adhesive taping on FPC and paste to human body; the electrical signal connection was through welding line between FPC and ECG module, as shown in Fig. 8. The character wave of ECG signal can be observed in this test. However, the signal to noise ratio was poor due to circuit impedance was not optimized and it would be addressed in integrated system.

D. Embedded Battery Solution

Power solution is the last mile to realize fully functioned wearable system as well as whole module size and persistent usage was limited to the battery size and energy capacity. In this study, Li-Po battery with thin (~ 1 mm) and flexible appearance was try to embedded into patch for rolling test. The low temperature encapsulation curing at 85°C can provide a feasible process to Li-Po integration. When rolling test upto 90,000 times, the embedded Li-Po battery became inflate and the output voltage was reduced, as in Fig. 9 where the rechargeable test monitoring as shown in Fig. 10.

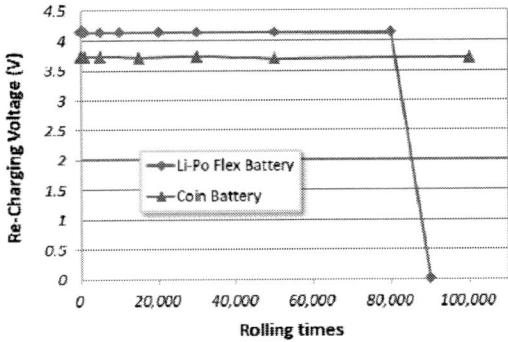

Fig. 10. Embedded battery rechargable under various rolling rounds.

IV. CONCLUSIONS

FHE for long time wearable in categories of iPatch (bendable) and iCloth (stretchable) were considered as shown in Fig. 11. Patch-type flex SiP, as first step demonstration, provided an attractive platform to reach the requirement. The capability criteria was highly correlated to the usage scenarios. ECG module for example, the radius R of patch bending curvature would be several tens mm for taping on chest but only few mm of R while access the bio-signal of finger. The durability criteria such as times and test mode (e.g. rolling, bending, twisting, etc.) be also considered. Moreover, different FPC deformation modes including bendable and stretchable should be defined before flex patch development for specific application. Patch-type flex SiP for ECG measurement could be realize by modifying current EP-Cu/PI FPC with low temperature solder for components mount as well as silicone encapsulation provide a module protection and flexible interface. The verification of integrated system would be scheduled for next study.

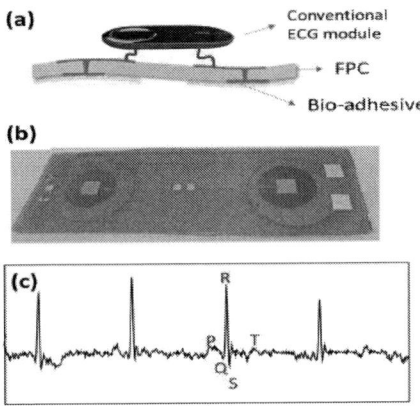

Fig. 8. Bio-adhesive functional test. Bio-adhesive on FPC form a new interaface between cnventional ECG module and human body for ECG signal reading.

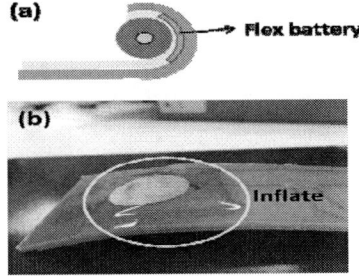

Fig. 9. Li-Po battery embedded in encapsulation of patch. After 80,000 rolling, the battery become inflate and voltage dowm.

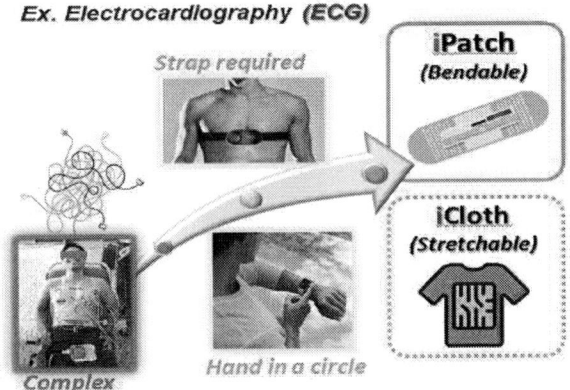

Fig. 11. FHE application for long-time wearables: ECG monitoring as eamaple where bendable (iPacth) and stretchable (iCloth) ideas were proposed.

REFERENCES

[1] C. Falconi, and S. Mandal, "Interface electronics: state-of-the-art, opportunities and needs," Sensors and Actuators A: Physical, vol. 296, pp. 24-30, 2019.

[2] D. Corzo, G. Tostado-Blázquez, and D. Baran, "Flexible electronics: status, challenges and opportunities," Frontiers in Electronics, vol. 1, 594003, 2020.

[3] S. P. Lee, G. Ha, D. E. Wright, Y. Ma, E. Sen-Gupta, N. R. Haubrich, and R. Ghaffari, "Highly flexible, wearable, and disposable cardiac biosensors for remote and ambulatory monitoring," NPJ digital medicine, vol.1 (1), pp. 2, 2018.

[4] Q. Li, J. Zhang, Q. Li, G. Li, X. Tian, Z. Luo, and J. Zhang, "Review of printed electrodes for flexible devices," Frontiers in Materials, vol. 5, pp.77, 2019.

[5] Y. Wang, C. Xu, X. Yu, H. Zhang, and M. Han, "Multilayer flexible electronics: Manufacturing approaches and applications," Materials Today Physics, 100647, 2022.

[6] W. Wu, "Stretchable electronics: functional materials, fabrication strategies and applications," Science and Technology of Advanced Materials, vol. 20(1), pp. 187-224, 2019.

Testing of electromigration resistance of copper and silver thick films

Jiri Hlina
Department of Materials and Technology
Faculty of Electrical Engineering
University of West Bohemia
Pilsen, Czech Republic
hlina@fel.zcu.cz

Jan Reboun
Department of Materials and Technology
Faculty of Electrical Engineering
University of West Bohemia
Pilsen, Czech Republic
jreboun@fel.zcu.cz

Marek Simonovsky
Elceram a.s.
Hradec Kralove, Czech Republic
simonovsky@elceram.cz

Ales Hamacek
Department of Materials and Technology
Faculty of Electrical Engineering
University of West Bohemia
Pilsen, Czech Republic
hamacek@fel.zcu.cz

Abstract—This paper is focused on the electromigration resistance of copper and silver thick films. Electromigration was tested by water drop test. Four common conductive thick film pastes were used for testing – silver paste with low platinum content, and two silver pastes with different content of palladium and copper paste. Electromigration was tested under voltage of 6 and 12 V and time to shortcut was measured. Results of testing including the comparison of electromigration resistance depending on thick film paste type and different electrode spacing are described in this paper.

Keywords—electromigration, copper, silver, thick film, water drop test

I. INTRODUCTION

Thick film technology is used in hybrid microelectronic applications for decades. This technology is intended for the realization of conductive, dielectric (insulation) or resistive films [1]. Thick films are printed in the form of paste (usually with screen printing) and fired in a belt furnace in an oxidative atmosphere. In general, thick film paste formulation contains functional component (metal particles in the case of conductive pastes), glass binder and organic component [2]. Commonly used conductive thick film pastes are based on silver (pure or in alloys with palladium or platinum), copper or gold [3].

Thick film technology can be also utilized for the realization of power hybrid modules. Special copper pastes have recently been used for these applications, but the use of silver pastes is also possible for whole or parts of circuits [4]. These modules usually integrate a power and control section and are exposed to harsh conditions. Due to the current trend of continuous miniaturization, control parts in particular are made up of high-density circuits. The use of copper and silver conductors in high-density circuits raises the issue of their long-term environmental reliability because mainly silver has a tendency to migrate under DC voltage in the presence of humidity and also in the presence of residues after soldering which is difficult to remove from the porous ceramic surface. Therefore, this paper is focused on testing of electromigration resistance of commonly used copper and silver pastes.

Electromigration is usually tested according to IPC standards or by water drop test when two electrodes (copper or silver) are subjected to droplets of distilled water and DC voltage [5,6]. Electromigration results in dendrite growth between these two electrodes and a causes short circuit.

Electromigration can be reduced by proper circuit design and with using of overglaze films which cover conductive film (mainly pure silver) and protects it against environmental effects. Another option is using silver pastes with palladium content. The addition of palladium reduces the tendency of silver to migrate. In the case of copper films, the tendency to migrate is low.

II. EXPERIMENT

Electromigration was firstly measured according to IPC standard 2.6.14.1 Electrochemical Migration Resistance Test [6]. The IPC-B-25A test pattern (Fig. 1) was used for the measurement and the electromigration was measured on part D. This pattern was printed with three commonly used thick film pastes – AgPt 99:1 Heraeus C1076SD, AgPd 3:1 Heraeus C2030 and Cu Heraeus C7403A on 96% Al₂O₃ substrates.

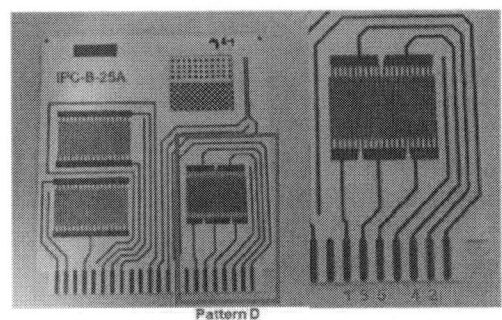

Fig. 1. Realized IPC-B-25A test pattern (left) and detail odf pattern D with marking of terminals (right).

According to the above-mentioned standard, the samples were tested in a climatic chamber at 85 °C and 88.5% RH. The test samples were first stabilized at these conditions for 96 hours. After stabilization, the insulation resistance was measured with an electrometer at 100 V between terminals 1-2, 3-2, 3-4 and 4-5 (Fig. 1). Then the samples were connected to a voltage of 10 V. T Limiting 1 MΩ resistors were connected in series in each measured path (at terminals 1, 3 and 5, to which the positive terminal of the source was connected - Fig. 1). The samples with the connected voltage were tested in the climatic chamber for 500 hours at the same conditions as during stabilization (85 °C and 88.5% RH). After 500 hours, the insulation resistance of the samples was measured again. The average insulation resistance was calculated using the following formula

$$IR_{avg} = 10^{\left[\frac{1}{N}\Sigma_1^N logIR_i\right]} \qquad (1)$$

where N is the number of test points, IR_i is individual insulation resistance measurements. Samples were observed under a microscope and evaluated for changes and growth of dendrites after the test.

Electromigration was further tested using a water drop test. Samples (Fig. 2 and Fig 3) were also manufactured from aluminium oxide (96% Al_2O_3) 0.635 mm thick. Similar conductive thick film pastes were used (AgPt 99:1 Heraeus C1076SD, Cu Heraeus C7403A, AgPd 3:1 Heraeus C2030) and in addition AgPd 6:1 Heraeus C2060. The electrode width was 1.5 mm. Spacing was chosen between 0.2 and 1.2 mm. Except for the testing area and contacts, samples were covered with a thick film overglaze to prevent water drops from spilling onto the surface. The description of materials used for each type of sample is mentioned in Table 1.

Fig. 2. CAD design of test samples.

Fig. 3. Realized test samples.

A drop of distilled water was applied to the test area. Samples were attached to the power source, the current was limited to 100 mA, the voltage was set from 6 V to 12 V and the time to 100 mA shortcut was measured. Dendrite growth was observed under a confocal microscope.

TABLE I. DESCRIPTION OF MATERIALS

Paste	Alloy	Overglaze
C1076SD	AgPt 99:1	IP9025ST
C2060	AgPd 6:1	IP9025ST
C2030	AgPd 3:1	IP9025ST
C7403A	Cu	IP7098

The third test was a water drop test with 5% NaCl solution which was proven only for samples with Cu films with spacing 0.4 mm under voltage of 20 V.

III. RESULTS

A. Test according to IPC standard 2.6.14.1

The results of insulation resistance measurement for all sample variants are summarized in Table 2. Samples were also observed under a microscope with bottom light for verification of dendrite growth and an example of the sample with AgPt 99:1 film is shown in Fig. 4.

TABLE II. RESULTS OF INSULATION RESISTANCE MEASUREMENT

Sample (paste)	Insulation resistance – IR_{avg} (MΩ)	
	After stabilization (96 h, 88.5 °C, 85 % RH)	After test (500 h, 10 V, 88.5 °C, 85 % RH)
AgPt 99:1	470,0	1,06
AgPd 3:1	558.6	2,73
Cu	541.6	2,48

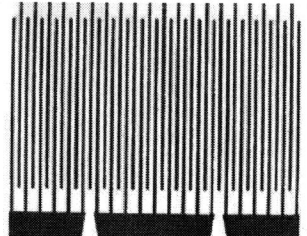

Fig. 4. Microscopic image of the sample with AgPt 99:1 film after the test.

It was verified that electromigration did not occur in the tested samples after testing according to IPC standard 2.6.14.1. No dendrite growth was observed in any type of sample. However, the insulation resistance decreased from values of ~500 MΩ to values of 1-3 MΩ.

B. Water drop test – distilled water

Considering that no electromigration occurred during the test according to the IPC standard, an additional test was chosen in which the samples are exposed to more harsh conditions – a water drop test.

The first measurement was performed under voltage 6 V. Results of electromigration testing under this voltage for different electrode spacing are shown in Fig. 5-8. Results of AgPt and AgPd films are described in graphs, Cu films shown current close to 0 mA.

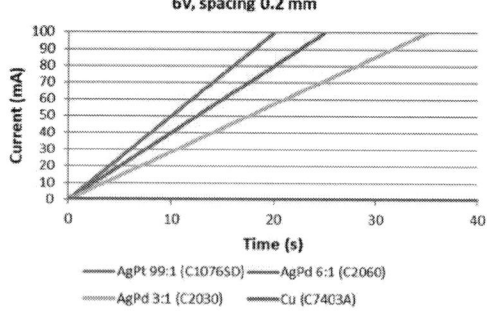

Fig. 5. Result of electromigration testing – test voltage 6 V, electrode spacing 0.2 mm.

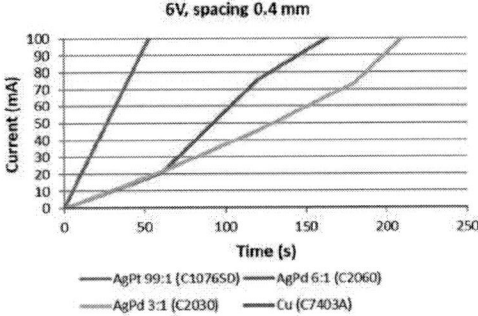

Fig. 6. Result of electromigration testing – test voltage 6 V, electrode spacing 0.4 mm.

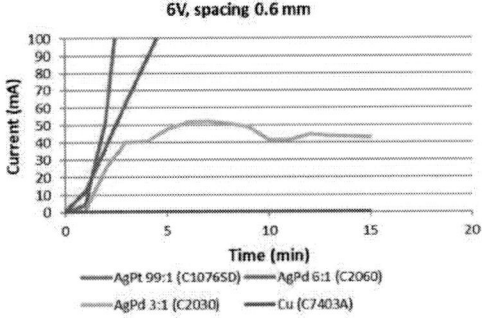

Fig. 7. Result of electromigration testing – test voltage 6 V, electrode spacing 0.6 mm.

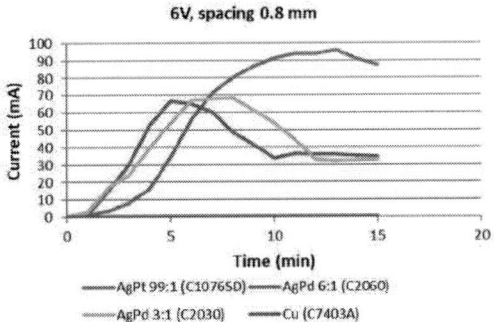

Fig. 8. Result of electromigration testing – test voltage 6 V, electrode spacing 0.8 mm.

It is obvious, that spacing (distance) between electrodes and also the material of thick film has a significant influence on the rate of electromigration. In the case of spacing 0.2 mm, all samples get a shortcut (shortcut current 100 mA) within 20-35 seconds. At two times higher spacing of 0.4 mm, a short circuit occurred at 50-210 s for all samples. For samples with a 0.6 mm spacing, a shortcut occurred only in the case of AgPt 99:1 and AgPd 6:1 films (within 2-5 minutes). For samples with 0.8 mm spacing, no shortcut was observed, however, the current was approaching 100 mA in the case of AgPt 99:1 film. In terms of the used thick film materials, AgPt 99:1 paste has the highest tendency to migrate compared to AgPd pastes. The tendency to migrate decreases with increasing palladium content. The neither high amount of palladium in the silver paste will not prevent electromigration completely.

The second measurement was performed under voltage 12 V. Results of electromigration testing under this voltage for different electrode spacing are shown in Fig. 9-13.

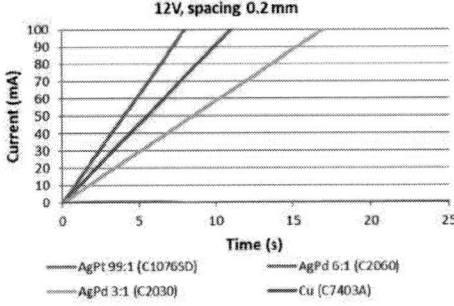

Fig. 9. Result of electromigration testing – test voltage 12 V, electrode spacing 0.2 mm.

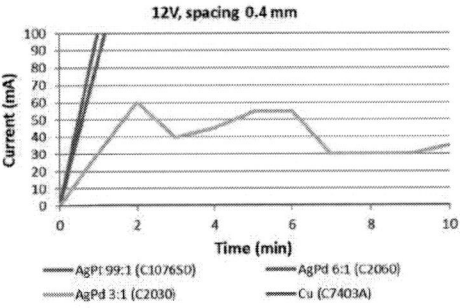

Fig. 10. Result of electromigration testing – test voltage 12 V, electrode spacing 0.4 mm.

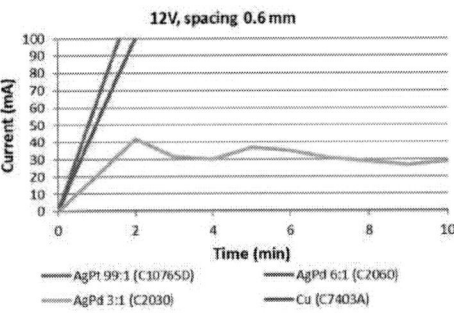

Fig. 11. Result of electromigration testing – test voltage 12 V, electrode spacing 0.6 mm.

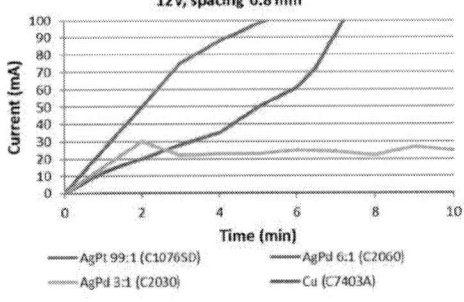

Fig. 12. Result of electromigration testing – test voltage 12 V, electrode spacing 0.8 mm.

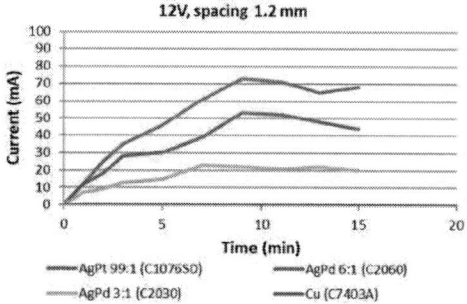

Fig. 13. Result of electromigration testing – test voltage 12 V, electrode spacing 1.2 mm.

The results show that migration is accelerated at the higher voltage. The behaviour of the tested materials is similar to that at lower voltages, only the shortcuts occurred in a shorter time – 7-17 s in the case of spacing 0.2 mm, around 1 minute in the case of spacing 0.4 mm for AgPt 99:1 and AgPd 6:1 films. For samples with 0.6 mm and 0.8 spacing, a shortcut occurred only in the case of AgPt 99:1 and AgPd 6:1 films (within 2 minutes for spacing 0.6 mm and within 7 minutes for spacing 0.8 mm). For samples with 1.2 mm spacing, no shortcut was observed. The copper film showed high resistance against electromigration, shortcut current was up to 1 mA for all variants of spacing.

In Fig. 14 are shown microscopic images of samples with spacing 0.4 mm after the electromigration test under voltage 12 V. There is visible dendrite growth for AgPt and AgPd films between electrodes, while copper dissolution on the anode electrode surface is apparent but no dendrite growth was observed for Cu film.

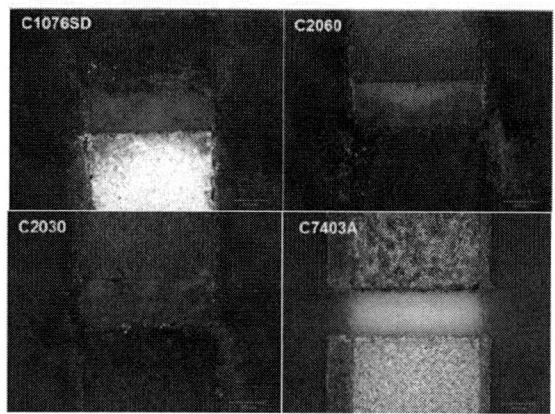

Fig. 14. Example of samples after testing of electromigration – voltage 12 V, spacing 0.4 mm.

C. Water drop test – 5% NaCl solution

No electromigration was observed in the samples with copper films during the water drop test. Therefore, in the case of these samples, the water drop test was performed with a 5% NaCl solution in distilled water. Samples with spacing 0.4 mm were tested under a voltage of 20 V. NaCl solution and higher voltage were chosen to promote the conditions for electromigration.

Results are shown in Fig. 15. There was no significant increase of the current between the electrodes even with the NaCl solution and higher voltage. The current was below 0.2

mA after 30 minutes of the test. Samples were also observed under the microscope before and after the test (Fig. 16). Similarly to the previous test at a lower voltage (12 V) and with distilled water, copper dissolution on the anode electrode surface is obvious but no dendrite growth was observed after the test.

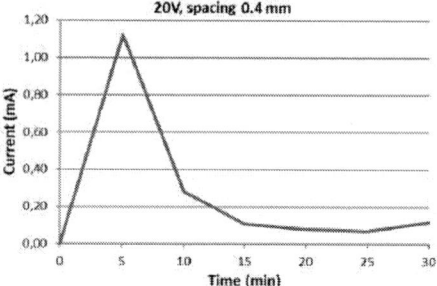

Fig. 15. Result of electromigration testing of sample with Cu film – test voltage 20 V, electrode spacing 0.4 mm, 5% NaCl solution.

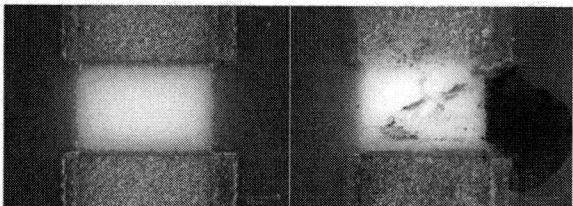

Fig. 16. Sample with Cu film before (left) and after water drop test in 5% NaCl solution (right).

IV. CONCLUSION

Experiments were carried out to verify electromigration in copper and silver thick films. Tests according to IPC standard 2.6.14.1 and water drop test with distilled water and 5% solution of NaCl in distilled water (only for samples with Cu films) were used.

Testing according to IPC standard 2.6.14.1. did not result in electromigration in samples with Ag and Cu films. Testing using water drop test under 6 V, 12 V and 20 V voltage showed far higher resistance of copper to migration compared to silver alloys. Copper dissolution on anode electrode surface is apparent but no dendrite growth was observed. Comparison between silver and silver-palladium alloys show detectable retardation of migration with increasing palladium content. While 6:1 alloy seems to be closer to pure silver, with 3:1 alloy retardation is quite apparent. Interestingly, with longer times of voltage applied, shortcut currents even decreased. The effect was most evident with 3:1 alloy. This is probably due to heavy oxidation/passivation by silver-palladium oxides limiting the availability of fresh metallic surfaces. Spacing is also important but a significant delay in shortcut time was observed only with spacing 0.8 and 1.2 mm.

From an economical point of view, given two times higher prices for AgPd 6:1 alloy and four times for AgPd 3:1 compared to silver, it is questionable where these alloys offer justifiable improvement. It seems more effective to use silver wherever possible (covered with low-cost overglaze) and apply AgPd 3:1 (or even 2:1) on most critical areas as soldering pads and contacts.

ACKNOWLEDGMENT

This research has been supported by the Technology Agency of the Czech Republic under the project POKER, No. FW01010067 " Advanced ceramic materials and technologies for power electronics" and by the Student Grant Agency of the University of West Bohemia in Pilsen, grant No. SGS-2021-003 "Materials, technologies and diagnostics in electrical engineering".

REFERENCE

[1] J. E. Sergent, Hybrid Microelectronics Handbook, 2nd ed. McGraw-Hill Inc., New York, 1995.

[2] T. K. Gupta, Thick- and Thin- Film MicroelectronicsHybrid Microelectronics Handbook, 1st ed. John Wiley & Sons Inc., New Jersey, 2003.

[3] R. Gee, M. V. Coleman, "A Thick Film Materials System for the Manufacture of Advanced Hybrid Microcircuits", Microelectron. Int., vol. 6, no. 2, pp. 23–28, 1989.

[4] J. Reboun, J. Hlina, P. Totzauer, A. Hamacek, "Effect of copper- and silver- based films on alumina substrate electrical properties", Ceramics International, Vol. 44, 2018, pp. 3497-3500.

[5] W. S Hong, B. Kang, B. Song, K. Kim, (2005). "A Study on the Metallic ion Migration Phenomena of PCB". Korean Journal of Materials Research. 15, 2005, pp. 54-60.

[6] IPC International, Inc., "IPC-TM-650 2.6.14.1 Electrochemical Migration Resistance Test. 2000.

Reliability Testing of Recycled SMD Components Reused in E-Textiles after Ageing by Washing Cycles

Martin Hirman
Faculty of Electrical Engineering
University of West Bohemia
Pilsen, Czech Republic
hirmanm@fel.zcu.cz
ORCID: 0000-0002-8481-8971

Jiří Navrátil
Faculty of Electrical Engineering
University of West Bohemia
Pilsen, Czech Republic
jirkanav@fel.zcu.cz

Andrea Benešová
Faculty of Electrical Engineering
University of West Bohemia
Pilsen, Czech Republic
benesov2@fel.zcu.cz
ORCID: 0000-0003-0879-7846

František Steiner
Faculty of Electrical Engineering
University of West Bohemia
Pilsen, Czech Republic
steiner@fel.zcu.cz
ORCID: 0000-0002-5702-7015

Abstract— This paper presents a method for recycling and reusing SMD chip components in e-textiles using a special contacting technique. The method involves using a UV-curable, non-conductive acrylic adhesive to connect SMD components onto electrically conductive textile stretchable ribbons. At the end of the product's life-cycle, the components are removed from the ribbon and reused to manufacture new products. The recycling procedure involves several steps, including disassembly, inspection, cleaning, repair, and reassembly of components. The results of an experiment testing the electrical resistance of new and reused components after washing and drying cycles showed that electrical resistance of reused joints did not significantly deteriorate compared to new joints. The reused components maintain their functionality and performance, suffering no significant impairment. They perform comparably to new components, even after repeated washing and drying cycles. The method offers several benefits, including conservation of raw materials, minimization of waste, and reduction of production costs. It can also help mitigate component shortages. However, it is found that the suitability of this remanufacturing method varies with different types of components. Certain component types may be more susceptible to damage or even unsuitable for this method. The method is particularly beneficial for higher-value or components experiencing market shortages, demonstrating its potential to address urgent industry challenges while contributing to environmental sustainability.

Keywords—e-waste, SMD chip components reuse, e-textiles, non-conductive adhesive, recycling.

I. INTRODUCTION

Recent reports from the European Commission and the World Health Organization indicate an exponential increase in the generation of electronic waste (e-waste) [1]. Data from a survey conducted by the Global E-waste Statistics Partnership corroborates this trend, revealing that 53.6 million metric tonnes of e-waste were generated in 2019 [2]. Projections suggest that this figure will continue to rise, with estimates indicating that up to 74 million metric tonnes of e-waste could be generated by 2030. Furthermore, it was determined that only approximately 17.4% of the total e-waste generated in 2019 was recycled. This implies that valuable materials such as gold, silver, and copper present in these devices were predominantly discarded or incinerated rather than reused.

Several factors have contributed to the rapid increase in e-waste observed in recent years. These include population growth leading to increased demand for and consumption of electronic products as well as the short life cycle and low repair rate of these products [3]. The constant development of new technologies has also played a role. Additionally, the Covid-19 pandemic has further increased interest in electronics among individuals who have had to work from home and businesses that have had to modernize their production significantly in order to remain competitive [4]. The amount and rate at which e-waste is produced poses a serious global issue with implications for both the environment and human health. As a result, an increasing number of countries are addressing this problem by implementing their own legislation and directives as well as adhering to supranational ones. Companies are now required to include environmental considerations in the development and design of new products. However, companies are currently facing another challenge: a shortage of electronic components such as chips and microcircuits necessary for producing these devices. This shortage has been caused by the closure of many factories dedicated to producing these components in 2020 due to efforts to combat the spread of Covid-19 [5].

It follows that the reduction of carbon footprint, saving of natural resources, reducing the amount of newly consumed materials and striving to reuse used materials and product parts are global trends. These trends are widely supported and they are in line with the overall ecologic ideas. Manufacturers in the electronics industry will have to adapt to these new conditions and look for innovative solutions for this trend. A possible way is using circular economy which is a sustainable model that aims to extend the life-cycle of products, reduce waste and minimize the use of new raw materials. One of its key principles is designing products with the potential for reuse. After a product reaches the end of its life-cycle, it can be recycled and some parts can be reused [6]. A central component of the circular economy is remanufacturing. This process involves disassembling, cleaning, inspecting, repairing, replacing and reassembling components in a product to extend its lifespan [7]. The remanufacturing process consists of several steps. First, products are collected from customers after their life-cycle ends. These products are then disassembled into individual parts. Next, these parts are inspected and tested to determine their current state and sorted into three groups: functional, repairable and destroyed. The functional and repairable components are then cleaned and repaired if necessary. The prepared components are reassembled into a new product and these products are thoroughly tested. Finally, the products are sold to new customers. This cycle can be repeated several times until the components are too worn and discarded [8]. The refurbishing is a form of remanufacturing where modules from returned products are disassembled, cleaned, replaced or restored, and reassembled into refurbished products. These products mainly come from secondhand goods, demo units, open-box products, those with shipping or exterior damage, or production defects. The refurbishment is a commonly used method in which whole used modules are used [9].

Also, e-textiles have been an important and growing market sector in recent years. In the paper, a procedure for the recycling and subsequent reuse of components contacted by a special method on e-textiles will be proposed and verified. Our long experience in this field (e.g. [10]) shows that e-textile product (in our case the conductive ribbon) or electrical connection itself are usually degraded, but the component remains intact. The important parameter to consider in the case of e-textiles is that the sweat from any activity is absorbed by the e-textile, and it has to be washed periodically for hygienic reasons. But the washing and drying can significantly affect the electrical parameters and reliability of e-textiles. They have to be also taken into consideration in the smart textile design [11], [12].

From the theoretical assumptions, two primary objectives of the investigation delineated in this manuscript were established. Firstly, to establish the feasibility of our method for reusing SMD (Surface Mount Device) chip components in e-textiles. Secondly, to evaluate the integrity of specimens containing reused components following a washing and drying cycle test.

The structure of this manuscript is as follows. In the preceding section, the significance of our topic matter and the fundamental principles of the circular economy were delineated. Also the aims of this paper are established in that section. Chapter II is devoted to the methodology and materials used in this research. Chapter III presents the outcomes of the experiment, which entailed 4 iterations of aging and reusing specimens. Each iteration consisted of 5 laundering and drying processes.

II. MATERIALS AND PROCEDURES

A. Textile ribbons

The experiment utilized electrically conductive textile stretchable ribbons protected by a national patent (308614) [13]. The ribbons are composed of PES threads (A) in warp and weft for strength and durability, rubber threads (B) in warp for stretchability, and hybrid conductive threads (C) in warp for electrical conductivity, see Figure 1. The hybrid threads contain PES monofilaments and 8 Ag plated Cu microwires per thread. The ribbons are woven in a stretched state and then shrunk, with the yarns inside the threads arranged in a horseshoe pattern to allow for repeated stretching without damage. The ribbon has conductive traces

with electrical resistance sufficient for standard e-textile applications.

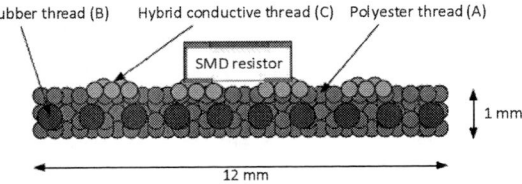

Fig. 1. Schematic cross-section of electrically conductive textile stretchable ribbons used in the experiment

B. SMD components

In the experiment, 15 SMD chip resistors with a case size of 1206 and a resistance of 0 ohms were used for their easy availability and the ability to evaluate the quality of the conductive joint by measuring the electrical resistance. However, our method of reusing components is especially suitable for more expensive or unavailable components on the market. For this reason, experiments were carried out with different components during which their functionality was evaluated.

C. Conductive connection technique

The method for the electrical connection of SMD components, specifically resistors with 0 ohm resistance and a case size of 1206, to conductive lines on ribbons is realized by our special contacting technique [10] utilizing a UV-curable, non-conductive acrylic adhesive (NCA) – specifically Loctite AA 3926 by Henkel Company. The procedure (see Figure 2) begins with dispensing the adhesive onto the space between the conductive lines on the ribbon. Carefully, the component is then placed into the adhesive, typically with the assistance of a tweezer for precision. The component is then pressed into the adhesive with a pressure around 24 MPa using a thorn. This pressure causes the adhesive to be squeezed outside the component pads or into the ribbon, and pushing the component into the ribbon, creating an intimate, direct electrical and mechanical contact, see Figure 3. Once this state of contact is achieved, the adhesive is cured by exposure to UV light for around 30 seconds, while still under pressure. This process ensures fixing the component in its connected position. Finally, the pressing thorn is carefully removed, and the hole left behind, along with the rest of the resistor, is fully encapsulated by the same adhesive for added stability and protection. The advantage of this process is that it provides a simple and reliable way for the mechanical and electrical connection of SMD components, realized through direct physical contact with the conductive pattern on the ribbon, facilitated by the UV-curable adhesive.

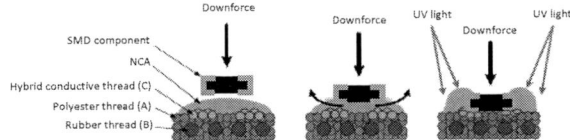

Fig. 2. The principle of SMD resistor electrical connection onto the ribbon by NCA technology.

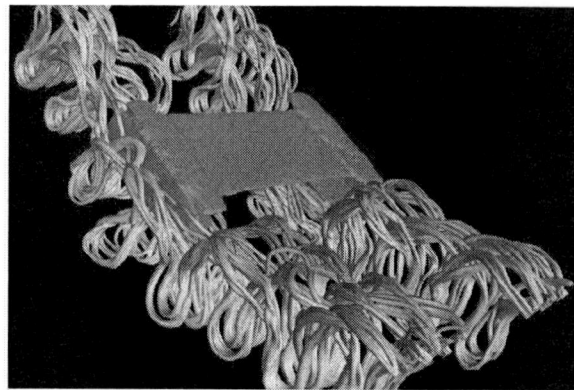

Fig. 3. An example of how an SMD component is connected to conductive ribbons using NCA.

D. Washing and drying cycles

To simulate the aging of the samples during their normal life cycle, washing and drying test according to the standard EN ISO 6330 was used. The washing cycles were chosen 4N (40±3)°C using a washing machine type A and 20 grams of standardized Non-Phosphate SDCE ECE detergent powder for each washing. Drying in an unfolded state, type C was realized (21°C, 63% RH, 24 hours). Overall, 5 cycles consisted of washing and drying process were realized for each reusing cycle. The whole testing procedure can be seen in Figure 4.

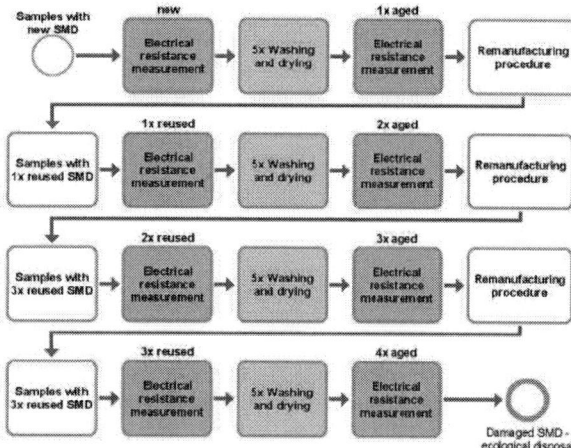

Fig. 4. Testing procedure

E. Recycling procedure

After significant deterioration in the characteristics of the ribbons in use, the product is collected according to the circular economy principle and the remanufacturing procedure is started. The remanufacturing procedure involves several steps, see Figure 5. First, the ribbons are removed from the collected e-textiles and placed onto a surface with acetone for 60 minutes with the component facing upwards. The component is then removed from the ribbon, either manually by a tweezer or automatically by an industrial robotic arm. The used ribbon and glue residues are then environmentally friendly disposed of. In the next step, the removed components undergo visual and functional control and are sorted into three groups: perfect components, components with minor mechanical damage without affecting their function, and

damaged components. The broken components are disposed of and the rest of the components are cleaned from residues by isopropyl alcohol. These components are then reused and contacted onto new ribbons using the same contacting technique as before. New e-textiles are created using these remanufactured samples. The remanufactured samples undergo functional testing and are sorted into two quality categories: perfect (perfect components with perfect function) and second quality (the product is functional but does not meet the demanding specifications set by the manufacturer for its products). In our experiment, the recycling procedure was realized three times, each after the five washing and drying cycles. It follows that four live cycles of components were realized.

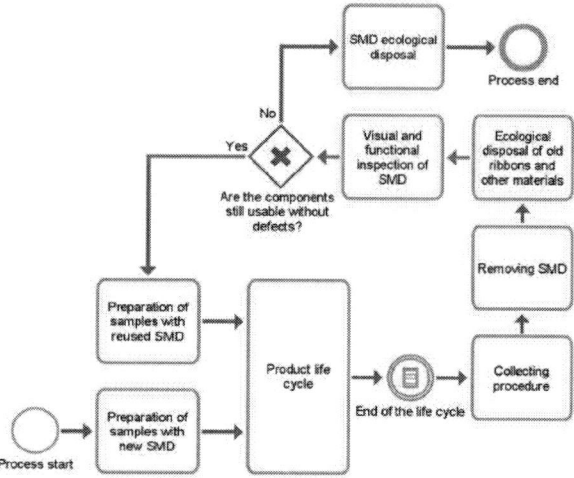

Fig. 5. Recycling procedure used in the experiment.

F. Electrical resistance measurement

The electrical resistance of tested joints was measured for new or reused samples and after aging using the four-point probe method with the Keithley DMM 7510 device. The conductive threads in ribbons were cut off in some places for resistance measurement using an electric discharge generated by a resistance spot welder from Sunstone company. The measured values consisted of the electrical resistance of two joints plus the electrical resistance of the resistor itself.

III. RESULTS AND DISCUSSION

The measured values of electrical resistance were statistically analyzed and presented in the boxplot diagram (see Figure 6). This graph type is important for understanding the distribution of values.

The results show that the electrical resistance of new samples is between 20 to 40 mΩ. The values for the new reused samples are fully comparable and show no degradation compared to the new components.

The results after five washing and drying cycles show that the electrical resistance of bonded joints increases as they are stressed by washing, confirming our previous research on the effect of washing on the reliability of bonded joints [14]. A comparative analysis of the electrical resistance of new and reused joints following washing and drying procedures indicates no significant difference or deterioration in the values of the reused joints.

136

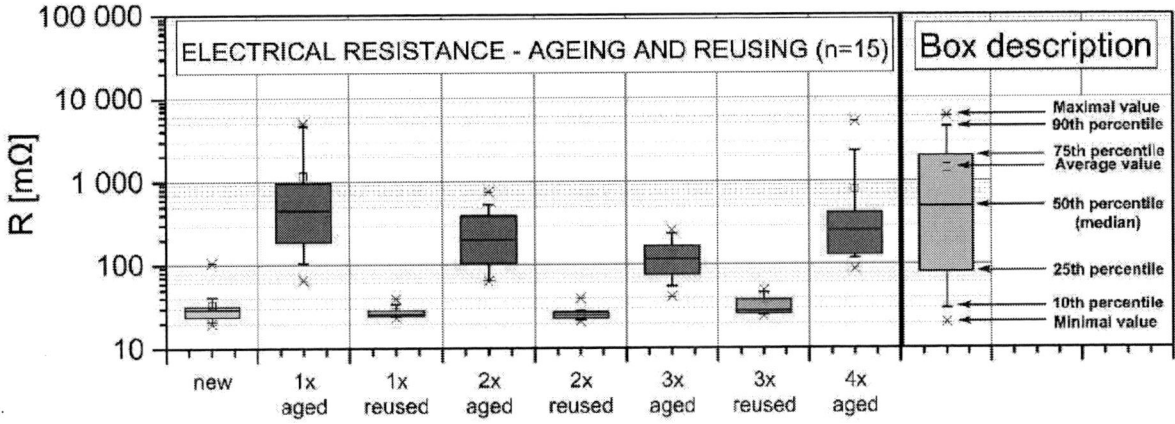

Fig. 6. Boxplot diagram of joints electrical resistance during the reusing after washing and drying cycle test.

However, our component reuse method is particularly suitable for more expensive or unavailable components. For this reason, initial experiments with various other components were also conducted and evaluation of their functionality was carried out. The results showed several generally valid findings. The non-conductive adhesive bonding method used is applicable provided that the components have sufficiently protruding pads compared to the rest of the component. If the pads are in the same plane as the body of the component or even embedded in the component, contacting them using this method is problematic or even impossible. The method of reusing components can be used if the components are sufficiently resistant to acetone and will not be damaged by it. In our experiments, for example, the power LED tested proved to be unsuitable because the transparent acrylic chip cover was degraded. However, it is possible to find LEDs that can withstand exposure to acetone without damage. Furthermore, the method of reusing components is not suitable for components that are too fragile to mechanically withstand the process of removing components from the old ribbon. Finally, it is important that there is a way to test the proper function of the component after it has been removed from the ribbon.

IV. Conclusion

In conclusion, this paper has successfully demonstrated the feasibility of our proposed method for reusing SMD chip components in e-textiles (first objective). As described in chapter II, our method involves the use of a special non-conductive adhesive bonding technique to connect SMD components onto electrically conductive stretchable textile ribbons during the production of a new product. At the end of the product's life-cycle, our recycling method is employed to manufacture a new product using the same (reused) components. The results indicate that our method is both possible and effective, with reused components remaining functional without any impairment to their performance (electrical resistance or functional test).

The second objective of this article was to assess the integrity of specimens containing reused components following a washing and drying cycle test. The results of the experiment demonstrated that our method is usable and that reused samples remain functional, even after undergoing washing and drying testing. Furthermore, the results indicated that our method can be used for multiple life-cycles (ideally no more than four). However, it was also observed that certain component types may be more susceptible to damage or

unsuitable for our method, necessitating slight variations in the procedure depending on the specific type, size, shape, and material of the component.

Our method offers several benefits, including the conservation of raw materials, minimization of waste, and reduction of production costs. Additionally, it can help to mitigate component shortages, which have become a significant issue in many manufacturing lines due to the SARS-CoV-2 pandemic and ongoing supply problems related to the Ukrainian conflict. While the applicability of our method may vary depending on various factors, it is particularly well-suited for use with higher-value or shortage components.

Acknowledgment

This research has been supported from the state budget by the Technology agency of the Czech Republic under the Future Electronics for Industry 4.0 and Medical 4.0 project No. TN02000067 and the Student Grant Agency of the University of West Bohemia in Pilsen, grant No. SGS-2021-003 "Materials, technologies and diagnostics in electrical engineering".

References

[1] Ankit, L. Saha, V. Kumar, J. Tiwari, Sweta, S. Rawat, J. Singh, and K. Bauddh, "Electronic waste and their leachates impact on human health and environment: Global ecological threat and management," *Environ. Technol. Innov.*, vol. 24, p. 102049, Nov. 2021, doi: 10.1016/j.eti.2021.102049.

[2] V. Forti, C. P. Baldé, and R. Kuehr, *The Global E-waste Monitor 2020. Quantities, flows, and the circular economy potential.* UNU/UNITAR SCYCLE, ITU, ISWA, 2020. [Online]. Available: https://www.researchgate.net/publication/342783104_The_Global_E-waste_Monitor_2020_Quantities_flows_and_the_circular_economy_potential

[3] J. C. Kerber, E. D. de Souza, M. Bouzon, R. M. Cruz, and K. Govindan, "Consumer behaviour aspects towards remanufactured electronic products in an emerging economy: Effects on demand and related risks," *Resour. Conserv. Recycl.*, vol. 170, p. 105572, Jul. 2021, doi: 10.1016/j.resconrec.2021.105572.

[4] E. R. Rene, M. Sethurajan, V. Kumar Ponnusamy, G. Kumar, T. N. Bao Dung, K. Brindhadevi, and A. Pugazhendhi, "Electronic waste generation, recycling and resource recovery: Technological perspectives and trends," *J. Hazard. Mater.*, vol. 416, p. 125664, Aug. 2021, doi: 10.1016/j.jhazmat.2021.125664.

[5] B. McCrea, "COVID-19 Disrupts the Electronic Parts Supply Chain," 2020. https://www.sourcetoday.com/industries/article/21131026/covid19-disrupts-the-electronic-parts-supply-chain

[6] European Parliament, "Circular economy: definition, importance and benefits," 2015. https://www.europarl.europa.eu/news/en/headlines/economy/2015120 1STO05603/circular-economy-definition-importance-and-benefits (accessed Nov. 04, 2022).

[7] N. Nasr and M. Thurston, "Remanufacturing: A Key Enabler to Sustainable Product Systems," in *13th CIRP International Conference on Life-Cycle Engineering*, 2006, pp. 15–18. [Online]. Available: https://www.researchgate.net/publication/255582393_Remanufacturin g_A_Key_Enabler_to_Sustainable_Product_Systems

[8] B. Salah, A. Ziout, M. Alkahtani, M. Alatefi, A. Abdelgawad, A. Badwelan, and U. Syarif, "A Qualitative and Quantitative Analysis of Remanufacturing Research," *Processes*, vol. 9, no. 10, p. 1766, Oct. 2021, doi: 10.3390/pr9101766.

[9] H. Liu, M. Lei, T. Huang, and G. K. Leong, "Refurbishing authorization strategy in the secondary market for electrical and electronic products," *Int. J. Prod. Econ.*, vol. 195, pp. 198–209, Jan. 2018, doi: 10.1016/j.ijpe.2017.10.012.

[10] M. Hirman, J. Navratil, F. Steiner, J. Reboun, R. Soukup, and A. Hamacek, "Study of low-temperature interconnection techniques for instant assembly of electronics on stretchable e-textile ribbons," *Text.*

Res. J., p. 004051752210847, May 2022, doi: 10.1177/00405175221084737.

[11] J. C. Lee, W. Liu, C. Lo, and C.-C. Chen, "Laundering Reliability of Electrically Conductive Fabrics for E-Textile Applications," in *2019 IEEE 69th Electronic Components and Technology Conference (ECTC)*, May 2019, pp. 1826–1832. doi: 10.1109/ECTC.2019.00281.

[12] W. Liu, D. Shangguan, and J. C. Lee, "Evaluation of Launderability of Electrically Conductive Fabrics for E-Textile Applications," *IEEE Trans. Components, Packag. Manuf. Technol.*, vol. 10, no. 5, pp. 763–769, May 2020, doi: 10.1109/TCPMT.2020.2981902.

[13] J. Řeboun, R. Soukup, T. Blecha, A. Hamáček, D. Moravcová, M. Ferkl, M. Hotmar, M. Pilíková, I. Hamanová, M. Tichý, and J. Benc, "Conductive elastic woven fabric, in particular conductive elastic woven ribbon," Patent number: 308 614, 2020 [Online]. Available: https://isdv.upv.cz/doc/FullFiles/Patents/FullDocuments/308/308614. pdf

[14] M. Hirman, J. Navratil, F. Steiner, and A. Hamacek, "Effect of Washing Cycles on Glued Conductive Joints Used on Stretchable Smart Textile Ribbons," in *2020 IEEE 8th Electronics System-Integration Technology Conference (ESTC)*, Sep. 2020, pp. 1–4. doi: 10.1109/ESTC48849.2020.9229797.

Process Window of Mini-LED Display Panel Packaging using Laser Assisted Bonding Technology

Yong-Sung EOM*, Gwang-Mun CHOI, Ki-Seok JANG, Ji-Ho JOO, Chan-Mi LEE, Jin-Heuk Oh, Seok-Hwan MOON and Kwang-Seong CHOI

Electronics and Telecommunications Research Institute, 138 Gajeongno, Yuseong-gu, Daejeon, 305-700, Korea

+82 42 860 5547, *yseom@etri.re.kr

Abstract

For high-resolution display panels using mini-LEDs, high alignment accuracy and high yield of LED transfer and bonding technology are very important parameters of packaging technologies. These parameters of packaging technologies are highly dependent on the interconnection material and bonding process. The use of laser assisted bonding (LAB) technology and epoxy-based interconnection materials has been introduced in mini-LED display panel packaging to improve bonding accuracy and yield. The bonding process of the entire display panel was performed by a tiling SITRAB process because the laser irradiation area is smaller than the LED bonding area on the substrate. For the SITRAB process, an epoxy-based SITRAB material to remove oxide on the surface of solder bump has been developed as a solvent free material. For good wetting of the solder bump during the SITRAB process, the temperature of the solder bumps must be higher than the melting temperature of solder. In general, the stage temperature of the laser process is determined to be around 100 ºC in order to minimize the laser power. However, if the stage temperature is high, the initial properties of the SITRAB material (such as low viscosity and deoxidation function) may be lost due to the high temperature. In this study, the process window of the SITRAB material was determined for a stable SITRAB process. The process window was identified according to three different pot lifes: room temperature pot life (RPL), stage pot life (SPL), and laser pot life (LPL).

Key words: simultaneous transfer and bonding, process window, room temperature pot life (RPL), stage pot life I(SPL), laser pot life (LPL), mini LED

Introduction

For high-resolution and high-brightness displays, mini- and micro-LEDs are attracting attention for large-area TV, laptop notebooks, smartphones, automotive head-up displays, virtual reality and augmented reality. The world's first micro-LED display was introduced by Sony at 2012 Consumer Electonics Show (CES), and Samsung demonstrated the world's first spliced 146-inch micro-LED TV, called "the Wall" in 2018 [1]. A 4K full color TV display is known to include 24,883,200 (3840 x 2160 x 3) LEDs. If the device bonding yield of a 4K display panel is 99.99%, there will still be 2,488 damaged pixels to be repaired. The yield and manufacturing cost are meanwhile highly dependent on the LED transfer and bonding technology. For high yield of mini- and micro-LED transfer and bonding, a new technology called "SITRAB" was introduced in SID 2021 [2]. Figure 1 shows a schematic of the simultaneous transfer and bonding (SITRAB) process. First, SITRAB paste or film is deposited on the display panel substrate through screen printing or dispensing or film lamination processes

at room temperature. The SITRAB paste and film consist of epoxy as a base resin, a reductant to remove oxide on the surface of solder bumps, a curing agent, a catalyst to control the rate of chemical reaction, and some addictives to optimize processability. The top surface of the mini- or micro- LED is aligned to the sticky PDMS on the glass interposer with a constant area in the form of an n x n matrix as shown in Figure 1(b). The glass interposer with LEDs held by a quartz chuck is aligned with the display panel substrate. After alignment between the LEDs and the display panel substrate, the solder bumps under the LEDs move down and contact the display panel substrate. An areal laser is then irradiated for a few seconds through a transparent quartz and glass interposer with PDMS, as shown in Figure 1(c). After penetrating the quartz and glass interposer with PDMS, the laser reaches the nontranspaent metal pad and solder bumps below the LEDs, and solder bumps and adjacent SITRAB material are heated to the melting temperature of the solder bumps. When the temperagture reachs the melting temperature of solder, the oxide on the solder surface is removed by the activated SITRAB material, and solder bumps are wetted to the metal pads on the display panel substrate. The LEDs are then separated from the PDMS of the glass interposer because the bonding strength of solder between the LEDs and the metal pads of the display panel substrate is greater than the sticky bonding strength between the LEDs and PDMS, as shown in Figure 1(d).

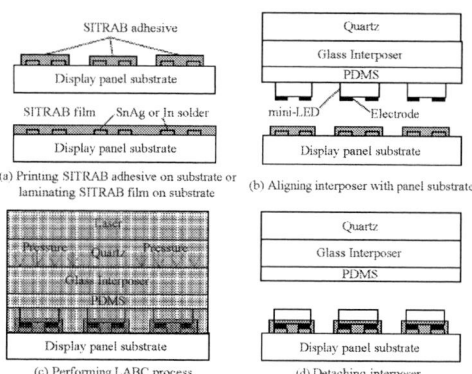

Figure 1: Schematic of SITRAB process [2]

In Figure 1(c), the display panel substrate is placed on a metal stage for the SITRAB process. The temperature of the metal stage is maintained at 80 ~ 100 °C to minimize laser power during the SITRAB process. Therefore, the epoxy-based SITRAB material must remain in a stable state without chemical reaction at a given stage temperature during the alignment process between the LEDs and the substrate. If the total area of the display panel is larger than the laser irradiation area, the SITRAB process for LED bonding must be repeatedly performed through tiling or laser line scanning processes. As shown in Figure 2, if there are three kinds of glass interposers performed with red, green, and blue LEDs, respectively, the SITRAB material is exposed three times in three SITRAB processes. Therefore, the SITRAB material can be exposed a maimum of six times by the SITRAB process due to the three different interposer applications and the tiling SITRAB process.

Table 1 shows three pot life definitions of SITRAB materials for the SITRAB process. Room temperature pot life (RPL) is determined by the time it takes for the SITRAB paste to reach 1.2 times its initial viscosity at 25 °C. Stage pot life (SPL) is defined as the time that the SITRAB paste can remain without undergoing a chemical reaction at a given stage temperature, such as 80, 90, and 100 °C. After the SPL time, the chemical performance of SITRAB paste is demonstrated by solder wetting tests. As illustrated in Figure 2, the SITRAB paste placed on a substrate with a given temperature of stage allows up to six laser irradiations due to the three interposers and the

tiling process. Therefore, laser pot fife (LPL) is defined as the number of times that laser is irradiated to the SITRAB paste at a given temperature of stage. After LPL, the SITRAB material is evaluated with solder wetting tests. In the current research, the RPL, SPL, and LPL of SITRAB paste are studied according to chemo-rheologically properties.

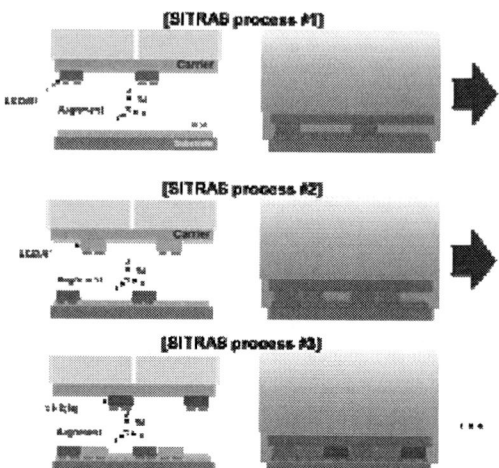

Figure 2: Schematic diagram of SITRAB process with three types of interposers [3]

Table 1: Definition of three different pot lifes of SITRAB material for SITRAB process

Items	Abbre viation	Environmental Temperature	Requirement
Room Temp. Pot Life	RPL	25°C	1.2 x Initial viscosity
Stage Pot Life	SPL	80, 90, 100°C	Solder wetting
Laser Pot Life	LPL	Laser Irradiations on 25, 80, 100°C Stage	Solder wetting

Materials and Experiment

Two types of SITRAB paste are prepared. One is the SITRAB paste with a viscosity of 37,000 cPs at 25 °C and 10 rpm. In Figure 1, if the solder bump is fabricated on the substrate, the SITRAB paste can chemically react by the catalytic activity of SAC305 solder [4]. The other material is SITRAB paste mixed with 1 wt.% SAC305 type 7 powder. When the solder bump is fabricated on the display panel, the surface area of the solder bumps of one pixel including three types of LEDs is similar to that of 1 wt.% type 7 solder.

The viscosity of SITRAB paste was measured at 25 °C and 10 rpm using high viscometer (BrookField Ltd.) for RPL. For SPL, the SITRAB paste was installed between a circular parlllel plate with a diameter of 20 mm. Its viscosity was measured at 1 Hz and a given temperature of stage, such as 80, 90, and 100 °C, using a rheometer (HAAKE, Mars 3). To evaluate solder wetting performance, after the SPL, a solder ball with a diameter of 0.5 mm was put into the SITRAB paste on a Cu plate, heated to 240 °C at a heating rate of 120 °C/min, and held for 10 minutes. To simulate the LPL, the SITRAB pastes placed on a silicon substrate was irradiated with an areal laser up to six times at 5-minute intervals at a given temperature of stage. After laser irradiations (LPL), the uncured glass transition temperature, conversion, and solder wetting performance were observed.

Results and Discussion

The viscosity and thixotropy index changes of the SITRAB paste were measured for six days at 25 °C, 10 rpm as shown in Figure 3. The initial viscosity and thixotropy index ere 37,000 cPs and 1.4, respectively, and the room temperature pot life (RPL) was six days, because the viscosity was increased to 1.2 times of its initial viscosity after six days.

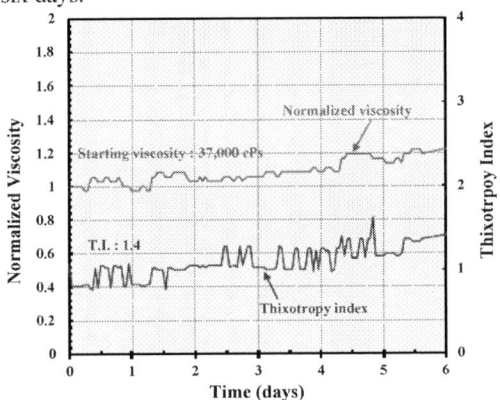

Figure 3: Measured room temperature pot life (RPL) of SITRAB paste.

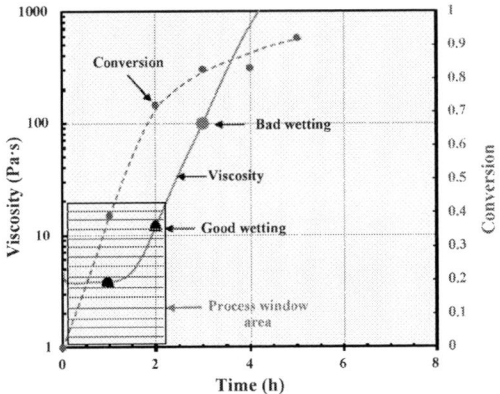

Figure 4: Satge pot life (SPL) of SITRAB paste at the stage temperature of 100 °C.

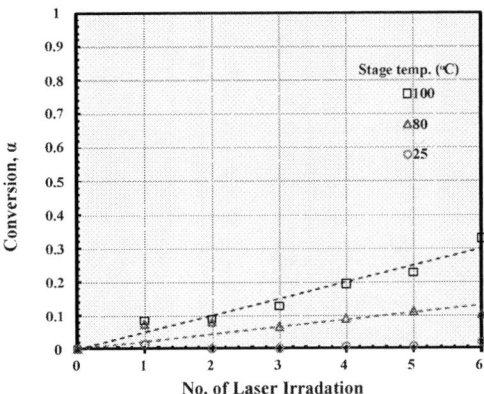

Figure 5: Laser pot life (LPL) of SITRAB paste at given stage temperatures.

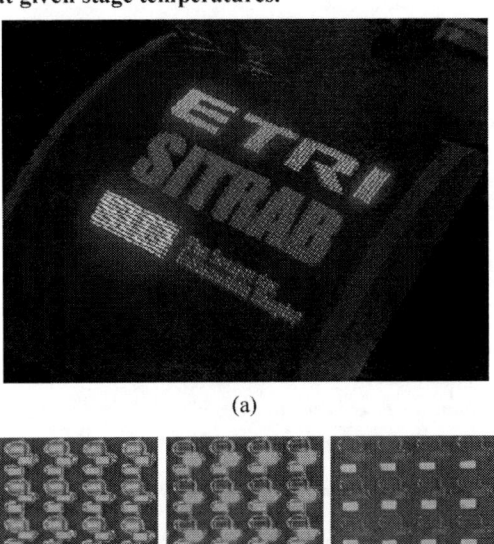

(a)

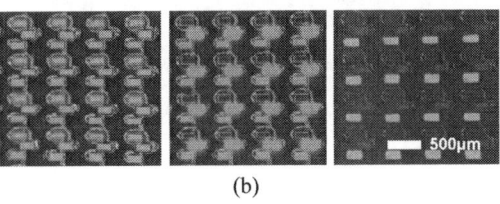

(b)

Figure 6: Fabricated full-color mini-LED display on flexible panel: (a) 64 x 64 color, and pixel pitch 0.576 mm display with mini-LED, (b) bonded red, green and blue LED by process window of SITRAB paste.

For stage pot life (SPL) of the SITRAB paste at a stage temperature of 100 °C, the viscosity, conversion, and solder wetting were measured. Good solder wetting was observed up to 2 hours, the conversion and viscosity of the SITRAB paste were approximately 0.7 and 20 Pa·s after two hours at 100 °C stage temperature, as shown in Figure 4. Thus, it was clearly shown that the process window for 100 °C SPL was two hours. Figure 5 shows the conversion changes with anincreasing number of laser irradiations at the stage temperature of 80, 90, 100 °C. After six iterations of laser irradiation at a stage temperature of 100 °C, the conversion of the

SITRAB paste was about 0.33, which was lower than the conversion after two hours at SPL in Figure 4. Therefore, it was observed that the process window of 100 °C LPL was higher than six iterations of laser irradiations.

Figure 6(a) shows a full-color mini-LED display fabricated with a 64 x 64 pixel array with 0.576 mm pixel pitch. The SITRAB bonding process using red, green and blue interposers are performed sequentially in the process window of SITRAB paste. In addition, the tiling SITRAB process was applied because the laser irradiation area was smaller than that of the display panel.

Conclusion

For mini- and micro-LED display panel packaging applications, a new SITRAB paste and process were introduced. An epoxy based SITRAB paste was developed for adoption in the SITRAB process using area laser irradiation at high stage temperatures. The process window of the SITRAB material was defined according to room temperarture pot life (RPL), stage pot life (SPL), and laser pot life (LPL). A full-color mini LED 64 x 64 pixel array display panel was successfully fabricated according to the process window of the tiling SITRAB process.

Acknowledgements

This research was supported by National R&D Program through the National Research Foundation of Korea (NRF) funded by Ministry of Science and ICT (NRF-2020M3H4A3081764, NRF-2020M3H4A3106383), a Korea Evaluation Institute of Industrial Technology (KEIT) grant by the Ministry of Trade, Industry and Energy (20010580), and an Electronics and Telecommunications Research Institute (ETRI) grant funded by the Korean government [22YB1110, Core technology for new microwave-reactive materials for low-carbon, high-quality semiconductor processing]. The authors would like to thank InSeok Gae and YoonHwan Moon for their support in the sample preparation and measurements.

References

[1] Y. Wu et al., Nanomaterials, Vol. 10, no. 2482, 2020.

[2] K.-S. Choi et al., SID Digest 2021, 30-4, pp. 436-439.

[3] J. Joo et al., SID Digest 2022, 74-4, pp. 1005-1008.

[4] Y.-S. Eom et al., ETRI J. 2014, Vol. 36, no. 3, pp. 343-351.

Effect of surface microstructure on joints using nanoporous Cu sheet for power devices

Hiroshi Nishikawa
Joing and Welding Research Institute
Osaka University
Osaka, Japan
nisikawa@jwri.osaka-u.ac.jp

Byungho Park
Graduate school of Engineering
Osaka University
Osaka, Japan

Mikiko Saito
Reseach Organization for Nano & Life
Inovation
Waseda University
Tokyo, Japan

Jun Mizuno
Reseach Organization for Nano & Life
Inovation
Waseda University
Tokyo, Japan

Abstract— **Wide-bandgap (WBG) semiconductors such as silicon carbide (SiC) and gallium nitride (GaN) are being developed as promising replacements for Si-based semiconductors. The operating temperature of SiC semiconductor is expected to be much higher. The replacement of soldering with new bonding process has been needed for WBG semiconductor devices. We have proposed a solid-state bonding process using nanoporous metal sheet, as a die-attach bonding method for WBG semiconductor devices. In this study, the bonding process using nanoporous Cu sheet as an insert metal was investigated to achieve a Cu disk/Cu disk bonding without solvent and flux. The shear strength of the joint using nanoporous Cu sheet at 350 and 300 °C in formic acid atmosphere exceeded the shear strength of conventional Pb–5Sn solder joints which was approximately 18 MPa.**

Keywords—Solid-state bonding, nanoporous Cu, dealloying, power module, shear strength

I. Introduction

Recently, there has been interest in improving the performance of electronic power modules that are used in electric vehicles and renewable energy sources for effective utilization of energy. And the electronic power modules are needed to endure harsh operating conditions such as more than 200°C. To assemble future power modules, high-thermostability packaging technologies are surely needed. Then, the requirement for a high-thermostability die attach material is also desirable for applications in areas such as aerospace and vehicles. Assembled joints in these power modules are required to exhibit good reliability against harsh environment. Although Pb-based solder has been used to join chips to base materials made of copper, a strong drive exists to find good Pb-free alternatives for power modules. In addition, alternative materials must be able to demonstrate high thermal and electrical conductivity; low coefficient of thermal expansion mismatch between the chip and the substrate; suitable mechanical properties which provide stress relaxation after bonding. Researchers have investigated alternative materials and bonding processes to withstand harsh conditions. A bonding process using nanoparticles have been proposed as a solder alternative [1, 2]. The sintering behavior of nanoparticles has attracted interest, because it is well known that nanoparticles of metals have lower sintering and melting temperatures than the bulk metal, which decouples the bonding temperature from the operating temperature. The other advantages of silver and copper are their high thermal and electrical conductivities in the range 6 to 8 times higher than Sn-Pb solder [3]. The sintering behavior of metal nanoparticles has exploited to join chips to substrates. However, there are some drawbacks of bonding process using metal nanoparticle pastes; for example, it is difficult to produce suitable nanoparticle pastes for the process conditions and the residual organic substances after the bonding process can induce unexpectedly large voids and insufficient densification in the joint [4]. These defects lead to decrease of the long-term reliability of the joint.

To solve these issues, we have proposed a novel solid-state bonding process suitable for application in future power modules. This process achieves nanoporous bonding (NPB) via nanoporous metal sheets in the absence of organic substances. Nanoporous metals are usually prepared by a chemical dealloying method, which involves the selective dissolution of less noble metal atoms into an acid solution from a precursor alloy. More noble metal atoms diffuse along the surface of the alloy to form nanoporous structures [5]. Kim et al. reported that solid-state NPB can be achieved using a nanoporous Ag sheet without the need for any solvent and organic substances [6]. Recently, we try to use nanoporous Cu sheet for solid-state NPB because Cu is less expensive than Ag and Au, while its electrical and thermal conductivities are excellent. Koga et al. reported that solid-state NPB using nanoporous Cu dealloyed from Mg-Cu precursor sheet achieved good shear strength at approximately 40 MPa [7]. Cu-based precursor alloys are generally fabricated by melt spinning method. However, our previous study revealed that the nanoporous Cu sheet by cold-rolled Mn-Cu precursor could achieve a homogeneous and large crack-free nanoporous structure using an adequate annealing process [8]. In this study, the bonding process using nanoporous Cu sheet formed from Mn-Cu precursor was investigated to achieve a Cu disk/Cu disk bonding without solvent and flux.

II. Experimetal

A. Fabrication of nanoporous Cu sheet

Nanoporous Cu (NPC) sheets were prepared using a chemical dealloying method, which involves the selective dissolution of Mn into 4% HCl from a Mn–30 at.% Cu precursor alloy. Pure Mn (99.99 mass %, Nilaco Co., Japan), and Cu (99.9 mass %, Nilaco Co., Japan) were used to prepare a parent alloy, and Mn–30 at.% Cu ingot was prepared by arc melting in an Ar atmosphere. Then, a cold-rolling process was adopted to fabricate Mn–Cu precursor alloy sheet with a thickness and width of 110–120 μm and 18–20 mm,

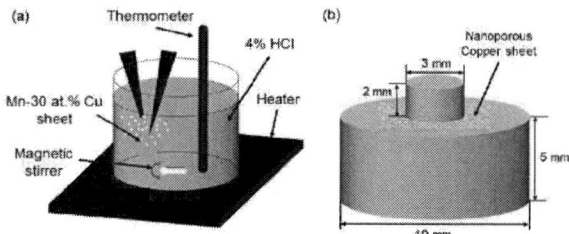

Fig. 1 Schematic diagram of (a) chemical dealloying method and (b) joint sample.

respectively. The cold-rolled sheets were immersed in a 4% HCl solution and dealloyed for 4 h at 50°C, as shown in Fig. 1(a). The dealloyed sheet samples were subsequently washed in distilled water and ethyl alcohol and dried in a vacuum desiccator. The sheet samples were examined using an X- ray diffractometer (Ultima IV, Rigaku, Japan) equipped with a Cu Kα radiation source.

Before a bonding experiment using a Cu disk, the nanoporous Cu sheet itself was heated at 350, 300, 250 °C for 10 min in a N2 atmosphere to clarify the sintering behavior of nanoporous structure on the surface of the sheet.

B. Bonding process

Two oxide-free Cu disks were bonded using the nanoporous Cu sheet as an insert material under various bonding conditions; the diameter and height of the top Cu disk were 3 mm and 2 mm, respectively, and of the bottom Cu disk were 10 mm and 5 mm, respectively, as shown in Fig. 1(b). To remove oxide layer on the Cu disks before bonding, they were first immersed in a 4% HCl solution using an ultrasonic cleaner for 5 min, and then immersed in ethyl alcohol using an ultrasonic cleaner for 5 min. Next, the nanoporous Cu sheet was put on the bottom Cu disk before the top Cu disk was attached. A thermo-compression bonding system (RB-100D, Ayumi Industry Co., Ltd, Japan) was used to achieve a Cu disk/ Cu disk joint in a N2 or formic acid atmosphere. The applied pressure was 10 MPa and the bonding temperature was 350, 300, 250 °C for 10 min.

C. Evaluation method

The shear strength of the joints was evaluated using a shear tester (STR-1001, Rhesca, Japan) at a strain rate of 1 mm/min, and the average shear strength of the five samples was obtained. The shear strength of a joint was calculated as the maximum fracture load divided by the area of upper disk. The surfaces of the NPC sheets after a shear test were observed using optical microscopy (OM, DM2700M, LEICA, Germany). The cross sections of the joints were observed using field emission scanning electron microscopy (FE-SEM, SU-70, Hitachi, Japan). The porosity ration in the joint layer was analyzed.

III. RESULTS AND DISCUSSION

The oxidation behavior of nanoporous Cu sheer just after dealloying was examined. Fig. 2 shows the XRD patterns of as-dealloyed nanoporous Cu sheet and stored nanoporous Cu sheet in air for 1 h. In the case of the as-dealloyed NPC sheet, diffraction peaks that are characteristic of face-centered cubic (fcc) Cu are observed, corresponding to its (111), (200), and (220) planes. In the case of the stored NPC sheet in air for 1h, low-intensity peaks near 36.6° and 61.3° were observed

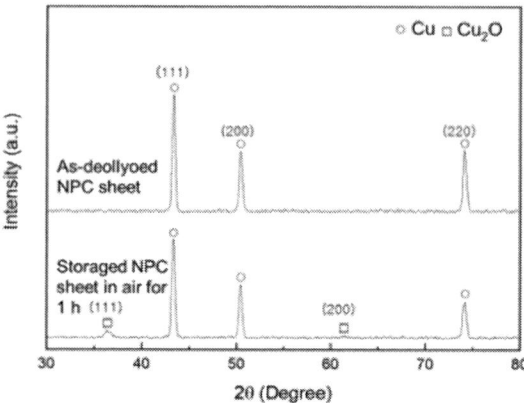

Fig. 2 XRD patterns of as-dealloyed NPC sheet and stored NPC sheet in air for 1 h.

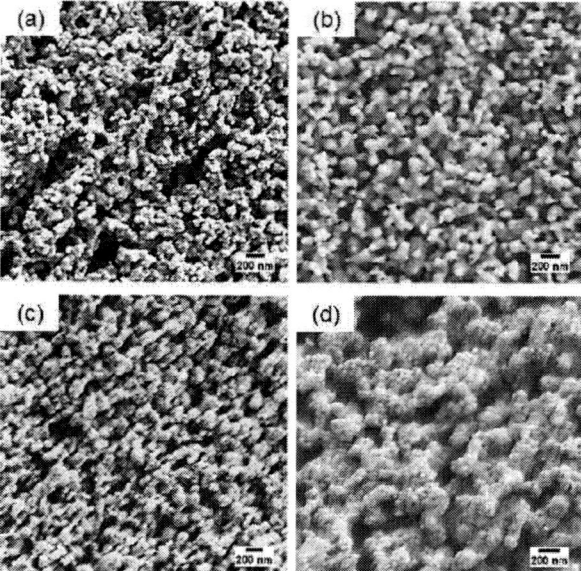

Fig. 3 SEM images of surface structure of (a) as-dealloyed NPC sheet and NPC sheets heated at (b) 250℃, (c) 300℃, (d) 350℃, (e) 350℃ for 10 min.

additionally, characteristic of Cu2O, indicating the presence of Cu2O layer on the stored NPC sheet in air for 1 h.

To clary the sintering behavior of nanostructure on the surface of NPC sheet at different temperatures, the microstructure of as-dealloyed NPC sheet and NPC sheets heated for 10 min at 350, 300 and 250 °C were observed by FE-SEM in Fig.3. As shown in Fig. 3, the heated NPC sheets showed three-dimensional porous structures similar to that of the as-dealloyed NPC sheet; however, their structures were coarsened, and the size of their ligaments increased with increasing the heating temperature for 10 min. The average ligament size of the as-dealloyed NPC sheet was 122 nm. The average ligament sizes of the heated NPC sheets for 10 min were 230 nm, 179 nm and 169 nm at 350 °C, 300 °C, and 250 °C respectively. The nanoporous structure was markedly denser. These results indicated that the ligaments were gradually coarsened by an increase of heat-treatment temperature because of accelerating surface diffusion of Cu atoms during heat treatment.

The nanoporous Cu sheet was applied as an insert metal for bonding between Cu disks. A thermo-compression

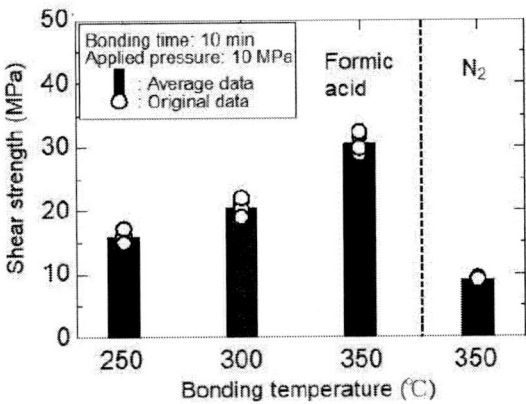

Fig. 4 Shear strength of NPC joints between Cu disks formed in N2 or formic acid atmosphere at various bonding temperatures for 10 min.

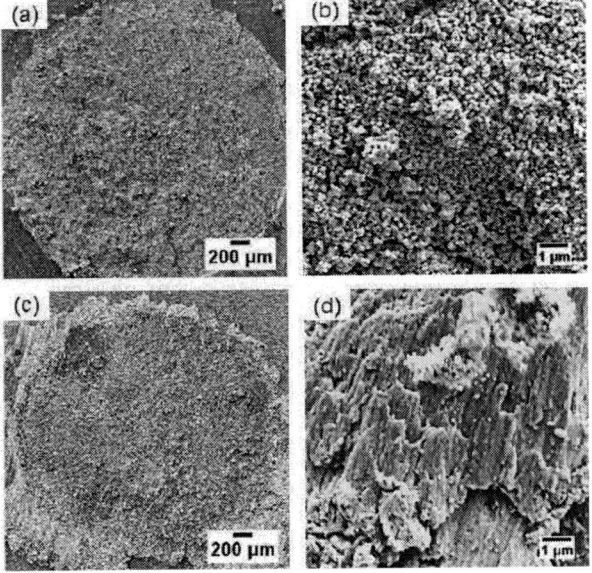

Fig. 5 Fracture morphologies after shear test for NPC joints on Cu disk at 350 °C for 10 min formed under (a and b) N2 and (c and d) formic acid atmospheres.

bonding system was used to achieve a joint at 350, 300 and 250 °C for 10 min in N2 or formic acid atmosphere. The applied pressure was 10 MPa. The shear strength of NPC joints is shown in Fig. 4. In the case of formic acid atmosphere, the shear strength of the NPC joint on Cu disk formed at 250 °C was approximately 16 MPa, and it gradually increased with increasing bonding temperature up to 30 MPa at 350 °C. The shear strength bonded at 350 and 300 °C in formic acid atmosphere exceeded the shear strength of conventional Pb–5Sn solder joints which was approximately 18 MPa. On the other hand, in the case of N2 atmosphere, the shear strength of the NPC joint on Cu disk formed at 350 °C was only around 9 MPa. The results revealed that the shear strength of NPC joints obtained in formic acid atmosphere is much higher than that of NPC joint obtained in N2 atmosphere. Fig. 5 shows fracture morphologies after the shear test of NPC joints under different atmospheres. Differences in the fracture morphologies between NPC joints

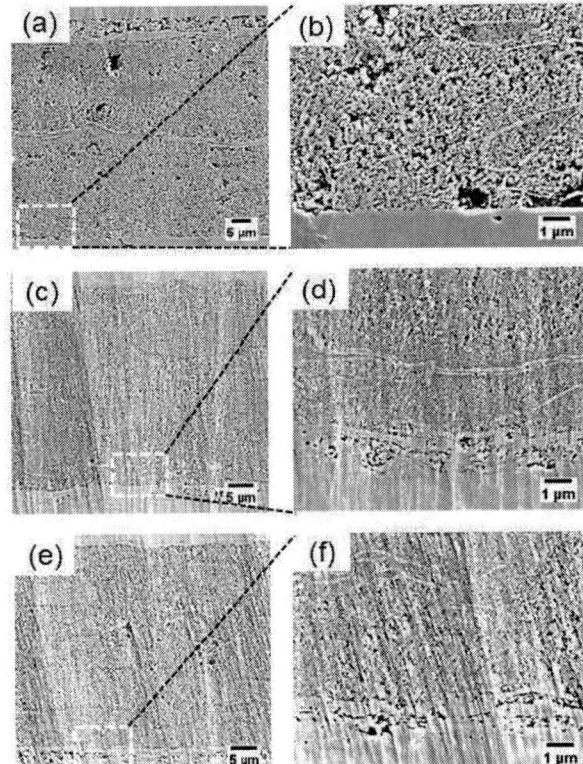

under different atmospheres were confirmed by SEM analysis. As shown in Fig. 5, the fracture

Fig.6 Cross-sectional SEM images of NPC joints formed under formic-acid atmosphere at (a) 250°C, (c) 300°C, (e) 350°C. And (b, d, f) High-magnification images marked by white squares from low-magnification images.

morphology of NPC joint under N2 atmosphere consisted of porous morphology after shear test. In contrast, the fracture morphology of NPC joint under formic acid atmosphere comprised dimple structure from the plastic deformation after shear test. These results indicated that the NPC joint under formic acid atmosphere was well bonded between NPC and Cu disk by reduction of oxidation layer of Cu.

The shear test results revealed that the shear strength of NPC joints under formic acid atmosphere increased clearly with increase of the bonding temperatures from 250 ºC to 350ºC. Therefore, different microstructures of NPC layer with various bonding temperatures were supposed to observe by SEM. Fig. 6 shows cross-sectional images of NPC joints on the Cu disk formed at various temperatures for 10 min under formic acid atmosphere. As can be seen in Fig. 6, the NPC joint on the Cu disk formed at 250 °C were incomplete, and some voids in the NPC layer and gaps between the NPC sheet and Cu disk were observed. The NPC joint on the Cu disk formed at 300°C contained denser structure than low-temperature range at 250 °C that were associated with neck growth that led to the densification of the microstructure. Moreover, in the NPC joint on the Cu disk formed at 350°C, significant denser structure could be observed. The bonding temperature is critical to the formation of the NPC joints on the Cu disk because an appropriate bonding temperature accelerates the densification reaction of NPC layer by applying pressure as well as increase of the diffusion rate of

Cu atoms between the NPC sheet and Cu disk to achieve a robust NPC joint.

IV. CONCLUSION

The bonding process using nanoporous Cu sheet formed from Mn-Cu precursor was investigated to achieve a Cu disk/Cu disk joint without solvent and flux. The morphology and XRD patterns of a self-oxidized NPC sheet confirmed the presence of a Cu_2O film on its surface. The shear strength of the NPC joints on the Cu disk ranged between 9 MPa and 30 MPa, depending on the bonding environment. The shear strength of the NPC joints increased under a formic acid atmosphere as the oxide layer on the surface of the NPC sheet was reduced by formic acid, facilitating the diffusion of Cu atoms. It means the surface nanostructure of nanoporous Cu sheet can enhances rapid surface diffusion of Cu atoms as well as the local plastic deformation.

REFERENCES

[1] E. Ide, A. Angata, A. Hirose and K.F. Kobayashi, "Metal-metal bonding using Ag metallo-organic nanoparticles," Acta Materialia, vol. 53, pp. 2358-2393, 2005.

[2] T.G. Lei, J.N. Calata and G-Q. Lu, "Low-temperature sintering of nanoscale silver paste for attaching large-are chips," IEEE Transactions on Components and Packaging Technology, vol. 33, pp. 98-104, 2010.

[3] F. Le Henaff, S. Azzopardi, J. Y. Deletage, E. Woirgard, S. Bontemps, J. Joguet, "A preliminary study on the thermal and mechanical performances of sintered nano-scale silver die-attach technology depending on the substrate metallization," Microelectron. Reliab. vol. 52, pp. 2321-2325, 2012.

[4] J. Yan, G. Zou, A. Wu, J. Ren, A. Hu and J.N. Zhou, "Improvement of bondability by depressing the inhomogeneous distribution of nanoparticles in a sintering bonding process with silver nanoparticles," Journal of Electronic Materials, vol. 41, pp. 1924-1930, 2012.

[5] J. Erlebacher, M.J. Aziz, A. Karma, N. Dimitrov, K. Sieradzki: "Evolution of nanoporosity in dealloying", Nature, vol. 410, pp. 450-453, 2001.

[6] M.S. Kim and H. Nishikawa, "Effects of bonding temperature on microstructure, fracture behavior and joint strength of Ag nanoporous bonding for high temperature die attach," Materials Science and Engineering: A, vol. 645, pp. 267-272, 2015.

[7] Shunichi Koga, Hiroshi Nishikawa, Mikiko Saito, Jun Mizuno: Fabrication of Nanoporous Cu Sheet and Application to Bonding for High-Temperature Applications, Journal of Electronic Materials, vol. 49, pp. 2151-2158, 2020.

[8] Byungho Park, Duy Le Han, Mikiko Saito, Jun Mizuno, Hiroshi Nishikawa: Fabrication and characterization of nanoporous copper through chemical dealloying of cold-rolled Mn–Cu alloy, Journal of porous materials, vol. 28, pp. 1823-1836, 2021

Investigation of Aluminum and Gold Flip-Chip Bonding for Quantum Device Integration

I. Cirulis[1], U. Zschenderlein[2], S. Braun[1], M. Radestock[1], R. Pantou[1], K. Vogel[1], F. Selbmann[1], S. Kurth[1], B. Wunderle[1,2], H. Kuhn[1,2]

[1]Fraunhofer Institute for Electronic Nano Systems, Chemnitz, Germany
[2]Technical University Chemnitz, Chemnitz, Germany

Abstract— The majority of qubit chip integration is realized in a two-dimensional (2D) architecture. Whereas 3D architecture enables more advantages like efficient interconnect routing, allowing for more compact qubit coupling geometries, reducing form factor, and increased connectivity beyond nearest-neighbor interactions. Flip-chip (FC) assembly has been demonstrated to enable 3D architecture connecting qubit chip, interposer, and readout in a sandwich-like structure. Moreover, 3D integration allows the fabrication of hosting chip circuitry without degrading the qubit performance [1–4]. Different material considerations have to be taken into account since qubit operational frequency is in the gigahertz range and operating temperature is in the mK range to avoid thermal excitation. Materials that possess superconducting characteristics like Indium (In), Titanium Nitride (TiN), Tantalum Nitride (TaN), and Niobium (Nb) have been discussed as potential interconnect between building blocks [3, 5]. The In bumping on Aluminum (Al) redistribution layers require under-bump metallization (UBM) layers thus introducing multiple fabrication steps before the bonding process. An alternative approach would be to use the existing Al surface to electrochemically grow Al bumps and bond the chip using thermosonic bonding (TSB) at below 150°C to form a homogeneous metal-metal interface [6, 7].

Keywords—Aluminum electrodeposition, Aluminum flip-chip bonding, Thermosonic bonding, Gold flip-chip bonding, Chip ultrasonic bonding, Thermal stress, Qubit chip integration, Finite Element analysis

I. INTRODUCTION

In ion trap-based quantum computers, the electrical, magnetic, thermal, and thermo-mechanical performance of the heterogeneously integrated package at cryogenic temperatures is crucial for the number of usable qubits per integrated area and thus ultimately determines the achievable computing power [4, 8–13]. Qubit chip integration using Flip-Chip bonding has been demonstrated to enable dense chip integration in 3D architecture [5, 14–19]. Wherein, traditional materials like copper (Cu), which is a standard metallization layer, are not superconducting due to quasiparticle excitation from heat formation thus reducing qubit performance. Also, silicon dioxide (SiO_2) and silicon nitride (SiN) dielectrics are reported to decrease the qubit lifetimes due to the qubit electric field interaction with defects [20]. Materials like Aluminum (Al), titanium nitride (TiN), and niobium (Nb), where substrate materials are silicon (Si) and sapphire (Al_2O_3), are discussed and used in the scientific community for qubit chip fabrication and integration [1, 3].

Different qubit chip integration methods can be used like wire bonding, soldering using In bumps as bumping material, or TSB using Al pillars to create a homogeneous metal-metal interconnect [1–3, 6, 7, 15, 21]. FC bumping using In and TCB have already been successfully demonstrated, however, In bumping requires UBM layers like Titanium (Ti), Platinum (Pt), Gold (Au), and Niobium nitride (NbN) to prevent the formation of Al-In intermetallic state, thereby introducing more fabrication steps and complex equipment to achieve the final layer before In electroplating process [1]. Also, considering CMOS processes the final metallization layer usually is Al, which can be utilized as a base for Al electrodeposition (ECD) [6, 7, 22, 23]. Finally, the Al-ECD-grown pillars could be bonded using the TSB method to bond the chip at below 150°C. Moreover, homogeneous metal-metal interconnects would benefit the overall system owing to reduced fabrication costs and good electrical contact between building blocks. The Au pillar chip bonding was used to compare the bond strength and cryogenic stress with Al pillar bonding.

II. EXPERIMENTAL

Aluminum and gold pillars were electrodeposited on 6-inch wafers with a thickness of 675 μm ± 25 μm. The substrate wafer had a similar wafer thickness with final Al and Au metallization layers.

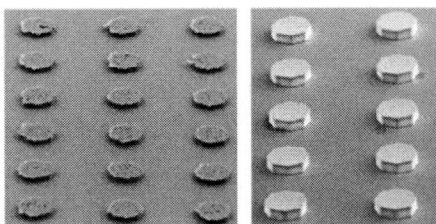

Figure 1. Aluminum (left) and Gold (right) chips with pillars

For the shear strength test, chips with electroplated Al and Au pillars were used to compare the bonding strength between the two materials. The average pillar thicknesses of Al and Au were 15.6 μm ± 0.5 μm and 15.2 ± 0.1 μm respectively (Figure 1). Each chip had 42 pillars with a pillar diameter of 100 μm and pillar distances of 120 μm in horizontal and 160 μm in vertical directions. Chips and substrates had sizes of 1.5mm x 1.5mm and 5mm x 5mm, respectively (Figure 2). The final bonding layer on the substrate was 1000 nm for Al and 250 nm for Au. The overall active bonding area of the pillars was 0.33 mm². The TSB of both chip materials was done using FINEPLACER® femto 2 from the company Finetech GmbH. The bonding machine is equipped with an ultrasonic module with a frequency of 42 kHz. An ultrasonic bonding tool with a flat surface was used for the experiments. The bonding parameters were previously experimentally tested.

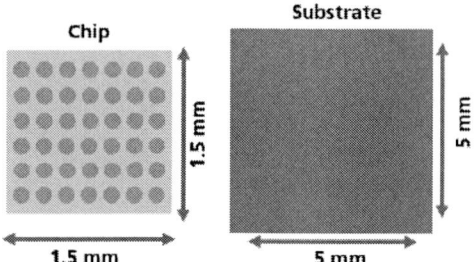

Figure 2. Flip-chip samples with chip dimensions of 1.5mm² and 42 pillars. The substrate with dimensions of 5mm² and PVD surface

A thermal stress test was conducted to measure bonded Al chip shear strength, with shear tester Condor Sigma from company XYZTEC, after immersing it in liquid nitrogen (-196°C) for the 60s and thawing it at +25°C for 30s (Figure 3). The first set of bonded Al and Au chips were sheared at room temperature (A) for comparison purposes and the second set of Al bonded chips was exposed to thermal cycling of 5x and 10x cycles (B) before the shear test.

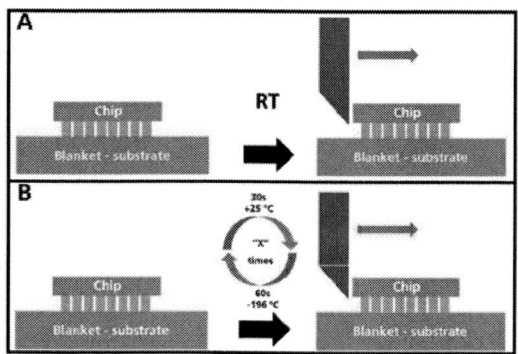

Figure 3. Bonded gold and Aluminum chip shear tests at A) Room temperature (25°C) and B) After thermal cycling

III. RESULTS & DISCUSSION

Results and discussion are separated into two parts – The Finite Element model (FE) and TSB using Al and Au samples.

A. Finite Element Analysis of the joint materials

The objective was to build FE models to study the thermo-mechanical performance of metallic joints on 3-layer setups (substrate, joint, chip) and to present an approach scalable to the package level. These models will further serve as a basis for future investigation of the electrical, magnetic, and thermal behavior of packages at cryogenic temperatures. The analysis was carried out using the simulation software Ansys.

The thermomechanical investigation considers a single joint since the chip and substrate are made of silicon and therefore exhibit no thermal mismatch. The metallization and joint are made of Al. The static analysis includes thermal shock cycling from 293 K down to 77 K and back to 293 K. Any visco-plastic behavior of the Al is neglected. The plastic deformation in the Al during cycling leads to fatigue crack growth. The accumulated equivalent plastic strain is quantitatively closely correlated with damage in ductile metals like Al and serves as a physical failure parameter.

Hence, this quantity was determined together with the *VON MISES* stress.

The FE model consists of a single joint with dimensions similar to the FC samples used in the shear tests. Figure 4 shows the quarter model of a joint with its mesh. The volumes are divided into chip, substrate, and joint together with seed layer (metallization) on chip and substrate. The boundary conditions include the fixation of the symmetry planes as well as the fixation of a point in the center in all 3 directions. The loading is applied at a uniform temperature of 293 or 77 K respectively.

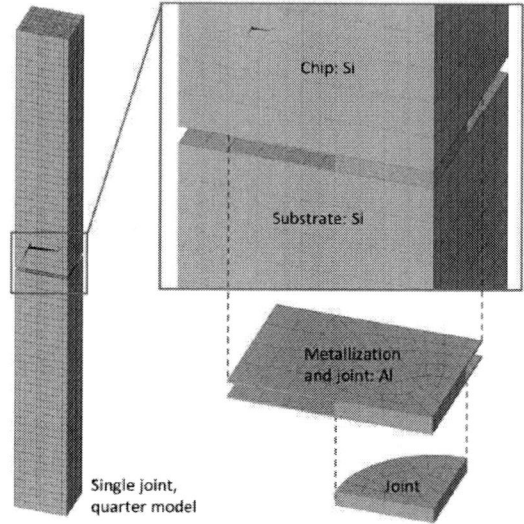

Figure 4. Quarter model of a single joint with its materials.

Temperature-dependent material data of Si and Al between 293 and 77 K is required for the modeling. The isotropic Young's modulus as well as coefficients of thermal expansions (CTE) are presented in Figure 5. The anisotropy of the mono-crystalline Si was neglected and isotropic Young's modulus and Poisson's ratio were used, computed from elastic constants considering the isostrain assumption (*VOIGT* average modulus) [24, 25].

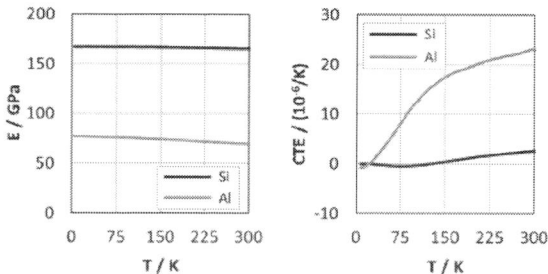

Figure 5. Anisotropic Young's modulus E (left) and coefficient of thermal expansion (right) of silicon and aluminum [24–27]

However, the actual plastic deformation of metals depends strongly on their microstructure. Therefore, that data must be obtained at a certain temperature from samples that feature the same microstructure as in the actual package. Since such equipment is not available at the author's facilities, the plastic behavior was estimated based on temperature-dependent offset yield strength (0.2% strain) and ultimate stress presented for a low alloy Al-1100 [27]. The ultimate strain was then chosen. Assuming a material behavior according to Ramberg-Osgood, temperature-dependent stress-strain curves could thus be determined. Their multilinear approximation is given in Figure 6. The Poisson's ratio was chosen temperature independent with 0.33.

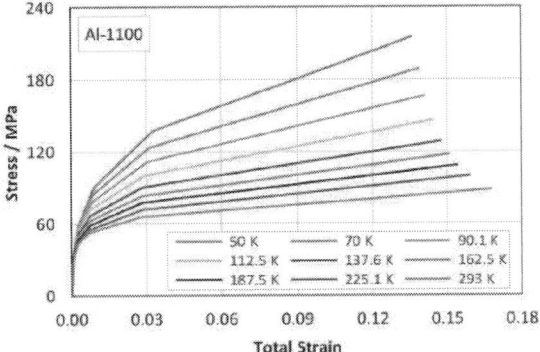

Figure 6. Estimated temperature-dependent multilinear stress-strain curves of alloy Al-1100. The model is based on yield stress and ultimate stress found in [27] assuming a material behavior according to Ramberg-Osgood.

Figure 7 shows the VON MISES stress after cooling down from 293 K to 77 K. The aluminum joint experiences a load of approximately 60 MPa to 80 MPa.

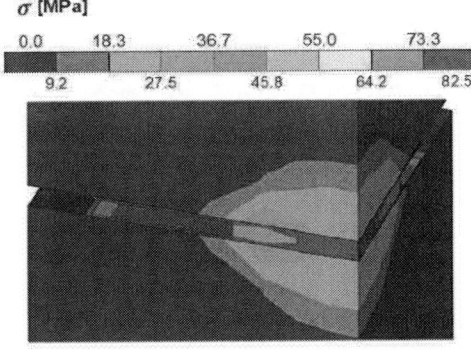

Figure 7. VON MISES stress after cooling down to 77 K.

The accumulated equivalent plastic strain as well as the VON MISES stress are averaged over the center layers of the joint for analysis of the thermal cycling. Stress concentrations at edges and material interfaces are therefore not included. The equivalent plastic strain is given as accumulation per cycle. Figure 8 presents the result after the first 3 cycles.

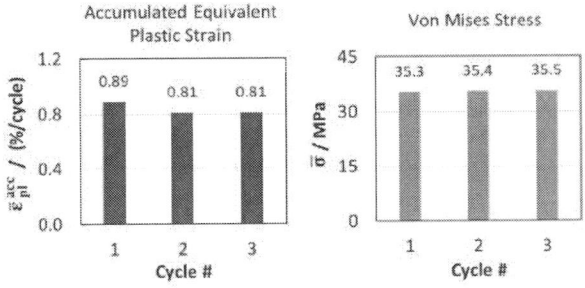

Figure 8. Accumulated equivalent plastic strain (left) and VON MISES stress (right) after the first 3 cycles.

The VAN MISES stress increases slightly for the first cycle due to strain hardening. It is only half as large at room temperature as at 77 K. The equivalent plastic strain accumulated per cycle settles after very few cycles to roughly 0.8%. These rather high values result from the large temperature step of 216 K and the large CTE mismatch between Si and Al, which produces a huge load of the joint. The relatively small joint height and its large diameter reinforce the behavior. Large shear components with large gradients develop towards the joint edges at Si-Al interfaces.

Caution is advised when analyzing absolute numbers since the elastic-plastic data of Al are subject to an unknown error. On one hand, tensile elongation values are estimated. And on the other hand, stress values are taken from a precipitation-hardening alloy (Al-1100) containing about 1% Si. Therefore, in the future, it is necessary to determine the material data of Al directly on the packages at different temperatures. The behavior of the bonded Al could in principle be more ductile so that the stress-strain curves are less steep. This would result in a lower VON MISES stress and higher plastic strain.

B. Aluminum and Gold shear strength

Chip TSB was performed at elevated temperatures of 100°C for Al-Al and 150°C for Au-Au. A bar chart shows the average shear strength of Al-Al and Au-Au samples (Figure 9). Shear strength was higher for Al-Al exhibiting values of 74 MPa ± 16 MPa compared to Au-Au with 40 MPa ± 15 MPa. The variation of bond strength can be attributed to the small pillar structure variation, which persists after Au-ECD and Al post-treatment using fly-cut planarization.

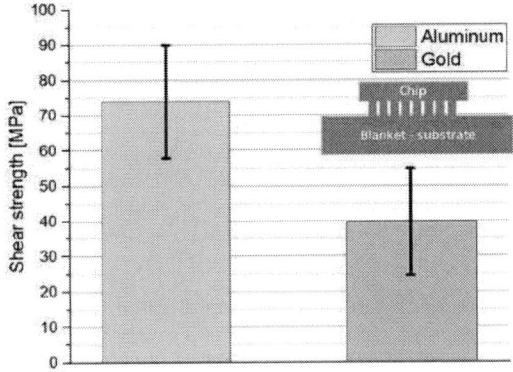

Figure 9. Comparison of Aluminum and Gold chip shear strength at room temperature

Shear tests of bonded Al samples were carried out after thermal stress tests of 5x and 10x cycles. Figure 10 shows shear strengths after 5x and 10x cycles at temperatures of +25°C and -196°C. The average shear strength value (63 MPa ± 8 MPa) is similar to the shear strength values without the thermal cycling test in Figure 9 for Al bonded samples. The difference between 5x and 10x cycles shows that the bonded chip was not or only minimally damaged after the thermal stress test.

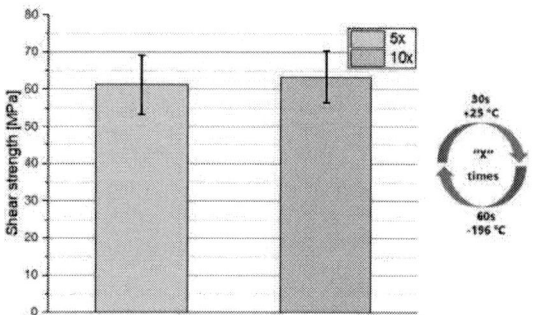

Figure 10. Shear strength of Aluminum chip after 5x and 10x cycles of +25°C and -196 °C

The cross-sectional images of Al-Al and Au-Au samples (Figure 11) show that both the pillar and substrate layer had made contact after TSB. From the measured pillar dimensions, the Au pillar remained close to its original thickness of 15.1 μm, but the Al pillar thickness decreased to 9.9 μm due to the high applied force on the chip. The width of the Au pillar was close to 100 μm, whereas the Al pillar's width increased by almost 15 μm.

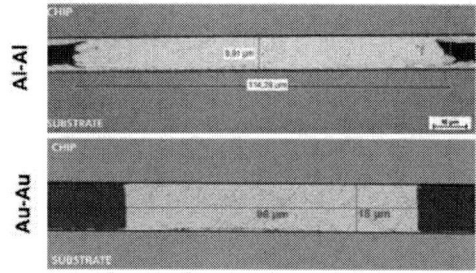

Figure 11. Cross section of gold and Aluminum chip after TSB bonding

Microscopic images (Figure 12) of different bond failures show that the metal-metal bond failed either at the chip-pillar interface (transferred pillar to the substrate) or the bond interface. Excessive damage to the Al pillar can be confirmed, showing that the pillar size changed along the pillar bonding direction. Therefore, adjustments to Al-Al bonding parameters should be made to reduce the excessive pillar damage.microscopic images of Au-Au samples confirm that the bonding mark and transferred pillars are regular in shape and correspond to initial dimensions.

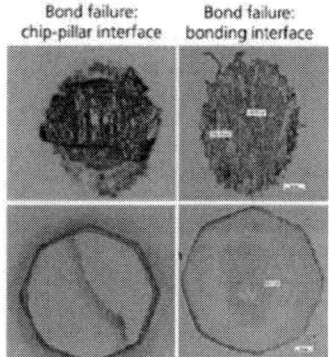

Figure 12. Light microscope image of bond failure of Aluminum and Gold chips after the shear test

IV. CONCLUSION

The Al-Al TSB was done at elevated temperatures showing the process's feasibility. The Al-Al samples showed higher average shear strength compared to the Au-Au samples. However, cross-sectional images showed that the Al pillar thickness changed by almost 5 μm in thickness from its initial size, which was not the case with Au pillars, where its pillar thickness did not change. It suggests that Al TSB parameters need to be adjusted to reduce the excessive deformation of the Al pillars for fine-pitch bonding applications. For further investigation, an optimized chip layout will be used to enable electrical test measurements and to examine the Al natural oxide at the interface of the bond. Additionally, different pitch sizes with varying numbers of pillars arranged in an array structure will be investigated to target fine-pitch applications.

The FE analysis provided an indication of a large load in the Al joint at 77 K. Changes in joint geometry to smaller diameters or larger heights can help to counter this large load. The FE model will serve as a starting point for analyses of possible ion trap layouts down to temperatures around 4 K. In this respect, the implementation of multiphysics in the model will help to gain system competence. A key task involves the determination of relevant temperature-dependent material data. To save computational time at the package level, simply meshed volumes with effective material properties could replace all but a few joints.

REFERENCES

[1] J. Yu *et al.*, "Indium-based Flip-chip Interconnection for Superconducting Quantum Computing Application," in *2022 23rd International Conference on Electronic Packaging Technology, ICEPT 2022*, Institute of Electrical and Electronics Engineers Inc., 2022. doi: 10.1109/ICEPT56209.2022.9873338.

[2] C. R. Conner *et al.*, "Superconducting qubits in a flip-chip architecture," *Appl Phys Lett*, vol. 118, no. 23, Jun. 2021, doi: 10.1063/5.0050173.

[3] S. Kosen *et al.*, "Building blocks of a flip-chip integrated superconducting quantum processor," *Quantum Sci Technol*, vol. 7, no. 3, Jul. 2022, doi: 10.1088/2058-9565/ac734b.

[4] K. R. Brown, J. Chiaverini, J. M. Sage, and H. Häffner, "Materials challenges for trapped-ion quantum computers," *Nat Rev Mater*, vol. 6, no. 10, pp. 892–905, Oct. 2021, doi: 10.1038/s41578-021-00292-1.

[5] Z. D. Romaszko *et al.*, "Engineering of microfabricated ion traps and integration of advanced on-chip features," *Nature Reviews Physics*, vol. 2, no. 6. Springer Nature, pp. 285–299, Jun. 01, 2020. doi: 10.1038/s42254-020-0182-8.

[6] I. Cirulis *et al.*, "Optimal design configuration for aluminum pillar fabrication towards fine pitch ultrasonic bonding applications," in *2022 IEEE 9th Electronics System-Integration Technology Conference, ESTC 2022 - Proceedings*, Institute of Electrical and Electronics Engineers Inc., 2022, pp. 6–10. doi: 10.1109/ESTC55720.2022.9939459.

[7] S. Braun, I. Cirulis, J. E. Liedtke, K. Hiller, M. Wiemer, and H. Kuhn, "Electroplated Aluminum Pillars for Ultrasonic Flip Chip Bonding."

[8] C. Monroe and J. Kim, "Scaling the Ion Trap Quantum Processor," 2011. [Online]. Available: https://www.science.org

[9] D. Rosenberg et al., "Solid-State Qubits: 3D Integration and Packaging," IEEE Microw Mag, vol. 21, no. 8, pp. 72–85, Aug. 2020, doi: 10.1109/MMM.2020.2993478.

[10] D. Kielpinski, C. Monroe, and D. J. Wineland, "Architecture for a large-scale ion-trap quantum computer," Nature, vol. 417, no. 6890, pp. 709–711, Jun. 2002, doi: 10.1038/nature00784.

[11] N. P. de Leon et al., "Materials challenges and opportunities for quantum computing hardware," Science, vol. 372, no. 6539. American Association for the Advancement of Science, Apr. 16, 2021. doi: 10.1126/science.abb2823.

[12] D. Stick, W. K. Hensinger, S. Olmschenk, M. J. Madsen, K. Schwab, and C. Monroe, "Ion trap in a semiconductor chip," Nat Phys, vol. 2, no. 1, pp. 36–39, Jan. 2006, doi: 10.1038/NPHYS171.

[13] C. D. Bruzewicz, J. Chiaverini, R. McConnell, and J. M. Sage, "Trapped-ion quantum computing: Progress and challenges," Appl Phys Rev, vol. 6, no. 2, Jun. 2019, doi: 10.1063/1.5088164.

[14] S. Tamate, Y. Tabuchi, and Y. Nakamura, "Toward Realization of Scalable Packaging and Wiring for Large-Scale Superconducting Quantum Computers," IEICE Transactions on Electronics, vol. 105, no. 6. Institute of Electronics Information Communication Engineers, pp. 290–295, Jun. 01, 2022. doi: 10.1587/TRANSELE.2021SEP0007.

[15] P. Zhao, Y. D. Lim, H. Y. Li, G. Luca, and C. S. Tan, "3D Integration Technologies for Various Quantum Computing Devices."

[16] R. Das et al., "Cryogenic Qubit Integration for Quantum Computing," in Proceedings - Electronic Components and Technology Conference, Institute of Electrical and Electronics Engineers Inc., Aug. 2018, pp. 504–514. doi: 10.1109/ECTC.2018.00080.

[17] B. Foxen et al., "Qubit compatible superconducting interconnects," Quantum Sci Technol, vol. 3, no. 1, Jan. 2018, doi: 10.1088/2058-9565/aa94fc.

[18] P. Zhao et al., "RF Performance Benchmarking of TSV Integrated Surface Electrode Ion Trap for Quantum Computing," IEEE Trans Compon Packaging Manuf Technol, vol. 11, no. 11, pp. 1856–1863, Nov. 2021, doi: 10.1109/TCPMT.2021.3114172.

[19] Z. D. Romaszko et al., "Engineering of microfabricated ion traps and integration of advanced on-chip features," Nature Reviews Physics, vol. 2, no. 6, pp. 285–299, Jun. 2020, doi: 10.1038/S42254-020-0182-8.

[20] D. Rosenberg et al., "3D integrated superconducting qubits," npj Quantum Inf, vol. 3, no. 1, 2017, doi: 10.1038/S41534-017-0044-0.

[21] N. D. Guise et al., "Ball-grid array architecture for microfabricated ion traps," J Appl Phys, vol. 117, no. 17, May 2015, doi: 10.1063/1.4917385.

[22] M. S. Al Farisi, S. Hertel, M. Wiemer, and T. Otto, "Aluminum patterned electroplating from AlCl3-[EMIm]Cl ionic liquid towards microsystems application," Micromachines (Basel), vol. 9, no. 11, Nov. 2018, doi: 10.3390/mi9110589.

[23] K. K. Mehta et al., "Ion traps fabricated in a CMOS foundry," Appl Phys Lett, vol. 105, no. 4, Jul. 2014, doi: 10.1063/1.4892061.

[24] M. A. Hopcroft, W. D. Nix, and T. W. Kenny, "What is the Young's modulus of silicon?" Journal of Microelectromechanical Systems, vol. 19, no. 2, pp. 229–238, Apr. 2010, doi: 10.1109/JMEMS.2009.2039697.

[25] Z. Liu, "Temperature-dependent elastic constants and Young's modulus of Silicon single," 2021, doi: 10.18429/JACoW-MEDSI2020-WEPC09.

[26] T. Middelmann, A. Walkov, G. Bartl, and R. Schödel, "Thermal expansion coefficient of single crystal silicon from 7 K to 293 K," Jul. 2015, doi: 10.1103/PhysRevB.92.174113.

[27] J. W. Ekin, "Experimental Techniques for Low-Temperature Measurements", Oct. 2006, Oxford Press, ISBN 0–19–857054–6

Adhesion Copper/Molding Compound: Modeling and Characterization

Samuele Zalaffi
STMicroelectronics
Agrate Brianza, Italy
samuele.zalaffi@st.com

Marco Rovitto
STMicroelectronics
Agrate Brianza, Italy
marco.rovitto@st.com

Carlo Passagrilli
STMicroelectronics
Agrate Brianza, Italy
carlo.passagrilli@st.com

Claudio Maria Villa
STMicroelectronics
Agrate Brianza, Italy
claudio-maria.villa@st.com

Luca Andena
Politecnico di Milano
Milan, Italy
luca.andena@polimi.it

Stefano Mariani
Politecnico di Milano
Milan, Italy
stefano.mariani@polimi.it

Abstract—Delamination at the resin-copper interface is fully characterized during operative conditions by adopting a combined experimental and numerical approach. Peeling and four-point bending fracture tests are used along with cohesive zone models to reach this purpose. The promising findings are beneficial to analyze the crack propagation at the interface during package-oriented reliability tests, and may therefore be employed to enhance these interfaces in microelectronic plastic packages.

Keywords— *Interfacial fracture, electronic package, peeling, four-point-bending, cohesive zone modeling*

I. INTRODUCTION

Electronic packages fulfil two important roles: to provide communication between the environment and the integrated circuits and to protect them from heat, corrosion, and mechanical damage [1]. The structural stability of a package is conferred by an epoxy molding compound (MC) encapsulation. MC is a composite material made of a polymeric cross-linked matrix and a very high ratio of silica filler [1]. In addition to mechanical stability, MC protects the die from any damage or contamination and provides a barrier to limit corrosion due to moisture absorption. On the other hand, the mismatch in the thermo-mechanical properties of the MC and the metallic lead frame may be responsible for a typical failure mechanism observed at the interface between the two materials, called decohesion. Therefore, a detailed thermo-mechanical assessment of the strength of this interface becomes crucial to identify the root cause for failure mechanism and improve package reliability. For this purpose, this work aims to address the study in three phases: macroscopic characterization of materials, MC analysis at the micro-scale, and investigation of interfacial delamination at the copper/MC interface. The latter phase is supported by experimental analysis as well as numerical modelling.

First, the macroscopic thermo-mechanical properties of copper and MC are identified by employing standard experimental methodologies. During this phase, digital image correlation (DIC) is employed as an auxiliary technique for the study of the heterogeneity of the MC. Due to the bi-component nature of the MC, a microscopic resin characterization is also necessary to understand if said heterogeneity can affect the adhesion with copper. Scratch tests are adopted for this purpose. The third step of this work provides an assessment of the copper/MC interfacial delamination rooted in fracture mechanics. The strength and toughness of the interface is quantitatively estimated through experiments, such as peeling and four-point bending (FPB)

tests, on dedicated copper/resin dual beam samples. Load versus displacement curves are collected during each experiment and for different configurations. The measured data serve as reference for the determination of the fracture energy at the interface. Then, finite element analysis (FEA) moves from this experimental estimation of fracture energy to find a set of model parameters able to reproduce the experimental curves. In this stage, the interface between copper and MC is modelled with traction-separation cohesive elements.

Once the interfacial delamination investigation has reached a sufficient maturity level, the validated methodology has the scope to endorse an approach able to optimize the accuracy and efficiency of both experimental characterization and numerical modeling.

II. MATERIALS AND SAMPLES

The interfacial adhesion between pure copper and two different MCs (A and B) is investigated. The choice of two MCs allows for a validation of the robustness of the method. Copper is provided as rolled sheets of two different thicknesses: 0.15mm and 0.25mm. It is cut in rectangular strips 10mm wide and 80mm long. MC samples are obtained through transfer molding process, in the shape of dumbbell specimens for characterization by uniaxial tensile tests. To test adhesion, a resin block is molded on copper strips to obtain a dual beam resin block 75mm long, 8mm wide, and 2mm thick.

III. EXPERIMENTAL METHODS

This section deals with the description of all experimental methodologies and procedures employed in the three phases of the study. For each phase, experimental techniques, test conditions, and post-processing analyses are fully detailed.

A. Macroscopic Material Characterization

The characterization of the tensile properties of the two MCs and copper is a fundamental step before moving on to study the MC/copper interface. MC dumbbell specimens are tested at room temperature, at a load rate of 1mm/min. During these tests DIC is also used to map the strain state of the whole sample to verify if there is an effect of the heterogeneous nature of the material on the deformation pattern. For this reason, samples are sprayed with silver paint as necessary preparation for DIC use. For copper characterization, rectangular strips of the two available thicknesses are tested at room temperature at a speed of 3mm/min, and strain measurement is performed using a video-extensometer.

B. Microscopic MC Characterization

A preliminary analysis on the heterogeneity of the MC is carried out to understand possible influences of the filler presence on the adhesion with copper. For this reason, in addition to the DIC test mentioned above, scratch tests are performed to measure the tip penetration and residual depth on the MC samples, to verify for possible correlation with the spatial distribution of filler into the MC itself. The scratch test leads to a controlled surface damage of the analyzed material. 3mm long scratches are applied on the MC surface at a speed of 5mm/min and a contact load of 10N. Data are analyzed to detect any periodicity which could be associated with the filler presence.

C. Interface Characterization

Two different types of tests are used for the characterization of the MC/copper interface. The dual beams with a copper thickness of 0.15mm are analyzed with a peeling configuration, while for the thicker ones a FPB test is chosen. The setup of these experiments are sketched in Fig. 1 and Fig. 2.

For what concerns the peeling tests, specimens are glued to a low friction roller and the copper strip is clamped to the movable part of the testing machine after a small preload is imposed to the system. Tests are performed at room temperature at a speed of 10mm/min. The same test is repeated for three different peeling angles: 45°, 90°, and 130°. Peeling of copper off the resin block occurs in a steady state way during the tests, with a nearly constant force measured during the analysis. Knowledge of the imposed angle, the peeling force, and the constitutive behavior of the peel arm material, it is possible to estimate the fracture energy of the interface [2,3,4,5,6,7,8]. The adhesion energy can then be computed by subtracting the contribution due to elastic/plastic deformation of the peel arms from the total energy input to the system [2,4].

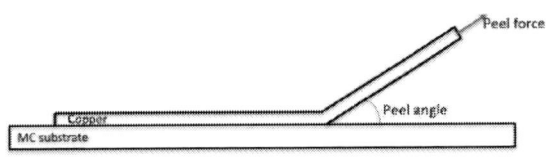

Fig. 1. Peeling test setup scheme.

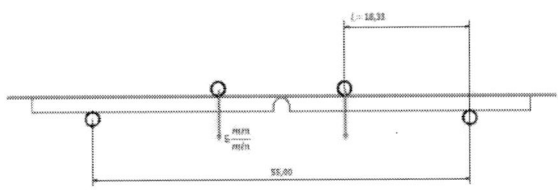

Fig. 2. FPB test setup.

FPB bending test requires a notch realized in the middle of the MC block of the dual beam, introduced using a diamond saw. Resin is removed by the sawing process to leave approximately 0.3mm of MC material covering the copper surface. The chosen setup has a lower span of 55mm and an upper span of one third that distance. Pins used as loading points have a diameter of 6mm. Tests are carried out at room temperature at a speed of 5mm/min. Thanks to the notch

realized in the middle of the MC block of the dual beam, twin cracks originate in the MC and then propagate to the MC/copper interface. Stable propagation occurs thanks to the constant bending moment applied in the region between the upper pins [9,10,11]. For this reason, the recorded force is also supposed to be constant, and its value can be related to the fracture energy G_c of the interface according to [9]:

$$G_c = \frac{(1-v_{Cu}^2)P^2L^2}{8E_{Cu}b^2}\left(\frac{1}{I_{Cu}} - \frac{\lambda}{I_c}\right) \tag{1}$$

where:

$$\lambda = \frac{E_{Cu}(1-v_{MC}^2)}{(1-v_{MC}^2)E_{MC}}, \tag{2}$$

$$I_c = \lambda I_{Cu} + \frac{t_{MC}^3}{12} + \frac{\lambda t_{Cu}t_{MC}(t_{Cu}+t_{MC})^2}{4(\lambda t_{Cu}+t_{MC})}, \tag{3}$$

and: v_{Cu} and E_{Cu} are the copper Poisson ratio and Young's modulus, respectively; P is the measured force; L is the distance between the upper and lower pins; b is the width of the beam; I_{Cu} and I_c are the moments of inertia of the copper strip and the entire composite beam, respectively;; v_{MC} and E_{MC} are the Poisson ratio and Young's modulus of MC, respectively; t_{MC} and t_{Cu} are the thicknesses of MC and copper. The specimens due to the different widths for the copper (b_{Cu}) and the MC (b_{MC}), so the following correction has been applied to the original formulation to account for the actual geometry:

$$G_c = \frac{(1-v_{Cu}^2)P^2L^2}{8E_{Cu}b_{MC}b_{Cu}}\left(\frac{1}{I_{Cu}} - \frac{\lambda'}{I_c}\right), \tag{4}$$

being:

$$\lambda' = \frac{E_{Cu}(1-v_{MC}^2)b_{Cu}}{(1-v_{MC}^2)E_{MC}b_{MC}}, \tag{5}$$

$$I_c = \lambda' I_{Cu} + \frac{t_{MC}^3}{12} + \frac{\lambda' t_{Cu}t_{MC}(t_{Cu}+t_{MC})^2}{4(\lambda' t_{Cu}+t_{MC})}. \tag{6}$$

IV. NUMERICAL MODELS

As mentioned before, the experimental analysis of this work is tightly connected to numerical modeling. FEA is employed for the investigation of delamination at the copper/MC interface. Numerical simulations of the performed experimental fracture tests (peeling and FPB) are run to validate the procedure and devise a prognostic procedure for this type of applications.

The cohesive zone model (CZM) is widely used for the numerical simulation of crack initiation and propagation in quasi-brittle materials [12]. In pure mode I, this fracture model requires two fundamental parameters: the interfacial strength $\sigma_{\max}$, and the fracture energy G_c. The displacement-based FEA framework also requires a (fictive) initial stiffness K at the interface to be introduced. According to CZM, the interface behaves elastically until the maximum stress reaches $\sigma_{\max}$. Afterwards, softening and damage ensue until the fracture energy G_c is dissipated [12]. In all the models described here, the interface is represented by a layer of cohesive elements between the two materials; the CZM law is applied as shown in Fig. 3, and delamination can therefore only occur there.

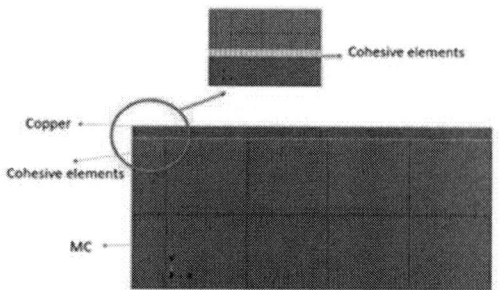

Fig. 3. Representation of dual beam structure with the copper/MC interface described by cohesive elements.

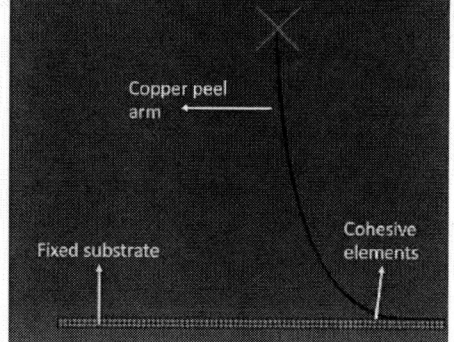

Fig. 4. Schematic of peeling test numerical model.

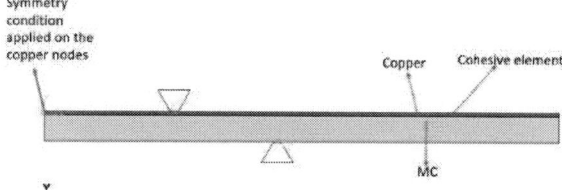

Fig. 5. Schematic of FPB test model.

A. Peeling Test

Due to the relatively large specimen width, 2D plane strain conditions are enforced to the simulation of the peeling test. The copper strip is modeled as an elastic-perfectly plastic material with a yielding stress of 337MPa and a Young's modulus of 124GPa. The MC substrate is instead assumed to be rigid, neglecting tensile strains across the thickness direction. Finally, a layer of cohesive elements is inserted between copper and MC, as depicted in Fig. 3. CZM parameters are identified directly from the experiments (G_c) or by searching for a good match with the experimental results (σ_{max}). During the investigation, the substrate is assumed fixed while the left end of the peel arm is displaced upwards in the peeling direction (see Fig. 4).

B. Four-Point Bending Test

FPB test is simulated in accordance with the former model assumptions. Further than that, the MC substrate is modeled as an elastic material. As per peeling test, a layer of cohesive elements is inserted between the two parts. To account for the symmetry conditions, only one half of the laminae is modeled [11], as depicted in Fig. 5.

V. RESULTS AND DISCUSSION

Having described the experimental techniques and the numerical modeling setup for all the phases of the present work, we next focus on the assessment of the findings coming out of the copper/MC adhesion investigation.

A. Macroscopic Material Characterization

The resulting stress-strain curve for the three materials tested are shown in Figs. 6 and 7. Copper displays an almost perfectly plastic behavior, with a yield stress of around 337MPa (Fig. 6). Fig. 7 depicts the mechanical behavior of the two MCs under uniaxial tensile test: results report similar Young's moduli for both A and B. The strain state on a MC A sample is depicted in Fig. 8, as obtained with the DIC: no strain concentration along the specimen is observed during the test. The longitudinal strain is plotted as a function of the length of the sample, along the path represented in Fig. 8, for different levels of average strain (0.002, 0.004, and 0.006). Results for both MCs show oscillations of the strain values which are not associated with the spatial periodicity of the filler.

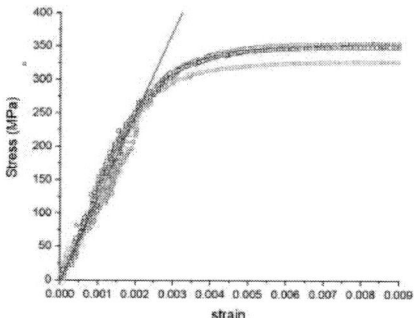

Fig. 6. Copper stress-strain curves.

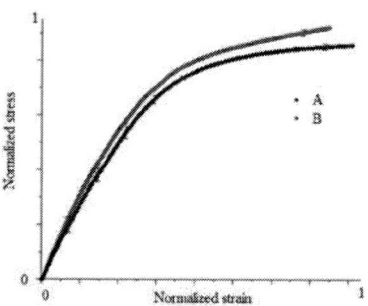

Fig. 7. Normalized MCs stress-strain curves.

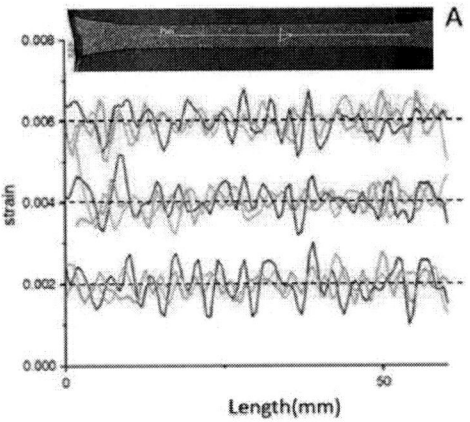

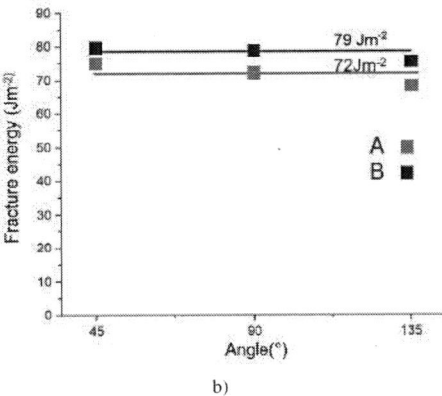

Fig. 10. a) Peeling forces and b) corresponding computed fracture energies for the two MCs.

Fig. 8. DIC results for MC A. Strain is plotted along path representing the length of the sample (57 mm).

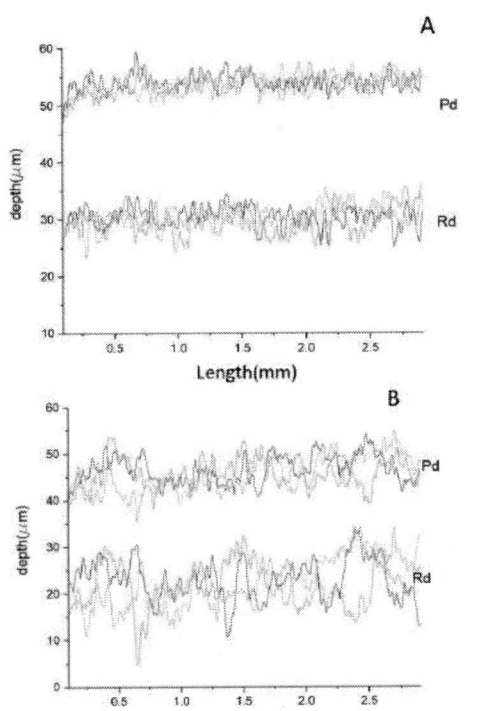

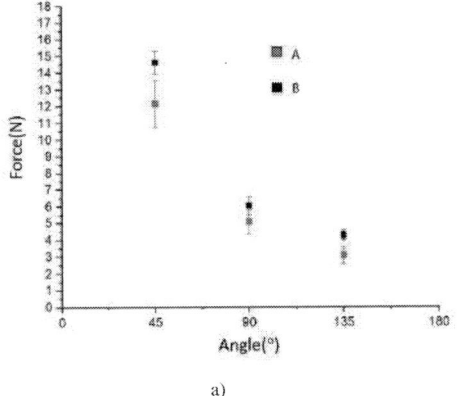

Fig. 9. Penetration and residual depth along 3mm scratch for MC A and B.

B. Microscopic MC Characterization

The results obtained with DIC didn't show any effect associated with the distribution of the MCs. A more comprehensive picture of the phenomenon can be provided by scratch tests, in which the following parameters are investigated: friction, penetration depth (Pd), residual depth (Rd), and acoustic emission. In fact, none of the above parameters show peaks at spatial frequencies associated with the filler dimensions. The results are represented in Fig. 9 for Pd and Rd along the scratch length on both MC A and B. Considering these findings, it is reasonable to assume that MC behaves as a homogeneous material for the purpose of its adhesion with copper at this scale. Other works which account for the presence of the filler into the polymer matrix could be helpful to understand the influence of filler particles on the MC/copper adhesion at the microscopic scale [13].

C. Copper/MC Interfacial Adhesion

Peeling and FPB tests on dedicated copper/resin dual beam samples are chosen as representative test cases.

As anticipated, peeling tests are carried out for three different angles. As shown in Fig. 10a the peeling force increases as the angle decreases for both the MCs investigated. Peeling angle is recorded during the test by employing a camera, to verify that its value remains constant during the test. The results are very similar for the two analyzed MCs, with a slightly higher value of peeling force for material B, for all the configurations. Combining peeling results with those of the copper tensile tests, the fracture energy of the copper/MC interface can be calculated. An open-source tool, called IC peel [14], is employed for this purpose. A bilinear function is selected to describe the almost perfectly plastic behavior of copper with a modulus of 124GPa and a yielding stress of 337MPa. The ratio between the stress-strain curve slopes before and after yielding for the bilinear function is 0.00337. To compute G_c, the width of the sample is considered as the one of the copper strips (10mm) so that the energy dissipated by the copper plastic bending, which is the most relevant contribution, is correctly computed. Given that the copper/MC interface has instead a 8mm width, the results obtained as G_c should be corrected by a factor of 1.25 equal to the ratio of the two widths. The same approach can be repeated for the estimation of G_c, which uses a 2D approximation of the specimen geometry. Fracture energy estimation is consistent among the different testing configurations and, as expected, the evaluated interfacial energy is almost angle independent. As for the peeling force, the values of fracture energy for the two materials A and B are very similar (see Fig. 10b).

For FPB tests the average measured forces and the fracture energies turn out to be again rather similar for both MCs investigated. The results are shown in Table 1, with the estimation of fracture energies obtained with (4).

TABLE I. MEASURED FORCES BY FPB TESTS AND FRACTURE ENERGIES ESTIMATION.

MC	Average F (N)	G_c (J/m²)
A	4.95	70.69
B	4.90	69.27

Starting from the estimated experimental values, different simulations are run by varying the parameters of the cohesive zone law to match at best the experimental forces measured. Finally, a good match is found for the three peeling angles configurations with these parameters: σ_{max} = 40MPa, G_c = 85Jm⁻², and K = 50000MPa/mm for MC A; σ_{max} =40MPa, G_c =95Jm⁻², and K =50000MPa/mm for MC B. Experimental peeling test results at 45° and 90° are very well matched with simulation results by employing the same set of values, while the 135° configuration shows forces slightly overestimated (Fig. 11). This trend is observed for both the investigated MCs and is repeatable meaning that best peeling angle configurations for G_c estimation are 45° and 90°.

FPB experiment is then simulated by employing a fracture energy of 70J/m², which is close to the one obtained analytically, and 90J/m², which is instead the one obtained as best estimate with the peeling numerical model. Given the 2D approximation of the numerical model, both the width of the MC beam (8mm) and the copper one (10mm) are considered to set the upper and lower limits of the simulated force. Results are presented in Fig.12.The procedure is repeated by employing the aforementioned values of the fracture energy of 70J/m² and 90J/m² (blue bars and red bars respectively in Fig.12). The value of the fracture energy tuned with the analytical solution seems to describe well the force trend obtained with the experimental results. Using instead the energy obtained with the peeling simulations, the results seem to slightly overestimate the experimental data if the 10mm thickness is considered, while they match them if the MC thickness is adopted in the analysis. Overall, the identified range of parameters offers a reasonably accurate description of the behaviour of the system under study, providing a very useful toolset to be implemented in more complex models of package delamination.

VI. CONCLUSION AND FUTURE DEVELOPMENTS

The main goal of this work has been the investigation of the interfacial adhesion between copper and MC, via materials characterization as well as numerical modeling. After a macroscopic experimental analysis of two MCs and copper, two delamination tests have been adopted to assess the properties of the interface between the two materials. Peeling and FPB tests prove to be efficient for the adhesion characterization with in-house produced samples and give results consistent with each other. Experimental tests are simulated to tune the CZM coefficients able to describe the response of the MC/copper interface. FPB and peeling tests give comparable results. The coefficients obtained by means

of numerical simulations are fully able to describe the experiments in different configurations.

Future developments of this research activity should be addressed to investigate the MC/copper interface under different conditions like temperature, thermal cycles, and fracture mode type. Finally, the obtained CZM coefficients can be used in a real package test case model to evaluate its ability to predict the failure mode in actual devices.

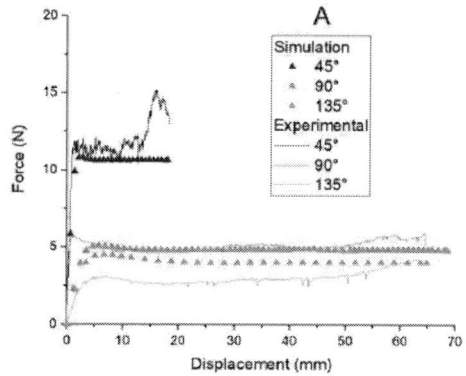

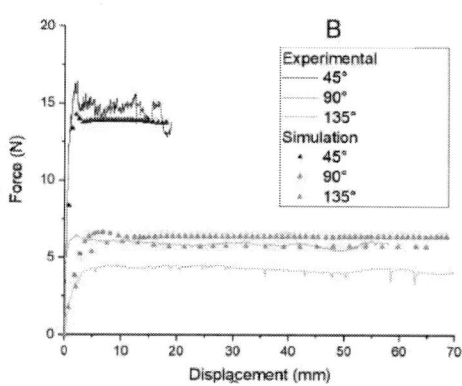

Fig. 11. Simulated versus experimental force-displacement curves for the peeling tests at different angles for MC A and B.

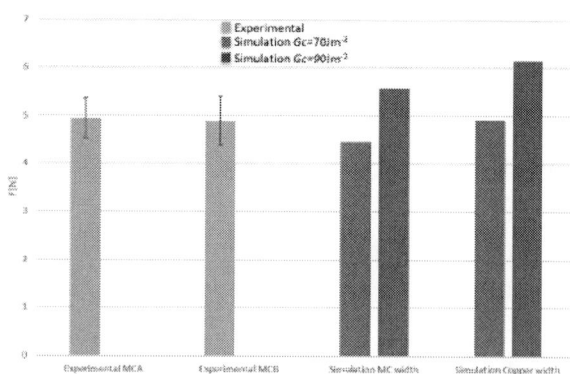

Fig. 12. Four-point bending tests: experimental data versus simulated results.

REFERENCES

[1] M.Linec and B.Music, "The Effects of Silica-Based Fillers on the Properties of Epoxy Molding Compounds" Materials, 12, 1811; 2019

[2] E.Simlissi, "Modeling of delamination and interface strength in printed circuit boards" Thesis for the Degree of Doctor of Philosophy of the Université de Lorraine, 2019.

[3] K.S. Kim and N.Aravas, "Elastoplastic analysis of the peel test" International Journal of Solids and Structures Vol. 24, No. 4, pp. 417-435, 1988

[4] L.F.Kawashita, "The peeling of adhesive joints" Thesis for the Degree of Doctor of Philosophy of the University of London and the Diploma of Imperial College,2006

[5] M. D. Thouless and Q. D. Yang, "A parametric study of the peel test" International journal of adhesion and adhesives Vol.28 pp.176-184,2008

[6] A.J.Kinloch, "The Application of Fracture Mechanics to the Peeling of Adhesive Joints" Conference: Annual Meeting of the Adhesion Society At: San Deigo, CA, USA, February 2014

[7] M.J. Loukis and N.Aravas, "The Effects of Viscoelasticity in the Peeling of Polymeric Films" J.Adhesion,Vol.35,pp.7-22,1991

[8] K.Kendall, "Thin-film peeling - the elastic term" J. Phys. D: Appl. Phys., Vol. 8,No.13, 1975.

[9] P.G. Charalambides, J. Lund,A.G. Evans and M.R. McMeeking, "A test specimen for determining the fracture resistance of bimaterial interfaces" Journal of applied Mechanics March vol. 56/79 pp.77-82, 1989

[10] P.G. Charalambides, H.C. Cao, J. Lund and A.G. Evans, "Development of a test method for measuring the mixed mode fracture resistance of bimaterial interfaces" Mechanics of Materials Vol.8, pp.269-283, 1990

[11] William E. R. Krieger, Sathyanarayanan Raghavan, and Suresh K. Sitaraman, "Experiments for Obtaining Cohesive-Zone Parameters for Copper-Mold Compound Interfacial Delamination" IEEE Transactions on components, packaging and manufacturing technology, vol. 6, no. 9, september 2016

[12] J.Jokinena,M. Kanerva, M. Wallinb , O. Saarelac, "The simulation of a double cantilever beam test using the virtual crack closure technique with the cohesive zone modelling" International Journal of Adhesion and Adhesives Vol.88 pp.50-58,2019

[13] A.Della Porta; S. Mariani; M. Rovitto; L. Andena; S.Zalaffi, "Leadframe-Epoxy Moulding Compound Adhesion: a Micromechanics-driven Investigation" 24th International Conference on Thermal, Mechanical and Multi-Physics Simulation and Experiments in Microelectronics and Microsystems (EuroSimE),Graz Austria, 2023

[14] Mechanical Engineering Department, Imperial College London. http://www3.imperial.ac.uk/meadhesion/testprotocols/peel

A High-Density Organic Package Solution to W-band SiGe Flip-Chip Applications

1st Fırat Altuntaş
Radar and Electronic Warfare Systems
Business Sector, Aselsan A.Ş.
Ankara, Turkey
faltuntas@aselsan.com.tr

2nd Nihan Öznazlı
Radar and Electronic Warfare Systems
Business Sector, Aselsan A.Ş.
Ankara, Turkey
ngokalp@aselsan.com.tr

3rd Olcay Kalkan
Radar and Electronic Warfare Systems
Business Sector, Aselsan A.Ş.
Ankara, Turkey
okalkan@aselsan.com.tr

4th Emrah Koç
Radar and Electronic Warfare Systems
Business Sector, Aselsan A.Ş.
Ankara, Turkey
emrahkoc@aselsan.com.tr

Abstract—**High-density packaging concepts need to be studied to meet the demands from telecommunication and radar applications which require large operational bandwidth and system compactness. In this work, package design of a transmitter SiGe flip-chip with C4 solder ball interface on a multilayer PCB operating around 94 GHz is presented.**

Keywords—W-band, multi-layer organic package, flip-chip, SiGe

I. INTRODUCTION

With the increase in demands from wireless communication in both civil and military applications, novel and progressive solutions are required in microwave field. For example, Fixed Wireless Access concept emerging with the 5G technology requires multi Gb\s data transmission [1]-[3]. In addition, automotive radars for the development of smart roads and vehicles, high resolution body scan and Foreign Object Debris Detection Radar (FODR) applications require transmitter and receiver systems with large operational bandwidth [4]-[7]. By taking advantage of millimeter-wave frequency band, minimization of systems and large operational bandwidths are achievable.

As a consequence of millimeter-wave signal utilization, spacing between elements in an antenna array would be decreased. At W-band frequencies, the spacing between antenna elements in an active electronically scanned array (AESA) would be comparable to the MMIC size. Considering this study, dimensions of the transmitter flip-chip is 5.9x6.7mm while the wavelength is in the order of millimeters at W-band. Hence, the most feasible and efficient solution is to combine MMIC flip-chip and antennas in the same structure such as antenna in package (AiP) [8]-[9].

The main purpose for an AiP concept is to obtain a fully functional package that contains T/R chip, antennas and necessary components such as DC connectors or passive SMD devices. Since measuring performance of an antenna integrated package requires complex test configuration such as an anechoic chamber system, it is required to verify electromagnetic design of the package and RF performance of vital discontinuity locations. For instance, SiGe flip-chip to package transition through C4 solder ball, microstrip and stripline via transitions should be designed carefully in W-band while regarding minimization of the structures to meet high density signal routing.

This study concentrates on the selection of the package technology and electromagnetic designs of the mentioned discontinuity regions. In the end, on-wafer and on-package RF measurements are to be compared.

II. SELECTION OF THE PACKAGE TECHNOLOGY

As related works are studied, ceramic based low temperature co-fired ceramics (LTCC) and organic substrate based printed circuit boards (PCB) shine out [10]-[11]. The main constraints determining the package technology are operating frequency and interface properties of the SiGe flip-chip. Regarding the operating frequency, substrates used in the package should have stable dielectric constant and low $\tan\delta$ at W-band. Furthermore, manufacturing capabilities of the selected technology should meet high density signal routing requirement.

In this study, a single transmitter MMIC element consists of approximately 200 C4 solder balls and 16 independent W-band front-ends. In order to obtain a scalable antenna array, the high frequency package could only be built up vertically since planar dimensions are determined by the array size. In other words, the technology should support vertical transitions with a multilayer structure. In addition, manufacturing limits and tolerances should provide fine resolution to the C4 interface properties. For example, distance between adjacent C4 solder balls are 225 μm. As solder and solder mask limits are taken into consideration, PCB technology provide better precision and repeatability for mass production than LTCC.

For the future consideration, antenna integration into the package is another difficulty. In addition to the structural compatibleness, antenna design and integration into the package should be analyzed in depth. The major and predominant factor is the relative dielectric constant of the material. Commercially available PCB and LTCC substrates that would support W-band signals have $\varepsilon r=3$ and $\varepsilon r=7.8$, respectively. Lower the dielectric constant is, higher the radiation efficiency would be. Therefore, PCB technology provide a better environment in terms of W-band signal transition, SiGe flip-chip with C4 solder ball interface and antenna integration.

Even though PCB manufacturing capabilities seem to be sufficient to design such a package, vertical via transitions in terms of W-band shielding and supply voltage connections could not be achieved in chip dimensions as common via technique is used. In Fig. 1, through via structure is shown. Despite having multilayer structure, vias are extended from bottom to top layer. Hence, efficiency of the area utilization in each layer is drastically reduced. Rather than extending vias unnecessarily, by using stacked via technique shown in Fig. 2, connection between each layer is provided individually.

Therefore, without occupying redundant layers, both RF and DC signal transmission is provided in an efficient way. In this study, 5+N+5 type high density interconnect (HDI) PCB is realized. That is, 5 prepreg materials are used above and below a core substrate resulting a total of 12 conductor layers with providing inter-layer connection using μvias.

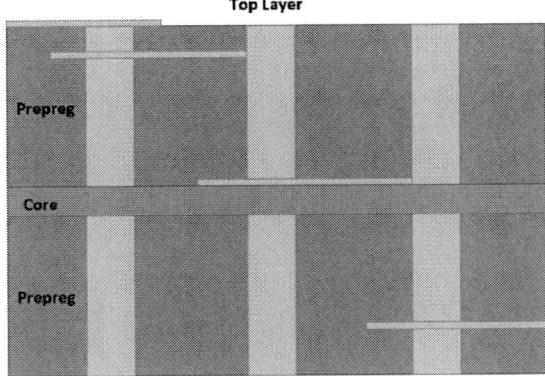

Fig. 1. Through via inter-layer connection technique

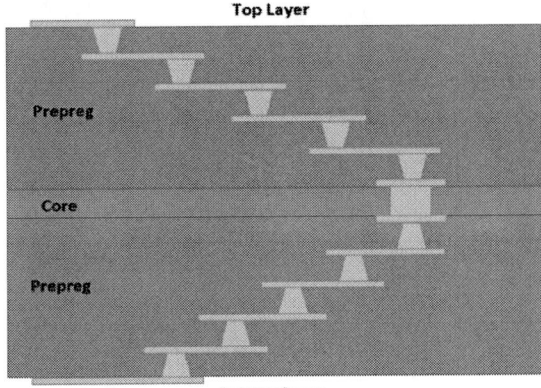

Fig. 2. Through via inter-layer connection technique

III. ELECTROMAGNETIC DESIGN OF THE ORGANIC PACKAGE AT W-BAND FREQUENCIES

High density packaging concept proposes a solution to system miniaturization regarding the RF and DC transmission of the SiGe flip-chip at W-band. However, electromagnetic design becomes more complex since it involves different considerations such as flip-chip to PCB transition and antenna array design. In addition to W-band signals, there are many different signals which requires special treatment such as DC supply voltages, TTL control voltages, intermediate frequency (IF) and reference signals. Altering any of these connections would require major revision in multilayer PCB structure. Hence, W-band signal transitions from flip-chip to inner layers of PCB are designed while optimizing required volume and return loss. By designing W-band transitions first, we make sure that W-band structures are kept unchanged during the package construction.

In this section of the study, package modelling and electromagnetic analysis conducted in CST Studio Suite is to be explained.

A. Flip-Chip to Stripline Transition

SiGe flip-chip contains 16 RF-front ends. Majority of the W-band channels located in the inner regions while remaining channels are close to edge of the flip-chip. Therefore, two different transition analysis should be conducted for interior and edge regions in W-band.

1) Corner Region Analysis

In Fig. 3, electromagnetic model of the W-band transition from flip-chip to stripline in PCB is shown including C4 solder balls. As SiGe flip-chip is inspected, it is seen that metal density on the solder side is high. Instead of modelling the surface of the flip-chip with μm resolution, it is accepted as a perfect electric conductor. The main reason is that order of μm is neglectable as wavelength of the 94 GHz center frequency is considered.

Shape of the C4 solder balls are shaped considering the inspection after assembly to the package. Solder balls are modelled by using lead material shown as purple color while flip-chip surface is shown as gray. Prepreg material is demonstrated as transparent blue while copper is yellow.

In Fig. 4, S-parameters are given after optimizing the matching while keeping size of the structure minimum. It is seen that between 80 – 110 GHz, return loss is lower than 18 dB while active operating region of the flip-chip is better than 20 dB centered around 94 GHz with 0.4 dB insertion loss.

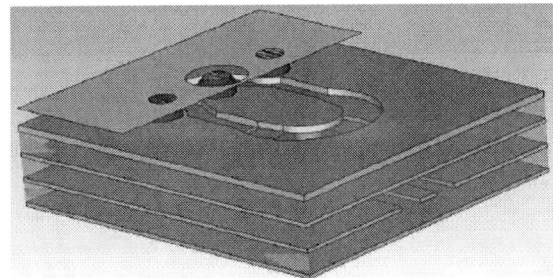

Fig. 3. Model of the flip-chip to stripline transition for edge part

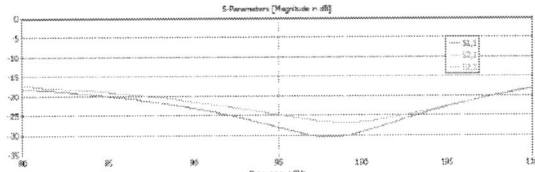

Fig. 4. RF matching performance for edge part analysis

2) Interion Region Analysis

As seen in Fig. 3, W-band signal is not directly routed through the PCB since placing vias under C4 solder balls are undesired in terms of matching performance and assembly reliability. Therefore, signal is routed in co-planar grounded wave (CPGW) structure. That is, signal matching would be distorted due to the flip-chip existence in the air. The affect is neglectable for edge part analysis, however, interior part analysis requires further interest.

In Fig. 5, interior part transition from flip-chip to stripline is shown. In Fig. 6, it is seen that length of the both CPGW and stripline structure are increased under the distortion of

SiGe, which is modelled as perfect electric conductor, to obtain fine matching performance.

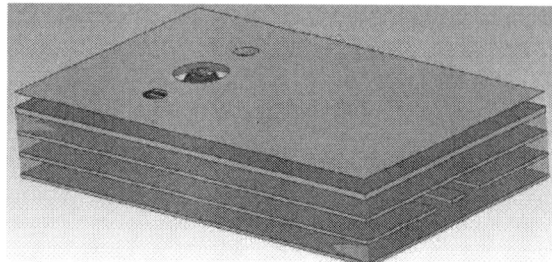

Fig. 5. Interior part of the flip-chip to stripline transition

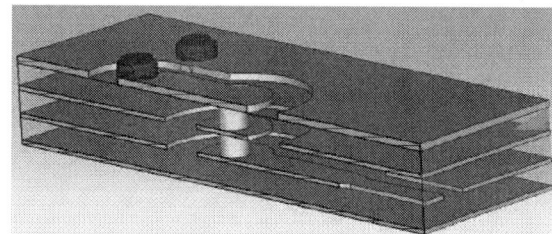

Fig. 6. Half-cut view of the interior part of the flip-chip to stripline transition

In Fig. 7, RF matching performance is given. Similar to previous analysis, below 20 dB return loss and 0.45 dB insertion loss are obtained throughout W-band and optimized around 94 GHz center frequency.

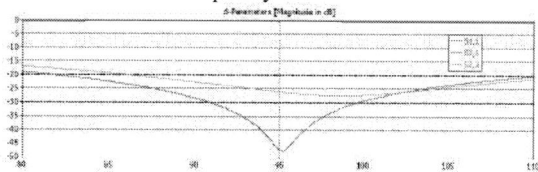

Fig. 7. RF matching performance for interior part analysis

B. RF Probe Transition

In Fig. 8, co-planar grounded wave structure to RF probe matching transition model is given. 50-ohm characteristic impedance of the CPGW line is increased so that RF probe could be landed without short circuiting through ground plane.

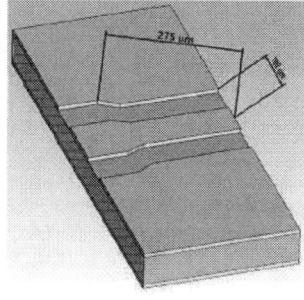

Fig. 8. 150 μm-pitch RF probe measurement structure on PCB

IV. HIGH DENSITY ORGANIC PACKAGE ASSEMBLY

By using the designed electromagnetic structures, 5+N+5 HDI PCB is manufactured. In Fig. 9, measurement configuration of the high-density organic package with assembled SiGe flip-chip is demonstrated. In addition, DC supply voltages and control signals are connected to bottom of the package through copper wires. Furthermore, relatively low frequency IF and reference signals are given through surface mount SMP type connectors. As seen from the left most part of the package, 150 μm-pitch probe measurement structures are placed. There are four different W-band outputs from flip-chip. Half of them are directly routed in co-planar grounded wave mode while other half is routed inside package in stripline mode. In other words, interior front-ends requires stripline routing while edge front-ends have only co-planar grounded wave structure.

Fig. 9. RF probe measurement package of W-band SiGe flip-chip with low frequency RF and DC connectors

In Fig. 10, RF probe landing on optimized CPGW structure view is shown under microscope.

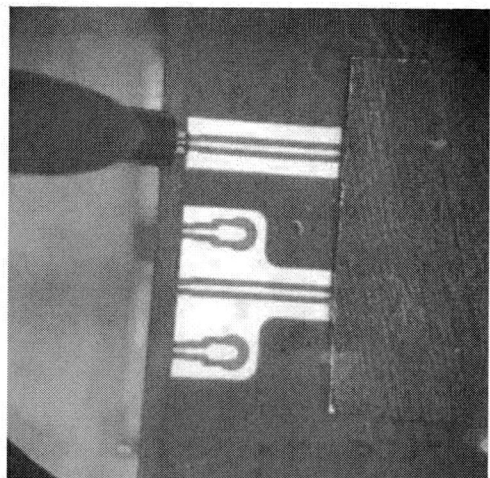

Fig. 10. RF probe landing on package under microscope during measurement

W-band measurement configuration includes two signal generators for 1 GHz reference and 10 GHz IF signals. Also, an FPGA is utilized to communicate with MMIC. Measured W-band signal is supported by waveguides through the W-band attenuator and down-converter connected to spectrum analyzer. RF cable, W-band probe, waveguide and mixer losses are added to spectrum analyzer result to calculate the 10 GHz IF to 94 GHz signal gain of the flip-chip.

V. RF Probe Measurement Configuration in W-Band

On-chip and on-package measurements are compared for a transmitter SiGe flip-chip. In Fig. 11, on-chip measurement view with custom-made RF and DC probes are shown. On-chip test results are:

- Small signal gain is between 30 – 35 dB in 92 – 96 GHz band
- 7 dBm saturation power at 94 GHz

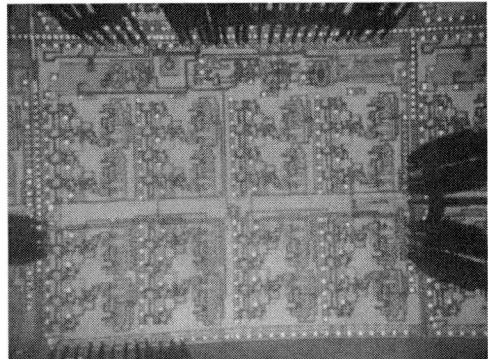

Fig. 11. On-chip measurement configuration

In Fig. 12, gain with respect to frequency graph is given. Different channels shown in Fig. 8 are measured. By taking package measurements, we observe how design is affected from manufacturing capabilities of different companies, effect of C4 solder balls, assembly process and electromagnetic modelling. It is seen that average gain 30 – 35 dB is achieved in active operation band.

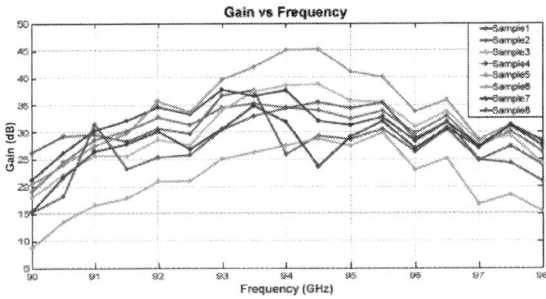

Fig. 12. On-package gain measurement in W-band

In Fig. 13, output power with respect to input power is given. Even though there exist a difference in small signal stage, output power is saturated around 7 dBm at 94 GHz. Only one sample separated from others in terms of saturated power is seen to have gain below 30 dB throughout the band in terms of gain seen in Fig. 12. The reason of deteriorated performance could be resulted from assembly process of flip-chip or manufacturing tolerances of the PCB.

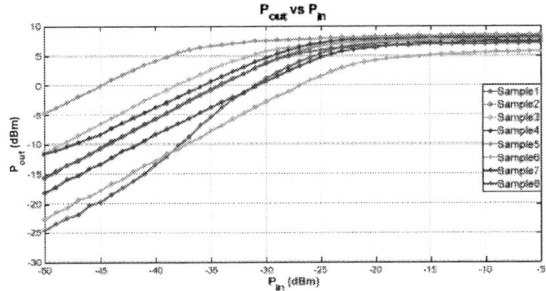

Fig. 13. On-package measurement P_{out} vs P_{in} at 94GHz

VI. Conclusion

In this paper, high density packaging concept of a SiGe flip-chip on a multilayer organic package operating around 94 GHz center frequency is explained. Electromagnetic modelling of interaction between flip-chip and PCB, C4 solder ball, via transitions considering minimum volume and W-band RF probe measurement structure are represented in detail. On-chip and on-package measurements show consistency which means that electromagnetic modelling via CST Studio in W-band verified. By using W-band transition structures conducted in this work, antenna implementation to the package becomes feasible for the antenna in package concept for the future work.

References

[1] S. Emami et al., "A 60GHz CMOS phased-array transceiver pair for multi-Gb/s wireless communications," 2011 IEEE International Solid-State Circuits Conference, 2011, pp. 164-166

[2] J. M. Gilbert, C. H. Doan, S. Emami and C. B. Shung, "A 4-Gbps Uncompressed Wireless HD A/V Transceiver Chipset," in IEEE Micro, vol. 28, no. 2, pp. 56-64, March-April 2008

[3] J. Thompson et al., "5G wireless communication systems: prospects and challenges [Guest Editorial]," in IEEE Communications Magazine, vol. 52, no. 2, pp. 62-64, February 2014

[4] W. Roh et al., "Millimeter-wave beamforming as an enabling technology for 5G cellular communications: theoretical feasibility and prototype results," in IEEE Communications Magazine, vol. 52, no. 2, pp. 106-113, February 2014

[5] F. Nsengiyumva, C. Migliaccio, L. Brochier, J. Lanteri, J. -Y. Dauvignac and C. Pichot, "90 GHz, 3-D Scattered Field Measurements for Investigation of Foreign Object Debris," in IEEE Transactions on Antennas and Propagation, vol. 67, no. 9, pp. 6217-6222, Sept. 2019

[6] P. Feil, W. Menzel, T. P. Nguyen, C. Pichot and C. Migliaccio, "Foreign Objects Debris Detection (FOD) on Airport Runways Using a Broadband 78 GHz Sensor," 2008 38th European Microwave Conference, 2008, pp. 1608-1611

[7] G. Mehdi and Jungang Miao, "Millimeter wave FMCW radar for Foreign object debris (FOD) detection at airport runways," Proceedings of 2012 9th International Bhurban Conference on Applied Sciences & Technology (IBCAST), 2012, pp. 407-412

[8] Y. Zhang and J. Mao, "An Overview of the Development of Antenna-in-Package Technology for Highly Integrated Wireless Devices," in Proceedings of the IEEE, vol. 107, no. 11, pp. 2265-2280, Nov. 2019

[9] T. -H. Lin et al., "Broadband and Miniaturized Antenna-in-Package (AiP) Design for 5G Applications," in IEEE Antennas and Wireless Propagation Letters, vol. 19, no. 11, pp. 1963-1967, Nov. 2020

[10] E. Cohen, M. Ruberto, M. Cohen, O. Degani, S. Ravid and D. Ritter, "A CMOS Bidirectional 32-Element Phased-Array Transceiver at 60 GHz With LTCC Antenna," in IEEE Transactions on Microwave Theory and Techniques, vol. 61, no. 3, pp. 1359-1375, March 2013

D. G. Kam, D. Liu, A. Natarajan, S. K. Reynolds and B. A. Floyd, "Organic Packages With Embedded Phased-Array Antennas for 60-GHz Wireless Chipsets," in IEEE Transactions on Components, Packaging and Manufacturing Technology, vol. 1, no. 11, pp. 1806-1814, Nov. 2011

Thermal design of stacked power modules for electric drive applications

Jianfeng Li
Zhuzhou CRRC Times Electric UK
Innovation Center
Solihull, United Kingdom
jianfeng.li@teic.crrczic.cc

Yuekang Du
Zhuzhou CRRC Times Electric UK
Innovation Center
Solihull, United Kingdom
yuekang.du@teic.crrczic.cc

Xingzhi Wang
Zhuzhou CRRC Times Electric UK
Innovation Center
Solihull, United Kingdom
xingzhi.wang@teic.crrczic.cc

Liangjie Liu
Zhuzhou CRRC Times Electric UK
Innovation Center
Solihull, United Kingdom
liangjie.liu@teic.crrczic.cc

Yong Pang
Zhuzhou CRRC Times Electric UK
Innovation Center
Solihull, United Kingdom
yong.pang@teic.crrczic.cc

Feixiang Liu
Zhuzhou CRRC Times Electric UK
Innovation Center
Solihull, United Kingdom
feixiang.liu@teic.crrczic.cc

Abstract—This paper reports the thermal design of a stacked arrangement inserting four IGBT modules between five heat sinks to achieve good cooling performance while enabling them to be installed into the limited space. Such stacked four IGBT modules can be used for a variable voltage converter (VVC), a three-phase motor inverter (M-INV) and a three-phase generator inverter (G-INV) of one electric control system in one electric vehicle or hybrid electric vehicle. Work presented in this paper are the thermally related structure design and thermal-electric and computational fluid dynamics coupled simulation, while the prototyping and thermal test of the stacked arrangement are still ongoing. The simulation results indicate that the stacked arrangement can be designed into a compact 121 mm × 94.7 mm × 90.8 mm structure while achieving the maximum virtual junction temperature below 150 °C under the coolant feed temperature of 70 °C, flow rate of 10 L/min and pressure drop of 0.16 bar for the VVC delivering 150 kW, the M-INV delivering 100 kW and the G-INV delivering 50 kW. If validated with the ongoing work, the stacked arrangement can be employed to increase power density and improve electrical and thermal performance of the power modules in the electric control systems with multiple converters/inverters.

Keywords—*IGBT module, heat sink, double side cooling, stacked structure, simulation, thermal test*

I. INTRODUCTION

There are generally the power semiconductor modules for three and more converters/inverters which are installed into the limited space in the electric control system of one electric vehicle (EV) or hybrid electric vehicle (HEV) [1,2]. In response to this demand, the stacked arrangements of power modules in which several power semiconductor modules and heat sinks are stacked together were developed [3,4]. However, in the previous stacked arrangements, each of the power modules comprises either single switch or half bridge switch which owns its independent structural, conductive paths and power terminals [1,3-5]. As a result, this may require additional footprint/space to connect the terminals of these modules to the power source, associated connectors, other passive components, and control/drive printed circuit boards (PCBs) for the three and more converters/inverters.

In the present work, a stacked arrangement of Si insulated gate bipolar transistor (IGBT) modules is developed where each IGBT module is integrated with two half bridge switches which share part of conductive tracks of substrates and share

part of power terminals. In particular, four such IGBT modules are inserted between five heat sink to form the stacked arrangement with further reduced volume and weight and with better layout of terminals to simplify the associated connections. The four IGBT modules can be used for constructing a variable voltage converter (VVC), a three-phase motor inverter (M-INV) and a three-phase generator inverter (G-INV), and thus increase the power density and improve the electrical and thermal performance of the electric control systems in the electric vehicle (EV) or hybrid electric vehicle (HEV) applications.

The aim of the present thermal design is to ensure the stacked four IGBT modules to achieve good cooling performance while enabling them to be installed into a specified space. The objectives of this paper are to: (i) report the compact structure stacking the modules and heat sinks together; and (ii) optimise the channel design of the heat sinks to meet the cooling requirement under an operating condition for EV and HEV applications. Despite that the stacked arrangement is formed with Si IGBT modules, the methodology and principles developed through the present work can be extended and applied to the development of the stacked arrangement comprising power modules with wide band gaps devices such as SiC and GaN power devices.

II. STACKED ARRANGEMENT

A. Requirement of Structure Design

The design of one electric control system in one EV or HEV requires the IGBT modules and heat sinks which can be installed in a limited space of 125 mm × 100 mm × 96 mm for a VVC, a three-phase M-INV and a three-phase G-INV. The VVC is specified with a maximum output voltage of 500 V and a maximum current of 300A. The M-INV and G-INV are specified both with a maximum direct current (DC) bus voltage of 500 V and respectively with the continuous output currents of 210 Arms and 100 Arms. To meet this requirement, the IGBT modules are designed with the 650V/300A IGBT and fast recovery diode (FRD) chips with the footprints of 12.2 mm × 12.2 mm × 0.07 mm and 12.0 mm × 6.1 mm × 0.07 mm. Each IGBT module is integrated with two half bridge switches which share part of conductive tracks of substrates and part of power terminals. Four such IGBT modules are inserted between five heat sinks to form the stacked arrangement for the VVC, M-INV G-INV.

B. IGBT Module

The IGBT module integrated with two half bridge switches is the same as the one presented in the previous paper [2]. As shown in Fig. 1, one half bridge switch is constructed with two paralleled IGBT chips in antiparallel to two paralleled FRD chips for each switching leg, and the other half bridge is constructed with one single IGBT chip in antiparallel to one single FRD chip for each switching leg. In each IGBT module, all the six IGBT chips, six FRD chips, one negative temperature coefficient (NTC) thermistor and the conductive shim interconnections are enclosed within two ceramic-based substrates to facilitate double side cooling. Both substrates are 0.3/0.32/0.3 mm thick Cu/ceramic/Cu substrates. With moulded epoxy compound encapsulant to provide additional support, the DC+ and DC- terminals are put at one side, the signal/control terminals are put at the opposite side, and the AC terminals are put at a third side of the IGBT module.

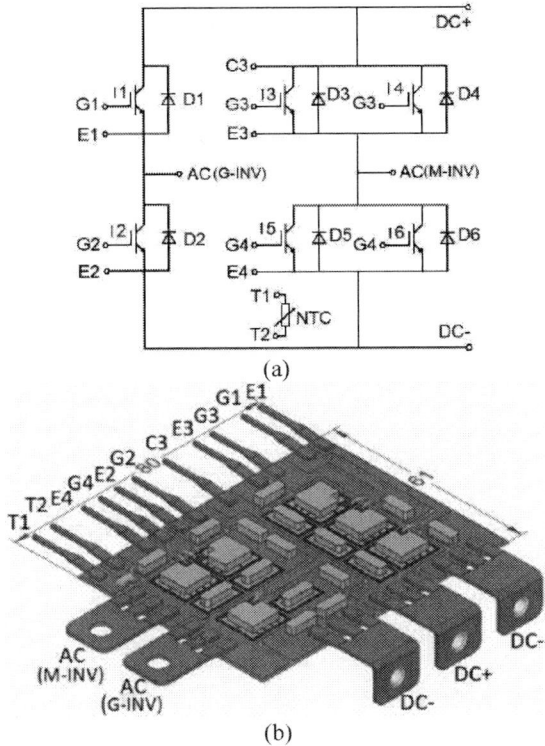

Fig. 1. The IGBT module integrated with 2 half bridges: (a) topology; (b) the designed structure without the top substrate and moulded encapsulant.

In such an IGBT module, the two half bridge switches share the DC+ and DC- terminals, the conductive track of one substrate directly connected to the collector sides of three IGBT chips and the cathode sides of three FRD chips, and the conductive track of the other substrate connected to the emitter sides of the other three IGBT chips and the anode sides of the other three FRD chips with 2 mm high conductive shims. The conducting paths are designed to achieve low parasitic inductances for the different commutation loops, and slots are added in the conductive tracks of one substrate to balance the parasitic inductances and resistances for the paralleled IGBT and FRD chips [6]. Ignoring the extended terminals, the dimensions of the IGBT module occupied by the substrates and the moulded encapsulant are 89.2 mm × 70 mm × 4.4 mm. The shared terminals and conductive tracks remarkably

reduce the footprint and weight of the present IGBT module.

C. Stacked Arrangement

Fig. 2 shows the stacked arrangement inserting four IGBT modules between five heat sinks. The two half bridge switches in the top IGBT module can be connected in parallel to form one half bridge switch. This half bridge switch, and inductor Lin, input capacitor C1, output capacitor C2 and resistor R1 are connected to form the VVC which is employed to step up the voltage of the power source battery. The other three IGBT modules are used as two three-phase half bridge switches. Here each half bridge switch with two paralleled IGBT/FRD chips per switching leg is used as one phase for the M-INV, and each half bridge switch with one single IGBT/FRD chip per switching leg is used as one phase for the G-INV. The shared DC+ and DC- terminals of both three-phase half bridge switches are connected to the output terminals of the VVC. The three AC terminals of the three-phase half bridge switches for the M-INV can be connected to a motor, and the three AC terminals of the three-phase half bridge switches for the G-INV can be connected to a generator.

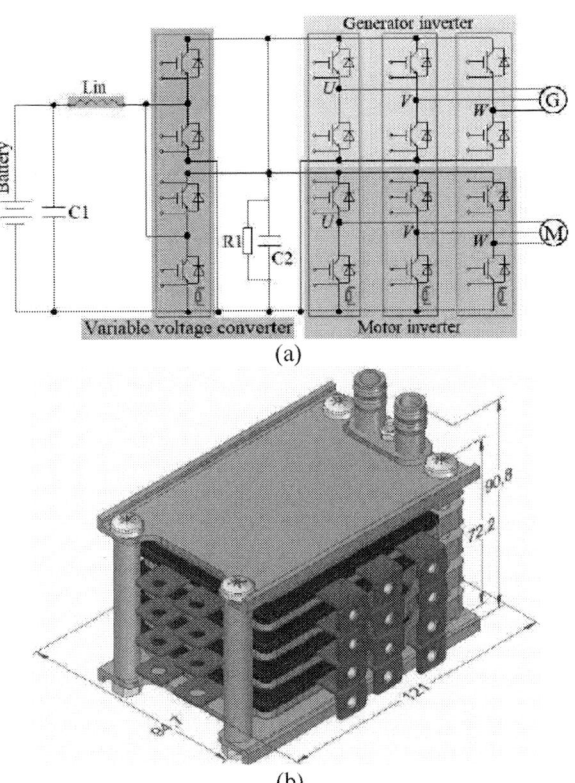

Fig. 2. The stacked arrangement inserting four IGBT modules between five heat sinks: (a) topology of the four IGBT modules in the four red boxes; (b) the designed strucutre.

To form the stacked arrangement, 0.2 mm thick thermal interface materials (TIMs) are placed between each of the four IGBT modules and its two adjacent heat sinks. All the five heat sinks are integrated with cooling channels and two-side cold plates. The inlets and outlets of these five heat sinks are connected in parallel and sealed with O-rings. Two 2.5 mm thick clamping plates attached with 1 mm thick compliant pads are further placed on the top heat sink and under the bottom heat sink. The stacking pressure is applied by bolting the top and bottom clamping plates with 4 pairs of M6 screws and nuts at a specified torque. Including the extended terminals and the extended inlet and outlet connectors of the coolant, the dimensions of the stacked arrangement are 121 mm × 94.7 mm × 90.8 mm. The shared conductive tracks and terminals in each module and the special layout of all the power and signal/control terminals lead to not only reduced volume and weight of the stacked arrangement but also simplified external connections of the four IGBT modules to the inductor, input capacitor, output capacitor, resistor, motor, generator and drive/control PCBs.

III. THERMAL DESIGN CONSIDERATION

A. Thermal-Electric Specifications

Thermal design of the stacked arrangement is to ensure good temperature distribution between the IGBT/FRD chips and the maximum junction temperature below a specified value, i.e. 150 °C, under the worst electro-thermal generation and the specified cooling condition. The latter are estimated following the electrical design and thermal management of the electric control system and the electrical and electro-thermal characteristics of the IGBT and FRD chips. Table I lists the specified parameters of electro-thermal generation and cooling condition for the thermal design of the stacked arrangement inserting the four IGBT modules between the five heat sinks. The coolant flow rate is the total value of the coolant flowing in the five paralleled heat sinks, and the maximum coolant pressure drop is the value across the two top connectors as inlet and outlet of the coolant. Due to the temperature-dependence of the electrical and electro-thermal characteristics of the IGBT and FRD chips, a few iterations between the design of the electric control system and the thermal design of the stacked arrangement should be done for estimating the power losses of the IGBT and FRD chips.

TABLE I.   SPECIFICATIONS OF ELECTRO-THERMAL GENERATION AND COOLING CONDITION FOR THE STACKED ARRANGEMENT

Parameter	VVC	M-INV	G-INV
Max current (A)	300		
Continuous output current (Arms)		210	100
Power loss per IGBT chip (W)	261.5	253.5	194
Power loss per FRD chip (W)	155.5	66.5	57
Coolant	50% glycol in water		
Coolant inlet temperature (°C)	70		
Coolant flow rate (L/min)	10		
Maximum coolant pressure drop (bar)	0.2		

B. Materials and Bonding Technologies

The two substrates in each of the four IGBT modules are 0.3/0.32/0.3 $Cu/Si_3N_4/Cu$ substrates because of sufficiently high thermal conductivity and good thermo-mechanical reliability [7]. The conductive shims are made of 1/4/1 Cu/Mo/Cu sandwiched structure, which has coefficients of thermal expansion compatible with IGBT and FRD chips [8]. All the power and signal/control terminals are made of oxygen-free copper (OFHC) with high thermal and electrical conductivities. All the IGBT/FRD chips, conductive shims and terminals may be bonded on or under the conductive tracks of the two substrates or the conductive shims may be bonded on the IGBT/FRD chips with near eutectic Sn-Ag-Cu

(SAC) solder joints and the conventional reflow soldering process. All the IGBT/FRD chips may be bonded on the conductive tracks of the bottom substrate, and all the conductive shims may be bonded under the conductive tracks of the top substrate with Ag sintering process for comparison. The five heat sinks and their connectors are made of aluminium alloy for low weight and cost. They are also assumed to be made of copper for comparison. The TIMs between each IGBT module and its two neighbouring heat sinks are 0.2 mm thick graphite sheets with the thermal conductivity higher than those of typical thermal greases.

C. Structure and Connection of Heat Sinks

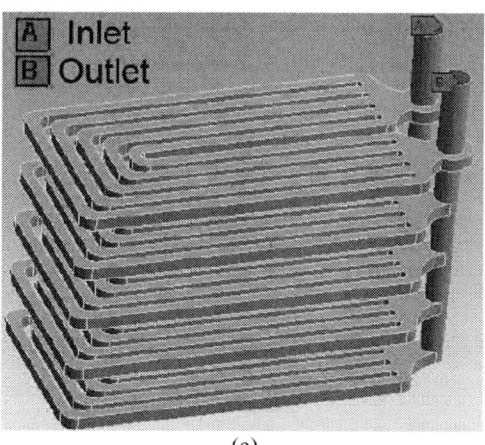

(a)

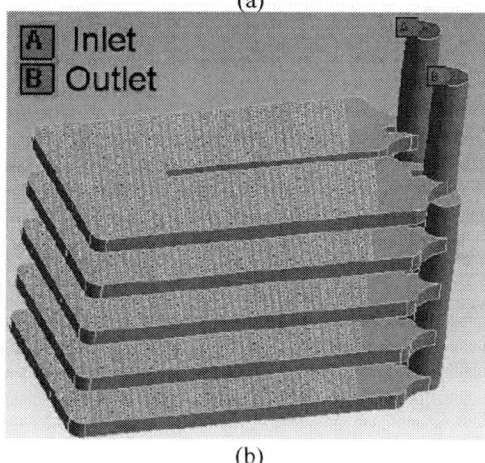

(b)

Fig. 3. The coolant domains in the stacked arrangement with: (a) U-shaped channel heat sinks; (b) pin fin heat sinks.

It is critical to design the structure and connection of the heat sinks for achieving good cooling performance while remaining low coolant pressure drop across and the inlet and outlet. For demonstrating this, the heat sinks with U-shaped channels and narrow connectors and the heat sinks with pin fins and wide connectors are considered and compared. Fig. 3 shows the coolant domains in the stacked arrangement with these two types of heat sinks. To meet the height requirement of the stacked arrangement, each of the five heat sinks is fixed to 8 mm in total thickness where the top and bottom cold plates are both 2 mm in thickness and the U-shaped channels or the fin fins are 2 mm in thickness or height. The simulation results from the U-shaped heat sinks are presented as one example of poor design of the heat sinks. For the pin fin heat sinks, the pin fins with cylindrical and diamond shapes, different size, angle and orientation were considered and compared. Only the simulation results from the optimised pin fin heat sinks with a specified size, angle and orientation are presented below.

IV. THERMAL-ELECTRIC AND CFD CO-SIMULATION

A. Modeling Description

Steady thermal-electric and computational fluid dynamics (CFD) coupled simulation was carried out using the system coupling method of Ansys Workbench. In the thermal-electric model, the entire stacked structure is considered, except for the two top and bottom clamping plates and the moulded encapsulants in the four IGBT modules. The thermal-electric loads include the power losses of the IGBT/FRD chips and the currents for the VVC, M-INV and G-INV as specified in Table I. The power losses are applied as surface heat sources on the corresponding IGBT and FRD chips, and the currents are applied as in and out currents on the corresponding DC+, DC- and AC terminals in the four IGBT modules.

In the CFD model, the coolant of 50% glycol in water occupies the flowing channels and connectors of the five heat sinks which are connected in parallel. Turbulent flow is assumed, and the most common K-epsilon (k-ε) turbulence model is employed. The inlet of the coolant is set to have a velocity corresponding to a total flow rate of 6, 8, 10 or 12 L/min entering the five heat sinks. The kinetic energy k, dissipation rate ε and the feed temperature of 70 °C are applied at the inlet boundary, and zero pressure is applied at the outlet boundary. The system coupling region is the interface between the solid structure and the coolant in the thermal-electric and CFD models.

B. Coupled Simulation Results

Figs. 4 to 12 present the representative coupled simulation results. Figs 4 to 11 compare the cooling performance of the stacked arrangements with the U-shaped channel and the optimised pin fin heats sinks where in both cases all the five heat sinks are made of Al alloy and all the four IGBT modules are assembled with SAC solder joints. Fig. 12 further compares the cooling performance of the stacked arrangements with the optimised Al alloy and Cu pin fin heat sinks where in both cases all the four IGBT modules are assembled with the sintered Ag joints to attach all the IGGBT and FRD chips on the bottom substrates and bond all the conductive shims under the top substrates.

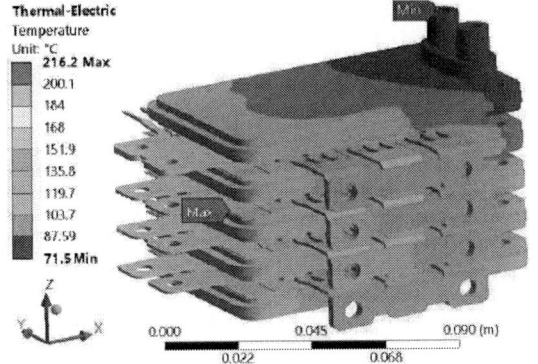

Fig. 4. The simulation result of the temperature filed in the stacked arrangement with the U-shaped channel heat sinks under the flow rate of 10 L/min for the coolant.

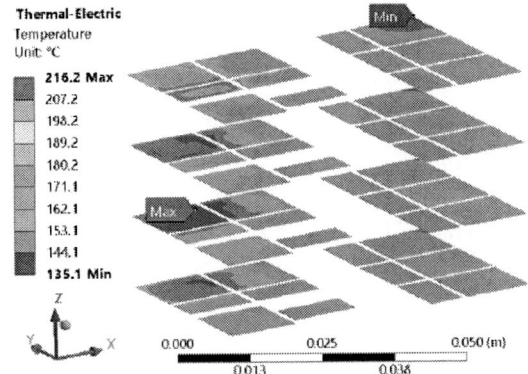

Fig. 5. The simulation result of the temperature filed in all the IGBT and FRD chips in the stacked arrangement with the U-shaped channel heat sinks under the flow rate of 10 L/min for the coolant.

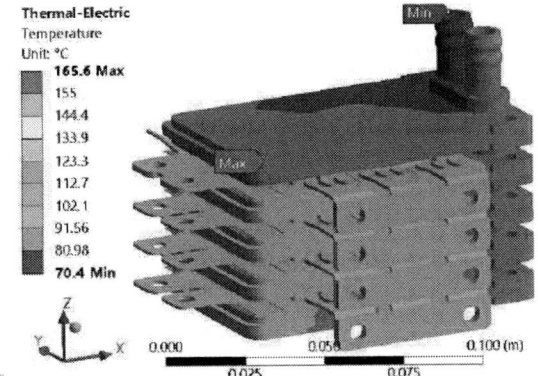

Fig. 6. The simulation result of the temperature filed in the stacked arrangement with the optimised pin fin heat sinks under the flow rate of 10 L/min for the coolant.

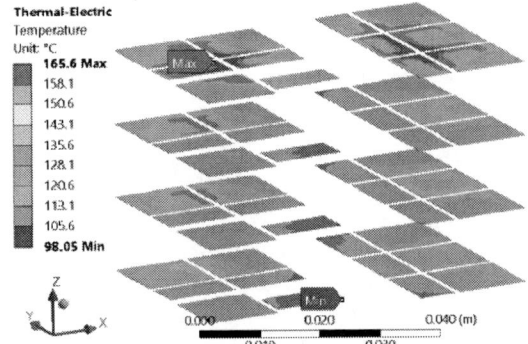

Fig. 7. The simulation result of the temperature filed in all the IGBT and FRD chips in the stacked arrangement with the optimised pin fin heat sinks under the flow rate of 10 L/min for the coolant.

In Figs 8 and 11, Tnode,max stands for the maximum node temperature, and Tvj,max stands for the maximum virtual junction temperature of the IGBT and FRD chips in the four IGBT modules of the corresponding stacked arrangement. The virtual junction temperatures were estimated from the average temperature on the active regions of the three paralleled IGBT or FRD chips in the half bridge switch for the VVC, the two paralleled IGBT or FRD chips in the three half bridge switches for the M-INV, or each IGBT or FRD chip in the three half bridge switches for the G-INV [9]. In Fig. 9, Tnode,max stands for the maximum node temperature in the coolant domain where the coolant flows to cool the corresponding stacked arrangement.

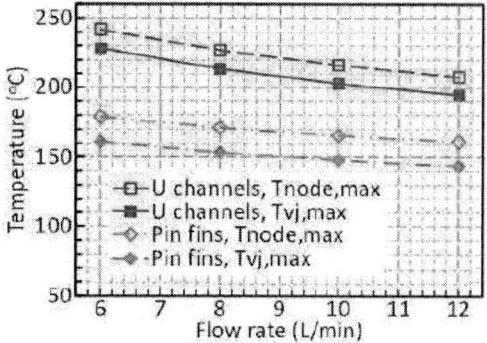

Fig. 8. Comparison of the maximum node and virtual junction temperatures with respect to the flow rate of the coolant in the stacked arrangements with the U-shaped channel and the optimised pin fin heat sinks.

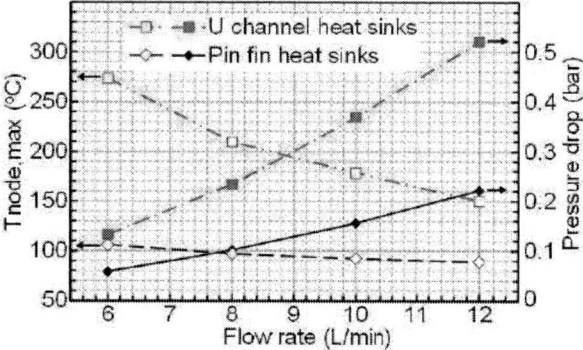

Fig. 9. The maximum node temperature in the coolant domain and the pressure drop across the inlet and outlet of the coolant in the stacked arrangements with the U-shaped channel and the optimised pin fin heat sinks.

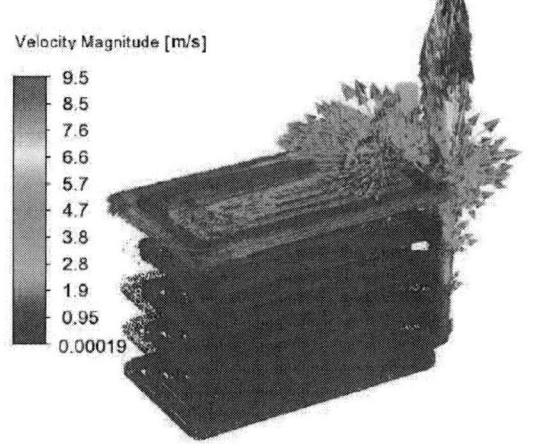

Fig. 10. The simulation result of the velocity magnitude in the coolant domain in the stacked arrangement with the U-shaped channel heat sinks under the flow rate of 10 L/min for the coolant.

The coupled simulation results reveal that the U-shaped channel heat sinks in the stacked arrangement have very poor cooling performance, leading to the maximum virtual junction temperature which is much higher than 150 °C and the pressure drop is remarkably higher than 0.2 bar under the flow rate of 10 L/min. In particular, the virtual junction

temperatures of the IGBT switches which have lower power losses in the three IGBT modules for the M-INV appear to be higher than the virtual junction temperatures of the two corresponding IGBT switches which higher power losses in the top IGBT modules for the VVC. This can be attributed to the fact that the channels in the connectors to connect the five U-shaped channel heat sinks in parallel are rather narrow and the flow resistance of the coolant through the top to the bottom heat sinks increases noticeably, resulting in reduced flowing velocities and thus reduced flow rate of the coolant through from the top to the bottom heat sinks, see Fig. 10.

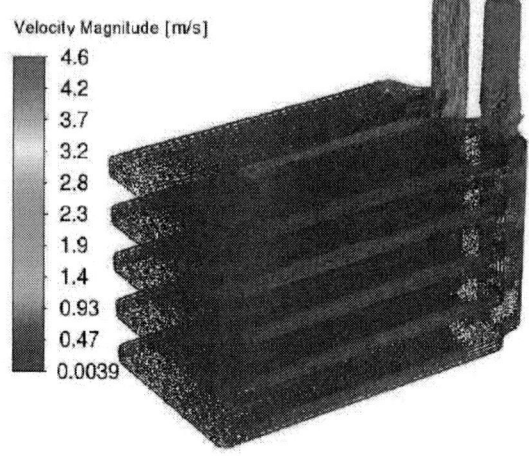

Fig. 11. The simulation result of the velocity magnitude in the coolant domain in the stacked arrangement with the optimised pin fin heat sinks under the flow rate of 10 L/min for the coolant.

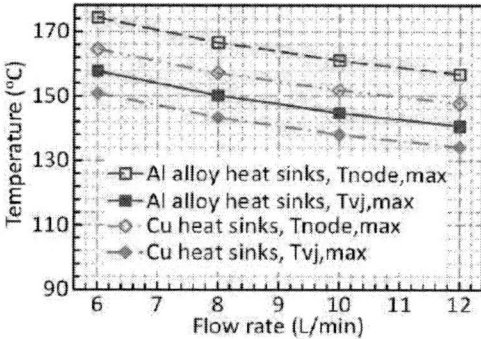

Fig. 12. The maximum node and virtual junction temperatures with respect to the flow rate of the coolant in the stacked arrangements with the IGBT modules assembled the sintered Ag joints and the optimised pin fin heat sinks made of Al alloy and Cu.

By modifying the channels in the connectors to connect the five paralleled heat sinks and optimising the orientation of pin fins in the pin fin heat sinks, the pressure drop across the inlet and outlet can significantly be reduced, and the flow rates of the coolant through the five heat sinks become quite consistent, see Figs. 9 and 11. As a result, the maximum virtual junction temperature occurs in the three paralleled low side FRD chips, followed by the virtual junction temperature in the three paralleled low side IGBT chips of the top IGBT module for the VVC which have higher power losses than the FRD and IGBT chips in the other three IGBT modules for the M-INV and G-INV. In particular, the coupled simulation result of the maximum virtual temperature in the stacked arrangement with the optimised pin fin heat sinks can be lower than 150 °C under the coolant feed temperature of 70 °C, flow rate of 10 L/min and pressure drop of 0.16 bar.

The cooling performance of the stacked arrangement under the different simulation cases may be compared with a thermal resistance defined as Rth=(Tvj,max-70)/P where P is the power loss which is the same value for all the cases. Comparing Figs. 8 and 12, it can be seen that replacing the SAC solder joints with sintered Ag joints in the four IGBT modules, the Rth of the stacked arrangement can be reduced about 2%, and further replacing the heat sink material of Al alloy with Cu, the Rth of the stacked arrangement can be reduced 6.3% to 6.6% under the flow rate from 6 to 12 L/min. The limited improvement in the cooling performance by the sintered Ag joints and the Cu heat sinks can be ascribed to the TIMs between the IGBT modules and their adjacent heat sinks which contribute relatively high thermal resistance.

V. EXPERIMENTAL VALIDATION

A. Prototyping processes

The designed IGBT modules have successfully been prepared with the following assembling processes: (i) a reflow soldering process to attach the IGBT/FRD chips and NTC thermistor on one bottom substrate, and bond all the conductive shims on the turned upside down top substrate for each IGBT module; (ii) a ultrasonic bonding process to bond the Al wires to interconnect the gate pads of the IGBT chips and the corresponding conductive tracks of the bottom substrate attaching the IGBT/FRD chips and NTC thermistor; (iii) a second reflow soldering process to bond all the power and signal/control terminals and the other sides of all the conductive shims on the conductive tracks of the bottom substrate attaching the IGBT/FRD chips and NTC thermistor or on the top sides of these IGBT/FRD chips; and (iv) a moulding process to inject and form the epoxy compound encapsulant for each IGBT module.

However, the delivery of the custom-made heat sinks has been delayed by the supplier. Once the heat sinks are received, each sample of the stacked arrangement will be produced with a mounting process to stack four IGBT module and five heat sinks together with eight pieces of graphite sheets, two clamping plates attaching the compliant pads and 4 pairs of M6 screws and nuts.

B. Thermal test

After the samples of the stacked arrangement are produced, transient thermal test will be carried out using a Mentor Graphics® 1500A power tester and a recirculating chiller to deliver the coolant of 50% glycol in water into the five paralleled heat sinks. The electrothermal generation in such thermal test cannot be adjusted to match the values specified in Table I for the above thermal-electric and CFD coupled simulation. This is because the values in Table I include both conducting losses and switching losses of the IGBT and FRD chips based on the operation of the electric control system where the VVC delivers 100 kW, the M-INV delivers 100 kW and the G-INV delivers 50 kW. In the transient thermal test using the Mentor Graphics® 1500A power tester which is equipped in our laboratory, only the conducting losses of the IGBT and FRD chips can be used as the heat sources.

Nevertheless, the transient thermal test results can still be compared with additional thermal-electric and CFD coupled simulation based on the testing condition and the electrothermal generation during the test. If the additional simulation results satisfactorily agree with the test results, this will demonstrate that the thermal-electric and CFD models and the electrical and thermophysical properties of all the materials in the models are sufficiently accurate for the electrothermal and cooling analysis, and thus validate the above thermal-electric and CFD coupled simulation results to support the thermal design of the stacked arrangement.

VI. CONCLUSIONS

The stacked arrangement inserting four IGBT modules between five heat sinks can be designed into a compact 121 mm × 94.7 mm × 90.8 mm structure for the variable voltage converter delivering 150 kW, the motor inverter delivering 100 kW and the generator inverter delivering 50 kW.

Thermal-electric and CFD coupled simulation results indicate the maximum virtual junction temperature in the four IGBT modules can be below 150 °C under the worst electrothermal generation, the coolant feed temperature of 70 °C, flow rate of 10 L/min and pressure drop of 0.16 bar.

Replacing the SAC solder joints with sintered Ag joints in the four IGBT modules and replacing the heat sink material of Al alloy with Cu can somewhat but not very effectively improve the cooling performance of the stacked arrangement which is limited by the thermal interface materials.

Experimental validation of the coupled simulation results has been delayed, and hopefully it will be obtained and presented at the conference. If validated, the stacked arrangement can be used to increase power density and improve electrical and thermal performance of the power modules in the electric control systems with multiple converters/inverters.

REFERENCES

[1] T. Tokuyama, K. Nakatsu, A. Mima, Y. Hattori, and T. Satoh, "Power semiconductor module," United States patent: US 2015/0214205 A1.

[2] J. F. Li, Y. Ma, F. Dong, Y. Du, J. C. Arcillas, and J. Yan, "An integrated IGBT module for dual inverter applications," in: CIPS 2022; 12th International Conference on Integrated Power Electronics Systems, 15-17 March 2022, Berlin, Germany.

[3] M. Sugita, "Stack unit," United States patent, US 2016/0192539 A1.

[4] I. Nakamura, and K. Matsuura, "Boost converter," United States patent, US 2018/0026533 A1.

[5] FF400R07A01E3_S6, "Double Side Cooled Module," Final Data Sheet, Published by Infineon Technologies AG, V3.4, 15 April, 2020.

[6] R. Wu, L. Smirnova, H. Wang, F. Iannuzzo, and F. Blaabjerg, "Comprehensive investigation on current imbalance among parallel chips inside MW-scale IGBT power modules," in Proceedings of the 2015 9th International Conference on Power Electronics and ECCE Asia, 1-5 June 2015, Seoul, South Korea.

[7] F. Lang, H. Yamaguchi, H. Nakagawa, and H. Sato, "Cyclic Thermal Stress-Induced Degradation of Cu Metallization on Si₃N₄ Substrate at −40°C to 300°C," Journal of Electronic Materials, vol. 44, pp. 482-489, Semptember 2015.

[8] J.F. Li, A. Castellazzi, T. Dai, M. Corfield, A.K. A.K. Solomon, and C.M. Johnson," Built-in reliability design of highly integrated solid-state power switches with metal bump interconnects," IEEE Transactions on Power Electronics, vol. 30, no. 5, pp. 2587–2600, May 2015.

[9] J.F. Li, K. Li; J.Dai, C.M Johnson, and X. Lin, "Thermal and thermo-mechanical design of an integrated IGBT module," In Proceedings of 19th International Conference on Electronic Packaging Technology, Shanghai, China, 8-11 Aug. 2018.

Lamination of capacitive micromachined ultrasonic transducer on a piezoelectric array: process and evaluation

Duy Hoang Le
Department of Microsystems
University of South-Eastern Norway
3184 Borre, Norway
hoang.d.le@usn.no

Tung Manh
Department of Microsystems
University of South-Eastern Norway
3184 Borre, Norway
tung.manh@usn.no

Lars Hoff
Department of Microsystems
University of South-Eastern Norway
3184 Borre, Norway
lars.hoff@usn.no

Abstract—Harmonic imaging plays a crucial role in medical ultrasound imaging, where the transducer probes are required to emit low-frequency ultrasonic waves into the tissues, and receive reflected signals at the higher, typically the second harmonic of the transmitted wave. We have designed a unique probe with a high-frequency capacitive micromachined ultrasonic transducer (CMUT) on top of a low-frequency piezoelectric array for harmonic imaging. This paper presents a lamination process between CMUT and piezoelectric array using epoxy. The bond-line between the two components was thoroughly investigated using mechanical and electrical methods. The resulting bond-line thickness was less than 2 µm, and shear strength fell within the range 4.8 – 5.1 MPa. Additionally, electrical impedance measurement of all elements in the low-frequency array exhibited resonance shifts before and after bonding, providing further insight into the bonding process.

Keywords—harmonic imaging, piezoelectric array, dual frequency, capacitive micromachined ultrasonic transducer.

I. INTRODUCTION

Harmonic imaging (HI) is a technique employed in medical ultrasound imaging to enhance image quality and diagnosis accuracy. The HI technique utilizes the non-linear properties of the tissues, which distort the transmitted wave as it propagates, generating higher harmonics of the transmitted fundamental frequency. Transducer probes for HI are mostly based on piezoelectric materials. These traditional transducers are operated by thickness vibration mode of the piezoelectric layer under an electrical excitation. A drawback of piezoelectric material is its high acoustic impedance, which is the product of density and longitudinal sound speed, around 35 MRayl, in contrast to 1.5 MRayl for water and human tissue. To compensate for the impedance mismatch, one or more quarter wavelength matching layers are added to the transducers. The matching layers secure a broad bandwidth for the piezoelectric transducers. The number of matching layers is normally limited to two due to production complexity and lower sensitivity. The matching layer is typically composed of particle-loaded epoxy with an optimal specific acoustic impedance. In this work, a single matching layer with an acoustic impedance 7.3 MRayl is utilized. The piezoelectric array is formed by dicing the piezoelectric plate, creating an array configuration to allow steering and focusing of the beam, necessary for imaging.

Over the last three decades, in parallel with piezoelectric technology, significant advancements have made in the development of CMUTs [1]. A CMUT cell consists of a thin membrane suspended on a cavity, two electrode layers, and a passivation base. The top membrane, which can range in thickness from nanometers to a few micrometers, is typically made of an elastic material, such as silicon nitride (Si_xN_y) or polymer. The CMUT operates based on the principle of capacitance. The two electrodes separated by the cavity, form a capacitor. In operation, a bias voltage is applied to maintain static charges on both electrodes. In transmission mode, when an electrical excitation is applied, the fluctuation of electrostatic force between the electrodes causes the membrane to vibrate. This vibration generates ultrasonic waves for transmission. In receiving mode, the membrane vibrates in response to incoming waves, resulting in a change in the capacitance between the electrodes. This vibration in capacitance generates an electrical signal that can be processed to extract information about the received waves. CMUTs offers high receiving sensitivity, flat bandwidth, high frequency operation and good integration with microelectronics. They can be designed in many shapes, which satisfy a wide range of applications from medical imaging to industrial inspection. However, one of the challenges in integrating CMUTs with conventional electronics is the requirement for high bias voltages, often in range of hundreds of volts. The fabrication process for CMUTs is complex and requires extremely well-controlled microfabrication techniques. In transmit, CMUTs typically generates lower acoustic intensity compared to piezoelectric probes. However, advancements in CMUT design and fabrication continue to address the challenges, making CMUTs an attractive technology for various ultrasonic applications.

In previous work, a unique single-element probe design was demonstrated, incorporating a high-frequency CMUT on top of a low-frequency piezoelectric element for harmonic imaging [2]. The piezoelectric transducer was responsible for transmitting acoustic signals at low frequencies, while the CMUT received the reflected waves at high frequencies. This design offered a combination of strong transmit sensitivity and linearity from the piezoelectric transducer, along with the broad bandwidth and configurability provided by the CMUT. The focus of this paper is on the development of a hybrid CMUT-piezoelectric transducer array, which introduces new challenges. To minimize crosstalk between elements of the piezoelectric array, air-filled kerfs were implemented. The lamination process between the piezoelectric array and the CMUT die was accomplished using the epoxy, Epotek 301-2 [3]. This epoxy is a biocompatible adhesive which can be cured within a wide range of conditions, from room temperature to 80°C. The epoxy layer was positioned between the matching layer of the low frequency array and the silicon nitride layer of the CMUT, without entering deeply into the kerfs of the array. The bond-line was characterized through cross-sectional inspection. The mechanical strength was tested using a shear tester. Additionally, electrical impedance measurements were performed on each piezoelectric element to investigate whether the bonding was successful.

II. METHOD

A. Piezoelectric array fabrication

The piezoelectric stack had dimensions 16.3 x 11.0 mm^2, and consisted of a PZT (Lead Zirconate Titanate) ceramic with an acoustic matching layer on top. The matching layer was composed of alumina and tungsten powder loaded epoxy. The construction of the stack involved a bonding process. Initially, the piezoelectric layer was bonded to the matching layer using glue DP 460 (3M, Minnesota, USA). Subsequently, the stack was bonded to a flex circuit using the same adhesive. The curing condition for the glue was set to 60°C for 2 hours. Alignment of the layers and flex during the bonding process was secured by employing a bonding tool. This tool helped to maintain precise alignment between the layers, ensuring the accurate bonding and integration of the stack and the flex circuit.

Fig. 1 Piezoelectric array on a flex. Active area, 16.3 x 11.0 mm^2, is diced with 160 μm pitch and 30 μm kerf.

The stack was diced into an 1D array using a 30 μm thick blade, as shown in Fig. 1. The resulting kerfs, measured less than 40 μm, and the elements were uniform, as depicted in Fig. 2. In addition, a dummy array was fabricated, which only contained the acoustic matching layer. This dummy array was specifically intended for destructive testing, such as cross-sectional inspection and shear testing. Prior to the lamination process, the top surface of the array, including dummy array, was activated using 200W oxygen plasma treatment for a duration of 60s right before the lamination process.

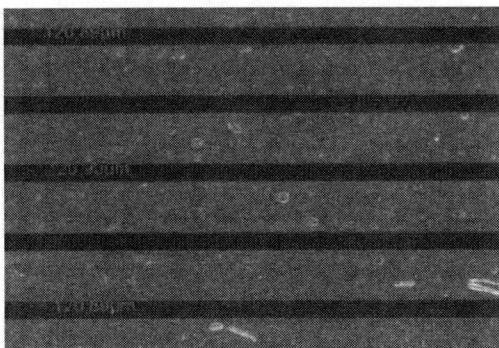

Fig. 2 Some elements of the piezoelectric array

B. Dummy CMUT die

The dummy CMUT dies with dimension 15.4 x 15.4 mm^2 were fabricated with Si$_x$N$_y$ structure on the top of a Si-handler, as illustrated in Fig. 3. The CMUT structure had a thickness of 8 μm, while the die had a thickness of 350 μm. Before the lamination process, the CMUT dies were cleaned using isopropanol and dried completely. Subsequently, the top surface of the dummy CMUT was subjected to a 200W oxygen plasma for 60s. This oxygen plasma process ensured the removal of any contaminants, provided an activated surface for next steps, and promoted adhesion during the lamination process.

Fig. 3 Dummy CMUT die, size 15.4x15.4 mm^2.

C. Lamination process of CMUT on a piezoelectric array

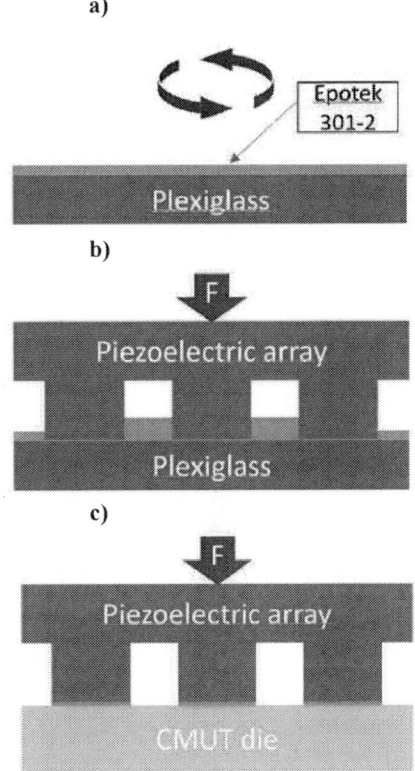

Fig. 4 Lamination process of a piezoelectric array on a CMUT die. a) Spin-coating of Epotek 301-2 on a plexiglass; b) The piezoelectric array is up-side-down on the epoxy film; c) The piezoelectric array is transferred to attach on a CMUT die.

The lamination process ensured the bonding and integration of the piezoelectric array with the CMUT die using the epoxy film as an adhesive layer. The process consisted of three main steps. Firstly, a plexiglass substrate was activated by oxygen plasma, at a power of 200W for a duration of 200s. This oxygen plasma was necessary to make the substrate hydrophilic before spin coating process. A thin film of Epotek 301-2 was deposited onto the plexiglass substrate using spin coating, as shown in Fig. 4a. To achieve 30 μm thick epoxy layer, the spinning speed was set to 1000 RPM, lasting for 60s. In the second step, the piezoelectric array was pressed onto the wet epoxy film to ensure proper wetting of the surface, as depicted in Fig. 4b. The pressing step lasted for 60s. Finally, the piezoelectric array, along with the wet epoxy on the top surface, was transferred onto a CMUT die, as shown in Fig. 4c. The array and the CMUT die were pressed together and left to cure at room temperature for a period of 3 days.

D. Evaluation of the bond-line

In the first prototype, which featured a real piezoelectric array, electrical impedance measurement was conducted. Electrical impedance of a PZT element is represented as a complex number, $Z = R + j*X$. The reactance, represented by X, encompasses the capacitance and inductance components of the transducer. The value of X influences the phase of electrical impedance, which indicates the phase shift between electrical input and output signals. At low frequencies, the impedance is predominantly capacitance, indicated by a large negative X. Resistance R should dominate around resonance. R indicates power dissipation, either within the transducer or as acoustic radiation. The value of electrical impedance depends on various factors, such as material properties, physical dimensions, layer structure and electrical connections. To access these electrical characteristics, the impedance of each active element in the array was measured both before and after the lamination process by an E4990A Impedance Analyzer (Keysight, California, USA).

For the second prototype, a dummy array without flex was utilized. The dummy array featured the same matching material on the surface as the first prototype to maintain consistency. The prototype was diced, allowing cross-sectional inspection under optical microscopy. Subsequently, a shear test was conducted using a 4000 Plus Bondtester (Nordson Dage, Ohio, USA). The purpose of the shear test to evaluate the strength of the structure.

III. RESULTS AND DISCUSSION

A. Bond-line thickness

A thin film of Epotek 301-2 coating, measuring 30 μm in thickness, was successfully achieved by a specific spinning schedule and the wetting properties of the plexiglass surface. Plexiglass was chosen as the substrate material because it facilitated the maintenance of a thin epoxy layer over extended periods of time. Si wafer and glass substrates were also tested, but they did not exhibit the same level of uniformity in maintaining a wet Epotek 301-2 epoxy film after spinning. Even when subjected to the same oxygen plasma activation, the epoxy would tend to shrink towards the center region on these substrates. Therefore, the use of plexiglass proved to be more suitable for obtaining the desired uniformity in the wet epoxy film.

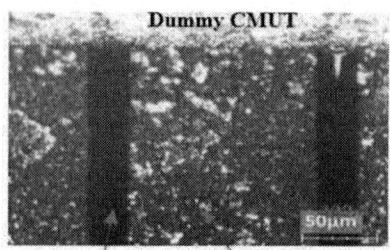

Air kerf   PA element

Fig. 5 Cross-sectional view of the bond-line between matching material and dummy CMUT

During cross-sectional inspection, it was observed that the air kerfs had uniform widths. There was no delamination after the dicing process, as depicted in Fig. 5. The Epotek 301-2 layer applied was of sufficient thickness to uniformly cover the entire sample surface without excessive epoxy infiltrating into the kerfs. The air kerfs in the piezoelectric array serve the purpose of reducing crosstalk effects between active elements.

The thickness of the bond-line was found to be less than 2 μm, which is much smaller than acoustic wavelength of 500 μm. Therefore, the acoustic effect of such a thin bond-line can be considered negligible.

To ensure the integrity of the active elements, the pressing force was carefully controlled to minimize the thickness of the bond-line. Applying high force could potentially cause damage to the active elements. The force exerted on the entire surface, equivalent to 10 bar pressure, remained consistent in both step 2 and step 3 of the lamination process. Traces of Epotek 301-2 on the dummy CMUT exhibited minimal variation in adhesion strength over a large area, as shown in Fig. 7.

B. Shear strength

The shear strength of the bond-line was measured within the range 4.8 – 5.1 MPa, as shown in Fig. 6. The surface roughness and flatness of the matching layer played a crucial role in achieving a strong adhesion and uniformity of the bond-line. The strong adhesion of the bond-line ensures good acoustic transmission through the interfaces.

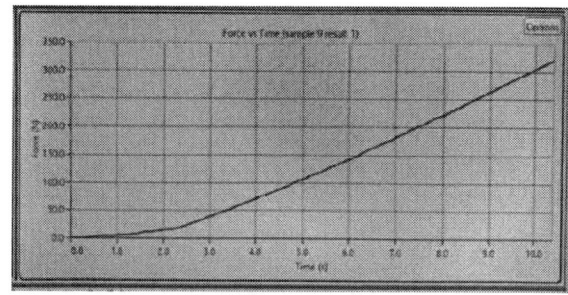

Fig. 6 Shear strength curve

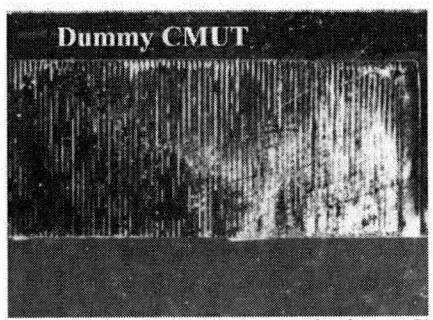

Fig. 7 Trace of Epotek 301-2 epoxy on the dummy CMUT after shear test

C. Electrical impedance measurement

The electrical impedance of each element on the array was measured within the range 1 – 10 MHz using connectors on the flex. The absolute value of electrical impedance spectrum exhibited a clear change after lamination with the CMUT die, as illustrated in Fig. 8. The impedance spectra show the resonance-antiresonance structures characteristic for a piezoelectric plate. Prior to lamination (blue curve), two such structures are seen in the impedance spectrum. The resonance frequencies are at the minima of these curves, i.e., at 3.6 MHz and 6.1 MHz. However, after lamination (red curve) a strong resonance is seen at 5.1 MHz, along with two weak resonances at 1.5 MHz and 6.5 MHz. These shifts in the resonance frequencies were a result of the presence of the CMUT die on the top of the matching layer. The Si-handler layer of the die creates acoustic mismatch with the matching layer and the tissue, which is undesirable. Consequently, wave reverberation and crosstalk between elements occur inside the transducer.

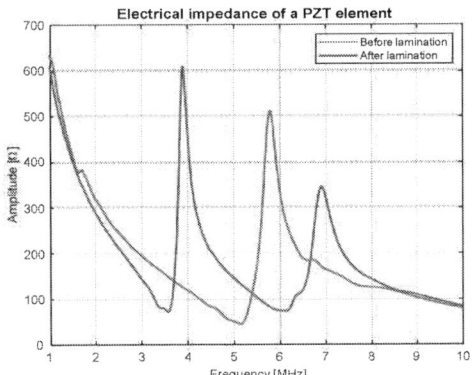

Fig. 8 Electrical impedance of a PZT element before and after lamination to the CMUT die

IV. Conclusion

A lamination process for a hybrid CMUT-piezoelectric ultrasound transducer was developed and successfully performed. The method uses traditional materials and equipment. Two different prototypes underwent shear testing, microscopic examination, and electrical impedance measurement. The strong and thin bond-line achieved ensured excellent acoustic transmission properties. Importantly, the epoxy did not enter the kerfs, thereby avoiding additional acoustic crosstalk in the low-frequency array.

The results validate the assembly process for constructing a new generation of ultrasonic transducer with the potential to improve image quality for medical diagnosis.

The resultes have also demonstrated the need for a thin silicon substrate layer, i.e., the necessity of removing or thinning down the Si handler layer.

Acknowledgment

The work described in this study is financially supported by The Research Council of Norway through project CIUS (Center for Innovative Ultrasound Solutions) and CFRON (Frontend and transducer technology for next generation cardiovascular ultrasound probes). The research team acknowledges the contribution of lab engineers at USN (University of South-Eastern Norway), and expertise provided by experts at GE (General Electric).

References

[1] M. I. Haller and B. T. Khuri-Yakub, "A surface micromachined electrostatic ultrasonic air transducer," in *1994 Proceedings of IEEE Ultrasonics Symposium*, Oct. 1994, pp. 1241–1244 vol.2. doi: 10.1109/ULTSYM.1994.401810.

[2] A. Stuart Savoia *et al.*, "Design, Fabrication and Characterization of a Hybrid Piezoelectric-CMUT Dual-Frequency Ultrasonic Transducer," in *2018 IEEE International Ultrasonics Symposium (IUS)*, Oct. 2018, pp. 1–4. doi: 10.1109/ULTSYM.2018.8580125.

[3] "MED-301-2 – Epoxy Technology." https://www.epotek.com/product/med-301-2/ (accessed Jun. 16, 2023).

Assembly of Ultra-thin MEMS Device on Driver Chip Using Anisotropic Conductive Film

Hoang-Vu Nguyen
Department of Microsystems
University of South-Eastern Norway
Borre, Norway
hoang.v.nguyen@usn.no

Knut Eilif Aasmundtveit
Department of Microsystems
University of South-Eastern Norway
Borre, Norway
Knut.Aasmundtveit@usn.no

Abstract—**Assembly solution for large-area, ultra-thin, MEMS-based transducer array dies on driver chips has been investigated using dummy Si dies and substrates. The dummy dies have a thickness of 50 μm, a bonding area on the order of 100 mm² populated with more than a thousand through-silicon vias connecting corresponding metal pads and redistribution layer (RDL) on both sides. The dummy substrates have conventional thickness as well as necessary RDL and pads for bonding and monitoring electrical resistance and stray capacitance of interconnects. Flip-chip interconnection technology based on anisotropic conductive film (ACF) was selected for bonding and characterization. Handling and ACF bonding of ultra-thin dies to rigid substrates were eased by means of glass carriers attached to the inactive side of the dies. Proper bonding parameters were identified, providing sufficiently low interconnect resistance with satisfactory yield. Stray capacitance of interconnects was found in acceptable range, though lower stray capacitance should be further pursued. The results from this work have demonstrated the feasibility of using ACF for assembling large-area, ultra-thin dies on rigid substrates.**

Keywords—ultra-thin dies, MEMS, assembly, ACA, ACF

I. INTRODUCTION

The demand for ultra-thin dies in applications such as stacked 3-dimensional (3D) packages, integrated circuit (IC) card, flexible hybrid electronics has been increasing considerably in the recent decade [1, 2]. Such ultra-thin dies, with a thickness of 50 μm and below, include IC chips and micro electro-mechanical systems (MEMS) sensors. The assembly of such dies on substrates is challenging in terms of handling, high-yield bonding, and reliable integration [3-5]. While soldering combined with underfilling has been the main solution for stacking 3D packages [1], anisotropic conductive adhesives, either in form of film (ACF) or paste (ACP), are common for assembling thin dies on polymer-based substrates in flexible electronics [2, 3, 6].

The present work investigates the assembly of MEMS-based capacitive micromachined ultrasound transducers (CMUT) for medical imaging applications. Such transducers are normally arranged in a 2D array with an extremely high number of elements (typically several thousand) on a large-area Si die in order to obtain high-resolution 3D images in real time. Since each transducer element needs a dedicated electrical connection to driver electronics, fanning out a high number of connections to the peripheral area of a sensor die for conventional wire bonds becomes impractical. Therefore, CMUT array with through-silicon vias (TSVs) connected to interconnection bumps/pads on the other side of a die for flip-chip integration is a proper solution [7-9]. Such sensor dies

normally have a thickness below 50 μm, to ensure short-distance electrical connections for sufficiently low stray capacitance and enhanced temporal resolution. Stray capacitance caused by individual interconnects, which is in parallel with transducer element capacitance, must be reduced as much as possible to achieve good receive sensitivity and preserve wide bandwidth, particularly for 2D array elements with very low device capacitance [9, 10].

A sensor die with a CMUT array needs to be integrated with a driver chip (ASIC – Application Specific Integrated Circuit) via short-distance interconnections. Flip-chip bonding a sensor die to an ASIC chip with extremely high number of I/Os is thus a proper solution. Since there are similarities between ultra-thin CMUT dies and Si interposers, the results obtained in this work could also be applied for bonding IC chips to Si interposers. Possible flip-chip bonding techniques include solder reflow, metal-metal thermocompression bonding, and anisotropic conductive film (ACF). ACF is of particular interest because it provides electrical connection, mechanical strength and sealing/underfilling in a quick bonding process. In addition, ACF bonds could be obtained at moderate bonding temperatures and pressures. The technology has also demonstrated high bond yield and reliable interconnects in relevant applications, such as chip-on-glass, chip-to-chip [11-14].

This work addresses ACF bonding process of large-area, ultra-thin transducer array dies to ASIC chips. Processes providing high bonding yield, sufficiently low electrical resistance and low stray capacitance for individual interconnects are of interests. Bonding and characterization are performed using dummy Si dies and substrates mimicking the transducer dies and ASIC chips. The dummy dies have ultra-low thickness, and TSVs connecting corresponding metal pads and redistribution layer (RDL) on both sides. The dummy substrates have conventional thickness as well as necessary RDL and pads for bonding and monitoring electrical resistance and stray capacitance of interconnects. While electrical resistance could be measured using conventional methods, stray capacitance of individual interconnects is defined as the capacitance between an ACF interconnect with its corresponding TSV and bulk Si.

II. EXPERIMENTAL

A. Test samples

Dummy transducer array dies were fabricated on highly doped Si wafers with SiO_2 as isolation layer on surface, targeting an active area on the order of 100 mm² and a thickness of 50 μm. One side of the dies are populated with Ni/Au-electroplated bumps, each electrically connects to Al

Funding agency: Research Council of Norway through the project "Frontend and transducer technology for next generation cardiovascular ultrasound probes" (project number: 317769)

RDL on the other side by means of a Cu-filled TSV. The bumps are 50 µm in diameter, 15 µm high, and distributed in form of m-by-n matrix with bump pitch in the range of 200–400 µm. Direct contacts between Al pads and bulk Si were realized at dedicated locations on the RDL side to enable stray capacitance measurements. Due to the ultra-low thickness, all dies were attached to individual glass carriers by means of a transparent adhesive, which would be detached after bonding.

Dummy substrates were fabricated on highly doped Si wafers with a thickness about 750 µm. SiO_2 was also used as isolation layer on the substrates' surface. Bond pads are composed of Ni/Au electroplated to dimensions of 80 µm x 80 µm x 15 µm, and distributed correspondingly to the bumps on dummy dies. In addition, the substrates also have probing pads for measurements of electrical resistance and capacitance. Direct contacts between Al pads and bulk Si were also found at dedicated locations on substrates to enable stray capacitance measurements.

Test samples comprise a dummy die bonded to a dummy substrate. The dies and substrates were designed and configured to form 57 daisy chains, each with 30 interconnects. Fig. 1 shows (a) the layout design of corresponding dies and substrates, as well as illustration of measurement concepts for (b) electrical resistance and (c) stray capacitance.

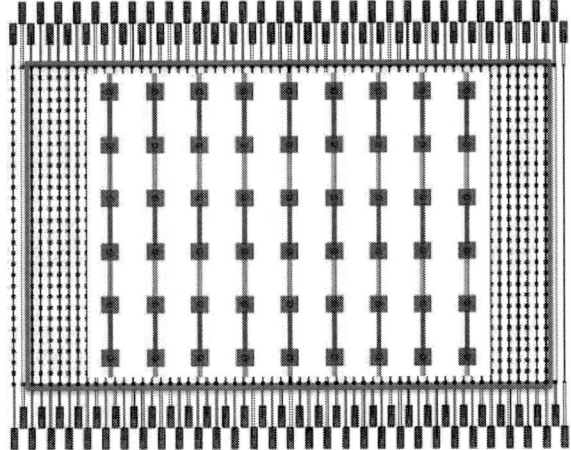

(a) Layout of a die (marked with a red rectangle) overlapping a substrate. Each column is a daisy chain of interconnects. The inset shows a zoom-in of circular die bumps and square substrate pads, in addition to Al RDL on both parts.

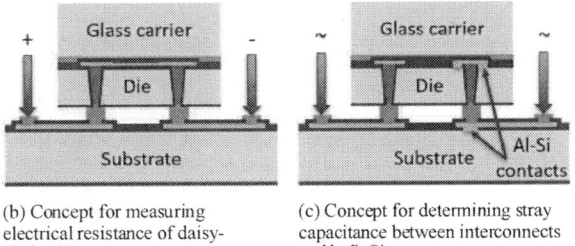

(b) Concept for measuring electrical resistance of daisy-chained interconnects

(c) Concept for determining stray capacitance between interconnects and bulk Si

Dimensions are not to the scale.

Fig. 1. Designs of dies and substrates for measurements of electrical resistance and stray capacitance of interconnects

B. Bonding and characterization

A commercial ACF product from H&S HighTech Corporation was used in this work. The ACF is single-layer film with a thickness of 35 µm and contains Ø5 µm Ni/Au coated monodisperse polymer spheres. ACF bonding of a dummy die to a dummy substrate was carried out using a flip-chip bonder Finetech FinePlacer pico. Bondline temperature was characterized using flat thermocouples inserted between a die and a substrate during a bonding process without ACF applied. A conventional ACF bonding process, with two main steps: i) pre-tack ACF on a substrate's surface, and ii) final bonding of a die to a substrate, was applied. Bonding pressure was the main varying parameter whereas bonding temperature and bonding time were selected as 180 °C and 30 seconds, according to recommendations of the ACF supplier.

Bonded samples were characterized by means of electrical resistance of daisy chains and bonding yield. The daisy-chain resistance was measured using two-point probe method with a Fluke 73 digital multimeter. The resistance of probe wires and needles was found negligible compared to the measured resistance of each daisy chain. For each bonded sample, there are a total of 57 daisy chains being measured. Bonding yield of each group of samples bonded at different pressures was determined by dividing the number of interconnects with sufficient conductance by the total interconnects of all samples in that group. As the highest resolution of probing is 30 interconnects in a daisy chain, all interconnects were claimed open if an open circuit happened to the chain.

Stray capacitance of individual daisy chains was obtained by measuring impedance of the entire chains. The impedance measurements were performed using a Keysight impedance analyzer E4990A, with one probe connecting to the entire chain and the other probe connecting to bulk Si thanks to the direct Al-Si contacts. Since the resistance of highly doped bulk Si is negligible as compared to the measured impedance, stray capacitance of individual daisy chains could be calculated from the impedance measurement results. Stray capacitance of individual ACF interconnects and its corresponding TSV was then estimated based on the values obtained for the daisy chains.

III. RESULTS

Test samples were bonded with bonding pressures varied from 2 MPa to 5 MPa. At each bonding pressure value, 5 samples were manufactured. The visual inspection of all bonded samples revealed no cracks on the ultra-thin dies after bonding. Fig. 2 shows a typical sample after ACF final bonding.

Fig. 3 shows effect of bonding pressure on electrical resistance of individual daisy chains, each comprising 30 interconnects. Considering even distribution of bonding pressure, and hence electrical resistance of individual interconnects in a daisy chain, the results in Fig. 3 also indicates impact of the bonding pressure on ACF interconnect resistance. The daisy-chain resistance decreases gradually with increasing bonding pressure from 2 MPa to 3 MPa. When bonding pressure is over 3 MPa, its impact on daisy-chain resistance is minor. Furthermore, the deviation of daisy-chain resistance increases considerably at bonding pressures higher than 3 MPa.

Bonding yield corresponding to different bonding pressures is shown in Table I. Better bonding yield is obtained when bonding pressure is increased over 2 MPa. However, the yield remains stable at about 85–87% for bonding pressures in the range of 3–5 MPa. Daisy chains with failed interconnects were observed at random locations from samples to samples. No systematic failures were recognized.

Impedance measurements were applied for a limited number of individual daisy chains on samples bonded at 3 MPa. Note that only daisy chains with proper electrical connection of all interconnects were of interests. The results showed relatively stable capacitive reactance of individual daisy chains of about 5.8 kΩ at 1 MHz. This corresponds to a capacitance of about 27.4 pF.

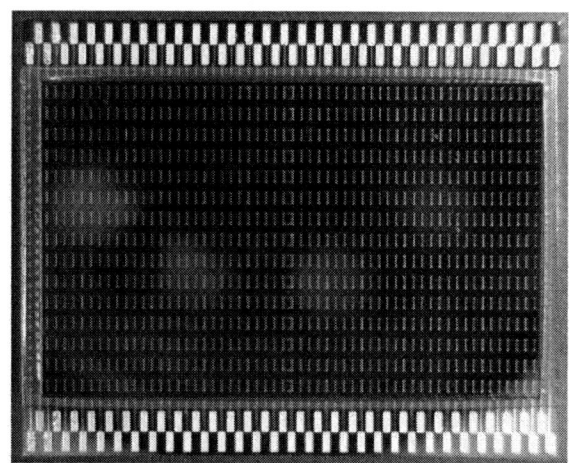

Fig. 2. Example of a test sample bonded at 3 MPa

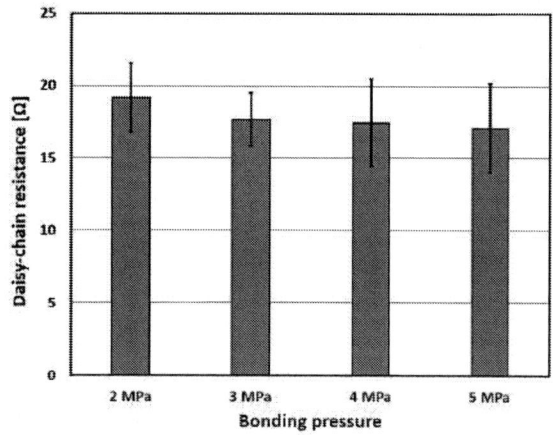

Fig. 3. Effect of bonding pressure on electrical resistance of individual daisy chains, each comprising 30 ACF interconnects

TABLE I.   BONDING YIELD OF SAMPLES BONDED AT DIFFERENT BONDING PRESSURES

	Bonding pressure			
	2 MPa	3 MPa	4 MPa	5 MPa
Bonding yield	73.7 %	85.1 %	87.3 %	86.8 %

IV. DISCUSSION

Electrical resistance of daisy chains presented in Fig. 3 includes resistance of Al RDL and Cu-filled TSVs that belong to individual chains. The resistance of all Al RDL in a chain was measured in practice by means of dummy chains designed for this purpose. The measurement results agree very well with the estimated values based on dimensions of Al tracks. By deducting the Al-RDL resistance, the correlation between electrical resistance of individual interconnects and bonding pressure were obtained, as shown in Fig. 4. Note that the resistance values in Fig. 4 are contributed from ACF interconnects and its corresponding Cu-filled TSV. By using TSV dimensions and electrical resistivity of pure Cu, the electrical resistance of individual TSVs was estimated in the range of 5–17 mΩ, being insignificant as compared to the resistance values presented in Fig. 4. These values thus represent ACF interconnect resistance.

Bonding yield shown in Table I is considered conservative. It is because all 30 interconnects in a daisy chain were considered as open circuit even if only 1 interconnect in that chain failed. In addition, failures in a daisy chain might stem from individual ACF interconnects and/or TSVs. The ACF bonding yield in practice is thus expected to be far better than the values presented in this work.

Failures were observed at random locations from samples to samples. This indicates a proper setup of bonding equipment. Failures seem to stem from the ACF bonding process as well as the fabrication of TSVs. Further work on failure analyses, such as cross-sectional microscopy, is crucial to identify the root causes of failed interconnects in bonded samples.

The interconnect resistance in Fig. 4 and bonding yield in Table I indicate 3 MPa as a proper bonding pressure within the pressure range tested in this work, providing sufficiently low interconnect resistance with moderate deviation as well as satisfactory bonding yield. ACF interconnect resistance obtained at this pressure value is in the range of 110–240 mΩ, in accordance with results from previous studies of ACF assemblies for chip-on-glass, chip-to-chip, chip-on-flex, chip-on-board, and several MEMS applications [12-16].

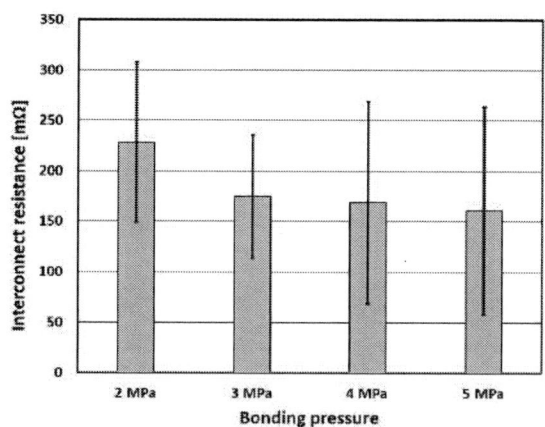

Fig. 4. Interconnect resistance versus bonding pressure

Capacitance of daisy chains with full connection was found to be about 27.4 pF. Since this result was obtained from measured impedance of a limited number of daisy chains on samples bonded at 3 MPa, the following interpretation is considered preliminary. Based on the dimensions of Al RDL as well as thickness and dielectric constant of SiO_2 layer, the capacitance of all Al RDL is estimated to be about 11.2 pF. Hence, the capacitance of 30 interconnects and TSVs in a chain is about 16.2 pF. That means stray capacitance of individual ACF interconnects and its corresponding TSV is about 0.5 pF. Whether such a stray capacitance is acceptable depends on the capacitance of individual CMUT elements used and its driver electronics. Previous studies specified element capacitance about 1 pF [10] or acceptable stray capacitance from electrical connections up to 1.7 pF [9]. This indicates that the interconnect stray capacitance obtained in this work might be acceptable. However, lower stray capacitance should be further pursued.

The assembly process employed in this work has demonstrated the ease of handling ultra-thin dies by means of glass carriers. No cracks were observed, even for samples bonded with as high as 5 MPa. Compared with handling of bare ultra-thin dies, which are common in assembly of flexible hybrid electronics [2, 3, 17, 18], the use of glass carrier indeed facilitated handling and ACF bonding of ultra-thin dies to rigid Si substrates. This solution also offered risk mitigation for squeezed-out ACF that might adhere to, and hence damage bonding tool during assembly process. No special treatment and additional steps are needed for protecting tool surface under ACF bonding. However, the use of glass carriers demands one extra step and instrument for removing the temporary adhesive used between the carriers and the ultra-thin dies after bonding. The observations in this work are in line with previous results reported by Huang and Lu [19], where ACA bonding of ultra-thin dies using rigid carriers provided considerably improved electrical performance and reliability of the assemblies, compared to the same process applied for bare ultra-thin dies.

V. CONCLUSION

ACF bonding process of large-area, ultra-thin Si dies to Si substrates was characterized. The dies have a thickness of 50 µm and a bonding area on the order of 100 mm² populated with more than a thousand TSVs connecting corresponding metal pads and RDL on both sides. The substrates have thickness of 750 µm as well as necessary RDL and pads for bonding and measuring electrical resistance and stray capacitance to bulk Si of interconnects.

Use of glass carriers attached to the inactive side of dies ensured proper handling and reliable ACF bonding of ultra-thin dies to rigid substrates. No cracks on dies were found on bonded samples, even with a bonding pressure as high as 5 MPa. No risk of bonding tool being damaged due to ACF squeezed out under bonding.

While bonding temperature and bonding time are kept fixed at 180 °C and 30 seconds, bonding pressures from 3 MPa and above are recommended. However, 3 MPa seems to be the proper bonding pressure in this work, providing sufficiently low interconnect resistance with moderate deviation (~110–240 mΩ) as well as satisfactory bonding yield. Stray capacitance of individual ACF interconnects and its corresponding TSV to bulk Si was found about 0.5 pF, which might be in acceptable range for MEMS-based

ultrasound transducer applications, though lower stray capacitance should be further pursued.

ACF flip-chip technology was found feasible for assembling large-area, ultra-thin dies on rigid substrates. This applies for not only bonding MEMS-based ultrasound transducer array dies to ASIC chips, but also bonding IC chips to rigid interposers.

ACKNOWLEDGMENT

The authors thank Hoang Duy Le and Lars Hoff, both with University of South-Eastern Norway, for their assistance in impedance measurements. The Research Council of Norway is gratefully acknowledged also for the support to the Norwegian Micro- and Nano Fabrication Facility (NorFab III, project number: 295864).

REFERENCES

[1] D. Liu and S. Park, "Three-Dimensional and 2.5 Dimensional Interconnection Technology: State of the Art " *Journal of Electronic Packaging,* vol. 136, no. 1, p. 014001, 2014.

[2] S. Gupta, W. T. Navaraj, L. Lorenzelli, and R. Dahiya, "Ultra-thin chips for high-performance flexible electronics," *npj Flexible Electronics* vol. 2, pp. 1-17, 2018.

[3] M. H. Malik, G. Grosso, H. Zangl, A. Binder, and A. Roshanghias, "Flip Chip integration of ultra-thinned dies in low-cost flexible printed electronics; the effects of die thickness, encapsulation and conductive adhesives," *Microelectronics Reliability,* vol. 123, pp. 1-7, 2021.

[4] J. J. M. Zaal, W. D. v. Driel, and G. Q. Zhang, "Challenges in the Assembly and Handling of Thin Film Capped MEMS Devices," *Sensors,* vol. 10, no. 4, pp. 3989-4001, 2010.

[5] W. Sun, W. H. Zhu, F. X. Che, C. K. Wang, A. Y. S. Sun, and H. B. Tan, "Ultra-thin Die Characterization for Stack-die Packaging," in *the 57th Electronic Components and Technology Conference (ECTC),* Sparks, NV, USA, 2007, pp. 1390-1396.

[6] C. Kallmayer *et al.,* "Stretchable Electronics for Large Area Sensor Skin," in *IMAPS Nordic's Annual Microelectronics and Packaging Conference and Exhibition (NordPac),* Oslo, Norway, 2023, pp. 1-6.

[7] K. Brenner, A. S. Ergun, K. Firouzi, M. F. Rasmussen, Q. Stedman, and B. P. Khuri-Yakub, "Advances in Capacitive Micromachined Ultrasonic Transducers," *Micromachines,* vol. 10, no. 2, pp. 1-27, 2019.

[8] A. S. Savoia *et al.,* "A 3D Packaging Technology for Acoustically Optimized Integration of 2D CMUT Arrays and Front End Circuits," in *IEEE International Ultrasonics Symposium (IUS),* Washington, DC, USA, 2017.

[9] I. O. Wygant *et al.,* "An Integrated Circuit With Transmit Beamforming Flip-Chip Bonded to a 2-D CMUT Array for 3-D Ultrasound Imaging," *IEEE Transactions on Ultrasonics, Ferroelectrics, and Frequency Control,* vol. 56, no. 10, pp. 2145-2156, 2009.

[10] Ö. Oralkan *et al.,* "Volumetric Ultrasound Imaging Using 2-D CMUT Arrays," *IEEE Transactions on Ultrasonics, Ferroelectrics, and Frequency Control,* vol. 50, no. 11, pp. 1581-1594, 2003.

[11] M. J. Yim, C.-K. Chung, and K. W. Paik, "Effect of Conductive Particle Properties on the Reliability of Anisotropic Conductive Film for Chip-on-Glass Applications," *IEEE Transactions on Electronics Packaging Manufacturing,* vol. 30, no. 4, pp. 306-312, 2007.

[12] M. J. Yim, J. Hwang, and K. W. Paik, "Anisotropic Conductive Films (ACFs) for Ultra-fine pitch Chip-On-Glass (COG) Applications," *International Journal of Adhesion and Adhesives,* vol. 27, no. 1, pp. 77-84, 2007.

[13] H.-V. Nguyen *et al.,* "Anisotropic Conductive Film Interconnects for Fine-pitch MEMS," in *the 4th Electronics System Integration Technology Conferences (ESTC),* Amsterdam, Netherlands, 2012, pp. 1-6.

[14] A. Larsson, F. Oldervoll, T. A. T. Seip, H.-V. Nguyen, H. Kristiansen, and Ø. Sløgedal, "Anisotropic Conductive Film for Flip-Chip Interconnection of A High I/O Silicon Based Finger Print Sensor," in *the 22nd Micromechanics and Micro systems Europe Workshop,* Tønsberg, Norway, 2011, pp. 186-189.

[15] H.-V. Nguyen, T. Eggen, and K. E. Aasmundtveit, "Assembly of Transducer Array Using Anisotropic Conductive Film for Medical Imaging Applications," in *the 20th European Microelectronics and Packaging Conference (EMPC)*, Friedrichshafen, Germany, 2015, pp. 1-6.

[16] H.-V. Nguyen, T. Eggen, B. Sten-Nilsen, K. Imenes, and K. E. Aasmundtveit, "Assembly of Multiple Chips on Flexible Substrate Using Anisotropic Conductive Film for Medical Imaging Applications," in *the 64th Electronic Components and Technology Conference (ECTC)*, Orlando, Florida, USA, 2014: IEEE, pp. 498-503.

[17] J. v. d. Brand, R. Kusters, M. Cauwe, D. v. d. Ende, and M. Erinc, "Flipchip bonding of thin Si dies onto PET foils: Possibilities and applications," in *the 18th European Microelectronics & Packaging Conference (EMPC)*, Brighton, UK, 2011, pp. 1-6.

[18] J.-H. Kim, T.-I. Lee, T.-S. Kim, and K.-W. Paik, "Effects of ACFs Modulus and Adhesion Strength on the Bending Reliability of CIF (Chip-in-Flex) Packages at Humid Environment," in *the 68th Electronic Components and Technology Conference (ECTC)*, San Diego, CA, USA, 2018.

[19] Y.-W. Huang and S.-T. Lu, "Development and Reliability of Ultra-Thin Chip on Plastic Bonding for Flexible Liquid Crystal Displays," in *the 60th Electronic Components and Technology Conference (ECTC)*, Las Vegas, NV, USA, 2010, pp. 575-580.

An Innovative Conformal Electronically Scanned Array Antenna for full 360° Steerability in the Ka-band

Peter Uhlig, Aline Friedrich, Markus Krengel, Winfried Simon, Oliver Litschke

IMST GmbH, 47475 Kamp-Lintfort, Germany, http://www.imst.com

Abstract— A novel approach to constructing an omnidirectional 5G/6G Ka-band Tx/Rx frontend module using an Active Electronically Steerable Array (AESA) will be presented. The 5G standard includes a number of mmWave frequency bands that provide the bandwidth for high data rates. However, up to now 5G still transmits primarily in the sub-6 GHz frequency range. Small cells like micro or pico cells operated in the mmWave range are capable to increase network capacity in densely populated areas. Efficient beam-based communications using directional antenna arrays can overcome the limitations imposed by propagation loss at high frequencies. These and a number of further requirements have been taken into account in the design of the antenna module presented here, making it a fundamental component for the cellular infra—structure of future 5G/6G mmWave networks. The dual-polarised module is using 2 x 96 elements to scan 360° in azimuth and ±45° in elevation. The operation frequency range is 24.25 GHz to 29.5 GHz, covering n258, n257, and n261 3GPP bands. Additionally, the module is capable to control up to four beams simultaneously.

I. INTRODUCTION

5G mmWave communications technology is being implemented in many commercial mobile phones today. Communication in the mmWave frequency bands of the 5G and 6G standards promises to meet the need for increased mobile bandwidth, high reliability and low latency data transmission [1][2], which is increasingly demanded by autonomous driving, robotics and telemedicine applications, among others. On the downside, mmWave frequency bands suffer from high propagating losses. Small cells and beam-based communications with an AESA are the answer to this challenge. The proposed front-end module concept provides an excellent solution for a compact 5G base station unit to be used, for example, in street lamps or in an indoor environment such as an exhibition centre or airport. Rather than combining a number of planar arrays, the antenna array in this concept is wrapped around a cylinder to provide omnidirectional coverage. A subset of the radiating elements is used to form the active antenna aperture for one beam.

II. FRONTEND MODULE CONCEPT

The omnidirectional RF front-end module covers the frequency band from 24.25 GHz to 29.5 GHz and supports both horizontal and vertical polarised RF signals. Up to four antenna beams can be controlled via a serial interface. 12 V and 10 A are required for power supply. The frontend module "5G CAN" is shown inFigure 1.

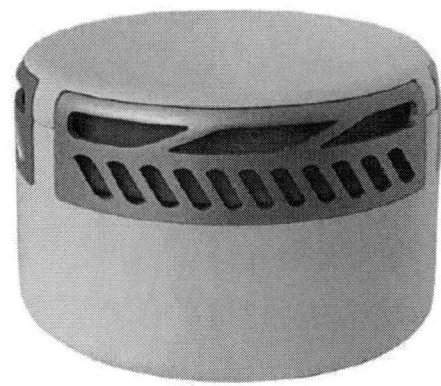

Figure 1: Omnidirectional mmWave 5G frontend module "5G CAN".

Today, many front-end modules are designed in tile architecture [3]. However, this design is based on a modified brick concept [4] where the PCBs are perpendicular to the antenna aperture. For the omnidirectional module, this architecture has several advantages, as explained below. The radiating elements are open-ended waveguides formed from two halves milled from aluminium and the via fence in the PCB sandwiched between the two metal plates. Each of the radiators is used for both polarisations, vertical and horizontal. The antenna aperture is the perimeter of a stack of five aluminium discs and four disc-shaped circuit boards. The organisation of the functional blocks on the different PCBs of the antenna module is shown in the block diagram in Figure 2.

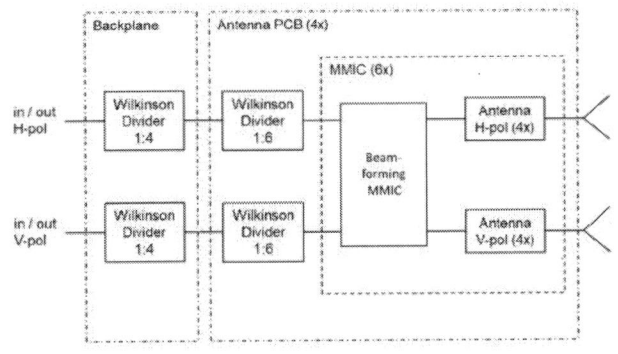

Figure 2: Schematic of the mmWave 5G/6G frontend module

The two bidirectional RF signals (input and output of horizontal and vertical polarisation) are distributed via the vertical backplane PCB to the four antenna PCBs to six beamforming MMICs. On the antenna PCB the signals for horizontal and vertical polarisation are routed separately on top and bottom layer to ensure good decoupling.

Each beamforming MMIC provides four horizontal and four vertical polarized output signals which can be controlled in amplitude and phase. The antenna element for the vertical polarisation is the end of a Substrate Integrated Waveguide (SIW) in the printed circuit board, while for the horizontal polarisation the open ended waveguide is used, fed by a stripline to waveguide transition in the same printed circuit board. Each RF board is sandwiched between two metal discs. The antenna array is designed in a sparse configuration for compactness. This means that only half of the grid positions are used in an alternating pattern. On the cylinder jacket, this is achieved by rotating the antenna board in the next layer by 7.5° (see Figure 3).

Figure 4: Connection of DC and SPI signals from the control board to the antenna board.

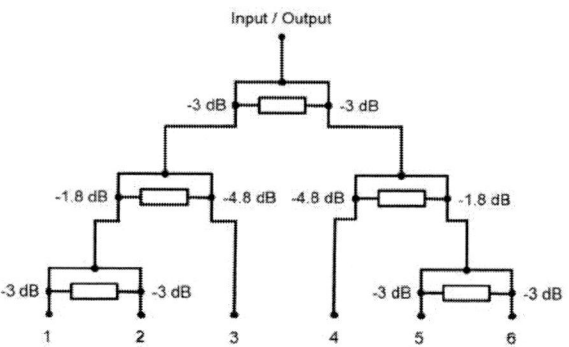

Figure 5: Circuit scheme of the 1:6 Wilkinson power divider / combiner.

In the block diagram of the antenna module (Figure 2) it can be seen that the signal for horizontal and vertical polarisation in transmit mode is distributed via a 1:4 Wilkinson divider [5] followed by a 1:6 Wilkinson divider network. In receive mode, the signal is summed in the opposite direction.

Figure 3: Antenna module for mmWave 5G/6G with the control board on top.

In addition to their function as partial waveguides and shielding between the antenna PCBs, the metal disks also serve as heat sinks for the MMICs. Passive cooling of these heat sinks is provided by air circulation flowing from the centre of the stack around the outside. The two RF input signals of the module are distributed to the different RF PCBs via a backplane PCB. The backplane PCB is placed in the polygon cut-out in the centre of the module. A 90° SIW transition routes the signal from the backplane PCB to the different RF PCBs.

The DC supply and SPI control board is shown on top of the antenna stack in Figure 3. DC voltages are converted from 12V to approximately 2V for the internal chip power supply. A microcontroller is used for SPI control. The power supply and SPI signals are routed to the antenna board below via connectors, as shown in Figure 4. A similar connection is used to connect the stacked antenna boards.

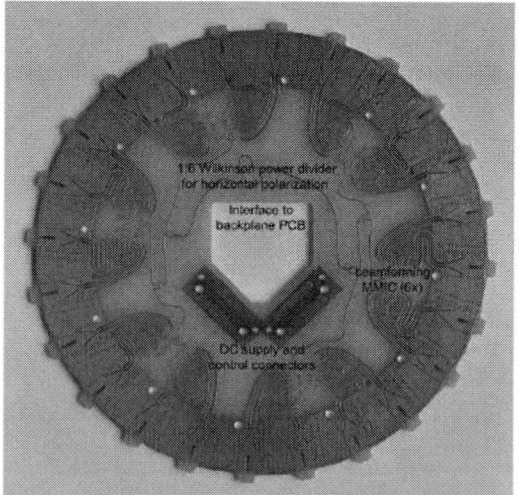

Figure 6: Component side of the antenna board with the footprint of six beamforming MMICs and the 1:6 power divider for horizontal polarisation.

Whereas the 1:4 divider on the backplane is a binary tree (two levels of 1:2 dividers), the 1:6 divider on the antenna board requires a more complex topology with unsymmetrical dividers in the second level, as shown in Figure 5. Note the extra loops in the microstrip line of the power divider. These are inserted to achieve equal transmission delay for all signal paths.

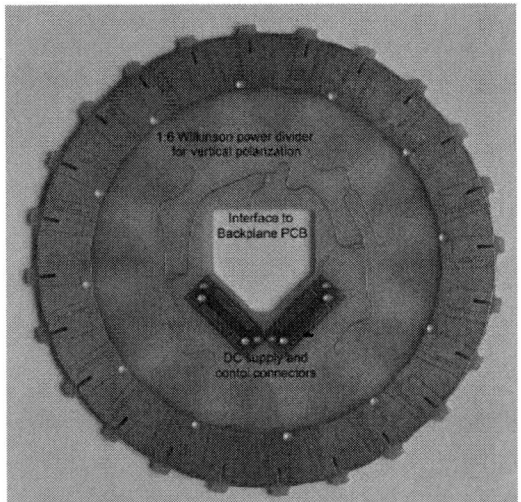

Figure 7: Bottom side of the antenna board with the 1:6 power divider for vertical polarisation.

Figure 6 and Figure 7 show both sides of the antenna board with the footprints of the six beamforming MMICs and the connectors for the DC and SPI signals. For better isolation, the Wilkinson power dividers for horizontal and vertical polarisation are on opposite sides of the board. The component side of the board (Figure 6) shows the microstrip to waveguide transitions for the horizontal polarisation of the 24 antenna elements in this layer.

The complete RF design of the module was performed using the 3D EM simulation software EMPIRE-XPU [6]. All the different building blocks (antenna, divider networks, RF transitions, backplane PCB, etc.) were designed independently in a first step. Step by step, more components were integrated and simulated (e.g. the beamforming MMIC with its eight corresponding antenna elements and the antenna feed network). Finally, the complete module, including all PCBs with components and antennas, was simulated together to ensure excellent RF performance. Thanks to efficient simulation techniques and software, the simulation of the entire module could be carried out in a few hours.

III. MODULE PERFORMANCE

The module achieves excellent performance over the entire frequency range from 24.25 GHz to 29.5 GHz. Omnidirectional coverage in azimuth is achieved with 96 antenna beams, 48 beams for vertical polarisation and 48 beams for horizontal polarisation.

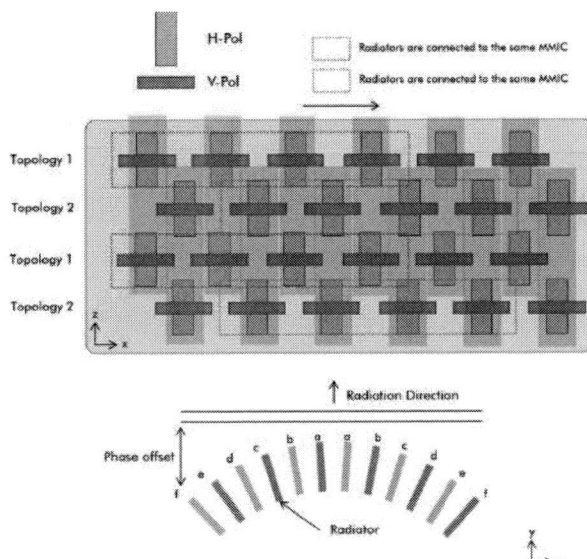

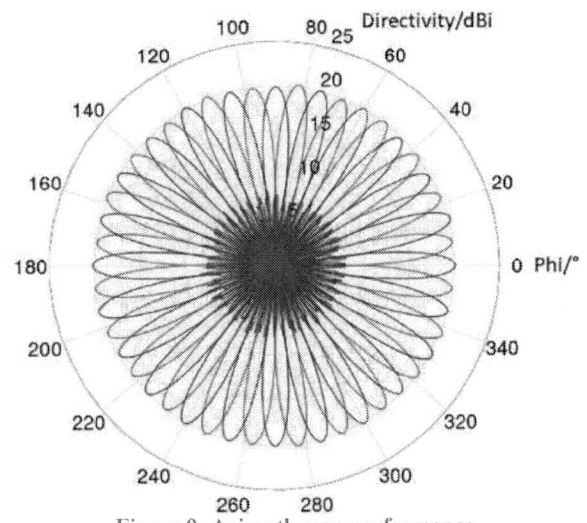

Figure 8: Antenna beam configuration.

Figure 8 depicts how each beam is realised by 4 x 6 antenna elements (using 4 antenna boards and 6 antenna elements per board). The resulting 3 dB beam width is approximately 8.5° in azimuth.

Figure 9: Azimuth scan performance for horizontal polarisation at 27 GHz

Figure 9 shows the simulated radiation pattern for the vertically polarised array (SIW) and Figure 10 for the horizontally polarised array (air waveguide) at 27 GHz. The achieved directivity is 20 dBi, resulting in an EIRP of 50 dBm at P1dB. The scan performance in elevation is ± 30° with a scan loss of less than 3 dB.

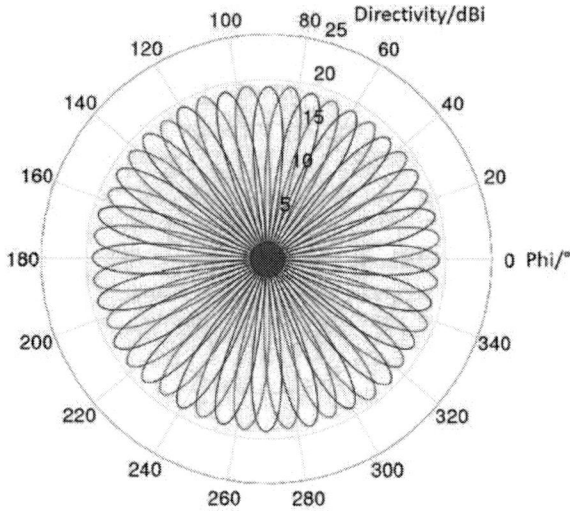

Figure 10: Azimuth scan performance
for vertical polarisation at 27 GHz

The simulated 3D radiation pattern for one vertically scanned beam is illustrated in Figure 11. Up to four simultaneous beams are supported, each emanating from a quarter of the circular module. Figure 12 shows the simulated 3D radiation pattern for two simultaneous beams in vertical polarisation (SIW) at 26 GHz.

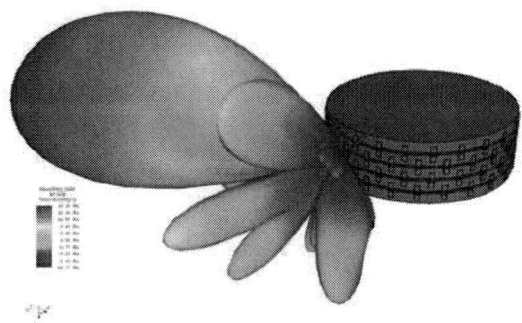

Figure 11: 3D radiation characteristic of one antenna beam
scanned in vertical direction.

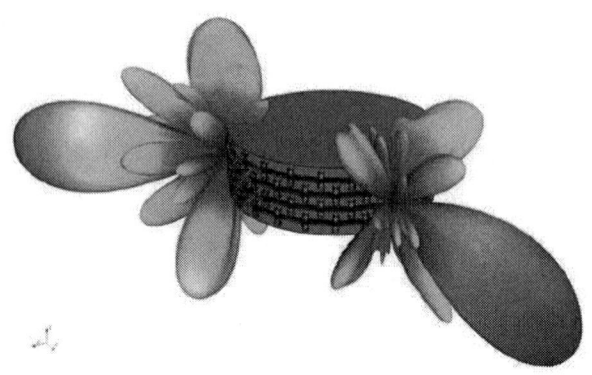

Figure 12: 3D radiation characteristic of two simultaneous
antenna beams.

IV. CONCLUSION

The 5G CAN front-end module is a compact, flexible and scalable phased array front-end that can be used in many 5G/6G scenarios. The module described is intended to be used as a development platform for 5G/6G systems. It is available as an evaluation kit for 5G/6G research teams and other customers. The described module covers frequencies from 24.25 GHz to 29.5 GHz (e.g. n258, n257 and n261 3GPP band), together with omnidirectional coverage in azimuth and ± 30° scanning in elevation. The design can be easily scaled in size (diameter) to increase both EIRP and number of beams. Upscaling in frequency up to 39 GHz (5G mmWave band n260) is also possible.

REFERENCES

[1] International Telecommunication Union, "IMT vision – Framework and overall objectives of the future development of IMT for 2020 and beyond," Recommendation ITU 2083, Electronic Publication Geneva, Switzerland, 2015. [Online]. Available: https://www.itu.int/rec/R-REC-M.2083.

[2] I. Ndip, "6G kann die Erwartungen erfüllen, die 5G geweckt hat," Produktion von Leiterplatten und Systemen (PLUS), no. 5, pp. 621–620, 2021

[3] W. Simon, D. Schäfer, S. Bruni, M.A. Campo, O.Litschke, "Highly Integrated Ka-Band Frontend Module for SATCOM and 5G," 2019 IEEE APMC.

[4] P. Uhlig, A. Friedrich, U. Lewark, and O. Litschke, "Brick or Tile? Evaluation of Integration Concepts for MicrowavePhased Array Antennas," in Proceedings of the Electronics System-Integration Technology Conference ESTC, Vestfold, Norway, 2020-09-17. doi: 10.1109/ESTC48849.2020.9229703.

[5] E. J. Wilkinson, "An N-way Hybrid Power Divider," IRE Trans. Microwave Theory Tech., vol. 8, no. 1, pp. 116–118, 1960-01, doi: 10.1109/tmtt.1960.1124668.

[6] A: Wien "EMPIRE XPU Reference manual", [Online]. Available: https://empire.de/resources/manual/ (assesed Jun, 2023)

Thermally conductive polymer composites with hexagonal boron nitride for medical device thermal management

Nu Bich Duyen Do
Department of Microsystems
University of South-Eastern Norway
Borre, Norway
nu.b.do@usn.no

Hoang-Vu Nguyen
Department of Microsystems
University of South-Eastern Norway
Borre, Norway
Hoang.V.Nguyen@usn.no

Kristin Imenes *
Department of Microsystems
University of South-Eastern Norway
Borre, Norway

* Correspondence:
Kristin.Imenes@usn.no

Erik Andreassen
Sintef Industry
Oslo, Norway
Department of Microsystems
University of South-Eastern Norway
Borre, Norway
Erik.Andreassen@sintef.no

Knut E. Aasmundtveit
Department of Microsystems
University of South-Eastern Norway
Borre, Norway
Knut.Aasmundtveit@usn.no

Abstract—Polymer composites with hexagonal boron nitride (hBN) have the potential to meet the heat dissipation and electrical insulation requirements of electronic and medical devices. In this study, hBN/polymer composites were fabricated with thermoplastic polyurethane (TPU) and epoxy as matrix materials. Three hBN powder types (BN1, BN2, BN3) with different platelet sizes and degrees of agglomeration were used. BN1 (with average size of 1 μm) and BN2 (with average size of 12 μm) mainly contained hBN platelets, while BN3 (with average size of 20 μm) contained both platelets and agglomerates of platelets. BN2 gave the highest thermal conductivity, while BN1 gave the lowest thermal conductivity.

Thermal simulations were performed for a medical device with a limit for the maximum surface temperature. For the encapsulation of this device, several material combinations were simulated, using anisotropic thermal conductivity data. This included encapsulations with inner layers of commercial thermally and electrically conductive materials and an outer layer of the best thermally conductive (and electrically insulating) hBN/TPU composite fabricated by the authors.

Keywords—*thermally conductive polymer composites, hexagonal boron nitride, thermoplastic polyurethane, epoxy, injection moulding, powder bed fusion, casting, thermal management.*

I. INTRODUCTION

For electronic devices, thermal management is vital for ensuring proper operation and preventing failures. In addition to efficient heat dissipation, electrical insulation is required for many applications, such as medical devices. Thermally conducting and electrically insulating polymer-based composites are therefore often used in electronics packaging [1]–[3]. Polymer composites have advantages of low density, low cost, and flexible processing via different techniques, such as injection moulding. Most polymers possess low thermal conductivity (0.1–0.5 W/mK) [4]. Incorporating inorganic fillers into polymer matrix is an effective solution for improving the thermal conductivity, while preserving the electrical insulation [1], [4], [5]. Common inorganic fillers are crystalline ceramic materials, either metal oxides (e.g.

alumina (Al_2O_3), quartz (crystalline SiO_2)), or non-oxides (e.g., aluminium nitride (AlN), boron nitride (BN), silicon nitride (Si_3N_4), silicon carbide (SiC)) [1], [4], [5].

Hexagonal boron nitride (hBN) has emerged as a promising filler for polymer composites due to its high intrinsic thermal conductivity, good electrical insulation, high thermal stability, good mechanical properties, and suitability for biomedical applications [6]–[10]. The filler hBN has been studied for enhancing the thermal conductivity of polymer-based composites, see examples in Table 1.

Hexagonal BN consists of B and N atoms arranged in a honeycomb configuration with a layer structure. Within the layers there are strong covalent bonds between B and N atoms, while between the layers there are weak van der Waals forces [6], [7]. The crystal structure of hBN results in platelet-shaped particles. The platelets have high in-plane thermal conductivity of about 300–600 W/mK, whereas the through-plane value is in the range of 2–30 W/mK [3], [6], [7]. Due to the strong B–N bonds, hBN is chemically stable, e.g., towards oxidation. However, this makes the functionalisation of hBN challenging [7].

Due to the hBN platelets' shape and anisotropic thermal conductivity, the orientation of the hBN platelets in the polymer matrix affects the thermal conductivity of the composite [4], [6]. Platelets can be oriented via processing or by using electric or magnetic fields [4], [6], [11]–[13]. Different processing methods have different impacts regarding distributing and orienting the platelets, as well as dispersing agglomerates and stacks of platelets into single platelets.

In scientific studies, thermally conductive polymer composite specimens are commonly fabricated by methods such as injection moulding (IM) and casting [1], [4], [6]. The hBN platelets are reported to be preferentially oriented with the platelet normal in the through-plane (thickness) direction of injection moulded parts, due to the flow in the IM process [4], [6]. Hence, the in-plane conductivity is higher than the through-plane conductivity, as shown in Table 1. In a casting process, composites are produced by mixing fillers into a

---

Funding: Research Council of Norway (269618 "Mechanical miniaturization in interventional medical instruments"); Norwegian Micro- and Nano-Fabrication Facility (295864 NORFAB III).

Table 1. Some studies of hBN/polymer composites. (wt% = weight percentage, vol% = volume percentage).

Materials	Filler loading (filler size)	Processing method	Thermal conductivity
hBN/epoxy [11]	95 wt%	Compression moulding	21.3 W/m·K (in-plane) 7 W/m·K (through-plane)
hBN/PU [14]	80 vol%	Freeze drying & hot pressing	39 W/m·K (in-plane) 11.5 W/m·K (through-plane)
hBN/polyimide [15]	60 vol%	Spin-cast (film)	17.5 W/m·K (in-plane) 5.4 W/m·K (through-plane)
hBN/epoxy [16]	57 vol% (5–11 μm)	Casting	5.27 W/m·K (through-plane)
hBN/PE [17]	50 vol% (4–5 μm)	Injection moulding	3.66 W/m·K (through-plane)
hBN/PEEK [18]	60 wt% (25–30 μm)	Injection moulding	12.45 W/m·K (in-plane) 2.34 W/m·K (through-plane)
hBN/TPU [19]	50 wt%	Solution mixing & hot pressing	3.06 W/mK (through-plane)
hBN/PA12 [20]	40 wt%	Powder bed fusion	0.55 W/m·K (through-plane, 77% higher than PA12)
hBN/Al₂O₃/PA12 [21]	15 wt% hBN and 35 wt% Al₂O₃	Powder bed fusion	1.05 W/m·K (through-plane, 275% higher than PA12)
hBN/AlN/TPU [22]	15 wt% hBN and 20 wt% AlN	Powder bed fusion	0.9 W/m·K (through-plane, 391% higher than TPU)
hBN/TPU [23]	30 wt%	Fused deposition modeling (material extrusion)	1.51 W/m·K (in-plane) 1.26 W/m·K (through-plane)

resin such as epoxy, followed by pouring or injecting the mixture into a mould for curing. The casting of hBN/polymer composites can result in almost randomly oriented platelets [4], [6], [16].

Powder bed fusion (PBF) is one of the most common 3D printing (additive manufacturing) processes for polymer materials. The feedstock is powder, and thermal energy (e.g., from a laser) selectively fuses regions of a powder bed, layer-by-layer, to fabricate 3D objects [24]. Compared to moulding processes, the principal advantages of PBF and other 3D printing techniques include fast prototyping and geometric freedom. Few polymer types are commercially available for PBF, but the most common are polyamide 12 (PA12) and thermoplastic polyurethane (TPU). There are several articles about polymer PBF with thermally and electrically conductive fillers (e.g., carbon-based fillers such as carbon fibres [25], graphite [26], CNT [27] and graphene [28], or metal fillers such as Cu [29] and Ag [30]). The literature on PBF with thermally conductive and electrically insulating polymer-based composites is sparse, and most studies use PA12 as the polymer matrix [24], [31], [32].

Encapsulation materials of medical devices used in the human body are required to have good heat transfer, electrical insulation, biocompatibility, and in some cases, a 'soft touch' [2], [33]. To achieve a significant increase in thermal conductivity in polymer-based composites, a high loading of inorganic fillers (e.g. hBN) is generally required. However, this normally results in increased hardness [4], [6] and reduced ductility. To compensate for this, a soft polymer can be used as the matrix material. The hBN loading also has a negative effect on the processability via higher viscosity.

Thermally conductive polymer composites can be used for the encapsulation of medical devices such as trans-esophageal echocardiography (TEE) probes. This device uses ultrasound waves to image the heart in 3D from inside the human esophagus. For the patient safety, the encapsulation of the device's scan head is required to provide good heat transfer (the maximum surface temperature of the scan head in contact with human tissue must be below 43 °C), electromagnetic interference (EMI) shielding, electrical insulation, and biocompatibility.

Thermoplastic polyurethane (TPU) is a soft thermoplastic elastomer, offering high elasticity over a broad temperature range and high wear resistance [24], [32], [34]. It is commonly used for PBF and IM, and its suitability for biomedical applications has recently been highlighted [31].

Therefore, the TPU "Ultrasint TPU 88A" (in the form of powder for PBF) was selected as the matrix for composites processed by PBF and IM in this study.

This study is a continuation of our previous research on the encapsulation of interventional medical devices [33] and thermally conductive polymer composites [35]. This paper adds results for composites with different hBN powder types, and how the hBN platelets' orientation in the injection moulding process can be simulated. Furthermore, thermal simulations for a TEE scan head are performed with various polymer composites, and with anisotropic thermal conductivity, representative of injection moulded polymer composites.

II. METHODOLOGY

A. Experiments

Materials

The polymer matrix for injection moulding and powder bed fusion (3D printing) was thermoplastic polyurethane (TPU), in the form of powder, with 88 shore A hardness (Ultrasint TPU 88A, BASF, Germany). The polymer matrix for casting was an epoxy containing 35 wt% unmodified bisphenol-F epoxy resin (Araldite GY 285-1), 35 wt% reactive diluent (Araldite DY 026), 30 wt% amine-based curing agent (Jeffamine D-230 Polyetheramine), from Huntsman. The epoxy system was formulated to have a low viscosity, suitable for preparing composites with high filler loading.

Three hBN powder types (BN1, BN2, BN3) with different platelet sizes and agglomeration degrees were used. BN1 had an average particle size of 1 μm (product no. 255475, Sigma Aldrich, Merck). BN2 and BN3 were supplied by Henze Boron Nitride Products AG, Germany. BN2 (HeBoFill LL-SP 120) [36] and BN3 (HeBoFill CL-ADH 020) [37] had an average particle size of 12 μm and 20 μm, respectively. BN1 and BN2 mainly contained hBN platelets, while BN3 had a partly agglomerated particle structure, claimed to provide good lubricating properties and low viscosity increase [37]. Platelets and spherical agglomerates of hBN are shown in Figure 1.

Specimen preparation by injection moulding, casting, and powder bed fusion

Injection moulded specimens were prepared using a 15 cm³ micro batch compounder (DSM Midi 2000) followed by injection moulding with a table-top machine (DSM). Moulded specimens for thermal conductivity measurement were 2 mm thick discs with diameters of 25 mm, see Figure 2. Specimens were prepared with 0, 15, 25, 35, 50 and 65 wt% BN3; 50 wt% BN2; and 50 wt% BN1.

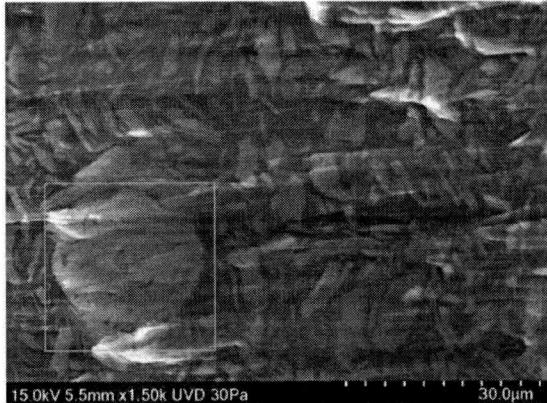

Figure 1: SEM micrograph of ion milled cross-section of the cast specimen with 55 wt% BN3. An agglomerate of hBN platelets is seen inside the yellow square.

Cast epoxy-hBN specimens were prepared by vacuum mixing, followed by casting into a Teflon mould (Figure 2) and then curing at 150 °C for 18 h. The cast specimens (2 mm thick discs) were grinded and polished on both sides. Specimens with 0, 35, 50 and 55 wt% BN3 were cast.

For powder bed fusion (PBF), a tabletop PBF 3D printer (Sharebot SnowWhite) was used to fabricate specimens of a (TPU and hBN) powder mixture. PBF specimen were prepared with 0, 35, 40 wt% BN3; 35 wt% BN2; and 35 wt% BN1.

Thermal conductivity measurements

The thermal conductivity of 2 mm thick specimens was indirectly determined by the non-contact, transient laser flash analysis (LFA) method. Details about the measurements are given in [35].

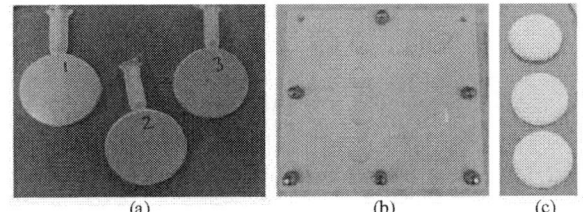

Figure 2. (a) Injection moulded specimens; (b) cast specimens in Teflon moulds; (c) PBF specimens.

B. Simulation of steady-state heat transfer

A TEE scan head with a two-layer encapsulation was used as a simulation model, which was previously reported in ref. [33], in order to evaluate the thermal performance when the outer layer is made of hBN/polymer composite. Previously, only isotropic thermal conductivity has been considered. However, hBN platelets typically induce anisotropic conductivity (see Table 1). Hence, anisotropic conductivity is included in the present simulation study.

The model represents the test sample used in experiments, see Figure 3. The two-layer encapsulation can be a metallized polymer composite (case 1 in Table 2), or it can consist of two polymer composites (case 2 in Table 2).

The metallized encapsulation (case 1) consists of a metal layer (0.15 mm) deposited on the inner surface of prefabricated polymer composite part (0.9 mm). The metal layer provides heat transfer and EMI shielding, while the outer layer provides electrical insulation, biocompatibility, and some heat transfer.

An encapsulation consisting of two composite layers (case 2) can be fabricated by methods such as two-component injection moulding. Both layers will contribute to the heat transfer. The outer layer should be thermally conductive, but electrically insulating and biocompatible. The inner layer should be thermally and electrically conductive, to take care of both heat transfer and EMI shielding. The thickness limit of the current encapsulation of the TEE scan head is ca. 1 mm. Therefore, the total encapsulation thickness was 1 mm in the simulations.

Table 2: Material data used in the thermal simulations.

Component	Material	Thermal conductivity k (Wm^{-1} K^{-1})
Heat source	Metal ceramic heater [38]	20
Heat sink	Al alloy 6082 [39]	180
Adhesive between heat source and heat sink	Thermally conductive adhesive [40]	1.3
Adhesive between heat sink and encapsulation	Thermally and electrically conductive adhesive [41]	2.5
Coverage of the hole for the electrical wires	Medical silicone	0.2*
Remaining volume inside the sample	Air [42]	k(T) via COMSOL's material library
Case 1: Metallized encapsulation (metallized polymer, or metallized polymer composite)		
Inner layer (0.15 mm)	Electroplated Cu [43]	380
Outer layer (0.9 mm)	Typical polymer material (case 1a)	0.2* (isotropic)
	hBN/TPU, assumed isotropic (case 1b)	2.1 (isotropic)
	hBN/TPU, anisotropic (case 1c)	2.1 (through-plane) ** (this work) 12 (in-plane) ** [18]
Case 2: Encapsulation consisting of two composite layers (total thickness 1.0 mm)		
Inner layer; 0.5 mm (case 2a) or 0.9 mm (case 2b)	Thermally and electrically conductive polymer composite from Celanese	4.5 (through-plane) *** 34 (in-plane) *** [44]
Outer layer; 0.5 mm (case 2a) or 0.1 mm (case 2b)	hBN/TPU (thermally conductive and electrically insulating)	2.1 (through-plane) ** (this work) 12 (in-plane) ** [18]

* Values from references [45], [46].

** The through-plane value (2.1 W/mK) is the highest obtained in our experimental study. The in-plane value (12 W/mK) is estimated based on ref. [18] which studied the anisotropic thermal conductivity of injection moulded hBN/PEEK composites. The highest in-plane and through-plane values in that study were 2 W/mK and 12 W/mK, respectively, for 60 wt% hBN (with size of 25 – 30 μm).

*** Values based on a commercial material (Celanese CoolPoly E5521) [44]. These values seem to be among the highest reported for commercial materials. Other materials were also simulated; one from Avient (with 5.5 W/mK through-plane and 19 W/mK in-plane) and one from Sabic (with 1.5 W/mK through-plane and 18 W/mK in-plane).

183

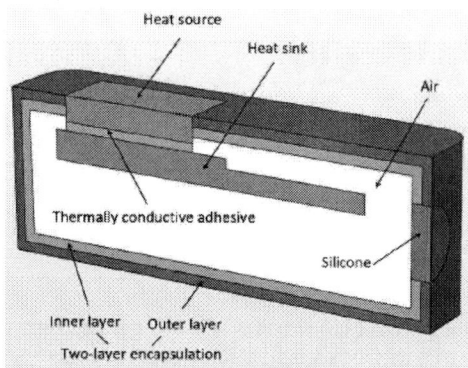

Figure 3: Schematic of cross-section of the test sample (not drawn to scale) used in experiments. See ref. [33] for details.

III. RESULTS AND DISCUSSIONS

A. Thermal conductivity of hBN/polymer composites

This section reports the effects of filler loading, filler type and processing method on the thermal conductivity. Figure 4 shows the thermal conductivity of hBN/polymer composites fabricated by injection moulding (IM), casting (C) and powder bed fusion (PBF) as a function of hBN loading. The highest conductivity (2.1 W/mK) was measured for an injection moulded specimen with the highest hBN loading in this study (65 wt% BN3). This conductivity is almost 900% higher than that of the pure TPU (injection moulded reference). For the cast composites, a 1313% higher conductivity was obtained with 55 wt% BN3. For a given hBN loading, the cast composites had the highest conductivity and the highest increase relative to the unfilled material. Among the PBF composites, the specimen with 40 wt% BN3, processed with the highest laser energy density, had the highest conductivity, which was 460% higher than that of the pure TPU (fabricated by PBF).

Figure 4 also shows that for IM composites with 50 wt% hBN, BN2 and BN1 gave the highest and lowest thermal conductivity, respectively. The same trend was observed for the PBF composites, although the effect was weaker.

With the same hBN loading, BN2 results in higher conductivity than BN3. Although some agglomerates in BN3 are broken up during processing, the lower conductivity with BN3 is probably due to platelets being less dispersed. Powders with hBN agglomerates, such as BN3, are claimed to have more isotropic properties and easier processing due to a lower viscosity from spherical fillers [4].

BN1 gives lower thermal conductivity than the two other powders at the same filler loading. This effect is observed for both IM and PBF specimen. The smaller platelets in BN1 gives a larger total surface area. This will reduce the conductivity of the composite if the matrix-filler interfacial thermal resistance is high [47]. The weak bonding between the polymer matrix and the hBN surface probably lead to high interfacial thermal resistance [48].

The surface of the hBN particles can be modified chemically in order to improve the dispersion of hBN particles and achieve a stronger hBN/polymer interface, thereby improving the thermal conductivity and the mechanical properties [7], [10], [49]–[51]. However, the chemical inertness and oxidation resistance of hBN make the functionalisation of hBN challenging.

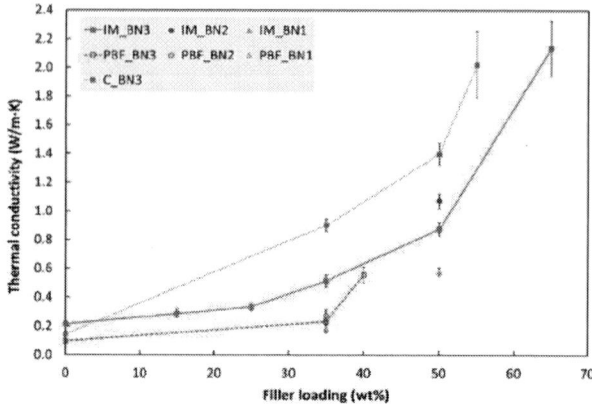

Figure 4. Thermal conductivity (at 30 °C) of composites fabricated by injection moulding ("IM"), powder bed fusion ("PBF") and casting ("C") as a function of hBN loading. The PBF specimen with 40 wt% BN3 was processed with a higher laser energy density than the other PBF specimens in this figure.

B. Anisotropy of thermal conductivity

The injection moulded composites in our study have lower conductivity than the cast composites. This is due to the preferred in-plane orientation of the hBN platelets in the former case, as shown by the XRD data in ref. [35]. Casting only resulted in a low preferred orientation, and powder bed fusion resulted in an almost random orientation.

In our previous work [35] we showed that injection moulding induces a preferred orientation of the platelets, with the platelet normal along the thickness direction of the specimens. The orientation varied through the thickness of the moulded specimens, and it increased with increasing hBN loading. Filler orientation by injection moulding was also reported by Grundler et al. [52] for their polymer composites with graphite flakes.

Numerical simulations of the injection moulding process were performed to illustrate how the platelets achieve their preferred orientation through the thickness of the injection moulded discs. The results agreed with the $<\cos^2\theta>$ values from the XRD measurements. The platelets have in-plane alignment (with high $<\cos^2\theta>$) in the 'shell' near the surface of the injection moulded disc, and less preferred alignment (with lower $<\cos^2\theta>$) in the core of the disc, see Figure 5.

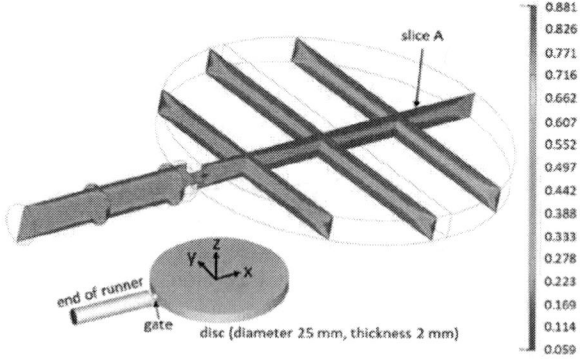

Figure 5: Simulated platelet orientation ($<\cos^2\theta>$) in an injection moulded 2 mm thick disc. The orientation data are shown in slices through the thickness of the disc, including gate and part of the runner.

C. Comparison with published thermal conductivity values

The thermal conductivity of our composites (hBN/TPU and hBN/epoxy) were compared with some commercially available polymer composites (Table 3) and published values for similar composites (Table 1 and Figure 6). However, such comparisons are challenging, due to differences in mixing/processing method, hBN loading, hBN particle size and matrix material.

The highest through-plane thermal conductivity of our hBN/TPU composites (2.1 W/mK) is higher than that of most commercial thermally conductive and electrically insulating materials, e.g. from Sabic, Covestro, Celanese and Kraiburg (Table 3). However, the commercial materials seem to have better mechanical properties.

Table 3: Through-plane thermal conductivity of some commercially available polymers, that are thermally conductive and electrically insulating and suitable for injection moulding.

Supplier – Materials	Through-plane thermal conductivity (W/mK)
Sabic – LNP Konduit Compounds [53] (with PA6, PPS or PC matrix)	0.6 – 1.5
Covestro - Makrolon TC [54] (with PC matrix)	0.2 – 0.3
Celanese – Coolpoly D Series [44] (with PP, PA6, PPS, LCP, or TPE matrix)	0.6 – 1.9
Kraiburg – Thermoplast K [55] (HTC1500/117, with TPE matrix)	1.5
TCPoly - Ice9 Flex [56] (TPE with two filler types)	2
Our hBN/TPU composites (65 wt% hBN)	2.1

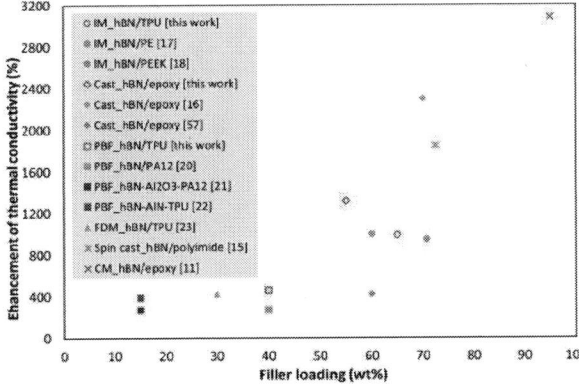

Figure 6: Comparison of the thermal conductivity enhancement (%) for the hBN/polymer composites prepared with different hBN loading (wt%) and processing methods [11], [15]–[18], [20]–[23], [57]. Abbreviations of processing methods in the legend: injection molding (IM), powder bed fusion (PBF), fused deposition modeling (FDM), compression molding (CM).

D. Using thermally conductive composites for the thermal management of medical devices – A simulation study

This section presents thermal simulations (steady-state heat transfer) for a TEE scan head. Cases and material data are given in Table 2. The scan head encapsulation must fulfil the requirement that the maximum surface temperature (T_{max}) of the scan head (in contact with human tissue) must be below 43 °C. T_{max} occurs at the surface of the heat source (which is not encapsulated). In most simulations the heat source power was 0.5 W (if nothing else is stated). In some cases, the power was varied to find the value corresponding to $T_{max} = 43$ °C.

Metallized encapsulation (case 1a-1c)

In our previous study [33], an isotropic metallized polymer encapsulation (case 1a in Table 2) was simulated. For a power of ca 0.5 W, T_{max} reached the limit of 43 °C. If the polymer is replaced by our composite and isotropy is assumed (case 1b, Figure 7a), T_{max} is reduced to 42.2 °C for the same power (0.5 W). $T_{max} = 43$ °C is reached for a power of 0.59 W. With the anisotropic material model (case 1c, Figure 7b), T_{max} is reduced to 42.1 °C for the same power (0.5 W). The anisotropic material provides better heat transfer, due to its higher in-plane conductivity.

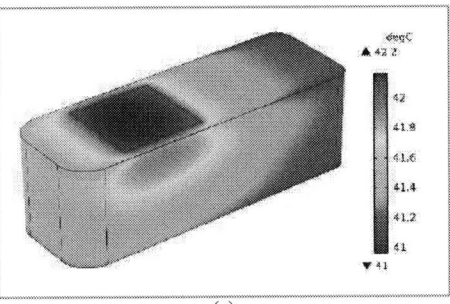

(a)

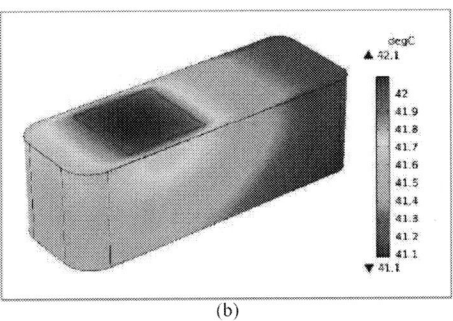

(b)

Figure 7: Simulated surface temperature of the TEE scan head with power 0.5 W. (a) Case 1b in Table 2. (b) Case 1c in Table 2.

Encapsulation consisting of two polymer composite layers (case 2a and 2b)

Two cases of two-layer polymer composite encapsulations were simulated (case 2a and 2b in Table 2). Both layers have fillers that increase the thermal conductivity. The inner layer is electrically conductive, while the outer layer is electrically insulating.

Such encapsulations perform similar to the metallized encapsulations (Figure 8 vs. Figure 7). With our hBN/TPU composite in the outer layer and the Celanese composite in the inner layer, the limit $T_{max} = 43$ °C is reached for a power level in the range 0.55-0.60 W.

Commercial polymer composites have different conductivities and degrees of thermal anisotropy. To investigate the overall effect on the surface temperature, simulations were performed with a range of different in-plane and through-plane thermal conductivities for the inner layer, see Figure 9. First, we notice that the difference in T_{max} between the four inner layer materials (white symbols in the figure) is small (T_{max} values within 1 °C, and below the 43 °C limit). The Celanese material (triangle) is the best, with the lowest T_{max} for both cases (0.5/0.5mm and 0.9/0.1mm).

185

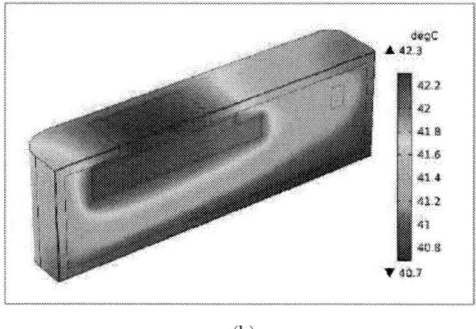

(a)

(b)

Figure 8: Heat transfer simulation of the TEE scan head (half model) with power 0.5 W. (a) Case 2a (0.5/0.5 mm thickness combination). (b) Case 2b (0.9/0.1 mm thickness combination).

For these four materials, the T_{max} values of the 0.9/0.1mm case and the 0.5/0.5mm case are quite similar. The 0.9/0.1mm case gives slightly lower T_{max} values for the best two materials (triangle and star symbols). When using our hBN/TPU material (square) in both layers of the encapsulation, there is obviously no difference between the two cases.

The difference in T_{max} between the two cases increases with decreasing conductivity values, since the 0.9 mm thick inner layer then gives a larger T_{max}, due to the insulating effect. For low conductivity values (e.g. $k_{through-plane} < 1$ W/mK), T_{max} is more sensitive to through-plane conductivity than to in-plane conductivity.

The minimum surface temperature T_{min} was also evaluated by simulations. For the 0.9/0.1mm case with low inner layer conductivities (< 0.5 W/mK), T_{min} is within 1 °C of the surrounding temperature (37 °C) due to low heat transfer to the coldest spot. T_{min} is more sensitive to $k_{in-plane}$ than to $k_{through-plane}$, as $k_{in-plane}$ contributes more to transferring heat to the coldest spot on the scan head surface.

The temperature difference ($\Delta T = T_{max} - T_{min}$) decreases with increasing conductivity values, in particular with in-creasing in-plane values, as more heat is then transferred to the coldest spot on the scan head surface. The material which gives the lowest T_{max} value also gives the lowest ΔT, since distributing heat over the surface lowers T_{max}. With the two best materials, (triangle and star symbols), ΔT is lower for the 0.9/0.1mm case than for the 0.5/0.5mm case.

In future studies, the local platelet orientation induced in the injection moulding process can be simulated (as in Figure 5). From this, the local anisotropic thermal conductivity can be calculated with an appropriate model (some discussed in our previous study [35]). Finally, a thermal simulation can be

performed, using the local anisotropic thermal conductivity as material input.

From a heat transfer point of view, the metallized encapsulation and the two-layer composite encapsulation studied in this paper can provide adequate thermal dissipation for the TEE scan head when the power is below 0.6 W. However, the EMI shielding of such encapsulations should be evaluated by experiments.

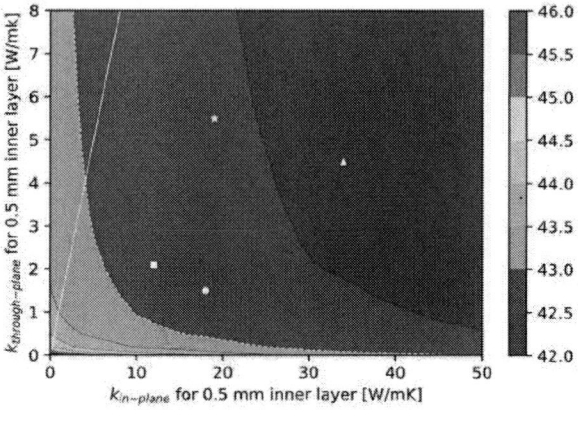

(a)

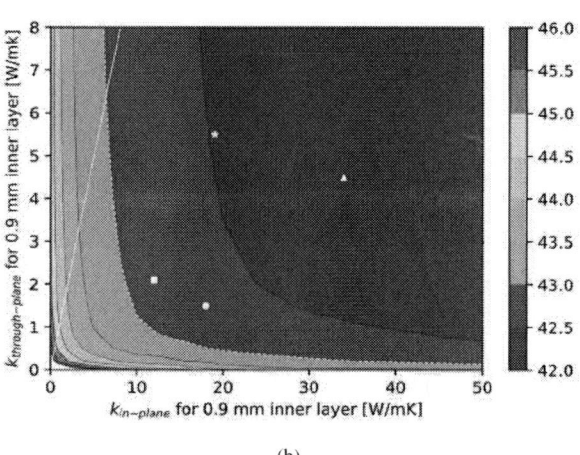

(b)

Figure 9: Contour plots of T_{max} vs. in-plane and through-plane thermal conductivity for the inner layer (case 2). The power is 0.5 W. The white dashed line represents the limit $T_{max} = 43$ °C, and the white solid line represents isotropic materials. The white symbols represent the four anisotropic materials in Table 2, case 2: The triangle, star and circle represent the materials from Celanese, Avient and Sabic, while the square is the best hBN/TPU composite in this study. (a) Case 2a (0.5/0.5 mm thickness combination). (b) Case 2b (0.9/0.1 mm thickness combination). In this plot, T_{max} values above 46 °C (for low conductivity values) are blank (white).

IV. CONCLUSIONS

This study has provided new insights on the effect of hBN powder type on the thermal conductivity of hBN/polymer composites. With the three powder types in this study, both the average particle size and the degree of agglomeration affect the conductivity. The conductivity of the composite is lowest for the smallest platelets. This is probably due to these giving the largest platelet-polymer interface area in combination with high interfacial thermal resistance. Powders with some platelet agglomerates seem to give lower conductivity than powders with only "free" platelets. The latter result in better dispersed platelets, which are more effective in transferring heat.

The highest through-plane thermal conductivity of the hBN/polymer composites obtained in this study was 2.1 W/mK. The applicability of hBN/polymer composites in a two-layer encapsulation of a TEE scan head was evaluated by thermal simulations. The maximum surface temperature of this scan head must be below 43 °C. A hBN/polymer composite can be used in a metallized encapsulation (with 0.15 mm of Cu and 0.9 mm of hBN/TPU), or in a 1.0 mm thick encapsulation consisting of two layers (an outer layer of hBN/TPU and an inner layer of a commercial thermally and electrically conductive polymer composite). Such encapsulations provide the TEE scan head with adequate heat transfer for power levels up to about 0.6 W.

Finally, we have demonstrated how thermal simulations with anisotropic material data can be used to evaluate and compare anisotropic materials.

ACKNOWLEDGMENT

The authors gratefully acknowledge Paul McMahon at SINTEF Industry; Jeyanthinath Mayandi at University of Oslo; Erik Kalland at Conpart AS; and Zekija Ramic, Svein Mindrebøe, Tayyib Muhammad, Chaoqun Cheng at University of South-Eastern Norway for their enthusiastic assistance with laboratory work. We sincerely thank Terje Finstad at University of Oslo and Helge Kristiansen at Conpart AS for valuable discussions.

REFERENCES

[1] R. Tummala, *Fundamentals of Microsystems Packaging*. McGraw-Hill Education, 2001. [Online]. Available: https://books.google.no/books?id=nf1vwR9VYGEC

[2] R. Bahru, A. A. Hamzah, and M. A. Mohamed, "Thermal management of wearable and implantable electronic healthcare devices: Perspective and measurement approach," *Int. J. Energy Res.*, 2020, doi: 10.1002/er.6031.

[3] C. Yuan, B. Duan, L. Li, B. Xie, M. Huang, and X. Luo, "Thermal Conductivity of Polymer-Based Composites with Magnetic Aligned Hexagonal Boron Nitride Platelets," *ACS Appl. Mater. Interfaces*, vol. 7, no. 23, pp. 13000–13006, 2015, doi: 10.1021/acsami.5b03007.

[4] H. Chen, V. V. Ginzburg, J. Yang, Y. Yang, W. Liu, Y. Huang, L. Du, and B. Chen, "Thermal conductivity of polymer-based composites: Fundamentals and applications," *Prog. Polym. Sci.*, vol. 59, pp. 41–85, 2016, doi: 10.1016/j.progpolymsci.2016.03.001.

[5] D. Li, D. Zeng, Q. Chen, M. Wei, L. Song, C. Xiao, and D. Pan, "Effect of different size complex fillers on thermal conductivity of PA6 thermal composites," *Plast. Rubber Compos.*, vol. 48, no. 8, pp. 347–355, Sep. 2019, doi: 10.1080/14658011.2019.1626596.

[6] C. Yu, J. Zhang, W. Tian, X. Fan, and Y. Yao, "Polymer composites based on hexagonal boron nitride and their application in thermally conductive composites," *RSC Adv.*, vol. 8, no. 39, pp. 21948–21967, 2018, doi: 10.1039/C8RA02685H.

[7] Z. Zheng, M. Cox, and B. Li, "Surface modification of hexagonal boron nitride nanomaterials: a review," *J. Mater. Sci.*, vol. 53, no. 1, pp. 66–99, Jan. 2018, doi: 10.1007/s10853-017-1472-0.

[8] Y. Zhang, W. Gao, Y. Li, D. Zhao, and H. Yin, "Hybrid fillers of hexagonal and cubic boron nitride in epoxy composites for thermal management applications," *RSC Adv.*, 2019, doi: 10.1039/c9ra00282k.

[9] A. Merlo, V. R. S. S. Mokkapati, S. Pandit, and I. Mijakovic, "Boron nitride nanomaterials: Biocompatibility and bio-applications," *Biomater. Sci.*, 2018, doi: 10.1039/c8bm00516h.

[10] G. Ciofani, G. G. Genchi, I. Liakos, A. Athanassiou, D. Dinucci, F. Chiellini, and V. Mattoli, "A simple approach to covalent functionalization of boron nitride nanotubes," *J. Colloid Interface Sci.*, vol. 374, no. 1, pp. 308–314, May 2012, doi: 10.1016/j.jcis.2012.01.049.

[11] Z. Zhu, P. Wang, P. Lv, T. Xu, J. Zheng, C. Ma, K. Yu, W. Feng, W. Wei, and L. Chen, "Densely packed polymer/boron nitride composite for superior anisotropic thermal conductivity," *Polym. Compos.*, vol. 39, no. S3, pp. E1653–E1658, Jun. 2018, doi: 10.1002/pc.24615.

[12] Y. Xue, X. Li, H. Wang, F. Zhao, D. Zhang, and Y. Chen, "Improvement in thermal conductivity of through-plane aligned boron nitride/silicone rubber composites," *Mater. Des.*, vol. 165, pp. 107580–107580, 2019, doi: 10.1016/j.matdes.2018.107580.

[13] Z. Lin, Y. Liu, S. Raghavan, K. S. Moon, S. K. Sitaraman, and C. P. Wong, "Magnetic alignment of hexagonal boron nitride platelets in polymer matrix: Toward high performance anisotropic polymer composites for electronic encapsulation," *ACS Appl. Mater. Interfaces*, vol. 5, no. 15, pp. 7633–7640, 2013, doi: 10.1021/am401939z.

[14] N. Zhao, J. Li, W. Wang, W. Gao, and H. Bai, "Isotropically Ultrahigh Thermal Conductive Polymer Composites by Assembling Anisotropic Boron Nitride Nanosheets into a Biaxially Oriented Network," *ACS Nano*, vol. 16, no. 11, pp. 18959–18967, Nov. 2022, doi: 10.1021/acsnano.2c07862.

[15] M. Tanimoto, T. Yamagata, K. Miyata, and S. Ando, "Anisotropic thermal diffusivity of hexagonal boron nitride-filled polyimide films: Effects of filler particle size, aggregation, orientation, and polymer chain rigidity," *ACS Appl. Mater. Interfaces*, vol. 5, no. 10, pp. 4374–4382, 2013, doi: 10.1021/am400615z.

[16] Y. Xu and D. D. L. Chung, "Increasing the thermal conductivity of boron nitride and aluminum nitride particle epoxy-matrix composites by particle surface treatments," *Compos. Interfaces*, vol. 7, no. 4, pp. 243–256, Jan. 2000, doi: 10.1163/156855400750244969.

[17] G.-W. Lee, M. Park, J. Kim, J. I. Lee, and H. G. Yoon, "Enhanced thermal conductivity of polymer composites filled with hybrid filler," *Compos. Part Appl. Sci. Manuf.*, vol. 37, no. 5, pp. 727–734, May 2006, doi: 10.1016/j.compositesa.2005.07.006.

[18] S. Gul, S. Arican, M. Cansever, B. Beylergil, M. Yildiz, and B. Saner Okan, "Design of Highly Thermally Conductive Hexagonal Boron Nitride-Reinforced PEEK Composites with Tailored Heat Conduction Through-Plane and Rheological Behaviors by a Scalable Extrusion," *ACS Appl. Polym. Mater.*, vol. 5, no. 1, pp. 329–341, Jan. 2023, doi: 10.1021/acsapm.2c01534.

[19] T. Fei, Y. Li, B. Liu, and C. Xia, "Flexible polyurethane/boron nitride composites with enhanced thermal conductivity," *High Perform. Polym.*, vol. 32, no. 3, pp. 324–333, Apr. 2020, doi: 10.1177/0954008319862044.

[20] L. Yang, L. Wang, and Y. Chen, "Solid-state shear milling method to prepare PA12/boron nitride thermal conductive composite powders and their selective laser sintering 3D-printing," *J. Appl. Polym. Sci.*, vol. 137, no. 23, pp. 1–13, 2020, doi: 10.1002/app.48766.

[21] Y. Yuan, W. Wu, H. Hu, D. Liu, H. Shen, and Z. Wang, "The combination of Al 2 O 3 and BN for enhancing the thermal conductivity of PA12 composites prepared by selective laser sintering," *RSC Adv.*, vol. 11, no. 4, pp. 1984–1991, 2021, doi: 10.1039/D0RA09775F.

[22] X. Zhang, W. Wu, T. Zhao, and J. Li, "The combination of AlN and h-BN for enhancing the thermal conductivity of thermoplastic polyurethane composites prepared by selective laser sintering," *J. Appl. Polym. Sci.*, Aug. 2022, doi: 10.1002/app.53051.

[23] J. Gao, M. Hao, Y. Wang, X. Kong, B. Yang, R. Wang, Y. Lu, L. Zhang, M. Gong, L. Zhang, D. Wang, and X. Lin, "3D printing boron nitride nanosheets filled thermoplastic polyurethane composites with enhanced mechanical and thermal conductive properties," *Addit. Manuf.*, vol. 56, pp. 102897–102897, Aug. 2022, doi: 10.1016/j.addma.2022.102897.

[24] S. C. Ligon, R. Liska, J. Stampfl, M. Gurr, and R. Mülhaupt, "Polymers for 3D Printing and Customized Additive Manufacturing," *Chem. Rev.*, vol. 117, no. 15, pp. 10212–10290, 2017, doi: 10.1021/acs.chemrev.7b00074.

[25] F. Sillani, F. de Gasparo, M. Schmid, and K. Wegener, "Influence of packing density and fillers on thermal conductivity of polymer powders for additive manufacturing," *Int. J. Adv. Manuf. Technol.*, vol. 117, no. 7–8, pp. 2049–2058, Dec. 2021, doi: 10.1007/s00170-021-07117-z.

[26] F. Lupone, E. Padovano, O. Ostrovskaya, A. Russo, and C. Badini, "Innovative approach to the development of conductive hybrid composites for Selective Laser Sintering," *Compos. Part Appl. Sci.*

Manuf., vol. 147, pp. 106429–106429, Aug. 2021, doi: 10.1016/j.compositesa.2021.106429.

[27] S. Yuan, Y. Zheng, C. K. Chua, Q. Yan, and K. Zhou, "Electrical and thermal conductivities of MWCNT/polymer composites fabricated by selective laser sintering," *Compos. Part Appl. Sci. Manuf.*, vol. 105, pp. 203–213, Feb. 2018, doi: 10.1016/j.compositesa.2017.11.007.

[28] A. Ronca, G. Rollo, P. Cerruti, G. Fei, X. Gan, G. Buonocore, M. Lavorgna, H. Xia, C. Silvestre, and L. Ambrosio, "Selective Laser Sintering Fabricated Thermoplastic Polyurethane/Graphene Cellular Structures with Tailorable Properties and High Strain Sensitivity," *Appl. Sci.*, vol. 9, no. 5, pp. 864–864, Feb. 2019, doi: 10.3390/app9050864.

[29] L. Lanzl, K. Wudy, S. Greiner, and D. Drummer, "Selective laser sintering of copper filled polyamide 12: Characterization of powder properties and process behavior," *Polym. Compos.*, vol. 40, no. 5, pp. 1801–1809, May 2019, doi: 10.1002/pc.24940.

[30] P. Gruber, G. Ziółkowski, M. Olejarczyk, E. Grochowska, V. Hoppe, P. Szymczyk-Ziółkowska, and T. Kurzynowski, "Influence of bioactive metal fillers on microstructural homogeneity of PA12 composites produced by polymer Laser Sintering," *Arch. Civ. Mech. Eng.*, vol. 22, no. 3, pp. 117–117, Jul. 2022, doi: 10.1007/s43452-022-00442-4.

[31] Shangqin Yuan, "Development and optimization of selective laser sintered-composites and structures for functional applications," 2018. doi: 10.32657/10356/73195.

[32] S. Yuan, F. Shen, C. K. Chua, and K. Zhou, "Polymeric composites for powder-based additive manufacturing: Materials and applications," *Prog. Polym. Sci.*, vol. 91, pp. 141–168, Apr. 2019, doi: 10.1016/j.progpolymsci.2018.11.001.

[33] N. B. D. Do, E. Andreassen, S. Edwardsen, A. Lifjeld, K. E. Aasmundtveit, H.-V. Nguyen, and K. Imenes, "Thermal management of an interventional medical device with double layer encapsulation," *Exp. Heat Transf.*, pp. 1–18, Jul. 2021, doi: 10.1080/08916152.2021.1946208.

[34] A. Frick and A. Rochman, "Characterization of TPU-elastomers by thermal analysis (DSC)," *Polym. Test.*, vol. 23, no. 4, pp. 413–417, Jun. 2004, doi: 10.1016/j.polymertesting.2003.09.013.

[35] N. B. D. Do, K. Imenes, K. E. Aasmundtveit, H.-V. Nguyen, and E. Andreassen, "Thermal Conductivity and Mechanical Properties of Polymer Composites with Hexagonal Boron Nitride—A Comparison of Three Processing Methods: Injection Moulding, Powder Bed Fusion and Casting," *Polymers*, vol. 15, no. 6, p. 1552, Mar. 2023, doi: 10.3390/polym15061552.

[36] Henze Boron Nitride Products AG, "HeBoFill ® LL-SP 120," *Tech. Datasheet*, pp. 8374–8374, 2020.

[37] Henze Boron Nitride Products AG, "HeBoFill ® CL-ADH 020," *Tech. Datasheet*, pp. 49–50, 2020.

[38] Heat Scientific, "Heat Scientific MCH Metal Ceramic Heater," *HS-PS101012Y Datasheet*, 2018.

[39] Aalco, "Aluminium alloy 6082," *Tech. Datasheet*, 2019.

[40] Epoxy Technology, "EPO-TEK T7109-19," *Tech. Datasheet*, 2019.

[41] Epoxy Technology, "EPO-TEK EJ2189-LV," *Tech. Datasheet*, 2017.

[42] COMSOL Multiphysics v. 5.3a, "COMSOL Material Library," *COMSOL AB Swden*, 2018.

[43] K. Jagannadham, "Thermal conductivity of copper-graphene composite films synthesized by electrochemical deposition with exfoliated graphene platelets," *Metall. Mater. Trans. B Process*

Metall. Mater. Process. Sci., 2012, doi: 10.1007/s11663-011-9597-z.

[44] Celanese, "Celanese Thermally conductive plastics CoolPoly TCP." Celanese corporation, 2023. [Online]. Available: https://www.celanese.com/products/coolpoly-thermal-conductivity-plastics/

[45] G. Wypych, *Handbook of Polymers: Second Edition.* 2016. doi: 10.1016/C2015-0-01462-9.

[46] A. J. T. Teo, A. Mishra, I. Park, Y. J. Kim, W. T. Park, and Y. J. Yoon, "Polymeric Biomaterials for Medical Implants and Devices," *ACS Biomaterials Science and Engineering.* 2016. doi: 10.1021/acsbiomaterials.5b00429.

[47] J. Ordonez-Miranda and J. J. Alvarado-Gil, "Thermal conductivity of nanocomposites with high volume fractions of particles," *Compos. Sci. Technol.*, vol. 72, no. 7, pp. 853–857, Apr. 2012, doi: 10.1016/j.compscitech.2012.02.016.

[48] K. Pietrak and T. Wiśniewski, "A review of models for effective thermal conductivity of composite materials," *J. Power Technol.*, vol. [Online] 95:1, 2014.

[49] S. Daneshmehr, F. Román, and J. M. Hutchinson, "The surface modification of boron nitride particles," *J. Therm. Anal. Calorim.*, 2019, doi: 10.1007/s10973-019-09160-1.

[50] Z. Cui, A. J. Oyer, A. J. Glover, H. C. Schniepp, and D. H. Adamson, "Large scale thermal exfoliation and functionalization of boron nitride," *Small*, 2014, doi: 10.1002/smll.201303236.

[51] S. Ryu, K. Kim, and J. Kim, "Silane surface treatment of boron nitride to improve the thermal conductivity of polyethylene naphthalate requiring high temperature molding," *Polym. Compos.*, vol. 39, no. S3, pp. E1692–E1700, Jun. 2018, doi: 10.1002/pc.24680.

[52] M. Grundler, T. Derieth, and A. Heinzel, "Polymer compounds with high thermal conductivity," presented at the PROCEEDINGS OF THE REGIONAL CONFERENCE GRAZ 2015 – POLYMER PROCESSING SOCIETY PPS: Conference Papers, Graz, Austria, 2016, p. 030015. doi: 10.1063/1.4965485.

[53] Sabic, "Rethinking your heat sink materials." Sabic, 2020. [Online]. Available: https://www.sabic.com/en/industries/electrical-and-electronics/lighting/thermal-conductive

[54] Covestro Deutschland AG - Engineering Plastics, "Whitepaper 'Power-Tools mit maximaler Leistung und Lebensdauer.'" Covestro, 2022. [Online]. Available: https://solutions.covestro.com/en/highlights/articles/theme/product-technology/makrolon-tc-thermal-conductive-plastics-next-generation-heat-management

[55] Kraiburg, "Technical datasheet - Kraiburg THERMOLAST K (HTC1500/117)." Kraiturg TPE, 2023. [Online]. Available: https://www.kraiburg-tpe.com/en/thermolast-k

[56] TCPoly, "Technical datasheet - Ice9 Flex (Thermally conductive, electrically non-conducting plastic TC-TPE1-145-000)." TCPoly, 2019. [Online]. Available: https://tcpoly.com/purchase-ice9-materials/

[57] J. Gu, Q. Zhang, J. Dang, and C. Xie, "Thermal conductivity epoxy resin composites filled with boron nitride: THERMAL CONDUCTIVITY EPOXY RESIN COMPOSITES," *Polym. Adv. Technol.*, vol. 23, no. 6, pp. 1025–1028, Jun. 2012, doi: 10.1002/pat.2063.

Microstructural based reliability investigation of water- and suspension free prepared integrated electronic packages

Sandy Klengel
Assessment of electronic system integration
Fraunhofer Institute for Microstructure of Materials and Systems
Halle, Germany
sandy.klengel@imws.fraunhofer.de

Robert Klengel
Assessment of electronic system integration
Fraunhofer Institute for Microstructure of Materials and Systems
Halle, Germany
robert.klengel@imws.fraunhofer.de

Tino Stephan
Assessment of electronic system integration
Fraunhofer Institute for Microstructure of Materials and Systems
Halle, Germany
tino.stephan@imws.fraunhofer.de

Abstract—In micro- and power electronics reliability limiting mechanical and (electro-) chemical processes are increasingly caused by the complex structure of the systems, the material compositions and often heterogeneous mixtures of metals, polymers, and ceramics. Additionally, this causes challenging requirements for microstructural analyses and reliability characterization of materials, interfaces, components, and systems. Standardized metallographic routines are not suitable for high resolution analyses of corrosion products and mechanically sensitive systems. Focused ion beam preparation is limited in terms of dimension of the cross sectioned area. Therefore, we developed new preparation concepts by complete water- and suspension free preparation methods using laser and broad ion beam techniques. Two case studies for analyses of power electronic systems potted by inorganic materials and embedded SiC dies show the great potential of these methods for following microstructural analyses.

Keywords—high resolution analyses, microstructure, laser preparation, ion beam preparation, reliability assessment, electronic packaging

I. INTRODUCTION

In microelectronics and power electronics, reliability-reducing mechanical and (electro-)chemical processes are increasingly occurring due to the complex structure of the systems, material compositions, and often heterogeneous mixtures of metals, polymers, and ceramics. In most cases, non-destructive analysis is not sufficient to understand the detailed causes of failures and subsequently decide on effective countermeasures. To enable the application of high-resolution microstructure techniques such as scanning electron microscopy (SEM) and X-ray spectroscopy (EDS), destructive target preparation methods must be used to gain direct access to the site of the failure. Furthermore, this poses demanding requirements for microstructural analysis and reliability characterization of materials, interfaces, components, and systems. Standardized metallographic routines are not suitable for high-resolution analysis of corrosion products and mechanically sensitive systems. The selection of the preparation method depends on the materials and samples under investigation, as well as the required precision and level of artifact risks. Especially in industrial applications, the analysis throughput and efficiency are mostly determined by the efficiency of the preparation. For the analysis of quality or reliability-critical interfaces of electronic components, precise cross-section preparation is required. The standard method is defined by metallographic techniques based on cutting, grinding, and polishing. For material boundaries between very hard and soft materials, such as the chip contact between a semiconductor chip, soft solder material or silver, used for sintering application, and a hard DCB ceramic substrate, metallographic cross-sectioning is not an easy task. Problems such as crack damage and the risk of smearing effects at the material interfaces must be avoided. Today, precise ion and laser-based preparation techniques are established, which can avoid mechanically induced preparation artifacts and improve both the quality and precision of preparation. For example, by ion beam cross-section polishing, where a broad ion beam is used to create relatively large cross-sections. Based on these methods, we have developed new preparation concepts that employ completely water- and suspension-free methods using laser and broad ion beam techniques. Two case studies on the analysis of power electronic systems using inorganic materials and embedded SiC chips demonstrate the great potential of these methods for subsequent microstructural analysis.

II. EXPERIMENTAL

A. Typical process flow for advanced material diagnostic

New routines have been developed for advanced material diagnostics to investigate the reliability of complex electronic systems. Especially in the investigation of corrosion and migration effects, new failure modes occur that can be strongly influenced by liquid-based preparation methods. Therefore, dry preparation methods are needed to minimize the influence of preparation artifacts to the origin failure mode. Pico- or femto-based laser techniques are a good alternative to water-/suspension cooled wire sawing methods for fast and efficient pre-preparation. Followed by high broad ion beam techniques a fast and efficient final preparation can be assured. Now, high-resolution analyses of interfaces and material interaction are possible to perform material-based reliability studies and failure analyses. Figure 1 shows a chart for a typical process workflow for advanced material diagnostics as described before.

189

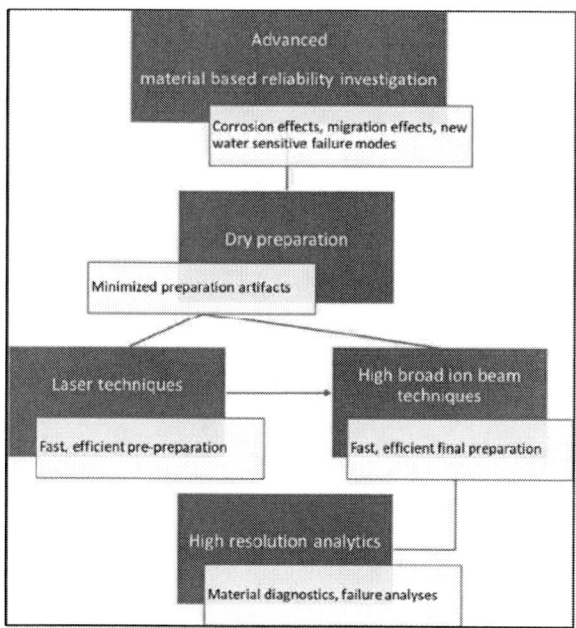

Fig. 1: Workflow for advanced material-based reliability investigation.

B. Laser Preparation

Laser have been commercially available for many decades and are used in various areas such as industrial manufacturing (welding, drilling, scribing, etc.), science (e.g., material characterization, measuring), and medical applications (invasive surgery techniques, LASIK). Laser radiation can ablate all kinds of materials when provided with sufficiently high power or fluence. By using ultrashort pulses and/or very high pulse energies, ablation is achieved through multiphoton absorption, allowing for the machining of transparent materials at the laser's wavelength. Lasers can be precisely delivered and focused using standard optical elements like galvanometer scanners. Laser micromachining is very clean in terms of contamination since laser radiation consists of photons, unlike corpuscular radiations like ion beams that may cause unwanted implantation effects. The ablation rate in laser micromachining is approximately six orders of magnitude higher than that of Ga+-FIB and roughly three orders of magnitude higher than the milling rate of Xe+-Plasma-FIB. These mentioned advantages make laser micromachining a highly interesting option for preparation applications related to analysis issues in electronics, such as connection technology and packaging. The continuous trend of miniaturization enables an increasing level of integration, including 3D systems in packages (3D-SiP). Consequently, the structure of electronic packages is becoming more complex, with material compositions often consisting of heterogeneous mixtures of metals, polymers, and ceramics. This poses challenging requirements for investigating and characterizing the reliability of materials, interfaces, components, and systems. The target positions are often buried or covered, and the cover and housing materials are designed to be resistant against chemical attack for harsh environment applications. These developments make the preparation for tasks such as local access for electrical measurements or artifact-free cross-sectioning increasingly complex. Focused ion beam preparation is limited in terms of ablation volume, chemical de-capping lacks selectivity and faces difficulties due to the resistivity of mold compounds or

alternative packaging materials like ceramics or inorganic potting against chemicals, and metallographic cross-sectioning may compromise the system's functionality or introduce artifacts that limit further investigations. Therefore, there is a growing demand for preparation techniques for microstructure diagnostics that are fast, reliable, cost-effective, artifact-free, and targeted at the micron scale or below. The MicroPrep Pro™ tool used, is a laser micromachining tool developed by 3D-Micromac AG (Chemnitz, Germany) with the intention of providing a fast, clean, and efficient development platform for laser preparation. By utilizing an ultrashort pulse laser, samples can be effectively and cost-efficiently prepared. This approach enables the creation of complex samples or 3D geometries. The flexibility of the laser process allows for comprehensive analysis of deep-lying structures, such as exposing contacts (e.g., through-silicon vias - TSVs) or targeted removal and exposure of traces in semiconductor structures (e.g., systems-in-packages - SiP). Furthermore, larger samples can be processed directly with micrometer-level accuracy. The integrated overview camera simplifies navigation on larger samples, and the high-resolution process camera ensures precise positioning. By using a picosecond laser, structural damage is significantly reduced, and material contamination is avoided. With the MicroPrep Pro™, much higher removal rates can be achieved compared to ion beam-based methods only [1]. Typical sample dimensions processable with this tool range from several hundred microns to several centimetres in size, with thicknesses up to several millimetres.

Fig 2.: MicroPrep Pro™, that can be used for efficient laser preparation.

C. Broad beam ion polishing

While Focused Ion Beams (FIB) have been successfully used for sample preparation in advanced CMOS device analysis, their relatively low milling rates (around 5-10 µm3/s for Silicon) do not make these systems well-suited for cross-sectioning package level interconnects, making large volume removal impractical. Some years ago, a new FIB technology has been developed that offers much higher material removal rates by utilizing an Inductively Coupled Plasma (ICP) Xenon ion source. This source can deliver beam currents 20-50 times higher than the traditional Ga Liquid Metal Ion source (LMIS). The Plasma FIB system, therefore, becomes an attractive tool for analysing interconnect structures by scanning electron microscopy (SEM) with relatively large cross sections. The Xenon ion beam can be used to deposit protective layers, mill cross sections with dimensions up to millimetres, and image the region of interest. This system achieves removal rates of approximately 20000 µm3/min for Silicon, allowing the removal of a volume of, for example, 200x200x100 µm3 in about 3 hours. The removal rate can be further improved by chemically induced FIB etching. We used a FEI Vion™ Plasma single-beam FIB system to evaluate its capability for high-quality, large-area cross-section preparation of selected interconnects in microelectronic devices as an alternative to traditional mechanical grinding, polishing steps, and ion beam bevel etch techniques for different packaging materials [2]. Furthermore, the Plasma FIB system incorporates charge neutralization based on low-energy electron flooding of the sample. The electron gun enables milling of non-conductive materials such as mold compounds or ceramic substrates, effectively preventing sample charging even at high ion beam currents. Thus, we have now developed ion-based preparation routines to investigate the reliability of new material compositions and their interaction with different contact materials (solder, wire bonds) and to assess the reliability of embedded SiC semiconductors in ceramic materials.

III. CASE STUDY 1: EMBEDDED SiC SEMICONDUCTOR

Silicon carbide (SiC) semiconductor devices offer significant economic potential for power electronic applications. To support the develop of a new packaging and integration technology of SiC devices into a monolithic ceramic package for high-power applications our work focuses on artifact minimized preparation and reliability investigation. This approach allows us to assess the inherent advantages of ceramics, such as high temperature stability, high insulation resistance, and high reliability, which are beneficial for SiC packaging concepts [3, 4]. The main challenge for this technology trend is a functional integration of non-shrinking SiC devices into a structured ceramic multilayer laminate, which undergoes shrinkage during the sintering process. This requires establishing mechanical, hermetic, and electrical connections between the SiC device and the ceramic prepackage during the sintering process. Cracks and delamination must be assessed after processing and reliability testing without the influence of mechanical artifacts. To perform sufficient failure analysis on such integrated power devices, laser-assisted target preparation was used in combination with ion beam polishing. The use of laser-assisted target preparation in conjunction with ion beam polishing enables the creation of an artifact-free surface for microscopic analysis. In comparison to conventional metallographic techniques for preparing SiC chips and ceramic structures, this approach allows the desired target plane to be reached much faster and without mechanical influences. The polished surfaces created without mechanical intervention allow a more targeted assessment of cracks and delamination at interfaces. This is demonstrated in Figure 3, which depicts a cross-section of a SiC chip prepared using laser-assisted preparation and ion beam polishing showing fine cracks in the ceramic near the buried silver via. Details of the crack are shown in Figure 4.

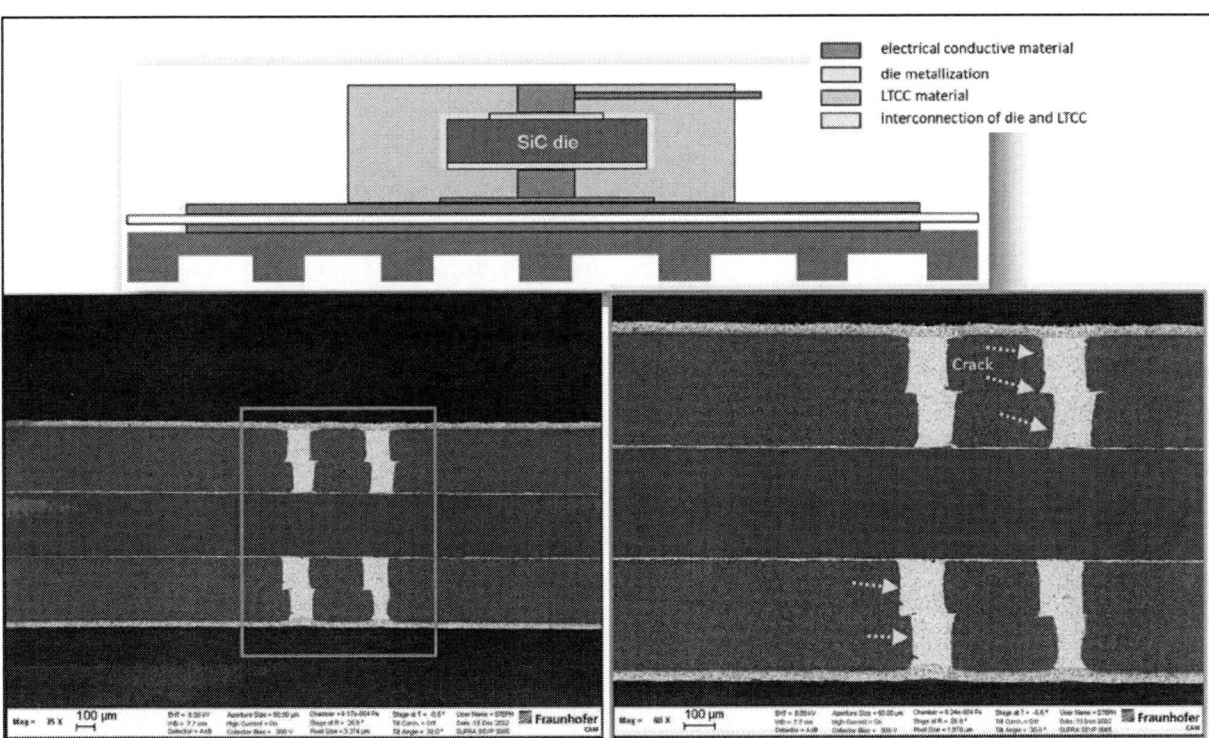

Fig 3.: Concept for an integrated SiC component and cross section prepared without mechanical influences.

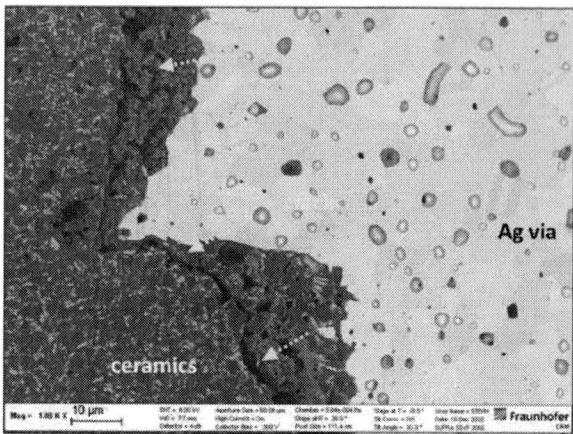

Fig 4.: Detail of cracked interface between buried silver via and sintered ceramic, cracks are marked by yellow arrows.

CASE STUDY 2: INTERFACE FAILURE MODES OF CEMENT ENCAPSULATION AND ADJACENT MATERIALS

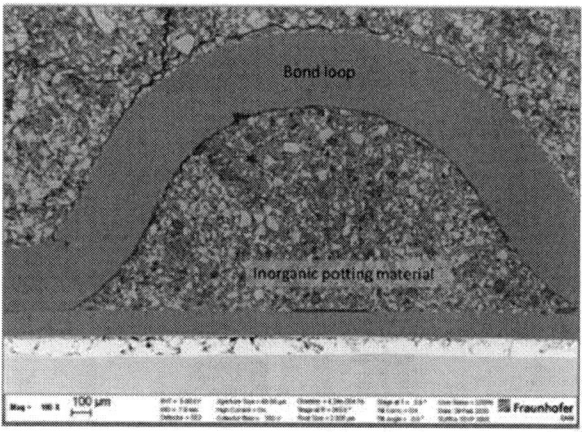

Fig. 5.: Aluminum bond wire with inorganic potting material shows interface delamination and cracking marked by red arrows [5]

In [5] we showed potential failure modes for electronic components potted by inorganic cement-based materials. As these materials become more and more popular, it is very important to develop preparation routines that can prevent smearing effects. Here, the combination of laser preparation and high-rate ion polishing is also suitable to ensure a proper interface preparation. For example, small cement cracks at the interface to the wire bonded contacts can lead to initial

weaking for delamination failure, see Figure **5Error! Reference source not found.**. If this interface is not prepared without smearing effects, misinterpretation is possible.

Figure 6 shows a direct comparison between a metallographic preparation and high broad ion beam polishing of a wire bond contact with mold encapsulation material. After broad beam ion polishing the wire bond shows a corroded area on top that appears porous with a void line below. As this layer has a thickness of about 1µm EDS analyses can be performed and the material composition can be analysed. In direct comparison, the metallographically prepared contact shows no clear layer of corrosion products. Only a brittle decomposed area is detectable on top of the stitch bond with a small gap to the mold compound. It is assumed that the corroded layer was dissolved in the liquids used for the metallographic preparation.

DISCUSSION & SUMMARY

In [6] we already showed the potential for combined laser preparation and broad ion beam polishing for 3D-integrated systems. It offers solutions to a wide range of problems and challenges, including the removal of potting compounds like mold compounds and gel, as well as the creation of large-scale cross-sections for ion beam-based target preparations. Laser pretreatment enables water-free or chemical-free preparation routines, or significantly reduces their usage, allowing for subsequent elemental analysis. Additionally, laser pre-preparation minimizes or eliminates the need for mechanical processing of sensitive samples, thereby avoiding or greatly limiting associated effects. This ensures artifact-free access that may not be possible with classical metallography, for example. Extensive investigations have been conducted to understand the impact of laser beam ablation on samples, such as the heat-affected zone, debris, and surface impairment, to ensure that subsequent analyses are not affected by the preparation process.

In this paper, additional case studies have demonstrated the versatility of the laser preparation method tool also for power electronics application. It has proven itself effective in processing large volume encapsulants and creating cross-sections of potted materials or integrated dies. Failure modes and material reactions can only be accurately determined and analysed if the preparation artifacts are minimised as far as possible. Therefore, it is crucial to adapt the preparation and analysis methods to the material and technological advancements.

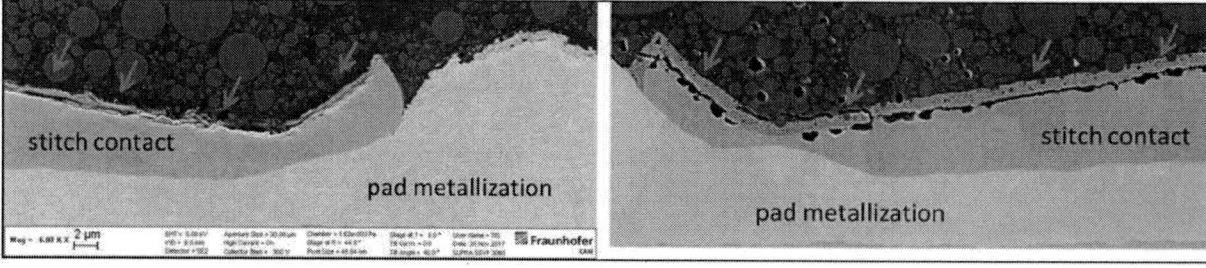

Fig. 6: Direct comparison between a metallographic preparation (left) and high broad ion beam polishing (right) of an encapsulated wire bond contact with encapsulation material showing corrosion products for the high-end prepared contact and a decomposed area for the metallographic prepared contact only.

ACKNOWLEDGMENT

This work was supported by the market-oriented strategic pre-competitive research (MAVO) from the Fraunhofer-Gesellschaft through grant "MESiC" (Use Case 1) and partial financial support (Use-Case 2) by the Federal Ministry for Economic Affairs and Climate Action with-in the AVEL project (Contract: 03ETE038B).

REFERENCES

[1] https://3d-micromac.de/laser-mikrobearbeitung/produkte/microprep/

[2] Altmann, F., Klengel, S., Schischka, J., & Petzold, M. (2013, May). Defect analysis using high throughput plasma FIB in packaging reliability investigations. In *2013 IEEE 63rd Electronic Components and Technology Conference* (pp. 1940-1945). IEEE.

[3] A L. Coppola, D. Huff, F. Wang, R. Burgos. D. Boroyevich, "Survey on High-Temperature Packaging Materials for SiC-Based Power Electronics Modules," *IEEE Power Electronics Specialists Conference*, June 2007, pp. 2234-2240.

[4] B B. Bayer, M. Groccia, H. L. Bach, C. F. Bayer, A. Schletz, C. Lenz, S. Ziesche, M. März, "LTCC Embedding of SiC Power Devices for High Temperature Applications over 400 °C," in *Electronics System-Integration Technology Conference (ESTC), 2020.*

[5] Naumann, F., Boettge, B., & Klengel, S. (2022, March). Potential failure modes of cement-based encapsulation concepts for reliable power electronics. *In CIPS 2022; 12th International Conference on Integrated Power Electronics Systems* (pp. 1-7). VDE.

[6] Klengel, R., Klengel, S., Schusser, G., & Krause, M. (2018, May). A New, Efficient Method for Preparation of 3D Integrated Systems by Laser Techniques. In *2018 IEEE 68th Electronic Components and Technology Conference (ECTC)* (pp. 2275-2279). IEEE.

Numerical Study on the Influence of Polyimide Thickness and Curing Temperature on Wafer Bow in Wafer Level Packaging

Prashant Kumar Singh
Advanced Packaging Development
GlobalFoundries Dresden Module One
LLC & Co.KG
Dresden, Germany
PrashantKumar.Singh@globalfoundries
.com

Patrick Rohlfs
Advanced Packaging Development
GlobalFoundries Dresden Module One
LLC & Co.KG
Dresden, Germany
Patrick.Rohlfs@globalfoundries.com

Gunther Sandmann
Advanced Packaging Development
GlobalFoundries Dresden Module One
LLC & Co.KG
Dresden, Germany
Gunther.Sandmann@globalfoundries.c
om

Kashi Vishwanath Machani
Advanced Packaging Development
GlobalFoundries Dresden Module One
LLC & Co.KG
Dresden, Germany
KashiVishwanath.Machani@globalfou
ndries.com

Dirk Breuer
Process Integration BEOL
GlobalFoundries Dresden Module One
LLC & Co.KG
Dresden, Germany
Dirk.Breuer@globalfoundries.com

Karsten Meier
Institute of Electronic Packaging
Technology
Technische Universität Dresden
Dresden, Germany
karsten.meier@tu-dresden.de

Frank Kuechenmeister
Global Reliability
GlobalFoundries Dresden Module One
LLC & Co.KG
Dresden, Germany
Frank.Kuechenmeister@globalfoundrie
s.com

Marcel Wieland
Advanced Packaging Development
GlobalFoundries Dresden Module One
LLC & Co.KG
Dresden, Germany
Marcel.Wieland@globalfoundries.com

Karlheinz Bock
Institute of Electronic Packaging
Technology
Technische Universität Dresden
Dresden, Germany
karlheinz.bock@tu-dresden.de

Abstract—In wafer level packaging, polyimide and electroplated copper are dielectric and conducting materials respectively in the so-called redistribution layers. During the wafer fabrication process large amount of stress is generated in those layers due to curing shrinkage of the polyimide and the coefficient of thermal expansion mismatch of both materials to silicon which can lead to severe wafer bow after a high temperature curing. In different applications polyimide can be used either only as passivation layer over electroplated copper, or polyimide is applied before the redistribution layer and then a second polyimide layer is applied as passivation. In the current study the effect of different polyimide integrations is investigated. In-situ wafer bow measurements and finite element method studies were conducted with samples, with and without polyimide below redistribution layer, exposed to different curing temperatures to understand the change in wafer bow and copper residual stress. It is observed that wafers with polyimide before redistribution layer show higher stress relaxation in copper with decreasing curing temperature, resulting therefore in a lower wafer bow. A good agreement is achieved between the experimentally measured and the simulated wafer bow.

Keywords— Wafer level packaging, redistribution layer, wafer bow, residual stress, finite element modeling

I. INTRODUCTION

Wafer level packaging (WLP) has been one of the fastest growing technologies in semiconductor packaging industry due to high demands, reduced fabrication cost and enhanced performance [1][2]. In a WLP, the redistribution layer (RDL) is used to re-route the I/O signal to the solder bumps. Polyimide (PI) as dielectric layer and electro-chemically deposited (ECD) copper as metallization is widely adapted in the RDL. Depending upon the I/O requirements the RDL can consist of five or more layers [3]. In RDL fabrication processes the wafer undergoes different high temperature

process steps that result in reliability risks due to stresses induced in the wafer during fabrication. For example, during the RDL fabrication PI is cured at high temperature that results in stresses induced in the wafer due to cure shrinkage of the PI [4]. Simultaneously, thermal stress is also induced in the wafer due to coefficient of thermal expansion (CTE) mismatch between the different materials used in the RDL processing. These stresses might be high and result in large amount of wafer bow that affect handling of the wafer and subsequent processing steps on the wafer that can lead to reliability risks like delamination and passivation cracks [4]. Therefore, it is important to predict the wafer bow and limit the stress evolution for different films in the RDL.

The thermal stress induced into the wafer due to CTE mismatch during the high temperature process step is at least partially recovered after the wafer is cooled down to room temperature (RT) again. To minimize the stresses a change in material used in RDL is usually not an option because most of the materials are optimized to have the best I/O response of the signal to the solder bumps and lower down the reliability risks. The stresses generated due to cure shrinkage of PI also cannot be avoided due to volume loss during the polymerization reaction. In past, different studies have been done to reduce warpage by deposition of an additional passivation layer on copper [5]. The effect of passivation on RDL copper still needs to be investigated, because the effect of passivation layer on stress relaxation behavior of copper is different for patterned structures deposited with ECD process than that of the copper film deposited with sputter method without patterning.

This study focuses on reducing the stresses which are generated due to the deformation of copper by using PI, for which the imidization can be achieved at lower curing temperatures. In some investigations the variation in mechanical response has been observed when PI is integrated only as passivation layer above the copper/PI patterned structures in comparison to PI that is integrated both as passivation over the copper/PI patterned structures and as a stress relief layer between the silicon substrate and the RDL [6]. In this paper, finite element method (FEM) based models with experimental verifications were used to predict the wafer bow for different PI integrations and to understand the effect of different curing temperatures on proposed PI integrations. Moreover, the work demonstrated measures to lower the wafer bow induced in the wafer during RDL fabrication process.

II. EXPERIMENT

A. Test Vehicle

In the experimental setup two sets of multi-layered thin films were fabricated on a 300 mm wafer. The variation in experimental setup on both the test vehicles is performed as follow:

(1) PI is used both as passivation over copper and as stress relief layer between substrate and RDL (see Figure 1 a). On the other hand, in the second setup PI is only used as passivation over the copper (see Figure 1 b).
(2) In addition, PI in both the cases have been cured at three different curing temperatures which are 250°C, 230°C and 210°C to understand the effect of curing temperatures on wafer bow.

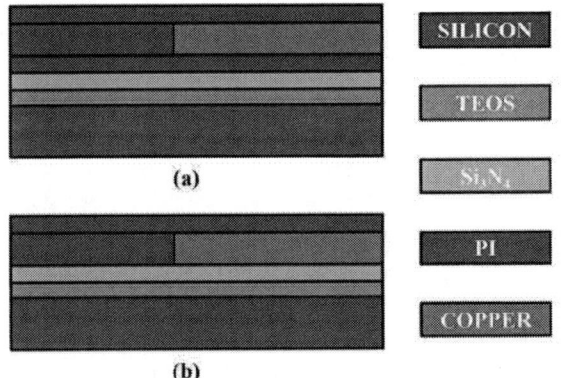

Figure 1: Schematic for fabricated RDL stacks: Sample with PI as stress relief layer (a) and sample without PI as stress relief layer (b)

In the wafer fabrication process for the samples with PI as stress relief layer (see Figure 1 a), TEOS is deposited on a silicon substrate followed by a Si₃N₄ passivation (see Figure 2). In the next step 1st PI layer (stress relief layer) is deposited followed by a curing at different temperatures, *i.e.* at 250°C for sample 1, 230°C for sample 2 and 210°C for sample 3. In the next step a 2nd PI layer is deposited on the wafer samples, the PI openings for the copper redistribution are exposed and a second curing is done at 250°C, 230°C and 210°C respectively for each sample. Before curing, the PI is structured using a lithography process and copper is deposited at RT. In the final step a 3rd PI layer is deposited as passivation on the patterned copper and again a third curing is done with 250°C, 230°C and 210°C respectively for each sample. The design of experiments (DOE) keeps all the process parameters and material thicknesses constant to investigate the change in wafer bow with the variation in

curing temperature. Moreover, after the deposition of 3rd PI layer in-situ measurements are conducted to measure the wafer bow of all three sample sets with the Patterned Wafer Geometry (PWG) Metrology [7]. The above steps are repeated to also fabricate the samples without the 1st PI layer (stress relief layer) as shown in Figure 1 b to investigate the influence of 1st PI layer on the stress evolution in copper for both the test vehicles.

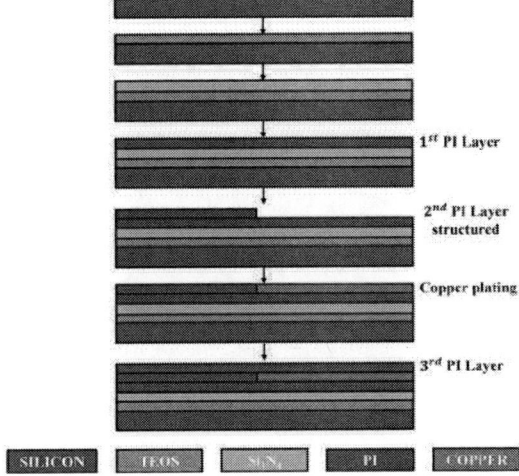

Figure 2: Schematic for steps involved in the fabrication process for test vehicles

III. FINITE ELEMENT METHOD MODELING

In addition to the experimental study, FEM based models were also developed to predict the wafer bow for different test vehicles processed at different curing temperatures. The FEM model setup involved a local to global modeling approach which was already presented in earlier work [7]. Firstly, the residual stress is calculated for different test structures in a local model. In the next step, both the material data and residual stress is added to a global full wafer level model to predict the wafer bow.

A. Residual stress calculation

The residual stress for the test vehicle (see Figure 1 a) is calculated by taking into consideration a die level model (see Figure 3 b) projected from a wafer reticle schematic (see Figure 3 a). The copper traces are also modelled in the local die level model embedded in the PI using GDS data [7]. In the next step, other layers are also modelled (see Figure 4 a) to complete the layer stack (see Figure 1 a).

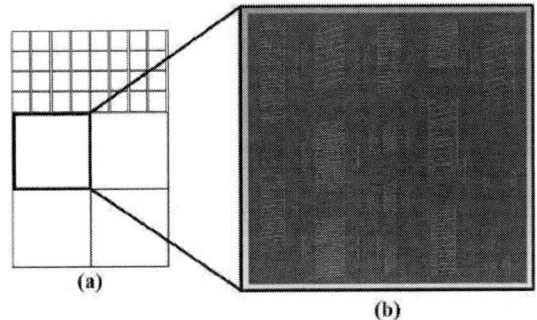

Figure 3: Wafer reticle schematic (a) and FEM model with copper traces embedded between the PI (b)

195

As the next step the residual stress is calculated by process modeling approach at die level model using element birth and death modeling technique. The test vehicle layer stack selected for this calculation is sample with PI as stress relief layer (see Figure 1 a). The curing temperature profile for each PI layer is shown in Figure 5.

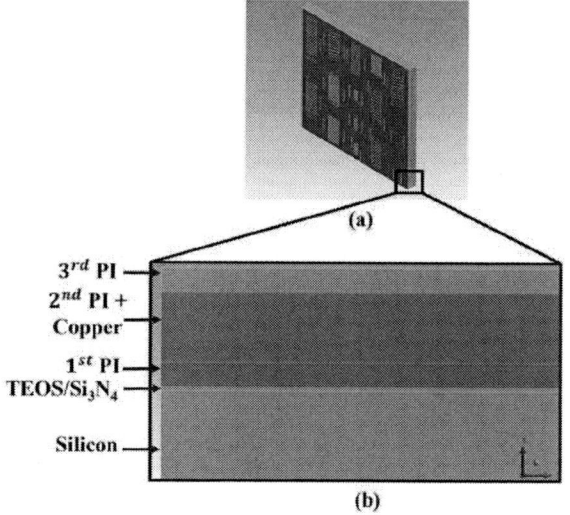

(a)

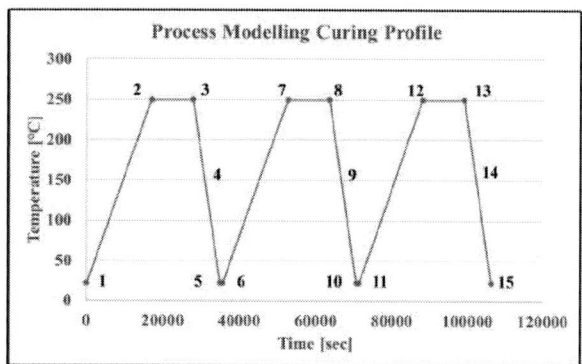

(b)

Figure 4: Full die level FEM model with copper traces (a) and cross-section of the model with complete test vehicle stack (b)

Process Modelling Curing Profile

Figure 5: Curing temperature profile at 250°C for test vehicle with stress relief layer (see Figure 1 a)

In setting up the residual stress simulation model linear elastic material data is selected for all the layers. In addition, bilinear isotropic hardening (BISO) plasticity model is used for copper and prony series viscoelastic model is used for PI. The process flow for the modeling step is shown in Figure 6. The simulation starts from step 1 and the model only has substrate, TEOS and Si_3N_4 layer activated. At step 2 1st PI layer is activated, and the curing is done at the peak temperature of 250°C for 3 hours. Once the model is cooled down to 150°C at step 4 the cure shrinkage strain is applied to the PI. Later the model is further cooled down to RT. The model is again heated to step 7 and the 2nd PI layer is activated with structured pattern (wherein copper is deposited in subsequent steps) and cured. Like 1st PI layer, cure shrinkage strain is again applied to 2nd PI layer at 150°C and the model is further cooled down to RT. At step 10 the copper is deposited in the structured pattern of

2nd PI layer, and the complete model is again heated to peak curing temperature. At step 12 the 3rd PI layer is activated and cured. At step 14 the cure shrinkage due to PI curing is applied and the model is cooled down to RT. The remaining stress at step 15 is considered as residual stress and used in further simulation setup for global wafer level model to calculate the wafer bow. Above steps are also repeated to calculate the residual stress for both the stacks used in this investigation (see Figure 1) at different curing temperatures. The actual non-uniform metallization pattern (see Figure 3 b) is also simulated to calculate the pattern dependent orthotropic material data [7] for the copper embedded between the 2nd PI layer. In addition to the calculated residual stress, the orthotropic material properties, *i.e.* Young's modulus, Poisson's ratio and CTE would be used as an input parameter for full wafer level model.

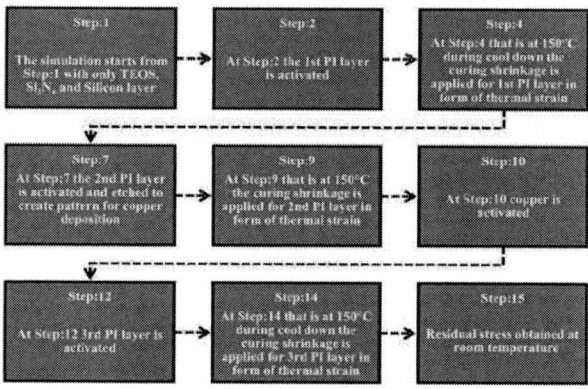

Figure 6: Process flow for the FEM modeling setup to calculate the residual stress

B. Full wafer level model

The full wafer level model is constructed based on the actual reticle layout (see Figure 3 a). Firstly, a reticle level model is constructed (see Figure 7 a) that is extended to a full wafer level model (see Figure 7 b). The full wafer level model also consists of all the thin film layers modelled with similar thickness as shown in Figure 4 b.

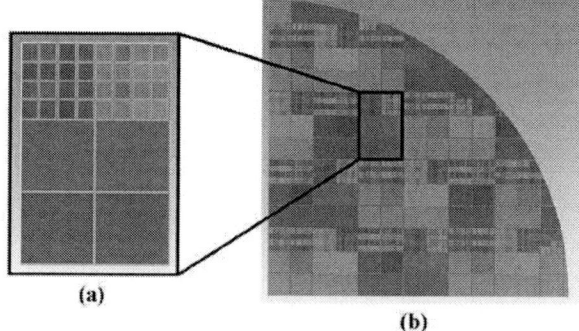

(a)

(b)

Figure 7: Reticle-level model (a) and quarter wafer level model (b)

The input parameter for the full wafer level model are the metallization pattern dependent orthotropic material data derived from the die level model. Also residual stress is applied for all the thin film layers shown in Figure 4 b calculated from local model discussed in section III A. For the residual stress calculation of PI layer embedded within

196

copper, an effective stress approach is used and input in the model [7]. To reduce the computational time a quarter model was established and the geometry is meshed with SOLSH190 mesh elements (see Figure 8 b) in Ansys with eight elements in the substrate thickness (see Figure 8 c). Symmetric boundary conditions were applied on the two symmetry planes and the center node is fixed in all the three directions to avoid rigid body motion (see Figure 8 a).

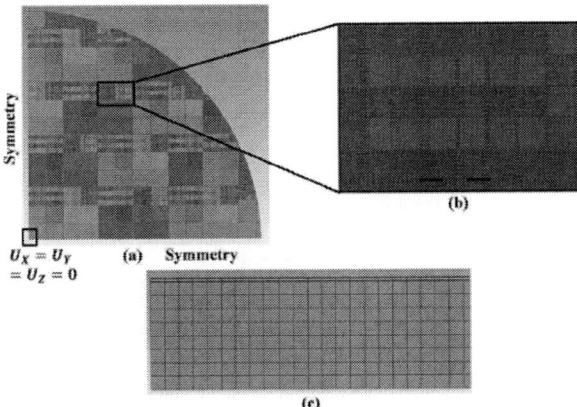

Figure 8: Boundary conditions applied in quarter wafer level model (a), mesh on wafer level model scoped to defined reticle location (b) and mesh structure in substrate thickness in the wafer level model (c)

IV. RESULTS AND DISCUSSION

A. Experimental Results

The wafer bow in each sample is measured after the curing of 3rd PI layer and cooled down to RT using the methodology discussed in section II A. Figure 9 shows the increase in wafer bow for the wafer samples without the 1st PI layer (stress relief layer).

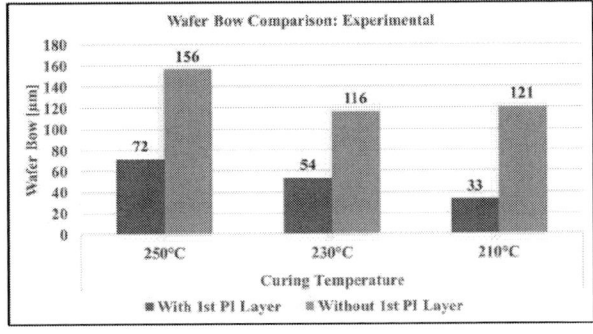

Figure 9: Experimental wafer bow comparison for stack with 1st PI layer (see Figure 1 a) and without 1st PI layer (see Figure 1 b) at different curing temperatures

At 250°C curing temperature wafer bow obtained for the samples with 1st PI layer is 72 μm and without 1st PI layer is 156 μm. The percentage deviation of wafer bow for wafer samples without 1st PI layer from sample with 1st PI layer is 118% (see Figure 10). At 230°C curing temperature wafer bow obtained for the samples with 1st PI layer is 54 μm and without 1st PI layer is 116 μm. The percentage deviation of wafer bow for wafer samples without 1st PI layer from

sample with 1st PI layer is 117%. Similarly, at 210°C curing temperature wafer bow obtained for the samples with 1st PI layer is 33 μm and without 1st PI layer is 121 μm. The percentage deviation of wafer bow for wafer samples without 1st PI layer from sample with 1st PI layer is 267%. The measured wafer bow was then compared to simulation results (see section IV C).

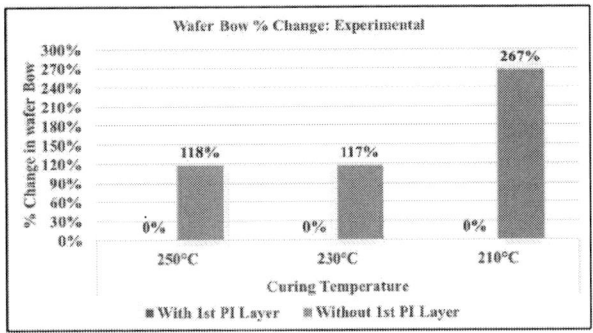

Figure 10: Percentage change in experimental wafer bow for wafer samples without 1st PI layer relative to wafer samples with 1st PI layer at different curing temperatures

B. Simulation Results

Wafer level models were setup as discussed in section III B and their input parameters including the residual stress for copper were derived from local die level models (see section III A). This was carried out for both the wafer samples configuration (see Figure 1) at different curing temperatures. Figure 11 shows the residual stress distribution in copper on both the samples discussed in Figure 1 at 250°C curing temperature. The samples without 1st PI layer show higher stresses within copper as compared to samples with 1st PI layer considered in the model.

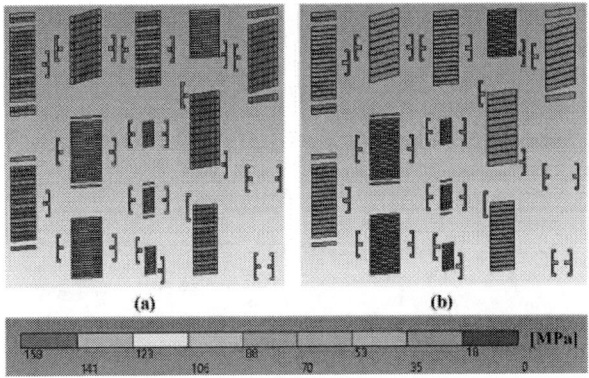

Figure 11: Simulated residual stress distribution: Sample without 1st PI layer (a) and sample with 1st PI layer (b)

Figure 12 shows the stress comparison at different curing temperatures for varied sample configuration (see Figure 1). The residual stress obtained at 250°C curing temperature in copper for sample without 1st PI layer and with 1st PI layer is 158 MPa and 78 MPa respectively. Residual stress obtained at 230°C curing temperature in copper for wafer sample without 1st PI layer and with 1st PI layer is 146 MPa and 58 MPa respectively. Similarly, the residual stress obtained in copper at 210°C curing temperature for wafer sample without 1st PI layer and with 1st PI layer is 138 MPa and 39 MPa respectively. The residual stress thus obtained was used in the wafer level simulation model to predict the wafer bow. Figure 13 shows the calculated wafer bow for the sample configurations (see Figure 1), at 250°C curing temperature. The results in the images are scaled up five times from true scale. Here again we see that the wafer bow is lower for samples with 1st PI layer.

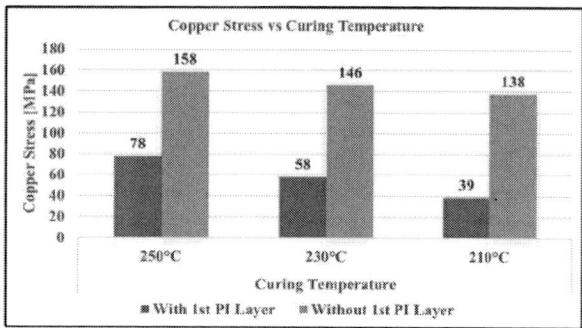

Figure 12: Residual stress calculated with simulation for copper at different curing temperatures for both sample configurations

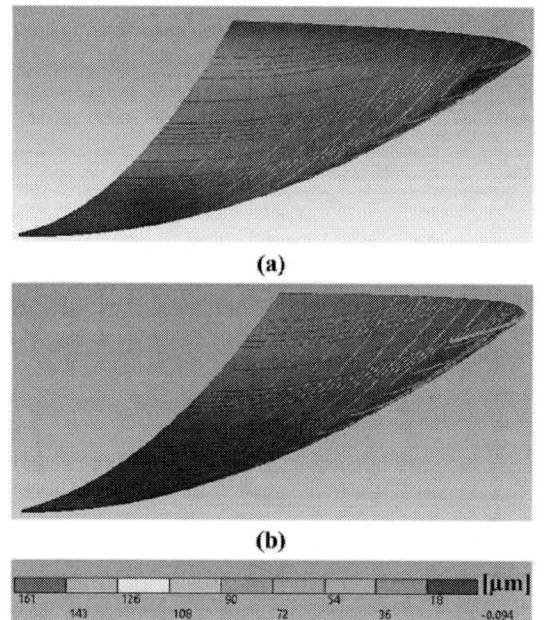

Figure 13: Wafer bow simulated results at 250°C curing temperature: Sample without 1st PI layer (a) and sample with 1st PI layer (b)

Figure 14 shows the wafer bow comparison at different curing temperatures for varied sample configuration (see Figure 1). The wafer bow obtained at 250°C curing temperature were 161 μm and 76 μm for samples without 1st PI layer and with 1st PI layer respectively. The percentage deviation in wafer bow for samples without 1st PI layer relative to sample with 1st PI layer is 122% (see Figure 15). At 230°C curing temperature wafer bow obtained for sample without 1st PI layer and with 1st PI layer were 122 μm and 58 μm respectively. The percentage deviation

in wafer bow for samples without 1st PI layer relative to sample with 1st PI layer is 109%. Similarly, at 210°C curing temperature wafer bow obtained for sample without 1st PI layer and with 1st PI layer were 115 μm and 36 μm respectively. The percentage deviation in wafer bow for samples without 1st PI layer relative to sample with 1st PI layer is 223%.

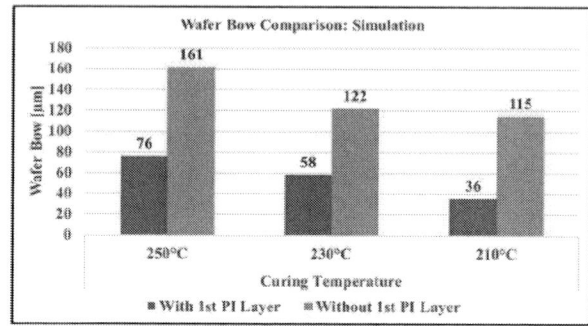

Figure 14: Simulated wafer bow comparison for stack with 1st PI layer (see Figure 1 a) and without 1st PI layer (see Figure 1 b) at different curing temperatures

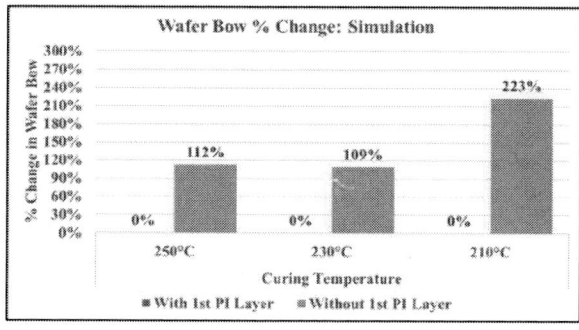

Figure 15: Percentage change in simulated wafer bow for wafer samples without 1st PI layer relative to wafer samples with 1st PI layer at different curing temperatures

C. Experimental vs. Simulation Results

The section IV A and IV B of this study shows the wafer bow development for different PI integrations schemes at varied curing temperatures. Both experimental and simulation shows dependence of wafer bow on PI integration methods and curing temperature. Due to the introduction of 1st PI layer the overall residual stress in the copper is relaxed (see Figure 12). The residual stress of copper obtained from simulation for samples with and without 1st PI layer are 78 MPa and 158 MPa respectively for a 250°C cure temperature, that shows higher stress relaxation in copper when the 1st PI layer is integrated within the sample. Similarly, as the PI acts as a stress relief layer, there is a reduction in the wafer bow observed with the introduction of 1st PI layer. Moreover, in copper films surface diffusion and vertical grain boundary diffusion are dominant stress relaxation mechanism. The traditional in-organic passivation on copper restricts this surface diffusion in vertical direction therefore restricting the grain growth which would lead to lower wafer bow. However, the PI as the passivation layer is softer than traditional passivation layers, hence there is a behavior change when it is integrated within copper RDL, *i.e.* the PI passivation (3rd PI layer) does not have influence on the wafer bow reduction efforts. Moreover, the 1st PI layer acting as a decoupling layer together with Si₃N₄ between substrate and copper can restrict the atomic migration and grain boundary diffusion that results in lower wafer bow.

In addition to the PI integration method, the curing temperature indicates a rather linear dependence on the wafer bow within the given temperature range. For instance, at 210°C curing temperature the wafer bow observed experimentally for samples with and without 1st PI layer are 33 μm and 121 μm (see Figure 9) respectively, which are lower than the wafer bow obtained at 250°C. Similarly, in simulation investigation at 210°C curing temperature the wafer bow observed for sample with and without 1st PI layer are 36 μm and 115 μm (see Figure 14) respectively, that are lower than the simulated wafer bow obtained at 250°C. These data lead to the conclusion, that there is a clear dependence of wafer bow on the curing temperature. Moreover, Figure 12 also confirms the indication of a linear reduction in simulated copper residual stress with reduction of curing temperature. Since the PI viscoelastic behavior did not contribute significantly to the residual stresses and deformations caused in the wafer. Also the plastic behavior of the copper was just not triggered because the residual stress did not exceed the yield stress of copper. Hence the linear dependency of wafer bow is well supported by simulation findings. Moreover, a lower curing temperature with appropriate imidization percentage of PI is proposed for wafer bow reduction efforts.

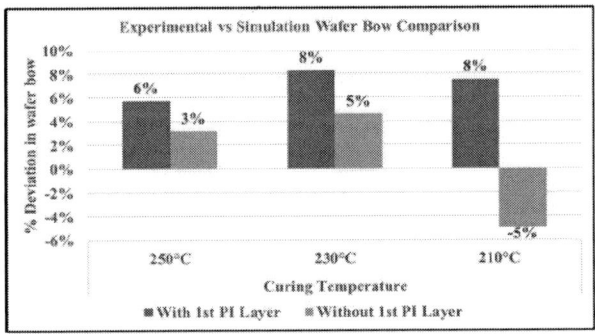

Figure 16: Percentage deviation between simulated and experimental wafer bow for stack with 1st PI layer (see Figure 1 a) and without 1st PI layer (see Figure 1 b) at different curing temperatures

The percentage deviation of simulated from experimentally generated wafer bow magnitude is 6% for wafer sample with 1st PI layer and 3% for wafer sample without 1st PI layer cured at 250°C (see Figure 16). At 230°C percentage deviation of simulated from experimentally generated wafer bow magnitude is 8% for wafer sample with 1st PI layer and 5% for wafer sample without 1st PI layer. Similarly, at 210°C percentage deviation of simulated from experimentally generated wafer bow magnitude is 8% for wafer sample with 1st PI layer and -5% for wafer sample without 1st PI layer. At 210°C experimental and simulation deviation for samples without 1st PI layer is -5% this can be due to the increase in experimental wafer bow at 210°C in comparison to 230°C. This increase in wafer bow at 210°C can be the effect of

averaging the results obtained at 210°C on three different wafer samples. Overall, the simulation results show a good correlation to the experimental data.

V. CONCLUSION

In this study the wafer bow development was studied for multi-layered structures. Experimental and simulation models were setup to understand the effect of different PI integration schemes and curing temperature on wafer bow. Since the modeling results are in a good correlation with the experimentally generated data, the FEM model had been validated. In next steps the validated FEM model can be used for further optimization of PI thickness and copper design for wafer bow reduction efforts. The experimental and simulation results show lower wafer bow obtained for test structures integrated with 1st PI layer (stress relief layer) when compared to test structures without 1st PI layer. Hence the PI-1 integration should be taken into account during the WLP fabrication phase to reduce the final wafer bow.

In addition to the PI integration schemes, curing temperature has a linear effect on the wafer bow. In both the test vehicles, the wafer bow decreases with the decrease in curing temperature. From this study, the authors propose to have a 1st PI layer integrated as decoupling layer and to have a curing temperature not to be more than 210°C during fabrication phase to reduce the final wafer bow.

REFERENCES

[1] Yong Liu, Trends of power semiconductor wafer level packaging, Microelectronics Reliability, Volume 50, Issue 4, 2010, Pages 514-521.

[2] Tong Yan Tee, Xuejun Fan, Yi-Shao Lai, Advances in Wafer Level Packaging (WLP), Microelectronics Reliability, Volume 50, Issue 4, 2010, Pages 479-480.

[3] P. Tumne, V. Venkatadri, S. Kudtarkar, M. Delaus, D. Santos, R. Havens, and K. Srihari, "Effect of Design Parameters on Drop Test Performance of Wafer Level Chip Scale Packages," Journal of Electronic Packaging, vol. 134, Jun. 2012, pp. 020905.

[4] S. S. Deng, S. J. Hwang, and H. H. Lee, "Warpage Prediction and Experiments of Fan-Out Wafer level Package During Encapsulation Process," IEEE Transactions on Components Packaging and Manufacturing Technology, vol. 3, no. 3, pp. 452-458, Mar, 2013.

[5] D. W. Gan, P. S. Ho, Y. Y. Pang, R. Huang, J. Leu, J. Maiz, and T. Scherban, "Effect of passivation on stress relaxation in electroplated copper films," Journal of Materials Research, vol. 21, no. 6, pp. 1512-1518, Jun, 2006.

[6] Chunsheng Zhu, Wenguo Ning, Gaowei Xu, Le Luo, Stress evolution during thermal cycling of copper/polyimide layered structures, Materials Science in Semiconductor Processing, Volume 27, 2014, Pages 819-826.

[7] P. K. Singh et al., "Finite Element Model for Prediction of Back-End-of-Line Process Induced Wafer Bow for Patterned Wafer," 2023 24th International Conference on Thermal, Mechanical and Multi-Physics Simulation and Experiments in Microelectronics and Microsystems (EuroSimE), Graz, Austria, 2023, pp. 1-10, doi: 10.1109/EuroSimE56861.2023.10100826.

Improvement of bonding strength and thermal shock reliability for Ag sinter joining direct on Al substrate

Chuantong Chen*
SANKEN
Osaka University
Osaka, Japan
chenchuantong@sanken.osaka-u.ac.jp

Ran Liu
SANKEN
Osaka University
Osaka, Japan
liu.ran@sanken.osaka-u.ac.jp

Koji Kobayashi
DOWA POWER DEVICE Co., Ltd
Nagano, Japan
kobayak2@dowa.co.jp

Hideyo Osanai
DOWA POWER DEVICE Co., Ltd
Nagano, Japan
osanaih@dowa.co.jp

Zheng Zhang
SANKEN
Osaka University
Osaka, Japan
zhangzheng@sanken.osaka-u.ac.jp

Katsuaki Suganuma
SANKEN
Osaka University
Osaka, Japan
suganuma@sanken.osaka-u.ac.jp

Abstract— Ag sinter paste joining as a proven die bonding technology have been used in the electric vehicle. However, for the Ag sinter paste joining, metallization layers on substrate are usually necessary such as Ni-P/Au or Ni-P/Ag to get a good bonding quality. The direct bonding on DBC (direct bonded copper) or DBA (Direct Bonded Aluminum) substrates without metallization layers are very attractive. In this study, we investigated the direct bonding on an aluminum (Al) substrate by Ag sinter paste and propose an abrasive blasting processes to treat the Al surface to increase the initial bonding quality. The result show that the shear strength of SiC/Al joint structure by Ag sinter paste can be improved by treating the Al surface. The shear strength of joint structure without surface treatment increased from 26.9 MPa to 32.2 MPa for a strong blasting treatment on Al surface. In addition, the results of thermal shock test (-40℃~150℃) also show that the blasting treatment on Al surface can improve the shear strength and reliability of SiC/Al joint structure.

Keywords—Direct bonding on Al, Ag sinter paste, thermal shock, blasting processes, SiC power modules

I. INTRODUCTION

With the development of wide bandgap semiconductors, such as silicon carbide (SiC) which present more potential properties for delivering improved power densities and frequencies, power module using SiC power chips attract more attention especially in the electric vehicle applications [1-4]. In power module structure, the SiC power chip was attached on a DBC (direct bonded copper) or DBA (Direct Bonded Aluminum) substrates by a die attached material, and finally connect to an aluminum (Al) heatsink for heat dispersal. The bonding quality and reliability of the power chips to substrates is one of the most important factors for the power module structure. Ag sinter paste joining as a proven bonding technology is one of the good selections because which possesses good thermal conductivity and can be sintered in low temperature low pressure air condition [5-8]. In addition, for the DBA and DBC substrate bonding to the Al heatsink, currently, some studies also focus on the Ag sinter paste to replace the traditional solder or TIM materials [9, 10]. The technology of bonding sintered Ag directly on Al substrate is very attractive, because it does not require a metallization layer on Al such as Ni/Au or Ni/Ag, which can save a lot of cost and time.

Recently, direct bonding of a SiC die to a bare DBA substrate via Ag sinter paste joining was achieved under a low-temperature, pressure-less condition [11]. However, a large deformation occurred on DBA for the direct bonding during a harsh thermal shock test due to the Al was soft. The deformation of Al surface lead to the cracks generation at the sintered Ag layer and thus lead to the possibility of reliability issue for the direct bonding case.

In this work, in order to improve the bonding strength and thermal shock reliability for Ag sinter joining on bare Al substrate, the Al substrate was treated by an abrasive blasting processes before bonding with Ag sinter paste. The abrasive blasting processes treatment was expected to increase the hardness of Al surface and decrease the grain size of Al, and thus to prevent the deformation during the thermal shock test after bonded with SiC chip by Ag sinter paste joining.

II. EXPERIMENTAL

A. Ag paste and Al substrate

In this study, micron-sized Ag flake particles were selected as the Ag precursor. The average size of the Ag particles is 2.5 μm, and cross section of the Ag flake particles was shown in Fig. 1(a). The CELTOL-IA, a kind of solvent was used to mix the Ag flake particles to fabricate the Ag paste with the wight ration of 1:13.

The Al substrate (Al 1050) with the dimension of 35x35x2 mm³ was used in this study. Two kinds of blasting processes treated on the Al surface called as weak treatment and strong treatment, respectively, that were implemented before joint by Ag sinter paste with SiC chip. Fig. 1(c) shows the printing process by Ag sinter paste. Firstly, the paste was printed on the Al substrate by a metal mask with the thickness of 100 μm, the size is 3x3 mm² (Fig. 1(d)). After printed the Ag sinter paste on the Al substrate, SiC with the sputtered Ti (100 nm) and Ag (1 μm) metallization on the bonding surface was attached to the paste and sintering at 300 °C with 2 MPa pressure (Fig. 1(e)).

B. Characterization

The shear strength SiC/Al joint structure was measured via a die shear tester (DAGE, XD-7500). In addition, to investigate the reliability on Al substrate with different surface treatment, SiC/Al joint structure was implemented with a thermal shock test from -40 ℃ to 150℃ for 500 cycles. The microstructure of SiC/ Al joint by Ag sinter paste was observed by field-emission scanning electron microscopy (FE-SEM, Hitachi SU8020, Hitachi), which equipped with energy dispersive spectroscopy (EDS). The of cross section of joint sample was mechanically polished and then fine polished by an ion milling process (IM 4000, HITACHI) before observation.

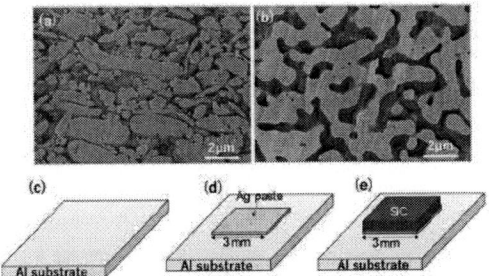

Fig.1 (a) the cross section of Ag flake particles, (b) the Ag porous structure sintering at 250°C for 5 min. (c), the Al substrate, (d)Ag paste printing and (e) SiC mounting process

III. RESULTS AND DISCUSSION

A. Cross section of SiC/Al joint

Fig. 2(a) shows the cross section of SiC/Al joint by Ag sinter paste, where Al substrate did not surface treatment. Sintered Ag layer shows a micron-porous structure without any large voids and interface cracks generation. The sintered Ag paste bonded well with the SiC and Al substrate via a strong interface necking growth as shown in Fig. 2 (b) and Fig. 2(c). By the EDS element analysis, the bonded interface just contains only Ag and Al metal element, not clear diffusion between the Ag and Al layers (Fig. 1(d-f)). The interface bonding mechanism of Ag and Al at low temperature have been discussed in our previous study [11], where the bonding interface between Ag and Al is mainly attributed to the Ag–O–Al bond, beginning from the self-generation of the Ag nanoparticles [12,13].

Fig.3(a) and Fig.3(c) show the cross section of SiC/Al joint structure with a weak and strong blasting processes treated Al, respectively. Sintered Ag layer also shows a micron-porous structure for the both case without any large voids and interface cracks generation. The magnified view of SiC/Al joint structure were shown in Fig.2(b) and Fig.2(d), respectively. It can be found that the sintered Ag paste bonded well with the Al even the Al have a rough surface after treatment.

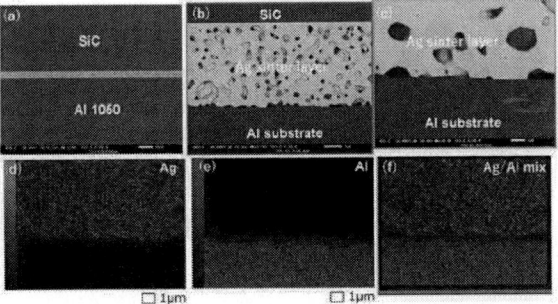

Fig.2 (a) The cross section of SiC/Al joint structure by Ag sinter paste, (b) the micron-sized sintered Ag direct on Al substrate, (c) the magnified view of the bonded interface between sintered Ag and Al substrate, (d-e) the EDS element mapping of Ag and Al, respectively, (f) the mix element mapping of Ag and Al.

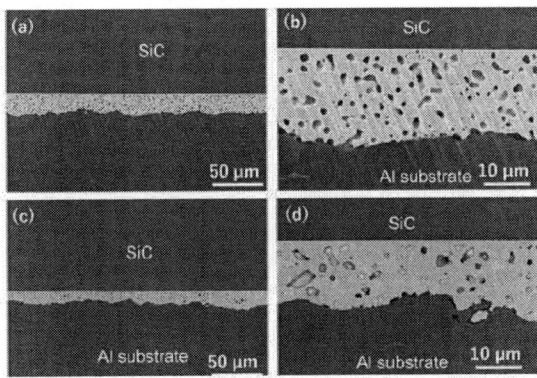

Fig.3 (a) the cross section of SiC/Al joint structure with weak blasting processes treated Al, (b) the magnified view of the bonded interface between sintered Ag and Al, (c) the SiC/Al joint structure with strong blasting processes treated Al and (d) the magnified view of interface between sintered Ag and Al substrate.

B. Fracture surface of SiC/Al joint structure

Fig. 4 (a) shows the fracture surface of SiC/Al joint structure by Ag sinter paste for the Al substrate without surface treatment after shear test. The fracture surface was observed on the Al substrate side, and the result suggested that the fracture occurred on sintered Ag necking interface nearly to the Al substrate as shown in Fig. 4 (b). The sintered Ag remains on the Al substrate with a ductile deformation. Fig. 4 (c) shows the fracture surface of SiC/Al joint structure by Ag sinter paste for the Al substrate with the weak blasting processes surface treatment. Similar with the Al substrate without surface treatment, the SiC/Al joint also fractured at the sintered Ag interface nearly to the Al substrate. The sintered Ag also remains on the Al substrate with a ductile deformation after shear test as shown in Fig. 4 (d). Fig. 4 (e) shows the fracture surface with the strong blasting processes treatment on Al surface. It was also found that some locations of Al substrate was broken with the strong blasting processes, that may be attributed to the large deformation of Al surface during the blasting processes. The sintered Ag still remains on the Al substrate with a ductile deformation even at the location where the Al surface was largely deformed as shown in Fig. 4 (f). These results imply that sintered Ag paste does not have much limitation on the surface roughness of the Al substrate for the direct bonding process with the Al.

C. Shear strength

Fig. 5 (a) shows the shear strength of SiC/Al joint structure by Ag sinter paste for the Al substrate with different surface treatment during shear test. The shear strength was achieved to 26.9 MPa for the directly bonding on Al substrate without any Al surface treatment. The shear strength was larger than that of the solder materials joint and also was able to compare to the Ag nano-paste bonding on an Ag metallized substrate [14, 15]. The shear strength was improved by the blasting surface treatment processes on Al substrate. The shear strength of SiC/Al joint structure separately increased to 29.1 MPa and 32.3 MPa in the case of the weak blasting surface treatment and the strong blasting surface treatment on Al substrates. Comparing with the Al substrate without surface treatment, the shear strength increased by 8.1% and 20.1%, respectively, for the weak and strong blasting surface treatment. The arithmetic mean roughness *Ra* and the maximum value of roughness *Rz* of the Al substrate without surface treatment are 0.49 μm and 1.98 μm, respectively. The values of *Ra* and *Rz* for the Al substrate with strong blasting surface treatment are 1.98 μm and 8.74 μm, respectively. The increased surface roughness may lead to an anchor effect for the interface bonding between sintered Ag and Al substrate [16] .

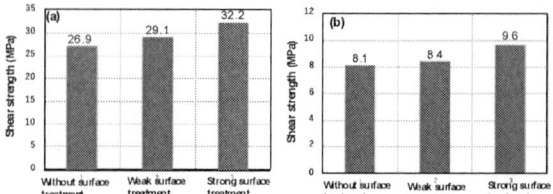

Fig.5 (a) the shear strength of SiC/Al joint structure by Ag sinter paste for the Al substrate with different surface treatment during shear test, (b) the shear strength after thermal shock test for 500 cycles.

IV. CONCLUSIONS

In this study, we investigated the direct bonding on an aluminum (Al) substrate by Ag sinter paste and propose an abrasive blasting processes to treat the Al surface to increase the initial bonding quality. The result show that the shear strength of SiC/Al joint structure by Ag sinter paste can be improved by treating the Al surface. The shear strength of joint structure without surface treatment increased from 26.9 MPa to 32.2 MPa for a strong blasting treatment on Al surface. The increased surface roughness may lead to an anchor effect for the interface bonding between sintered Ag and Al substrate In addition, the results of thermal shock test also show that the blasting treatment on Al surface can improve the shear strength and reliability of SiC/Al joint structure. This study provides a new method to increase bonding strength and structure reliability of Ag sinter paste direct on Al substrate .

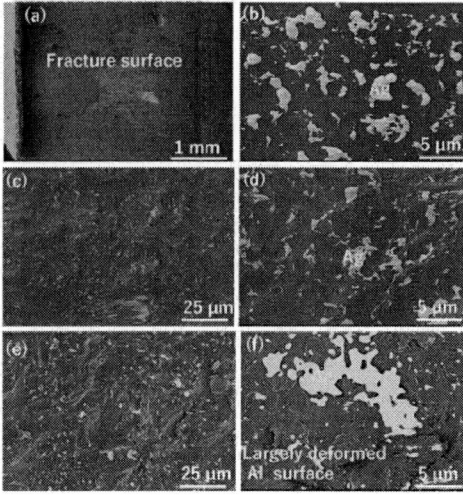

Fig. 4(a) The fracture surface of SiC/Al joint structure for the Al substrate without surface treatment, (b) the magnified view of fracture surface, (c) (e) the fracture surface of SiC/Al joint structure for the Al substrate with weak blasting treatment and strong blasting treatment, respectively, (d) (f) the magnified view of the fracture surface corresponding to the (c) and (e), respectively.

ACKNOWLEDGMENT

This work was supported in part by the New Energy and Industrial Technology Development Organization NEDO under Grant Number JPNP20004, and JSPS KAKENHI Grant Number JP22K04243. The author acknowledges both of financial support and material supply from DOWA Holdings Co., Ltd.

REFERENCES

[1] J. Millan, P. Godignon, X. Perpina, A. Perez-Tomas, and J. Rebollo, A survey of wide bandgap power semiconductor devices, IEEE Trans.Power Electron., vol. 29, no. 5, pp. 2155–2163, May 2014.

[2] L. Zhang, X. Yuan, X. Wu, C. Shi, J. Zhang, and Y. Zhang, "Performance evaluation of high-power SiCMOSFET modules in comparison to Si IGBT modules," IEEE Trans. Power Electron., vol. 34, no. 2, pp. 1181–1196, Feb. 2019 .

[3] T. Funaki, J. C. Balda, J. Junghans, A. S. Kashyap, and H. A. Manto, Power conversion with SiC devices at extremely high ambient temperatures," IEEE Trans. Power Electron., vol. 22, no. 4, pp. 1321–1329, Jul. 2007 .

[4] D. Kim et al., "Online Thermal Resistance and Reliability Characteristic Monitoring of Power Modules With Ag Sinter Joining and Pb, Pb-Free Solders During Power Cycling Test by SiC TEG Chip," in IEEE Transactions on Power Electronics, vol. 36, no. 5, pp. 4977-4990, May 2021.

[5] Z. Zhang and G. Q. Lu, "Pressure-assisted low-temperature sintering of silver paste as an alternative die-attach solution to solder reflow", IEEE Trans. Electron. Packag., vol. 25, no. 4, pp. 279-283, Oct. 2002.

[6] F. Le Henaff et al., "A preliminary study on the thermal and mechanical performances of sintered nano-scale silver die-attach technology depending on the substrate metallization", Microelectron. Rel., vol. 52, pp. 2321-2325, 2012.

D. Thermal shock test

Fig.5 (b) shows the shear strength of SiC/Al joint structure by Ag sinter paste for the Al substrate with different surface treatment after thermal shock test for 500 cycles. Comparing with the initial shear strength, the shear strength after thermals shock test decreased a lot for the both Al surface treatment and without surface treatment. The decrease in shear strength may be due to delamination at the interface between sintered Ag and Al substrate, and due to the appearance of cracks in the sintered Ag layer. On the other hand, it is found that the shear strength for the Al substrate with strong blasting treatment was higher than that Al substrate without surface treatment. The result indicated that the thermal shock reliability of SiC /Al joint structure could be improved by the blasting surface treatment process. We evaluated the Vickers hardness of the Al surface without blasting treatment and after the strong blasting treatment as 37.2 HV and 43.1 HV , respectively. The Vickers hardness increase after strong blasting treatment may be attribute to the decrease of grain size of Al . The Vickers hardness increase may prohibit the Al surface deformation, and thus influence and improve the structure reliability of SiC/Al direct bonding during thermal shock test.

[7] D. Kim et al., "Development of high-strength and superior thermal shock-resistant GaN/DBA die attach structure with Ag sinter joining by thick Ni metallization", Microelectron. Reliab., vol. 100–101, May 2019.

[8] C. Chen, Y. Gao, Z-Q. Liu and K. Suganuma, "3-D pyramid-shape Ag plating assisted interface connection growth of sinter micron-sized Ag paste", Scripta. Mater., vol. 179, pp. 36-39, 2020.

[9] B Zhang, C Chen, T Sekiguchi, Y Liu, C Li, K Suganuma, Development of anti-oxidation Ag salt paste for large-area (35× 35 mm2) Cu-Cu bonding with ultra-high bonding strength, Journal of Materials Science & Technology 113, 261-270, 2020.

[10] Y. -S. Tan, X. Li, X. Chen, G. -Q. Lu and Y. -H. Mei, "Low-Pressure-Assisted Large-Area (>800 mm2) Sintered-Silver Bonding for High-Power Electronic Packaging," in IEEE Transactions on Components, Packaging and Manufacturing Technology, vol. 8, no. 2, pp. 202-209, Feb. 2018.

[11] C Chen, D Kim, Z Zhang, N Wakasugi, Y Liu, MC Hsieh, S Zhao, A .Suetake, K. Suganuma, Interface-Mechanical and Thermal Characteristics of Ag Sinter Joining on Bare DBA Substrate During Aging, Thermal Shock and 1200 W/cm² Power Cycling Tests, IEEE Transactions on Power Electronics 37 (6), 6647-6659,2022.

[12] J Yeom, S Nagao, C Chen, T Sugahara, H Zhang, C Choe, CF Li, K Suganuma, Ag particles for sinter bonding: Flakes or spheres? Applied Physics Letters 114 (25), 253103.

[13] C Chen, Z Zhang, B Zhang, K Suganuma, Micron-sized Ag flake particles direct die bonding on electroless Ni–P-finished DBC substrate: low-temperature pressure-free sintering, bonding mechanism and high-temperature aging reliability, Journal of Materials Science: Materials in Electronics 31, 1247-1256.

[14] A. Sharif, C. L. Gan, and Z. Chen, "Transient liquid phase Ag-based solder technology for high-temperature packaging applications," J. Alloys Compounds, vol. 587, pp. 365–368, 2014.

[15] O. Mokhtari and H. Nishikawa, "The shear strength of transient liquid phase bonded Sn–Bi solder joint with added Cu particles," Adv. Powder Technol., vol. 27, no. 3, pp. 1000–1005, 2016.

[16] C Chen, Y Gao, ZQ Liu, K Suganuma, 3D pyramid-shape Ag plating assisted interface connection growth of sinter micron-sized Ag paste, Scripta Materialia 179, 36-39,2020.

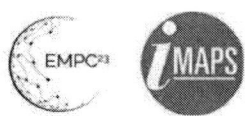

24th European Microelectronics Packaging Conference

Interconnect Stress Testing as a tool for assessment of reliability of modern PCB's.

Marek Koscielski, Krzysztof Gliński, Dariusz Ostaszewski, Jan Oklej, Tomasz Klej, Aneta Cholaj, Wojciech
Steplewski, Stefan Galinski, Janusz Borecki
Łukasiewicz Research Network - Tele and Radio Research Institute, Ratuszowa 11, 03-450 Warsaw, Poland
marek.koscielski@itr.lukasiewicz.gov.pl

Abstract (Word count: 327)

1. Background

The of miniaturization can be seen in electronics for some time already. The pinout of active components grows enabling them to have additional functions. The trend in usage of System in a Package (SiP) solutions is also evident thus routing of the active and passive elements became also challenging. To overcome this challenge boards with High Density Interconnection (HDI) where not only through hole, but also different kinds of microvias e.g. blind, buried, stacked, staggered via are used. The standard procedure to test reliability of boards and assemblies is very resource-hungry especially in time and money, as the tests have to be carried out in climatic chambers which lasts for weeks, followed by additional destructive testing like cross-sections. On the other hand IST (Interconnect Stress Testing) is a quick test, that enables to have assessment of the used technology especially the quality of the through holes of all types. Although the coupons are destroyed during the test, it is not necessary to test the manufactured circuits. This saves both materials and time. The technology optimizations and the key findings will be presented within the paper.

2. Method

The IST is a technique where a current is passed through specially designed PCB where "power" and "sensing" traces overlay. The passage of the current leads to heating of the board which is monitored through sensing side where resistance change is recorded. The described methodology is based on IPC test method where own device was built and software was written to enable testing. IST test coupons is a specialized tool that is assumed to act as "representatives" of printed circuit boards produced. Such coupons are placed in the production panels, the same on which the ultimately manufactured boards are located. This allows one to easily control the production process almost in real time. If any discrepancies are diagnosed on the test coupon, a given batch of products can be eliminated from further production or subjected to a more thorough than standard inspection to determine whether the product is still suitable for repair and restoration of its 100% functionality, or whether it must be definitively rejected. The proposed diagnostic technique includes implemented operational aspects already at the production stage, which allows for verification of the quality of printed circuit boards produced not only in terms of compliance with the design, but also in relation to the possible occurrence of hidden defects and material compatibility. Different sets of technological parameters were altered during the production of the PCBs. Exemplary PCB layout used for tests is presented in Figure 1.

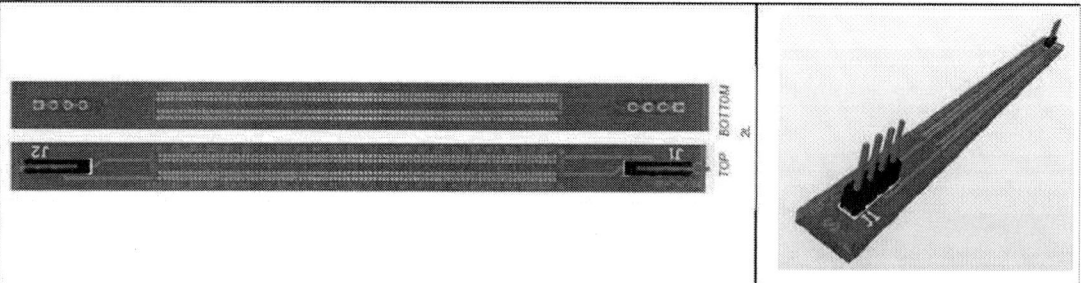

Figure 1. Top and bottom view of an exemplary test board (left), 3D model of the tested board (right).

IST works by taking the test sample through a quick thermal cycle between ambient temperature and 150°C followed by forced air cooling back to ambient temperature. As the coupon thermal cycles, changes in circuit resistance are monitored. An increase of 10% is considered as a failure and testing is aborted. So IST stops testing (stress) at the exact moment of failure. Testing temperatures can be raised to 260°C (for polyimde boards). The test is dependent on the coupon design, which reflects the board's attributes, including critical hole size, copper weight, number of layers, connection types, etc. The coupon is designed with two discrete circuits - POWER and SENSE. The power circuit is used to heat the coupon and check the integrity of the post. The sensing circuit is measured to monitor resistance changes in the PTH or PTV (through shell). The sensor circuit is not receiving significant power. The power circuit heats the coupon to 150°C using DC current. At the same time the resistance of the sense circuit is monitored. Typically, the test is runned until resistance in either circuit increases by greater than 10% or till a certain number of cycles.

These tests determine the physical resistance of representative samples of printed circuit boards to cyclic heating to high temperatures. Temperature changes are intended to create thermomechanical stresses on the sample. Generally, during the test, the sample is heated by a current flowing through it to bring the temperature of the copper to a certain value. Turning the electric current on and off at precisely defined intervals causes thermal cycles between the ambient temperature and the designated test temperature. The laminate and surrounding materials are heated to varying degrees depending on their thermal conductivity. Thermal cycling is designed to accelerate the detection of hidden anomalies. The number of cycles to failure allows for a quantitative assessment of the "technology quality" and predicts the durability of the electronic packages produced. It is expected that for consumer electronics devices, test packages should withstand at least 100 IST cycles. The IST test is also used to assess the quality of manufactured printed circuits intended for operation in space conditions. For some applications, there are requirements for strength of 1500 cycles and an allowable change in the electrical resistance of the sensory circuit not more than 5% of the output value.

The standard diagnostics of potential hidden defects, which mainly appear only during operation, requires the use of long-term, often months-long, fatigue tests. The proposed method of verifying the quality of printed circuit boards radically reduces this time to less than 2 days. Cyclically forced changes in the temperature of the test coupon, through thermal expansion of the construction materials used, cause cyclical changes in the geometric dimensions of the holes, and these in turn cause cyclical stretching and compression of the copper layer covering the walls of the holes. This phenomenon has the greatest impact along the axis of the holes, causing cyclic stretching and compression of the copper plated walls of the holes, due to CTE mismatch. The observed significant changes in the electrical resistance of interlayer connections may be caused by the low quality of metallization of the walls of the holes. This is particularly noticeable in the case of coupons with relatively small diameter holes, i.e. a large aspect ratio. This may indicate serious problems related to the penetration and exchange of metallising chemical baths inside holes with a large aspect ratio.

3. Results

Based on the literature data and the results of research work, the following design assumptions for the test boards were formulated:

1. Two separate POWER and SENSE circuits.

2. Comparable electrical resistance of circuits in the range - 0.3 ÷ 1.2 Ω.

3. Pins: two times four pins arranged in a row with a spacing of 2.54 mm for soldering GOLDPIN type connectors. The pins for the POWER and SENSE circuits can be located on opposite or one side of the test coupon.

4. Test coupon should contain current and signal traces with holes, in accordance with the production design:

- Through Holes,

- Blind Vias,

- Buried Vias.

5. The size of the holes - adequate to the diameter of the holes on the production boards.

6. Number of layers / thickness of the laminate - adequate to the construction of production boards.

7. Test coupon dimensions: The tester can test coupons up to 121 x 26 mm. In addition, it is important that the GOLDPIN type connectors are not mounted close to the edge of the test coupon.

In the test system, it is necessary to provide a small clearance (min. 1-2 mm) for linear elongations of the laminate related to its thermal expansion. During the test, cyclic changes in the temperature of the test coupon are caused in the range from room temperature, approx. 25 ° C, to 150 ° C, which cause its elongation.

8. Coupons should be designed in a program that allows saving in the IPC-2581 Rev. format. B Best or

ODB++ v8, e.g. Altium Designer.

During the work, software for automatic IST measurements was developed. Proprietary software allows for accurate measurements of each production batch, each panel separately. Instructions for carrying out tests on a test stand have also been prepared, thanks to which measurements can be performed easily by a trained employee. This allows for quick verification of the quality and reliability of the PCBs made.

One of the basic needs of diagnostics conducted with this method is to multiply the number of tests performed at one time. As part of the research work, an eight-channel stand for testing IST coupons was developed, Fig. 2-3.

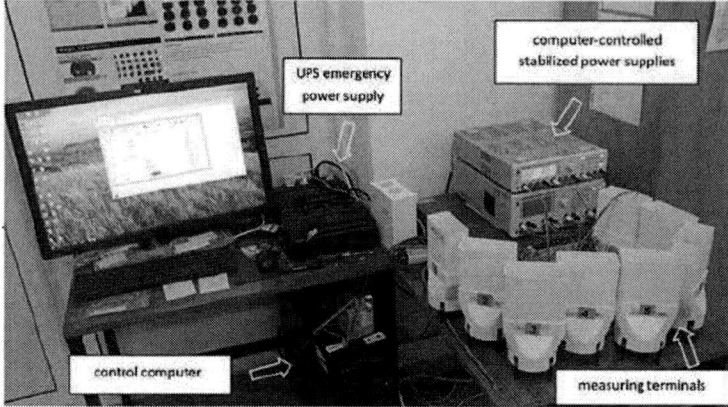

Fig.2. IST test stand.

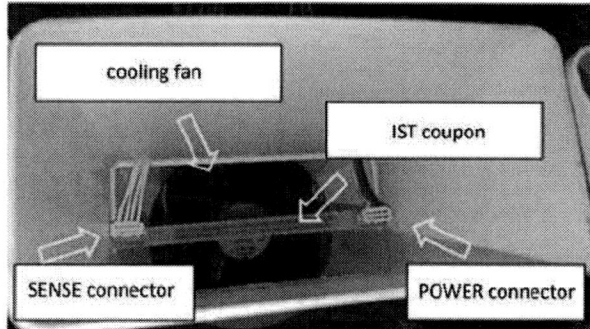

Fig. 3. IST test coupon placed inside the measuring terminal (top view).

The system can displays graphs of resistance activity in real time, showing individual thermal cycles and cumulative cycle data. Data analysis is simplified with automated graphical tools in each cycle and coupons. The data collection takes place in the set time period. The data collected during the test includes the number of cycles to failure or termination, the resistance of each circuit, the resistance at both high and low temperatures. Data evaluation can indicate failure mode and circuit integrity. The results are compared with baseline results for similar products. This gives you a quantification of how performance compares to industry, customer or internal requirements.

The diagrams in Fig. 5 and 6 show exemplary waveforms of changes in the resistance of the POWER circuit of the test coupuns (designations 13 and 14) as a function of the number of exposure cycles. As can be seen from the graph in Fig. 35, the permissible resistance change ($\pm$10% of the output resistance) was exceeded after a little more than 380 cycles. On the other hand, the graph in Fig. 36 shows the complete failure of the circuit after a few more than 270 cycles. Until the circuit was damaged, changes in resistance did not exceed 5÷6%.

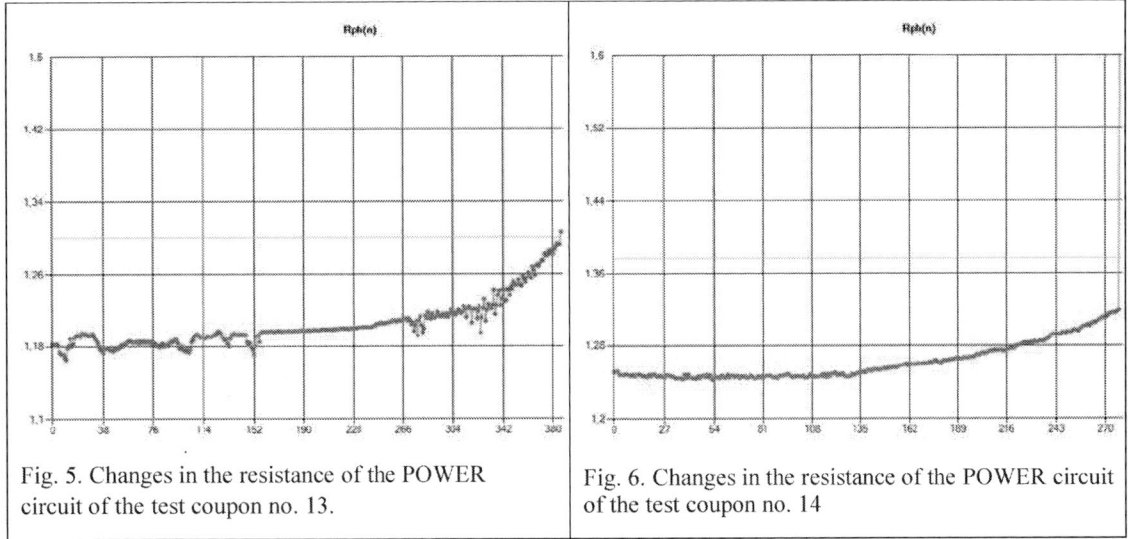

Fig. 5. Changes in the resistance of the POWER circuit of the test coupon no. 13.

Fig. 6. Changes in the resistance of the POWER circuit of the test coupon no. 14

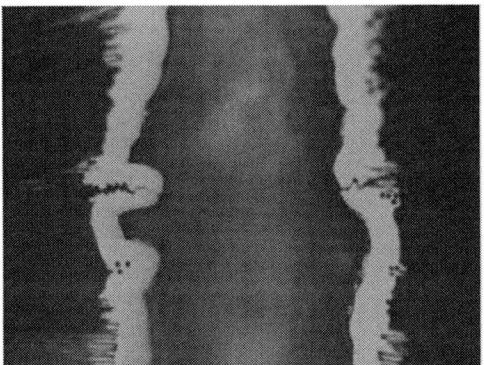

Fig. 7. Metallization damage in the through hole of coupon no. 13.

The failure of the metallization occurred as a result of a rapid increase in temperature.

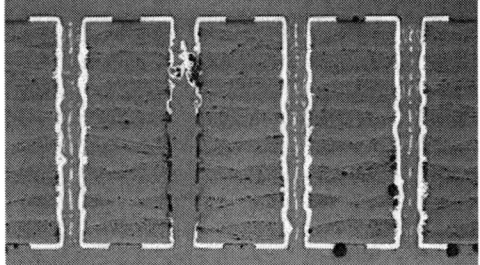

Fig. 8. Metallization crack in the through hole of coupon no. 14.

The failure of the metallization occurred as a result of barrel cracking. This is a typical defect observed for through holes, due to the higher thermal expansion of the system in the Z axis than in the XY axis.

4. Conclusion

The main reason why PCB manufacturers use IST is to save costs and time. IST costs less than full accelerated standard thermal tests while being more comprehensive. IST tests hundreds of holes and connections simultaneously, so statistically IST tests are more representative of PCB quality. The IST test ends before or after the PCB is fully damaged according to requirements. The combination of IST and thermography greatly improves the ability to find and assess failure causes. The use of thermal imaging to locate the damaged area allows for precise metallographic examination and quick determination of the cause of the failure. Other test methods have significant limitations. Analysis based solely on metallographic microsections is laborious and requires qualified employee preparation and subjective assessment. Standard thermal tests are more expensive and time consuming and do not distinguish whether a failure occurs for PTH or interlayer connections. IST is cost effective test method that enables swift analysis of the PCB reliability. This method is successfully introduced to the scope of research in the Laboratory. It allowed to quickly assess the quality of the production panels and optimize the production process. The outcomes helped to identify failure in early fabrication stages and lead to further optimization of the PCB manufacturing process.

5. References

[1] IPC TM 650 2.6.26 DC Current Induced Thermal Cycling Test.

[2] ECSS-Q-ST-70-60c (1june2018) Space product assurance - Qualification and procurement of printed circuit boards.

[3] Reliability Testing of PWB Plated Through Holes in Air-to-Air Thermal Cycling and Interconnect Stress Testing after Pb-free Reflow Preconditioning, Joe Smetana at. all, Apex Expo IPC, 2011

[4] Reliabilities and Failure Analysis of Printed Circuit Boards Interconnect Stress Test, Ying Yang, 2018 19th International Conference on Electronic Packaging Technology (ICEPT)

[5] Challenges in introducing high-density interconnect technology in printed circuit boards for space applications, Maarten Cauwe at. all, CEAS Space Journal, **15**, 101–112 (2023), Published 24 November 2021

High Speed Transmission Characteristics on Glass based interposers

Satoru Kuramochi
Dai Nippon Printing
Chiba, Japan
Kuramochi-S2@mail.dnp.co.jp

Masaya Tanaka
Dai Nippon Printing
Chiba, Japan
Tanaka-M32@mail.dnp.co.jp

Takahiro Tai
Dai Nippon Printing
Chiba, Japan
Tai-T@mail.dnp.co.jp

Abstract— **Interposers for system in package will became more and more important for advanced electronic systems. As the industry moves toward HPC (High Performance Computing) for huge data transmission with low power consumption. The major engineering requirements for HPC application are high-density, high speed data transmission, low loss, precision manufacturing and low cost. On the 2.5D heterogenous package structure, key technology is fine wiring connected with CPU chip and HBM chips. The fine wires needed high speed transmission.**

This paper presents the demonstration of Glass-based twin type interposers, Glass interposer with fine pitch metalized through via and RDL interposer with low loss dielectrics on Glass carrier. We compare twin types of Interposes fabrication process capability with panel size format 300x400mm. Finally, we compare the transmission characteristics using eye diagram for heterogenous integration application

Glass Interposer has low loss glass via, it is effective vertical interconnection. RDL interposer has low loss dielectric layers, excellent transmission characteristics obtained on fine pitch trace area.

Keywords— Glass Interposer RDL Interposer

I. INTRODUCTION

As electronic product becomes smaller and lighter with an increasing number of functions, the demand for high density and high integration becomes stronger. As the industry moves toward HPC (High Performance Computing) for huge data transmission with low power consumption. The requirements for PKG structure have become more challenging. The major engineering requirements for HPC application are high-density, high speed data transmission, low loss, precision manufacturing and low cost. On the 2.5D heterogenous package structure, key technology is fine wiring connected with CPU chip and HBM chips. The fine wires needed high speed transmission. On the other hand, Fanout panel level packaging (FOPLP) technology spread out 2.5D system integration. FOPLP has no core substrate like silicon TSV or glass TGV, it has advantage on manufacturing cost and panel scale productivity.

This paper presents first the demonstration of glass interposers with fine pitch metalized through via and low loss dielectric layers.

Firstly we demonstrate TGV process capability with panel size format 300x400mm. We model high frequency characteristics for high-speed transmission applications using basic high frequency characteristics.

Secondary we show process flow and demonstrate RDL interposer with low loss dielectrics layers. High speed transmission characteristics for RDL interposer modeled with eye diagram. Excellent transmission obtained on fine pitch trace area using low loss dielectrics.

II. GLASS BASED INTERPOSERS

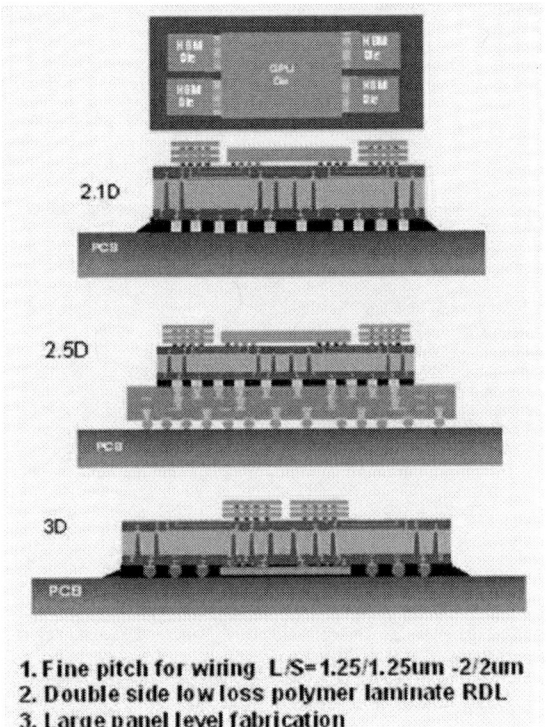

1. **Fine pitch for wiring L/S= 1.25/1.25um -2/2um**
2. **Double side low loss polymer laminate RDL**
3. **Large panel level fabrication**

Figure 1. Schematic structure of Glass based Interposer

III. GLASS INTERPOSERS

Figure 2 shows process flow of Glass interposer.

First of all, TGVs of 80µm in diameter and 200µm in pitch were formed on 400µm thick alkali-free glass. Ti/Cu seed layer deposition. Then the via was deposited with Cu by conformal electroplating. Thick dielectric polymer layer was laminated on the wafer as RDL passivation film. Redistribution lines were patterned with photo resist. Cu RDL line of 5 µm thickness was deposited by Cu electroplating followed by photo resist and Cu seed layer removal

209

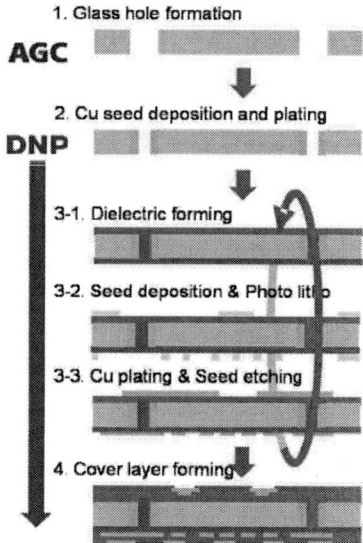

Figure 2　Cross sectional view of via formation

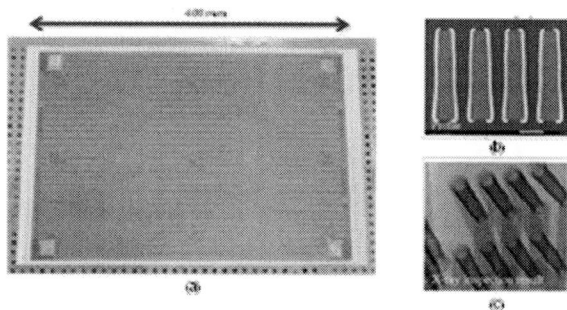

Figure 3　Cross sectional view of via formation

The conformal plating method has great advantage of process time of plating. After depositing the seed layer, plasma ashing was done from the both sides of the wafer to improve the hydrophilic property of the surface of Cu seed layer. Conformal TGV with 300x400mm panel demonstrated shown in Figure 3.

A. High frequency characteristics

High frequency electrical characteristics of interposers were evaluated with three type metalized method. We have selected a combination of a transmission wiring with coplanar waveguide structure (CPW) and TGV in order to study the influence of TGV on high frequency region. The measurement of TGV only itself is difficult because measurement TEG needs outer pad for probing. It is not ignored influence of the outer pad. We choice the method compared CPW and CPW+TGV show in Figure 4. The data are corrected from TEG which have Line and space are 30μm and 15μm for impedance 50ohm matching. Length of CPW is 10mm.

Double side metal layer test vehicle containing co-planer waveguides (CPW) with TGV transmission were designed. Electrical measurement was performed after SOLT

calibration and the CPW transmission characterized up to 40GHz. Figure 5(a) shows test vehicle containing co-planer waveguides (CPW) with filled types of metalized method. Figure 5(b) shows test vehicle containing co-planer waveguides (CPW) with conformal types of metalized method. The network analyzer S-parameter measurements indicate that CPW has lower loss than CPW+TGV in transmission shown in Figure 5.

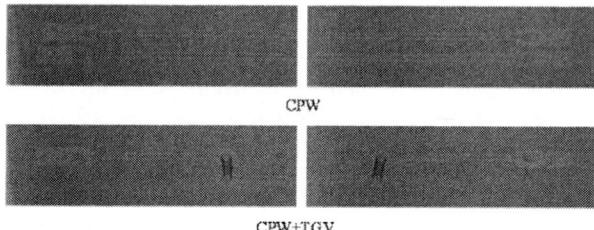

Figure 4　CPW transmission line TEG

Both types of metalized TGV have excellent transmission. Filled via has slightly lower loss than Conformal via in transmission shown in Figure 5.

It causes reflection on the joint with trace and TGV. Highly bulk resistance of TGV resulted in an insertion loss less than -0.5dB at 30GHz. It causes reflection on the joint with trace and TGV.

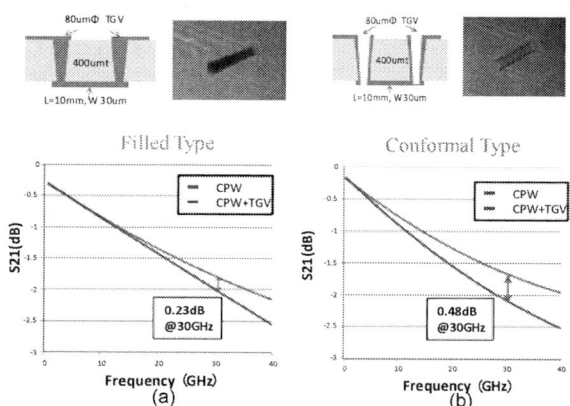

Figure 5　Co-planer waveguide test vehicle

The new types of partial filling plating method applied for measurement. Partial filling types have great advantage of fine pitch arrangement due to half diameter of conventional ones. [1] AGC developed high aspect small diameter TGV. Figure 6 shows test vehicle containing co-planer waveguides (CPW) with partial filling types of metalized method using small diameter TGV. Figure 7 shows measurement result CPW and CPW with TGV. The differential of S21 summarized in Table 1. Three types of TGV were all very low loss.

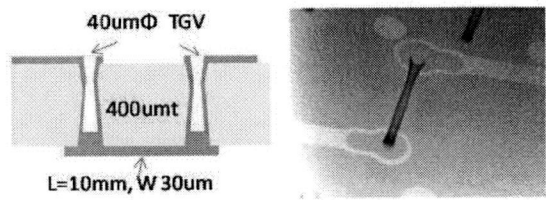

Figure 6 Co-planer waveguide test vehicle

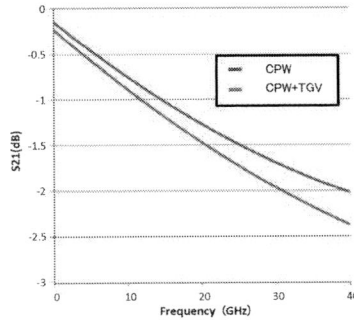

Figure 7   Measurement result of partial filling TGV

Table 1

Types of TGV plating method	Difference of S21
Filled type	0.23 @30GHz
Conformal type	0.48 @30GHz
Half fill type	0.34 @30GHz

Fused quartz has an amorphous structure which is composed of Si-O-Si network and no periodic atomic configuration. Therefore, the dielectric characteristics of fused quartz, especially its loss tangent, is lower than other typical materials. [2]

Quartz glass and conventional glass demonstrated TEG for high frequency measurement as core glass.

Table2

Core Glass		Er	Tanδ
AQ	Synthetic fused silica Glass	3.8	0.0002 @10GHz
EN-A1	Conventional Glass	5.8	0.006 @10GHz

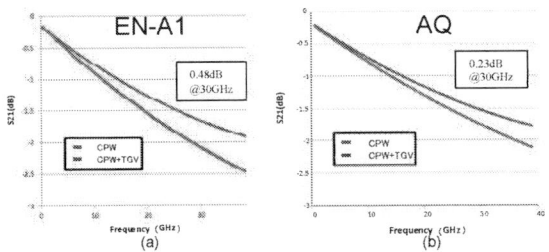

Figure 8(a) and 8(b) shows comparison of insertion loss.

Figure 8 Measurement result of AQ fused silica glass

AQ fused silica glass indicates lower less than ENA1 conventional glass.

Differential of S21 with TQV also smaller than TGV are shown in Table 3. AQ fused silica glass has lower dielectric tangent than Conventional glass. Extremely low dielectric tangent of Quartz glass resulted in an insertion loss of less than -0.23dB at 30GHz.

Table 3

Core Glass		Differential of S21(dB)
AQ	Synthetic fused silica Glass	0.23dB @30GHz
EN-A1	Conventional Glass	0.48dB @30GHz

Figure 9 shows modeling method of Eye diagram. Measurement result of insertion loss was converted to Eye diagram of transmission 10Gbbs input pulse signals.

Figure 9(a)(b) and 9(c)(d) shows comparison of Eye diagram. AQ fused silica glass indicates lower rise time less than ENA1 conventional glass. TGV(TQV) is very low influence for the rise time.

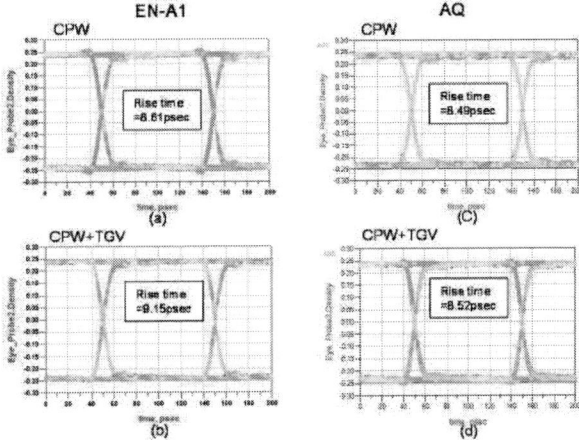

Figure 9 Modeling result of Eye diagram

3D 3D-TGV-IP

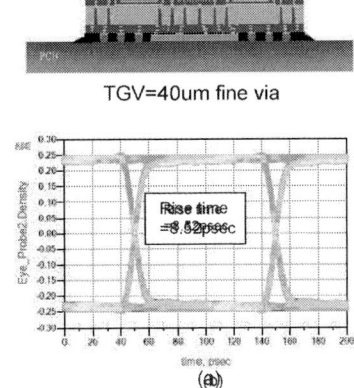

TGV=40um fine via

Figure 10 schematic structure of heterogenous 3D -TGV PKG

Figure 10 show high speed transmission characteristics for 3D-TGV PKG modeled with eye diagram. Excellent transmission obtained on fine pitch trace area.

IV. RDL INTERPOSERS

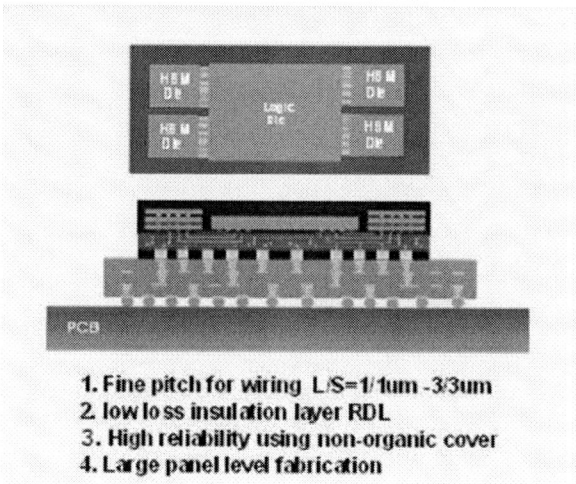

1. Fine pitch for wiring L/S=1/1um -3/3um
2. low loss insulation layer RDL
3. High reliability using non-organic cover
4. Large panel level fabrication

Figure 11 (a)Schematic structure and feature of RDL Interposer.

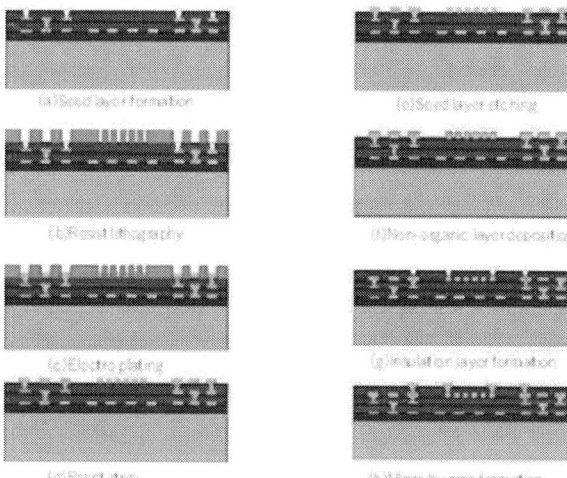

Figure 12 Process flow of RDL interposer

Figure 12 shows Process flow of RDL layers. At first, seed layer formation on lower steps on the insulation layer. (a) Secondary, Photo resist formation and open the wiring area with developer. Thirdly, electro-plating applied open area of photo resist. (c) After resist striped(d), seed layer metal was etched. (e) And then non-organic layer deposit on the Cu trace(f). It is original process for high reliability of the RDL layer. After insulation layer formation, Nonorganic layer was dry-etched through opening area of insulation layer. Finally micro bumps formation on the top layer. [3]
Figure13. shows fine wiring layers on 300x400mm Glass panel format. 2-μm-pitch Cu trace fabricated on the glass.

Figure14 show L/S=1/1um trace with protective layer. Non-organic layer enhance highly reliability on the narrow pitch under 4um [3].

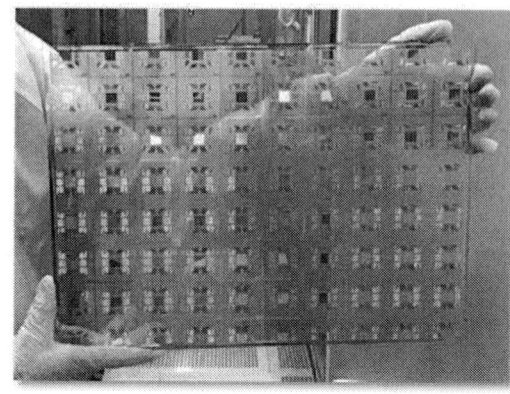

Figure 13. Fine wiring layers on 300x400mm Glass panel format.

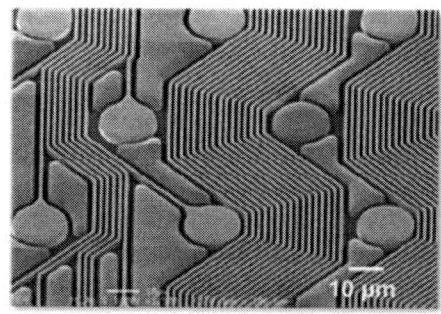

(a)

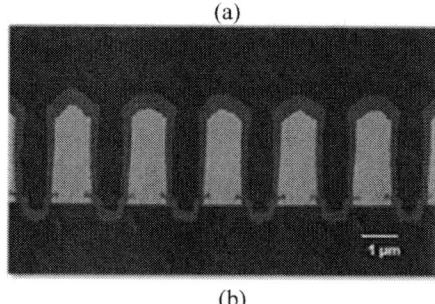

(b)

Figure 14(a) L/S=1/1um trace
(b)trace profile with protective layer

Insulation layer materials is important role of RDL properties. Insulation layer materials needed small size micro via, Low stress for reduction warping and low loss tangent for high-speed signal transmission.
Figure 15 Shows result of via diameter reproduction PI-A. Via shapes is taper forms, minimum diameter was top 5.5um and bottom 2.4um with H line LDI.
Figure 16. shows multi-layer stack of RDL. 5metal RDL layer include fine pitch layer demonstrated stacking on carrier glass. Fine wiring layer fabricated on flat intermediate RDL surface.

212

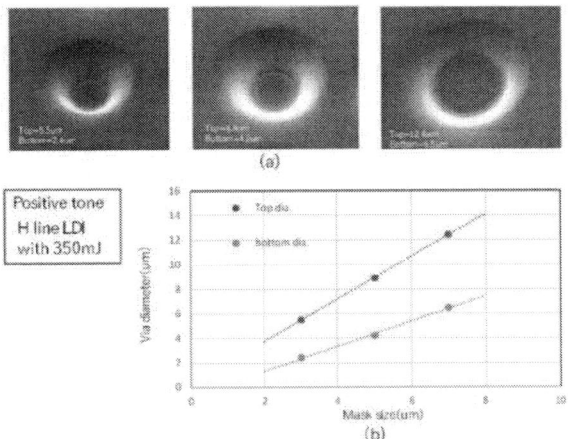

Figure 15. Via diameter of insulation materials PID

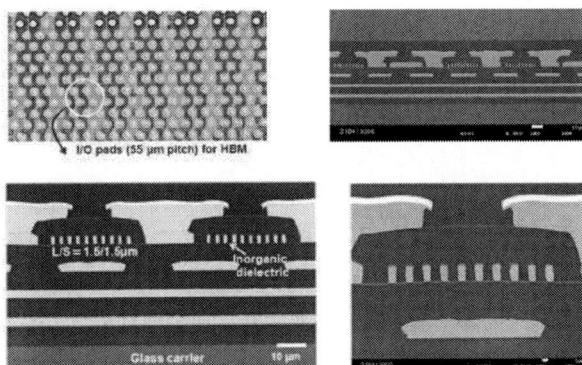

Figure 16. Profile of 5metal layer stack

We demonstrate CPW transmission line with fine wiring layers. TEG fabricated on core glass to measure high frequency characteristics. We calculated insertion loss with fine line. Figure 24 shows real measurement result of fine wiring consist of CPW structure. Insertion loss has large value with fine wiring pitch due to the conductive loss will be large.[4]

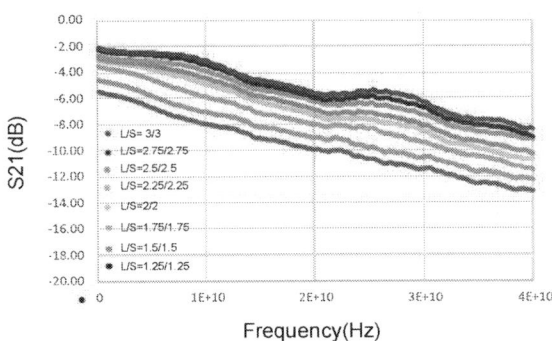

Figure 17 Measurement result of fine wiring consist with CPW structure.

Figure 18 shows simulation result of analysis on 10Gbps. CPW 10mm length wiring with dielectric polymer calculated for conversion Eye diagrams.   Wiring width is small, eye opening is small with wiring width.

Eye diagram of L/S=0.5/0.5um show signal delay is large with rising time. It indicates that typical BEOL damacine process same factor as L/S=0.5/0.5um. 10Gbps will next standard after HBM3. In the next generation, It needed more critical signal integrity.

10Gbps

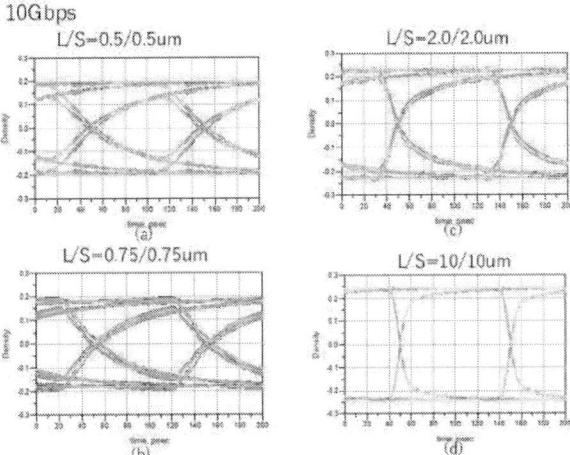

Figure 19. Simulation result of CPW on RDL

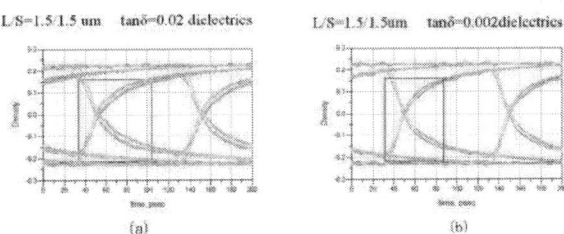

Figure 20. Eye diagram analysis with insulation layer

Figure 20(a) show the ED (eye diagram) of 10Gbps, L/S=1.5/1.5um CPW with tanδ=0.02 dielectrics.

 Figure 20(b) show the ED (eye diagram) of 10Gbps, L/S=1.5/1.5um CPW with tanδ=0.002 dielectrics.

Table 4 summarized result of low loss tangent of dielectric layers. Rising time of the eye diagram using low loss dielectrics 30% smaller than other.

Table 4

Model	Eye height(V)	Rising time(psec)
CPW tanδ=0.002 dielectrics	0.30	25.7
CPW tanδ=0.02 dielectrics	0.27	33.8

It indicates that trace width will be optimized as wide as possible to route between the chips with low loss dielectric insulator.

The specification opened in UCIe 1.0 is expected to provide standardization of chiplet-based communication for advanced packages. In the specification of advanced packaging, the required processing speed is ranged from 4 to 32 Gbps/line, which corresponds to the bandwidth density of from 1.0 to 10.5 Tbs/mm. On the basis of the specification of the UCIe1.0, we set signal line length between chiplets to 1.0 mm in the I/O sections (1.0 mm x 2) and 0.4 mm in the middle section, so the total length was 2.4 mm (Fig. 21). This is 20% longer that of the required signal line length/channel length between chiplets (2.0 mm) in the specification of UCIe 1.0.

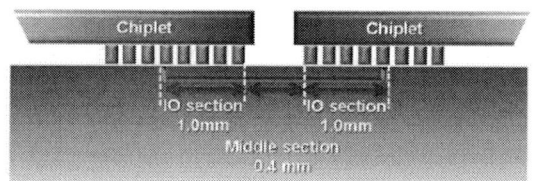

Figure21   Length of signal line consisting of both IO sections and middle section for specification of UCIe 1.0.

Cross-sectional dimensions used in the simulation for the topologies of GSG, SSG and SSS are shown Figure. 21 (a), (b) and (c).

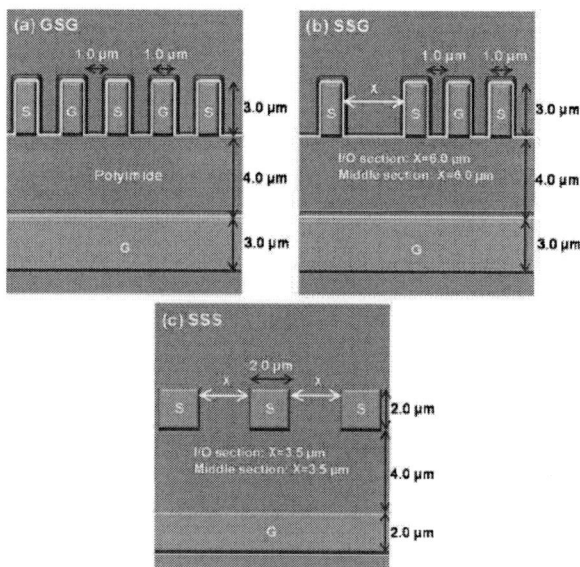

Figure 22   Cross-sectional dimension of signal line for topologies of GSG (a), SSG (b) and SSS (C).

The wave form for the topology of SSG did not cross the keep-out area, indicating a shield effect created by the ground line neighboring the signal line reduced the cross-talk (Figure.23 (b)). The wave form for the topology of GSG was away from the keep-out area enough (Fig.23 (a)). Both side ground lines neighboring the signal line greatly removed electromagnetic force generated around the signal lines. Both topologies of GSG and SSG met the specification of UCIe 1.0 at 32 Gbps.

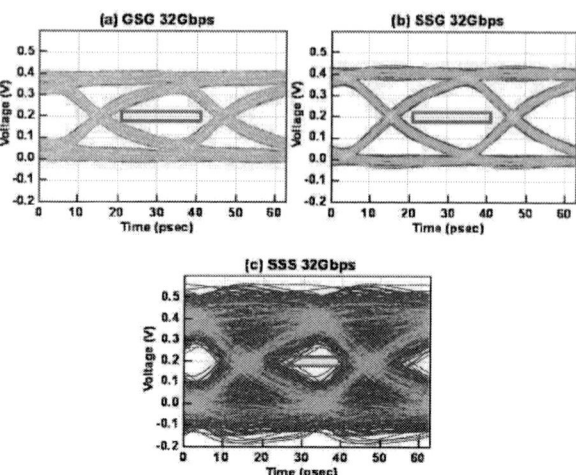

Figure23 Simulated eye diagram of topology GSG (a), SSG (b) and SSS (c) at 32 Gbps in specification of UCIe 1.0.

V.   SUMMARY

Interposers with TGV interconnects was demonstrated with fine pitch through via. High speed transmission characteristics for 3D-TGV interposer modeled with eye diagram. Excellent transmission obtained on fine pitch trace area.

Fine wiring L/S=1/1um demonstrated using Semi-Additive-Process on the 300x400 panel format. Excellent uniformity of resistance obtained for HPC application.

Modeling transmission structure, Excellent transmission characteristics obtain fine pitch area using low loss dielectrics.

ACKNOWLEDGMENT

The authors would like to thank Asahi glass corporate for supplied glass substrate with TGVs

REFERENCES

[1]   Hiroshi Kudo et al., "A chracterized redistribution layer atchitecture for Advanced Packaging technology" Proceeding of 66th Electronic Components and Technology Conference, 2021, pp. 1829-1833

[2]   Yoichiro Sato, Nobutaka Kidera., "Demonstrationof 28GHz Band Pass Filter toward 5G using Ultra Low Loss and High Accuracy Through Quartz Vias," Proceedings 68rd Electronic Components and Technology Conference, 2018, pp. 2237-2241

[3]   Hiroshi Kudo et al., "Panel-Based Large scale interposer Fabricatied using 2-um pitch semi additive process for chiplet-based integration" Proceeding of 67th Electronic Components and Technology Conference, 2022, pp. 836-844

[4]   Satoru Kuramochi and Masaya Tanaka., "High speed Data Transmission characteristics on Glass Based Interposer" Proceedings of SMTA International Nov.15-Dec. 31. 2021, pp. 49-pp.57

Design, Fabrication, and Characterization of a 4H-SiC CMOS Readout Circuit for Monolithic Integration with SiC Sensors

Romina Sattari
Department of Microelectronics
Delft University of Technology
Delft, The Netherlands
R.Sattari@tudelft.nl

Henk van Zeijl
Department of Microelectronics
Delft University of Technology
Delft, The Netherlands
h.w.vanzeijl@tudelft.nl

Guoqi Zhang
Department of Microelectronics
Delft University of Technology
Delft, The Netherlands
g.q.zhang@tudelft.nl

Abstract— **This paper reports the design and fabrication of a 4H-SiC CMOS readout circuit enabling monolithic integration of silicon carbide (SiC) sensors and circuits. Compared to conventional Si electronics, 4H-SiC integrated circuits can sustain operation in harsh conditions such as higher temperatures and radiation levels. The proposed amplifier performance is well balanced through the temperature range of 25 ℃ to 400 ℃. Compared to state-of-the-art, the proposed SiC readout circuit does not include any off-chip components. The amplifier is fully differential, and hence shows improved common-mode rejection and signal-to-noise ratio (SNR). It can be monolithically integrated with SiC sensors in a scalable SiC technology.**

Keywords—SiC technology, monolithic integration, readout circuit, off-chip component, fully differential, common-mode rejection, signal-to-noise ratio (SNR).

I. INTRODUCTION

SiC CMOS (Silicon Carbide Complementary Metal-Oxide-Semiconductor) circuits are a type of integrated circuit technology that utilizes silicon carbide as the semiconductor material. SiC is a wide-bandgap semiconductor material with several advantages over traditional silicon-based technologies, including higher breakdown voltage, faster switching speed, and higher temperature tolerance.

SiC CMOS circuits are designed using a complementary metal-oxide-semiconductor (CMOS) process, which is a widely used technology for fabricating integrated circuits. CMOS technology uses both p-type and n-type transistors, which allows for low power consumption and high speed operation. The use of SiC as the semiconductor material in CMOS circuits allows for higher operating frequencies and temperatures compared to traditional silicon-based technologies.

Recently, the importance of high-temperature electronics and sensors is expanding for harsh environment applications such as aerospace and automotive, where reliability and durability are critical factors. Additionally, SiC CMOS circuits are being developed for use in emerging applications such as 5G wireless communication, Internet of Things (IoT) devices, and renewable energy systems. SiC is one of the available wide bandgap materials explicitly used for these purposes. It can tolerate extremely high temperatures up to 600 °C and above, while silicon CMOS, the most common semiconductor electronics platform, is generally limited to operations below 200 °C because of junction leakage.

In [1], the recessed channel SiC CMOS process is proposed to implement high-temperature inverters and ring oscillators. This process does not need ion implantation. However, the PMOS transistors indicated inadequate performance in this study. Furthermore, the transistor-transistor logic (TTL)-based process design kit (PDK) has been established recently for SiC bipolar junction transistor (BJT) process [2]. SiC TTL circuits can be employed to develop logic gates. However, one of the main challenges in SiC technology is providing analog readout circuits to be used in harsh environments monolithically integrated with SiC sensors.

This study presents a novel 4H-SiC fully differential readout circuit to address the constraints mentioned above. The proposed circuitry can be integrated with multi-functional sensors for harsh environments applications.

II. DESIGN AND METHODOLOGY

A. Design Process

The proposed amplifier is designed using the process design kit (PDK) provided by Fraunhofer IISB. The SiC wafer fabrication is performed at IISB, a schematic cross section of the NMOS/PMOS is given in Fig. 1. This manufacturing route has been proven successfully in earlier work [3]. To start the design, single transistors were simulated. Fig. 2 shows that the measurement results from the processed wafer coincide with the simulated data based on the FhG PDK. The results show much lower driving capability, especially in PMOS transistors, compared to mature silicon technology. Therefore, to reach a robust output signal headroom and high output impedance, a fully differential folded cascode structure is proposed. Furthermore, common mode feedback (CMFB) is employed to properly adjust the output and input operating points and guarantee the stability and performance of the amplifier. The layout of the circuit is illustrated in Fig. 7.

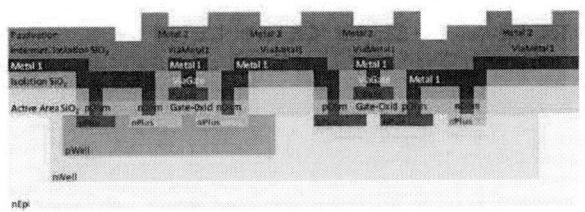

Fig. 1. 4H-SiC CMOS technology cross section.

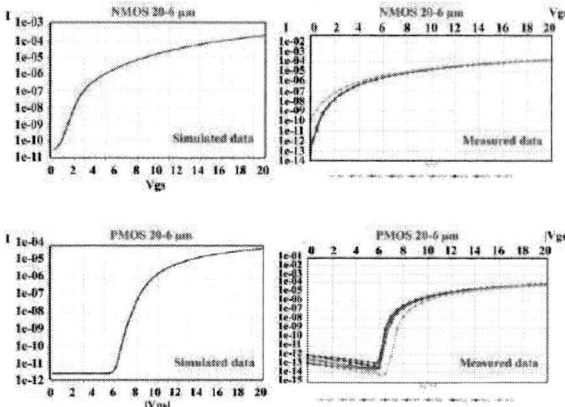

Fig. 2. 4H-SiC NMOS/PMOS characterization.

B. Fabrication and Manufacturing

The SiC wafer fabrication is performed at IISB, where the process is carried out using their in-house SiC CMOS technology. This process involves a series of steps, including substrate preparation, epitaxial growth, ion implantation, metallization, and annealing. The entire process is performed in a clean room environment to ensure high product quality and minimize contamination.

During the fabrication process, special metallization and solicitation techniques are employed to ensure stable and robust n-type and p-type ohmic contacts. Ohmic contacts are the electrical connections between the metal electrodes and the doped SiC semiconductor material, which is critical for the device's electrical performance. The stability and durability of the ohmic contacts are essential to ensure reliable circuit performance. Therefore, the development of new techniques to ensure stable and robust ohmic contacts is an essential aspect of the SiC wafer fabrication process.

The successful verification of the IISB process in previous fabrication runs and the employment of the same procedure in the current fabrication run confirm the process's effectiveness. Furthermore, the use of novel metallization and solicitation techniques highlights the commitment to innovation and continuous improvement in the SiC wafer fabrication process. The resulting SiC wafers are of high quality and are suitable for the production of SiC-based devices with reliable and consistent performance.

III. RESULTS AND DISCUSSION

This section describes the simulation and verification of a proposed amplifier circuit that is designed for high-temperature applications using SiC technology. The simulation results showed that the amplifier circuit has an open-loop gain of 51 dB and a closed-loop gain of 20 dB, with a bandwidth of 10 kHz and a phase margin of 61°. The closed-loop and open-loop gain results are reported in Fig. 3 and Fig. 4, respectively. These parameters ensure proper SiC sensor readout. Fig. 5 and Fig. 6 show the transient responses of the circuit, confirming the amplification gain of 20 dB.

The circuit's performance was found to be well balanced over a temperature range of 25 °C to 400 °C, which is crucial for high-temperature applications. The amplifier's layout is shown in Fig. 7, and a processed 4H-SiC wafer is illustrated in Fig. 8.

Compared to traditional silicon-based readouts, this SiC-based amplifier supports high-temperature applications, which is a significant improvement. It also eliminates the need for off-chip components required by previously proposed SiC readouts, such as relaxation oscillator-based readouts. Additionally, the amplifier is fully differential, which improves common-mode rejection and signal-to-noise ratio (SNR). Furthermore, the output signal is a differential voltage, which eliminates the need for frequency to voltage translation and any complex interfaces for detection, which is beneficial for high-speed and reliable detection.

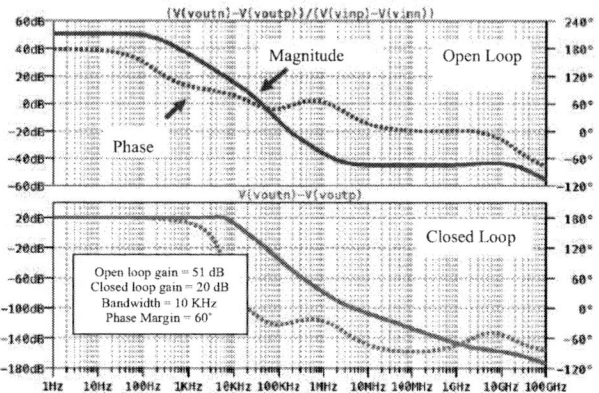

Fig. 3. The simulated AC frequency response.

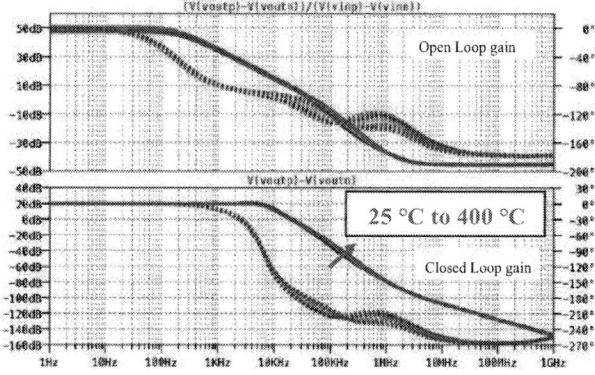

Fig. 4. Performance at high temperatures.

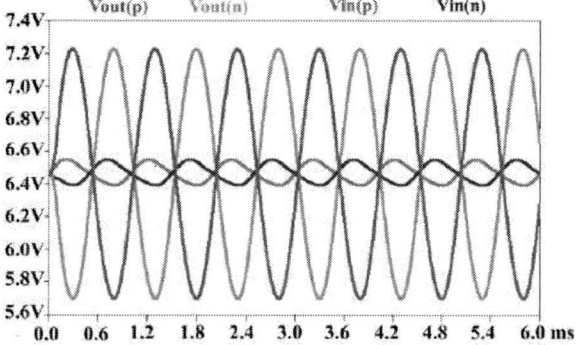

Fig. 5. Transient response simulated at 1KHz frequency.

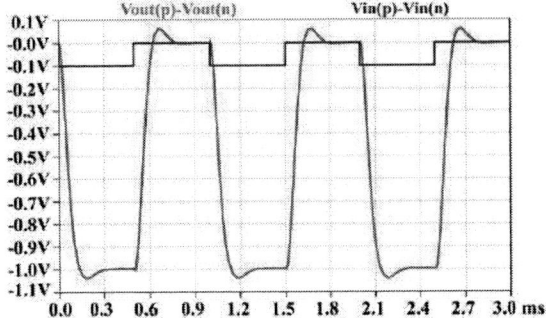

Fig. 6. Transient step response at 1KHz frequency.

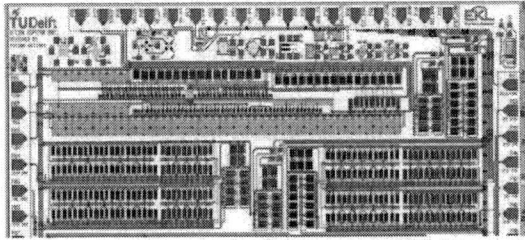

Fig. 7. The layout of the readout circuit.

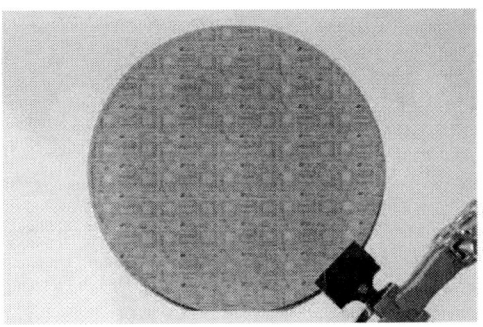

Fig. 8. The processed 6-inch 4H-SiC multi-project wafer.

IV. CONCLUSION

Overall, the proposed SiC amplifier circuit shows promising results for high-temperature applications and has several advantages over traditional silicon-based technologies and previously proposed SiC readouts. The SiC wafer fabrication is performed at Fraunhofer IISB, where the process is carried out using their in-house SiC CMOS technology. The open-loop gain of 51 dB, and closed-loop gain of 20 dB ensure reliable performance of the circuit. The results have been simulated and verified in a temperature range of 25 °C to 400 °C, using the PDK provided by IISB. Additionally, the amplifier is fully differential, which improves common-mode rejection and signal-to-noise ratio (SNR). Furthermore, the output signal is a differential voltage, which eliminates the need for frequency to voltage translation and any complex interfaces for detection, which is beneficial for high-speed and reliable detection. The resulting SiC wafers are of high quality and are suitable for the production of SiC-based devices with reliable and consistent performance.

ACKNOWLEDGMENT

The author would like to thank Tobias Erlbacher and Alexander May from Fraunhofer IISB for their invaluable contribution and efforts in process fabrication of 4H-SiC multi-project wafers.

REFERENCES

[1] Ekström, Mattias & Malm, Bengt & Zetterling, Carl-Mikael. (2019). High Temperature Recessed Channel SiC CMOS Inverters and Ring Oscillators. IEEE Electron Device Letters. PP. 1-1. 10.1109/LED.2019.2903184.

[2] Shakir, M., et al. (2019). "Towards Silicon Carbide VLSI Circuits for Extreme Environment Applications." Electronics 8(5): 496.

[3] J. Romijn et al., "Integrated Digital and Analog Circuit Blocks in a Scalable Silicon Carbide CMOS Technology," in IEEE Transactions on Electron Devices, vol. 69, no. 1, pp. 4-10, Jan. 2022, doi: 10.1109/TED.2021.3125279.

[4] Middelburg, Luke. (2020), From Silicon toward Silicon Carbide Smart Integrated Sensors.

Reliability of Copper Sintered Interconnects under Extreme Thermal Shock Conditions

Sri Krishna Bhogaraju[1]*, Francesco Ugolini[2], Federico Belponer[2], Alessio Greci[2] and Gordon Elger[1]

[1]*Institute of Innovative Mobility, Technische Hochschule Ingolstadt, Esplanade 10, 85049, Ingolstadt, Germany

[2]AMX Automatrix srl, Località Bolina, 1D/1E, 25085 Gavardo BS, Italy

*Email: srikrishna.bhogaraju@thi.de Tel: +49-841-9348-6449/6416

Abstract— Copper flakes-based sinter paste showed high reliability under thermal shock conditions. A dense and homogeneous interconnect was realized while sintering for 5 min at 275°C under 15 MPa bonding pressure, with the flakes showing the unique behavior of stacking over each other to realize a dense and homogeneous interconnect. Porosity under 10% could be achieved. The paste could be sintered under a constant nitrogen flow and did not need a fully inert atmosphere during sintering. Under high stress thermal shock cycling of +175/-65°C, the sintered interconnects showed a drop of 25% in the shear strength values after 1000 TST, consistent with observations made with Ag sintering. Thermal conductivity more than 200 W/mK and electrical conductivity of 20 Ms/m is achieved using the Cu sinter paste.

Keywords — sintering; copper flakes; reliability; SiC, die-attach

I. INTRODUCTION

New power electronics applications, mainly driven by the rapid pace of electrification in the automobile industry has necessitated the need for novel, sustainable materials with improved reliability [1], [2]. Conventional lead free-based solders have a low melting point (220-230 °C), thereby limiting the reliable temperature use to below 150 °C.

Among the alternatives, solid state sintering has shown a high rate of adaptability by the industry. Ag sintering under pressure is industrialized and has proven to be a reliable interconnect material [3]–[6]. But, Ag is expensive and prone to electromigration [7]. Cu offers the next best alternative to Ag and has been the focus of research and development.

However, Cu is prone to oxidation. Organic capping agents, sintering under formic acid enriched nitrogen/forming gas or pure H_2, or the use of a binder with the in-situ capability to reduce copper oxides are some of the common approaches to tackle the issue of oxidation. For Cu sintering to be easily adopted by the industry, it is essential to design a paste capable of operating in the similar process window as is established for Ag sintering in the industry today. Therefore, for the overall cost of ownership to be attractive, not only should the material cost be low, the handling and operating costs of the Cu sinter material should also be comparable, if not better than commercial Ag sinter materials.

Ag sintering under pressure is an industrialized process in the power electronics packaging industry with numerous publications on the reliability of Ag sintered interconnects for die-attach applications. Flakes based Ag sinter pastes have been subject to harsh thermal cycling of -50/+250 °C with a dwell time of 30min. While initially sintered interconnects showed a high shear strength of 45 MPa, it degraded drastically to 11 MPa after 750 cycles. The degradation is attributed to a combination of degradation of the sintered microstructure and the delamination between the die and the sputtering layer [8]. Typically, a crack propagates from the edges to the center of the die. However, recent publications in the field of Ag sintering, comparing nanoparticle based Ag sinter paste and bimodal (nano + microparticle) Ag sinter pastes have shown that delamination in the bimodal paste was faster than in case of the nanoparticle based paste when performing TST at -40/+150°C. Two different failure mechanisms were also reported, namely, edge crack mode and void growth mode [9]. Recent studies on Cu nanoparticle sintering have studied the effect of oxidation on the strength of the sintered interconnect after thermal ageing in air [10]. It is reported that the oxidation initially leads to the increase in the density of the sintered microstructure as well as the shear strength by filling the pores, eventually however leading to a drop in shear strength due to coarsening by the growth of copper oxides around the sintered nanoparticles. Grain boundaries are observed to act as nucleating sites for oxide, leading to formation of nanovoids. However, by achieving a dense and homogeneously sintered microstructure (~10% porosity), the oxidation is reported to be significantly limited [10]. Therefore, working with flake type particles offers a substantial advantage in realizing a dense sintered microstructure, which is discussed later in the paper.

II. MATERIALS AND METHODS

A. Sinter paste & characterization

Microscale copper particles-based paste, commercially available as Cuprum 81 from CuNex GmbH, capable of sintering at 275°C under constant nitrogen flow was evaluated. The unique chemistry of the paste allowed for sintering under N_2 flux, eliminating the need for a vacuum step prior to introducing N_2 into the sintering chamber. Therefore, an inert atmosphere during sintering was not necessary. Non-functional SiC devices with Ag end metallization and a footprint of 24mm² were used as test devices in the study. These were sintered onto Si_3N_4 based AMB substrates, also with Ag end metallization. A two-step sintering process is designed. The paste was printed using a 75 µm stencil on a Uniprint PBT Go3v semi-automatic stencil printer using a squeeze pressure of 20 N, printing speed of 13mm/s and a stencil separation speed of 2.3mm/s. A double stroke printing was performed. After printing, the paste is pre-dried in air at 100°C for 5 min. Finally, sintering was carried out by the application of 15 MPa bonding pressure at 275 °C for 5 min on a AMX P100 sinter press at the application lab of AMX Automatrix in Gavardo (BS), Italy (Fig. 1).

Cross sections were prepared with a JEOL SM-09010 Ar-Ion beam miller. Optical microscopy imaging was performed on a Keyence Digital Microscope. Profilometery is performed using a Nanofocus profilometer. Scanning acoustic microscopy (SAM) imaging was performed on a Nordscan Sonoscan D9600 with 50MHz transducer. Shear tests were performed on a XYZ-Condor Sigma Lite shear tester under a shear speed of 250 μm/s and a shear height of 20 μm. Scanning Electron Microscopy (SEM) analysis was performed using a Zeiss Auriga 40 Crossbeam microscope. The thermal conductivity of the sintered samples was measured by In-Plane Thermal Material Analyzer (LaTIMA) and the electrical conductivity was measured using the electrical conductivity analyzer (ECLA). Both LaTIMA and ECLA measurements were performed by Berliner Nanotest und Design GmbH [11]. Profilometer measurements of the printed samples was completed using the Nanofocus profilometer. Thermal shock test (TST) was performed on a Thermotec shock chamber with a profile +175/-65°C, air-air with a dwell time of 30 min.

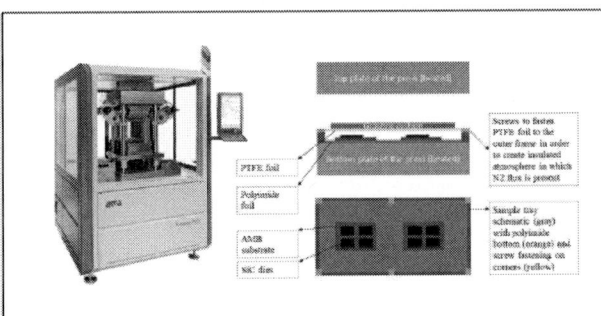

Fig. 1 – AMX P100 sinter press and the schematic of the assembly of samples prior to sintering.

III. RESULTS AND DISCUSSIONS

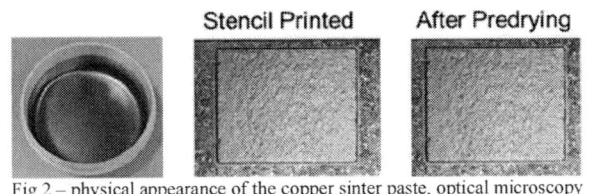

Fig 2 – physical appearance of the copper sinter paste, optical microscopy image after printing & after pre-drying in air for 5min at 100°C.

The paste shows very good form adherence and printability as shown in Fig 3. No bleed out is observed. Hot-die placement as is the case with many Ag sinter pastes was not required. Die placement could be performed at room temperature with a placement force of 0.1N. The paste retained its tackiness even

after pre-drying. Therefore, the SiC die could be placed without the need for any tacking agent to hold it in place.

The freshly printed paste had a measured height of 64 ± 3 μm. After pre-drying, the height of the printed paste was 51 ± 3. The measurements were averaged across 16 samples. After sintering, the bondline thickness (BLT) measured on the cross-sectioned samples across 4 different dies and 10 positions for each die (40 positions in total), averaged to be 12 ± 2 μm.

In conclusion, the paste showed a reduction of ~20% in print height after pre-drying and a reduction of 82% from the original printed paste to the final sintered BLT. This is attributed to the stacking nature of the flakes. As seen in Fig 3, a clear indication of this phenomenon is observed under low bonding pressure sintering (5 MPa), indicating the flake type morphology and the ability of the flakes to orient parallel to the interfaces. With the application of a higher bonding force, a dense and homogeneous interconnect with uniformly distributed porosity is obtained as seen in Fig. 4.

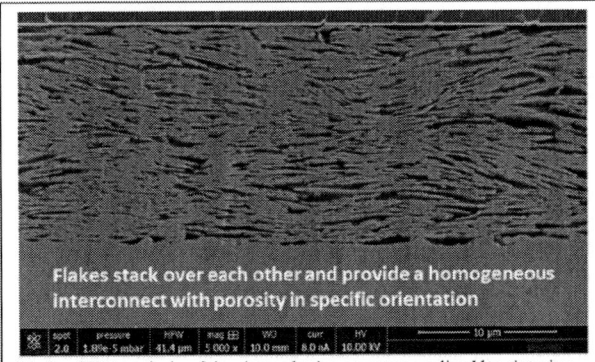

Fig 3 – SEM analysis of the sintered microstructure realized by sintering with 5 MPa bonding pressure at 275°C for 5 min, showing the unique capability of the flakes to stack over each other.

To understand the thermal and electrical properties of the sintered microstructure, stripes measuring 5mm x 50 mm, were manually stencil printed onto stainless tell substrates using a 150 μm stencil. These stripes were sintered at 275°C for 5 min, resulting in a dense sintered microstructure (Fig. 5). Three sample stripes were measured by LaTIMA and ECLA. The average width and thickness of the samples are measured to be 5.23 ± 0.15 and 50 ± 4 μm respectively, resulting in a thermal conductivity of 205 ± 27 W/mK and an electrical conductivity of 20.45 ± 1.76 MS/m.

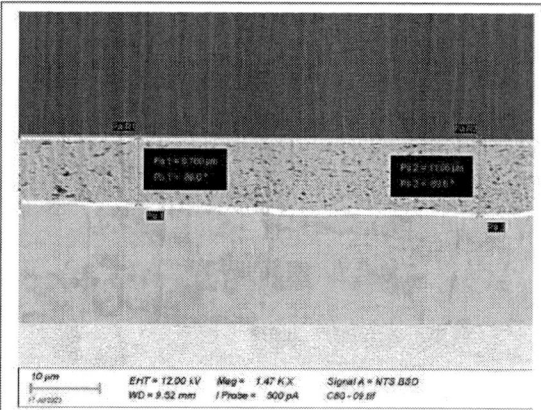

Fig. 4 – SEM analysis of the sintered microstructure.

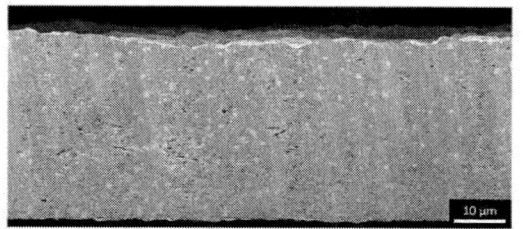

Fig. 5 – SEM analysis of the stripes used for LaTIMA and ECLA analysis.

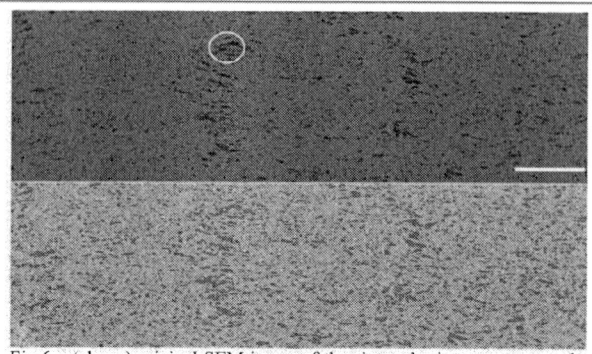

Fig 6 – (above) original SEM image of the sintered microstructure, scale bar referring to 5 µm and (below) ImageJ threshold image used to calculate porosity of the interconnect. The red highlights showing the pores.

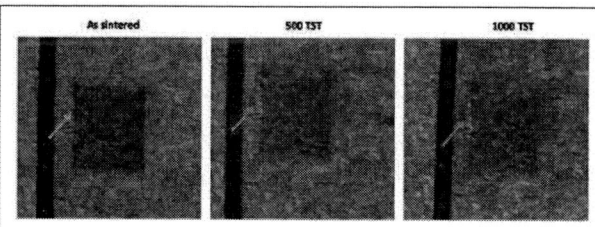

Fig. 7 – Backside SAM analysis of the sintered interconnect after TST at +175/-65°C, air to air, 30min dwell time. Red arrow shows the crack.

The porosity of the sintered interconnect as measured using the open-source ImageJ software resulted in a porosity of 8.6% (area %), a pore circularity of 0.9 ± 0.2 (a value of 1 indicating the pores are circular) and pores with an average width of 0.23 ± 0.1 µm and height of 0.22 ± 0.09 µm. The measurements indicated that the interconnect is largely dominated by nearly circular pores, apart from some long, interconnected pores as highlighted by the yellow marking in Fig 6. All values were measured across 4800 elements in a SEM cross-section of the sintered interconnect as shown in Fig 6.

The shear test results between the as-sintered samples and ones after 1000 TST are compared. As sintered interconnects showed a shear strength of 60.5 ± 9.5 MPa which reduced to 45.4 ± 5.3 MPa after 1000 TST, indicating ~ 25% drop in shear strength, but still considerably higher than the reported values for Ag sintering [4][12]. It must also be noted that SiC is a very stiff material, with Young's modulus that is approximately 3 times that of Si [13]. Therefore, under high thermo-mechanical stress conditions, while Si shows some flexure, reducing the total strain in the die-attach layer, SiC due to its high stiffness results in the die-attach layer carrying the bulk of the strain from the test conditions. Therefore, it is important for the die-attach material to be able to carry the high strain.

SAM analysis (backside scan) of the sintered interconnects is as shown in Fig. 7. A 50MHz transducer was used for the imaging. The TST was performed on the substrates as sintered, without encapsulating them. As is evident from the image, the first delamination was observed after 500 cycles along the left edge of the chip as shown by the red arrow. However, it is observed that this defect did not propagate or grow substantially in the next 500 cycles as can be seen in the 1000 TST image. But, although scanned under the same settings, the image quality is considerably compromised between the scans compared to the as sintered interconnect. The degradation within the AMB substrate as well as the impact of oxidation on AMB substrate impact the SAM scans. Further, as is evident from the images, the reflection of the substrate features (uniform white texture) could not be eliminated. A top side scan through the structure of the SiC chip did not yield the desired results due to the structuring on top of the chip and the corresponding reflections as can be seen in Fig. 8. Further, the dummy SiC devices used in this study had a polyimide coating on the top surface which degraded partially during sintering at 275°C for 5 min and also during thermal shock tests, further compromising the SAM top scan possibilities. Therefore, top scan SAM imaging was not pursued further.

Fig. 8 – SAM analysis (top side scan) of the SiC devices after sintering, showing the impact of the structuring on the top surface of the chip and "stain" like features/spotting attributed to degradation of polyimide layer on die surface.

The subsequent cross-sectioning and SEM analysis of the sample showed interesting defect phenomena. The SEM image is shown in Fig. 9. The delamination at the chip edge and the crack further into the interconnect is observed on the edge as marked with the red arrow (top image in Fig. 9). On the opposite edge, consistent with the SAM image does not show any crack or delamination as shown by the green arrow. However, the SAM images do not reveal the defects at the substrate interface as observed in the bottom image of Fig. 9, shown by the red arrow. Although the large void exists, the microstructure above the void is dense and uniformly sintered. With no such defect observed under the SAM, it could be a possibility that this is a defect during the sample preparation. However, the defect with the substrate metallization is clearly observed.

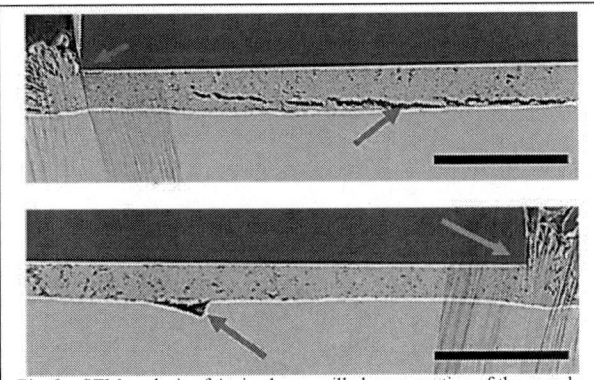

Fig. 9 – SEM analysis of Ar-ion beam milled cross-section of the sample after 1000 TST. Red arrows indicate the defects. Scale bar refers to 50 μm.

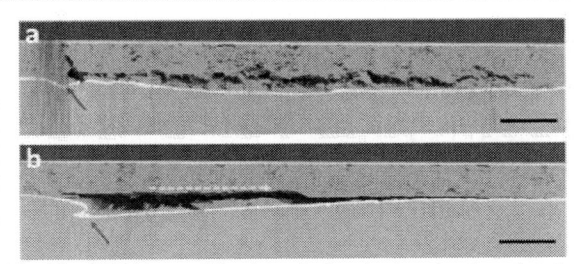

Fig. 10 – SEM images of the sintered microstructure showing substrate metallization defects and the corresponding cracks in the sintered microstructure. Scale bar refers to 20 μm.

In order to ascertain if this is a recurring effect, further samples from the same batch were cross-sections and the resulting analysis of the sintered microstructure among other samples indicated cracks to have predominantly initiated in areas where defects on substrates were also noticed. The origin of the cracks is consistent with the position of the defect on the substrate and hence localized. As a result, the typical edge to center crack propagation is not observed as reported earlier [14]. This can be possibly explained by the phenomena that the sintered interconnect is characterized by long pores which are nearly parallel to the interfaces as shown earlier in Fig 3. Therefore, the stress concentration formulation that gives the maximum stress at crack tip can be applied in terms of the long pores in case of the stacked Cu flakes Fig. 11. This implies, the longer the pore and sharper the pore tip radius, higher the stress concentration. Therefore, the stress concentration at these locations is of several magnitude higher than the applied stress, thereby causing delamination by crack initiation and propagation. Therefore, while the stacking of the flakes over each other is beneficial for sintering, it is important to ensure a low porosity interconnect on sintering, i.e., a densely packed structure to avoid the above-mentioned phenomena.

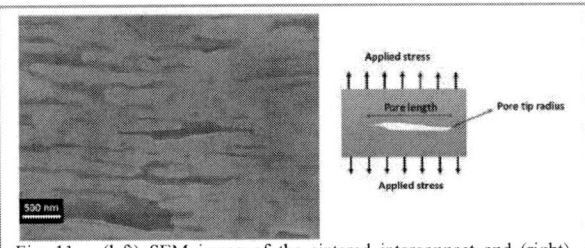

Fig. 11 – (left) SEM image of the sintered interconnect and (right) schematic representation of the impact of applied stress.

IV. CONCLUSIONS

Cu flakes-based sinter pastes provide a promising approach to die-attach bonding of SiC devices. The paste shows good workability, stencil printing and no bleed out. The compatibility to Ag metallization is also proven. The performance of the copper sinter paste is comparable with state of the art in Ag sintering. A thermal conductivity > 200 W/mK could be achieved, with as sintered interconnected resulting in shear strength over 60 MPa and an interconnect with < 10% porosity. Even after 1000 TST under harsh testing conditions of +175/-65°C, non-encapsulated, an average shear strength of 45 MPa is recorded. The unique stacking behavior of the flakes resulted in a dense and homogeneous interconnect with porosity < 10% (area %). Typical delamination failures as known from Ag sintering were not observed. Even with a thin BLT of ~12 μm, the sintered microstructure could withstand the extreme thermal shock conditions with SiC devices bonded onto AMB substrates. Localized cracks were observed which initiated at surface anomalies observed on the substrate. However, the typical failure mode with cracks initiation in the corners, leading to the center and complete delamination is not observed. In conclusion, the flakes-based paste offers a promising and reliable approach to die-attach bonding.

REFERENCES

[1]   J. Fabian, M. Hirz, and K. Krischan, "State of the Art and Future

Trends of Electric Drives and Power Electronics for Automotive Engineering," *SAE Int. J. Passeng. Cars - Electron. Electr. Syst.*, vol. 7, no. 1, pp. 293–303, 2014, doi: 10.4271/2014-01-1888.

[2] J. Watson and G. Castro, "A review of high-temperature electronics technology and applications," *J. Mater. Sci. Mater. Electron.*, vol. 26, no. 12, pp. 9226–9235, 2015, doi: 10.1007/s10854-015-3459-4.

[3] M. Schaal, M. Klingler, and B. Wunderle, "Silver Sintering in Power Electronics: The State of the Art in Material Characterization and Reliability Testing," *2018 7th Electron. Syst. Technol. Conf.*, pp. 1–18.

[4] S. A. Paknejad and S. H. Mannan, "Review of silver nanoparticle based die attach materials for high power/temperature applications," *Microelectron. Reliab.*, vol. 70, pp. 1–11, Mar. 2017, doi: 10.1016/j.microrel.2017.01.010.

[5] Y. Tan, X. Li, G. Chen, Q. Gao, G.-Q. Lu, and X. Chen, "Effects of thermal aging on long-term reliability and failure modes of nano-silver sintered lap-shear joint," *Int. J. Adhes. Adhes.*, p. 102488, Nov. 2019, doi: 10.1016/j.ijadhadh.2019.102488.

[6] F. Qin *et al.*, "Crack Effect on the Equivalent Thermal Conductivity of Porously Sintered Silver," *J. Electron. Mater.*, vol. 49, no. 10, pp. 5994–6008, 2020, doi: 10.1007/s11664-020-08325-1.

[7] J. Dryzek, "Migration of vacancies in deformed silver studied by positron annihilation," *Mater. Sci. Forum*, vol. 255–257, pp. 533–535, 1997, doi: 10.4028/www.scientific.net/msf.255-257.533.

[8] Z. Zhang, C. Chen, A. Suetake, M.-C. Hsieh, and K. Suganuma, "Reliability of Ag Sinter-Joining Die Attach Under Harsh Thermal Cycling and Power Cycling Tests," *J. Electron. Mater.*, vol. 50, no. 12, pp. 6597–6606, Dec. 2021, doi: 10.1007/s11664-021-09221-y.

[9] K. Wakamoto, T. Otsuka, K. Nakahara, and T. Namazu, "Degradation Mechanism of Pressure-Assisted Sintered Silver by Thermal Shock Test," *Energies*, vol. 14, no. 17, p. 5532, Sep. 2021, doi: 10.3390/en14175532.

[10] Y. Zuo, A. Robador, M. Wickham, and S. H. Mannan, "Unraveling the complex oxidation effect in sintered Cu nanoparticle interconnects during high temperature aging," *Corros. Sci.*, vol. 209, no. July, p. 110713, Dec. 2022, doi: 10.1016/j.corsci.2022.110713.

[11] M. A. Ras *et al.*, "'LaTIMA' an innovative test stand for thermal and electrical characterization of highly conductive metals, die attach, and substrate materials," in *2015 21st International Workshop on Thermal Investigations of ICs and Systems (THERMINIC)*, Sep. 2015, pp. 1–6. doi: 10.1109/THERMINIC.2015.7389630.

[12] T. F. Chen and K. S. Siow, "Comparing the mechanical and thermal-electrical properties of sintered copper (Cu) and sintered silver (Ag) joints," *J. Alloys Compd.*, vol. 866, p. 158783, Jun. 2021, doi: 10.1016/j.jallcom.2021.158783.

[13] K. Järrendahl and R. F. Davis, "Chapter 1 Materials Properties and Characterization of SiC," in *Semiconductors and Semimetals*, vol. 52, no. C, 1998, pp. 1–20. doi: 10.1016/S0080-8784(08)62843-4.

[14] S. K. Bhogaraju, M. Schmid, H. R. Kotadia, F. Conti, and G. Elger, "Highly reliable die-attach bonding with etched brass flakes," in *2021 23rd European Microelectronics and Packaging Conference & Exhibition (EMPC)*, Sep. 2021, pp. 1–6. doi: 10.23919/EMPC53418.2021.9584967.

Rapid sintering of inkjet printed Cu complex inks using Laser in air

Nihesh Mohan
Institute of Innovative Mobility (IIMo)
Technische Hochschule Ingolstadt
Ingolstadt, Germany
Nihesh.Mohan@thi.de

Juan Ignacio Ahuir-Torres
School of Engineering
Liverpool John Moores University
Liverpool, United Kingdom
J.I.AhuirTorres@ljmu.ac.uk

Sri Krishna Bhogaraju
Institute of Innovative Mobility (IIMo)
Technische Hochschule Ingolstadt
Ingolstadt, Germany
SriKrishna.Bhogaraju@thi.de

Hiren Kotadia
School of Engineering
Liverpool John Moores University
Liverpool, United Kingdom
H.R.Kotadia@ljmu.ac.uk

Gordon Elger
Institute of Innovative Mobility (IIMo)
Technische Hochschule Ingolstadt
Ingolstadt, Germany
Gordon.Elger@thi.de

Abstract—Cu complex inks offer a facile and cost-effective alternative to commercial nanoparticle inks for printed electronics application. Cu complex inks are prepared by mixing Cu formate tetrahydrate (metal precursor) with amino-2-propanol (complexing agent) in a molar ratio of 1:2. A carrier solvent consisting of ethanol and ethylene glycol is then added to help adjust the rheological properties of the ink for printing purpose. The Cu complex ink having a viscosity of 12 mPa.s is then inkjet printed onto a polyimide substrate (thickness: 125 μm). After printing 10 layers, the ink is pre-dried under air at 100°C for 5 minutes followed by sintering using an IR laser (942 nm) to enable rapid thermal decomposition of the Cu formate. After sintering using a 100 W line beam laser scanned at 3 mm/sec under air; a contiguous and homogenous Cu metallic trace is formed with a minimum bulk resistivity of 4.8 μΩcm (2.84 times of bulk Cu). The laser sintered trace exhibited a highly compact and homogenous sintered structure after thermal decomposition compared to the conventional oven sintering process. Th result is remarkable and proves the importance of faster decomposition process. In contrast, sintering the same trace using slow ramp rate within the traditional oven sintering pr process, it is observed that a Cu metallic trace is formed with particles in the range of 2-5 μm that are sparsely distributed due to low metal content of Cu complex inks (< 10 metal wt.%). Due to sparse distribution of particles, a low bulk resistivity value of 102 μΩcm is obtained.

Keywords—Cu complex inks, Laser sintering, Flexible substrate, Printed electronics

I. INTRODUCTION

Printed electronics provides an innovative approach using principles of traditional printing process to fabricate electronic devices on flexible and unconventional substrates using functional materials [1-5]. These functional materials mainly consist of conductive inks that can be printed onto suitable substrates by utilizing various printing techniques. These printing techniques can be classified as contact (flexographic printing, screen printing and soft lithography) and non-contact (inkjet printing, aerosol jet printing and direct writing technologies) patterning processes [5-7]. Among these printing technologies, non-contact patterning process provides advantages in terms of flexibility in design, higher patterning accuracy and lower risk of contamination [5-7]. Using this emerging technology, industries have the potential to produce low-cost, lightweight, and flexible electronic devices with applications such as wearable sensors, RFID components,

energy harvesting and storage, thin film transistors, health monitoring devices and many more [8-17].

Over the last decade, there has been rapid development and commercialization of conductive Ag, Cu and Au nanoparticles (NPs) inks for printed electronics applications [18-22]. Noble metals like Ag and Au offer high conductivity and stability in terms of oxidation, however owing to high cost they are not suitable for large volume applications. On the other hand, Cu offers a good alternative due to its high electrical conductivity (58 MS/m, second only to Ag), higher abundance (1000 times more than Ag) and low cost (nearly 100 times cheaper than Ag) [23]. The biggest challenge however is addressing it's high affinity to oxidation at elevated temperatures (formation of Cu_2O at 150°C) [24]. Cu NPs are prone to oxidation due to its high surface energy, which leads to formation of surface oxide layers that hinder the sintering [25]. To tackle this issue, the sintering process is required to be done under an inert (N_2) or reducing atmosphere (formic acid enriched nitrogen, H_2 or forming gas). Aside from this, capping agents (such as Polyvinylpyrrolidone) are used not only to prevent oxidation of Cu NPs at lower temperatures, but also to prevent them from agglomerating within the ink formulation [26-27]. This leads to a complex synthesis route of Cu NPs ink. And typically, higher sintering temperature (>250°C) remove these capping agents to provide good sinterability [28]. Other issues pertaining to Cu NP inks are storage (refrigeration or deep freeze) and shorter shelf life (< 6 months).

In recent years, there has been a growing interest in development of Cu particle free inks [29-30]. It offers a facile and low-cost synthesis route, where nanoparticles are generated *in-situ* upon thermal decomposition. These inks are usually composed of Cu metal salt (mainly Cu formate tetrahydrate) and an amine based solvent that coordinates with formate ion in a monodentate or bidentate fashion to form a metal salt-amine complex [31]. Further, suitable solvents are added to improve the properties (viscosity and wettability) of the ink related to printing purpose. Since, the Cu exists in an ionic form, it prevents oxidation, particle agglomeration and provides a longer shelf life at room temperature. Additionally, these inks have lower decomposition temperature (<160°C) making it suitable for sintering on some polymeric substrates.

Based on these advantages, it also provides flexibility in terms of choosing a wide range of sintering processes for fabrication of conductive traces. Conventional oven sintering of Cu complex inks faces many challenges. It requires use of an inert/reducing atmosphere because the formed NPs due to their reactivity are prone to oxidation. Also, efficient degassing of organics is required for fabrication of contiguous conductive traces [32]. Therefore, slower ramp rate (typically 5°C/min) is required for controlled evaporation of solvents present in the ink which in turn lead to formation of larger aggregates ranging from 2-5 μm [32]. This leads to sparse distribution of metallic particles resulting in poor conductivity of the metallic traces. Apart from conventional sintering processes under inert/reducing atmosphere, there have been several studies carried out using laser sintering, photonic sintering, microwave sintering and electrical sintering processes [33-37]. Among them Laser sintering provides several advantages in terms of high speed localized sintering preventing substrate damage to low melting point substrates (such as PET) and integration of patterning and sintering process for large scale production [33-34].

In this work, a comparison is made between the conventional oven sintering process and laser sintering process. Laser sintering of Cu complex inks is performed under air on a flexible polyimide substrate using two different types of laser optics (spot laser and line beam laser). Focus is made on laser sintering of fine printed structure (trace thickness after sintering < 1-2 μm and trace width < 100 μm) using inkjet printer. The challenges related to process optimization of laser sinter parameters will also be briefly discussed.

II. EXPERIMENTAL

A. Material

Cu (II) formate tetrahydrate (Cuf), (98%) is purchased from Thermo Fischer Scientific. Amino-2-propanol (A2P) (93%) is purchased from Sigma-Aldrich. Ethanol (>96%), and Ethylene glycol (EG) (>99%) are purchased from Carl Roth. Polyimide films (Kapton® HPPST-125 μm) from DuPont are used for printing. Apart from Cu (II) formate tetrahydrate and polyimide film, other chemicals are used as received. Additionally, propylene glycol methyl ether acetate (PGMEA) is used for cleaning of the ink cartridges and isopropanol is used to clean the substrates.

B. Synthesis

A Cuf-A2P complex is prepared by mixing in a molar ratio of 1:2. The mixture is then stirred for 15 minutes at 500 rpm in a planetary-rotary solder paste mixer (Thinky SR500). Then, an ink formulation for inkjet printing is prepared using ethanol and EG (mixed together in 75:25 wt. ratio). 6 g of Cuf-A2P complex is added to 6 g of ethanol-EG solvent combination to prepare a low viscosity ink of 12 mPa.s, essential for ink-jet printing with a Cu metal content of 7wt.% Cu. The viscosity of the inks was measured using a viscometer (Thermo Scientific Haake TM Viscotester TM) and the contact angle of the inks formed on polyimide substrate were measured using a drop shape analyser (Kruss DSA 30M). After the ink synthesis, spectral absorbance of the Cu complex ink is measured under near infrared (NIR) until 1050 nm using a spectrometer (PMA-12 Photonic multichannel analyser from Hamamatsu) to check the compatibility of the laser for performing the sintering experiments.

C. Printing and processing

The prepared Cu complex ink is printed using a Dimatix Materials Printer (DMP-2850, Fujifilm Dimatix Inc.) onto the polyimide substrates. Before printing, the polyimide substrates were Ar-plasma treated for 5 minutes for activation of the surface and removing of contaminations. The Ar-plasma treatment is done using Zepto PLS from Diener electronic under UV in a manually controlled Ar atmosphere with chamber pressure maintained between 0.4-0.45 mbar. After printing, a conventional oven sintering, and laser sintering process is conducted. A two-step process, pre-drying and sintering is conducted in a reflow oven (UniTemp RSS 160-S) under a constant N_2 flow of 5L/min. The pre-drying is done at 100°C for 5 min followed by sintering at a slow ramp rate of 5°C/min upto 160°C and then to 250°C for an isothermal holding time of 5 min. A predrying step is also conducted before the laser sintering process at 100°C for 5 min in air. Laser sintering of printed Cu complex ink traces are done using a spot (200W T-SMILS, optic size: 1.6 mm in diameter, wavelength: 940 nm) and line beam (360 W T-SPOLD, optic size: 2mm x 80mm, wavelength: 940 nm) laser from Hamamatsu Photonics Deutschland GmbH.

D. Characterization

The particle morphology and sinterability of fabricated traces after sintering are analyzed and measured using an optical microscope(Keyence VHX-900F 3D microscope). They are evaluated using a scanning electron microscope (Verse FEI dual beam) with following parameters (Electron High Tension: 10 KV, Working Distance: 10 mm, Current: 8 nA, Mode: Secondary Electrons). The thickness (t) of the trace was measured using an optical light profilometer (Nanofocus μ-surf custom) and sheet resistance (R_s) measured using a 4-point probe (Keithley 2461). The bulk or volume resistivity (ρ), thereafter is calculated using the following equation [38]:

$$\rho = t * R_s \tag{1}$$

Additionally, the bulk resistivity (ρ) can also be calculated using a standard formula as shown in equation 2 [38]:

$$\rho = R * \frac{w.t}{L} \tag{1}$$

where, R is the measured resistance and L, w, t are length, width, and thickness of the sintered trace.

III. RESULTS AND DISCUSSION

A. Ink design and printability

Cuf upon thermal decomposition (>200°C) forms following by-products as shown in equation 2 and 3 [39]:

$$Cu(HCOO)_{2(s)} + heat \rightarrow Cu_{(s)} + 2CO_{2(g)} + H_{2(g)} \tag{2}$$

$$Cu(HCOO)_{2(s)} + heat \rightarrow Cu_{(s)} + CO_{2(g)} + CO_{(g)} + H_2O_{(g)} \tag{3}$$

A Cuf-A2P complex is prepared which decreases the thermal decomposition temperature to 160°C. The A2P in the complex helps in stabilizing the ink formulation preventing agglomeration and increasing the shelf life. The OH⁻ bond in the alkanolamine also promotes solubility with low-boiling point alcohol-based solvents (such as ethanol) that may help in lowering the viscosity of prepared complex for inkjet printing purpose. In order to prepare the ink for printing purpose using inkjet printing, it must have a viscosity in the range of 10-12 mPa.s. A solvent combination is prepared using ethanol and EG in 75:25 wt. ratio, which then added to the complex in a 50:50 wt.% resulting in a viscosity of 12 mPa.s. In the prepared ink, ethanol is responsible for controlling the viscosity of the ink. EG is responsible for providing good stability by decreasing the rate of evaporation during printing as well as storage and improving patterning onto polyimide substrate by increasing the contact angle from 10° to 45°. This prevents uneven spreading and formation of single droplets instead of contiguous trace. Fig. 1a shows a schematic of drop deposition based on ink-substrate interaction, where γ_T and γ_E are surface tension of the ink and surface energy of the substrate, respectively. When surface tension of the Cu complex ink is very less than the surface energy of the polyimide substrate, it causes uneven spreading whereas when it is greater it causes droplet formation on the substrate [40]. In this case, the surface tension of the designed Cu complex ink is 35 mN/m and surface energy of the polyimide substrate is 47 mJ/m² [41]. Fig. 1b shows the contact angle measurement of the Cu complex ink with and without the presence of EG on polyimide substrate. In addition, printing parameters related to inkjet printing also play a key role in determining the uniformity and resolution of the printed pattern. Printing parameters such as waveform design, jetting voltage and drop spacing plays a significant role in controlling the drop volume, drop velocity and accurate placement of each drop in concurrence with the print design, respectively. Fig. 2 shows the inkjet printing parameters used in this study. More details related to optimization of printing parameters related to inkjet printing can be found in our previous publications [40].

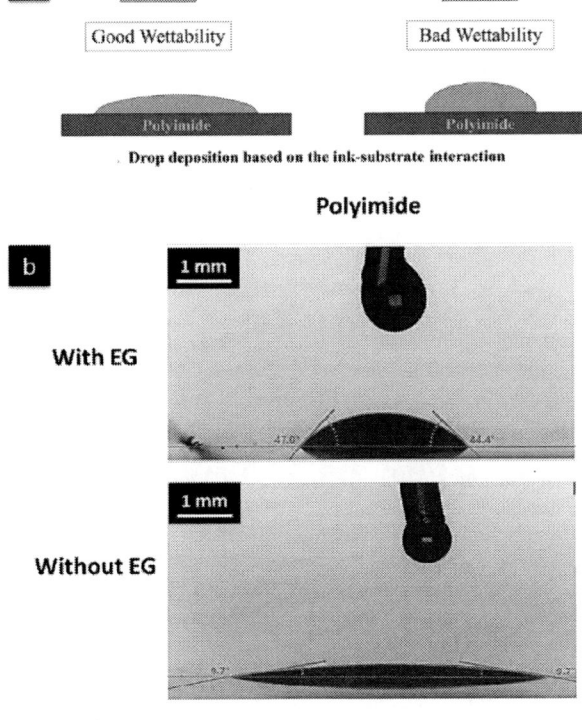

Fig. 1 Contact angle measurement of the Cu complex ink with and without the presence of EG on polyimide substrate.

Printing Parameters	Values/ Ranges
Printhead type	Samba™
Drop size (in volume, pL)	2.4
Number of nozzles in use	4-8
Jetting frequency (kHz)	50
Jetting voltage (V)	40
Cartridge temperature	Room temperature
Print head height (mm)	0.75
Drop spacing (µm)	15

Fig. 2 Inkjet printing parameters for Cu complex ink and the design after printing.

B. Conventional oven sintering process

Firstly, the Cu complex ink traces after inkjet printing are subjected to conventional oven sintering. Cu complex inks have high volume shrinkage due to low solids content (<10 wt.% metallic Cu), therefore a slower ramp rate is necessary for efficient degassing of organics. The first step involves predrying of the printed Cu complex ink at 100°C for 5 mins followed by a slow ramp rate process upto 160°C at 5°C/min and then sintering at 250°C for 5 min. From the previous investigations, it was observed that using the slow ramp rate upto 160°C (decomposition temperature for A2P), a controlled evaporation of ink byproducts and better patterning post-sintering was achieved [32]. The *in-situ* realized NPs were then further sintered at 250°C for 5 mins. This method helped in fabricating contiguous traces, however showed a poor conductivity. Fig. 3a shows the differences in formation of Cu sintered trace when sintered using a slow ramp rate process and fast ramp rate process. Fig. 3b shows the SEM images of the Cu trace sintered using the slow ramp rate process (5°C/min). It was observed that large Cu aggregates (2-5 µm) were formed resulting in sparse distribution of particles across the sintered trace. The bulk resistivity of the Cu traces obtained was 102 µΩcm. It should be noted that unlike screen printing or other deposition methods, the thickness of printed trace is < 20 µm where the complex ink has a Cu metal content of < 7 wt.%. Aggregation of NPs is a tightly bound collection, where NPs are held together by strong mechanical forces that are difficult to breakup [42]. Usually as NPs form, they have high surface energy, and they tend to reduce this surface energy either by routes of agglomeration or aggregation [42]. This can however be thwarted via surface engineering (use of surfactants) or by use of rapid sintering methods (the nucleation and growth rate of NPs are increased with heating temperature and time) [42-43]. Therefore, rapid sintering processes using laser are required to produce a fine and homogenous sintered structure, where formed NPs are <100 nm and the resulting Cu metallic trace has high electrical conductivity.

C. Laser sintering process

In order to find a laser with suitable wavelength for laser sintering, the spectral absorbance of the Cu complex ink is measured. Fig. 4a shows the setup for measuring spectral absorbance using the spectrometer. It is observed that stronger transmission is observed in the blue region of visible spectrum with a peak value of 477.8 nm and stronger absorbance is observed in the red region of visible spectrum with a peak value of 700.64 nm. Fig. 4b shows the spectral absorbance of Cu complex ink. Based on this measurement, it is clear that a laser in the red or NIR spectrum should be selected for the laser sintering process. From the available lasers, a laser of wavelength 940 nm was selected for the sintering process. Another important parameter to consider is selection of a suitable laser optic (spot laser and line beam laser).

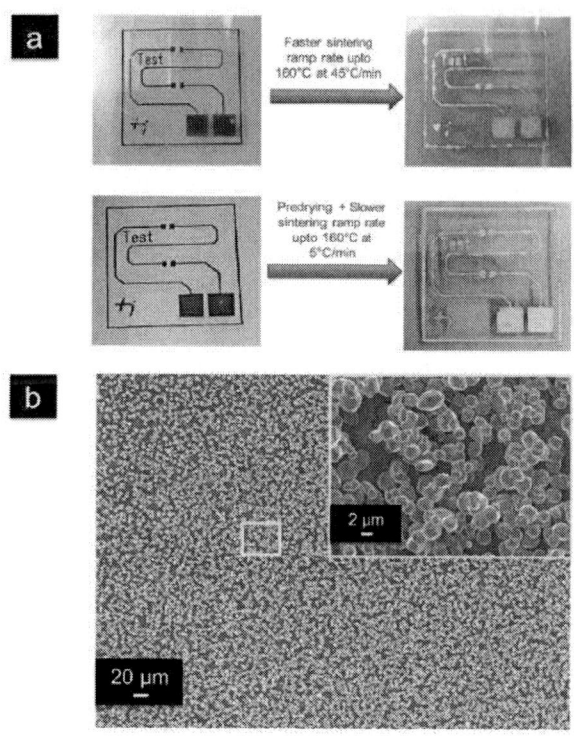

Fig. 3 (a) Inkjet printed pattern before and after sintering processes, faster ramp rate (top) and predrying plus slower ramp rate (bottom), (b) SEM image of the sintered sample after predrying followed by slower sintering ramp rate upto 160°C at 5°C/ min and then sintered at 250°C for 5 mins.

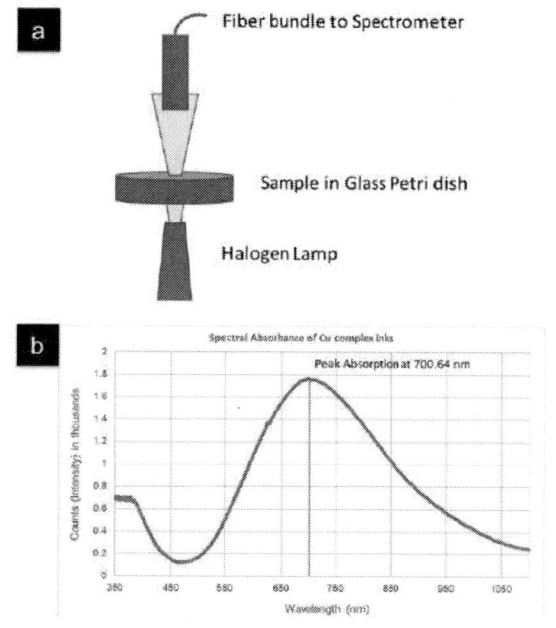

Fig. 4 (a) Setup for measuring spectral absorbance of Cu complex ink using the spectrometer, (b) Spectral absorbance of Cu complex ink.

After inkjet printing 10 layers of Cu complex ink, the printed traces are then predried at 100°C for 5 mins under air to avoid rapid evaporation of ethanol present in the ink. After process optimization, a spot laser of 1.6 mm diameter with a laser power of 10 W was used to sinter three identical printed traces using three different scan rates (30 mm/s, 20 mm/s and 10 mm/s). Using laser sintering parameters such as laser power (P), scan rate (v) and laser spot diameter (d) energy density (E_d) was calculated. Equation 4 shows the formula for calculation of energy density.

$$E_d = \frac{P}{v.d} \tag{4}$$

It was found that when scan rate was 30 mm/s, the calculated energy density was 20J/cm² which resulted in a partially sintered sample with large voiding as seen on the trace resulting in an inhomogeneous trace. Similarly, for a scan rate of 10 mm/s, the energy density was 62.5 J/cm² which caused substrate damage and subsequent oxidation of the trace. Using a scan rate of 20 mm/s, a sintered conductive trace having a sheet resistance of 3.2 Ω/□ under air was obtained, where energy density was calculated to be 31.25 J/cm². The trials with spot laser showed good sinterability between the Cu nanoparticles, however there was trace inhomogeneity and large voiding due to uneven heat distribution across the printed trace (see Fig. 5). From the experiments, the energy density is calculated to be in the range of 20-50 J/cm² for well sintered samples without causing substrate damage.

Subsequently, laser sintering was performed using a line beam laser with optic size of 2 mm x 80 mm. The operational advantages of using a line beam laser are larger scanning area which enables sintering of intricate designs without the need for retracing the printed path and its suitability for large scale production. They require however higher power to meet the required energy density for sintering the same printed trace resulting in higher operational costs. After process optimization, using a laser power of 100W and scanning rates of 12mm/s and 3mm/s sheet resistances of 1.2 Ω/□ and 0.8 Ω/□ were obtained respectively under air. Fig 5 shows the SEM images of laser sintered Cu complex traces using laser scan rate of 3mm/s. From SEM images, a well sintered homogenous structure is observed. The energy density calculated was found to be 30 J/cm², which aligns well with the good sinterability range as discussed earlier. Compared to spot laser, line beam laser provided an easy laser processing of the sintered traces without requiring any retracing of printed trace using g-codes. It also enabled complete sintering of printed traces without leaving behind any residual ink in the sintered traces. This can however be adjusted using a spot laser with larger spot diameter. Based on the profilometer results, the sintered trace formed using spot laser (for scan rate 20 mm/s) had an average thickness of 1.5 ± 0.8 μm, however the sintered trace formed using a line beam laser (for scan rate 3 mm/s) had an average thickness of 1.2 ± 0.2 μm. This resulted in a bulk resistivity (measured over a trace length of 20mm and average trace width of 1mm) of 24 μΩcm and 4.8 μΩcm for spot laser and line beam laser respectively. Using line beam laser, sintered Cu traces had bulk resistivity 2.84 times that of bulk Cu.

When compared with conventional oven sintering process, it is found that laser sintering in general produces a finer and densely packed structure. Further, in case of laser sintering the *in-situ* generated particles are exposed for a very short time (<100 ms) preventing the coalescence of the particles into larger aggregates [34]. In Fig. 5 (SEM images), it can be inferred that the particles formed tend to coalesce together in form of rod like structure in that short process time generating a densely packed web like structure [34]. Further it can be seen that after sintering using a line beam laser, this structure has a more uniform and packed structure. This can be attributed to the homogenous heat distribution. Further process optimization related to spot laser trials are ongoing, as it provides an advantage in terms of integration with the printing process and to understand the physical mechanism of laser sintering of Cu complex inks.

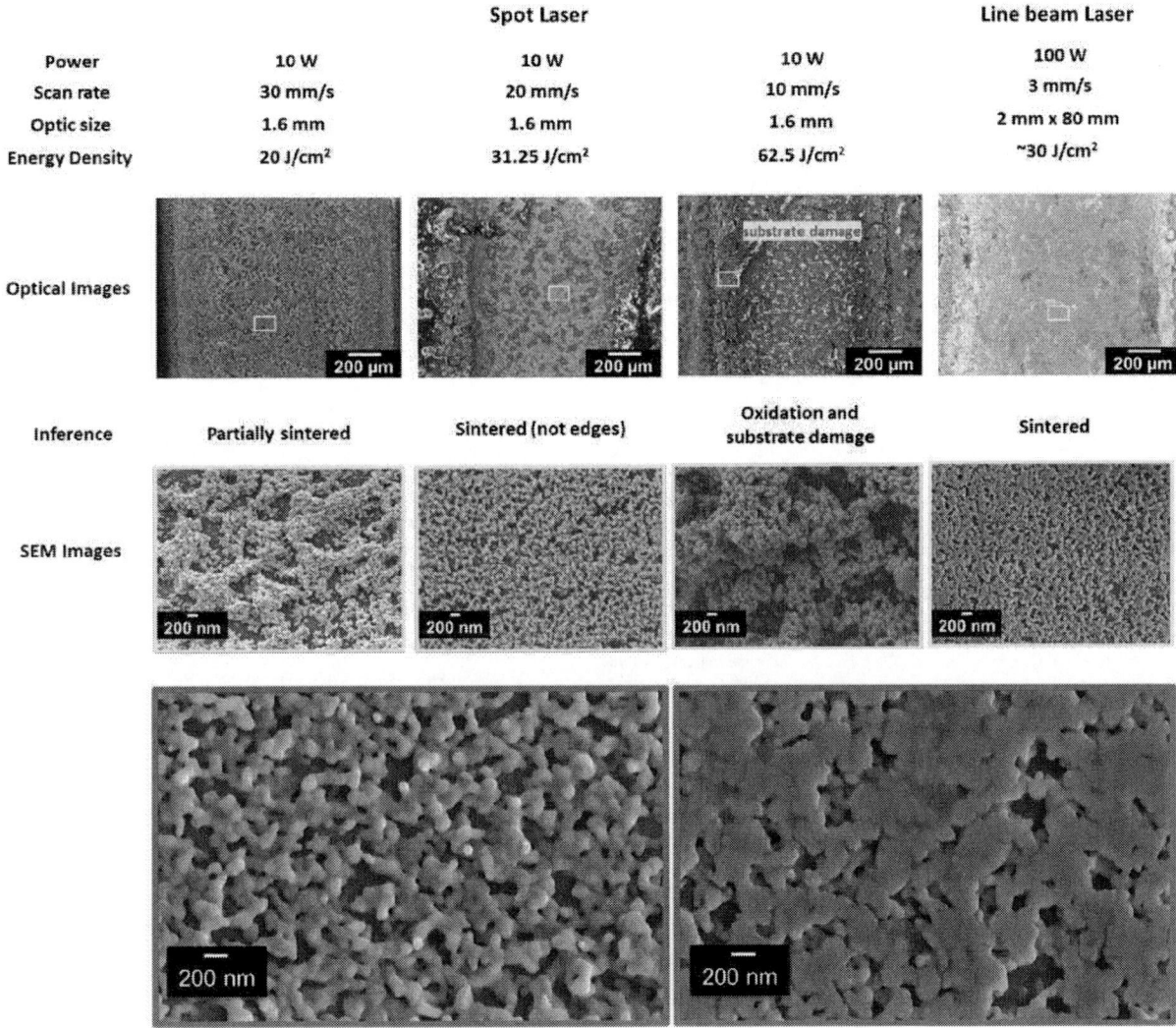

Fig. 5 Laser parameters, optical and SEM images of laser sintered sample using a spot laser of 1.6 mm diameter and line beam laser of optic size 2 mm x 80 mm.

IV. CONCLUSION

Cu complex inks provide a facile, low-cost alternative to nanoparticles-based inks for printed electronics application. Since, they exist in ionic form, they don't exhibit issues of agglomeration during storage and oxidation at ambient temperature. Using conventional reflow oven process, they can be sintered at temperature less than 160°C. However, post printing these inks require efficient degassing to prevent discontinuity of sintered traces post sintering process. Sintering using conventional process tend to form larger aggregates (2-5 µm) contributing to sparse distribution of particles within the sintered trace affecting the overall bulk resistivity (102 µΩcm). One method to solve this problem is using rapid sintering method such as laser sintering. Initial trials with spot laser showed good sinterability between the nanoparticles however trace inhomogeneity was an issue due to uneven heat distribution caused by the spot laser. Using

227

line beam optic laser, the morphology of the sintered trace was more homogeneous compared to the one obtained by spot laser as the scanning area is larger and helps in uniform heat distribution. The measured bulk resistivity was 24 μΩcm and 4.8 μΩcm for spot laser and line beam laser respectively. Using line beam laser, sintered Cu traces had bulk resistivity 2.84 times that of bulk Cu. In addition, the laser sintering was performed under air, and it also gives a possibility for sintering onto low melting temperature substrates such as PET.

V. Acknowledgment

The authors gratefully acknowledge the experimental support provided by external partners especially Hamamatsu Photonics Deutschland GmbH (Laser sintering) and Conti Temic microelectronic GmbH (Contact angle measurements). This research was funded by the German Ministry of Education and Science within the scope of project 'ADDIRA'.

References

[1] Khan, Y., Thielens, A., Muin, S., Ting, J., Baumbauer, C., & Arias, A. C. (2020). A new frontier of printed electronics: flexible hybrid electronics. Advanced Materials, 32(15), 1905279.

[2] Suganuma, K. (2014). Introduction to printed electronics (Vol. 74). Springer Science & Business Media.

[3] Kamyshny, A., & Magdassi, S. (2014). Conductive nanomaterials for printed electronics. Small, 10(17), 3515-3535.

[4] Cui, Z. (2016). Printed electronics: materials, technologies and applications. John Wiley & Sons.

[5] Beedasy, V., & Smith, P. J. (2020). Printed electronics as prepared by inkjet printing. Materials, 13(3), 704.

[6] Cruz, S. M. F., Rocha, L. A., & Viana, J. C. (2018). Printing technologies on flexible substrates for printed electronics. In Flexible electronics. Intech Open.

[7] Zhang, H., Moon, S. K., & Ngo, T. H. (2020). 3D printed electronics of non-contact ink writing techniques: status and promise. International Journal of Precision Engineering and Manufacturing-Green Technology, 7, 511-524.

[8] Sreenilayam, S. P., Ahad, I. U., Nicolosi, V., Garzon, V. A., & Brabazon, D. (2020). Advanced materials of printed wearables for physiological parameter monitoring. Materials Today, 32, 147-177.

[9] Kim, J., Kumar, R., Bandodkar, A. J., & Wang, J. (2017). Advanced materials for printed wearable electrochemical devices: A review. Advanced Electronic Materials, 3(1), 1600260.

[10] Kim, S. (2020). Inkjet-printed electronics on paper for RF identification (RFID) and sensing. Electronics, 9(10), 1636.

[11] Subramanian, V., Chang, P. C., Lee, J. B., Molesa, S. E., & Volkman, S. K. (2005). Printed organic transistors for ultra-low-cost RFID applications. IEEE transactions on components and packaging technologies, 28(4), 742-747.

[12] Subramanian, V., Fréchet, J. M., Chang, P. C., Huang, D. C., Lee, J. B., Molesa, S. E., ... & Volkman, S. K. (2005). Progress toward development of all-printed RFID tags: materials, processes, and devices. Proceedings of the IEEE, 93(7), 1330-1338.

[13] Grande, L., Chundi, V. T., Wei, D., Bower, C., Andrew, P., & Ryhänen, T. (2012). Graphene for energy harvesting/storage devices and printed electronics. Particuology, 10(1), 1-8.

[14] Kim, S., Tentzeris, M. M., & Georgiadis, A. (2019). Hybrid printed energy harvesting technology for self-sustainable autonomous sensor application. Sensors, 19(3), 728.

[15] Mirshojaeian Hosseini, M. J., & Nawrocki, R. A. (2021). A review of the progress of thin-film transistors and their technologies for flexible electronics. Micromachines, 12(6), 655.

[16] He, Y., Wang, X., Gao, Y., Hou, Y., & Wan, Q. (2018). Oxide-based thin film transistors for flexible electronics. Journal of Semiconductors, 39(1), 011005.

[17] Ma, L. Y., & Soin, N. (2022). Recent progress in printed physical sensing electronics for wearable health-monitoring devices: A review. IEEE Sensors Journal, 22(5), 3844-3859.

[18] Jung, I., Jo, Y. H., Kim, I., & Lee, H. M. (2012). A simple process for synthesis of Ag nanoparticles and sintering of conductive ink for use in printed electronics. Journal of electronic materials, 41, 115-121.

[19] Dang, M. C., Dang, T. M. D., & Fribourg-Blanc, E. (2014). Silver nanoparticles ink synthesis for conductive patterns fabrication using inkjet printing technology. Advances in Natural Sciences: Nanoscience and Nanotechnology, 6(1), 015003.

[20] Jeong, S., Song, H. C., Lee, W. W., Lee, S. S., Choi, Y., Son, W., ... & Ryu, B. H. (2011). Stable aqueous based Cu nanoparticle ink for printing well-defined highly conductive features on a plastic substrate. Langmuir, 27(6), 3144-3149.

[21] Dai, X., Xu, W., Zhang, T., & Wang, T. (2018). Self-reducible Cu nanoparticles for conductive inks. Industrial & Engineering Chemistry Research, 57(7), 2508-2516.

[22] Deng, M., Zhang, X., Zhang, Z., Xin, Z., & Song, Y. (2014). A gold nanoparticle ink suitable for the fabrication of electrochemical electrode by inkjet printing. Journal of nanoscience and nanotechnology, 14(7), 5114-5119.

[23] W. Li, Q. Sun, L. Li, J. Jiu, X.-Y. Liu und M. Kanehara, „The rise of conductive copper inks: challenges and perspective," Applied Materials Today, Bd. 18, Nr. 100451, 2019.

[24] Castrejón-Sánchez, V. H., Solís, A. C., López, R., Encarnación-Gomez, C., Morales, F. M., Vargas, O. S., ... & Sánchez, G. V. (2019). Thermal oxidation of copper over a broad temperature range: Towards the formation of cupric oxide (CuO). Materials Research Express, 6(7), 075909.

[25] Roy, N. K., Foong, C. S., & Cullinan, M. A. (2018). Effect of size, morphology, and synthesis method on the thermal and sintering properties of copper nanoparticles for use in microscale additive manufacturing processes. Additive Manufacturing, 21, 17-29.

[26] Li, W., Li, W., Wei, J., Tan, J., & Chen, M. (2014). Preparation of conductive Cu patterns by directly writing using nano-Cu ink. Materials Chemistry and Physics, 146(1-2), 82-87.

[27] Li, W., Chen, M., Wei, J., Li, W., & You, C. (2013). Synthesis and characterization of air-stable Cu nanoparticles for conductive pattern drawing directly on paper substrates. Journal of nanoparticle research, 15, 1-10.

[28] Deng, D., Jin, Y., Cheng, Y., Qi, T., & Xiao, F. (2013). Copper nanoparticles: aqueous phase synthesis and conductive films fabrication at low sintering temperature. ACS applied materials & interfaces, 5(9), 3839-3846.

[29] Shin, D. H., Woo, S., Yem, H., Cha, M., Cho, S., Kang, M., ... & Piao, Y. (2014). A self-reducible and alcohol-soluble copper-based metal-organic decomposition ink for printed electronics. ACS applied materials & interfaces, 6(5), 3312-3319.

[30] Yabuki, A., Tachibana, Y., & Fathona, I. W. (2014). Synthesis of copper conductive film by low-temperature thermal decomposition of copper–aminediol complexes under an air atmosphere. Materials Chemistry and Physics, 148(1-2), 299-304.

[31] Farraj, Y., Grouchko, M., & Magdassi, S. (2015). Self-reduction of a copper complex MOD ink for inkjet printing conductive patterns on plastics. Chemical Communications, 51(9), 1587-1590.

[32] Mohan, N., Saccon, R., Bhogaraju, S. K., & Elger, G. (2022, November). Cu complex inks for printed electronics application-challenges and solutions. In Mikro-Nano-Integration; 9. GMM-Workshop (pp. 1-5). VDE.

[33] Min, H., Lee, B., Jeong, S., & Lee, M. (2017). Fabrication of 10 μm-scale conductive Cu patterns by selective laser sintering of Cu complex ink. Optics & Laser Technology, 88, 128-133.

[34] Lee, J., Lee, B., Jeong, S., Kim, Y., & Lee, M. (2014). Microstructure and electrical property of laser-sintered Cu complex ink. Applied surface science, 307, 42-45.

[35] Araki, T., Sugahara, T., Jiu, J., Nagao, S., Nogi, M., Koga, H., ... & Suganuma, K. (2013). Cu salt ink formulation for printed electronics using photonic sintering. Langmuir, 29(35), 11192-11197.

[36] Kwon, Y. T., Kim, Y. S., Lee, Y., Kwon, S., Lim, M., Song, Y., ... & Yeo, W. H. (2018). Ultrahigh conductivity and superior interfacial adhesion of a nanostructured, photonic-sintered copper membrane for printed flexible hybrid electronics. ACS applied materials & interfaces, 10(50), 44071-44079.

[37] Zhang, J., Yuan, Y., Liang, G., Arshad, M. N., Albar, H. A., Sobahi, T. R., & Yu, S. H. (2015). A microwave-facilitated rapid synthesis of gold nanoclusters with tunable optical properties for sensing ions and fluorescent ink. Chemical Communications, 51(52), 10539-10542.

[38] Apostolakis, A., Barmpakos, D., Pilatis, A., Patsis, G., Pagonis, D. N., Belessi, V., & Kaltsas, G. (2022). Resistivity study of inkjet-printed structures and electrical interfacing on flexible substrates. Micro and Nano Engineering, 15, 100129.

[39] Bhogaraju, S. K., Mokhtari, O., Conti, F., & Elger, G. (2020). Die-attach bonding for high temperature applications using thermal decomposition of copper (II) formate with polyethylene glycol. Scripta Materialia, 182, 74-80.

[40] Mohan, N., Bhogaraju, S. K., Łysień, M., Schneider, L., Granek, F., Lux, K., & Elger, G. (2021, September). Drop feature optimization for fine trace inkjet printing. In 2021 23rd European Microelectronics and Packaging Conference & Exhibition (EMPC) (pp. 1-6). IEEE.

[41] Akram, M., Bhowmik, S., Jansen, K. M. B., & Ernst Leo, J. (2010). Surface modification of polyimide by atmospheric pressure plasma for adhesive bonding with titanium and its application to aviation and space. materials of SAMPE, 326-335.

[42] Shrestha, S., Wang, B., & Dutta, P. (2020). Nanoparticle processing: Understanding and controlling aggregation. Advances in colloid and interface science, 279, 102162.

[43] Liu, S., Reed, S. N., Higgins, M. J., Titus, M. S., & Kramer-Bottiglio, R. (2019). Oxide rupture-induced conductivity in liquid metal nanoparticles by laser and thermal sintering. Nanoscale, 11(38), 17615-17629.

Improving Semiconductor Reliability of Silver Sintering Die-Attach Adhesives for Large Die on Copper Lead Frames

Henry Martin
Chip Integration Technology Center
Nijmegen, The Netherlands
henry.martin@citc.org

Xinpei Cao
Henkel Electronic Materials
Irvine CA, USA
xinpei.cao@henkel.com

Jan Wijgaerts
Henkel Electronic Materials
Benelux, EU
jan.wijgaerts@henkel.com

Edsger Smits
Chip Integration Technology Center
Nijmegen, The Netherlands
edsger.smits@citc.org

Bo Xia
Henkel Electronic Materials
Irvine CA, USA
bo.xia@henkel.com

Ruud de Wit
Henkel Electronic Materials
Benelux, EU
ruud.dewit@henkel.com

Abstract—**Emerging trends like 6G Telecom and Electric Vehicles (EVs) are driving advancements in semiconductor packaging, specifically in back-end assembly. These enabling technologies are focused on achieving higher functionality, improved connectivity at higher frequencies, smaller form factors, efficient heat dissipation, and reduced power consumption. The instrumental role of Wide-Band Gap semiconductors to enable high-frequency RF transmissions, and efficient power switching necessitates advancements in thermal dissipation and the ability to withstand higher thermo-mechanical stresses in next-generation power devices. Concurrently, the semiconductor industry has been seeking a lead (Pb) solder replacement with improved automotive reliability for over two decades. The lead-free alternatives need to be compatible with larger die sizes on metal lead frames and must align with the high productivity demands of semiconductor back-end processing. However, finding a suitable, drop-in lead-free solution that can address these challenges has proven to be highly challenging [1]. This paper describes the ongoing product development and comprehensive testing of a new hybrid silver sintering die-attach development with stress-absorbing additives specifically designed for large die on lead frame based applications following close collaboration between Henkel and CITC. This development work has been the result of the dire need to meet the industry's evolving trends and requirements towards higher power and sustainable products.**

Index Terms—**Silver sintering; Hybrid sintering; Power semiconductors; Automotive reliability; Conductive die-attach; Pb solder replacement; Pb free.**

I. INTRODUCTION

Given the complex challenges encountered during the package assembly processes and the imperative to fulfill application-specific requirements, it is vital to assess the compatibility of individual components, processes, and materials [2–4]. Therefore, this study places specific emphasis on evaluating the die-attach layer, encompassing the assembly process and its impact on the reliability of the end product.

The choice of an appropriate die-attach material plays a critical role in package assembly. A review of die-attach materials for high-temperature applications can be found in [5–7]. Previously, high-lead (Pb)-based solders were mostly preferred due to their high melting point and relatively low stiffness. However, concerns regarding lead toxicity have led to restrictions on their usage. This has resulted in market demand for lead-free die-attach materials that offer superior thermal properties and improved thermo-mechanical stability. Addressing the challenge of reducing thermo-mechanical stresses between a large die with a coefficient of thermal expansion (CTE) of ~3ppm/°C and copper lead frames (~17ppm/°C) require a multi-disciplinary approach, encompassing thermal, electrical, mechanical, and material science domains. Traditional lead-free solders are not viable alternatives for semiconductor die-attach due to their low melting point. The reliability of lead-free solders with high-melting temperatures for harsh environmental applications has been a subject of comprehensive studies on its own [8].

Sintering technology presents a promising solution that offers several advantages compared to traditional solders. The sintering process involves the fusion of metal (micro/nano) particle precursors in paste form under heat and optionally pressure. One of the main benefits of sintering is its ability to sinter at relatively low temperatures, typically between 200–300°C, and achieving the high-melting-point characteristics of a bulk metal after sintering. Sintered joints exhibit improved thermal stability and enhanced thermal performance, making them highly suitable for high-temperature and high-power applications. However, the success of sintering heavily relies on the sintering paste chemistry and precise processing conditions. In electronic packaging, sintering materials are typically composed of silver or copper base metals. Copper is favored in terms of process compatibility, but sintering remains challenging due to the reactivity and sensitivity of copper precursors (risk of Cu oxidation/corrosion). As a result, silver sintering has emerged as the most promising and reliable lead-free die-attach solution [9–16].

Pressure-less silver sintering materials provide an excellent basis to improve performance, both thermal and electrical, as well as thermo-mechanical reliability compared to solders. Furthermore, from a practical point of view, pressure-less silver sintering is expected to enable increased production output as compared to solders and pressure-assisted sintering. However, a significant challenge with pure solvent-based silver sintering (pressure-less) materials is their inherent porosity, which leads to microscopic stress concentrations within the die-attach material and allows moisture ingress. This, in turn, reduces the Moisture Sensitivity Level (MSL) of the finished component and increases its brittleness compared to bulk silver material. To address these concerns, an effective approach has been the incorporation of an organic resin phase into the sintered joint, referred to as 'hybrid silver sintering.' This method minimizes porosity through filling and sealing, thereby improving the overall flexibility and toughness of the sintered bulk material. As a result, the thermo-mechanical performance of the sintered die-attach joint during stressful Moisture Sensitivity Level (MSL) and Thermo-Mechanical Cycling Lifetime (TMCL) reliability testing is enhanced. However, it should be noted that pressure-less hybrid silver sintering materials often exhibit lower thermal performance as compared to pure silver or pressure-assisted silver, which was previously investigated on similar silver sintering materials with metal-organic and resin-reinforced technology [17–19].

In this study, a new class of experimental 'stress absorbing' hybrid silver sintering die-attach materials ('EXP3' and 'EXP5') with ~150–200 W/m-K effective bulk thermal conductivity were investigated. The fundamental difference between these two new die-attach materials is that EXP5 has a higher bulk sintering density and consequently higher modulus than EXP3. These materials were developed to pass severe Automotive Electronics Council (AEC) Q100 and Q101 (respectively for Integrated Circuits and Discretes) reliability test requirements (up to 2000 thermal cycles −55°C/+150°C for highest Grade 0). This is achieved by optimizing the silver paste formulation with special filler packages and stress-absorbing additives, to increase the bulk sintering network and interfacial sintering to die and lead frame, next to improving MSL and thermal cycling performance. The thermo-mechanical reliability of these stress-absorbing hybrid silver sintering materials was benchmarked against a commercially available hybrid silver sintering material ('8068TI') with ~165 W/m-K bulk conductivity without the stress-absorbing feature [1] (for reference, LOCTITE® ABLESTIK ABP 8068TI is successfully released in automotive power semiconductors with die sizes up to ~3×3mm, but facing thermal cycling limitations on a copper lead frame with larger die sizes).

This work focuses on examining the package assembly process and thermo-mechanical reliability testing. The following section II provides a detailed description of the various steps involved in the test package assembly. The subsequent section III discusses the hot die-shear strength experiments and the experimental findings from the Thermo-Mechanical Cycling Lifetime (TMCL) testing. This section includes the functional device's electrical measurement (changes in electrical resistance R_{DS}(on)) during thermal cycling for the new class of hybrid silver sintering die-attach materials with stress-absorbing technology versus commercially available reference material. This paper concludes with section IV including remarks on the potentially broader use of the hybrid silver sintering die-attach adhesives towards larger die sizes on copper-based lead frame applications.

II. EXPERIMENTAL METHODS

A. Sample preparation

Silicon-based N-channel enhancement mode Field Effect Transistor (FET) with TrenchMOS technology was chosen as a test vehicle in this study. These devices are commercially used in high-performance automotive systems with a maximum power rating of up to 333W. The MOSFETs were assembled in Power Quad Flat No-Lead (PQFN) surface mount packages based on the following steps.

(a) The package assembly process commences with the preparation of the lead frame. As depicted in Fig. 1a, the package substrate (lead frame) is made of fully hardened C194 copper with 500μm thickness. To enhance adhesion, the lead frame surface was plated. Two distinct lead frame finishing options were examined: (i) Nickel-Palladium-Gold (NiPdAu, also known as 'PPF') and (ii) Silver. Prior to processing, the lead frames were treated with Argon plasma to remove surface contamination.

(b) Next, the commercial and experimental die-attach materials (8068TI, EXP3, EXP5) were dispensed in a snowflake pattern (Fig. 1b) using a Musashi Image Master 350PC Smart Dispense Robot equipped with a 25-gauge needle and applying 100–200KPa of pressure.

(c) Die placement was performed on a Finetech Sigma Fine-placer equipped with a position-controlled z-axis. The Silicon MOSFETs dies with 4.5×5.5×0.17 mm dimensions and silver backside metallization (Ag BSM) were carefully picked and placed with a placement force of approximately 0.3N (Fig. 1c). Once the die is properly positioned, the sintering process was carried out in a Budatec VS160 vacuum oven under Nitrogen. The first step involves staging at 130 °C for two hours, followed by pressure-less sintering at 200 °C for one hour. After sintering, all samples were inspected with a confocal microscope to measure the die-warpage.

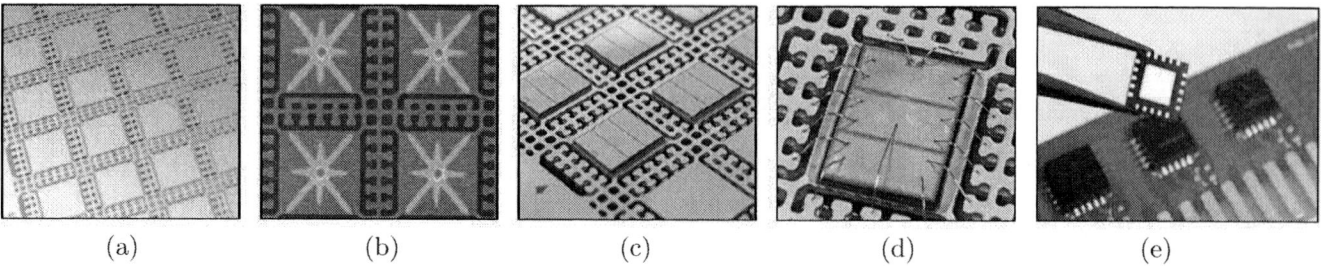

Fig. 1. A pictorial representation of the package assembly processes is shown. The various stages of the assembly process are explained in steps (a) to (e) in section II. A.

(d) Electrical connections were established to the lead frame through wire bonding (Fig. 1d). A 4-point probe configuration was used for all terminals; Source, Gate, and Drain.

(e) The final step involves transfer molding. This was performed with Sumitomo EME-G700LA epoxy molding compound at 175°C. The molded devices were further laser marked, singulated, and soldered with SAC305 alloy onto a Printed Circuit Board (Fig. 1e). These soldered devices were further subjected to Moisture Sensitivity Level (MSL3) and TMCL tests as described in the following subsection.

In this study, six types of functional test packages (3 die-attach materials and 2 lead frame metallizations) were prepared with over 32 packages per leg. The Bond-Line Thickness (BLT) of the sintered die-attach layer was evaluated using confocal optical microscopy. The typical wet BLT after die placement was ~45–50um and the average dry (sintered) bond line thickness was ~28–37um. In addition to the functional test packages, a series of test samples were prepared in a similar process to evaluate the Hot Die-Shear Strength (HDSS) at typical solder reflow temperature (260°C) for both commercial and experimental die-attach materials. A dummy silicon die of 4.9×4.6×0.52mm with silver backside metallization was sintered on a 2mm copper flange with both NiPdAu and Silver metallization using the same staging and sintering profile. The die-shear experimental results were evaluated in comparison with the die-warpage measurements after sintering.

B. Thermo-Mechanical Cycling Lifetime Experiments

To assess the reliability of the PQFN test packages prepared with commercial and experimental hybrid silver sintering die-attach adhesives, Thermo-Mechanical Cycling Lifetime (TMCL) experiments were conducted. Before commencing the TMCL testing, all packages underwent the Moisture Sensitivity Level 3 (MSL3) pre-conditioning step. The MSL classification system aids to assess the moisture sensitivity of an electronic component during storage before assembly, particularly surface mount devices (SMDs). This MSL3 step

involved drying the PQFN packages at 125°C for 24 hours, followed by exposure to 60% relative humidity at 60°C for 42 hours. Subsequently, the PQFN packages were subjected to three consecutive reflow soldering cycles (260°C) within one week.

The TMCL experiments were conducted based on the guidelines provided by AEC Q100 and Q101 as mentioned earlier. These stress tests are recommended for qualifying automotive-grade Integrated Circuits (ICs) and discrete semiconductors involving temperature cycling from −55°C (compression state) to 150°C (expansion state), as depicted in Fig. 2. The temperature cycling rate was maintained at approximately 1–2 cycles per hour. Throughout the testing process, the device's electrical performance was monitored intermittently.

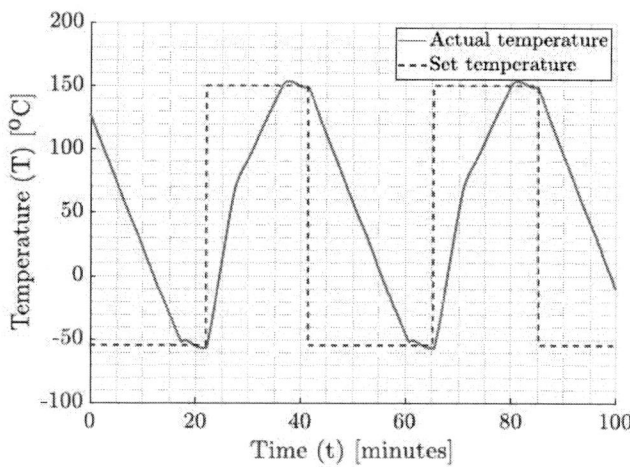

Fig. 2. The temperature cycling profile −55°C to 150°C shown is based on the automotive AEC norms for ICs & discrete semiconductors and the JEDEC temperature cycling standards.

The on-state electrical resistance (R_{DS}(on)) of the N-channel MOSFET was measured at specific thermal cycle intervals: 0, 100, 500, 1000, 1500, and 2000 cycles. The PQFN device was operated under a forward source-drain bias with a positive gate-source voltage (V_{GS}) of 20V and a drain-current (I_D) of 2A for a duration of 5ms. The reliability of the die-attach joint was monitored effectively during thermal cycling based on intermittent electrical measurements. This was possible since the device's drain terminal is electrically connected to the package substrate through the die-attach interface. The subsequent experiment results section discusses in detail the hot die-shear strength test results in comparison to the die-warpage measured after sintering, and the in-package electrical measurements on the PQFN test vehicle during TMCL testing.

III. EXPERIMENTAL RESULTS

A. Hot Die-Shear Strength Experiments

The die-shear strength of both commercial and experimental die-attach materials was evaluated at the typical lead-free reflow soldering temperature of 260°C to investigate and benchmark the adhesive and the sintering performance. The experiments were conducted using a Universal Bond Tester (Royce 650). The evaluation encompassed all three die-attach materials 8068TI, EXP3, and EXP5 on both copper lead frame metallizations (NiPdAu and Ag). The results of the HDSS experiments are presented in Table I. Both experimental hybrid silver sinter materials exhibited good adhesion/sintering on NiPdAu and Ag finish lead frames, yielding HDSS in the range of approximately 14 MPa ($\sim$1.4 kgf/mm^2 which is in line with typical application requirements). Cohesive failure modes within the sintered die-attach bond line were observed in all cases. However, for commercial 8068TI on NiPdAu metallization, a significantly lower HDSS of approximately 5 MPa was observed, with adhesive failure occurring at the NiPdAu interface. On the other hand, when 8068TI was used with the silver finish, failure occurred at the interface of the silver backside of the die with an HDSS of approximately 14 MPa (like EXP3 and EXP5).

TABLE I

HOT DIE-SHEAR STRENGTH (HDSS) OF THREE HYBRID SILVER SINTERING DIE-ATTACH ADHESIVES AT 260°C ON TWO DIFFERENT LEAD FRAME METALLIZATIONS.

HDSS [MPa]	8068TI			EXP3			EXP5		
NiPdAu	3.2	7.1	5.4	13.8	14.7	15.6	12.8	15.2	11.8
	3.5	2.9	4.8	12.2	13.2	14.6	13.7	16.1	16.2
	Avg. 4.48 MPa			*Avg. 14.02 MPa*			*Avg. 14.3 MPa*		
Silver	15.0	17.2	17.4	11.1	15.7	13.5	15.2	11.5	13.7
	10.9	10.1	12.5	11.2	11.5	13.6	13.5	12.4	14.6
	Avg. 13.85 MPa			*Avg. 12.76 MPa*			*Avg. 13.48 MPa*		

The difference in thermal expansion CTE of the copper lead frame ($\sim$17 ppm/°C) and the silicon die ($\sim$3 ppm/°C), causes measurable warpage at room temperature after sintering at $\sim$200°C. This is a direct result of the residual stress within the sintered die-attach bond line after cooling down. It is driven by the stiffness, CTE, and the temperature at which sintering occurs, and the die-attach bond line solidifies (which was found

experimentally to be around 150°C). Lower-than-expected warpage values are reminiscent of relaxation within the die-attach bond line layer. The die-warpage was experimentally measured using a Keyence laser confocal surface profiler. The results are tabulated in Table. II. In this configuration, Finite Element Method (FEM) simulations confirmed that the material properties (modulus, CTE, and thickness) of the die-attach layer have a neglectable influence on the die-warpage itself. Rather it is fully defined by the mechanical properties of the lead frame and die.

TABLE II

DIE-WARPAGE FOR ALL THREE DIE-ATTACH MATERIALS ONTO TWO LEAD FRAME FINISHES MEASURED AFTER THE SINTERING PROCESS OVER TWO AXES OF THE DIE (SHORT AND LONG).

Warpage [μm]	Die short side - 4.5mm			Die long side - 5.5mm		
	8068TI	EXP3	EXP5	8068TI	EXP3	EXP5
NiPdAu	*0.5*	5.8	6.6	*0*	9	9.3
Silver	*2.3*	5.2	5.5	*4.3*	8.6	10.2

The experimental 'stress-absorbing' hybrid silver sintering materials (EXP3 and EXP5) demonstrate a curvature of approximately 10μm along the long side of the die and around 6μm along the short side on both NiPdAu and Silver metalized lead frames. These results confirm a favorable bonding/sintering between the die and lead frame, contributing to the observed warpage of the complete stack. The die-warpage with EXP3 and EXP5 aligns within 30% of that obtained from FEM simulations. In contrast, the commercial silver sintering material (8068TI) on NiPdAu exhibits no warpage indicating poor adhesion and delamination at the die edges. Lower-than-expected warpage was observed for 8068TI material on Silver metalized lead frames. The poor adhesion of 8068TI to NiPdAu is in good alignment with the results from the Hot Die-Shear Strength (HDSS). Contradicting evidence was observed for 8068TI on Silver metalized lead frames. The HDSS on Silver indicates good adhesion, however, the die-warpage measured suggests some relaxation within the 8068TI material.

In summary,

- The new 'stress-absorbing' hybrid silver sintering materials (EXP3 and EXP5) demonstrate high die-shear strength at 260°C on both NiPdAu and Silver metalized lead frames. These findings are further supported by die-warpage measurements conducted on functional devices after sintering.

- The 8068TI reference material shows lower adhesion to these specific NiPdAu finished PQFN lead frames. This was confirmed through hot die-shear strength and die-warpage measurements. 8068TI adhesion to these specific Silver metalized lead frames indicates higher HDSS values, however die-warpage measurements indicate some form of mechanical relaxation.

B. Electrical Measurements

The electrical on-state resistance (R_{DS}(on)) refers to the resistance measured across the drain-source terminals of the packaged device under forward bias. It is a combination of the electrical resistances of the die (R_{Die}), die-attach (R_{DA}), lead frame (R_{LF}), and the package substrate solder joint interface (R_S). It's assumed that the solder joint interface remains relatively stable during the TMCL test. This assumption is based on the matched coefficient of thermal expansion (CTE) between the copper lead frame and the copper metallic pad on the PCB substrate. Hence, changes in the electrical resistance R_{DS}(on) are expected to correspond to degradation, specifically in the die-attach bond line which is more prone to failure as compared to other materials during cyclic loading. The packaged devices were subjected to thermal cycling from $-55°C$ to $150°C$, and the electrical resistance (R_{DS}(on)) was measured intermittently at room temperature ($\sim 25°C$). The test results are presented in Fig. 3.

The measurement results were analyzed based on Weibull probability distribution. Weibull probability analysis refers to a statistical method of analyzing and interpreting the failure data based on the Weibull distribution. It involves fitting the observed failure data to the Weibull distribution, thereby estimating the shape and scale parameters of the distribution. The shape parameter β represents the slope of the fitted line and the scale parameter η is related to the intercept (electrical shift or drift). Accordingly, the electrical resistance (R_{DS}(on)) of the packages measured at fixed intervals (0, 100, 500, 1000, 1500, and 2000 thermal cycles) were plotted against the percentage distribution (Fig. 3). For each sample variation, the data from every cycle was fitted with a straight line to determine the shape and scale parameters. The evolution of the shape β and scale η parameters over thermal cycling up to 2000 cycles can be visually observed in Fig. 3. From a physical context, the scale parameter (η) signifies the shift or drift in electrical performance (typically degradation by higher resistance values), where the shape parameter (β) reflects the distribution of the degraded samples after thermal cycling.

For the PQFN packages with NiPdAu finish, a noticeable drift in R_{DS}(on) is observed with increasing thermal cycles. In particular, 8068TI and EXP5 show more shifts to the right

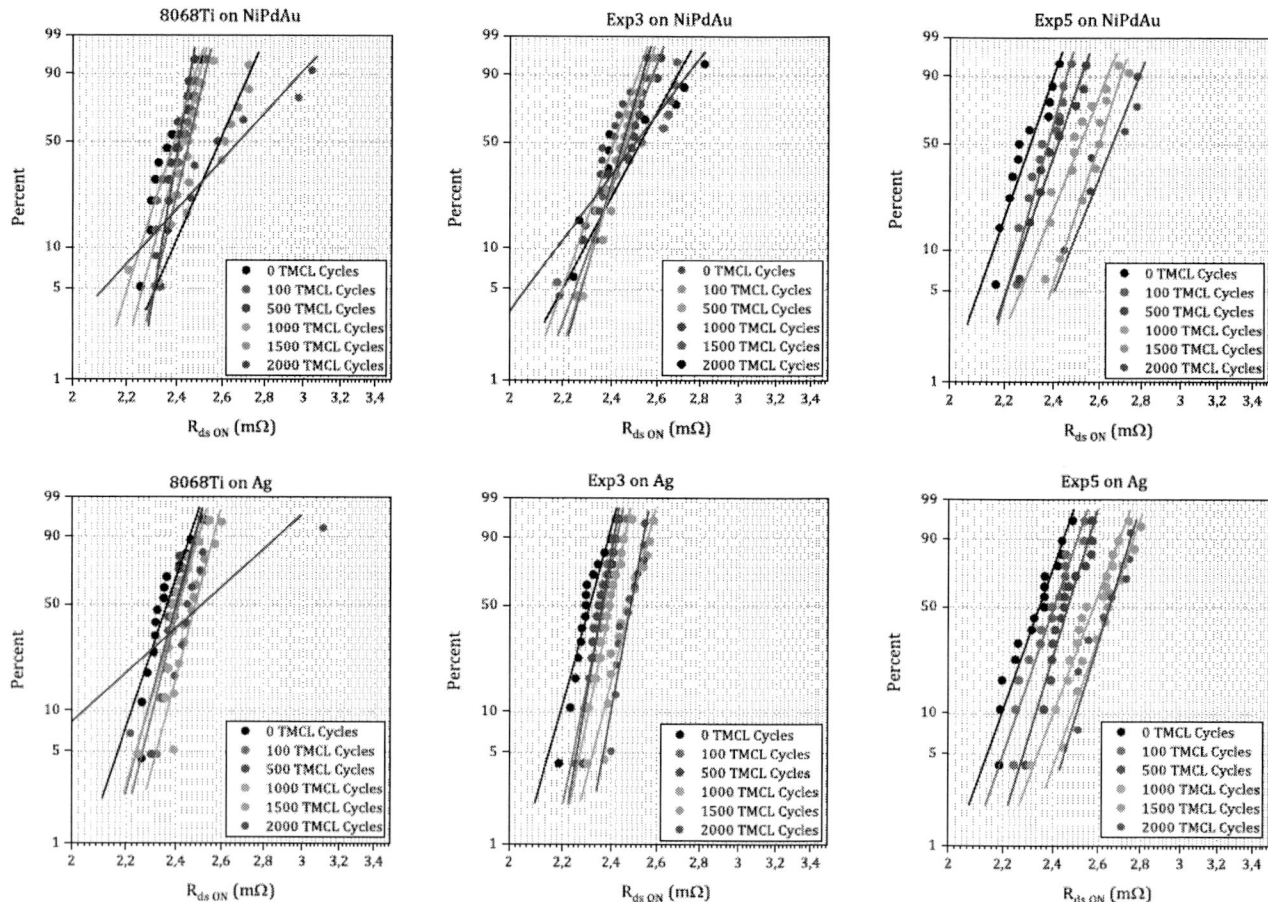

Fig. 3. Weibull probability plots showing the electrical on-state resistance (R_{DS}(on)) measured on all three tested die-attach materials (8068TI, EXP3, and EXP5) on two PQFN lead frame metallizations (NiPdAu and Ag) up to 2000 thermal cycles ($-55°$ C to $150°$C).

indicating a higher degree of degradation compared to EXP3, leading to considerably higher electrical resistance as depicted in Fig. 3. Conversely, the slope of the line decreases with increasing thermal cycling in the case of 8068TI and EXP3. This suggests that the degree of performance diverges among the samples, implying the presence of a possible stochastic degradation mechanism. In the case of the die-attach materials (8068TI, EXP3, and EXP5) on Silver (Ag) finish, a relatively limited drift in R_{DS}(on) is observed as compared to NiPdAu finish, suggesting a more gradual and slower degradation process during thermal cycling. Besides, the measurement data does not exhibit significant variations in comparison to the NiPdAu test samples. This indicates that the degree of degradation is relatively consistent among all samples with a Silver (Ag) finish, with EXP3 showing the most stable electrical performance with the lowest electrical drift after 2000 thermal cycles.

Another statistical aspect of interpreting the experimental measurements is to understand the dispersion present within the datasets. The interquartile range (IQR) is a robust statistical method to evaluate the dispersion. The IQR method provides a summary of data spread by focusing on the middle 50% of observations between the 25th and 75th percentiles. Based on the experimental measurement results shown in Fig. 3, the IQR was determined and visually represented in Fig. 4. These box plots effectively illustrate the central 50%

data spread, while the whiskers indicate the 1.5IQR range. Additionally, Fig. 4 presents the mean, median, and outliers present in the measurement data, providing a comprehensive analysis of the thermal cycling datasets' characteristics.

The IQR data as presented in Fig. 4 clearly indicates that all three die-attach materials (8068TI, EXP3, and EXP5), when used with both NiPdAu and Ag finish lead frames on this specific PQFN test package, exhibit a shift of less than 20% in the electrical on-state resistance over 2000 thermal cycles. Notably, the EXP3 die-attach with stress-absorbing additive stands out, demonstrating a deviation of less than 10% on both lead frame metallizations. Meeting the stringent long-term reliability requirement of less than 10% variation in electrical resistance (R_{DS}(on)) is of utmost importance, especially for automotive grade semiconductor packages. In this regard, EXP3 convincingly meets this criterion with both NiPdAu and Ag finish PQFN copper lead frames on relatively large die size ($\sim$4.5$\times$5.5 mm).

In summary,

- A noticeable difference in the electrical performance during thermal cycling is observed between the 8068TI, EXP3, and EXP5 test samples on both lead frame metalization types; NiPdAu and Ag finish lead frames (Fig. 3 & 4).

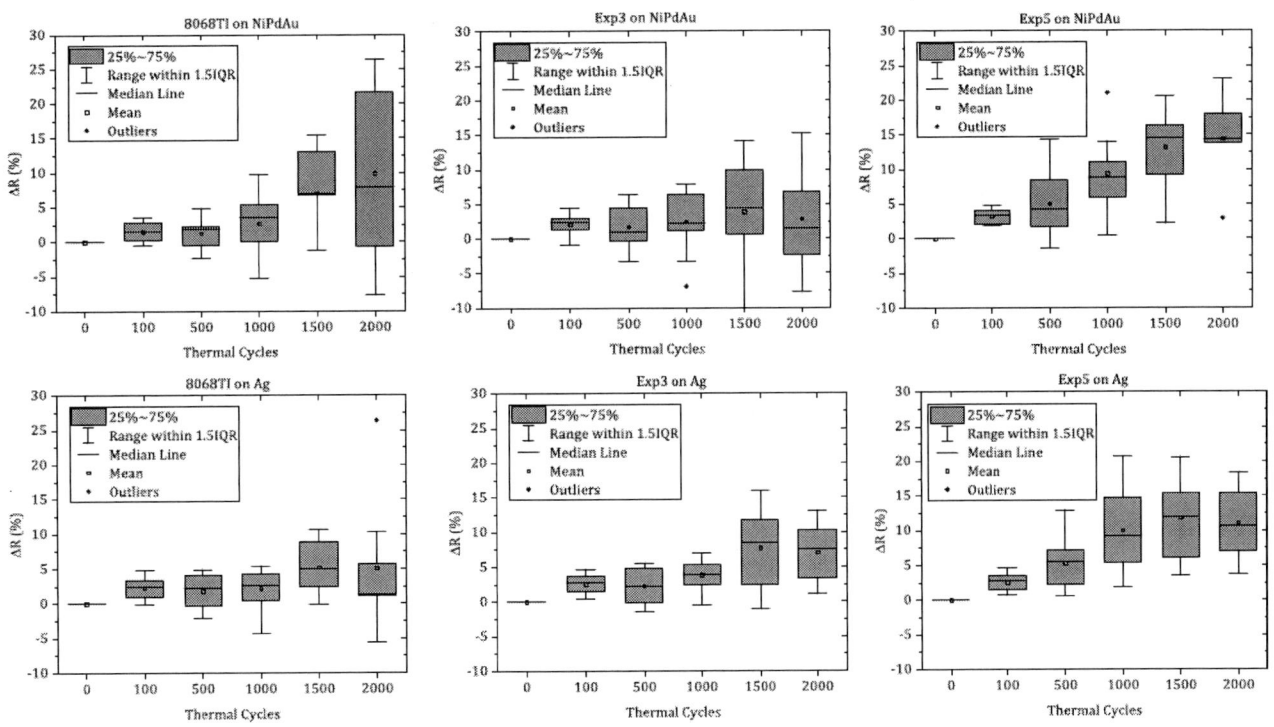

Fig. 4. Interquartile range (IQR) plots showing the percentage change in the electrical on-state resistance (R_{DS}(on)) measured on all three tested die-attach materials (8068TI, EXP3, and EXP5) on two PQFN lead frame metallizations (NiPdAu and Ag) up to 2000 thermal cycles (-55°C to 150°C).

- The new experimental hybrid silver sintering material EXP3 with stress-absorbing additives, exhibits the best resilience to thermo-mechanical loading. EXP3 demonstrates less than 10% deviation in $R_{DS}(on)$ over 2000 thermal cycles on both copper lead frame types (with Ag finish giving the best test results versus NiPdAu).

- Comparatively, 8068TI reference material without 'stress-absorber' performs similarly well when used with Ag finish lead frames, but shows large variety in $R_{DS}(on)$ test results on NiPdAu.

- The other new EXP5 material with 'stress-absorber' displays an $R_{DS}(on)$ drift of $\sim$10% when used with Ag lead frames and $\sim$15% when used on NiPdAu finish lead frames.

IV. DISCUSSIONS AND CONCLUSIONS

This comprehensive test work on PQFN test packages with $4.5\times5.5\times0.17$ mm die size demonstrates that all three hybrid silver sintering die-attach materials exhibit reasonably good to very good thermo-mechanical reliability performance based on electrical measurements. It became evident that the adhesion, sintering, and reliability performance of these three die-attach formulations on Ag finish is overall better as compared to NiPdAu finish lead frames. Notably, the experimental hybrid silver sintering die-attach materials (EXP3 and EXP5) with 'stress absorbing technology' demonstrate improvements compared to the commercially available (8068TI) reference material without 'stress-absorber' in terms of adhesion and sintering performance. The EXP3 die-attach formulation stands out in this specific PQFN test work with superior performance on both NiPdAu and Ag finish lead frames, as confirmed by hot Die-Shear experiments and automotive thermal cycling (TMCL) testing up to 2000 cycles ($-55/+150$°C) with a maximum of 10% drift in $R_{DS}(on)$. Similarly, the EXP5 die-attach formulation with higher bulk sintering density and consequently higher modulus than EXP3, shows improved adhesion in Die-Shear experiments compared to 8068TI reference material. However, the thermal cycling reliability performance of EXP5 falls somewhere between that of 8068TI and EXP3.

It is essential to mention, realize, and acknowledge that the outcome of all test work described in this study is based on the specific PQFN test package used and that the test results can be different on other test devices, such as different geometries (die size and thickness), die types (Si/SiC/GaN), backside metals (Ag/Au), lead frame finishes, surface roughness, etc. This understanding is crucial when selecting the most suitable die-attach material for specific applications, as the choice can significantly impact the overall performance and reliability of the package. For instance, for applications with a thicker lead frame and thinner die, and vice versa, EXP5 with higher bulk sintering density and higher modulus may perform better than EXP3 having lower modulus.

To summarize, the new experimental hybrid silver sintering materials with stress-absorbing feature exhibit commendable die-shear strength at elevated temperatures (260°C), which are further supported by die-warpage measurements. Furthermore, the comprehensive thermo-mechanical cycling test results strongly indicate that the stress-absorbing hybrid silver sintering technology offers suitable die-attach solutions with enhanced thermo-mechanical reliability for larger die sizes on copper lead frame applications. With the introduction of stress-absorbing additives, the application space of this 'next generation' of hybrid silver sintering die-attach adhesives, developed for copper-based lead frames, can be expanded from typical die sizes of $\sim$3$\times$3mm in the past, towards $\sim$5$\times$5mm and above, by qualifying the severe automotive AEC Q100 and Q101 reliability test requirements (2000 thermal cycles $-55/+150$°C for highest Grade 0). Besides, this pressure-less hybrid silver sintering technology offers other benefits such as high thermal conductivity, sustainable lead-free composition, and well-established dispensing application using existing die-attach equipment and processes proven in high-volume IC and Discrete production.

ACKNOWLEDGMENT

The authors extend their sincere gratitude and recognition to all other Henkel Product Development and Application Engineering colleagues, as well as Dave Reijs and other CITC colleagues involved for their precise and comprehensive test work, next to their valuable support and input throughout this extensive investigation. We also would like to acknowledge the appreciated assistance of SENCIO BV in Nijmegen (NL), providing their transfer molding knowledge and capabilities to encapsulate the PQFN test parts prior to testing.

REFERENCES

[1] R. d. Wit, "Silver sintering die attach developments for rf, power and automotive applications," in *2021 23rd European Microelectronics and Packaging Conference & Exhibition (EMPC)*, 2021, pp. 1–4. DOI: 10.23919/EMPC53418.2021.9584971.

[2] Z. Chen, J. Zhang, S. Wang, and C.-P. Wong, "Challenges and prospects for advanced packaging," *Fundamental Research*, 2023, ISSN: 2667-3258. DOI: https://doi.org/10.1016/j.fmre.2023.04.014. [Online]. Available: https://www.sciencedirect.com/science/article/pii/S2667325823001334.

[3] D. R. Frear, L. N. Ramanathan, J.-W. Jang, and N. L. Owens, "Emerging reliability challenges in electronic packaging," in *2008 IEEE International Reliability Physics Symposium*, 2008, pp. 450–454. DOI: 10.1109/RELPHY.2008.4558927.

[4] H. Wang et al., "Transitioning to physics-of-failure as a reliability driver in power electronics," *IEEE Journal of Emerging and Selected Topics in Power Electronics*, vol. 2, no. 1, pp. 97–114, 2014. DOI: 10.1109/JESTPE.2013.2290282.

[5] K. Siow, *Die-attach materials for high temperature applications in microelectronics packaging*, Oct. 2018.

[6] V. R. Manikam and K. Y. Cheong, "Die attach materials for high temperature applications: A review," *IEEE Transactions on Components, Packaging and Manufacturing Technology*, vol. 1, no. 4, pp. 457–478, 2011. DOI: 10.1109/TCPMT.2010.2100432.

[7] H. S. Chin, K. Y. Cheong, and A. B. Ismail, "A review on die attach materials for sic-based high-temperature power devices," *Metallurgical and Materials Transactions B*, vol. 41, pp. 824–832, 2010.

[8] J. Wang, S. Xue, P. Zhang, P. Zhai, and Y. Tao, "The reliability of lead-free solder joint subjected to special environment: A review," *Journal of Materials Science: Materials in Electronics*, vol. 30, May 2019. DOI: 10.1007/s10854-019-01333-w.

[9] S. A. Paknejad and S. H. Mannan, "Review of silver nanoparticle based die attach materials for high power/temperature applications," *Microelectronics Reliability*, vol. 70, pp. 1–11, 2017, ISSN: 0026-2714. DOI: https://doi.org/10.1016/j.microrel.2017.01.010. [Online]. Available: https://www.sciencedirect.com/science/article/pii/S0026271417300161.

[10] S. Fu, Y. Mei, G.-Q. Lu, X. Li, G. Chen, and X. Chen, "Pressureless sintering of nanosilver paste at low temperature to join large area ($\geq$100mm2) power chips for electronic packaging," *Materials Letters*, vol. 128, pp. 42–45, 2014, ISSN: 0167-577X. DOI: https://doi.org/10.1016/j.matlet.2014.04.127. [Online]. Available: https://www.sciencedirect.com/science/article/pii/S0167577X14007046.

[11] W. Liu *et al.*, "Recent progress in rapid sintering of nanosilver for electronics applications," *Micromachines*, vol. 9, p. 346, Jul. 2018. DOI: 10.3390/mi9070346.

[12] H. Yan, P. Liang, Y. Mei, and Z. Feng, "Brief review of silver sinter-bonding processing for packaging high-temperature power devices," *Chinese Journal of Electrical Engineering*, vol. 6, no. 3, pp. 25–34, 2020. DOI: 10.23919/CJEE.2020.000016.

[13] Y. Liu, H. Zhang, L. Wang, X. Fan, G. Zhang, and F. Sun, "Effect of sintering pressure on the porosity and the shear strength of the pressure-assisted silver sintering bonding," *IEEE Transactions on Device and Materials Reliability*, vol. 18, no. 2, pp. 240–246, 2018. DOI: 10.1109/TDMR.2018.2819431.

[14] W. Schmitt, L. M. Chew, and R. Miller, "Pressure-less sintering on large dies using infrared radiation and optimized silver sinter paste," in *2018 IEEE 68th Electronic Components and Technology Conference (ECTC)*, 2018, pp. 539–544. DOI: 10.1109/ECTC.2018.00085.

[15] K. Sugiura *et al.*, "Reliability evaluation of sic power module with sintered ag die attach and stress-relaxation structure," *IEEE Transactions on Components, Packaging and Manufacturing Technology*, vol. 9, no. 4, pp. 609–615, 2019. DOI: 10.1109/TCPMT.2019.2901543.

[16] L. A. Navarro *et al.*, "Thermomechanical assessment of die-attach materials for wide bandgap semiconductor devices and harsh environment applications," *IEEE Transactions on Power Electronics*, vol. 29, no. 5, pp. 2261–2271, 2014. DOI: 10.1109/TPEL.2013.2279607.

[17] N. Mizumura and K. Sasaki, "Development of low-temperature sintered nano-silver pastes using mo technology and resin reinforcing technology," in *2014 International Conference on Electronics Packaging (ICEP)*, 2014, pp. 526–531. DOI: 10.1109/ICEP.2014.6826735.

[18] K. Sasaki, N. Mizumura, A. Tsuno, S. Yagci, and G. Kopp, "Development of low-temperature sintering nano-silver die attach materials for bare cu application," in *2017 21st European Microelectronics and Packaging Conference (EMPC) & Exhibition*, 2017, pp. 1–5. DOI: 10.23919/EMPC.2017.8346875.

[19] X. Hu *et al.*, "Microstructure analysis based on 3d reconstruction model and transient thermal impedance measurement of resin-reinforced sintered ag layer for high power rf device," in *2023 24th International Conference on Thermal, Mechanical and Multi-Physics Simulation and Experiments in Microelectronics and Microsystems (EuroSimE)*, 2023, pp. 1–7. DOI: 10.1109/EuroSimE56861.2023.10100799.

Pick and Place of sensitive chips with vacuum-free Gecomer® tools

Lukas Lorenz
Fraunhofer Institute for Photonic
Microsystems
Dresden, Germany
lukas.lorenz@ipms.fraunhofer.de
ORCID 0000-0003-4145-0342

Thomas Ludewig
Fraunhofer Institute for Photonic
Microsystems
Dresden, Germany
thomas.ludewig@ipms.fraunhofer.de

Kai Swiecinski
Fraunhofer Institute for Photonic
Microsystems
Dresden, Germany
kai.swiecinski@ipms.fraunhofer.de

Henrik Ollmann
Innocise GmbH
Saarbrücken, Germany
h.ollmann@innocise.com

Amirabbas Razkordanisharahi
Innocise GmbH
Saarbrücken, Germany
a.razkordani@innocise.com

Volker Bock
Fraunhofer Institute for Photonic
Microsystems
Dresden, Germany
volker.bock@ipms.fraunhofer.de

Abstract—To cope the lack of suitable pick and place methods for sensitive chips, we present a new process using a gripper tool from the company INNOCISE called Gecomer®. The tool, made of polymer pillars, is using Van der Waals forces to create adhesion between the gripping tool and the chip surface. Hence, the tool works without any vacuum. While placing, the Van der Waals interactions are reduced by buckling those pillars resulting in a decreased contact area. The challenge is to find the right parameters to avoid a re-gripping of the chip during the Gecomer®-movement away from the chip. In this paper we present results on the placement accuracy including the reproducibility compared to a standard vacuum tool for different process parameters. For that, we used a pick and place machine with built-in inspection camera to examine the deviation from the target position. It is revealed that both vacuum and polymer gripper have a similar deviation of approx. ±15 μm. Hence, the placing accuracy is not negatively influenced by the Gecomer® tool and depends only on the machine parameters. Furthermore, we analyze the surfaces of the chips, whether there are any residues of the gripping tool. In a first optical inspection, no damages or residues could be found. The new vacuum-free pick and place process has the potential to open new opportunities in the assembly of sensitive (MEMS)-chips regarding reproducibility, accuracy, and process throughput.

Keywords—pick and place, MEMS assembly, die bond, chip bond, gripping tool, chip handling, die sorting

I. INTRODUCTION (*HEADING 1*)

A. Motivation

MEMS chips are an emerging market. Especially the packaging of these chips is considered crucial, because 60% to 80% of the MEMS device costs are packaging [1]. Hence, efficient packaging processes are required. One example in this context is the MEMS handling and pick and place. Spatial light modulators (SLM) are a subsegment of MEMS chips, using micromirror arrays for lithography and imaging applications. Large SLM-chips have mirror counts in the millions with mirror edge lengths in the micrometer range. For these chips, conventional pick and place processes with tools using vacuum are difficult to use, due to the risk of damaging the movable mirror parts on the chip by getting close to the vacuum. Hence, the chip must be designed with enough space between the mirror and vacuum tool touching area. The alternative is a manual assembly with tweezers or special grippers. However, this is not reproducible, cannot be used for large chips and holds the risk of damaging the edges of the chip.

B. Challenges for MEMS packaging and handling

As already mentioned, pick and place and assembly are a difficult tasks for MEMS and need special attention [2]. Packaging of MEMS is linked to the specific application and in contrast to standard IC packaging, MEMS most often need an interface to the environment [3]. TABLE I shows the challenges for MEMS during the entire packaging workflow and possible solutions. In this paper, we focus on the challenges, which arise during die handling.

TABLE I. OVERVIEW OF THE PACKAGING CHALLENGES FOR MEMS CHIPS [3]

	Challenge	Possible solution
Release	Damaging the movable parts	
Dicing	Risk of contamination or destruction	Release after dicing
Die handling	Device failure, due to a very sensitive die top face to contact	Handling at the side faces
Stress	Performance degradation or device failure	Low modulus die attach, compatible CTE match-ups, cooling
Outgassing	Corrosion, performance degradation	Low outgassing adhesives
Testing	Possible device failure after many assembly steps	Pre-package testing and/or post package testing

For chips, which do not need a dedicated contact to the environment, the standard packaging includes a cap, to cover the movable MEMS parts. This is typically done in wafer-level packaging processes [1] and is used for example for gyroscope MEMS [4]. For MEMS, vacuum tools can be used, if the movable parts are robust enough and/or the sensitive surfaces are far away from the vacuum point or the MEMS part is encapsulated [5, 6]. If the chip is properly protected, it can be handled by standard vacuum pick and place tools [7].

However, sensitive MEMS chips cannot be moved by vacuum pick tools [3]. As pointed out in TABLE I, the state-of-the-art handling method for more sensitive MEMS dies is a gripping at their side faces [8]. Especially, dies where the active area is not protective covered, as it is often the case for SLM [9, 10], need special assembly processes. Since edge grippers are mostly not compatible with standard (vacuum) pick and place tool heads, separate infrastructure and machines are required. However, a handling at the edges has the risk of damaging the chip and more space in the package is required next to the chip for the gripper. Manual assembly with special tweezers is an option as well, but not suitable for a reproducible packaging of higher chip quantities. In every case, chip-package co-design is crucial for MEMS-chips [6] and especially for the die handling. This includes space for the gripping or distances between gripping-/vacuum surfaces and sensitive structures.

In this paper we introduce a new way of die handling, to overcome the lack of proper pick and place tools for sensitive chips. The Gecomer ® gripper works without any vacuum and uses the top surface of the chip for gripping. This is possible due to the use of Van-der-Waals forces. The basic principle of the Gecomer® tool is briefly introduced in the next section, followed by the results of the pick and place experiments covering the test setup, the placing accuracy, and the comparison to standard vacuum tools. We finish with a short conclusion and an outlook.

II. BASIC PRINCIPLE OF MICROSTRUCTURED ADHESIVE SURFACES

The Gecomer® technology is a sustainable bio-inspired gripping solution suitable for pick and place applications without need of any external energy. It is inspired by hairy hierarchical structures on gecko's feet that enables it to move rapidly on vertical surfaces and the ceiling. These grippers utilize a phenomenon called dry adhesion, which is attained by arranging specific microstructures on the gripping surface.

Millions of fine hairs called setae equipped with spatula-shaped ends on gecko's feet, as depicted in Fig. 1, enables it to strongly attach their feet to different surfaces and quickly detach them. No vacuum tool, glue, or sticky material exists on their feet, and the studies show this strong temporary and reversible adhesion relies on intermolecular forces, i.e., Van der Waals forces between the setae and the surface [11, 12].

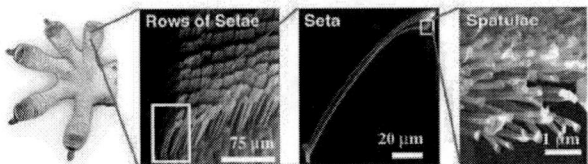

Fig. 1. Gecko's feet comprising of hierarchical structure of fine hairs which is an inspiration for dry adhesives [13].

In the nature, adhesive forces between large surfaces are usually not caused by van der Waals forces, since the real contact area is normally way smaller than the apparent one. Meanwhile, a higher number of smaller sub-contacts can adapt and conform better with the roughness of the counter surface and therefore provide a larger real contact area, which increases the adhesion.

To grasp the mechanism of dry adhesive structures, it is helpful to have a quick review on the contact mechanics. In the Johnson, Kendall, and Roberts theory, adhesive contact of elastic bodies, adhesive force refers exactly to the energy required to create new surfaces through detachment [14]. It

also suggests that when a cylindrical geometry is attached to a rigid substrate, the stress distribution through the interface is not uniform and there is a stress concentration at the edge, as it is shown in Fig. 2. Maugis et al. [15] considered a cohesive zone that supports the stress concentration applied to the edge. As long as the work required to overcome the work of adhesion attributed to the cohesive zone is done, the interfacial detachment initiates, grows and detachment occurs [15, 16].

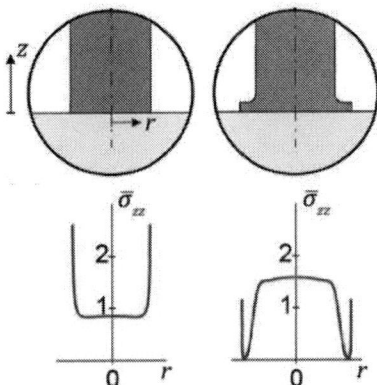

Fig. 2. A comparison of stress distribution at the tip of a flat punch and a mushroom-shaped pillar [17].

For a continuous contact surface, once the detachment starts to propagate, it advances along the whole interface, while in case of a higher number of smaller contacts, e.g., setae of gecko's feet, the detachment initiation results in detachment of just one sub contact, and for the next one there should be another detachment initiation induced [18]. Furthermore, for smaller contacts, the stored elastic energy is reduced [16], so that its dissipation does not provide the required energy to initiate or propagate a detachment in the next sub contact [18]. Also, at an interface with a high number of small sub contacts the stress distribution is more uniform, and defects are smaller and less destructive compared to a large continuous contact.

For handling microscopic parts, the challenge is mostly not to pick up the part but to release it again. Microstructures with cylindrical tips, flat punches, show a higher switching between a high and low adhesive state, therefore more suitable for a pick and place cycle in this application, as it is shown in Fig. 3.

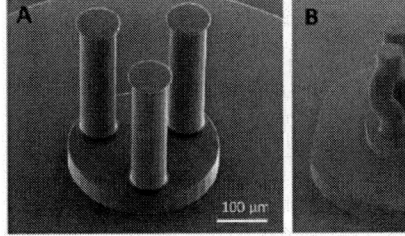
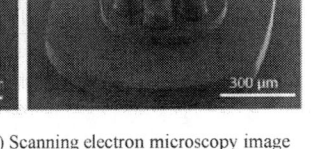

Fig. 3. Micro handling grippers, A) Scanning electron microscopy image (SEM) of a micro handling gripper with three mushroom pillars, and B) SEM image of a micro handling gripper comprising of 5 flat punch pillars.

III. PICK AND PLACE EXPERIMENTS

A. Test setup

For the pick and place experiments, a Gecomer® tool is used, as it is shown in Fig. 4. There are five pillars with a length of 315 µm and a diameter of 75 µm each. To evaluate the suitability of the tool for chip sorting and placing, a test is performed with a standard pick and place machine. The machine setup is shown in Fig. 5. For the test, blank Silicon dummy chips with edge dimensions of 2.5 x 2.5 mm² and a thickness of 1.4 mm are used. They are picked from a waffle pack (1) to either a Gel Pak® or a second waffle pack (2). The target position is defined in relation to the lines and edges in the Gel Pak® or waffle pack. The Gecomer® tool is mounted in the pick and place module (3), as it is shown in Fig. 6. To ensure precise alignment between gripper and chip, the gripper position is thought to the machine by two cameras (4,5). After placing the chip, an inspection camera (6) controls the resulting position and measures the deviation in x,y direction and the rotation angle around the z-axis. The manufacturer of the machine indicates the accuracy for placing with better than ±10 µm, which represents the goal for the Gecomer® tool. If the placing accuracy with the gripper reaches ±10 µm, the precision is only limited by the machine itself and not by the Gecomer® tool. To proof that, a comparative measurement with a standard vacuum tool is performed as well.

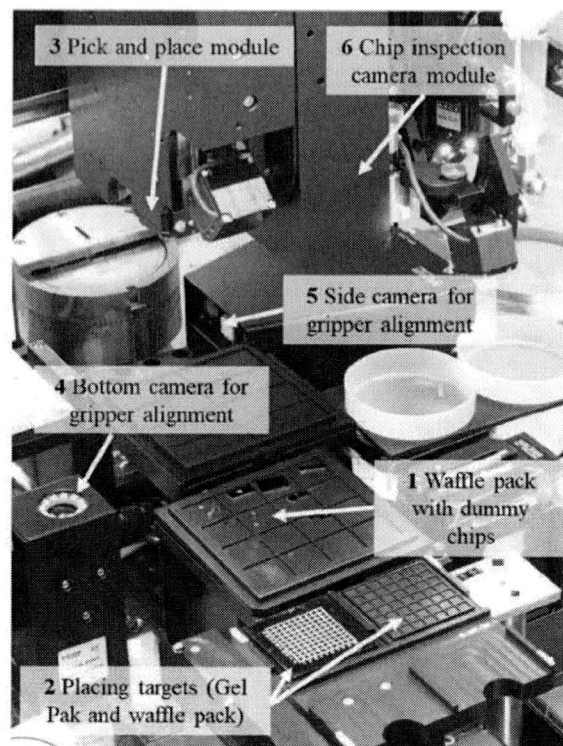

Fig. 5. Setup of the pick and place machine with the waffle pack in the middle for providing the dummy chips and the Gel Pak® and waffle pack at the bottom as placing target

Fig. 4. Gecomer® tool with five pillars, used for the pick and place experiments.

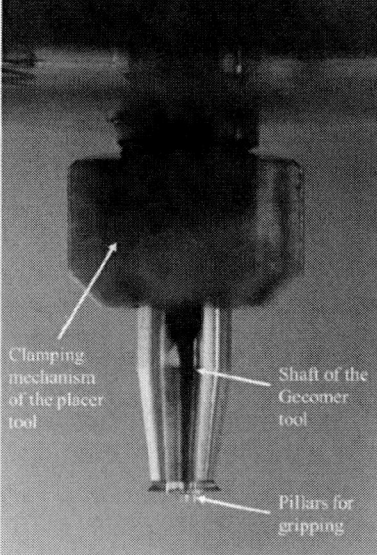

Fig. 6. Detailed picture of the tool head of the machine with the Gecomer® gripper.

The first challenge is to pick the chip from the initial waffle pack. For that, the gripper is positioned above the chip, which is to be picked. After that the pick and place head is moved down until 0.5 mm above the chip surface. From this point the head's moving speed is reduced to 0.1 mm·s⁻¹, to allow a soft touch. For a proper touching the pick and place head is programmed to move 50 µm below the chip surface level. Because of the elastic behavior of the Gecomer® tool, the pillars and the chip surface huddle against each other and the gripping mechanism, described in section II, starts. After that, the head moves slowly upwards (again 0.1 mm·s⁻¹) until he reaches a height of 0.5 mm again. This ensures, that the chip sticks to the gripper. If the movement is too fast, the chip may drop. The successful lifting of a dummy chip is shown in Fig. 7. It also illustrates the small area, which is effectively used for gripping, showing the potential for chips with large active areas and only few space for gripping.

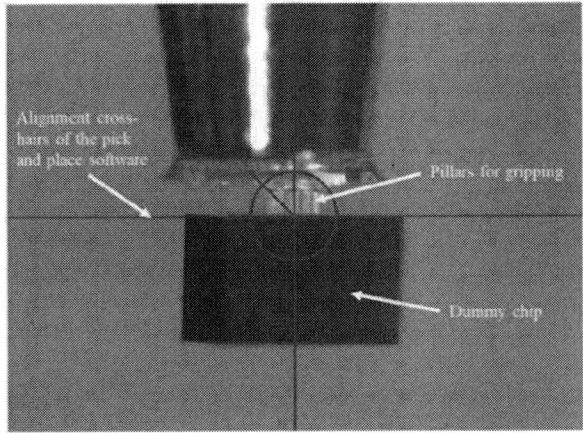

Fig. 7. Image of a dummy chip attached to the pillars of the Gecomer® tool captured with the monitoring camera built-in in the pick and place machine.

B. Pick and place from waffle pack to gel-pak

After lifting the chip with the Gecomer® gripper, the next challenge is to find the right parameters for placing. The chip must detach from the gripper while it must not change its position on the placing target. This could be the case if the chip briefly stays attached to the gripper and drops after lifted again.

To overcome this challenge, we used Gel Paks® as a first placing target. Since the Gel Pak® has a sticky surface, the force between chip and Gel Pak® is higher than between chip and gripper. Hence, the chip is easily released from the Gecomer® tool. This effect could also be used for a placing in wet adhesives for die bond procedures.

During the test, we varied the movement speed of the pick and place head to see, if there is any influence. For every different speed, 20 chips were picked and as mentioned above, we inspected the placing result visually and measured the accuracy. Fig. 8 shows the deviation in x and y direction for three different movement speeds.

The test reveals that most of the chips are placed with an accuracy within the machine specs of ±10 µm. Only a total of six chips does not fulfill this requirement but are still in the range of ±15 µm. Furthermore, the movement speed has no significant influence on the placing accuracy. As a result, it can be summarized, that the pick and place from waffle pack to Gel Pak® with a Gecomer® tool is working properly.

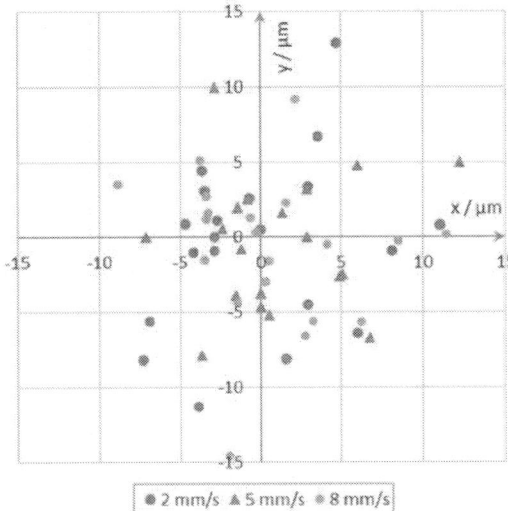

Fig. 8. Lateral deviation of placed chips for a WP-to-GP picking compared between different movement speeds of the tool head.

C. Pick and place from waffle pack to waffle pack

For the next test, we want to pick from a waffle pack to another waffle pack, to see whether it is possible to detach the chip from the gripper on a non-sticking surface. To release the chip, the pillars of the Gecomer® tool need to be buckled, until their surface no longer attaches to the chip surface. This is achieved by a movement of the pick and place head in z-direction towards the chip surface. For these tests we moved the tool head 150 µm further downwards after the chip touches the waffle pack, which corresponds to roughly the half of the pillar's height.

After buckling the pillars, the chip is detached from the Gecomer® tool. Nevertheless, the challenge is to move up the pick and place head without re-grip the chip again. For the 5 mm·s⁻¹ speed used in the previous test, we observed "jumping" chips. These chips only detach when the pick and place head was already back in the air. Although that was only the case for the minority of the chips, we increased the speed to 50 mm·s⁻¹ and 100 mm·s⁻¹ to overcome this issue for all of the chips.

Again, we measured the accuracy of the placed chips. The results are shown in Fig. 9 and Fig. 10. At a first glance, it seems all placed chips are again in the range of ±10 µm for the x and y deviation, even for the 5 mm·s⁻¹ speed. Only a look at an enlarged axis division (Fig. 10) shows the outliers. While for the two higher speeds all data points lie in the ±10 range, at 5 mm·s⁻¹ three chips "jumped" off the Gecomer® tool, resulting in deviations up to 450 µm. Nevertheless, for 50 mm·s⁻¹ and 100 mm·s⁻¹ the placing is very accurate and within the machine's specs.

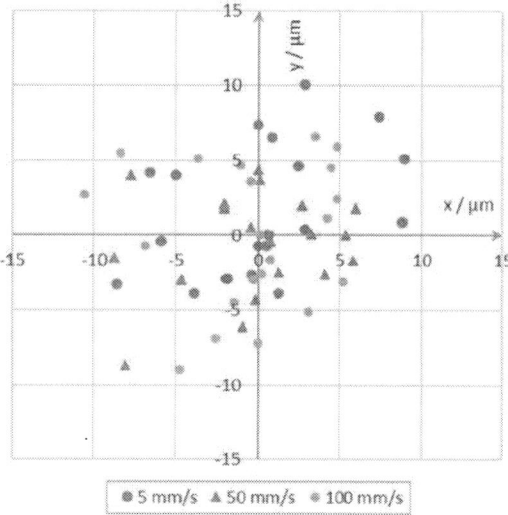

Fig. 9. Lateral deviation of placed chips for a WP-to-WP picking compared between different movement speeds of the tool head.

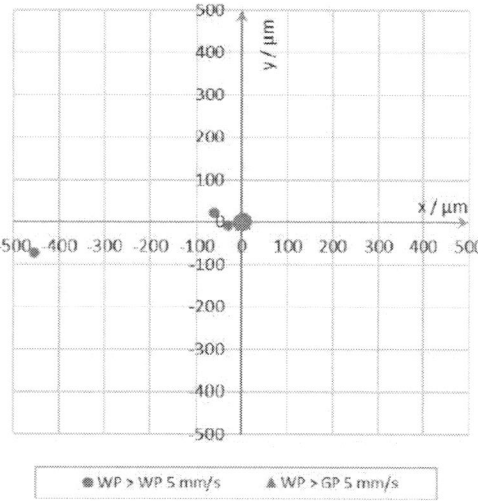

Fig. 10. Lateral deviation of placed chips compared between waffle pack (WP) and Gel Pak® (GP) as placing target; The greater axes scale reveals three outliers, which were not placed properly.

D. Comparison to standard vacuum tools

In the last experiment, we compare the placing accuracy of the Gecomer® tool to a standard pick and place tool, which uses vacuum to hold the chip. The advantage of the latter one is the easier release of the chip. Simply interrupt the vacuum and the chip is no longer attached to the tool. Since this avoids chip movement after placing, this might be beneficial for the accuracy.

Fig. 11 shows the lateral deviation of 20 chips placed with Gecomer® tool and standard vacuum tool respectively. It is revealed that there is no significant difference between the two tools, proving the pick and place process is only influenced by the machine itself and not by the used gripper.

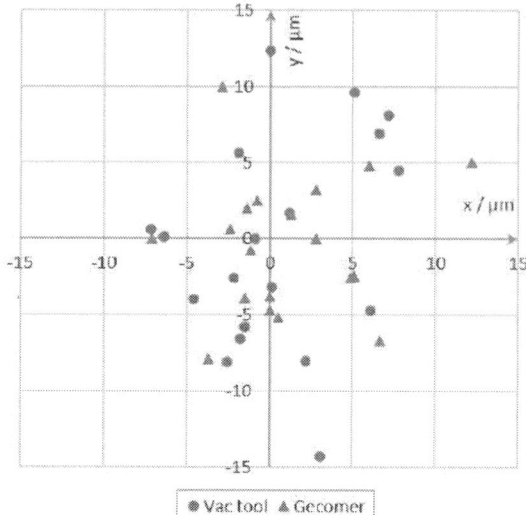

Fig. 11. Lateral deviation of placed chips compared between standard vacuum tool and Gecomer® gripper for pick and place from waffle pack and gel-Pak at 5 mm/s

E. Inspection on residues on the chip surface

The last question to be answered in this paper is, whether there are any residues or damages on the chip surface induced by the Gecomer® gripper. For that, we inspected the chip by microscope before and after the pick and place process. As shown in Fig. 12, no changes or residues on the surface are visible. Hence, the gripper is suitable to touch, e.g. on bond pads without affecting the pads.

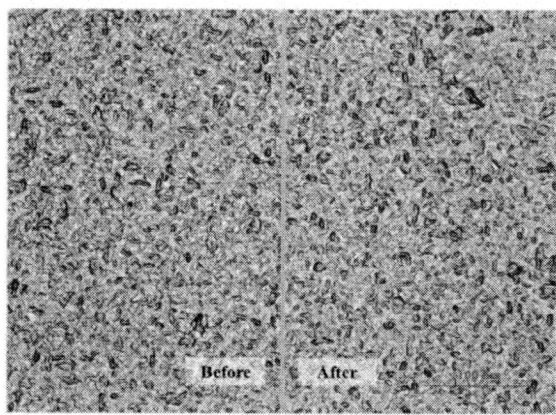

Fig. 12. Surface of the dummy chip, which can be picked with the Gecomer® gripper.

IV. CONCLUSION

In this paper, we introduced a vacuum-free technology for pick and place of silicon dies. To our best knowledge, it is the first time this is demonstrated without using edge-grippers. The Gecomer® tool allows for safe handling of chips and a placing accuracy at the same level as standard vacuum tools. The achieved precision of ±10 μm is only determined by the xyz-portal of the machine and is not negatively influenced by the new gripping technology. Furthermore, an advantage compared to side grippers is the compatibility to standard pick and place machines. This allows for easy changeover on established equipment.

However, the limit of the gripping technology is the dependance on the chip surface. An example can be seen in Fig. 13, which shows the surface of a dummy chip with oxide coating. The coarse-grained finish prevents a sufficient contact of the Gecomer® pillars, which is why the Van-der-Waals interactions are not developed.

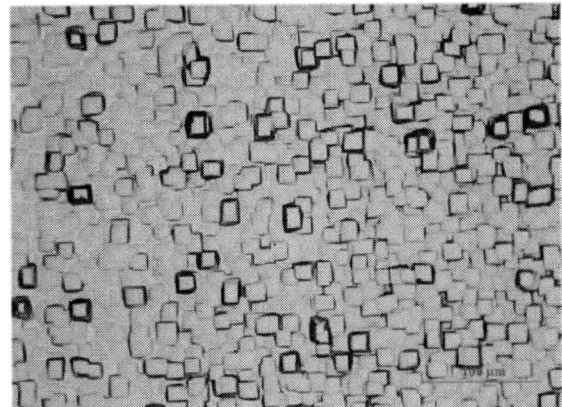

Fig. 13. Surface of a dummy chip with oxide coating, which was not pickable. The rough surface prevents a proper adhesion to the gripper.

The next step is to design a chip specific Gecomer® gripper, which is adopted to the chip layout. The idea is to align the pillars to the bond pads. Hence, the gripping is done only on the robust surface of the pads, while the sensitive MEMS parts are spared. The exact alignment is challenging and needs to be tested as well as the machine parameters for the new tool. Furthermore, the boundaries of the gripping regarding surface of the chip need to be identified.

REFERENCES

[1] J. H. Lau, *3D IC integration and packaging*. New York: McGraw-Hill Education, 2016.

[2] T.-R. Hsu, *MEMS packaging*. London: INSPEC, 2004.

[3] C. B. O'Neal, A. P. Malshe, S. B. Singh, W. D. Brown, and W. P. Eaton, "Challenges in the packaging of MEMS," in *Proceedings International Symposium on Advanced Packaging Materials. Processes, Properties and Interfaces (IEEE Cat. No.99TH8405)*, 1999.

[4] J.-W. Joo and S.-H. Choa, "Deformation Behavior of MEMS Gyroscope Sensor Package Subjected to Temperature Change," *IEEE Transactions on Components and Packaging Technologies*, vol. 30, no. 2, pp. 346–354, 2007, doi: 10.1109/tcapt.2007.897948.

[5] J. J. M. Zaal, W. D. van Driel, and G. Q. Zhang, "Challenges in the assembly and handling of thin film capped MEMS devices," *Sensors (Basel, Switzerland)*, vol. 10, no. 4, pp. 3989–4001, 2010, doi: 10.3390/s100403989.

[6] H. Tilmans *et al.*, "MEMS packaging and reliability: An undividable couple," *Microelectronics Reliability*, vol. 52, 9-10, pp. 2228–2234, 2012, doi: 10.1016/j.microrel.2012.06.029.

[7] F. Janek *et al.*, "Feasibility Study of an Automated Assembly Process for Ultrathin Chips," *Micromachines*, vol. 11, no. 7, p. 654, 2020, doi: 10.3390/mi11070654.

[8] S. Parrish, "Non-Surface Contact Approach for Device Flip," in *International Wafer Level Packaging Conference (IWLPC)*, San Jose, CA, USA, 2020, pp. 1–6.

[9] Jan-Uwe Schmidt, Peter Duerr, and Michael Wagner, "Customized micro mirror array for highly parallel industrial laser direct processing on the micro-and nanoscale," in *MikroSystemTechnik Congress*, 2017.

[10] P. Dürr, A. Neudert, M. Nitzsche, C. Hohle, H. Stolle, and J. Pleikies, "MEMS piston mirror arrays for computer generated holography," in *MOEMS and Miniaturized Systems XXI*, San Francisco, California, United States, 2022, pp. 47–54.

[11] E. Arzt, S. Gorb, and R. Spolenak, "From micro to nano contacts in biological attachment devices," *PNAS*, vol. 100, no. 19, pp. 10603–10606, 2003, doi: 10.1073/pnas.1534701100.

[12] A. K. Geim, S. V. Dubonos, I. V. Grigorieva, K. S. Novoselov, A. A. Zhukov, and S. Y. Shapoval, "Microfabricated adhesive mimicking gecko foot-hair," *Nature Materials*, vol. 2, no. 7, p. 461, 2003.

[13] A. K. Kasar, R. Ramachandran, and P. L. Menezes, "Natural Adhesion System Leads to Synthetic Adhesives," *J Bio Tribo Corros*, vol. 4, no. 3, p. 43, 2018, doi: 10.1007/s40735-018-0160-1.

[14] E. Barthel, "Adhesive elastic contacts: JKR and more," *J. Phys. D: Appl. Phys.*, vol. 41, no. 16, p. 163001, 2008, doi: 10.1088/0022-3727/41/16/163001.

[15] D. Maugis, "Adhesion of spheres: The JKR-DMT transition using a dugdale model," *Journal of Colloid and Interface Science*, vol. 150, no. 1, pp. 243–269, 1992, doi: 10.1016/0021-9797(92)90285-T.

[16] R. Hensel, K. Moh, and E. Arzt, "Engineering Micropatterned Dry Adhesives: From Contact Theory to Handling Applications," *Adv. Funct. Mater.*, vol. 28, no. 28, p. 1800865, 2018, doi: 10.1002/adfm.201800865.

[17] E. Arzt, H. Quan, R. M. McMeeking, and R. Hensel, "Functional surface microstructures inspired by nature – From adhesion and wetting principles to sustainable new devices," *Progress in Materials Science*, vol. 120, p. 100823, 2021, doi: 10.1016/j.pmatsci.2021.100823.

[18] M. Kamperman, E. Kroner, A. Del Campo, R. M. McMeeking, and E. Arzt, "Functional Adhesive Surfaces with "Gecko" Effect: The Concept of Contact Splitting," *Adv. Eng. Mater.*, vol. 12, no. 5, pp. 335–348, 2010, doi: 10.1002/adem.201000104.

Research of Chip Placement Accuracy for Fan-Out WLP using A Novel Self-Assembly Stage

Tadatomo Yamada, Ken Takano, Toshiaki Menjo, Shinya Takyu
Next Generation Innovation Group
LINTEC Corporation
1-1-1 Koishikawa, Bunkyo-ku, Tokyo, 112-0002, Japan
t-yamada@post.lintec.co.jp

Abstract—This paper discusses the alignment accuracy after self-assembly using a novel self-assembly method with a uniquely developed porous stage. In a preceding paper, we have introduced our process that combines tape expansion technology and self-assembly technology. By using this process and machine device, it is possible to significantly improve the throughput of pick and place process for FO-WLP. Furthermore, it is indicated that our self-assembly method using the porous chuck stage enables achievement of high precision chip alignment and high throughput. In this study, two factors are evaluated to clarify the impact on the chip accuracy after self-assembly; 1) the influence of chip position before self-assembly, 2) the relationship between stage size of self-assembly area and chip size. As a result of evaluation, if the chip size is the same as the porous stage size, the alignment accuracy after self-assembly is within 5 μm or less even with varied chip positions before self-assembly. On the other hand, when the porous stage size and chip size are different, the chip position before self-assembly has an impact on the alignment accuracy after self-assembly. Based on the results obtained in this study, further development of machine device and process will be pursued to improve the alignment accuracy after self-assembly.

Keywords- FO-WLP; pick-up and place; tape expansion; self-assembly;

I. INTRODUCTION

Die attach is also commonly known in the semiconductor industry as die bonding or die mount. It is the process of attaching a silicon chip to a substrate or package. This process is an essential for creating a reliable electrical and mechanical connection between the chip and the rest of the electronic devices. Various precision die attach equipment are used to accurately pick and place the chip onto the substrate. These devices are an essential tool in semiconductor backend process, but it faces challenges in terms of throughput. In order to solve this problem, the self-assembly technology has the potential to address the throughput problem associated with traditional pick and place assembly process. Self-assembly technology is known for higher placement accuracy process using liquid surface tension and it has long been reported in the field of semiconductor devices. In 2005, Fukushima *et al.*, demonstrated self-assembly technique for the first time to apply for 3D integration [1]. They have employed self-

assembly technologies for advanced chip-to-wafer 3D integration [2] [3]. CEA-LETI has been working since several years on self-assembly process. Recently, they have demonstrated collective D2W direct bonding self-assembly process [4] [5]. These previous papers mainly reported on the method of self-assembly technique. Although there have been previous papers on self-assembly technique, there have been no papers on process or devices specifically applying self-assembly technique. In 2021, we have introduced for the first time our novel process that combines tape expansion technology and self-assembly technology [6]. Our proposal process flow is illustrated in Figure 1. There are two parts basically, one is tape expansion and the other is self-assembly. Figure 2 shows the photograph of tape expansion process. We have developed the tape expansion machine device with a function allowing that tape can be expanded in four directions individually [7]. To achieve precise alignment after tape expansion, we have developed our own self-assembly technique using a novel porous stage. Our unique self-assembly process is illustrated in Figure 3. Generally, self-assembly technique uses a supporting wafer which surface is formed with hydrophilic and hydrophobic area. This process enables high precision alignment, but throughput is a challenge due to the required steps of liquid dropping before self-assembly and liquid drying after self-assembly. On the other hand, our method performs liquid supply and recovery through the porous stage. Due to the elimination of liquid dropping step and liquid drying step, our method enables high throughput. We have confirmed that it is possible to achieve high precision chip alignment through self-assembly using our novel porous stage (see Figure 4).

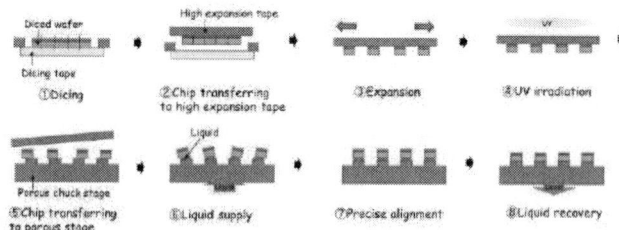

Figure 1. A novel chip placement process using tape expansion technology and self-assembly technology.

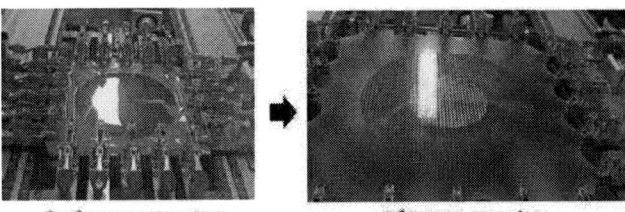

Before expansion   **After expansion**

Figure 2. Before and after photos of tape expansion.

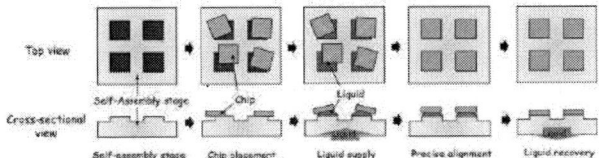

Figure 3. Self-assembly process using a novel porous stage.

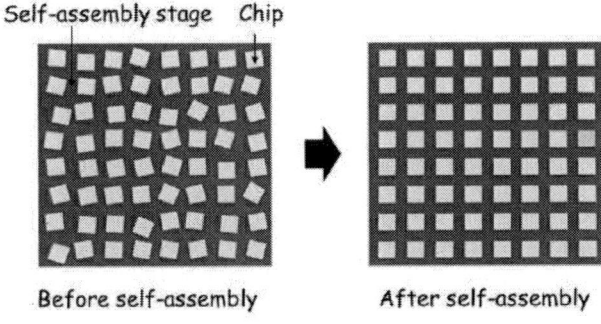

Before self-assembly   **After self-assembly**

Figure 4. Before and after photos of self-assembly using a novel porous stage.

We are initially proposing this process for FO-WLP applications. One of the highest cost manufacturing process for FO-WLP is pick and place process. To reduce the process cost, pick and place equipment try to provide the high productivity machines [8]. However, since the conventional pick and place process is a one chip at a time process, it is difficult to improve the throughput dramatically. Figure 5 shows the comparison of throughput between our novel chip placement device and conventional pick and place device. Throughput comparison is performed under following conditions; WPH of our device is 15 wafer which size is 8 inch, CPH of pick and place device is 20K chips. As shown in Figure 5, a significant advantage of our device is its remarkably higher throughput compared to conventional device. Since our process is designed to manufacture based on wafer size, it is possible to produce 15 wafer per hour regardless of chip size. On the other hand, since the porous chuck stage is still in the prototype stage, there are remaining challenges regarding alignment accuracy. In this study, two factors are evaluated to clarify the impact on the alignment accuracy after self-assembly, 1) the influence of chip position before self-assembly, 2) the relationship between stage size of self-assembly area and chip size.

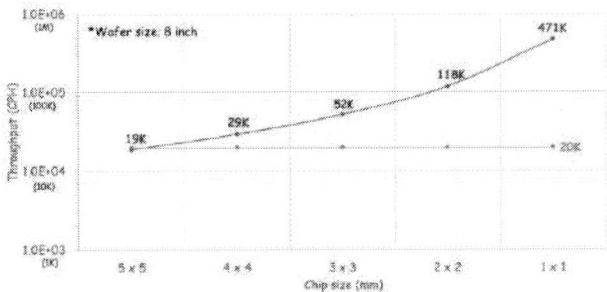

Figure 5. Comparison of throughput between a novel chip placement device and conventional pick and place device.

II. EXPERIMENTAL AND RESULTS

A. Evaluation method of self-assembly

First, a porous stage with the following specifications is prepared for this study; self-assembly area size of 3 mm square and the distance between self-assembly areas is 2 mm. Evaluation method of self-assembly is illustrated in Figure 6. Chip is randomly placed on the porous stage, allowing it to extend beyond its edges. Then, the distance between the center of chip and the center of stage is measured with CNC vision measuring system (Mitutoyo Corporation; Quick Vision ACCEL) and its distance is defined as the misalignment distance. Liquid is supplied through the porous stage until it reaches the surface of the stage and chip is precisely aligned by liquid surface tension. Water is used as the liquid for self-assembly in this study. After chip is aligned, liquid is recovered through the porous stage. Then, the distance between the center of chip and the center of stage is measured with CNC vision measuring system again.

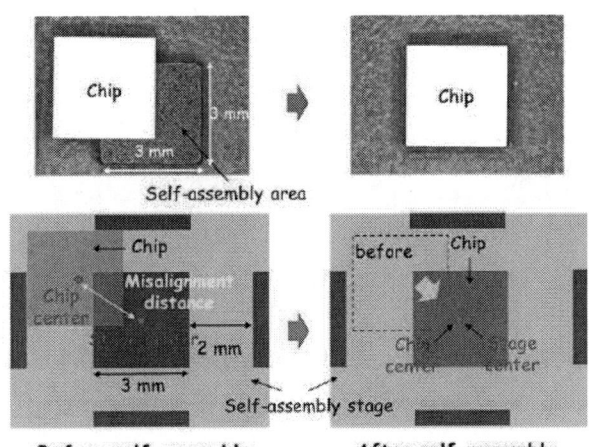

Before self-assembly   **After self-assembly**

Figure 6. Evaluation method of self-assembly.

B. Influence of chip position before self-assembly

In this section, we evaluated the influence of the chip position before self-assembly on the alignment accuracy after self-assembly. We prepared chips with size of 3 mm

245

square and thickness of 350 μm for this evaluation. As shown in Figure 7, the chip positions before self-assembly are divided into four areas; 1) 501~1,000 μm, 2) 1,001~1,500 μm, 3) 1,501~2,000 μm, 4) 2,001~2,500 μm. Each chip in four areas are aligned by self-assembly and the alignment accuracy is measured as described in paragraph A. Figure 8 shows the result of alignment accuracy after self-assembly. Chip alignment is possible even when the chip is placed in other area, as long as the chip has a large contact area with target location of self-assembly. Furthermore, regardless of the initial chip position, the alignment accuracy after self-assembly shows a median value of 5 μm or less. As indicated by the evaluation results, high precision chip alignment is achievable regardless of the chip position before self-assembly, as long as the chip size is the same as the porous stage size.

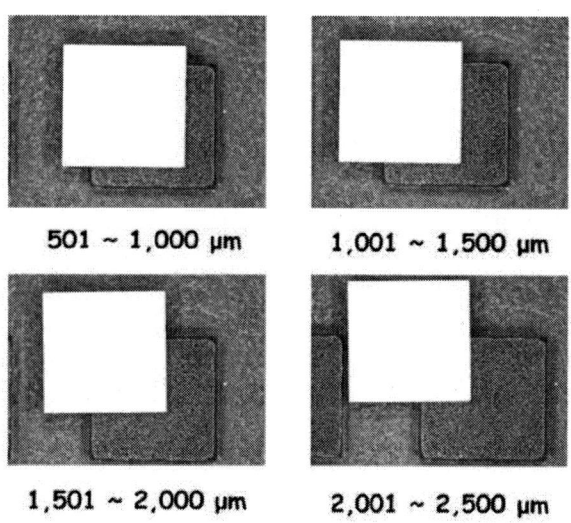

Figure 7. Photos of chips with different positions before self-assembly.

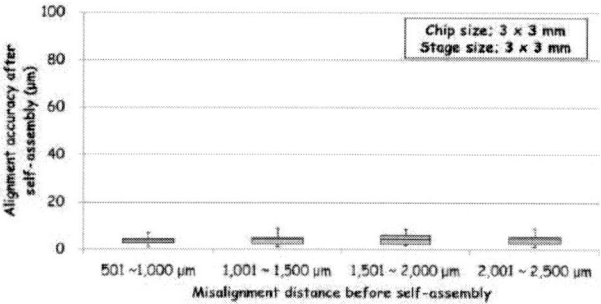

Figure 8. Alignment accuracy after self-assembly with different misalignment distances.

C. Relationship between stage size and chip size

In this section, we evaluated the influence of relationship between stage size of self-assembly area and chip size. We prepared seven different chip sizes; 2.7 mm, 2,8 mm, 2,9 mm, 3.0 mm, 3.1mm, 3.2 mm and 3.3 mm, all chips were square in shape. Before self-assembly, the chips are randomly positioned to surround the stage in 360-degree arrangement. Misalignment distance before self-assembly is below 1,200 μm for this evaluation. Evaluation result shows in Figure 9 and 10. When the chip size is 3 mm square, the alignment accuracy shows a median value of 3 μm. As the chip size becomes smaller, the alignment accuracy becomes lower. For the chip size of 2.7 mm square, the alignment accuracy shows a median value of 120 μm. As the chip size becomes larger, the alignment accuracy becomes lower as well. However, the alignment accuracy is still better compared to when the chip size is smaller. For the chip size of 3.3 mm square, the alignment accuracy shows a median value of 37 μm. From these results, it is indicated that when the stage size and chip size are different, the alignment accuracy becomes lower. Particularly, when the chip size becomes smaller, this tendency becomes more pronounced.

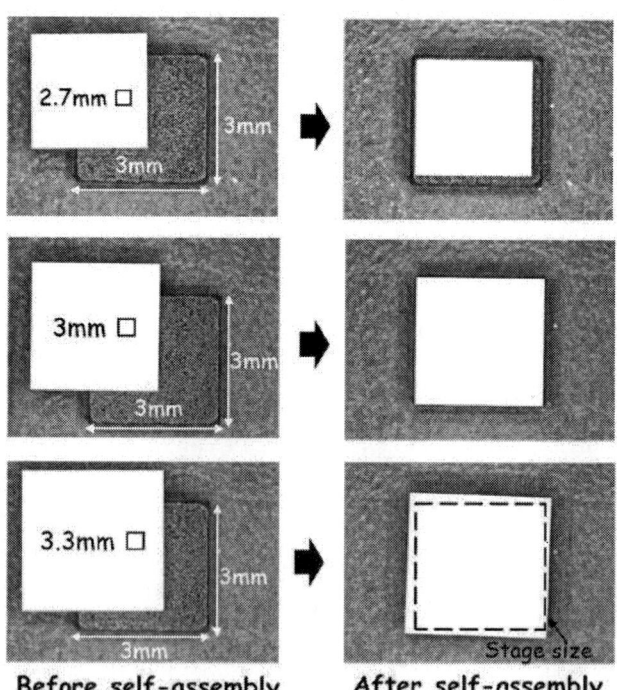

Figure 9. Alignment accuracy after self-assembly with different chip sizes.

Figure 10. Before and after photos of self-assembly with different chip sizes.

III. DISCUSSION

As I stated above, when the stage size and the chip size are different, the alignment accuracy becomes lower. This tendency becomes more pronounced when the chip size is smaller. The question we have to consider is why the chip size affects the alignment accuracy after self-assembly. To clarify the reason, the positions of chips before and after self-assembly are plotted in Figure 11. The chips after self-assembly align around the center of stage when the chip size is 3 mm square. On the other hand, the chips after self-assembly tend to move away from the center of stage when the chip size differs from the stage size. Furthermore, as the chip size becomes smaller, the positions of the chips move even further away from the center. For further analysis, the positions of chips before and after self-assembly are compared. Although the distance from the center may vary before and after self-assembly, the overall trend of chip placement remains the same. Based on these results, it can be concluded that when the stage size and chip size are different, the initial position of the chip before self-assembly has an impact on the alignment accuracy after self-assembly.

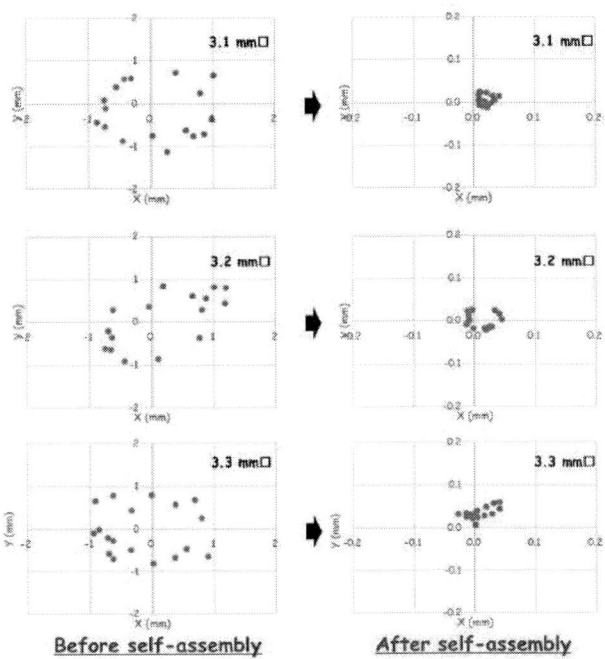

Figure 11. Chip positions before and after self-assembly with different chip sizes.

To discuss the reason why the position of the chip before self-assembly has an impact on the alignment accuracy, the movement of chips during self-assembly is illustrated in Figure 12. Chips are capable of moving due to the surface tension of liquid even when the chip size is changed. However, when the stage size and chip size are different, the chips come into contact with the corner of the stage during movement and then unable to continue moving. The evaluation results for chip size of 2.7 mm square and 3.3 mm square with different initial chip positions before self-assembly are presented in Figure 13 and 14. Looking at the photographs after self-assembly, it can be observed that both sizes of chips are stopped at the corners of the stage. Additionally, it can be observed that the corners of the stage where the chips come to stop vary depending on the initial positions of the chips before self-assembly. This tendency is particularly strong in chip size of 2.7 mm square. With 3.3 mm square chip, a similar tendency can be observed where the chip stops at the corners of the stage. However, there is a possibility of chip tilting after self-assembly due to the chip being larger than the stage. Based on these results, the following measures are necessary; 1) adjusting the dimensions and shape of the stage to accommodate different chip size, 2) to further enhance the tape expansion technique for controlling the initial chip positions of self-assembly.

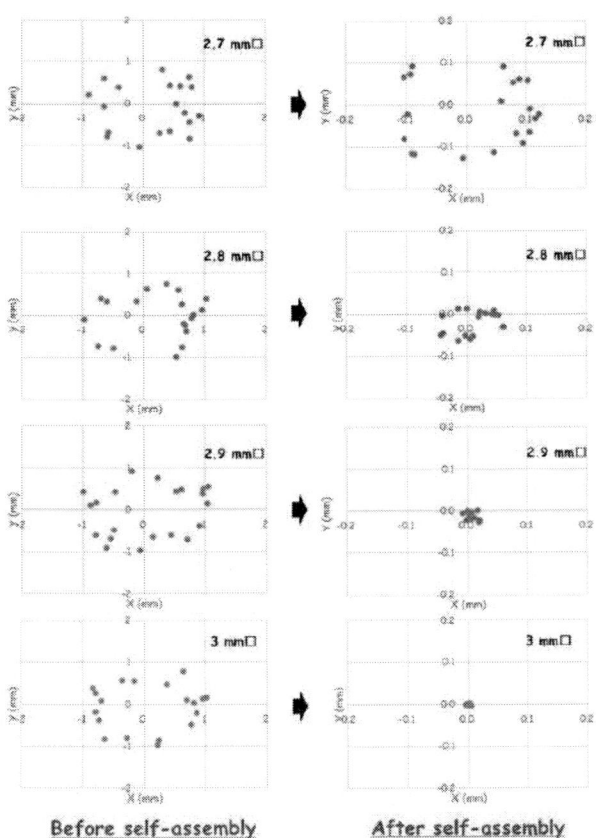

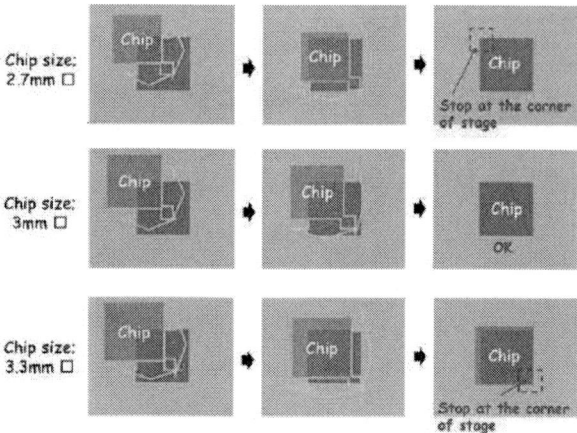

Figure 12. Movement of chips during self-assembly with different chip sizes.

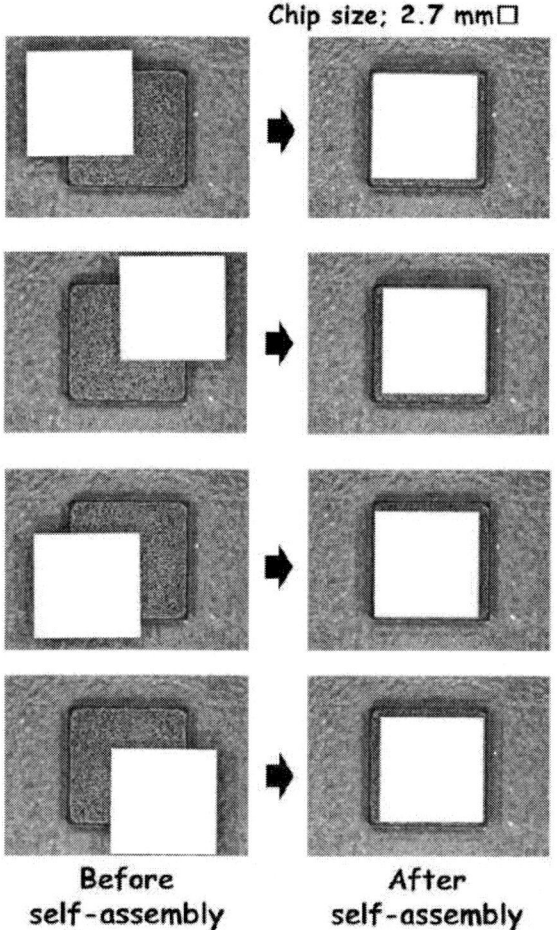

Figure 13. Before and after photos of self-assembly with different initial chip positions with the chip size of 2.7 mm square.

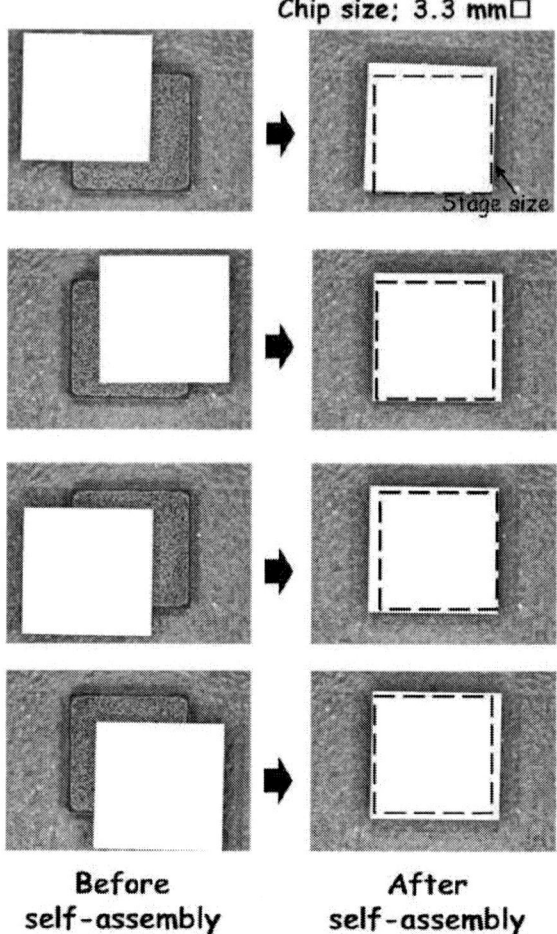

Figure 14. Before and after photos of self-assembly with different initial chip positions with the chip size of 3.3 mm square.

IV. CONCLUSION

In order to improve the throughput of Fan-out WLP process, we have developed the machine device that incorporates a unique self-assembly method. In this study, two factors are evaluated to clarify the impact on the alignment accuracy after self-assembly; 1) the influence of chip position before self-assembly, 2) the relationship between stage size of self-assembly area and chip size. As a result of evaluation, if the chip size is the same as the porous stage size, the alignment accuracy after self-assembly is within 5 μm or less even with varied chip positions before self-assembly. On the other hand, when the porous stage size and chip size are different, the chip position before self-assembly has an impact on the alignment accuracy after self-assembly. Based on the results obtained in this study, further development of machine device and process will be pursued to improve the alignment accuracy after self-assembly.

ACKNOWLEDGMENT

The authors would like to thank Professor Koyanagi and Professor Fukushima at Tohoku University for useful discussions.

REFERENCES

[1] Fukushima, T., Yamada, Y., Kikuchi, H., and Koyanagi, M., "New Three-Dimensional Integration Technology Using Self-Assemlby Technique", IEEE International Electron Devices Meeting (IEDM) Technical Digetst (2005) pp. 359-362.

[2] T. Fukushima, Y. Yamada, H. Kikuchi, T. Tanaka, M. Koyanagi, "Self-Assembly Process for Chip-to-Wafer Three-Dimensional Integraion" Proc. 57th Electronic Components and Technology Conerence, Reno, NV, May 2007, pp. 836-841

[3] T. Fukushima, T. Konno, T. Tanaka, M. Koyanagi, "Multichip Self-Assembly Technique on Flexible Polymeric Substrate", Proc. 58th Electronic Components and Technology Conference, Orland, FL, May 2008, pp. 1532-1537.

[4] A. Jouve et al., "Self-Assembly Process for 3D Die-to-Wafer using Direct Bondin: A Step Forward Toward Process Automatisation", 69th Electronic Components and Technology Conference (ECTC), 2019, pp225-234.

[5] A. Bond et al., "Collective Die-to-Wafer Self-Assembly for High Alignment Accuracy and High Throughput 3D Integration", 72nd Electronic Components and Technology Conference (ECTC), 2022, pp168-176.

[6] T. Yamada, K. Takano, T. Menjo, S. Takyu, "A Novel Chip Placement Technology for Fan-out WLP using Self-Assembly Technique with Porous Chuck Table" 2021 IEEE 70th Electronic Components and Technology Conference.

[7] S. Takyu, N. Okamoto, T. Yamada, T. Menjo, "A Novel Pick-up and Place Process for FO-WLP Using Tape Expansion Machine Device", 2017 IEEE 67th Electronic Components and Technology Conference, pp.364-370.

[8] H. Selhofer, A. Mayr, H. Pristauz, "Large Panel Size Bonder with High Performance and High Accuracy", 2019 IEEE 69th Electonic Components and Technology Conference, pp.1492-1497, 2019.

One-probe nanoprobing of power devices and electronic packages

Chengliang Huang[1], Vignesh Viswanathan[2], Andreas Rummel[3], Greg M. Johnson[4], Heiko Stegmann[4], Elliott Andrew[5], Allen Gu[4], Masako Terada[4], Thomas Rodgers[4]

[1] Carl Zeiss Co. Ltd, Shanghai, China
[2] Carl Zeiss Pte Ltd, Singapore
[3] Kleindiek Nanotechnik, Ruetlingen, Germany
[4] Carl Zeiss Microscopy GmbH, Oberkocken, Germany
[5] Carl Zeiss Limited, Cambridge, United Kingdom

Abstract— Results involving the use of a single nanoprobing contact are presented, on both a SiC planar MOSFET device, and a small system-in package. In both cases, mechanical polished samples were prepared, and then probed with a single contact. For the SiC MOSFET, the depletion zones were imaged while the samples were mounted on a 45 ° stub. Discussion of the different signals generated from Passive Voltage Contrast (PVC) and Electron Beam Induced Current (EBIC), as well as possible artifacts of sample grounding, are provided. For the package, Electron Beam Absorbed Current (EBAC) provided indication of connected features within the sample. The experiment showed the capability of measuring features across three orders of magnitude in size.

Keywords—EBAC, EBIC, package, SiC, nanoprobing, PVC

I. INTRODUCTION

A semiconductor failure analysis lab may have to work with a wide variety of failure types and form factors, especially if that were one working with packaging. That is because a failure may occur anywhere from the relatively large-scale wiring of the package to intricate devices or transistors deep inside the chip. For this work, two types of form factors were considered for study, a planar SiC MOSFET, and a system in package.

Power semiconductors are increasingly important to a wide range of industrial applications. SiC valued for its low resistance and high breakdown voltage. Thus, it is seeing increasing applications [1] in power conversion, elevators, [2], and in the drivetrain and charging systems of electric vehicles. [3] As such, SiC devices may suffer from a number of failure modes, including those such as body diode degradation [4], or involve hot carriers in the depletion zone. [5] [6] . Because these properties will depend on the health of the *p/n* junctions in a sample, techniques which enable the evaluation of junction health and placement will be of great interest in yield monitoring and failure analysis.

At the same time, on a much larger scale, packaging FA is becoming more challenging. Concepts like Heterogeneous Integration (HI) [7] will introduce complexities in layers, stacking, materials, etc. which will only make it more important to be able to perform localization on as large and as intact a sample as possible. For this experiment, the ADAQ23875 µModule® Data Acquisition Solution is a System-in-Package was used as a model test vehicle for the evaluation of the technique, with the use of a randomly selected, good device.

II. TECHNIQUES

A. Electron Beam Induced Current and Passive Voltage Contrast

Two techniques are available for the examination of p/n junctions, EBIC [8] and PVC [9]. The theoretical backgrounds for these two are discernable by consulting the energy band diagram for a p/n junction, as shown in Figure 1. The most simple technique that most engineers would have exposure to in electron microscopy is passive voltage contrast, which will often occur naturally due to different properties within the area examined. Across the p/n junction there is a difference in electron affinity, as depicted by the sloping shape of the conduction band of the junction. As the sample is scanned with an electron beam, any electrons seeking to escape from the N side will incur an energy penalty relative to the P side. Thus, there will be a lower secondary electron yield, and a lower energy among those electrons that do escape. Detection of such secondary electron energy distributions would require an electron microscope sensitive to low-energy secondary electrons [10]. It should be noted that the PVC effect is constant across the entire implant region.

The second technique, also informed by Figure 2, is Electron Beam Induced Current for evaluation of *p/n* junctions has been widely studied [11] [8] [12] [13] . Here, any charge that is introduced to a region with sloped energy profiles—the depletion region—will see an electric field that will push the carriers, and this movement of carriers may be detected as a current if a probe is opportunistically placed on the sample. These carriers may be due to the incident electron beam, secondary electrons, or any electron-hole pairs created by the beam. It should be noted that this effect only occurs at the depletion zone. Charge introduced at other places is either recombined or lost to ground.

A third probing technique is Electron Beam Absorbed Current, which simply maps regions of connectivity in the sample. The probe acts as either a path to ground for the electrons impinging upon the sample, or re-supplies electrons lost due to secondary electron emission in the sample.

These two techniques in this system were conducted by use of a Kleindiek MM3E prober and amplification system, and imaging in a SigmaSEM 300.

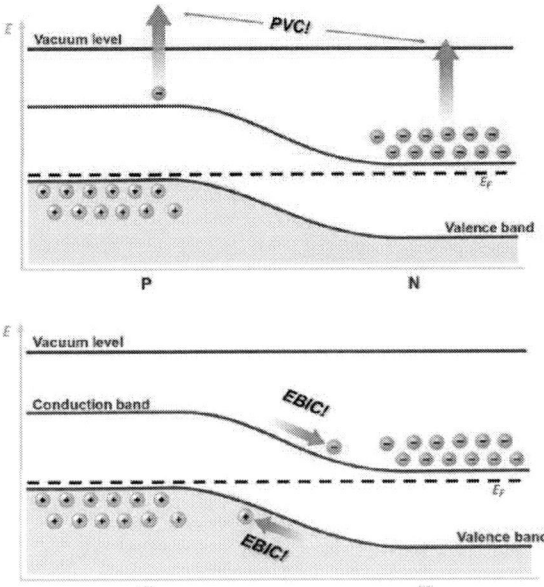

Figure 1. Band diagram for a p/n junction, indicating how these electrical properties influence microscopy techniques at operate on junctions. The top image explains how the P and N sides of a junction have a different electron affinity, and this imposes an energy penalty on any secondary electrons that might want to escape the N side. The depicts how the electric field at a p/n junction's depletion zone will cause EBIC currents at the junction, and not over the full body of the device.

III. PROCEDURE

A. Sample selection and preparation

For the first section, a commercially available SiC MOSFET chip was selected, and a mechanically polished cross section was prepared, perpendicular to the gates. The samples were attached by carbon tape to 45-degree stubs which were placed in a 9-stub carousel. The microprobes then came in from a roof mounting to approach and touch down on the samples The samples were then inserted in a Sigma 300 ZEISS SEM. The EBIC analysis was performed with a MM3A nanoprobing system provided by Kleindiek Nanotechnik. Several other samples were also mounted similarly on a 9-up carousel and examined at the same time. The setup is depicted in Figure 2, as seen from the chamberscope of the SEM.

Other samples were cleaved and examined at 90 ° in a GeminiSEM 300.

Figure 2. Chamber scope view of the experimental setup for the study. Several polished cross sections of chips were mounted on 45° stubs and placed on the same carousel. The probe needle is raised above the sample and is just visible at the red arrow

For the second section, a commercially available ADAQ23875 system-in-package (SiP) device was purchased using a standard commercial channel. The sample then was imaged by 3D X-ray microscopy in a Versa XRM system. at 10 μm/voxel resolution to capture full field of view (FOV) of the package. 120 kV X-ray energy was used to image this medium density sample. Then the package was mounted in epoxy inside a standard 2-inch metallographic mount, and polished with mechanical polishing and alumina slurry. The resulting "puck" was then examined with a GeminiSEM 300. A depiction of this setup is shown in the optical photo of Figure 3.

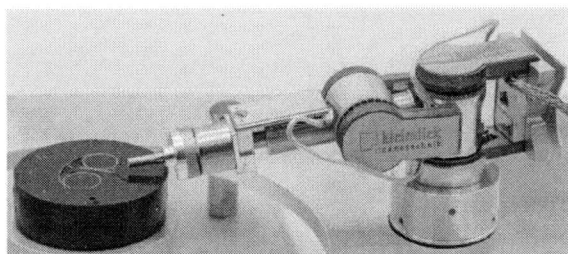

Figure 3. Demonstration of the placement of a nanoprobing probe on a polished cross section (optical photo, outside of chamber).

IV. RESULTS

A. SiC MOSFET

Figure 4 depicts the SiC MOSFET used in this analysis. The cartoon on the left depicts the typical implant regions used in the top half of the chip, and the image at right is from the cleaved surface. One can easily see the implant regions of the N+ source, P-well, and the N regions below.

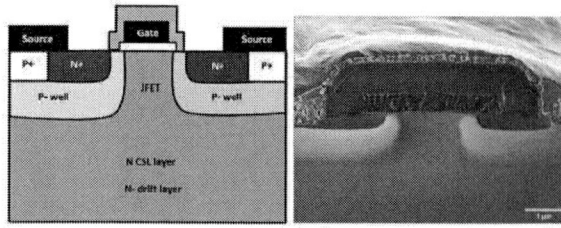

Figure 4. Diagram and cross-sectional image of the implant regions around a gate in a SiC MOSFET. Left: cartoon of implant regions and names (not to scale). Right: 2 kV Inlens image of device from the same area.

Figure 5 now compares the same general view of two different chips: the polished section, viewed on a 45° stub, with 60 um aperture, and the image from Figure 2 taken with a lower aperture, on a cleaved surface, at 90°. The greater current and angle provide for more sensitivity [14] in some voltage contrast, such that we can now distinguish (left image) among the N-doped regions of the JFET, CSL layer, and N-drift regions.

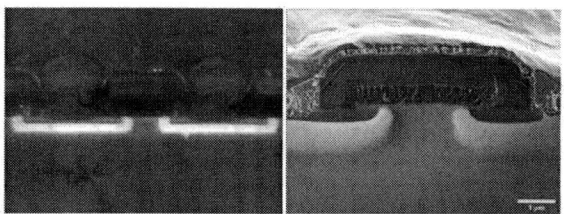

Figure 5. Diagram and cross-sectional image of the implant regions around a gate in a SiC MOSFET. Left: cartoon of implant regions and names (not to scale). Right: 2 kV Inlens image of device from the same area.

For the next set of experiments, with a 60 μm aperture, EBIC images were taken from the cleave, as shown in Figure 6, at three different beam acceleration potentials. In all images, one can see a small white zone beneath the main EBIC signal. It is believed that these may be an artifact of the degree of grounding possible with the use of a single probe. It is also clear that as the beam energy gets above 5 kV, the electron scattering cloud becomes unusually large, and therefore one is sampling from too large an area to get a meaningful characterization of the depletion zone.

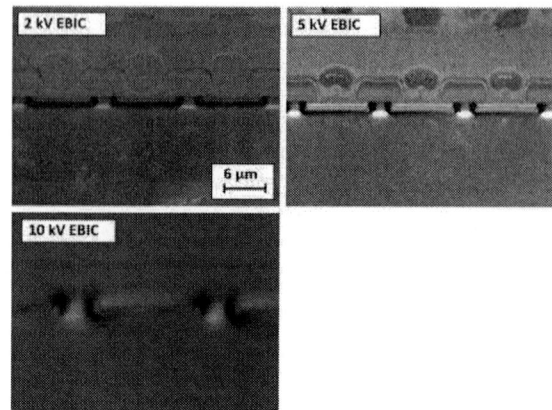

Figure 6. EBIC images at 2 kV, 5 kV, and 10 kV for the devices

Next the PVC and EBIC images were compared side by side, as seen in Figure 7. It is clear that these zones highlight different features in the p/n junctions.

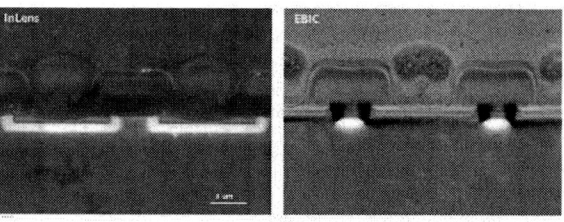

Figure 7. InLens and 5 kV EBIC images generated from the SiC MOSFET. In the InLens image, the P- well region stands out in bright contrast against the surrounding N-type regions, due to PVC. In the EBIC images, the depletion zone between this P-well and surrounding N- drift regions is highlighted in black. An additional white spot is seen, which may be a combination of the grounding problem introduced by use of a single probe, and the resistance at the JFET region of the device.

B. ADAQ23875 section

Three views of the System-in-Package sample are shown in Figure 8. At left is a light microscopy image. In the middle and right, respectively, are two different virtual reconstructions from XRM data of the top and middle planes in the sample. From the 3D reconstruction is it possible to examine multiple planes in the sample, choose a feature of interest, and as desired, to note top surface features that will enable one to target for section and thus find buried features underneath.

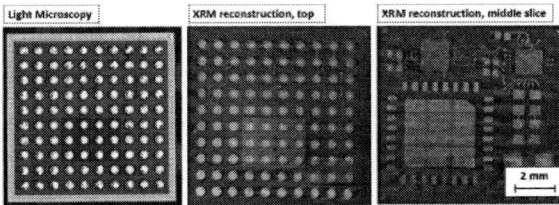

Figure 8. Multiple views of the ADAQ23875 package, including optical, and two different reconstructions from XRM data, taken at different virtual slices in the sample.

Next the polished cross section of the package was examined in SEM. It is seen in Figure 9.

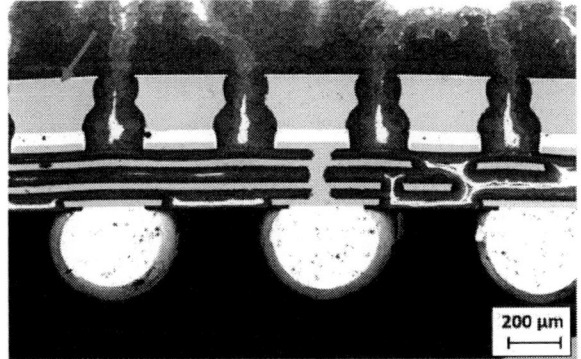

Figure 9. SEM micrograph at 15 kV of a random, polished, mechanical cross section of the package. The feature tested for continuity is at the left in the image.

Figure 10 shows the EBAC results undertaken at 20 kV. Only a single metal feature is highlighted, indicating that the section has intersected a feature that was disconnected from its net by the plane of section. It nevertheless demonstrates the utility of the technique.

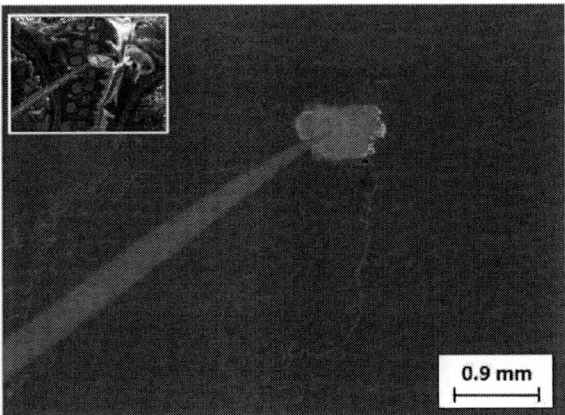

Figure 10. Results from nanoprobing a specific metallic feature in the polished cross section. Inset: 20 kV SEM image with color overlay of the EBAC signal. Main image: EBAC image showing the one feature and that there are no features next to it, indicating that the cross section has cut off this wire from its surroundings.

V. SUMMARY

Results were presented from two different types of analysis, showing nanoprobing characterization of sections of either a power device or a package. In the SiC MOSFET, the depletion zones of the sample were clearly imaged, as well as clear PVC images of the implant regions. It is likely one spurious feature was seen that may be due to the degree of grounding possible with a single probe.

For the System-in-Package, XRM data was combined with a cross section to demonstrate the ability to probe one net in a sample.

VI. CONCLUSIONS

This study has shown that nanoprobing with a single, manipulatable probe offers versatility in everything from EBIC of p/n junctions, to accommodating a large polished structure. Effective results across three orders of magnitude in magnification.

A sample preparation workflow is shown for examination of p/n junctions. These results show useful properties of the junction even in cases where there may be some debris on the surface. Multiple samples can be examined each within a five minute window.

The work demonstrates what passive voltage contrast can do. It highlights levels of N-doping between the JFET, CSL, and N- drift region.

By virtue of being on 45 degree mounted samples, the work demonstrates the fundamental difference between EBIC and PVC. PVC is fundamentally a surface effect, related to the energy penalty that escaping secondary electrons undergo between differently doped regions, those with different electron affinities. EBIC meanwhile is more of a volume sampling effect, related to the introduction of charge into a region, which naturally spreads into the sample according to Monte Carlo physics into a teardrop-shaped bulb. In this experiment, higher kV was particularly unhelpful in that it was likely that the bulb was so large as to sample from regions of different implant regions.

The work also showed that grounding is key to eliminating artifacts that may influence the interpretation of the background level.

This work also shows a workflow where one may use XRM to find a net in question, mount a sample on a 2 inch puck, polish to a net in question, based on top surface marks, and then use EBAC to trace nets on the surface of the sample.

For advanced logic nodes with, say, 20 nm features, one may want a more sophisticated system, but whether looking at a planar compound semiconductor or a package, these systems appear to work well for the application

It takes less than 5 mins to land one probe on the target sample structure. This setup is capable of quick and streamlined in-situ electrical chacterization in SEM.

VII. FUTURE WORK

These results have strong merit for the power devices examined, devices in for a future paper after further optimization of the technique.

VIII. REFERENCES

[1] V. Veliadis, "SiC chip cost, the impact of defects and the case of price parity with Si at the system level," SEMI webinar series, June 6, 2023.

[2] T. Kimoto, "SiC Power Devices: Overview, Defect Electronics, and Reliability," in *IRPS Tutorial*, 2019.

[3] J. Palmour and J. Edmond, "The golden age of silicon carbide: 25 years of innovation," *Compound Semiconductor*, no. 6, pp. 38-42, 2020.

[4] B. J. Nel and S. Perinpanayagam, "A brief overview of SiC MOSFET failure modes and design reliability," in *The 5th International Conference on Through-life Engineering Services*, 2017.

[5] B. Kok, M. Looi and H. Goh, " Insulated Gate Bipolar Transistor Failure Analysis in Overvoltage Condition," in *ICREPQ'12*.

[6] N. Valentine, D. Das and M. M. Pecht, "Failure Mechanisms of Insulated Gate Bipolar Transistors (IGBTs)," in *2015 NREL Photovoltaic Reliability Workshop*.

[7] G. Kenyon, "Heterogeneous Integration and the evoluation of IC packaging," EE Times Europe, 6 April 2021.

[8] J. R. Beall and L. C. Hamiter, "EBIC - A Valuable Tool for Semiconductor Evaluation and Failure Analysis.," *15th International Reliability Physics Symposium*, pp. 61-69, 1977.

[9] R. Rosenkranz, ""Failure localization with active and passive voltage contrast in FIB and SEM"," *Journal of Materials Science-materials in Electronics*, 2011.

[10] J. Jatzkowski, M. Simon-Najasek and F. Altmann, "Novel techniques for dopant contrast analysis on real IC structures," *Microelectronics Reliability*, 2012.

[11] E. Cole, ""Beam-Based Defect Localization Techniques"," *ISTFA*, vol. Tutorials, 2007.

[12] K. Nikawa, "Method and system for testing an interconnection in a semiconductor integrated circuit," *US Patent number 5,804,980*.

[13] A. Rummel, G. Johnson and M. Kemmler, "Low-voltage EBIC investigation of EBIC fails," *IPFA*, 2021.

[14] H. Seiler, "Secondary electron emission in the scanning electron microscope," *Journal of Applied Physics*, vol. 54, no. doi: 10.1063/1.332840, 1983.

Study of spatial distortion in InP nanophotonic membranes on different carrier substrates

Salim Abdi
Eindhoven Hendrik Casimir Institute
(EHCI)
Eindhoven University of Technology
Eindhoven, the Netherlands
s.a.abdi@tue.nl

Aleksandr Zozulia
Eindhoven Hendrik Casimir Institute
(EHCI)
Eindhoven University of Technology
Eindhoven, the Netherlands
a.zozulia@tue.nl

Jeroen Bolk
Nanolab@tu/e
Eindhoven University of Technology
Eindhoven, the Netherlands
j.bolk@tue.nl

Erik Jan Geluk
Nanolab@tu/e
Eindhoven University of Technology
Eindhoven, the Netherlands
e.j.geluk@tue.nl

Kevin Williams
Eindhoven Hendrik Casimir Institute
(EHCI)
Eindhoven University of Technology
Eindhoven, the Netherlands
k.a.williams@tue.nl

Yuqing Jiao
Eindhoven Hendrik Casimir Institute
(EHCI)
Eindhoven University of Technology
Eindhoven, the Netherlands
y.jiao@tue.nl

Abstract—Electron Beam Lithography (EBL) metrology and least-square estimation of wafer-scale distortions is used to determine InP membrane deformation as a result of bonding to different substrate materials. First, the accuracy of EBL as a metrology tool for this particular application was assessed. Next, modelling of distortions was tested on unbonded InP wafers as a reference for extracting post-bonding distortions. Then we investigated the effect of substrate material choice on InP membrane deformation after bonding. We found residual expansion factors of 4.53 ± 1, 312.4 ± 1, and 317 ± 1 ppm of the InP membrane bonded to InP, Si, and 3C-SiC carriers, respectively. For the SiO_2 carrier, the 3inch InP membrane split into smaller membranes to reduce the stress, highlighting the importance of substrate choice.

Keywords—*Adhesive bonding, metrology, 3D integration, overlay lithography, heterogeneous integration*

I. INTRODUCTION

Nanophotonics has emerged as a groundbreaking field that enables the generation and guiding of light in sub-micron sized planar structures, with applications ranging from telecommunications to biomedical sensing [1]. A promising approach to nanophotonic design is the utilization of InP membranes that are bonded onto a substrate, like Silicon (Si), using some dielectric low refractive index material, such as SiO_2 or benzocyclobutene (BCB) [1]. This method is also purposed for vertical integration of photonics with electronic drivers for applications requiring high bandwidth or high integration density [2]. While membrane technology shows potential for the miniaturization of photonic integrated circuits (PICs), the fabrication of membrane devices poses challenges as a result of a complex process flow, involving double-side membrane processing before and after bonding. In particular, it becomes a challenge to perform wafer-scale alignment for overlay lithography before and after bonding due to non-uniform distortion of the membrane [3].

To gain quantitative insight into membrane distortions, one method employed is EBL metrology [3], [4]. This technique involves the utilization of a set of EBL markers that are patterned on the wafer as the initial step of the fabrication process to allow for overlay alignment. These are subsequently read by EBL after each process step such as adhesive bonding or hard mask deposition. Such processing steps can introduce distortions in the form of expansion, non-orthogonality, and higher-order distortions [5], [6]. By comparing the positions of markers found by EBL before and after processing, it is possible to determine the magnitude and direction of distortions and decompose it into systematic and non-systematic errors. This reveals how different physical processes influence the membrane expansion [5]. This method could be used as an in-line metrology tool during fabrication of nanophotonics, and it also allows for guiding future work to improve overlay lithography using industrial tools for membrane photonics [4], [7].

In this paper, we analyzed the influence of the EBL tool's beam current and beam drift on the writing and reading positional accuracy of markers. We used a 6-parameter model to fit the distortions and quantify systematic and non-systematic errors. This model corrects for in-plane translation, rotation, asymmetric scaling, and non-orthogonality on the wafer scale. We first validated the possibility to use the model with InP wafers before bonding. Then we applied the model to investigate how bonding parameters can influence membrane distortion.

For the effect of adhesive bonding on membrane distortion, the bonding is carried out at 280°C. So we investigated bonding a 300nm-thick InP membrane to different substrates to vary the coefficient of thermal expansion (CTE) mismatch between the membrane and the carrier substrate. We fixed the BCB thickness at 2μm and chose InP, Si, 3C-SiC (SiC), and Pyrex glass (SiO_2) carrier substrates to achieve a wide range of CTE mismatch of 0 to 4.27×10^{-6}/°C, as will be discussed later on. We observed that the CTE mismatch, *i.e.* choice of substrate, plays a crucial role on membrane expansion, namely scaling. Moreover, our findings suggest that such deformations are dynamic and can be further reduced by extra fabrication steps or tailored bonding parameters.

This work was supported by the H2020 ICT TWILIGHT Project.

II. EXPERIMENTAL DETAILS

The study was conducted on 3" wafers. A simplified process flow is shown in Fig.1. Herein, we used a dedicated epitaxial stack on InP wafers and bonded it to different substrates to assess the effect of substrate carrier type on membrane distortion. The chosen substrates are: InP with identical CTE to the membrane of $4.75\times10^{-6}/°C$ [8], Si and 3C-SiC with CTE of $2.55\times10^{-6}/°C$ [9] and $2.77\times10^{-6}/°C$ [10], and finally Pyrex glass with low CTE of $0.48\times10^{-6}/°C$ [11]. To fabricate overlay markers, we grew a basic epitaxial stack on InP consisting of 300 nm InGaAs etch stop layer and 300 nm InP used as the membrane for device fabrication.

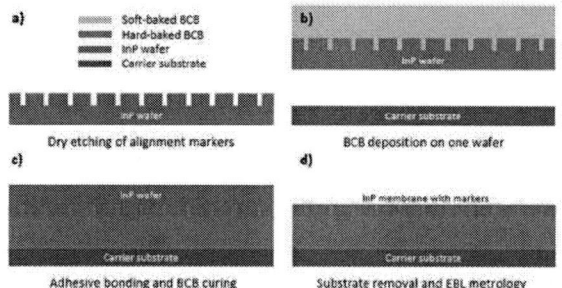

Fig. 1. Simplified illustration of the general process flow used in this study

We begin marker fabrication by depositing a SiN hard mask and patterning it with EBL using a positive resist. The pattern is then transferred to the hard mask with RIE etching, and subsequently to the InP wafer using ICP RIE plasma ensuring an etch depth of 450nm (Fig.1.a). Next in Fig.1.b, We prepare the InP wafer by depositing SiO_2. We then load it into EBL and read the markers. After that, we coat a thin layer of AP3000 for better adhesion to BCB and follow it with coating a 2 µm-thick BCB and evaporation of solvents. The carrier substrate is also prepared by depositing the same thickness of SiO_2 followed by coating with AP3000. The substrates are then aligned using the orientation of the major flats and locked in a cassette holder, which we load into EVG bonder for bonding. This operation is carried out at 280°C for 1h. We use a low ramp-up rate of 5 °C/min and 700N of force. During bonding, the temperatures of the top and bottom parts of the bonder are controlled separately within 0.1°C accuracy. Bonding results in full crosslinking of BCB, turning it to a solid dielectric layer that permanently joins the two substrates (Fig1.c). We note that the BCB thickness non-uniformity is below 5% before bonding, but it can increase after bonding because of BCB reflow before full-crosslinking [12]. After bonding, the bulk substrate and etch-stop layers are wet etched. For the InP substrate carrier case, we use protective multi-coatings to preserve the carrier during wet etching of the other InP wafer, and these are removed afterwards. After this stage, the markers are visible and ready for EBL reading from the backside(Fig1.d).

In this study, we use a Raith EBPG5150 EBL system. Both pattern writing and reading require placing the InP substrate into a 3" holder. For this, the substrate is clamped into the holder using three pins from the top side and pushed from the backside by the holder's backplate using a spring mechanism. Here, the substrate's bow is not fully neutralized. The locations of the 3 pins are shown in Fig.2.b. After loading the holder and reaching a vacuum level of ~10^{-7} mbar, marker reading or writing is carried out. We use square markers with an area of 20×20 µm^2 for marker fabrication with our system. The markers are uniformly distributed across the entire 3" wafer. We designed 3 maps on the same wafer with different marker pitches to investigate the impact of the spatial resolution on our analysis. The chosen pitches are 1.25, 2.5, and 5 mm in the (x,y) directions, where x and y are the directions perpendicular and parallel to the major flat, respectively. A beam current of 100 nA was chosen to ensure fast writing of the markers.

For marker reading, EBL logs the design and absolute positions of found markers. These are extracted and fitted using a wafer-scale 6-parameter model based on the least-squares estimation method. More details on the model are discussed next. The fitted maps can either be pre-bonding or post-bonding maps based on the desired information to be extracted. Experimentally, we explored a beam currents range of 5-190nA and investigated its effect on the residual errors. The effect of drift was also investigated as it can be a major source of reading errors [13].

For modeling our results, we used a 6-parameter model to correct for various systematic distortions of the studied maps. The model transforms given (x.y) coordinates of a map to new coordinates (x_{opt}, y_{opt}) according to the following equations [5]:

$$x_{opt} = x.\cos(P_1) - y.\sin(P_1) + P_2 + x.(1 + P_4) + y.\tan(P_6) \tag{1}$$

$$y_{opt} = x.\sin(P_1) + y.\cos(P_1) + P_3 + x.(1 + P_5) + x.\tan(P_6) \tag{2}$$

Here, P_1 is the rotation in (rad). P_2 and P_3 are the translations in the x- and y-direction in (µm) perpendicular and parallel to the flat, respectively. P_4 and P_5 are the scaling factors in parts-per-million (ppm) in x- and y-directions, respectively. P_6 is the non-orthogonality factor in (rad).

Optimal values of P_1-P_6 are obtained using a least-squares estimation method to match (x_{opt}, y_{opt}) with a given (x_0, y_0). The given coordinates can be the design coordinates or coordinates read by EBL at a certain stage during fabrication. In the post-bonding case, the coordinates are flipped in the y-direction to match pre-bond observed coordinates, *i.e.* we match (x,-y) with pre-bond (x_0, y_0). For optimal results, we use a null initial guess for all parameters and then fit the data with a suitable algorithm for minimization. Throughout this study, we use the standard deviation (StDev) of residual errors (x_{opt}- x_0, y_{opt}-y_0) to assess the quality of the acquired results. Hence, to benchmark results, we first calculated with Eq.1 and Eq.2 how the StDev grows when each of the variables P_1-P_6 is varied individually. For this, we used wafer-scale marker positions that are identical to the ones used in our experiments dedicated to this study. We matched our modelling estimation errors using this information with experimental data revealing experimental systematic errors from EBL, namely from beam drift. Given that translation and rotation depend on the initial wafer position and are related to wafer placement in the EBL holder, we do not include P_1- P_3 in our results. For P_4 and P_5 the errors are within ±1ppm, and for P_6 it is within $\pm2\times10^{-6}$rad.

III. RESULTS AND DISCUSSION

A. Evaluation of the effect of EBL settings and model accuracy on residual errors

In this section, we investigate the effect of chosen EBL settings during marker reading/writing on the standard deviation. We also test our model on unbonded wafers. The goal is to decouple systematic and non-systematic errors before bonding to ensure the accuracy of results in this paper. In metrology, the main non-systematic EBL errors arise from beam drift and the used beam current. Beam drift is caused by temperature fluctuations <0.1°C of the chamber, and it affects both marker fabrication and reading [13]. It can particularly affect results for large writing areas or as a result of temperature instability of the holder, for instance during loading. Beam drift is corrected periodically by EBL each 1h in this study, and these corrections if present are extracted and subtracted from results in this paper for better modelling accuracy. Also, increasing the beam current leads to a higher beam diameter, which affects marker reading time, image contrast, and edge sharpness, thereby affecting the registered marker positions.

First, we assessed the effect of beam drift and current on marker fabrication and reading without using the model. For that, we used 100 wafer-scale markers spanning the full wafers. These required a fabrication time of 10 minutes to expose. An identical map with a 675µm x-shift was also fabricated and designed to take 60 minutes to expose. The maps were then read multiple times with multiple beam currents without loading/unloading the holder from the main chamber. Next, marker positions were extracted from different reading times and subtracted from each other. We calculated the StDev from the residuals and the results are shown in Fig.2.a. The residuals map extracted using 5nA beam is shown in Fig2.b. The standard deviation is below 5nm for beam currents below 100 nA and around 8.5 nm for a beam current of 190 nA. This could be caused by the increase in beam diameter that lowers the accuracy of observed marker positions, especially since the effect of charging is likely mitigated by the high pitch between markers. The increase in markers' exposure time from 10 to 60 minutes during fabrication does not increase the StDev above 1nm. The map in Fig.2.b shows the movement as vector directions and lengths. The displacement appear to be locally defined, with no clear global wafer-level trends.

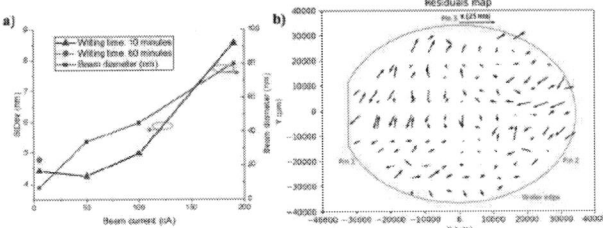

Fig. 2. a) Standard deviation of residuals and beam diameter *vs* beam current. b) Residuals map showing the positional differences between markers read at two different times highlighting the effect of drift.

Next, we assessed the accuracy of our 6-parameter model based on these findings. Here, we compared the observed marker positions with the designed positions and extracted the six distortion parameters based on Eq.1 and Eq.2. The fitted map using 5nA beam and averaged for >10 datapoints per marker

is shown in Fig.3. The inset of Fig.3 shows bell plots of residuals from averaged data *vs* different beam currents. Here, we observe that using smaller beam currents yield slightly higher accuracy based on lower residuals. However, the overall standard deviation values are similar. Moreover, the obtained values for P_4-P_5 for different currents are all within ± 1 ppm. For non-orthogonality all obtained values of P_6 are below 1.10^{-6} Rad lower than the error range, indicating that the effect of non-orthogonality is minimal. Based on this, and to ensure the highest accuracy of our marker reading, we use 5nA beam and averaged data per marker in the rest of the study.

For the map in Fig.3, the largest vectors lie on the edges of the wafer. We note that the pattern near pin 1 is consistent in all of our extracted maps, suggesting that such deformations are affected by the metal pins used to secure the wafer on the holder. Since the wafer was not yet bonded, the overall observed distortions might come from the effect of removing the SiN hard mask used initially to fabricate the markers. Indeed, residual errors in the order of tens of nm can be the result of localized wafer shape changes induced by residual stress from thin films [14]. Such distortions would require both localized and wafer-scale correction models to account for the full distortion, for instance with a 10-parameter model based on the used lithography tool, depending on the neutralization of the wafer bow by the holder [14]. It also might be possible to better model the patterns near the edge with only wafer-scale corrections by using higher-order corrections like cubic distortion. However, as will be shown in the next sections, the post-bonding distortions are mainly dominated by scale (P_4 and P_5), so we do not pursue this option.

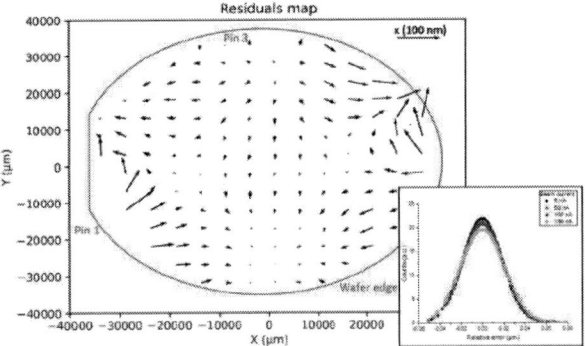

Fig. 3. Residuals map showing the difference between design and observed marker coordinates after correction of the data with fitted model parameters. Inset: bell plots of residuals from fitted maps for different beam currents

B. Effect of substrate choice on post-bonding membrane distortion:

The choice of the substrate directly affects the CTE mismatch between the InP wafer and the carrier substrate during bonding. It also affects the InP membrane after substrate removal in a similar manner. For isotropic materials in CTE, the equation to describe ideal values for scaling based on CTE is given as follows:

$$P = \Delta T . \Delta\alpha \qquad (3)$$

Where P is the scaling coefficient, ΔT is the difference between bonding and room temperatures, and $\Delta\alpha$ is the CTE mismatch. Using Eq.3, we first calculated the theoretical thermal expansion values of the InP membrane on different substrates for bonding temperatures of 250 and 280°C, which are the most often used in the literature [15]. We also plotted the average of P_4 and P_5 extracted by our model from our dedicated experiments. Results are shown in Fig.4.a. We note that the CTE of InP is higher than the used carrier substrates here, so the scaling is limited to expansion in this study as InP expands more than other substrates here by the effect of temperature.

For the InP carrier case, the CTE mismatch is 0. However, an average scaling factor of 4.53±1 ppm was found from our experiment. This indicates the existence of some expansion in the InP membrane. It is likely caused by the partial relaxation of residual stresses that might be present in the BCB, given that the temperature of both InP wafers is the same during bonding within 0.1°C. This is because the thermal expansion of BCB as a polymer is an order of magnitude higher than that of InP or other semiconductors, causing the accumulation of residual stresses in the BCB layer. The latter might affect the InP membrane more than the InP substrate underneath, since the substrate thickness is three orders of magnitude thicker than the membrane.

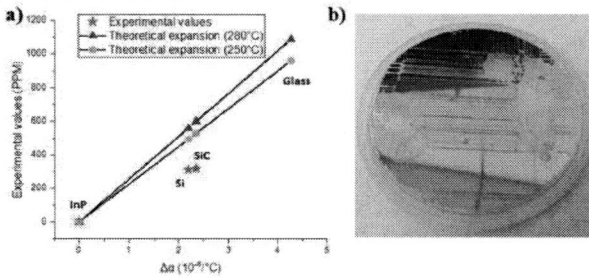

Fig. 4. a): Theoretical and experimental values of the InP membrane expansion *vs* CTE mismatch, b) image of an InP bonded membrane on top of a SiO₂ substrate before InGaAs etch-stop removal showing InP memebrane splitting

The measured scaling (expansion) factors on Si and SiC wafers are 312.4±1 and 317±1 ppm, respectively. These values are lower by 248.6±1 and 282.2±1 ppm from the expected expansion values for Si and SiC, respectively. Hence this part of the expansion could have been released after bonding as a residual stress by increasing the wafer bow. Based on this, we decompose the membrane expansion into two parts. We refer to the experimentally obtained values of P_4 and P_5 as residual expansion, while the difference between experimental and theoretical values is referred to as released expansion. It might be that the released expansion is only possible by the relatively higher elastic deformation of BCB compared to the InP semiconductor membrane on top [15]. Moreover, in all of our measurements P_4 is slightly higher than P_5. This difference is 11.22±2 ppm for Si, 7.19±2 ppm for SiC, and 5.14±2 ppm for InP carrier substrates. This might point to the existence of an anisotropic behavior in CTE or the mechanical properties of the substrate carriers. Another possibility to explain why the experimental values are much lower than theoretical values is that as the temperature ramps up during bonding, the crosslinking of BCB increases across the two wafers. A point where BCB permanently fastens the two wafers could then occur before full crosslinking. A calculation of this temperature based on experimental values yields a temperature of 170°C. The degree of crosslinking at this temperature is below 50% during the slow ramp-up [15]. Therefore, this possibility if present might only coexist with the aforementioned mechanism. This will be further investigated to determine if the membrane deformation is permanent or dynamic and can be further relieved.

For bonding to the SiO₂ wafer, an image of the bonding result is shown in Fig.4.b. Apart from the large defects where the membrane is detached, vertical and horizontal lines are apparent. These defects are only visible after substrate removal. Theoretically, the membrane is expected to expand by 1088.9 ppm given the large CTE mismatch of $4.27\times10^{-6}/°C$. The presence of such lines where the InP membrane split indicates that the released expansion is higher than the values previously found for Si as it required plastic deformation of the membrane. Although it is not possible to determine the values of residual expansion in this case, *i.e.* P_4 and P_5, the released expansion is higher than the highest found value in this study. Hence, the residual expansion of the InP membrane after breaking is lower than 806.7±1 ppm. Consequently, a more tailored bonding process is needed to reduce the risk of membrane breaking.

IV. CONCLUSIONS

We used EBL as a metrology tool to determine InP membrane deformation as a result of bonding to different substrates. We found residual expansion factors of 4.53±1, 312.4±1, and 317±1 ppm of the InP membrane to InP, Si, and SiC carriers, respectively. For the SiO₂ carrier, the 3" InP membrane split into smaller membranes to further release the stress as the CTE mismatch is high. These results highlight the importance of substrate choice and/or tailoring the bonding parameters for improved bonding yield and lower risks during post-bonding processing.

ACKNOWLEDGMENT

This research was performed in the NanoLab@TU/e cleanroom facility. This work was supported by the H2020 ICT TWILIGHT Project (contract No. 781471) under the Photonics PPP.

REFERENCES

[1] Y. Jiao *et al.*, "InP membrane integrated photonics research," *Semicond. Sci. Technol.*, vol. 36, no. 1, p. 013001, Jan. 2020, doi: 10.1088/1361-6641/abcadd.

[2] M. Spyropoulou *et al.*, "The future of multi-terabit datacenter interconnects based on tight co-integration of photonics and electronics technologies," in *2023 Optical Fiber Communications Conference and Exhibition (OFC)*, Mar. 2023, pp. 1–3. doi: 10.1364/OFC.2023.Tu3I.3.

[3] A. Sakanas, E. Semenova, L. Ottaviano, J. Mørk, and K. Yvind, "Comparison of processing-induced deformations of InP bonded to Si determined by e-beam metrology: Direct vs. adhesive bonding," *Microelectron. Eng.*, vol. 214, pp. 93–99, Jun. 2019, doi: 10.1016/j.mee.2019.05.001.

[4] S.-C. Horng, "Compensating Modeling Overlay Errors Using the Weighted Least-Squares Estimation," *IEEE Trans. Semicond. Manuf.*, vol. 27, no. 1, pp. 60–70, Feb. 2014, doi: 10.1109/TSM.2013.2243925.

[5] J. D. Armitage, Jr. and J. P. Kirk, "Analysis Of Overlay Distortion Patterns," presented at the 1988 Microlithography Conferences, K. M. Monahan, Ed., Santa Clara, CA, United States, Jan. 1988, p. 207. doi: 10.1117/12.968368.

[6]   S. E. Steen, D. LaTulipe, A. W. Topol, D. J. Frank, K. Belote, and D. Posillico, "Overlay as the key to drive wafer scale 3D integration," *Microelectron. Eng.*, vol. 84, no. 5, pp. 1412–1415, May 2007, doi: 10.1016/j.mee.2007.01.231.

[7]   C.-F. Chien, K.-H. Chang, and C.-P. Chen, "Modeling Overlay Errors and Sampling Strategies to Improve Yield," *J. Chin. Inst. Ind. Eng.*, vol. 18, no. 3, pp. 95–103, Jan. 2001, doi: 10.1080/10170660109509462.

[8]   I. Kudman and R. J. Paff, "Thermal expansion of InxGa1−xP alloys," *J. Appl. Phys.*, vol. 43, no. 9, pp. 3760–3762, Nov. 2003, doi: 10.1063/1.1661805.

[9]   Y. Okada and Y. Tokumaru, "Precise determination of lattice parameter and thermal expansion coefficient of silicon between 300 and 1500 K," *J. Appl. Phys.*, vol. 56, no. 2, pp. 314–320, Jul. 1984, doi: 10.1063/1.333965.

[10]   N. M. Sultan, T. M. B. Albarody, H. K. M. Al-Jothery, M. A. Abdullah, H. G. Mohammed, and K. O. Obodo, "Thermal Expansion of 3C-SiC Obtained from In-Situ X-ray Diffraction at High Temperature and First-Principal Calculations," *Materials*, vol. 15, no. 18, Art. no. 18, Jan. 2022, doi: 10.3390/ma15186229.

[11]   R. Roy, D. K. Agrawal, and H. A. McKinstry, "Very Low Thermal Expansion Coefficient Materials," *Annu. Rev. Mater. Sci.*, vol. 19, no. 1, pp. 59–81, 1989, doi: 10.1146/annurev.ms.19.080189.000423.

[12]   S. Abdi, T. de Vries, M. Spiegelberg, K. Williams, and Y. Jiao, "Novel wafer-scale adhesive bonding with improved alignment accuracy and bond uniformity," *Microelectron. Eng.*, vol. 270, p. 111936, Feb. 2023, doi: 10.1016/j.mee.2023.111936.

[13]   C. K. Chen, "Electron-Beam Lithography Error Sources," in *Electron-Beam, X-Ray, and Ion-Beam Techniques for Submicrometer Lithographies III*, SPIE, Jun. 1984, pp. 2–7. doi: 10.1117/12.942306.

[14]   K. T. Turner, S. Veeraraghavan, and J. K. Sinha, "Relationship between localized wafer shape changes induced by residual stress and overlay errors," *J. MicroNanolithography MEMS MOEMS*, vol. 11, no. 1, p. 013001, Mar. 2012, doi: 10.1117/1.JMM.11.1.013001.

[15]   X. Wang and F. Niklaus, "Polymer Bonding," in *3D and Circuit Integration of MEMS*, John Wiley & Sons, Ltd, 2021, pp. 331–359. doi: 10.1002/9783527823239.ch15.

Laser-Assisted Bonding Approach for Photonic Integration Processes

Aleksandr A. Vlasov[1], Topi Uusitalo,
Evgenii Lepukhov, Heikki Virtanen,
Samu-Pekka Ojanen, Jukka Viheriälä,
and Mircea Guina.
Optoelectronic Research Centre,
Physics Unit,
Tampere University,
Tampere, Finland, FI-33720
[1]aleksandr.vlasov@tuni.fi
[1]https://orcid.org/0000-0001-8839-6721

Abstract—Current development trends concerning miniaturizing of electronics and photonics systems are aiming at assembly and 3D co-integration of a broad range of technologies including MEMS, microfluidics, wafer level optics, and silicon photonics. To this end, on-chip integration using silicon-photonics platform offers a wide range of possibilities addressing passive optics functionality, active optoelectronic devices, and compatibility with CMOS fabrication. On the other hand, the hybrid technology enabling volume manufacturing of such system-on-chip components it is still in an early development stage. In this work, the original LAB setup with a coaxial bottom irradiation architecture was developed. The setup allows a rapid and energy-effective process with the ability to produce successful electrical bonds with negligible thermal-induced stress and warpage to bonded surfaces. The proposed machine-vision based temperature sensor is validated for photonic integration assembly processes.

Keywords—assembly processes, laser-assisted bonding, photonic integration, silicon devices, silicon photonics, through-silicon vision, infrared imaging, infrared microscopy.

I. INTRODUCTION

Photonic integrated circuits (PICs) have gained growing attention motivated by their ability to combine an increasing variety of optical functions on a chip-level, allowing to reduce cost, and drastically miniaturize optical systems [1]. PICs will inevitably introduce a paradigm shift in the use of photonics in applications such as wearable electronics, low-dissipation power datacom, and advanced sensing. Despite many important advances, photonic integration technology requires significant development to attain the yield, the throughput, and versatility required to engage with emerging applications. To this end, one of the most promising solutions to co-integrate complimentary photonic chips is laser-assisted bonding (LAB), owing to low thermal-induced stress and warpage to bonded surfaces [2].

In LAB processes, used in microelectronic integration, silicon chips are usually bonded to printed-circuit boards (PCBs), so the laser usually is placed above the assembly (Fig. 1, a), and could be combined with vacuum pick-up-tool [3]. As for photonic integration processes, where semiconductor components are bonded to silicon photonic integration circuits or wafers, laser is usually placed below the assembly (Fig. 1, b) [4]. Laser beam delivery system is usually combined with a through-silicon microscope, but it does not allow the simultaneous observation, alignment, and irradiation, requiring switching between projecting and

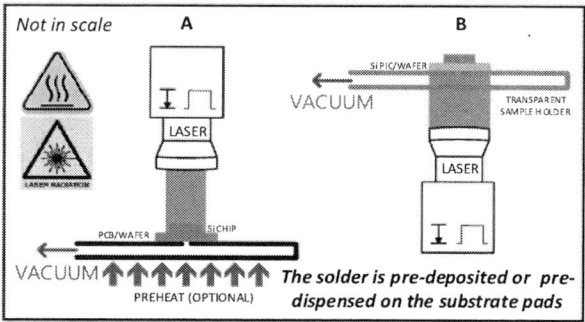

Fig. 1. LAB processes: a – uniform beam profile (top irradiation), b – uniform beam profile (bottom irradiation).

imaging optical channels [4].

For temperature (t°) control and feedback, a thermocouple [5] - [6], a pyrometer [7] - [8], or an IR camera (so-called thermal imager) [9] - [11] could be used. Thermal imager allows to control the temperature distribution among several points or the whole bonding surface.

In this work, we introduce the developed LAB setup with a bottom coaxial irradiation architecture [12], combined with through-silicon machine vision system [12], for obtaining an electrical interconnects (bonds) between pads on a silicon PIC, using pre-dispensed solder paste.

II. EXPERIMENTAL SETUP AND METHOD

The experimental setup consists of a through-silicon machine vision system [12], optically combined with LAB 980 nm laser bottom irradiation beam delivery system (Fig. 2). Both systems have common optical path and objective/projector, whereas vision and irradiation optical signals are divided using Wavelength Division Multiplexing (WDM) principle (Fig. 2). This design allows simultaneous observation, alignment, and irradiation of the silicon sample, placed on the custom transparent sample holder [13] during LAB process, which is highly beneficial for scientific and R&D applications. To suppress the environmental noise and vibrations impacts, the sample holder and the setup itself are mounted on a solid rigid basis with the use of properly designed vibration isolation system [14].

The test vehicle consists of a 15x10x0.65 mm silicon PIC [15] - [16] with 100x100 µm *Au*-plated pads, covered with a pre-dispensed type 619D solder paste (Sn62/Pb36/Ag2, metal content 86%, mesh type 3), (Fig. 3). The target temperature of 300°C was chosen to ensure the melting of solder and the

---

This work was supported by Business Finland under project PICAP (decision 44761/31/2020) and Academy of Finland through Photonics Flagship program PREIN #320168.

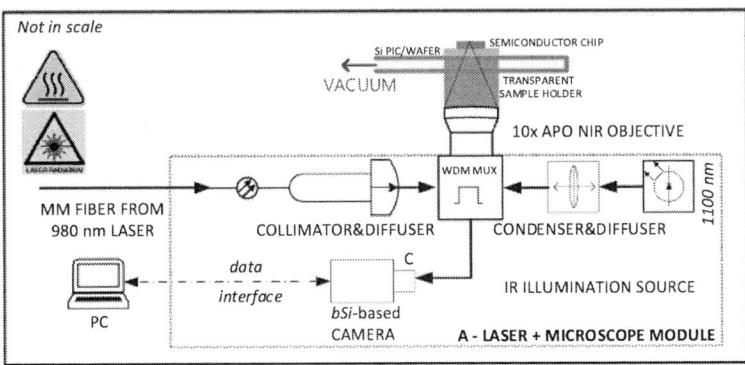

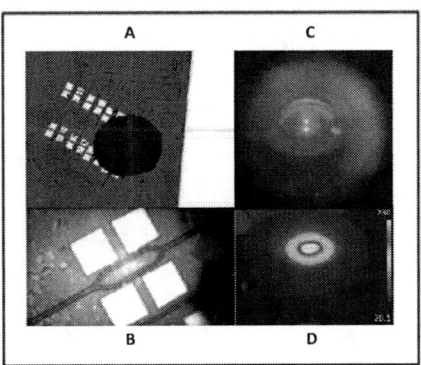

Fig. 2. The proposed design of the developed LAB setup, combined with through-silicon machine vision system [12].

Fig. 3. Test vehicle: A and B – top and bottom through-silicon [12] view of a PIC with dispensed solder paste, respectively; C and D – IR visualizer and thermal imager pictures taken during LAB process, respectively.

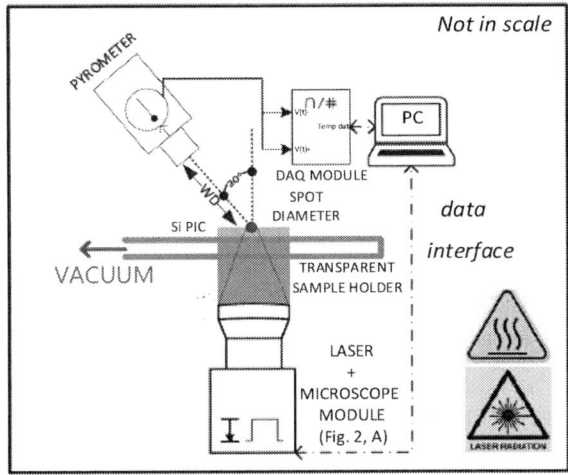

Fig. 4. Simultaneous thermal measurements using the machine-vision based sensor and a pyrometer sensor.

wetting of a bonding surfaces. Depending on an assembly process requirements (e.g., to satisfy *RoHS* directives), a *Pb*-free soldering materials could be used (with necessary temperature adjustments).

Two irradiation modes with the power densities of 110 and 240 W/cm^2 at 980 nm are investigated. A machine vision-based sensor is employed for non-contact temperature control of the silicon substrate. The temperature data obtained using a through-silicon vision system [12] is compared with data obtained with a pyrometer. For the best temporal resolution, the measurements are taken using a power density of 40 W/cm^2. The operation principle of the vision-based temperature control system relies on the transparency vs temperature dependence of silicon, due to which a through-silicon image undergoes darkening with temperature increase; thus, the change in image brightness carries information about temperature change [17] - [19]. For the temperature measurements validation, the setup is equipped with a pyrometer with a 300 mm working distance (WD, Fig. 4).

III. RESULTS AND DISCUSSION

The time-temperature diagrams (so-called "shark fins") of a Ø3mm central spot on the PIC top surface and a photo of bonds are given in Fig. 5. Both LAB operation modes resulted in the heating of the PIC above the solder melting temperature, which in turns led to the successful bond formation with electrical impedance in the range of hundredths of an Ohm, which is consistent with the published data [20]. The Pearson correlation coefficient (PCC) of the developed vision-based temperature sensor and pyrometer signals was 0.9522, which demonstrates their high mutual correlation. The vision-based approach to temperature measurements is highly beneficial as a non-contact threshold sensor for *in situ* LAB photonic integration assembly process control.

The results obtained are of great interest for the needs of heterogeneous photonics integration.

IV. CONCLUSION

The original LAB setup with a coaxial bottom irradiation architecture was developed. The setup allows a rapid and energy-effective process with the ability to produce successful electrical bonds with negligible thermal-induced stress and warpage to bonded surfaces. The proposed machine-vision based temperature sensor is validated for photonic integration assembly processes. The detailed review on LAB in the fields of microelectronics and photonics is presented in [13].

ACKNOWLEDGMENT

The authors are sincerely grateful to FabLab Tampere team and personally to Mika Kiirikki, Ilkka Hirvonen, Antti Tukiainen, Bengt-Olof Holmström, Tuomas Seppä, Tommi Salo from Tampere University, and to Andrei Chumachenko and Andrei Gurovich from Ampliconyx Oy (Tampere, Finland) for the invaluable help with prototyping of proposed design.

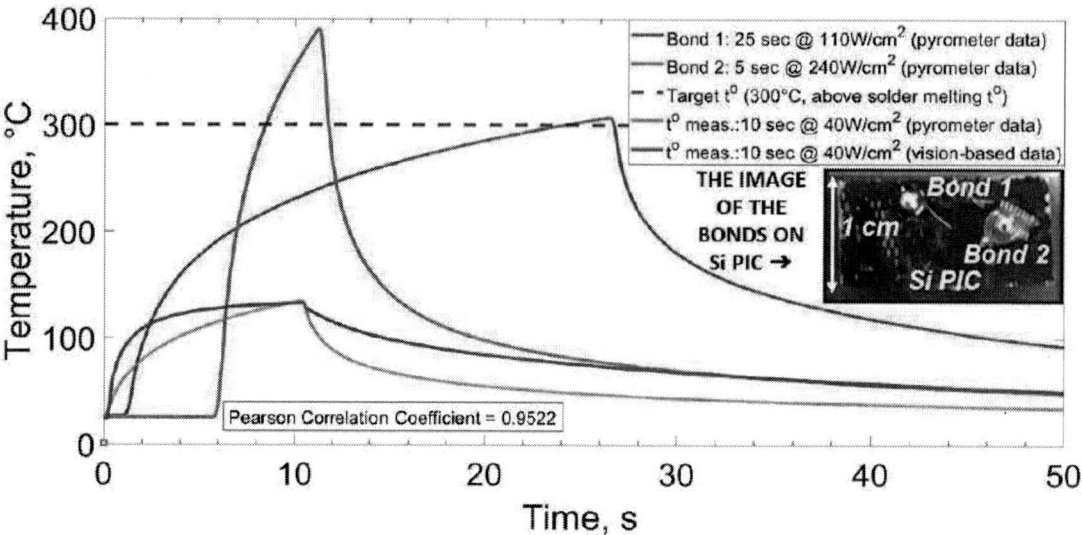

Fig. 5. The LAB processes research results: time-temperature diagrams ("shark fins") and a photo of obtained solder joints (bonds).

REFERENCES

[1] N. Margalit et. al., "Perspective on the future of silicon photonics and electronics," *Applied Physics Letters,* vol. 118, no. 22, p. 220501, May 2021, doi: 10.1063/5.0050117.

[2] M. Gim et. al., "High-Performance Flip Chip Bonding Mechanism Study with Laser Assisted Bonding," *2020 IEEE 70th Electronic Components and Technology Conference (ECTC),* Jun. 2020, doi: 10.1109/ectc32862.2020.00166.I.

[3] A. Kolbasow, T. Kubsch, M. Fettke, G. Friedrich and T. Teutsch, "Vertical Laser Assisted Bonding for Advanced "3.5D" Chip Packaging," *2019 IEEE 69th Electronic Components and Technology Conference (ECTC),* 2019, pp. 210-217, doi: 10.1109/ECTC.2019.00039.

[4] "ficonTEC Webinars – ficonTEC Service (Sub-μm Through-Silicon Alignment)" *ficonTEC.* https://www.ficontec.com/webinar/ (accessed May 31, 2023).

[5] W. A. Braganca, K. KyungOe and K. YoungCheol, "Development of a laser-assisted bonding process for a flip-chip die with backside metallization," *2020 IEEE 22nd Electronics Packaging Technology Conference (EPTC),* 2020, pp. 68-72, doi: 10.1109/EPTC50525.2020.9315012.

[6] W. Alves Braganca, Y. Eom, K. Jang, S. H. Moon, H. Bae, and K. Choi, "Collective laser-assisted bonding process for 3D TSV integration with NCP," *ETRI Journal,* vol. 41, no. 3, pp. 396–407, Apr. 2019, doi: 10.4218/etrij.2018-0171.

[7] M. Fettke, T. Kubsch, A. Kolbasow, V. Bejugam, A. Frick and T. Teutsch, "Laser-assisted bonding (LAB) and de-bonding (LAdB) as an advanced process solution for selective repair of 3D and multi-die chip packages," *2020 IEEE 70th Electronic Components and Technology Conference (ECTC),* 2020, pp. 1016-1024, doi: 10.1109/ECTC32862.2020.00165.

[8] M. Fettke et al., "A study on laser-assisted bonding (LAB) and its influence on luminescence characteristics of blue and *YAG* phosphor encapsulated *InGaN* LEDs," *2020 IEEE 70th Electronic Components and Technology Conference (ECTC),* Jun. 2020, doi: 10.1109/ectc32862.2020.00301.

[9] Y. Jung et al., "Development of Next Generation Flip Chip Interconnection Technology Using Homogenized Laser-Assisted Bonding," *2016 IEEE 66th Electronic Components and Technology Conference (ECTC),* 2016, pp. 88-94, doi: 10.1109/ECTC.2016.76.

[10] C. Kim et al., "Development of extremely thin profile flip chip CSP using laser assisted bonding technology," *2017 IEEE CPMT Symposium Japan (ICSJ),* 2017, pp. 45-49, doi: 10.1109/ICSJ.2017.8240085.

[11] Y. M. Jang, Y. Kim, and S.-H. Choa, "Development and optimization of the laser-assisted bonding process for a flip chip package," *Microsystem Technologies,* vol. 26, no. 3, pp. 1043–1054, Sep. 2019, doi: 10.1007/s00542-019-04624-8.

[12] A. A. Vlasov et al., "Machine Vision System Utilizing Black Silicon CMOS Camera for Through-Silicon Alignment," *IEEE Transactions on Components, Packaging and Manufacturing Technology,* 2022, doi: 10.1109/TCPMT.2022.3225051.

[13] A. A. Vlasov et al., "Reconfigurable Vacuum Sample Holder for Through-Silicon Microscopy and Laser-Assisted Bonding," *in IEEE Photonics Journal,* doi: 10.1109/JPHOT.2023.3302404.

[14] A. A. Vlasov et al., "Development of the passive vibroacoustic isolation system for the path matched differential interferometry based fiber-optic sensors," *Optical Fiber Technology,* vol. 57, p. 102241, Jul. 2020, doi: https://doi.org/10.1016/j.yofte.2020.102241.

[15] T. Aalto et al., "Open-Access 3-μm SOI Waveguide Platform for Dense Photonic Integrated Circuits," *IEEE Journal of Selected Topics in Quantum Electronics,* vol. 25, no. 5, pp. 1–9, Sep. 2019, doi: 10.1109/jstqe.2019.2908551.

[16] S.-P. Ojanen et al., "*GaSb* diode lasers tunable around 2.6 μm using silicon photonics resonators or external diffractive gratings," *Applied Physics Letters,* vol. 116, no. 8, p. 081105, Feb. 2020, doi: 10.1063/1.5140062.

[17] Boyd, I. W., et al. "Absorption of infrared radiation in silicon." *Journal of Applied Physics,* 55.8 (1984): 3061-3063.

[18] Bergmann, Joachim, et al. "Temperature dependent optical properties of amorphous silicon for diode laser crystallization." *Optics express,* 20.106 (2012): A856-A863.

[19] E.-J. Sin, C. K. Ong, and H. Tan, "Temperature Dependence of Interband Optical Absorption of Silicon at 1152,1064, 750, and 694 nm," *Physica status solidi,* vol. 85, no. 1, pp. 199–204, Sep. 1984, doi: https://doi.org/10.1002/pssa.2210850124.

[20] J. Joo et al., "Development of bonding process for flexible devices with fine-pitch interconnection using Anisotropic Solder Paste and Laser-Assisted Bonding Technology," *2020 IEEE 70th Electronic Components and Technology Conference (ECTC),* 2020, pp. 1309-1314, doi: 10.1109/ECTC32862.2020.00207.

Aleksandr Andreevich Vlasov was born in Starychi, Lviv region. He received the B.Eng. in Telecommunications (2014), M.Eng. in Optical Systems and Communication Networks (2016) and Ph.D. in Engineering (Optical and optoelectronic devices and complexes, 2020) degrees from ITMO University, Saint-Petersburg. His dissertation research was related to various methods of fiber-optical interferometers and fiber lasers cavities desensitization to environmental noise and vibration impacts. Having overall diverse R&D experience of more than 10 years on positions from laboratory assistant to research engineer at ITMO University, he joined the Optoelectronics Research Centre of Tampere University in March 2021 as Postdoc Research Fellow. His current research interests include semiconductor packaging and bonding technologies, silicon photonics and optoelectronic instrumentation. He has authored or co-authored more than 15 scientific papers.

Topi Uusitalo received the Ph.D. degree in engineering physics from the Tampere University of Technology in 2018. He is currently working as a postdoctoral researcher at Optoelectronics Research Centre of Tampere University in research areas in high power edge emitting laser diodes, VCSELs, from device design to characterization. He has authored or co-authored more than 20 scientific works.

Evgenii Lepukhov is a Doctoral Researcher at Optoelectronics Research Centre, Tampere University. He graduated from the Technical Physics master's degree program of Lappeenranta University in 2022, the topic of his master's thesis was the electronics protection from ionizing radiation impact. In the summer of 2022 Evgenii completed an internship at the CERN, Switzerland. His major tasks were the design and test of semiconductor detectors. His current research interests include semiconductor packaging and bonding technologies, silicon photonics and optoelectronic instrumentation.

Heikki Virtanen obtained his Doctor of Science and Technology in 2017 from Tampere University of Technology. He is currently working as a researcher in Optoelectronic Research Centre at Tampere University. His research is focused on the modelling, design, fabrication, and characterization of active optoelectronic devices such as laser diodes, semiconductor amplifiers, photodiodes, and superluminescent diodes. Moreover, his research is involving the development of active micro-optics assemblies. He has authored or co-authored more than 40 scientific works.

Samu-Pekka Ojanen is a Doctoral Researcher at Optoelectronics Research Centre, Tampere University. He received his M.Sc. (Tech) in engineering physics from Tampere University of Technology. His research focuses on the development of tunable hybrid lasers for sensing applications. With expertise in semiconductor laser processing, packaging, and characterization, he also specializes in photonic integrated circuit design and simulation, as well as tunable hybrid laser cavity design and characterization.

Jukka Viheriälä (Dr.Sc., Adj. Prof.) is Senior Scientist at Tampere University responsible for micro- and nanofabrication in Faculty of Engineering and Natural Sciences. Dr. Viheriälä's research interests are linked to development of novel optoelectronic devices at Optoelectronics Research Centre, Tampere. Dr. Viheriälä has more than 15 years of experience as a researcher, with particular focus on optoelectronics device fabrication, testing and characterization. Most recently he was responsible for the nano-imprint lithography capabilities for laser diode processing and contributed to obtaining several pioneering results in this area. He has also developed optoelectronics devices tailored for hybrid integration to silicon photonics technology and contributes to progress in high power single frequency diode lasers and mid-IR laser and gain chips. He has authored or co-authored more than 100 scientific papers. Besides his academic work, Dr. Viheriälä has been active in commercialization of research at Optoelectronics Research Centre in various roles in technical and business development.

Mircea Guina (Member, IEEE) received the Ph.D. degree in physics from the Tampere University of Technology, Tampere, Finland, in 2002. He is currently a Professor of semiconductor technology (optoelectronics), since 2008 and leads the Optoelectronics Research Centre team, a research group part of the Faculty of Engineering and Natural Sciences, Tampere University, Tampere. He conducts research on several major topics including molecular beam epitaxy of novel optoelectronic compounds, development of semiconductor lasers and high-efficiency solar cells, photonic integration, and use of lasers in medicine, light detection and ranging (LIDAR), and sensing. He has authored more than 200 journal articles, nine book chapters, has given more than 40 invited talks at major international conferences, and holds four international patents. Dr. Guina has an outstanding track record in initiating and leading large-scale research projects extending from basic science to technology transfer. He is a current recipient of an European Research Council (ERC) Advanced Grant for the development of high efficiency solar cell technology (AMETIST). He is also Cofounder and Chairman of three start-ups related to laser technologies (Vexlum Oy, Reflekron Oy, and Picophotonics Oy). He is a Topical Editor for the Optics Letters journal and the Journal of European Optical Society and was awarded the Optical Society (OSA) Fellow and the Society of Photo-Optical Instrumentation Engineers (SPIE) Fellow distinctions for his work on optoelectronics and laser technologies.

Integration of Multi-Lithography Technologies for the Fabrication of Flexible Optical Link

Akash Mistry
Institute of Electronic Packaging Technology
Technical University of Dresden
Dresden, Germany
akash_sunilkumar.mistry@tu-dresden

Krzysztof Nieweglowski
Institute of Electronic Packaging Technology
Technical University of Dresden
Dresden, Germany
krzysztof.nieweglowski@tu-dresden.de

Karlheinz Bock
Institute of Electronic Packaging Technology
Technical University of Dresden
Dresden, Germany
karlheinz.bock@tu-dresden.de

Abstract—The advancement in demand for high bandwidth energy-efficient communication in the data centre and edge cloud servers needs a viable optical interconnection solution to cope with the demands. Therefore, the presented study describes the concept of flexible multi-mode waveguides (MM-WGs) as an optical link for co-packaged optics. It evaluates three lithography technologies; UV-lithography, 2 photon polymerization direct laser writing process (2PP-DLW), and nano-imprint lithography (NIL) for the fabrication of flexible MM-WGs. The UV-lithography and 2PP-DLW process were evaluated for the fabrication of MM-WGs and micro-mirrors, respectively, for the master pattern of the NIL stamp. The NIL evaluates the imprinting of the MM-WGs with micro-mirrors at either end on the flexible and transparent PEN substrates with a low-loss OrmoClad lower cladding layer. There are five different cross-sections from 10x10 μm^2 to 50x50 μm^2 of MM-WGs with micro-mirrors were imprinted. Additionally, it presents the importance of integrating multi-lithography technologies to fabricate flexible optical links where a 2PP-DLW process shows the best results for printing μm-scale optical components. On the other side, UV-lithography with SU-8 gives the foremost definition of the master for the polymeric MM-WGs. Furthermore, NIL offers the industrial mass-production option alongside prototyping.

Index Terms—MM-WGs, micro-mirror, NIL, flexible substrate, multi-lithography

I. INTRODUCTION

The increasing demand for the Internet of Things (IoT), cloud computing, cloud storage, and energy-efficient communication in data centres necessitates an alternative solution to electrical interconnects. However, the miniaturization of semiconductor industries, driven by Moore's Law [1], has made it challenging to fabricate high-speed and energy-efficient electrical interconnects using existing technologies. Furthermore, scaling down copper interconnects has resulted in increased resistivity compared to bulk counterparts due to electron migration at the grain boundaries [2]. Nevertheless, significant advancements have been made in optical interconnects over the past decades, thanks to their potential for high bandwidth and energy-efficient communication. Optical cables have already demonstrated potential for long-distance communications. Additionally, a previous study [3] highlighted the application of single-mode waveguides (SM-WGs) for short-distance communications at the interposer level, utilizing a hybrid-lithography approach that combines UV-lithography and 2PP-DLW processes to fabricate SM-WGs with micro-mirrors for out-of-plane communication. However, the slow nature of the 2PP-DLW process makes mass production of micro-mirrors challenging without advancements in processing tools. Nevertheless, to counter the problem, [4] shows the application of NIL for the fabrication of SM-WGs with micro-mirrors.

Flexible optical links present promising applications for high-bandwidth and energy-efficient communication in co-packaged optics. Therefore, this work discusses the concept of multi-lithography for fabricating flexible optical links. The concept mainly focuses on fabricating MM-WGs with micro-mirrors at each end on a flexible PEN substrate using UV-NIL to transfer the signal out-of-plane. Multi-lithography refers to three lithography processes used to fabricate these flexible optical links: UV-lithography, 2PP-DLW process, and UV-NIL technology. The integration of these technologies aims to leverage their specific capabilities with suitable materials to create customized and efficient structures for flexible optical links. To replicate the MM-WGs with micro-mirrors onto the low-loss OrmoCore polymer, a master stamp needed to be fabricated. As described in a previous study [5], UV-lithography with SU-8 polymer was employed for the fabrication of the master stamp, providing precise 90° sidewall angles and smooth structures with nanoscale roughness. Conversely, the 2PP-DLW process with IP-DIP resin, explicitly designed for printing applications, enables better structural definition with sharp edges for the micro-mirrors and faster printing of these structures with lower surface roughness.

In this study, five different cross-sections ranging from $10 \times 10 \ \mu m^2$ to $50 \times 50 \ \mu m^2$ of optical links were successfully imprinted on the OrmoCore polymer, which serves as the core layer for MM-WGs over a low-loss OrmoClad lower cladding layer on a flexible substrate. Section II provides an introduction to the concept of integrating multi-lithography, while section III discusses the experimental results. Sections IV cover the conclusion and future outlook of the work.

II. INTRODUCTION TO THE CONCEPT OF MULTI-LITHOGRAPHY TECHNOLOGIES FOR FLEXIBLE OPTICAL LINK

As previously mentioned, the concept of integrating multi-lithography technologies aims to maximize the performance of each technology with the most suitable materials. Fig. 1 depicts the processing steps of fabricating flexible optical links using multi-lithography technologies. Firstly, as shown in Fig. 1(a), SU-8 waveguides were fabricated using a specially designed mask that ensured uniform pressure distribution for achieving a nearly uniform residual layer at the end of the UV-NIL process. The fabrication of SU-8 waveguides followed a standard procedure described in reference [5], employing UV-lithography. After fabricating the waveguides, micro-mirrors were created at each end using the 2PP-DLW process with IP-DIP resin, as illustrated in Fig. 1(b). The alignment accuracy during micro-mirror fabrication was within ± 1 μm, and the roughness was below 0.1 λ.

Once the master with MM-WGs and micro-mirrors was prepared, an anti-sticking layer was applied using the provided silanization solution. Subsequently, the stamp was fabricated by drop-casting OrmoStamp polymer onto the pattern master, as shown in Fig. 1(c). Due to the low viscosity of OrmoStamp at 25 °C, which is 0.5 Pa·s, it settled automatically through the openings under the weight of the glass substrate. For the subsequent steps illustrated in Fig. 1(d) and 1(e), substrates were prepared by bonding a flexible PEN film onto the silicon wafer, followed by the coating of the lower cladding layer.

The next steps involved the drop-casting method, where small droplets of 0.01-0.02 gm of OrmoCore-OrmoThin polymer were deposited on the prepared substrate to form the core layer, as depicted in Fig. 1(f). Before contacting the stamp with the droplet, it was coated with the anti-sticking coating trichlor-(1H,1H,2H, 2H-perfluoroctyl)-silane provided by Sigma-Aldrich. After curing, as shown in Fig. 1(g), the stamp was de-bonded from the substrate. Finally, the flexible PEN film, along with the layer of MM-WGs, was also de-bonded from the silicon substrate to obtain the flexible optical link.

III. EXPERIMENTAL RESULTS AND DISCUSSION

In the initial stage, as outlined in reference [5], a master for the NIL stamp was prepared. The fabrication of the master patterns involved two steps. In the first step, SU-8 MM-WGs were fabricated using UV lithography. To achieve MM-WGs with straight 90° sidewalls and square cross-sections, different combinations of power and timing were experimented with, and successful results were obtained at 225 mJ/cm^2 for all cross-sections of WGs. Fig. 2 [5] illustrates the results

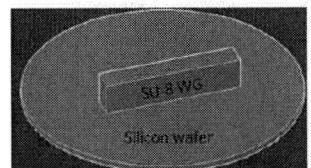 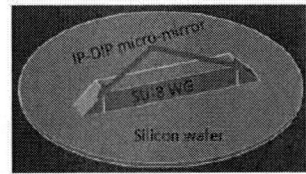

(a) UV-lithography for SU-8 MM-WGs

(b) 2PP-DLW process for IP-DIP micro-mirrors (master)

 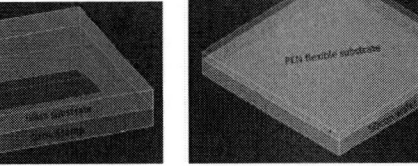

(c) Drop casting process for stamp

(d) PEN flexible substrate on silicon wafer

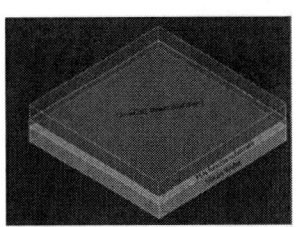

(e) Lower cladding over PEN

(f) Drop casting process for core layer fabrication

(g) De-bonding of the stamp after UV-curing

(h) De-bonding of the PEN from the silicon wafer

Fig. 1. Processing steps for the fabrication of flexible optical links using multi-lithography technologies

for 30×30 μm^2 of cross-section of the MM-WGs. Additionally, [5] shows the results for all other cross-sections of MM-WGs. One advantage of using SU-8 is its compatibility with contact mode lithography, which ensures the formation of 90° sidewalls and smooth MM-WGs surfaces, facilitating easy demolding during NIL and enabling better optical signal transmission.

In the second step, the micro-mirrors were printed using IP-DIP resin and the 2PP-DLW process, achieving an alignment accuracy of ± 1 μm. The IP-DIP resin, developed by Nano-scribe GmbH & Co. KG, is specifically designed to enable faster printing with sharp structural contours. The experimental procedure and processing parameters for this step can be found in reference [5]. Fig. 3 [5] displays the printed mirrors located

265

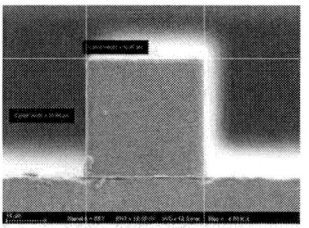 

(a) 30×30 μm^2 cross-section   (b) 30×30 μm^2 tilted view

Fig. 2. Patterned 30×30 μm^2 cross-section of SU-8 MM-WGs by UV-lithography [5]

at the ends of the SU-8 MM-WGs for 30×30 μm^2 cross-section, serving as the pattern master for the NIL process.

Fig. 3. 2PP-DLW printed 30×30 μm^2 micro-mirrors with IP-DIP resin at the end of the SU- 8 MM-WGs [5]

In the subsequent step, the pattern master was coated with an anti-sticking layer using Silanization Solution I, n-Hexane, and I-Octane from Sigma-Aldrich. The silanization solution reacted with the silicon surface, forming a solid covalent bond through a condensation reaction and resulting in the presence of an organosilane molecule on the surface. This organosilane layer provided a highly hydrophobic surface. The pattern master was immersed in the silanization liquid for 20 minutes to achieve this hydrophobic surface. Subsequently, it was immersed in n-Hexane and I-Octane for 15 minutes each to remove any residual contamination or unreacted silane molecules. These non-polar solvents, n-Hexane and I-Octane, did not alter the hydrophobic properties of the organosilane surface on the pattern master.

For the next step, a glass substrate measuring 50x50 mm and 3.8 mm thick was prepared. It underwent a standard cleaning process, followed by plasma treatment at 200 W for 3 minutes. An OrmoPrime layer was then spin-coated onto the glass substrate at 4000 RPM for 60 seconds, resulting in a thickness of approximately 150 nm. This OrmoPrime layer improved the adhesion of the OrmoStamp resin to the glass surface. The glass substrate was then baked on a hot plate at 150 °C for 5 minutes.

Next, the stamp was fabricated using a drop-casting method by using the master pattern. A precise amount (0.01 gm in this case) of OrmoStamp was dropped onto the pattern master using a syringe. Degassing was performed before or after dropping the resin to remove any trapped air bubbles, and a pointed syringe tip also be used to burst the bubbles. Subsequently, the glass substrate, treated with OrmoPrime,

was gently placed upside down to make contact with the OrmoStamp drop on the pattern master. Due to the lower viscosity of the OrmoStamp resin, it spread across the pattern master's surface under the weight of the glass substrate. It took approximately 10 minutes for the resin to completely fill all the patterns on the pattern master. The thickness of the resulting OrmoStamp master depended on the amount of OrmoStamp resin applied initially as a droplet. The manufacturer, micro resist technology GmbH, provides data in [7] on the recommended amount of OrmoStamp resin for a specific substrate size. Once the OrmoStamp was settled, a full exposure at 20 mW/cm^2 for 50 seconds was performed. To de-bond the pattern master from the glass substrate with the cured OrmoStamp resin, a small force was applied at one of the corners using a sharp blade, resulting in easy demolding. The stamp was further hardened through a 30-minute hard baking process at 130 °C. Fig. 4 illustrates the fabricated stamp result. The middle section of Fig. 4 depicts the MM-WGs with micro-mirrors at the ends, while different stamps were fabricated for each of the various cross-sections of the MM-WGs.

Fig. 4. Fabricated glass-OrmoStamp stamp

In the subsequent process steps, a substrate was prepared by applying double-sided tape to a clean 4-inch silicon wafer. A flexible PEN substrate was then applied to the tape, followed by spin-coating and curing 40 μm thick lower-clad layer using OrmoClad polymer.

To facilitate easy demolding after NIL curing, an anti-sticking layer was coated on the glass stamp with OrmoStamp. Trichlor-(1H,1H,2H,2H-perfluorooctyl)-silane (PFOTS) was used for this purpose. A drop of PFOTS was placed beside the glass stamp and the setup was placed in a desiccator, where the pressure was reduced to 0.06 bar to allow for the evaporation of PFOTS over a period of 20 minutes. Subsequently, the glass stamp was hard-baked on a hot plate at 150 °C for 15 minutes. During this process, a reaction occurred between the silanol (Si-OH) group in the OrmoStamp and the trichlorosilane (Si-Cl) group of PFOTS. Hydrolysis resulted in the formation of PFOTS-OH, followed by a condensation reaction that formed PFOTS-O-Si-OrmoStamp bonds, providing hydrophobic and oleophobic properties to the stamp.

To minimize the residue layer at the end of the NIL process, a lower thickness of the OrmoCore polymer was required for the core layer. However, OrmoCore alone had a higher viscosity, making it unsuitable for NIL. To address this, it was diluted with OrmoThin in a 1:1 ratio to reduce the viscosity. Since the substrates used in this work were sized at 0.8x50 mm, the spin-coating process was not applicable. Therefore, a drop-casting method was employed. Where, a drop of the diluted OrmoCore-OrmoThin mixture, weighing around 0.01 to 0.02 grams, was gently placed on the prepared substrate. However, covering the entire surface with the polymer is a problem as the viscosity of the OrmoCore is regained after prebaking, which limits the flow of the OrmoCore resulting in an incomplete coating of the substrate while imprinting. Therefore, manually spreading the liquid over the surface was done with a slightly more quantity of the Ormothin-OrmoCore mixture. After manually spreading the liquid over the surface, a pre-baking step was performed at 85 °C for 2 minutes to evaporate the OrmoThin from the substrate surface. Subsequently, an anti-sticking layer-coated glass stamp was placed upside down, with the treated surface in contact with the spread OrmoCore on the substrate, applying a pressure of 35N for 30 minutes to ensure complete filling of all patterns by the OrmoCore polymer. UV curing was then conducted at 3000 mJ/cm^2 (20 mW/cm^2 for 150 seconds). To demold the glass stamp from the substrate, a slight force was applied at the edge of the substrate surface using a sharp blade. Finally, the recommended hard-baking step at 130 °C for 10 minutes was performed to harden the OrmoCore patterns containing MM-WGs with micro-mirrors at either end. Fig. 5 illustrates the imprinted results for the five different cross-sections of MM-WGs with micro-mirrors.

In Fig. 5(a), the OrmoCore MM-WG cross-section of 10×10 μm^2 is depicted, while a tilted view with the micro-mirror is shown in Fig. 5(b). In Fig. 5(b), a desired square cross-section is observed for the 10×10 μm^2 MM-WGs, indicating that the shape was successfully preserved throughout the processing steps. However, a residue layer of 980 nm can also be seen. It should be noted that the pressure applied during NIL was limited to 35N, so the main focus was on successfully imprinting the pattern. Some visible dust particles can also be observed, likely from the dicing saw used during the MM-WG cross-section characterization. Nevertheless, the surface of the MM-WGs and micro-mirror appears smooth. Similarly, the imprinted cross-sections of MM-WGs with micro-mirrors, including 20×20 μm^2, 30×30 μm^2, 40×40 μm^2, and 50×50 μm^2 are shown in Fig. 5(c) and 5(d), Fig. 5(e) and 5(f), Fig. 5(g) and 5(h), Fig. 5(i) and 5(j), respectively. Some defects can be observed around the micro-mirror structures, but they can be easily mitigated by slightly adjusting the processing parameters. However, the shape and roughness values were preserved at the end of the process.

IV. CONCLUSION AND FUTURE OUTLOOK

In this work, the integration of three lithography technologies, namely UV-lithography, 2PP-DLW process, and nanoim-

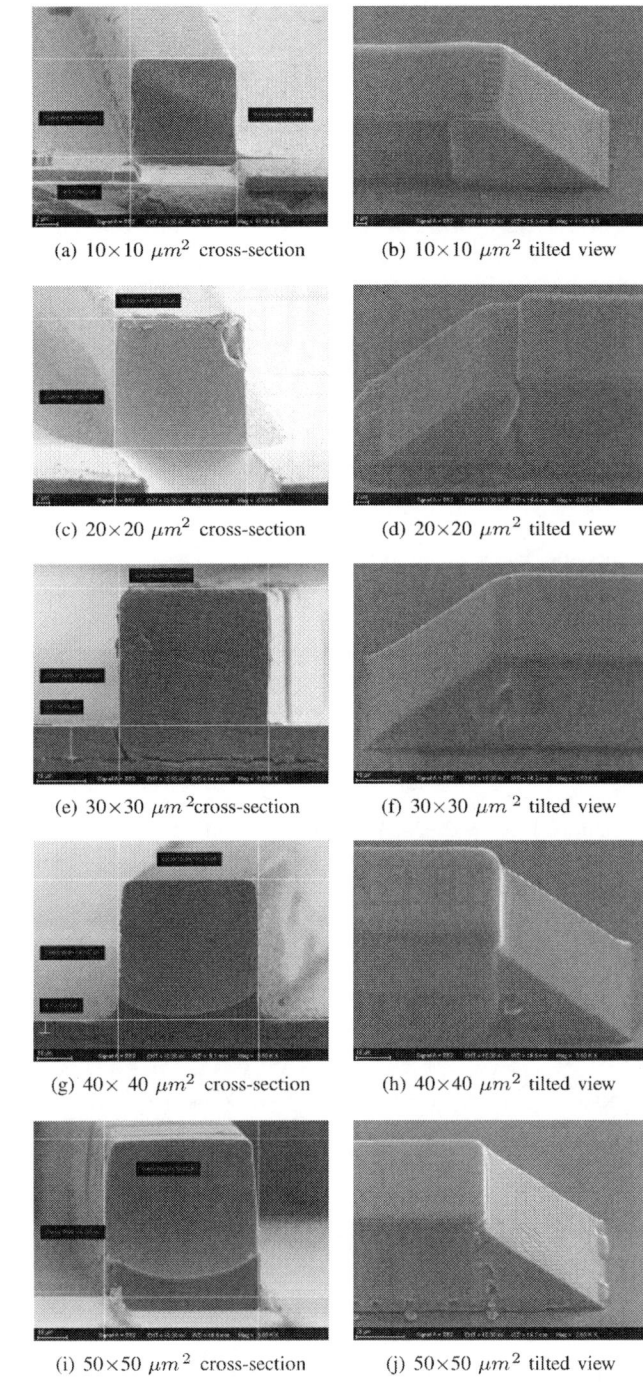

(a) 10×10 μm^2 cross-section   (b) 10×10 μm^2 tilted view

(c) 20×20 μm^2 cross-section   (d) 20×20 μm^2 tilted view

(e) 30×30 μm^2 cross-section   (f) 30×30 μm^2 tilted view

(g) 40×40 μm^2 cross-section   (h) 40×40 μm^2 tilted view

(i) 50×50 μm^2 cross-section   (j) 50×50 μm^2 tilted view

Fig. 5. NIL processed OrmoCore MM-WGs with their cross-sections and micro-mirrors at the end

print lithography, was demonstrated for fabricating MM-WGs with micro-mirrors at both ends to enable out-of-plane signal direction for high-speed co-packaged optics applications. A previously fabricated master with SU-8 MM-WGs and micro-mirrors, created using UV-lithography and 2PP-DLW process, respectively, was utilized in this study. The processing steps involved applying an anti-sticking layer to the previously fabricated master, fabricating a glass stamp from OrmoStamp material, coating the glass stamp with an anti-sticking layer, preparing the substrate, and successfully imprinting the patterns containing MM-WGs and micro-mirrors onto the low-loss OrmoCore hybrid polymer through nanoimprint lithography. The roughness and shape of the structures were preserved throughout the process, with no noticeable effects on their quality. However, some defects were observed around the micro-mirror area, likely arising from the final nanoimprint lithography steps. These defects can be mitigated by making slight adjustments to the process parameters.

For future work, the process parameters will be further optimized to achieve the best possible results from the employed technologies. With improved processing steps in higher-pressure facilities, efforts will be made to minimize the residue layer to an optimum level, enabling clean imprinting with a better pattern-to-defect ratio. Then an etching will be performed to completely remove the residue layer. Subsequently, the optical performance of the MM-WGs with micro-mirrors, along with their mechanical performance, will be characterized.

ACKNOWLEDGMENT

The research work presented within the E4C project has been supported by Germany's Federal Ministry of Education and Research. The author would like to thank I. Sotiriou and M. Göhler, the technical support staff at IAVT, TU Dresden, for their invaluable contribution in preparing the working samples of SU-8 MM-WGs, imprinting substrates with flexible PEN substrate on the lower clad layer, and dicing samples for SEM characterization.

REFERENCES

[1] R. R. Schaller, "Moore's law: past, present and future," in IEEE Spectrum, vol. 34, no. 6, pp. 52-59, June 1997, doi: 10.1109/6.591665.

[2] A. A. Vyas, C. Zhou and C. Y. Yang, "On-Chip Interconnect Conductor Materials for End-of-Roadmap Technology Nodes," in IEEE Transactions on Nanotechnology, vol. 17, no. 1, pp. 4-10, Jan. 2018, doi: 10.1109/TNANO.2016.2635583.

[3] D. Weyers, A. Mistry, K. Nieweglowski and K. Bock, "Hybrid lithography approach for single mode polymeric waveguides and out-of-plane coupling mirrors," 2022 IEEE 72nd Electronic Components and Technology Conference (ECTC), San Diego, CA, USA, 2022, pp. 1919-1926, doi: 10.1109/ECTC51906.2022.00301.

[4] Mikko Karppinen, Noora Salminen, Tia Korhonen, Teemu Alajoki, Jarno Petäjä, Erwin Bosman, Geert Van Steenberge, John Justice, Umar Khan, Brian Corbett, Arjen Boersma, "Optical coupling structure made by imprinting between single-mode polymer waveguide and embedded VCSEL," Proc. SPIE 9368, Optical Interconnects XV, 93680H (3 April 2015); https://doi.org/10.1117/12.2082935.

[5] A. Mistry, K. Nieweglowski and K. Bock, "Hybrid-Lithography for the Master of Multi-ModeWaveguides NIL Stamp," 2023 46th International Spring Seminar on Electronics Technology (ISSE), Timisoara, Romania, 2023, pp. 1-5, doi: 10.1109/ISSE57496.2023.10168370.

[6] del Campo, Aránzazu, and Christian Greiner. "SU-8: a photoresist for high-aspect-ratio and 3D submicron lithography." Journal of micromechanics and microengineering 17.6 (2007): R81.

[7] Mühlberger, M., Bergmair, I., Klukowska, A., Kolander, A., Leichtfried, H., Platzgummer, E., Loeschner, H., Ebm, C., Grützner, G., Schöftner, R. (2009). UV-NIL with working stamps made from Ormostamp. Microelectronic Engineering, 86(4–6), 691–693. https://doi.org/10.1016/j.mee.2008.11.020.

Hybrid lithography fabrication of single mode optics for signal redistribution and coupling

David Weyers, Krszystof Nieweglowski and Karlheinz Bock

Institute of Electronic Packaging Technology
Technical University of Dresden, D-01062 Dresden, Germany
Email: david.weyers@tu-dresden.de

Abstract—This paper describes advances in hybrid-lithography process, combining UV-lithography for planar, single mode redistribution layer (RDL) and 2-photon-polymerization direct-laser-writing (2PP-DLW) for micro-mirrors inside RDL-opening. Improvements to multi-layer direct patterning of OrmoCore/-Clad material system using UV-lithography and need for broadband UV-LED source are presented. Near square core cross sections and smooth sidewalls are achieved. Openings in full stack with steep sidewalls without residual layer are patterned. To optimize 2PP-DLW-process processing window for both OrmoComp and IP-DIP is thoroughly characterized. Roughness measurements prove feasibility even of coarsely printed structure as reflective μ-mirror for 1550 nm wavelength. Finally these results are applied to periscope probe for wafer-level-testing of edge emitting lasers and proof of concept is shown. Outlook to further research on UV-lithography of multi-layer waveguide stack and alignment with μ-mirror printing is given.

Keywords—optical interconnects, single mode, ORMOCER®, hybrid lithography, direct patterning, UV-lithography, 2-photon-polymerization direct-laser-writing, micro-mirrors

I. INTRODUCTION

The ongoing increase in performance and density in high performance computing (HPC) culminating in so-called co-package optics approach where photonic processing units (PPU) or receiver and transmitter (Rx/Tx) are integrated on the same interposer with central processing unit (CPU) and high-bandwidth memory (HBM) is the biggest driver for interposer level optical interconnects and also heavily pushed by industry [1]–[3]. Even more so increasing research and fabrication volume in photonic integrated circuits (PICs) is demanding adequate packaging solutions.

Often butt-coupling to single mode fibers or fiber arrays aligned in v-grooves is used. Whilst this is well established and mechanically stable [4], we believe that it also poses as a limit in integration density and scaling as well as costs, due to the needed chip area for v-grooves, peripheral arrangement on chip/interposer edges and limitation to 1D-arrays. In addition it only provides chip-to-fiber and no chip-to-chip interconnects.

The capability of polymer waveguides (WG) to provide interposer and module level interconnects has been largely demonstrated [5]–[10]. In [11] we proposed a combination of polymer WGs patterned in to an optical RDL via UV-lithography and μ-mirrors structured with 2PP-DLW to provide surface coupling. With the combination of two processes that we termed hybrid lithography we combine the advantages of both. UV-lithography gives parallel, high throughput 2.5D-structuring for the optical RDL. Whilst 2PP-DLW allows for arbitrary 3D-shaped μ-mirrors changing both optical axis and beam shape. This allows for a higher IO-count and integration density compared to fiber butt-coupling. The so-called optical back end of line (BEOL) shown in Fig. 1 can be added to an existing electrical interposer to add optical function.

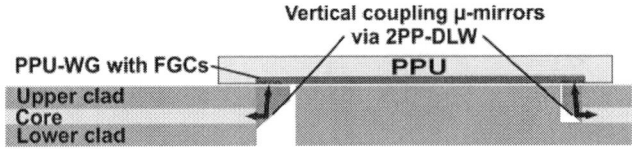

Figure 1. Schematic of the optical BEOL consisting of polymeric RDL with vertical coupling micro-mirrors used to connect to a PPU with integrated WGs and FGCs.

This paper details our recent advances in the hybrid lithography process. The processing section will begin with a brief description of the flow. We demonstrate that bottom cladding can be used to planarize topography from metal RDL and pads forming the electrical interconnects on the interposer. Thereafter measures taken to improve the UV-lithographic patterning, with the final one being the upgrade to a new UV-LED source, will be described. The second part of processing will detail a method used for semi automated resin characterization and comparison for the 2PP-DLW process. This will then be applied to a periscope probe for wafer level testing of edge emitting lasers. Thereafter we will characterize UV-lithographic patterning results using scanning electron microscopy (SEM) and explain why the UV-LED source is necessary for reproducible patterning. We will present the 2PP-DLW process window for IP-DIP resin and conclude its suitability for our application. SEM images of the printed structures will be undermined with tactile profilometry measurements to prove that even a visible roughness in SEM can be well below $\lambda/10$. Thereafter printing results of the periscope probe and its application in the measurement of edge emitting lasers are shown. Finally we conclude our work and provide an outlook to future research.

II. PROCESSING

We previously demonstrated light transmission across two μ-mirrors and a WG inside the optical BEOL [12]. However patterning quality was not ideal, thus this chapter will focus on upgrades to the individual steps of hybrid lithography process to improve structure quality, thus bare silicon is used as a substrate. Fig. 2 gives an overview on the full process flow to transform an electrical Si-interposer into an hybrid. Second focus will be on ensuring that all process development is compatible to the Si-interposer substrate.

A. UV-Lithography

In [10]–[12] we improved our UV-lithographic direct patterning of WGs from multi to single mode dimensions. With the reduction in layer height and feature size the formation of an inhibition layer of 5 μm with the used ORMOCER®-material system demanded inert atmosphere during exposure. Additionally it has to be exposed with a proximity gap, since it is sticky after spin-coating.

Previously we set the proximity gap by referencing contact of wafer and mask before coating the layer, however the used mask aligner EVG640 does not allow for reproducible setting of low proximity gap. Since the resin is still sticky after pre-bake the mask will be fully wetted even from a small contact. To ensure a low and reproducible gap we

structure SU8 spacers around the wafer edge. First the substrate is spin cleaned with acetone und isopropanol, thereafter SU8-3025 is coated at 1500 RPM and soft baked for 15 min at 95 °C. Since resolution is uncritical for spacers a foil mask and contact exposure with $400\,\mathrm{mJ\,cm^{-2}}$ at 365 nm (i-line) is used to structure 1 mm long and 400 μm wide, triangular spacers that are 2 mm from wafer edge and pointing inwards. Post exposure bake is carried out for 1 min at 65 °C and 5 min at 95 °C before the spacers are developed for 10 min in Mr-Dev600 and rinsed with IPA. Any residues of SU8 will later cause holes in the BEOL, thus full development of the layer must be achieved. Spacers are fully inside the edge bead and thus the height of 90 μm is much higher then data sheet value of 40 μm. They are high enough to be used for the full 50 μm optical BEOL and accompany for some wafer bow and thickness non uniformity but still small enough to allow patterning of the 7 μm WG-cores.

To achieve a numerical aperture (NA) of 0.14, matched to typical single mode fiber (SMF), a mixture of 70% OrmoCore and 30% OrmoClad is used as bottom and top cladding and will be referred to as OrmoCladXP. By ordering this premixed in syringes directly from micro resist technology, we were both able to solve problems with small bubbles after spin coating and achieve good control over applied amount of resin by using a pneumatic dispenser. Before applying the resin O_2-plasma activation is performed for best adhesion [11]. Undiluted resin and a spin speed of 3000 RPM yielding a thickness of 27 μm was derived from planarization tests on 6 μm topography from pads and RDL on Si-interposer, that will be explained later in the characterization chapter. edge beat removal (EBR) is performed by spraying OrmoThin onto the wafer edge during the last 10 s of spin coating. This removes most of the resin from the spacers, small amounts might touch the mask, but the wafer can not be pulled towards the mask due to the spacers. Since the mask is opaque at the wafer edge, those residues will not be exposed and can be easily washed away using the development process. Thirdly pre-bake for 4 min at 85 °C is carried out. Even though this is only necessary for thinned resin to evaporate OrmoThin, we see improvement to layer uniformity during the pre-bake. At forth exposure is performed. To ensure a good N_2-atmosphere and thus reduce inhibition we use a custom chuck with N_2-inlets surrounding the wafer and rinse for 2 min with 400 μm separation before bringing spacers into contact with the mask. Exposure is then performed in this quasi-contact-mode, initially with a i-line Hg-lamp. Puddle development in spin coater is used. Sample is exposed to OrmoDev for 30 seconds and spin cleaned twice with OrmoDev and once with IPA. Those steps are then repeated for core and top cladding. 3 mL of OrmoCore diluted 4:1 with OrmoThin are pipetted for the core layer to yield a thickness of 7 μm at 3000 RPM. After printing the μ-mirrors using 2PP-DLW, full exposure with $1000\,\mathrm{mJ\,cm^{-2}}$ to deplete residual iniator and a 3 h hard bake with 30 min up-ramp is carried out to improve stability and

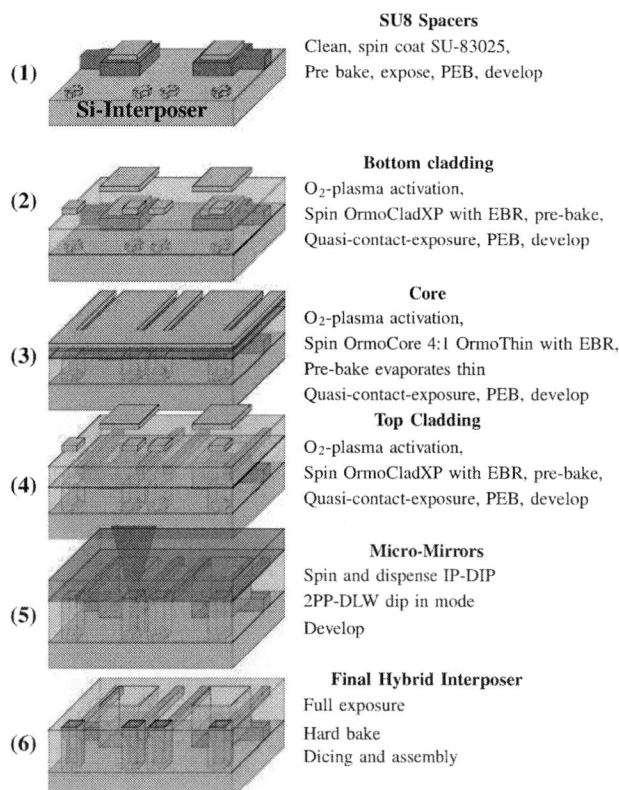

SU8 Spacers
Clean, spin coat SU-83025,
Pre bake, expose, PEB, develop

Bottom cladding
O₂-plasma activation,
Spin OrmoCladXP with EBR, pre-bake,
Quasi-contact-exposure, PEB, develop

Core
O₂-plasma activation,
Spin OrmoCore 4:1 OrmoThin with EBR,
Pre-bake evaporates thin
Quasi-contact-exposure, PEB, develop

Top Cladding
O₂-plasma activation,
Spin OrmoCladXP with EBR, pre-bake,
Quasi-contact-exposure, PEB, develop

Micro-Mirrors
Spin and dispense IP-DIP
2PP-DLW dip in mode
Develop

Final Hybrid Interposer
Full exposure
Hard bake
Dicing and assembly

Figure 2. Schematic hybrid lithography process flow for polymeric optical BEOL on Si-interposer with Cu-RDL, passivation and pads.

adhesion.

Main focus of the process development is to improve patterning quality by working on the exposure. In [12] we explained, that our i-line Hg-lamp non-ideal, since it has to be moved away from its dedicated operation point to reduce lamp power, so that needed doses of a few $mJ\,cm^{-2}$ can be delivered at reliable shutter times $> 0.8\,s$. Nevertheless we tried to improve the patterning both by the spacers and using optical proximity structures (OPC) patterns in the mask. To further improve patterning we upgraded to a UV-LED exposure source with two broadband spectra at 365 nm (i-line) and 405 nm (h-line), separately tunable in power, and pulse time as short as 100 ms and up to $65\,mW\,cm^{-2}$, allowing for short, high power exposure to further reduce inhibition. Patterning results for bottom cladding and core with the two setups and different doses will be compared in the characterization section.

B. Micro-Mirror Fabrication

The 2PP-DLW-process is based on two-photon-absorption (TPA) of negative tone resins inside the focal volume, so-called voxel (volumetric pixel) of a highly focused 780 nm laser. Regular single photon absorption is proportional to intensity whereas TPA is to intensity square, thus resin contrast is much larger and voxel can be tuned down 300-400 nm diameter and 800-1000 nm height [13], [14]. Laser beam is moved along a 3D-trajectory with the overlapping voxels forming an arbitrary 3D-structure. This capability makes it an ideal process for μ-mirrors.

Nanoscribe photonic GT2 machine is used for printing the μ-mirrors. A resin droplet is dispensed onto the substrate, for larger substrates a thin layer has to be spin coated in addition. Since the substrate is opaque we use so called dip-in mode, where the objective is then moved into the droplet. The surface is referenced by its reflection, thus only the interface between substrate and polymeric BEOL can be found, since all other interfaces have low refractive index step and thus reflection. This makes good process stability in lithography critical so that the mirrors can be accurately placed in front of WG-cores in z-direction. Since the linear stage used to move laterally across the wafer is only accurate to a few μm, local alignment marks in the pad layer are used for every mirror. They can be referenced using the in-line camera, however only with the 63x objective resolution is high enough for sub-μm alignment. Therefore we use it, even though 25x would allow for faster printing with still sufficient resolution for our 20 μm wide mirrors. To form the structure the beam is steered in concentric trajectories in plane using a galvanic mirror system and the planes are stacked by moving the sample away from the objective with a piezo. In-plane trajectory distance is called hatch (H) and plane distance slice (S). Together with laser power and scan speed those are the most important process parameters. After printing the objective is moved out of the droplet and sample is developed in PGMEA bath for 10 min and rinsed in isopropanol for 2 min.

Since this is a well established process our focus is on reducing processing time and variation without degrading structure and thus optical quality. In [15] we achieved good results with OrmoComp resin, which is related to used WG material system. However OrmoComp has a strong tendency to overexposure causing the structures to look inflated, which causes additional lens effects. Thus we present a comparison to IP-DIP resin in this paper. Since the path of beam through the mirror is very short, small differences in absorption only have a minor impact, thus the patterning conformity, process stability and mechanical stability are the most important aspects. To ensure the later, adhesion tests similar to those for optical RDL in [11] were carried out on the contact interfaces of the μ-mirrors: Si, SiO$_2$ and bottom clad. For characterizing the processing window a large matrix of identical structures with different hatch and slice, scan speed and laser power was printed. Since we are aiming at roughly 45° mirrors identical hatch and slice from 100-700 nm was used. All structures were lifted on a 15 μm-high coarse written base to get away from the silicon reflection, which alters power density. Laser power was varied in 7 levels between 6.5 mW and the maximum of 65 mW. Maximum scan speed of the machine is $200\,mm\,s^{-1}$. Minimum scan speed for each combination of laser power, hatch and slice was set using the equation 1 and at least to $100\,\mu m\,s^{-1}$, below the line starts to wobble [11]. Logarithmic sweep with 10 values per decade and at least 2 decades is applied.

$$SS_{min} = 0.1 * \frac{(P/[\mathrm{mW}])^2}{h/[\mu m] * s/[\mu m]} \qquad (1)$$

Characterization of these samples will be presented later in the respective section. For roughness characterization 40 μm wide, 50 μm high trapezoid structures with 50 μm long, 45° up and down ramp are printed. All parameter sets within processing window that allow for feasible printing time are used. To ensure compatibility of μ-mirrors to proposed interposer adhesion tests as for polymeric RDL in [11] were carried out. Pillars of IP-DIP with 50 μm diameter and 20 μm were printed on Si, SiO$_2$ and OrmoCladXP. For printing on substrate materials coarse parameters as for bases in process window experiment and on OrmoCladXP fine mirror parameters were used. For substrate materials tested adhesion treatments were: none, dehydration bake for 5 min at 200 °C, oxygen plasma, OrmoPrime and silanization. Finally results were already applied to so called periscope probe in Fig. 3.

A project partner in VE-Silhouette presented us with the challenge of probing edge emitting lasers on wafer level. Since the facet is inside a small trench periscope probe was proposed. Centerpiece are the vertical and horizontal WGs connected via μ-mirrors, coupling emission from the laser to a multi mode fiber placed underneath the support structure. Probe tip is extending 60 μm above, thus can be lowered into the laser's trench without touching the substrate surface. It is tilted so the tip can be found when looking through the assembly under prober microscope.

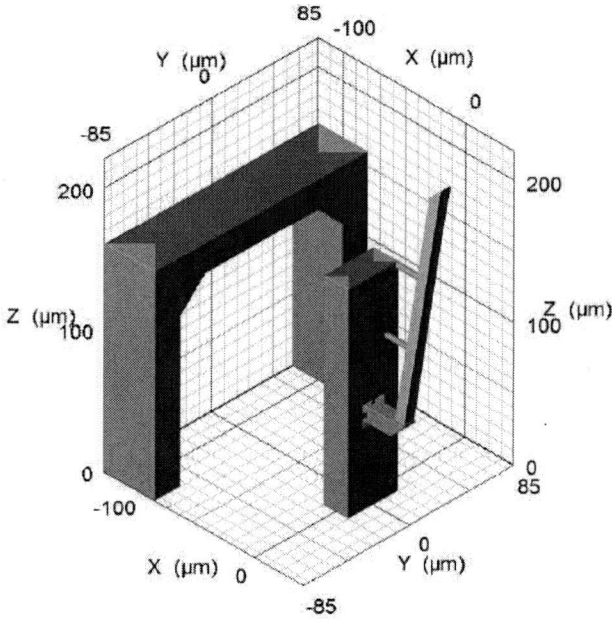

Figure 3. Design of the periscope probe with tilted 8 μm square WG with mirror on top, coupling into 10 μm horizontal WG via mirror. Horizontal WG height is set to match multi mode fiber core held with support structure. WGs are held with 2 μm studs.

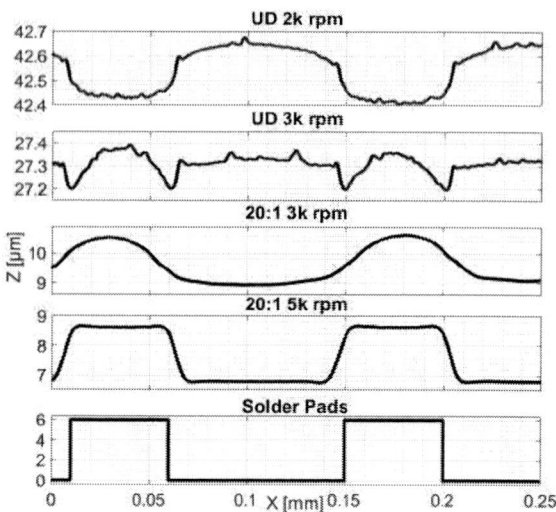

Figure 4. Surface profile measured using tactile profilometry of bottom clad on 6 μm solder pad topography pretreated with oxygen plasma. Undiluted (UD) OrmoCladXP and dilution 20:1 with OrmoThin at different spin speeds are compared. Best planarization is achieved for undiluted OrmoCladXP spun on at 3000 RPM.

Laser is emitting at short visible wavelength, thus hatch and slice of 100 nm with laser power and scan speed at 20 mW and 10 mm were used for the light path. This gives very low roughness and corner rounding with acceptable process time. For the support structures hatch and slice of 200 nm with laser power and scan speed at 30 mW and 20 mm were used to decrease processing time but still give a transparent structure, so it does not obscure tip. This leads to a print time of 30 min. Print is placed roughly 200 μm from edge of 5 mm by 25 mm fused silica substrate, again so that the diced substrate edge does not obscure tip. Fused silica is used since it is transparent and refractive index is low enough to reference interface to IP-DIP resin. After printing the substrate is glued into an stereo lithographically printed base allowing for mechanical fixation to probe manipulators and the fiber. A 50 μm core, 0.2 NA fiber is stripped at its end and cleaved. The fiber facet is carefully maneuvered under the support structure and in front of horizontal WG facet. It is then glued to both the substrate and base.

III. CHARACTERIZATION

In this section we characterize produced samples both geometrically and optically. We highlight and explain deviations as well as possible solutions.

A. Optical RDL

For fast process development Optical RDL test samples are first characterized using optical microscopy, tactile profilometry and SEM imaging. Only if process parameters are deemed,

feasible functional WGs are manufactured and optically tested. For the polymeric RDL to function on a Si-interposer bottom clad has to planarize topography from 6 μm solder pads so that WG-cores do not get extra loss from roughness, waviness and z-position is accurately aligned to μ-mirrors. Different dilutions and spin speeds were tested with OrmoCladXP on Si-interposer samples treated with either oxygen plasma or OrmoPrime. After spin coating bottom clad was fully exposed and topography was measured using tactile profilomety. Ormo-Prime lead to extended waviness above pads, needing bottom clad thickness exceeding 50 μm, that can no longer be bridged by the electrical interconnect. Fig. 4 shows selected results for oxygen plasma. With increasing layer thickness sharp pad topography gets smoothed out, until at 27 μm only ±100 nm topography from pad and edge is left and the section between pads, where WG-cores will be in functional layout, is flat with roughness below $\lambda/10$.

With the introduction of SU8 spacers we achieved a lower proximity gap yielding close to square core cross sections in Fig. 5 with improved end facets, compared to what we reported in [11]. Additionally 0.5 μm wide scattering bars in the mask present for all but the right most WG give a more perpendicular side wall. Best result is achieved for second right with lowest distance between scattering bar and WG mask opening. However root like structures are still clearly visible, even though extend is reduced by lower proximity gap. Additionally sidewalls have rough, standing wave pattern in Fig. 6a and correct exposure dose is only found for parts of wafer due to big inhomogeneity. Thus WGs with fine 32 μm pitch showed losses above $1.5\,\mathrm{dB\,cm^{-1}}$, exceeding what we reported in [11], and strong cross coupling shown in Fig. 7.

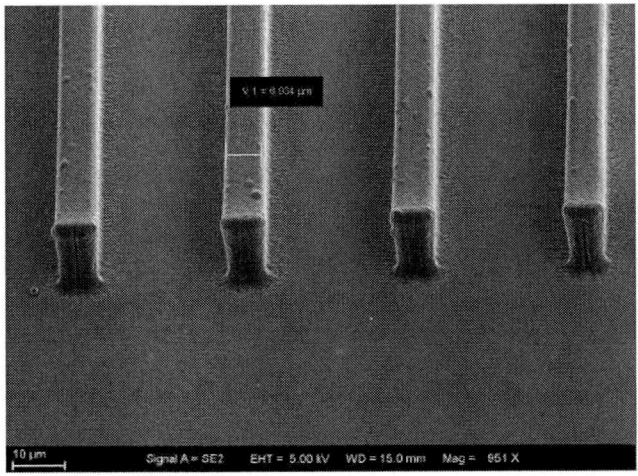

Figure 5. Dimetric SEM-image of WG cores processed with Hg-lamp and SU8-spacers. right most has no scattering bar, others have $0.5\,\mu m$ bar with increasing distance to mask opening. Square cross section but root structures and rough sidewalls.

Similarly for the clad openings we were able to improve structure quality shown in Fig. 6c, however we could not achieve open bottom with full layer thickness for $20\,\mu m$ openings. OPC structures only improved the hole shape but not sidewall angle or residual layer at the bottom.

Thus we concluded that our setup with Hg-lamp moved out of dedicated working point is not feasible, mainly due to bad parallelization of light and standing wave effects. However with the recent upgrade to dual wavelength UV-LED source we were able to achieve core cross section in Fig. 6b and clad opening in Fig. 6d. Core has smooth sidewalls and

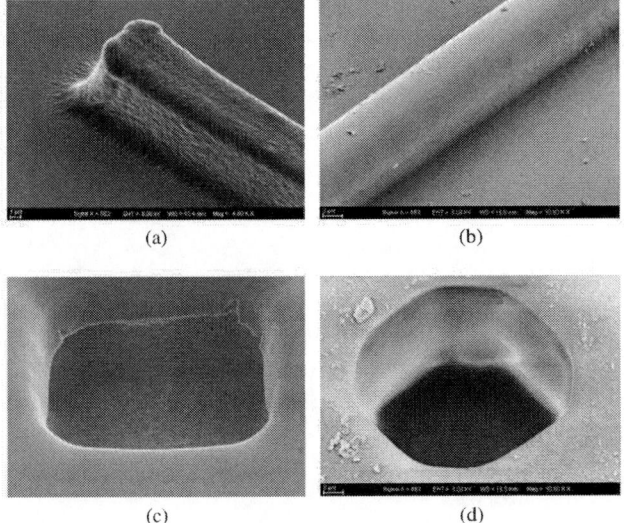

(a)   (b)

(c)   (d)

Figure 6. Best results for WG-cores (top) and clad openings (bottom) patterned with Hg-lamp (left), 365 nm and 405 nm dual wavelength UV-LED source (right).

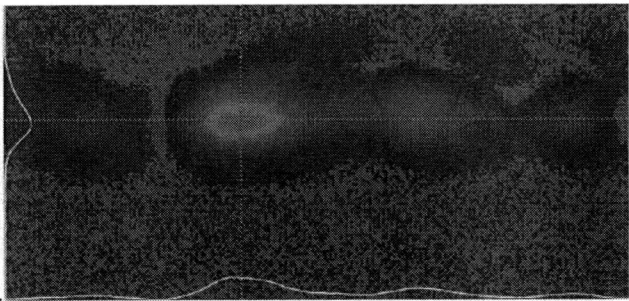

Figure 7. Power distribution at output facet of 4 WGs with $32\,\mu m$-pitch and light only being coupled into the second from left, measured using InGaAs beamprofiler showing strong cross coupling.

top without root like structure. We attribute this to the dual wavelength source preventing standing wave effects and much more parallel light source. Attenuation and cross coupling are yet to be measured but expected to be improved. Exposure result is uniform across wafer and identical for $32\,\mu m$ and $64\,\mu m$ pitch. Similarly sidewall of opening in Fig. 6d is steeper and smooth, there is no residual layer at the bottom and full layer thickness of $27\,\mu m$ is achieved.

B. Micro-Mirrors

To characterize processing of different resins with 2PP-DLW more than 1000 structures per resin and optic we're printed. Due to serial nature of the process this can be done in a single run. To extract full information from those parameter sweeps and not just find one working process point, automatic characterization method had to be developed. Therefore we automatically took isometric images of every structure using SEM. Every parameter set could then be graded qualitatively in 8 different ranks shown in Fig. 8.

Since absorption is proportional to intensity square it can be so high that resin evaporates locally and forms a bubble, if this is the case all along the structure it ends up being defect as in Fig. 8a. If this is only reached due to lower printing speed when changing direction in corners or at a particle in resin a bubble only takes up part of the structure as in Fig. 8b. If dose is decreased but still too high the structure appears inflated in Fig. 8c giving a certain deviation from the designed shape. Goal is to produce a smooth, according to design structure, as in Fig. 8d. As the dose is further decreased, voxel overlap reduces and thus the individual lines can be distinguished under SEM in Fig. 8e, impact on structure roughness will be detailed later. Further decrease in dose makes the structure shrink (fig. 8f) and become unstable (fig. 8g). In addition low dose corresponds to high printing speed, for used structure size only speeds up to roughly $40\,\mathrm{mm\,s^{-1}}$ can be used, above corner deviation dominates as shown in Fig. 8g. If power inside voxel is below absorption threshold no structure is formed (fig. 8h).

Comparing the results in Fig. 9 gives different tendency of resins. While Ormocomp shows no bubbling, it has strong overexposure and processing window is thin but ranges across

273

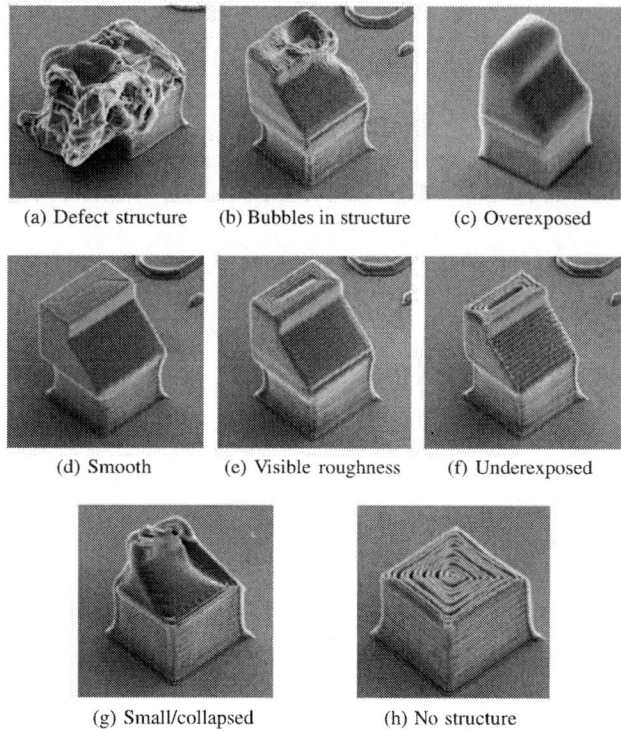

(a) Defect structure   (b) Bubbles in structure   (c) Overexposed

(d) Smooth   (e) Visible roughness   (f) Underexposed

(g) Small/collapsed   (h) No structure

Figure 8. Typical structures for each qualitative grade used to define processing window and compare resins. Exposure dose decreases from (a) to (h). Usable process window from (c) to (f), however ideally only (d) and (e) if roughness is below $\lambda/10$ are used.

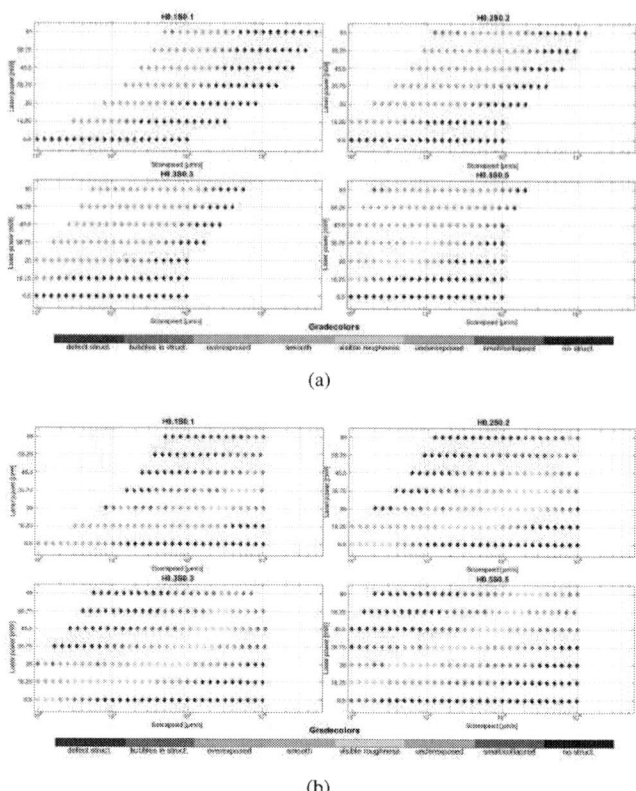

Figure 9. Process window graph for resins OrmoComp (a) and IP-DIP (b) showing bigger usable process window for IP-DIP and small overexposure effect, but higher tendency to form bubbles. OrmoComp however has strong overexposure causing structures to appear inflated with higher dose.

most powers. In comparison IP-DIP shows strong bubbling but low tendency to overexposure. Goal is to produce a structure with below $\lambda/10$ roughness, low shape deviation and processing time. From a time standpoint hatch and slice should be chosen as high as feasible. Since printing speed is reduced during direction change in corner, time is not directly proportional to printing speed and a sweet spot is roughly at $10\,\mathrm{mm\,s^{-1}}$. In addition process variation mainly from resin temperature and aging as well as laser power variation and drifting must be taken into account. Thus it is preferable to use a parameter set in the middle of a bigger processing window. IP-DIP meets these criteria for both hatch and slice of $0.2\,\mu\mathrm{m}$ and $0.3\,\mu\mathrm{m}$ and was thus used for further experiments. However within this processing window structure quality changes from smooth to visible roughness. For further investigation tactile profilometry with $8.3\,\mathrm{nm}$ lateral and $0.1\,\mathrm{nm}$ vertical resolution using $50\,\mathrm{nm}$ tip radius stylus was carried out. Since the structure is only $140\,\mu\mathrm{m}$ long, 30 parallel lines are measured to increase data points.

For roughness characterization on 3D structures filters are used to separate profile and roughness. For the used structures a cut-off of $1\,\mu\mathrm{m}$ should be used, however this would also partially filter roughness signal from voxel lines with $500\,\mathrm{nm}$ hatch. Thus each line was separated into 3 segments up, top and down and linearly flattened. For all samples a waviness

with $5\text{-}10\,\mu\mathrm{m}$ period and up to $1\,\mu\mathrm{m}$ amplitude was observed. This should be visible in iso- and dimetric view under SEM, but is not. Thus is likely introduced by the measurement system. To further investigate spatial line scan data was transferred to a periodic spectral density (PSD) graph using Welch's method. Fig. 10 shows this for coarse hatch and slice

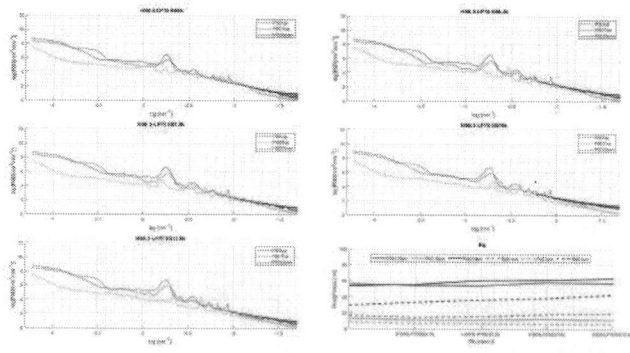

Figure 10. PSD-graph from tactile profilometry with $8.3\,\mathrm{nm}$ lateral and $0.1\,\mathrm{nm}$ vertical resolution using $50\,\mathrm{nm}$ tip radius stylus. Peak from $500\,\mathrm{nm}$ hatch is increasing with scan speed, since lower dose decreases voxel size and thus separation. However roughness in bottom right graph is still below $\lambda/10$ and only mildly increasing.

of 500 nm. For all scan speeds the peak corresponding to hatch is clearly distinguishable and is increasing in height with scan speed, since voxel outlines are becoming more separated. This is stronger in up and down ramp, since here scan direction is always perpendicular to print trajectory however on top it is mostly parallel. For lower length scale all three segments give similar reads in PSD-graph. In contrast for higher length scale of few µm there is a gap of two decades in spectral density between the two graphs, given by the previously explained measurement error. This leads to the fact that R_q with $10\,\mu m$ cut-off, given by the solid lines, is only minor higher then for $1\,\mu m$ cut-off for the top section but much higher for the sloped sides. We conclude from that, that the $1\,\mu m$ cut-off can be used. As shown in the R_q-graph it is higher for the slopes and slightly increasing due to the increase of voxel separation. However it has to be noted, that even for coarse printing it is still below $\lambda/10$ and thus still feasible as µ-mirror.

For lower hatch the increase in roughness suggested by transition from smooth to visible roughness in Fig. 9b could not be measured. Within the processing window R_q is between 10-15 nm for hatch of 100 nm and 200 nm and increases to around 20 nm for 300 nm hatch.

Functional feasibility for µ-mirrors is shown, for compatibility to hybrid interposer adhesion tests were performed. Each pair of surface treatment and substrate 20 pillars were sheared before and after thermal treatment. Adhesion for both Si and SiO₂ was by far the best with OrmoPrime in the 30-40 MPa range and with only minor impact of thermal treatment up to 300 °C, simulating reflow soldering. All other treatments yielded inferior adhesion that was dropped to below 5 MPa already by thermal treatment with 270 °C. For OrmoCladXP no pretreatment was compared only to oxygen plasma and bake. Here no pretreatment yielded the best adhesion of arround 40 MPa and is stable with temperature exposure.

Since the mirror has large contact area to OrmoCladXP in opening of polymeric RDL, low adhesion to substrate should not present a problem. Since OrmoPrime is made to bond acrylates to inorganic substrates it can not be used on Ormo-CladXP. The only way would be to use it for polymeric RDL already, were it yielded similar adhesion as oxygen plasma [11]. However it performed worse in planarization tests and it's unclear whether it would still be present on substrate surface after polymeric RDL processing.

As one application of µ-mirrors periscope probe was tested with edge emitting laser still on wafer. Fig.11a shows that the goal of a transparent assembly allowing for coarse placement of the probe to laser facet was met. In addition assembly is small enough so that it does not block the DC-probes supplying laser with current. This is to our knowledge the first demonstration of such a setup, previously similar probes were printed on fiber tip, however here the fiber obscures tip view [16]. During testing this assembly automatic manipulation with sub-µm resolution as for testing of optical RDL is necessary to find maximum coupling without crashing the fiber into trench bottom. Fig. 11b shows the tip after crashing it. However with this tip we were still able to measure 600 µW from laser

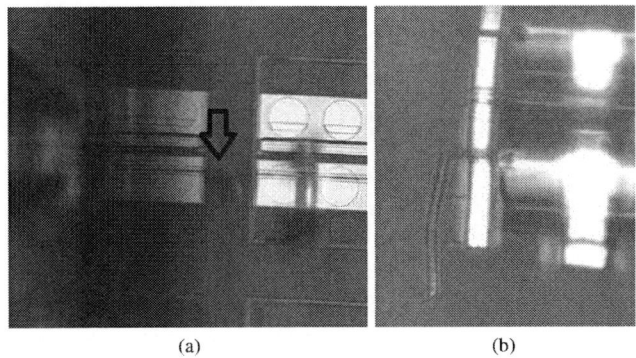

Figure 11. (a) View through periscope probe aligned to facet of edge emitting laser supplied with current via DC-probes. (b) Side view of crashed probe with vertical WG detached from studs.

emission that should be between 10-20 mW, before it finally broke. Probably due to thermal degradation since heat can not dissipate from the small tip. This shows a proof of concept, however for further tests lower emission power have to be used.

IV. CONCLUSION AND OUTLOOK

We presented our current state in hybrid lithography processing polymeric optical RDL via UV-lithography and µ-mirrors via 2PP-DLW. We demonstrated that SU8 spacers allow for setting of a low, reproducible proximity gap with stick resin, improving patterning results and making batch processing more easy. Comparison of Hg-lamp and dual wavelength LED-source showed, that the second gives improved patterning results both for WG-cores and clad openings, especially with respect to sidewall roughness. Next step is to measure attenuation of those greatly improved core cross sections. Thereafter the full polymeric RDL has to be put together, with the main challenge being good alignment and homogeneity of core thickness on topographic bottom clad. However the latter can be solved by having a nearly full bottom clad with only the necessary openings.

For µ-mirror structuring we introduced a process flow for automatic resin characterization yielding the to our knowledge most detailed published processing window characterization on OrmoComp and IP-DIP. Comparison showed that IP-DIP is more feasible due to bigger processing window. Thereafter roughness characterization showed that also with coarse writing hatch of 500 nm below 50 nm roughness can be achieved, even if single lines are visible in SEM and tactile profilometry measurement. In the future we will optically characterize µ-mirrors made with those parameters, to see if impact on performance is as low as expected. Additionally we showed that adhesion of IP-DIP to polymeric RDL is very good and no issues for proposed interposer. Finally we applied this to so-called periscope probe, providing a solution for wafer level testing of edge emitting lasers.

275

ACKNOWLEDGMENT

This Work has been supported by Germany's Federal Ministry of Education and Research (BMBF) within the project VE-Silhouette. The authors would like to thank I. Sotiriou, M. Göhler and N. Gohla for their extensive processing work.

.

REFERENCES

[1] M.-J. Lu, S.-Y. Mu, C.-S. Cheng, and J. Chen, "Advanced packaging technologies for co-packaged optics," in *Proc. of IEEE 72nd Electronic Components and Technology Conf. (ECTC)*. San Diego, CA, USA: IEEE, 2022, pp. 38–42.

[2] B. Chou, B. M. Sawyer*, G. Lyu, E. Timurdugan, C. Minkenberg, A. J. Zilkie, and D. McCann, "Demonstration of fan-out silicon photonics module for next generation co-packaged optics(cpo) application," in *Proc. of IEEE 72nd Electronic Components and Technology Conf. (ECTC)*. San Diego, CA, USA: IEEE, 2022, pp. 394–402.

[3] S. B. N. Gourikutty, M. C. Jong, C. V. Kanna, D. S. W. Ho, S. W. Wei, S. L. P. Siang, J. Wu, T. G. Lim, R. Mandal, J. T.-Y. Liow, and S. B. 1, "A novel packaging platform for high-performance optical engines in hyperscale data center applications," in *Proc. of IEEE 72nd Electronic Components and Technology Conf. (ECTC)*. San Diego, CA, USA: IEEE, 2022, pp. 422–427.

[4] A. Janta-Polczynski and M. Robitaille, "Optical fiber pigtails integration in co-package," in *Proc. of IEEE 72nd Electronic Components and Technology Conf. (ECTC)*. San Diego, CA, USA: IEEE, 2022, pp. 410–422.

[5] Z. Zhang, D. Felipe, V. Katopodis, P. Groumas, C. Kouloumentas, H. Avramopoulos, J.-Y. Dupuy, A. Konczykowska, A. Dede, A. Beretta, A. Vannucci, G. Cangini, R. Dinu, D. Schmidt, M. Moehrle, P. Runge, J.-H. Choi, H.-G. Bach, N. Grote, N. Keil, and M. Schell, "Hybrid photonic integration on a polymer platform," *Photonics*, vol. 2, no. 3, pp. 1005–1026, 9 2015. [Online]. Available: http://www.mdpi.com/2304-6732/2/3/1005

[6] E. Bosman, G. V. Steenberge, A. Boersma, S. Wiegersma, P. Harmsma, M. Karppinen, T. Korhonen, B. J. Offrein, R. Dangel, A. Daly, M. Ortsiefer, J. Justice, B. Corbett, S. Dorrestein, and J. Duis, "Scalable electrophotonic integration concept based on polymer waveguides," in *Optical Interconnects XVI*, H. Schröder and R. T. Chen, Eds. SPIE, mar 2016.

[7] L. Ma, X. Xu, and Z. He, "Single-mode polymer waveguides and devices for high-speed on-board optical interconnect application," in *Optical Interconnects XIX*, H. Schröder and R. T. Chen, Eds. SPIE, mar 2019.

[8] T. Barwicz, A. Janta-Polczynski, S. Takenobu, K. Watanabe, R. Langlois, Y. Taira, K. Suematsu, H. Numata, B. Peng, S. Kamlapurkar, S. Engelmann, P. Fortier, and N. Boyer, "Advances in interfacing optical fibers to nanophotonic waveguides via mechanically compliant polymer waveguides," *IEEE Journal of Selected Topics in Quantum Electronics*, vol. 26, no. 2, pp. 1–12, mar 2020.

[9] M. Hiltunen, M. T. Harjanne, T. Vehmas, B. Wälchli, P. Heimala, and T. Aalto, "Polymer interposer for efficient light coupling into 3 μm silicon-on-insulator waveguides," in *Optical Interconnects XX*, H. Schröder and R. T. Chen, Eds. SPIE, feb 2020.

[10] D. Weyers, K. Nieweglowski, L. Lorenz, and K. Bock, "Analysis of polymeric singlemode waveguides for inter-system communication," in *2021 23rd European Microelectronics and Packaging Conference & Exhibition (EMPC)*. IEEE, sep 2021.

[11] D. Weyers, A. Mistry, K. Nieweglowski, and K. Bock, "Hybrid lithography approach for single mode polymeric waveguides and out-of-plane coupling mirrors," in *Proc. of IEEE 72nd Electronic Components and Technology Conf. (ECTC)*. San Diego, CA, USA: IEEE, 2022, pp. 1919–1926.

[12] D. Weyers, K. Nieweglowski, and K. Bock, "Advances in UV-lithographic patterning of multi-layer waveguide stack for single mode polymeric RDL," in *2022 IEEE 9th Electronics System-Integration Technology Conference (ESTC)*. IEEE, sep 2022.

[13] J. Fischer and M. Wegener, "Three-dimensional optical laser lithography beyond the diffraction limit," *Laser & Photonics Reviews*, vol. 7, no. 1, pp. 22–44, mar 2012.

[14] X. Zhou, Y. Hou, and J. Lin, "A review on the processing accuracy of two-photon polymerization," *AIP Advances*, vol. 5, no. 3, p. 030701, mar 2015.

[15] A. Mistry, D. Weyers, K. Nieweglowski, and K. Bock, "Out-of-plane mirrors for single-mode polymeric rdl using direct laser writing," in *Proc. of 9th Electronic System-Integration Technology Conference (ESTC)*. Sibiu, Romania: IEEE, 2022.

[16] P.-I. Dietrich, M. Blaicher, I. Reuter, M. Billah, T. Hoose, A. Hofmann, C. Caer, R. Dangel, B. Offrein, U. Troppenz, M. Moehrle, W. Freude, and C. Koos, "In situ 3d nanoprinting of free-form coupling elements for hybrid photonic integration," *Nature Photonics*, vol. 12, no. 4, pp. 241–247, mar 2018.

Comparing the solderability of different SAC0307 composite solder pastes

Balázs Illés
Department of Electronics Technology
Budapest University of Technology and
Economics
Budapest, Hungary
illes.balazs@vik.bme.hu

Halim Choi
Department of Electronics Technology
Budapest University of Technology and
Economics
Budapest, Hungary
inertia9192@gmail.com

Agata Skwarek
LTCC Technology Research Group
Łukasiewicz Research Network - IMiF
Krakow, Poland
agata.skwarek@imif.lukasiewicz.gov.pl

Abstract— In the present study, the effect of four oxide ceramics (TiO_2, ZnO, ZrO_2, and CuO) as reinforcements were investigated on the soldering parameters of the SAC0307 solder alloy. The oxide ceramics were used in nano-powder format. Composite solder pastes were composed by the standard ball milling process of the nano-particles into the SAC0307 solder paste in 0.25wt% weight fraction. The wettability, shear strength, and microstructure of the composite solder joints were studied and compared. The ceramic nano-particles did not have a major effect on the wettability of the solder. Shear strength increases were observed at the composite joints, except in the case of ZnO. Generally, Sn grain refinement was found in the Sn-matrix which could be the reason for the shear strength increase.

Keywords— composite solder alloy; oxide ceramics; nanoparticles; shear strength; reflow soldering.

I. INTRODUCTION

The transition to lead-free soldering technology resulted in the change of classical SnPb to SnAgCu alloys. Recently, the so-called low Ag content ones got the most research attention, like Sn99Ag0.3Cu0.7 (SAC0307) and Sn98.5Ag1Cu0.5 (SAC105). The low Ag content is advantageous from the point of the alloys' price as well as from the point of decreased Ag_3Sn formation in the solder bulk, which can cause mechanical problems [1]. Further mechanical improvement of the solder joints can be achieved with the application of ceramic reinforcements in the solder alloys. The reinforcements are usually nano-particles (NPs) (mostly nano-powders, sometimes nano-fibers) and applied in 0.1¯1 wt%. A wide range of oxide ceramics was tried in composite solder joints: Al_2O_3, SiC, ZnO, TiO_2, ZrO_2, Si_3Ni_4, La_2O_3, etc [2].

The ceramic NPs are non-soluble in Sn, so they are dispersed at the grain boundaries in the solder matrix (at the intermetallic (IMC) and Sn and grains). During the solidification of the solder alloy, NPs act as incubation Kseeds_ for grain formation, which results in considerable grain refinement in composite solder joints as well as blocks the dislocation movements [3]. The previously discussed effects of the NPs usually improve the basic mechanical parameters (microhardness, tensile and yield strength) of the composite solder alloys [4]. However, the optimization of the composite solder pastes for industrial use is still a complex challenge.

For our study, we selected four oxide ceramics: TiO_2, ZnO, ZrO_2, and CuO. TiO_2 is the most researched material in composite soldering. It adapts very well to the soldering environment, not sensitive neither to the applied weight fraction (between 0.1¯1 wt%) nor to the primary particle size (between 20¯200 nm). Below 0.5 wt%, TiO_2 can even imporve the wetting ability of the composite solder alloys [5],

which is not typical with other ceramics. According to the latest results, TiO_2 can bond to Sn atoms, which could improve the reliability of the composite solder joints in a corrosive environment [6]. However, it can increase the liquidus temperature of the composite solder alloy even with 6éC [7], which is not advantageous in the case of low Ag content solder alloy with already higher liquidus temperatures (around 227éC).

From the melting temperature point, ZnO could be the best candidate, El-Daly et al. [8] found that ZnO increased the liquidus temperature only with 1-2K. Furthermore, adding ZnO NPs increased the creep resistance and the ultimate tensile strength of the composite solder joints [9]. Unfortunately, ZnO is much more sensitive to applied weight fraction than TiO_2. It was reported by several studies that over 0.5 wt%, ZnO NPs prone to agglomerate at the grain boundaries, which could ruin the structural integrity of the solder bulk [10]. In addition, ZnO is a hydrophobic material that fastly dries out the flux from the solder paste, which decreases the wettability and increases the void formation [11].

ZrO_2 is a much less researched oxide ceramic in composite soldering than TiO_2 or ZnO. Gain et al. [12] reported that ZrO_2 NPs increased the shear strength of Sn96.5Ag3Cu0.5 solder joints by a second-phase dispersion strengthening mechanism. Rajendran et al. [13] studied the reliability of Sn96.5Ag3Cu0.5 (SAC305) solder with ZrO_2 NPs during isothermal aging. They observed that the ZrO_2 NPs at the IMC layer region suppressed the diffusion of Cu into the solder bulk so, slowed the growth of the IMC layer in the solid phase. Furthermore, they observed an increase in shear strength as well. Gain et al. [14] found the same, namely that the growth rate of IMC layer in SAC305 composite solder joints is lower.

The application of CuO ceramic in composite soldering is an untouched area, but it could be promising since in other applications, CuO shows similar properties as TiO_2. Ismail [15] put 1wt% CuO and TiO_2 NPs into Sn90Zn10 solder alloy and found that both composite solder joints presented very similar microstructural and mechanical properties. Bahadur and Sunkara [16] increased the wear resistance of polyphenylene sulfide (PPS) by applying TiO_2 and CuO as fillers.

In this study, the effects of four oxide ceramics, TiO_2, ZnO, ZrO_2, and CuO, have been investigated on the wetting and shear properties of SAC0307 alloy. Furthermore, the microstructures of the composite solder joints were compared.

II. MATERIALS AND METHODS

TiO_2, ZnO, ZrO_2, and CuO NPs (produced by Sigma Aldrich) in 0.25wt% were put into SAC0307 solder paste (produced by Alpha). The primary particle sizes of the NPs were the following: 21nm - TiO_2, <50nm - ZnO, <100nm ZrO_2, and <50nm CuO. The NPs were homogeneously mixed into the SAC0307 solder paste by a YX paste mixer with 400rpm for 12 mins. 4 different composite solder pastes and the reference SAC0307 were investigated (Table 1).

TABLE I.   SOLDER ALLOYS

Name	Composition
SAC	SAC0307
SAC-TiO$_2$	SAC0307-TiO$_2$ (0.25wt%)
SAC-ZnO	SAC0307-ZnO (0.25wt%)
SAC-ZrO$_2$	SAC0307-ZrO$_2$ (0.25wt%)
SAC-CuO	SAC0307-CuO (0.25wt%)

The first investigation was the spreading test to study the wetting properties of the solder alloys (Tab. 1). Test surfaces (50x50mm) were prepared on FR4 substrates with Cu wiring with imm-Ag surface finishing. On each test boards four circular solder deposits were screen printed with 5mm diameter and 125 ı m thickness. The solder pastes were reflowed in a vapour phase soldering (VPS) oven. A linear thermal profile was applied: preheating till 170ιC from 0 to 120s, keeping a soak between 170-210ιC from 120 to 280s, and ramp-up till 255ιC from 280 to 360s. The wetted surface area was measured by an Olympus BX 50 optical-microscope. Four test boards (with 16 solder deposites) were prepared from each sample type.

During the second experiment, solder joints were fabricated by SMT on one-layer standard FR4 PCBs with Cu wiring. Solder pads were covered by imm-Ag surface finishing. Solder pastes were printed with a 125ı m thick stainless steel stencil. 0603 sized chip resistors were placed into the solder paste by manual pick & place. The test board had places for 50 chip resistors. The solder pastes were reflowed in a vapour phase soldering (VPS) oven used the linear thermal profile presented above. Fig. 1 shows a part of a test board with the chip resistors.

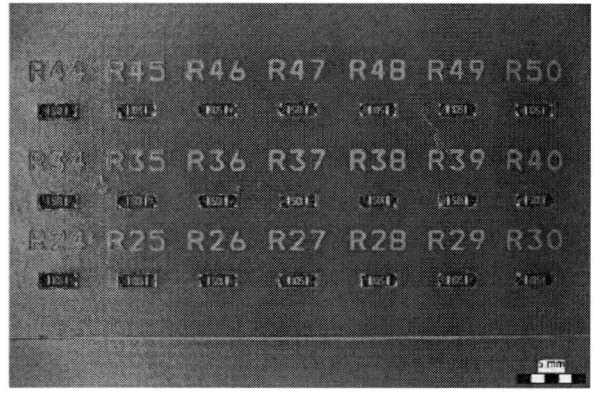

Fig 1.   Test board.

A DAGE 2400 shear tester has been used to measure the shear force of the solder joints. 20 resistors were pushed off from each sample type. Metallographic cross-sections were also prepared from the solder joints to compare their microstructures. The metallogrpahic cross-sections were investigated with an FEI Inspect-S50 SEM.

III. RESULTS AND DISCUSSION

Typical examples of the spreading tests are presented in Fig. 2. Generally, the wettability of the composite solder alloys decreased a bit compared to the ref. SAC. In the case of the reference SAC0307 alloy, the wetted area usually even increased a bit (with 5-6%) over 5mm (Fig. 2a), while in the case of the composite alloys usually decreased below 5mm (Fig. 2 b-d). However, the decrease was marginal, between 1-2%. No difference was observed according to the type of the applied NPs. It can be stated that the addition of the NPs in 0.25wt% did not have a significant effect of the wettability of the composite solder alloy.

Fig. 3 shows the shear force statistics of the different solder alloys in a box plot diagram. The dashed line marks the average (mean) shear force of the reference SAC0307 solder joints, the squares mark the averages, and the lines mark the medians. The deviations of the results were generally high in all cases. However, it is clearly visible that the TiO$_2$, ZrO$_2$, and CuO NPs slightly increased the average shear force of the composite joints compared to the ref. SAC, with 10%, 13%, and 9%, respectively.

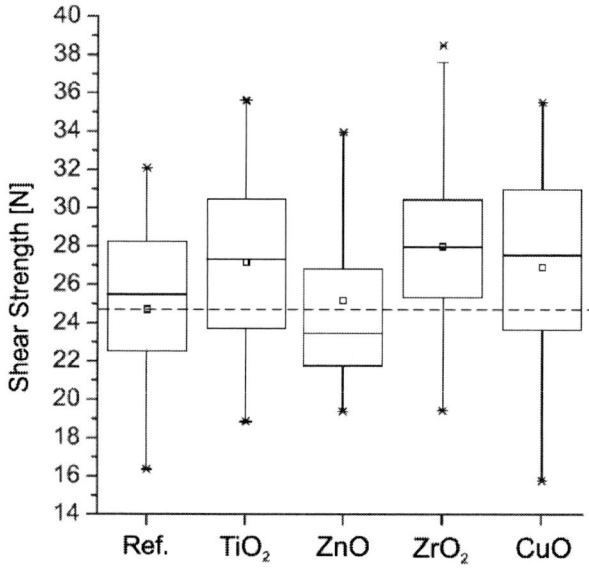

Fig. 3   Shear strength statistics of the solder alloys.

In the case of the ZnO NPs the average shear force of the composite solder joints did not change, but the median decreased. It was observed only in the case of ZnO samples. If the median is smaller than the average, it means that the distribution of the data is skewed to the left, which means that the values below the average were in the majority of the data set. In the case of TiO$_2$ and ZrO$_2$ the median was almost identical with the average. In the case of ref. SAC and CuO solder joints, the median was a bit higher than the average. It means that the distribution of the data skewed to the right, namely that the values above the average were in the majority in the data set. From the point of the shear strength, the higher mean (the lastly discussed occasion) is favorable. It needs to be highlighted that the ₖunknown_ CuO NPs performed as well in the shear strength tests as the TiO$_2$ and ZrO$_2$ NPs.

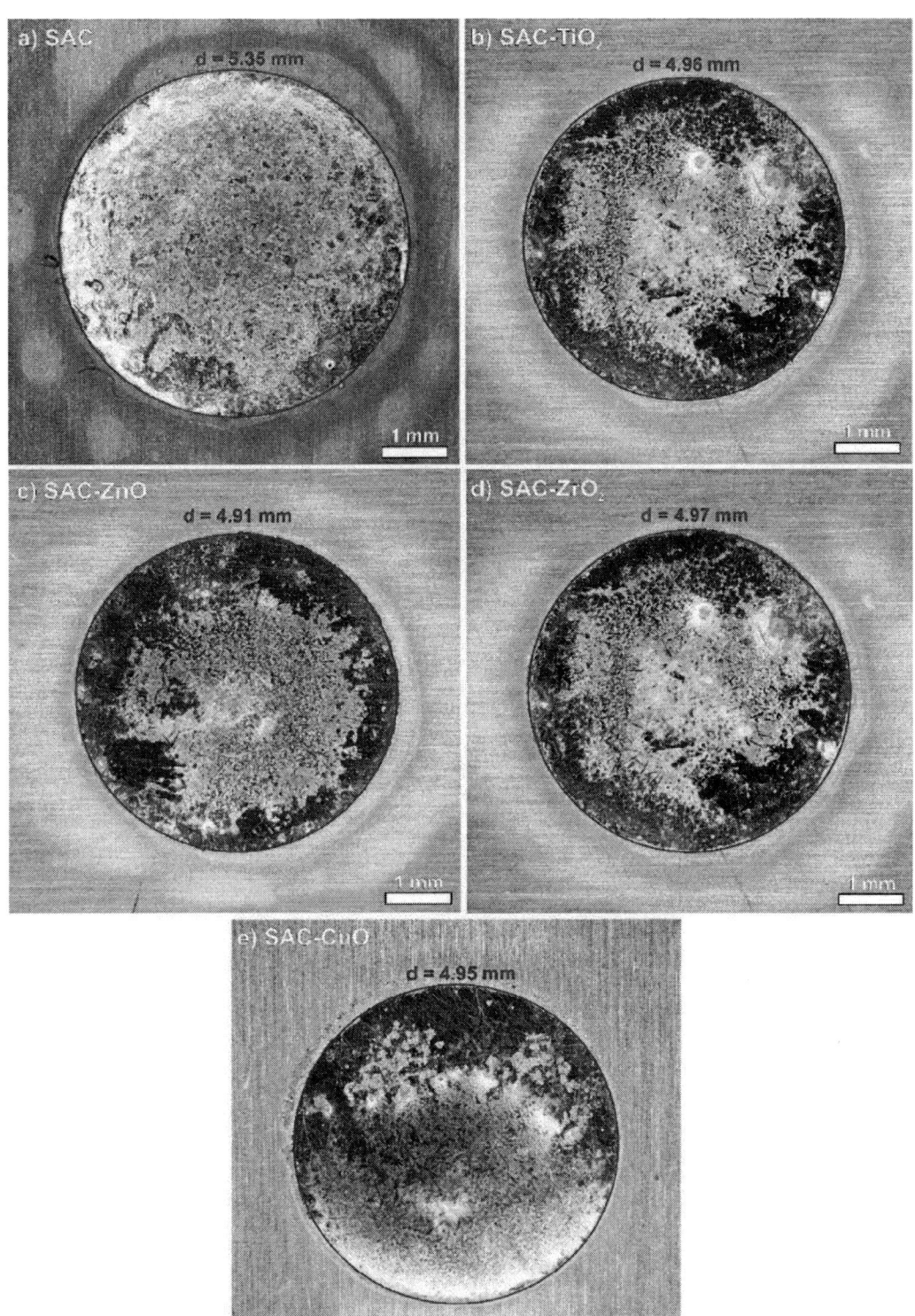

Fig. 2 Example results of the spreading test.

Metallographic cross-sections were fabricated from ref. SAC, SAC-TiO₂, and SAC-ZnO solder joints to study the microstructural differences. The cross-sections were observed with SEM. Example results done by backscattered electron detector (BSE) can be seen in Fig. 4. The addition of the oxide ceramic NPs resulted in considerable microstructural changes in the composite joints (Fig. 4b and c) compared to the ref. SAC (Fig. 4a). General microstructural refinement occurred in the composite joints. The different phases (Sn and different

IMCs) in the Sn-matrix can be easily distinguished due to the elemental contrast of the BSE micrographs.

The Cu_6Sn_5 layer thickness in the composite joints decreased by 20-40% compared to the reference SAC. The size of the dispersed Cu_6Sn_5 and Ag_3Sn IMCs in the solder bulk also decreased considerably. In the reference SAC sample (Fig. 4a), even 5-10 µm long Cu_6Sn_5 IMC grains can be found, while in the composite joints (Fig. 4b and c) the size of the same particles are 1-2 µm. In the case of the Ag_3Sn IMC grains, next to the size decrease, the most visible change was the lack of their agglomeration (Fig. 4b and c), which was typical in the reference joints (Fig. 4a).

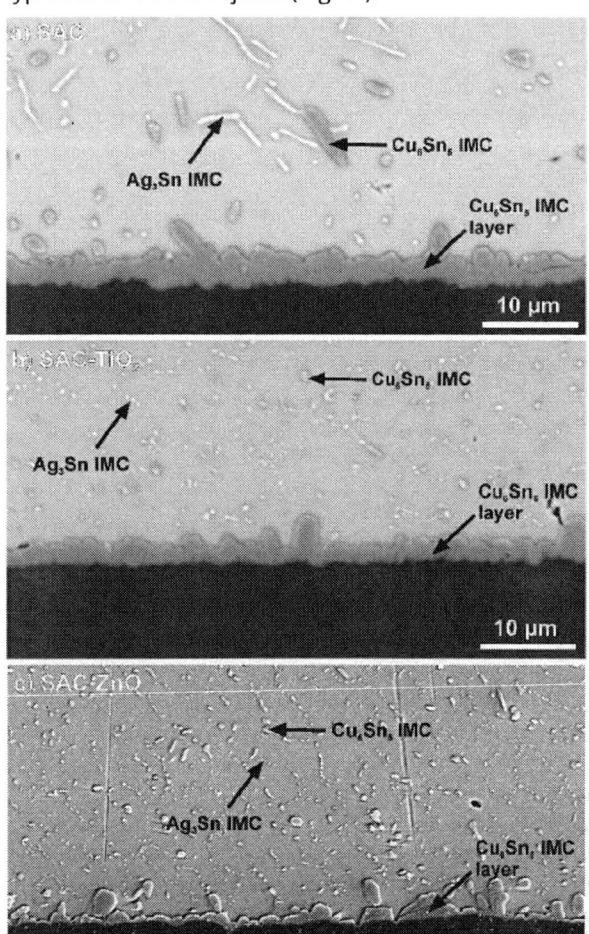

Fig. 4   SEM-BSE micrograph about the microstructure of the solder joints: a) SAC; b) SAC-TiO₂; c) SAC-ZnO.

The finer Ag_3Sn network in the composite joints indicated the Sn grain refinement as well. The Ag_3Sn IMCs are always stuck between the Sn grains, so they are good indicators for calculating the Sn grain size in the solder bulk. According to the Ag_3Sn networks, Sn grains with 20-25 µm size are visible in the reference SAC solder joint (Fig. 4a), which grain size decreased to 4-6 µm in the composite joints (Fig. 4b and c).

IV. Conclusions

The effect of TiO₂, ZnO, ZrO₂, and CuO oxide ceramics as reinforcements were investigated on the soldering parameters of the SAC0307 solder alloy. The main conclusions are the followings:

¿   The oxide ceramics slightly decreased the wettability of composite solder alloys. Considerable differences were not observed between the different ceramics.

¿   The increased shear strength of the composite joints showed that the second phase dispersion strengthening mechanism worked in the case of TiO₂, ZrO₂, and CuO NPs.

¿   ZnO NPs did not reach mechanical improvements, which is interesting since the microstructural refinement occurred in these composite solder joints as well. However, the SAC-ZnO solders performed the worst spreading/wetting, which could have a negative effect on the mechanical improvements.

¿   The unknown CuO NPs performed very similar and positive results as TiO₂ NPs, so it is suggested for application in soldering technology just like ZrO₂ NPs. According to our results, ZnO NPs are not favorable for soldering.

Acknowledgment

This work was executed in the frame of the NUMmic project financed by the National Science Center Poland (NCN) within the framework KOPUS-24_ project no. UMO-2022/47/B/ST5/00997 and supported by the National Research, Development and Innovation Office´ NKFIH, FK 132186.

References

[1]   O Krammer, T Garami, B Horváth, T Hurtony, B Medgyes, L Jakab, Investigating the thermomechanical properties and intermetallic layer formation of Bi micro-alloyed low-Ag content solders, J. Alloys Compnd. 634 (2015) pp. 156-162.

[2]   J. Wu, S. Xue, J. Wang, S. Liu, Y. Han, L. Wang, Recent progress of Sn¯Ag¯Cu lead-free solders bearing alloy elements and nanoparticles in electronic packaging, J Mater Sci: Mater Electron 27 (2016) pp. 12729¯12763

[3]   A. Skwarek, P. Ptak, K. GÆrecki, T. Hurtony, B. Illés, Microstructure Influence of SACX0307-TiO2 Composite Solder Joints on Thermal Properties o Power LED Assemblies, Materials 13 (2020) 1563.

[4]   Y. Tang, G. Li, Y. Pan, Effects of TiO₂ nanoparticles addition on microstructure, microhardness and tensile properties of Sn¯3.0Ag¯0.5Cu¯xTiO2 composite solder, Mater. Des. 55 (2014), pp. 574¯582.

[5]   Y. Tang, Y.C. Pan, G.Y. Li, Influence of TiO2 nanoparticles on thermal property, wettability and interfacial reaction in Sn¯3.0Ag¯0.5Cu¯xTiO2 composite solder, J. Mater. Sci. Mater. Electron. 24 (2013), pp. 1587¯1594.

[6]   B. Illés, H. Choi, J. Byun, K. Dusek, D. Busek, A. Skwarek, Incorporation and corrosion protection mechanism of TiO2 nano-particles in SnAgCu composite solder alloys: experimental and density functional theory study, Ceram. Inter. (2023) in press.

[7]   L. Tsao, S. Chang, Effects of Nano-TiO2 additions on thermal analysis, microstructure and tensile properties of Sn3.5Ag0.25Cu solder, Mater. Des. 31 (2010), pp. 990¯993.

[8]   A.A. El-Daly, T.A. Elmosalami, W.M. Desoky, M.G. El-Shaarawy, A.M. Abdraboh, Tensile deformation behavior and melting property of nano-sized ZnO particles reinforced Sn¯3.0Ag¯0.5Cu lead-free solder, Mater. Sci. Eng. A 618 (2014), pp. 389¯397.

[9]   Fawzy, A.; Fayek, S.A.; Sobhy, M.; Nassr, E.; Mousa, M.M.; Saad, G. Tensile creep characteristics of Sn¯3.5Ag¯0.5Cu (SAC355) solder reinforced with nano-metric ZnO particles, Mater. Sci. Eng. A 603 (2014), pp. 1¯10.

[10] H. Peng, G. Chen, L. Mo, Y.C. Chan, F. Wu, H. Liu, H. An investigation on the ZnO retained ratio, microstructural evolution, and mechanical properties of ZnO doped Sn3.0Ag0.5Cu composite solder joint, J. Mater. Sci.: Mater. Electron. 27 (2016), pp. 9083¯9093.

[11] A. Skwarek, O. Krammer, T. Hurtony, P. Ptak, K. Gǽecki, S. Wrovski, D. Straubinger, K. Witek, B. Ill¶, Application of ZnO nanoparticles in Sn99Ag0.3Cu0.7 based composite solder alloys, Nanomaterials 11 (2021), 1545.

[12] A.K. Gain, Y.C. Chan, W.K.C. Yung, Effect of additions of ZrO_2 nanoparticles on the microstructure and shear strength of Sn-Ag-Cu solder on Au/Ni metallized Cu pads. Microelectron. Reliab. 51 (2011), pp. 2306~2313.

[13] S.H. Rajendran, S.J. Hwang, J.P. Jung, Shear Strength and Aging Characteristics of Sn-3.0Ag-0.5Cu/Cu Solder Joint Reinforced with ZrO_2 Nanoparticles, Metals 10 (2020), 1295.

[14] A.K. Gain, Y.C. Chan, Growth mechanism of intermetallic compounds and damping properties of Sn¯Ag¯Cu-1 wt% nano-ZrO2 composite solders. Microelectron. Reliab. 54v (2014) 945-955.

[15] R.A. Ismail, An Investigation of Microstructure and Mechanical Properties of Different Nano-Particles Doped Sn-Zn Lead-Free Solder Alloys, Arab J. Nucl. Sci. Appl., 53/1 (2020), pp. 191-199.

[16] S. Bahadur, C. Sunkara, Effect of transfer film structure, composition and bonding on the tribological behavior of polyphenylene sulfide filled with nano particles of TiO2, ZnO, CuO and SiC, Wear 258 (2005), pp. 1411-1421.

Anisotropic Solder Paste (ASP) Material Solution for Laser Assisted Bonding (LAB) Process

Ki-Seok JANG
Superintelligence Creative Research Laboratory
ETRI
Daejeon, Republic of KOREA
sroka80@etri.re.kr

Yong-Sung EOM
Superintelligence Creative Research Laboratory
ETRI
Daejeon, Republic of KOREA
yseom@etri.re.kr

Gwang-Mun CHOI
Superintelligence Creative Research Laboratory
ETRI
Daejeon, Republic of KOREA
gmchoi@etri.re.kr

Jiho JOO
Superintelligence Creative Research Laboratory
ETRI
Daejeon, Republic of KOREA
jihojoo@etri.re.kr

Jin-Hyuk OH
Superintelligence Creative Research Laboratory
ETRI
Daejeon, Republic of KOREA
jing5king@etri.re.kr

Chan-Mi LEE
Superintelligence Creative Research Laboratory
ETRI
Daejeon, South KOREA
hichanmi@etri.re.kr

Yoon-Hwan MOON
Superintelligence Creative Research Laboratory
ETRI
Daejeon, Republic of KOREA
sssmoon@etri.re.kr

Seok-Hwan MOON
Superintelligence Creative Research Laboratory
ETRI
Daejeon, Republic of KOREA
shmoon@etri.re.kr

Kwang-Seong CHOI
Superintelligence Creative Research Laboratory
ETRI
Daejeon, Republic of KOREA
kschoi@etri.re.kr

Abstract—Anisotropic conductive paste (ACP) or anisotropic conductive film (ACF) consisting of a polymer matrix and a metal-coated polymer bead are introduced as interconnection materials in mini-LED display panel packaging through a thermal compression bonding process. However, this packaging solution has been leading to important problems such as alignment difficulties due to high bonding temperature and pressure, and its repair issues for bad LED devices after the bonding process. Recently, laser-assisted bonding (LAB) process for mini-LED display packaging has been introduced to solve these problems[Ref. 1-2]. This study introduces, an anisotropic solder paste (ASP) using conductive solder particles and applies it to the mini-LED packaging through the LAB process.

Keywords— Anisotropic Solder Paste (ASP), Anisotropic Conductive Film (ACF), Laser Assisted Bonding (LAB), mini LED

I. INTRODUCTION

Recently, there has been a growing demand for chips and high-resolution packages such as Mini and Micro LED displays. Therefore, the number of I/O(Input/Output) is increasing and the pattern pitch needs to decrease. Anisotropic conductive paste (ACP) or anisotropic conductive film (ACF) are generally used for electrical interconnection between metal pads of the upper and bottom substrates, as the high density interconnection materials, and the conductive adhesives have many advantages such as fine pitch. There are generally used for chip on glass (COG) and chip on flex (COF) packaging technologies[Ref. 3-8].

However, the transfer method and bonding technology of Mini & Micro LED devices to substrates still has many challenges like lack of productivity, high bonding force and high electrical resistance by ACP interconnection. Anisotropic Solder Paste (ASP) material for Laser Assisted Bonding (LAB) technology may be an excellent solution to

high bonding pressure and electrical resistance. Also, a LAB process using ASP material can prevent electrical short by controlling solder volume and polymer bead size.

This study newly introduced ASP using conductive solder particles and applied it to the mini-LED packaging through the LAB process which was developed for good electrical interconnection and low bonding pressure.

II. MATERIALS AND EXPERIMENT

A. Anisotropic Solder Paste (ASP)

The resin of Anisotropic Solder Paste (ASP) consists of an epoxy-based solvent free resin, solder particles for electrical interconnection, and non-conductive polymer beads(spacer) for the standoff height of solder joints between the device and the substrate. The base resin was composed of an epoxy, a reductant to remove oxide from the metal surfaces, curing agent and catalyst for controlling the chemical reaction. A residue removal process of thermosetting based resin was not necessary because unreacted chemical components did not remain after the bonding process[Ref. 9]. We developed a novel ASP material for fine pattern pitch using solder particles with a Sn/58Bi type6 size and 2vol.% of polymer beads with a 10μm diameter. ASP material was placed between the metal pads of the device and the substrate, and the bond process was performed. The viscosity of the ASP with a 6%vol. of solder particles was 130,000 cPs@10rpm at Room Temperature in a Brookfiled viscometer. Since it did not depend on the ASP pattern size and pitch, both dispensing and screen printing processes had the advantage of being easy.

B. Laser Assisted Bonding (LAB) Process

The traditional flip chip bonding thermal compression bonding (TCB) technology with a long process time and a substrate fully heat path process can warpage of flexible substrates. The Laser-Assisted Bonding (LAB) is next generation flip chip bonding technology that can avoid warpage of flexible substrates by controlling the temperature

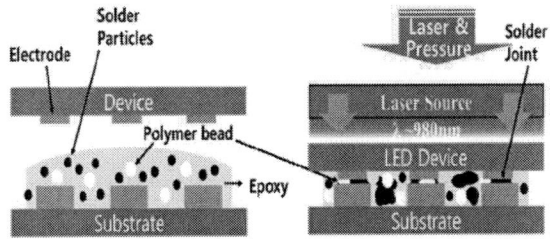

Figure 1. Schematics diagram of ASP material solution for LAB Process

and time of the bonding process.

Figure 1 shows a schematics diagram of the LAB process with ASP. ASP printing on substrate was applied by a dispensing method. The device of the upper and bottom substrates were aligned, pressured with a quartz head, and laser irradiated. The heating mechanism of the LAB process was based on the laser absorption of a device with in 980nm wavelength homogenized laser beam. The ASP consisted of a solvent-free resin, solder particles for electrical interconnection, and non-conductive polymer beads for the stand-off height of the solder joints between the device and the substrate. After the LAB process, we observed that the interconnection was completely achieved by the solder wetting between the device and the bottom metal pads. The post curing process was 120℃ for two hours in a convection oven after the LAB process.

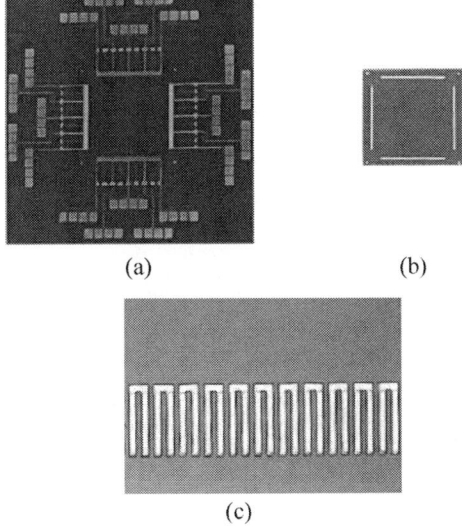

Figure 2. Pictures of Si test vehicles (a) Si substrate and (b) Si device of a 40μm PAD pitch

III. RESULTS AND DISCUSSION

The chemo-rheological properties of ASP were measured using rheometer and differential scanning calorimetry (DSC).

For the rheometer measurement, the ASP material was placed between parallel plates with a 20mm diameter gap into and 1 mm thickness. The applied frequency and heating rate were 1 Hz and 10°C/min, respectively.

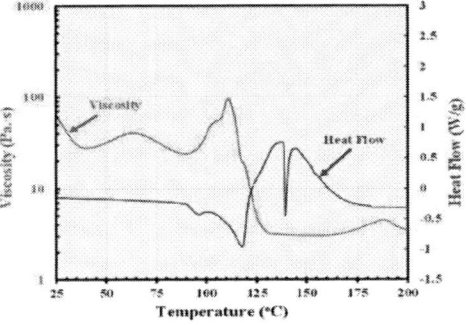

Figure 3. Viscosity and heat flow of ASP with a heating rate up 10°C/min

The measured viscosity and heat flow by DSC and Rheometer are shown in Figure 3, where the red line indicates the viscosity and the blue line indicates the heat flow as the temperature increases. An exothermic peak is observed around 135°C, indicating an exothermic reaction caused by chemical reaction. The endothermic peak observed at around 139°C refers to the melting of Sn/58Bi solder. The viscosity was maintained while the chemical reaction was in progress and the viscosity gradually increased from 170°C near the end point of the chemical reaction. As a result of the viscosity and heat flow, it was observed that the processability characteristics were retained because enough low-viscosity of ASP viscosity with 3Pa·s at the process temperature (melting point of solder).

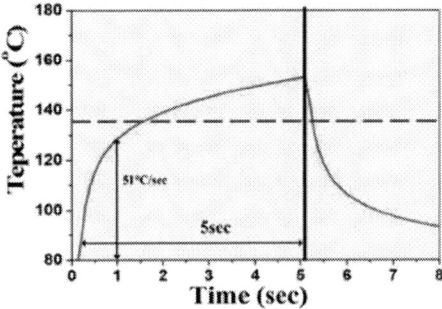

Figure 4. Temperature vs. Laser Irradiation time Curve

The LAB process was performed by a demonstrated with a homogenized laser power of 200W for 5 seconds irradiation to the chip using a wavelength of 980nm at a stage temperature of 80°C. To measure the process temperature during the laser irradiation, a thermo-couple was placed between the device to the substrate. Figure 4 shows the process temperature profile measured using the thermo-couple which was over the melting temperature of the solder when laser irradiation at 200W for 5sec. At this time, the measured heating rate maximum, that is 51°C/sec.

The Scanning electron microscope (SEM) image in figure 5 is of a conductive polymer bead trapped between the electrodes, and the remainder of the gap is filled with resin. A sample for cross-section observation was produced by polishing process the bonded sample. The interconnection, polymer bead shape and solder joint analysis of the polished cross-section of the composite was observed using SEM.

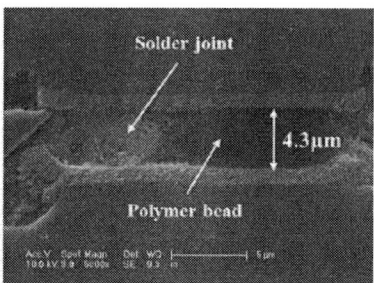

Figure 5. Cross-Sectional SEM image of the device on a bonded substrate

The solder joint area in this case refers to an electrical contact area that forms an electrical path from the device to the substrate electrode pads. The interconnections were partially contacted by melted volume content 6% of Type-6 solder particles. In this study, the optimized composition was 6 vol.% of solder type-6 particle size and 2 vol.% of polymer beads with a diameter of 10 um. If there is a difference in volume ratio or size of the solder particles and polymer beads, the electrical interconnection to open or short expected. After the LAB process, we observed that the interconnection was completely achieved by the solder wetting between the device and bottom metal pads. A Bonding Line Thickness(BLT) of 4.3μm was observed after the LAB process with 5sec at 200W with a bonding force of 0.41 Mpa. Since the 10um diameter polymer bead was compressed by pressure, it became a constant gap of 4.3um. The ASP material composition for LAB optimized the non-conductive polymer bead and solder particle size and LAB process pressure. The electrical interconnections were also optimized by size and volume of polymer beads and solder particles. The solder particles formed electrical connections, while the polymer beads achieved uniform dispersion of the solder particles and prevented electrical shorts.

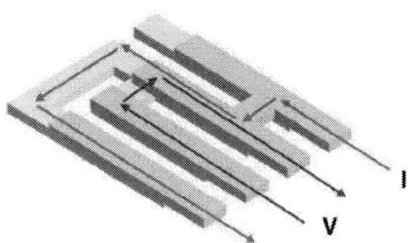

Figure 6. Schematic circuit to measure the electrical resistance by ASP interconnection using the four-point probe method

The electrical resistance by ASP interconnection was measured by employing the four-point probe method and configuring the circuit according to the schematic shown in Figure 6.

ACF is metal-coated conductive particles form mechanical contacts, which creates a relatively small electrical path compared to solder joints, so it has a higher electrical resistance. As shown in Table 1, the ASP and ACF process bonding force was 0.18MPa and 3.27MPa, and ASP was about 18 times less than the ACF. The ASP needed a low bonding force because lowering the viscosity and forming electrical contact while the solder particles melt. On the other hand, ACF requires a high bonding force as it contacts the metal-coated conductive beads.

Material	ASP	ACF
Conductive particle	SnBi Solder Type-6	Au/Ni coated polymer
Polymer bead size [μm]	10	10
Electrode pattern (W x L) [mm]	0.02 x 0.2	0.125 x 3.0
Bonding force [MPa]	0.18 (~ 18 times less)	3.27
Gap between device and sub. [μm]	4.3	5.91
Electrical resistance by ASP interconnection [mΩ/mm²]	62.5 (~ 4 times less)	289

Table 1. Comparison of bonding process conditions and results of ASP and ACF [Ref. 10].

After the bonding process of the ASP and ACF test vehicles, the electrical resistance of each test vehicle was measured by a four-point probe method. The electrical resistances of the ASP and ACF were each 62.5 mΩ/mm² and 289 mΩ/mm², respectively. The electoral resistance of the ACF is about 4-times higher than that of the ASP because of the small electrical path.

IV. CONCLUSION

The chemical reaction during processing was analyzed using Differential Scanning Calorimetry (DSC). A novel Anisotropic Solder Paste (ASP) consists of a solvent-free resin, solder particles for the electrical interconnection, and non-conductive polymer beads for the stand-off height of the solder joints between the device and the substrate. The Laser Assisted Bonding Compression (LABC) process was successfully demonstrated with a laser power of 200W for 5 seconds applied to a chip using a wavelength of 980nm with a homogenized laser beam. The post curing process for the ASP was carried out at 120°C for 120min and was optimized by controlling the isothermal processing temperature. Moreover, ASP can be observed that the array patterns are interconnected without electrical shorts. The epoxy-based resin in the ASP is expected to be an underfill between the device and the substrate to improve the bonding shear strength. Also, since the solder joint forms the electrical connection, it has a lower electrical resistance by the interconnection and the advantage of a low bonding pressure. According to our experimental results, we believe that ASP for LAB will greatly influence the commercialization of Micro/Mini LED display packaging.

ACKNOWLEDGMENT

This research was supported by the National R&D Program through the National Research Foundation of Korea (NRF) funded by the Ministry of Science and ICT (NRF-2020M3H4A3106383, NRF-2020M3H4A3081764), a Korea Evaluation Institute of Industrial Technology (KEIT) grant by the Ministry of Trade, Industry and Energy (20010580), and an Electronics and Telecommunications Research Institute (ETRI) grant funded by the Korean government [22YB1100, Core technology for new microwave-reactive materials for low-carbon, high-quality semiconductor processing].

REFERENCES

[1] Kwang-Seong Choi, Jiho Joo, "Simultaneous Transfer and Bonding (SITRAB) Process for Mini - LED Display", SID, Vol. 52, 841-844, 2021

[2] Jiho Joo, Gwang-Mun Choi, "Development of Flexible Full - Color Mini - LED Display Using Simultaneous Transfer and Bonding (SITRAB) Technology" , SID, Vol. 53, 1005-1008, 2022

[3] H. Nishida, K. Sakamoto, H. Ogawa, "Micropitch connection using anisotropic conductive materials for driver IC attachment to a liquid crystal display", IBM J. Research and Development, Vol.42, No.3/4, pp. 517-525, 1998

[4] Yim M-J, Paik K-W. "The electrical resistance by ASP interconnection and reliability of anisotropically conductive film (ACF)," IEEE Transactions on Advanced Packaging, Vol. 22, No. 2, pp.166-167, 1999

[5] I. Watanabe et al., "Packaging Technologies using Anisotropic Conductive Adhesive Films in FPDs", In Proc. Asia Display/IDW, pp. 553~556, 2001

[6] J. Liu, A. Tolvgard, J. Malmodin, and Z. Lai, "A Reliable and Environmentally Friendly Packaging Technology-Flip Chip Joining Using Anisotropically Conductive Adhesive", IEEE Trans. Comp. Packag., Manufact. Technol. Vol. 22, No. 2, pp.186~190, 1999

[7] JM Kim, K Yasuda, M Rito, K Fujimoto,"New electrically conductive adhesives filled with low-melting-point alloy fillers", J. Electron. Mater. 33(11) 1331–1337, 2004

[8] JM Kim, K Yasuda, M Rito, K Fujimoto,"Novel interconnection method using electrically conductive paste with fusible fille", J. Electron. Mater. 34(5) 600–604, 20057

[9] Yong-Sung Eom, Gwang-Mun Choi, "Process window of simultaneous transfer and bonding materials using laser - assisted bonding for mini - and micro - LED display panel packaging", ETRI Journal.2023, 1–13

[10] Haksun Lee, Kwang-Seong Choi, "Sn58Bi Solder Interconnection for Low-Temperature Flex-on-Flex Bonding", ETRI Journal, Vol 38, 2003, 1163-1171

Durability of Lead-Free Solder Interconnections for Printed Circuit Board Applications: Comparing Energy-Based Thermo-Mechanical Fatigue Models

Chien-Ming Huang
Mechanical Engineering
University of Maryland
College Park, USA
cmhuang@umd.edu

Jeffrey W. Herrmann
Mechanical Engineering
University of Maryland
College Park, USA
jwh2@umd.edu

Abstract—Fatigue models for predicting the cycles to failure of solder interconnections under temperature cycling situations have been discussed and developed for decades. However, most models were developed for different solder materials, components, and printed circuit boards. No previous work has systematically compared these models. Therefore, the variability of their durability predictions is unknown. This study compared nine existing low-cycle energy-based fatigue models for different solder materials and components and then analyzed the differences among them. Each fatigue model had a specific combination of the factors that affect the strain energy density accumulation. Therefore, we adjusted the strain energy density (input) and the predicted cycles to failure (output) in a consistent way to compare the selected fatigue models on the same basis. The differences among the predictions on cycles to failure from the fatigue models was significantly reduced after applying the adjustments, and they exhibited excellent consistency around 1 mJ/mm^3 strain energy density. In the end, factors that can affect the prediction consistency of fatigue mode were provided, including the number of data points while building the fatigue model, the range of strain energy density while selecting the electronic components, and the application of volume-weighted averaging technique on the critical solder joint.

Keywords—lead-free solders, durability prediction, fatigue life, temperature cycling, printed circuit boards

I. INTRODUCTION

The durability of solder interconnections is critical in modern electronic products because the solder joints provide mechanical and electrical connections between chips, components, and printed circuit boards (PCBs). Solder interconnections can fail due to various loading conditions, including temperature cycling, vibration, or drop/shock impact [1]. Among these loading conditions, temperature cycling induced low-cycle fatigue is considered as a one of the major failure mechanisms for the electronics.

The durability depends, in part, upon the choice of solder. Many lead-free solders have been developed during the past 20 years because of various considerations, including cost and reliability, and the component-level and board-level reliability and durability of different lead-free solders has been reported in the literature [2]–[4]. These results, which provide some insights about the performance of different solders, are based on experiments and finite element analysis (FEA) simulations with specific lead-free solders and assemblies.

An engineer who is developing a new product and wishes to use a specific solder material should conduct reliability tests with prototypes of the new system. Performing the entire series of reliability tests can be time-consuming and costly, however. Therefore, FEA simulation tools, such as ANSYS, have become a popular technique to estimate the durability of solder interconnections in the electronics industry [5],[6] in order to obtain preliminary results without running costly tests. The results from such simulations depend upon multiple parameter values and modeling choices, however, and we are not aware of any systematic study of how these factors affect reliability estimates. This paper describes studies that we conducted to provide such information.

We considered how the solder material, the fatigue life model parameters, the use of volume-weighted averaging technique, and the parameters in the stress-strain model affect the expected number of cycles to failure (CTF) when using commercially available FEA software such as ANSYS. Our results show the inelastic strain energy density can be affected by the preferences of simulation methodology. Moreover, the limitations of some fatigue models are pointed out and the recommendations are provided for the engineers. These results will help engineers make better decisions when using modeling to support durability evaluation.

The remainder of this paper is organized as follows. Section II reviews related work and the selected fatigue models. Section III presents the analysis approach, and Section IV presents the simulation results, and Section V discusses the adjustment of durability predictions. Section VI concludes the paper.

II. RELATED WORK

For the fatigue models with cyclic strain energy input, both viscoplastic [7] and creep-only [8] energy models have been proposed. (Viscoplastic models include both creep and plastic energies.) In general, a damage model with an inverse power law form describes the relationship between the damage metric and cycles to failure. Damage metrics include accumulated viscoplastic and creep-only strain energy density, and the cycles to failure can be median life (50% failures) or characteristic life (63.2% failures in a Weibull distribution). Let N_f be the cycles to failure. Let D be the damage metric, and let A and n be the material constants, which can be affected by many factors, including solder materials, temperature cycling profile, and assembly structure. Equation (1) is a generalized fatigue life model for solder joint under temperature cycling.

$$N_f = (A \times D)^n \qquad (1)$$

The cycles to failure of solder joint can be estimated by applying the power law (1) to the fatigue coefficient *A*, fatigue exponent *n*, and damage metric *D*. Many solder fatigue life constants have been proposed for SAC solder joint under various temperature cycling conditions [9],[10]. Table I lists the fatigue coefficients and exponents for solder joint fatigue life models from prior studies that determined these values by fitting the power law fatigue model to experimental data and simulation results. As shown in Table I, the damage metric in the fatigue model was categorized into two groups, which were viscoplastic strain energy density and creep strain energy density. This paper will discuss the prediction difference between these two groups later.

TABLE I. ENERGY-BASED LOW-CYCLE FATIGUE MODELS FOR SOLDER JOINTS

Prior Studies	Strain Energy Density	Fatigue Coefficient	Fatigue Exponent	Fatigue Life Prediction
Chen et al. [9]	Viscoplastic	1.05×10^{-5}	−0.59	N50%
	Creep	4.75×10^{-6}	−0.55	N50%
Syed [10]	Creep	1.90×10^{-3}	−1.00	N50%
Schubert et al. [11]	Viscoplastic	3.30×10^{-3}	−1.02	N63.2%
Hsieh [12]	Creep	4.15×10^{-6}	−0.60	N50%
Hsieh and Tzeng [13]	Creep	1.09×10^{-2}	−2.26	N63.2%
Ghorbani and Spelt [14]	Creep	2.70×10^{-2}	−2.27	N50%
Sun et al. [15]	Creep	4.38×10^{-8}	−0.39	N63.2%
Zhang et al. [16]	Viscoplastic	1.69×10^{-4}	−0.77	N63.2%

Many studies have worked on the fatigue life prediction of solder joints with strain energy density. Some models focused on viscoplastic strain energy density, whereas other models focused on creep strain energy density. When fitting the constants of fatigue model, utilizing different types of strain energy density would generate different numbers of the constants. Moreover, employing an established fatigue model without understanding its underlying assumptions can make significant deviation on the prediction of fatigue life.

In this study, the multilinear isotropic hardening model [9] was utilized to handle the plastic strain energy density and the Garofalo-Arrhenius creep model [17] was applied to estimate the creep strain energy density. The Garofalo creep model is composed of a hyperbolic sine equation with stress-dependent term and an Arrhenius equation with temperature-dependent term, as shown in (2), where *C1* and *C2* are coefficients related to the material, σ is applied stress, ε_{cr} is creep strain, *C3* is stress exponent, *C4* is activation energy divided by Boltzmann's constant, and *T* is temperature in Kelvin scale. Many studies have been published to establish data sets of Garofalo-Arrhenius parameters for lead-free solder materials [9],[18]–[20], and are summarized in Table II. For the convenience of discussion in this paper, a code name was used instead of the original name for each study.

$$d\varepsilon_{cr}/dt = C1 \times [\sinh(C2 \times \sigma)]^{C3} \times \exp(-C4/T) \quad (2)$$

TABLE II. SUMMARY OF GAROFALO-ARRHENIUS CREEP CONSTANTS FOR SOLDER MATERIAL

Solder Alloy	Code Name	Garofalo-Arrhenius Parameters			
		C1 (s⁻¹)	C2 (MPa⁻¹)	C3	C4 (K)
SAC305 [9]	GA01	6.07	0.18	2.3	6,710
SAC387 [11]	GA02	277,984	0.02447	6.41	6,500
SAC405 [12]	GA03	1.15×10^{6}	0.0335	7.5	8,703.4
SAC105 [13]	GA04	2.31×10^{6}	0.026	6.5	6,962.7
SAC387 [14]	GA05	441,000	0.005	4.2	5,412
SAC387 [16]	GA06	1.5×10^{3}	0.19	4	8,575.9

The multilinear isotropic hardening property was utilized to model the time-independent plastic strain (ε_{pl}). The relationship between the applied stress and strain was addressed in (3) to (5) [9], where σ is the instantaneous stress, *K* and n_{pl} are temperature-dependent instantaneous plasticity material constants, and *T* is temperature in Celsius scale.

$$\varepsilon_{pl} = (\sigma / K)^{1/n_{pl}} \quad (3)$$

$$K = 121.6 - 0.4 \times T \quad (4)$$

$$n_{pl} = 0.29 - 0.00046 \times T \quad (5)$$

Because the contribution of creep strain energy density is much higher than plastic one (around 10 times higher), this paper employed only one prior study for the multilinear isotropic hardening property, as addressed in (3) to (5), to simplify the comparison process.

III. APPROACH

To support the reproducibility of our work, we used a PCB design from IPC-2581, the online digital product model exchange that is a generic standard for the printed circuit board and assembly manufacturing description data. The test case for our study was IPC-2581 B Test Case 3 board design [21] with BGA168 component from ANSYS Sherlock part library. ANSYS Sherlock is a physics-based reliability tool that provides life predictions for electronics at the component, board, and system levels. Details about the dimensions and material properties are listed in Table III and Table IV. The values of Poisson's ratio for SAC305, BGA168, and the PCB, which were considered as isotropic, were 0.36, 0.20, and 0.15, respectively.

In our approach, we created the FEA model with ANSYS Mechanical APDL because it provided the flexibility to adjust the mesh of solder joints. In the FEA model, the geometry of the solder joint was built in a ball-shape. We then imported the FEA models into ANSYS Workbench for the FEA simulations. In order to consider the impact of the volume-weighted averaging technique, the number of temperature cycles, and the constitutive equations, we conducted multiple simulation runs using different combinations of values for these factors. Each simulation run yielded a value for the strain energy density ΔW with corresponding solder joint volume fraction, number of temperature cycles, and constitutive constants. In the end, we analyzed the variation of the cycles to failure prediction among the fatigue models. The results are given in the next section.

TABLE III. PARAMETERS OF THE BGA168 COMPONENT [21]

BGA168 Size	Ball Matrix	Number of Balls	Ball Diameter	Ball Height	Ball Pitch
13.5 × 13.5 mm	13 × 13 Full	168	0.645 mm	0.3 mm	1 mm

TABLE IV. MECHANICAL PROPERTIES OF MATERIALS IN BGA168 ASSEMBLY [21]

	Young's Modulus E (GPa)	CTE α (ppm/°C)	Shear Modulus G (GPa)
SAC 305	39.99	22.28	14.70
BGA168	25.08	9.70	10.43
PCB (x-y direction)	25.95	18.32	11.28
PCB (z direction)	7.05	57.53	11.28

Due to the nature of symmetry on the in-plane dimensions of the component, solder joints, and PCB, a quarter model of the FEA geometry was built to reduce the computational costs; the corresponding meshes are shown in Fig. 1.

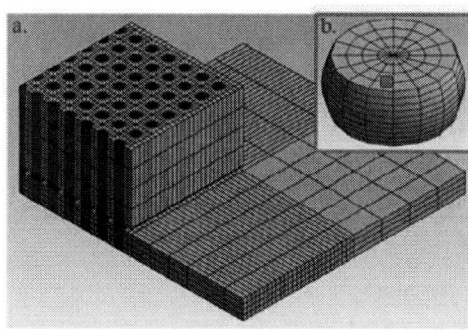

Fig. 1. Finite element meshes of BGA168 component and the PCB: (a) isometric view of the quarter model; (b) Solder ball

The strain energy density was determined with the volume-weighted averaging technique [22], which can use only a fraction of the elements in the calculation. We conducted simulations with different values for this fraction: 0.23% (critical element), 9%, 18%, 50%, and 100%. The corresponding volumes are shown in Fig. 2. The information of the volumes was obtained from ANSYS Workbench by inserting a Volume result object in the Solution.

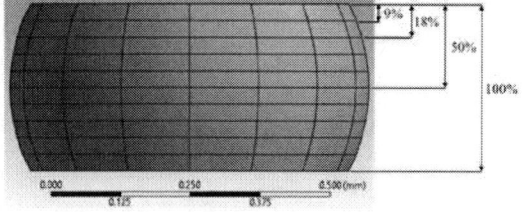

Fig. 2. A schematic of the number of elements selected for volume-weighted averaging technique

The cycles chosen for the accumulation of strain energy density can also affect the fitting of the constants of fatigue model. Fig. 3 showed the profile of temperature cycling used in this study. The value of this strain energy density accumulation (ΔW) can vary from cycle to cycle. Therefore, we examined the impact of using Cycle 1 to Cycle 4 accumulation to determine ΔW.

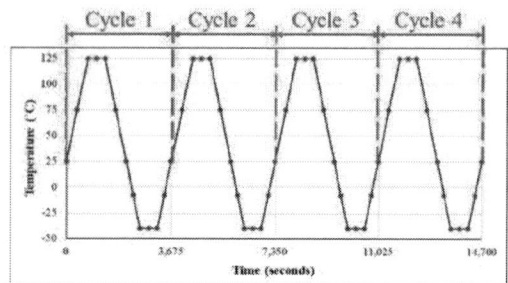

Fig. 3. Temperature cycling profile used in this study

Combing the abovementioned factors, the conditions of running simulations for the ball-shape solder joint were summarized in Table V. Our experiments employed a full factorial design, so there were $6 \times 5 \times 4 = 120$ combinations for the subsequent analysis.

TABLE V. CONDITIONS FOR THE FACTORS OF EXPERIMENT DESIGN

Factors	Level 1	Level 2	Level 3	Level 4	Level 5	Level 6
Constitutive model	GA01	GA02	GA03	GA04	GA05	GA06
Volume-weighted %	0.23%	9%	18%	50%	100%	---
Cycle accumulation	Cycle 1	Cycle 2	Cycle 3	Cycle 4	---	---

IV. RESULTS

The simulation results were summarized according to the solder volume and the simulated cycle, as addressed in Section III, in order to determine the necessary information for comparing the constitutive models and fatigue models. We used SAS JMP software to create variability charts (Fig. 4) to summarize the strain energy density data from all the considered situations and to compare the data by constitutive models, strain energy density types, cycles, and solder volumes.

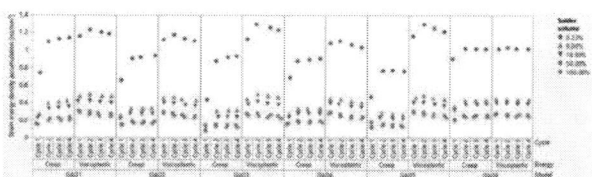

Fig. 4. Variability chart of strain energy density with the selected Garofalo-Arrhenius models

Fig. 4 can be interpreted from various perspectives. First, there were significant differences on the group means between 0.23%, 9% and 18%, as well as 50% and 100% solder volumes. Second, the strain energy density accumulation exhibited different behaviors for viscoplastic and creep energies. For the creep energy type, the energy accumulation of Cycle 1 was much lower than that of Cycles 2 to 4. On the contrary, the viscoplastic energy type displayed similar strain energy density values from Cycle 1 to 4. Third, the strain energy densities of viscoplastic energy type was higher than the ones of creep energy type for all constitutive models. However, the difference between the viscoplastic and creep energy type was not consistent among the models. Fourth, Student's t tests were conducted to compare the group means of each pair of the constitutive models for both viscoplastic and creep energy type, there were no significant differences discovered among the constitutive models.

V. DISCUSSION

From Fig. 4, the effect of cycle accumulation on the strain energy density variation could be ignored for 9% to 100% volume-weighted conditions. For the 0.23% volume (maximum element value) application, the strain energy density variation was around 3.7%. Fortunately, the fatigue models that employed the maximum element value took the cycle accumulation after Cycle 4 except Schubert et al. [11], who didn't address the cycle accumulation, as shown in Table VI. Therefore, when comparing the fatigue models in this paper, Cycle 4 accumulation was employed for the greater than three cycles and not specified situations.

TABLE VI. STRAIN ENERGY DENSITY (ΔW) METHOD FROM THE SELECTED FATIGUE LIFE MODEL STUDIES

Constitute Model	Volume-Weighted Amount	Cycle Accumulation	ΔW from Our Simulations (mJ/mm³)
GA01	8%	Cycle 4	0.47 (viscoplastic) 0.42 (creep)
GA02	5.6%	Cycle 1 or 2	0.33
GA02	100%	Not specified	0.22
GA03	Maximum value (element)	Cycle 5	0.93
GA04	Maximum value (element)	Cycle 5	0.90
GA05	Maximum value (element)	Cycle 10	0.76
GA02	9.3%	Not specified (performed 3 cycles in their simulation)	0.33
GA06	5%	Cycle 3	0.44

In Table VI, five of the fatigue models directly described the amount of volume-weighted for the critical solder or the maximum value of the critical element (no volume-weighted) in their studies [11]–[14],[16]. The remaining fatigue models [9],[10],[15] described only the thickness or the layer(s) of elements used for the volume-weighted technique. In this situation, the values in Table VI were determined by dividing the thickness of the layer(s) over the critical solder volume, which was calculated from the figure in each study. The volume of the critical element in each study was not possible to be approximated due to many reasons. Therefore, the 0.23% solder volume data was employed for the maximum element value. When the volume-weighted amount was less than 10%, the 9% solder volume data was applied for the subsequent comparisons because there should not exist a significant difference. Moreover, there were significant differences between the group means of 0.23%, 9%, and 100% solder volumes. For instance, the strain energy density of 0.23% volume was about 2.5 times higher than the one of 9% volume. Hence, the data from selected solder volume was critical while substituting into the fatigue models.

There was a significant difference on the group means between the viscoplastic energy group and creep energy group from the six studied Garofalo-Arrhenius models. Moreover, the difference between the viscoplastic energy and creep energy groups was not consistent among the six models. Therefore, it is important to be consistent with the energy type while fitting the constants of the Garofalo-Arrhenius model and to be careful when referring to the available Garofalo-

Arrhenius models. Based on all of these considerations, we determined that the strain energy density value needed to be adjusted before substituting it into the fatigue model. The strain energy density data from our simulations for each fatigue model situation was listed in Table VI.

Each fatigue model had a specific combination of factors (as shown in Table V) that affects the strain energy density calculation. For a given combination set, we found the ΔW by taking the accumulation of strain energy density from the cyclic strain energy density curve. Fig. 5 presents the process flow to calculate the adjustment ratio of strain energy density for each fatigue model. We first chose one of the fatigue life models, which was Chen et al. [9] with viscoplastic strain energy density, as the reference fatigue model A, ΔW_A. Then, for any of the other fatigue life models in Table VI, as the target fatigue model B. We repeated the simulation with the same temperature cycling profile, component, and PCB to get the target strain energy density, ΔW_B. For the creep constants of solder material, however, we used the values that were used in that study. Moreover, for each target fatigue life model, different settings of volume-weighted averaging amount and cycle accumulation was applied to determine the ΔW_B. Therefore, each fatigue life model had its own ΔW_B value. We then calculated the ratio R_B between each ΔW_B and the ΔW_A, as shown in (6). For the comparison purpose, these ratios would be employed for the subsequent discussions of the predictions of cycles to failure.

$$R_B = \Delta W_B / \Delta W_A \qquad (6)$$

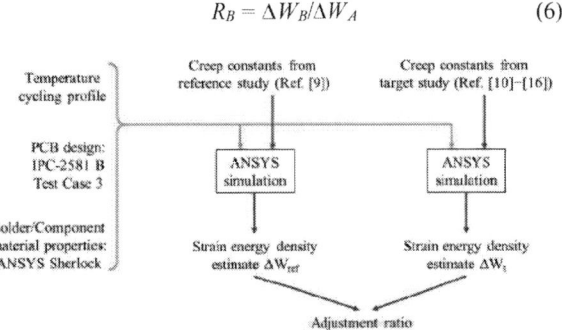

Fig. 5. Strain energy density adjustment flow for the target studies

Fig. 6 plots the published values of strain energy density (ΔW) and cycles to failure (N_f) on a log-log scale. The range of ΔW and the fatigue coefficients and exponents were taken directly from the literature (listed in Table I and Table VI), and then the cycles to failures were determined by using them with the inverse power law in (1) for each model from the literature. Most of the reported ΔW values covered only a small range, and the range was less than 1 mJ/mm³. However, this phenomenon could also reflect this small range might be a practical situation for modern electronic devices. For ΔW was greater than 2 mJ/mm³, fewer studies are available. Note that the strain energy density values are the results of simulations that used different sets of parameter values, and the cycles to failure values are different points (percentiles) from the empirical distributions collected by the different studies.

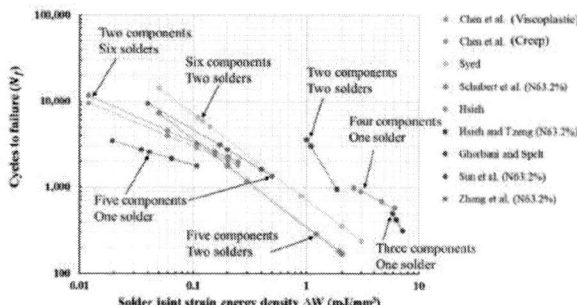

Fig. 6. Distributions of strain energy density from the selected fatigue life models and their predictions on the cycles to failure

In order to have a consistent comparison for all the selected fatigue models, five strain energy density values were chosen based on all of the combinations of our simulation results, which were $\Delta W_i = \{0.09, 0.575, 1.06, 1.545, 2.03\}$ mJ/mm³ for $i = 1, 2, 3, 4, 5$. These five values were used as the inputs of the fatigue models. Therefore, the predicted cycles to failure can be compared within the same strain energy density scope.

The strain energy density value, ΔW_i, was multiplied by that model's adjustment ratio, R_B, before applying the inverse power law, as shown in (7), to get the adjusted strain energy density, $\Delta W_i'$, for each target fatigue life model, where i was from 1 to 8 due to total eight target models. After that, the cycles to failure (N_f) can be predicted with the adjusted strain energy density and the inverse power law with the corresponding coefficient and exponent from each target model, as presented in (8).

$$\Delta W_i' = \Delta W_i \times R_B \tag{7}$$

$$N_f = \left(A_B \times \Delta W_i'\right)^{\eta_B} \tag{8}$$

To have a more straightforward understanding of the predictions of all the selected fatigue models, the adjustment of life percentile was applied, as shown in (9) and (10). N_{fi}' was referred to values that were derived from different ΔW_i. If the target model predicted median life (N50%), N_{fi}' was the same as the N_f, which was the outcome of the inverse power law. On the other hand, if the target model predicted characteristic life (N63.2%), N_f was multiplied by $0.6931^{1/\beta_B}$ to get the adjusted N_{fi}'. β_B was the shape parameter of the two-parameter Weibull distribution from the field failure data in each target fatigue life model. We averaged that study's published values of the shape parameter β to determine an aggregate β_B for that study. The ratio $0.6931^{1/\beta_B}$ is the scaling factor from N63.2% to N50% in the Weibull distribution. The entire adjustment process flow for the cycles to failure prediction is shown in Fig. 7.

$$N_{fi}' = N_f \quad \text{(if predicting N50\%)} \tag{9}$$

$$N_{fi}' = 0.6931^{1/\beta_B} \times N_f \quad \text{(if predicting N63.2\%)} \tag{10}$$

The adjusted cycles to failure predictions were shown in Fig. 8. The differences of the predictions on cycles to failure among the fatigue models was significantly reduced after applying the adjustments and exhibited excellent consistency around 1 mJ/mm³ strain energy density.

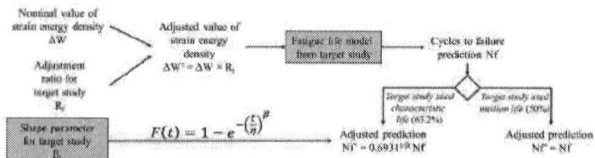

Fig. 7. Process flow of the adjusted prediction of fatigue model

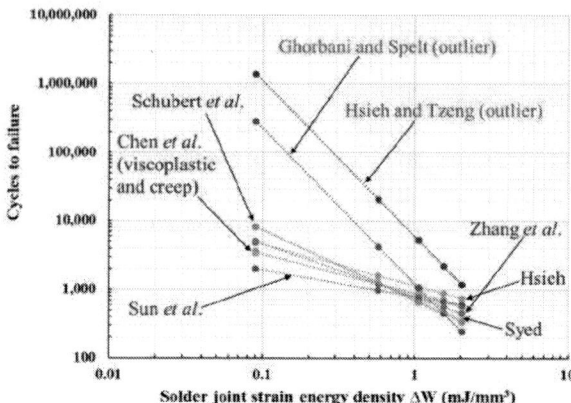

Fig. 8. Distributions of strain energy density with selected ΔW range: With ΔW and characteristic life (N63.2%) to median life (N50%) adjustment

Two of the fatigue life models, Hsieh and Tzeng [13] and Ghorbani and Spelt [14], exhibited divergences from the other seven models (see Fig. 8). Three factors were found that were responsible for the significant divergence. First, both models did not apply volume-weighted averaging technique for determining the strain energy density. They employed the maximum element value as the strain energy density (see Table VI). This could induce large variation on the strain energy density while changing the mesh size of the FEA model. Therefore, the adjustment method for the strain energy density (see Fig. 5) may not be appropriate for these two models because their mesh sizes were obviously smaller than our mesh size. Second, these two models only used three data points while curve fitting the coefficient and exponent of the inverse power law. This was the minimum number of data points for curve fitting an inverse power law, so predictions that were based on values of ΔW outside the range of their data can deviate in a certain amount. Third, the ΔW range selected by these two models was around 1 mJ/mm³, which was too small comparing to the reported ΔW range (from 0.01 to 8 mJ/mm³). Combining with the second point, the prediction of the cycles to failure with smaller ΔW values can produce a huge deviation from the other seven models. Due to these factors, these two models appear to be outliers among the total nine fatigue models.

VI. SUMMARY

Meteorologists use computer models to forecast the weather, and they use multiple models to estimate the range of what might happen. In the same way, engineers might use a set of fatigue models to get a range of durability predictions. If used inappropriately, however, the range of predictions might be misleading. This paper describes a study that provides some insights on how to use existing fatigue models.

This paper presented the results of nine selected low-cycle energy-based fatigue models that were built with their own solder materials, assemblies, simulation methodologies, and field failure data to predict the median life or characteristic life of the solder joint.

The differences between the selected Garofalo-Arrhenius constitutive relationships from different solder materials were not significant on the accumulation of strain energy density. Other factors, including the amount of the critical solder for volume-weighted averaging technique, the selection of temperature cycle for strain energy density calculation, and the type of strain energy density (viscoplastic or creep), did make significant differences on the strain energy density accumulation. The percentile of predicted cycles to failure of the selected fatigue models were also not consistent. Therefore, adjusting the cycles to failure predictions was necessary after using the strain energy density as the input to the inverse power law equation.

For the adjustments, the strain energy density range (0.09 to 2.03 mJ/mm^3) with five data points was chosen as the inputs of the selected fatigue models. The distribution of the predictions on cycles to failure among the fatigue models was significantly reduced after applying the adjustments, and exhibited excellent consistency around 1 mJ/mm^3 strain energy density. Two of the fatigue models still showed large deviations from the other models, especially for small strain energy density (less than 1 mJ/mm^3).

For engineers who wish to use multiple fatigue models to get a range of durability predictions, our results suggest the following: First, choose fatigue models that were constructed using at least five data points. Second, select fatigue models that cover more than 1 mJ/mm^3 of strain energy density. Third, use fatigue models that apply a volume-weighted averaging technique on the critical solder to avoid extreme values of the strain energy density.

REFERENCES

[1] Kim, Y.K. and Hwang, D.S., 2015. PBGA Packaging Reliability Assessments under Random Vibrations for Space Applications. Microelectronics Reliability, 55(1), pp.172-179.

[2] Singh, B., Menezes, G., McCann, S., Jayaram, V., Ray, U., Sundaram, V., Pulugurtha, R., Smet, V. and Tummala, R., 2017. Board-Level Thermal Cycling and Drop-Test Reliability of Large, Ultrathin Glass BGA Packages for Smart Mobile Applications. IEEE Transactions on Components, Packaging and Manufacturing Technology, 7(5), pp.726-733.

[3] Osterman, M., 2018, September. Modeling Temperature Cycle Fatigue Life of Select SAC Solders. In Proceedings of SMTA International. Online available: https://web.calce.umd.edu/lead-free/SMTAI2018-Osterman.pdf

[4] Coyle, R., Johnson, C., Hillman, D., Pearson, T., Osterman, M., Smetana, J., Howell, K., Zhang, H., Silk, J., Geng, J. and Daily, D., Enhancing Thermal Fatigue Reliability of Pb-Free Solder Alloys with Additions of Bismuth and Antimony. Proceedings of SMTA International, Sep. 28 - Oct. 23, 2020. pp. 339-354.

[5] Schubert, A., Dudek, R., Walter, H., Jung, E., Gollhardt, A., Michel, B. and Reichl, H., 2002, May. Reliability Assessment of Flip-Chip Assemblies with Lead-Free Solder Joints. In 52nd Electronic Components and Technology Conference 2002. (Cat. No. 02CH37345) (pp. 1246-1255). IEEE.

[6] Tee, T.Y., Ng, H.S. and Zhong, Z., 2006. Board Level Solder Joint Reliability Analysis of Stacked Die Mixed Flip-Chip and Wirebond BGA. Microelectronics Reliability, 46(12), pp.2131-2138.

[7] Morrow, J., 1965. Cyclic plastic strain energy and fatigue of metals. In Internal friction, damping, and cyclic plasticity. ASTM International. Lazan, B.J. (edited). West Conshohocken, PA, USA.

[8] Dasgupta, A., Oyan, C., Barker, D., and Pecht, M. (June 1, 1992). Solder Creep-Fatigue Analysis by an Energy-Partitioning Approach. ASME. Journal of Electronic Packaging, June 1992; 114(2): 152–160.

[9] Chen, D.Y., Osterman, M. and Dasgupta, A., 2020. Energy based modeling for temperature cycling induced tin silver copper solder interconnect fatigue life. Microelectronics Reliability, 109, p.113651.

[10] Syed, A., 2004, June. Accumulated creep strain and energy density based thermal fatigue life prediction models for SnAgCu solder joints. In 2004 Proceedings. 54th electronic components and technology conference (IEEE Cat. No. 04CH37546) (Vol. 1, pp. 737-746). IEEE.

[11] Schubert, A., Dudek, R., Auerswald, E., Gollhardt, A., Michel, B. and Reichl, H., 2003, May. Fatigue life models for SnAgCu and SnPb solder joints evaluated by experiments and simulation. In Electronic components and technology conference (pp. 603-610). IEEE; 1999.

[12] Hsieh, M.C., 2015, October. Modeling correlation for solder joint fatigue life estimation in wafer-level chip scale packages. In 2015 10th International Microsystems, Packaging, Assembly and Circuits Technology Conference (IMPACT) (pp. 65-68). IEEE.

[13] Hsieh, M.C. and Tzeng, S.L., 2014, August. Solder joint fatigue life prediction in large size and low cost wafer-level chip scale packages. In 2014 15th International Conference on Electronic Packaging Technology (pp. 496-501). IEEE.

[14] Ghorbani, H.R. and Spelt, J.K., 2007. An analytical elasto-creep model of solder joints in leadless chip resistors: part 2—applications in fatigue reliability predictions for SnPb and lead-free solders. IEEE Transactions on Advanced Packaging, 30(4), pp.695-704.

[15] Sun, W., Zhu, W.H., Danny, R., Che, F.X., Wang, C.K., Sun, A.Y. and Tan, H.B., 2007, April. Study on the board-level SMT assembly and solder joint reliability of different QFN packages. In 2007 International Conference on Thermal, Mechanical and Multi-Physics Simulation Experiments in Microelectronics and Micro-Systems. EuroSime 2007 (pp. 1-6). IEEE.

[16] Zhang, Q., Dasgupta, A., Nelson, D., and Pallavicini, H. (January 6, 2005). "Systematic Study on Thermo-Mechanical Durability of Pb-Free Assemblies: Experiments and FE Analysis." ASME. Journal of Electronic Packaging, December 2005; 127(4): 415–429.

[17] Garofalo, F., 1965. Fundamentals of creep and creep-rupture in metals (Creep and creep rupture in metals and alloys, fundamental information for instruction and reference). NEW YORK, MACMILLAN CO., LONDON, COLLIER- MACMILLAN, LTD., 1965. 258 P.

[18] Clech, J.P., 2003. Review and Analysis of Pb-Free Solder Properties. report to the NEMI Pb-Free Solder Project.

[19] Cuddalorepatta, G. and Dasgupta, A., 2008, April. Effect of primary creep behavior on fatigue damage accumulation rates in accelerated thermal cycling of Sn3.0Ag0.5Cu Pb-free interconnects. In EuroSimE 2008-International Conference on Thermal, Mechanical and Multi-Physics Simulation and Experiments in Microelectronics and Micro-Systems (pp. 1-8). IEEE.

[20] Jiang, Q., Deshpande, A. and Dasgupta, A., 2022. Multi-scale Crystal Viscoplasticity Approach for Estimating Anisotropic Steady-State Creep Properties of Single-Crystal SnAgCu Alloys. International Journal of Plasticity, 153, p.103271.

[21] IPC-2581 B Test Case 3, IPC-DPMX (IPC-2581) Consortium. Online available: http://www.ipc2581.com/b-test-cases/

[22] Akay, H.U., Paydar, N.H. and Bilgic, A., 1997. Fatigue Life Predictions for Thermally Loaded Solder Joints using a Volume-Weighted Averaging Technique. Journal of Electronic Packaging, 119(4), pp. 228-235.

An experimental investigation of a flexible sintered silver joint for micro-joining based on a design of experiments

Laurent VIVET
VALEO
THS Material Laboratory
78321 La Verrière, France
laurent.vivet@valeo.com

Lahouari BENABOU
LISV
Université de Versailles SQY-Paris Saclay
78321 Vélizy, France
lahouari.benabou@uvsq.fr

Olivier SIMON
LISV
Université de Versailles SQY-Paris Saclay
78321 Vélizy, France
olivier .simon@uvsq.fr

Abstract— **The increasing electrification of vehicles requires the use of ever higher voltages and electrical power. The thermal stresses to which the assembly joints of electronic chips are subjected are becoming increasingly severe. As a result, thermal stresses acting on the electronic assembly, and particularly on the electrical interconnections, are becoming increasingly severe. The objective of this study is to create, characterize and optimize the structure and properties of a porous silver sintered joint using the design of experiments (DoE) method. Joints consisting of a porous network of silver particles are obtained, showing remarkable flexibility allowing for thermal stresses accommodation and having sufficient tensile strength to ensure a correct bonding of the components. Depending on the selected levels for the sintering process parameters, the microscopy analysis shows a transition of the mode of rupture from the silver layer to the intermetallic compounds forming at the interface with the substrate.**

Keywords— *sintering; nano-silver paste; mechanical testing; microstructure*

I. INTRODUCTION

With the development of power electronics systems, there is an increasing need of having interconnection materials with properties that can ensure both a good mechanical strength of the bonding and high thermal/electrical conductivities in the assembly. So far, the lead-free solders have been considered as an attractive solution for die attach [1-3] but sintered joints based on nano-silver paste [4-6] are now considered as a replacement solution in the most demanding applications. The sintering technique proves indeed to be a promising solution, offering a thermal conductivity increased by many folds for the electronic packaging [7]. In addition, the sintered interconnect material can be manufactured with sufficient mechanical flexibility to accommodate the mismatch in coefficient of thermal expansion between the different components of the assembly. In power optics applications, the mechanical stresses can cause geometrical distortion in the active laser medium, and thus degrade the optical beam quality. It is then necessary to realize a flexible joint in such cases to solve the issue, which can be achieved by using low-pressure and moderate temperature conditions during the sintering process. However, evaluating quantitatively the effects of the process on the joint properties remains a challenge. It would of great use for the electronic manufacturing industry to have a systematic means to select precisely the process parameters for the desired properties.

The objective of this study is to fabricate sintered Ag joints on gold-plated copper and investigate, in relation with the chosen sintering conditions, their physical and mechanical properties. The analysis of the measurements is done by following a design of experiments (DoE), making it possible to estimate effects of process parameters, such as the heating rate and the applied pressure, on the mechanical properties of the fabricated samples, as well as on their microstructural features.

II. SAMPLE PREPARATION

A. Assembly silver precursor

Silver oxalates used in this study were obtained by the following process [7,8]. The Ag oxalates are decomposed into metallic silver below 200°C under controlled atmosphere, which allows for reduction of the metallic oxides. The chemical decomposition reaction occurs as follows:

$$Ag_2C_2O_4(s) \rightarrow 2Ag(s) + 2CO_2(g) \qquad (1)$$

During this reaction, the Ag oxalates reduce to highly reactive and pure nano-Ag particles after the organic content evaporates due to heat. At this stage, the spongy metallic nanometric grains replace the initial oxalate particles and are partially linked to each other by small metallic bridges. In order to make the process easier, a silver paste is made by mixing the decomposed silver oxalate powder with pure ethylene glycol in an ultrasonic bath. The obtained solder paste can then be deposited more easily on the surfaces to bond. The bonding process is finally realized by heating the assembly at a moderate temperature ($\leq 300°C$) as illustrated in Fig. 1a. The nanometric silver grains formed during the early stages of the oxalate decomposition possess a high propensity for sintering and, thus, only a relatively small bonding pressure is sufficient to enhance further the growth of the particles and the inter-particle bonds.

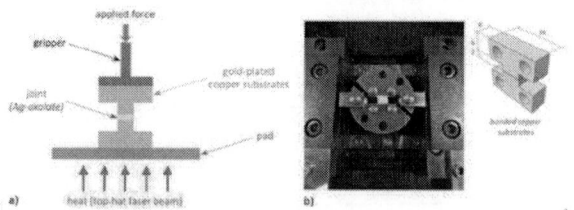

Fig. 1. (a) Schematic of the sintering setup for bonding copper parts, and (b) shear loading of a sintered sample under quasi-static conditions at room temperature.

B. Assembly substrate

This process is used to assemble several pairs of copper substrates. To ensure a good adhesion with the silver joint, the surfaces of the substrates to be bonded are gold-metallized beforehand [9]. The small-sized copper substrates have been fabricated with comparable dimensions to those of components used in microelectronics or optics packaging (bonded surfaces of 4×4 mm^2).

C. Assembly process

The sintering dwell time is fixed at 30 min and only two main controlling factors are made vary in the experimental design, namely the heating rate and the applied pressure. The experimental design consists of 6 different configurations as reported in Table 1 (D-optimal RSM DoE). Each configuration is replicated 5 times to have statistically representative results. All the specimens are tested under shear loading until rupture using a micro-tensile machine (Fig. 1b).

TABLE I.

Values of the controlling process parameters for the 6 configurations of the DoE		
Configuration	Applied pressure (MPa)	Heating rate (°C/min)
C1	0.5	20
C2	10	90
C3	7.5	20
C4	0.5	90
C5	4.5	60
C6	10	50

3. RESULTS AND DISCUSSION

A. Mechanical characterization

From the load-displacement curves of the shear tests, the following mechanical properties have been extracted: maximal force or strength (N), elastic stiffness of the assembly (N/mm), critical elongation or displacement at rupture (mm) and work of fracture (mJ). It is observed that the values of all these properties increase significantly with the increase of the sintering pressure (Fig. 2). This effect is reinforced at higher values of the heating rate.

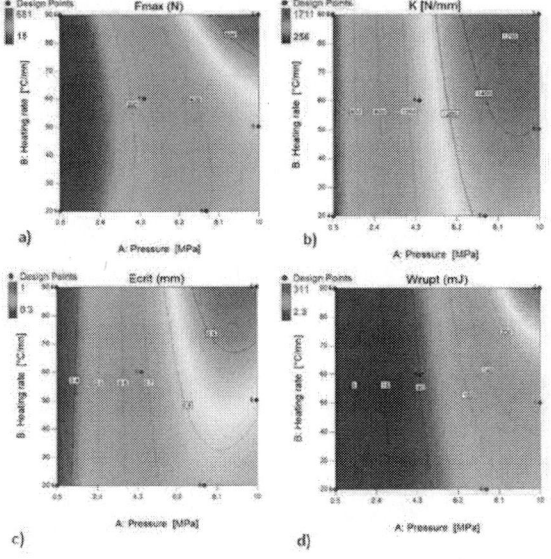

Fig. 2. Effects of the applied pressure and the heating rate on (a) maximal force, (b) elastic stiffness, (c) displacement at rupture, and (d) work of fracture.

The correlation coefficients between the different measured quantities have been estimated and are represented in Fig. 3a. It appears that all the mechanical properties are strongly correlated with each other. Having a tool to find the optimal process parameters leading to the best compromise in terms of mechanical properties is of great interest. For example, a high flexibility can be desired to accommodate the thermo-mechanical stresses under service, together with a good bonding strength of the microelectronic assembly. A specific relationship relating the maximal force and the elastic stiffness of the sintered silver joint is thus derived from the experimental data as represented in Fig. 3b. This diagram helps find the optimum corresponding to the lowest value of the elastic stiffness for compliance associated with a necessary and sufficient strength (maximal force) for the assembly.

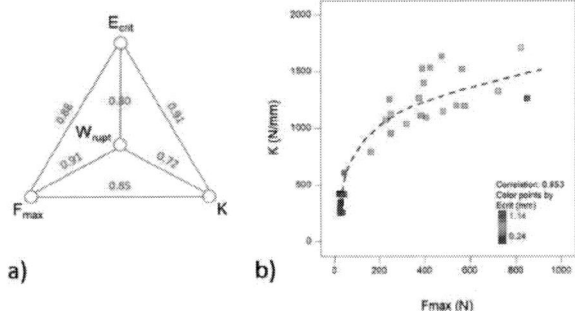

Fig. 3. a) Estimated correlations between the different mechanical properties, and (b) relationship between the elastic stiffness of the assembly and the strength of the sintered joint.

B. Microstructure of the bulk sintered Ag

The fracture surfaces are now investigated and the microstructural features of the sintered joints are compared. The effects of the process parameters are discussed by considering the two extreme configurations of the experimental design, associated respectively with the lowest sintering conditions C1 (20°C/min at 0.5 MPa) and the highest sintering conditions C2 (90°C/min at 10 MPa). In Figs. 4a and c, global views of the sintered joints after debonding are shown for the two configurations, showing a porous metallic network with typical micrometric porosities distributed more or less uniformly throughout the material. In Figs. 4b and d, close views of the same samples clearly exhibit that the porosity strongly differs depending on the sintering condition. The formation of necks (grain boundaries) is still at work in the case of the weak configuration (C1), which results in a network of large interconnected pores (open porosity). In the case of the strong configuration (C2), the reduction of porosity is far more important, indicating that an advanced stage of sintering has been reached due to a superior applied pressure combined with a higher heating rate.

293

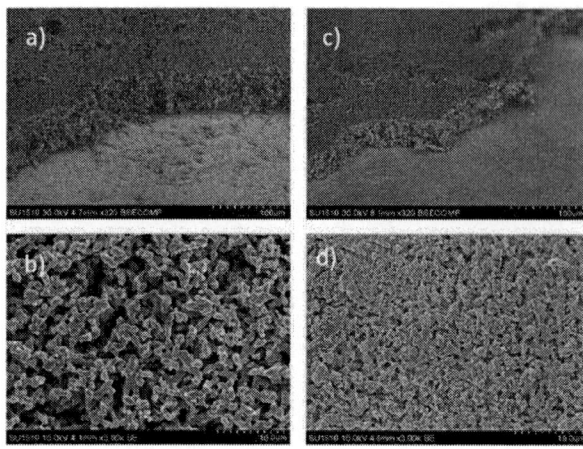

Fig. 4. Fig. 4: SEM Global and close views of the sintered silver joints after shear test achievement with (a)-(b) sintering conditions C1, and (c)-(d) sintering conditions C2.

Concerning the characteristics pertaining to particles morphology, the sintered joint in conditions C1 exhibits submicron-sized particles (~500 nm) with a good proportion of cubic-shaped particles which are typically associated with the earlier stages of the decomposition phase of silver oxalate precursors (Fig. 5a). It also appears that the material in the fracture zone does not display any plastic deformation. The sintered joint in conditions C2 displays, on the contrary, a deformed microstructure with particles elongated and plastically deformed after the shear loading. The particles in this case are also larger, over 1 micron in size (Fig. 5b).

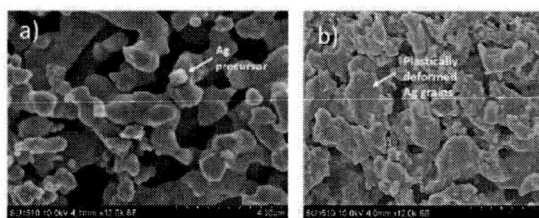

Fig. 5. SEM magnified views of the Ag particles after shear deformation for (a) sintering conditions C1, and (b) sintering conditions C2.

C. Analysis of the Ag/substrate junction

Analysis of the rupture surfaces shows that fracture occurs either inside the bulk sintered-Ag layer or at the interface with the gold-plated copper substrate, depending on the strength of the sintered material reached in the process. For the sintering conditions C1 (weakest configuration), large and thick chunks of the silver layer are found across the fracture zone (Figs. 6a-b), which indicates that the cohesive strength of the joint (strength of the bond between the sintered Ag particles) is weaker than the adhesion strength of the joint on the gold-plated copper substrate in these regions. Areas of the interface where silver was removed reveals that the junction layer that is expected to form by inter-diffusion of species (Au-Ag compound) is absent or hardly present. For the sintering conditions C2 (strongest configuration), no thick residues of the silver layer are observed on the fracture surface. The rupture appears to have occurred very close to the junction layer (formed from the gold metallization and the silver of the joint) or in the junction layer itself (Figs. 6c-d).

The Ag particles are evenly spread over the fracture surface, exhibiting an elongated shape due to plastic shear deformation.

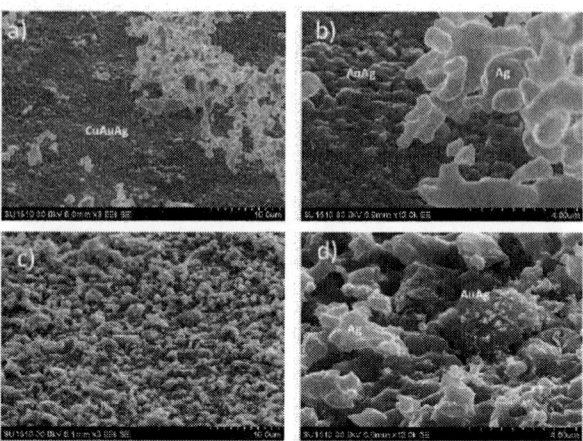

Fig. 6. SEM global and close views of the junction area between the sintered joint and the substrate for (a)-(b) sintering conditions C1, and (c)-(d) sintering conditions C2.

Cross-sections of the samples were also analyzed to have another view on the compounds forming at the sintered Ag/Au-plated Cu interface during the sintering process. Solid solution of the elements is achieved since gold is miscible in any proportions with both silver and copper, resulting in homogeneous phases [10]. The growth and final size of these compounds are dependent on the chosen levels of the process parameters during sintering, i.e. pressure and heat rate. The gold-silver (AuAg) solid solution, making the junction layer previously mentioned, is shown for both conditions C1 and C2 in Figs. 7a and 7b. The measured thicknesses associated with the two configurations are approximately 1.7 μm and 2.7 μm, respectively.

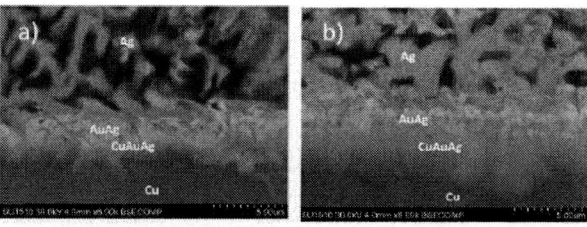

Fig. 7. Cross-section SEM views of the interface between the sintered silver layer and the gold-plated substrate for (a) sintering conditions C1, and (b) sintering conditions C2.

D. Microstructural measurements

Thicknesses of the joints in the different configurations were measured using the cross-sectional images. It can be seen in Fig. 8 that the obtained joint thickness is primarily determined by the sintering pressure. The heating rate has little or no effect and only a small decrease of thickness can be observed when the heating rate is changed from 20°C/min to 90°C/min. For the lowest applied pressure (0.5 MPa), the average thickness of the joints is 60±9 μm, while it is 36±5 μm for the highest pressure (10 MPa).

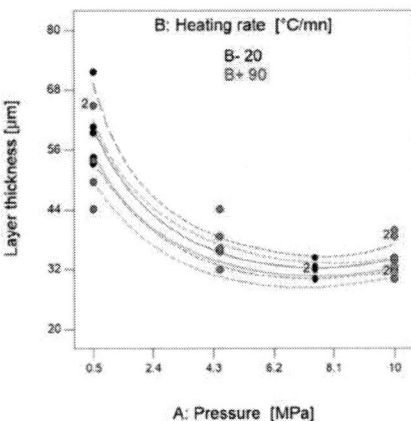

Fig. 8. Evolution of the sintered joint thickness with respect to the applied pressure and heating rate.

Using the ImageJ software [11], the relative density and the particle size distribution are measured for the sintered joints with sintering conditions C1 and C2 by processing their microscope images. The lowest sintering conditions (C1) leads to a highly porous sintered joint with open porosity. The measured value of the relative porosity is as high as 53%, and this is associated with relatively small grain sizes exhibiting a log-normal distribution (Figs. 9a-b). The grain sizes lie in the range between 0.45 µm and 1.25 µm, with an average size of 0.75 µm. The smallest silver grains obtained under these conditions are about the size of the silver precursors resulting from the silver oxalate decomposition (400 to 500 nm), which indicates that the sintering process remained in its first stages where the necks forms but grain growth is not significant. In the case of the highest sintering conditions (C2), image processing reveals a close porosity network with relative porosity dropping to 33% (Figs. 9c-d). The grain sizes, ranging from 0.9 µm to 2.5 µm, follow a Gaussian distribution and the average grain size with a value of 1.5 µm has doubled compared to that for the weak configuration.

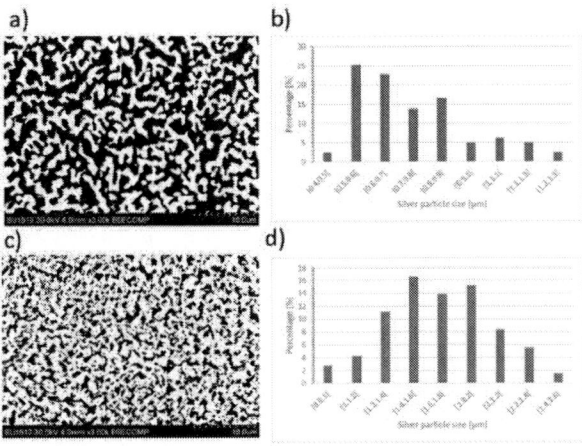

Fig. 9. Relative density and particle size distribution based on cross-section SEM images analysis, for (a)-(b) sintering conditions C1, and (c)-(d) sintering conditions C2.

Silver grain size is the measured physical characteristic that is the most dependent on the rate of temperature rise (Fig.10a), while porosity decreases mainly with pressure

(Fig.10b). This effect is enhanced at higher values of the temperature rise rate, meaning that porosity can be decreased when the rate of temperature rise is increased under a high pressure, but it is little or not affected by the rate of temperature rise if a low pressure is applied.

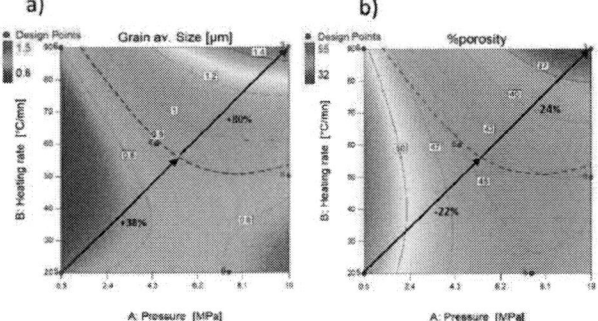

Fig. 10. Effects of the applied pressure and the heating rate on (a) the silver grain average size, and (b) the relative porosity of the sintered joint.

It can be established that the study area is clearly divided into two regions, following the 0.9µm grain size iso-line. Above this iso-value line, the increase in grain size with temperature rise rate and sintering pressure is twice as strong as that observed in the area below this iso value line. There is therefore a very strong acceleration of the growth of silver grains when this threshold is exceeded. The needed pressure for transitioning from the first regime of low grain size growth to the second regime of high grain size growth decreases considerably as the rate of temperature rise increases.

In terms of effects on the porosity rate, it is noticeable that, for both regions of the study area associated with the two grain size growth regimes, the decrease of the porosity rate with the rate of temperature rise and the sintering pressure remains approximately constant (Fig.10b). The silver particles get closer in the first regime (shrinkage), and they increase in size during the second regime (grain growth). These two regimes of transformation of the microstructure of the porous joints correspond typically to the first two stages of the sintering process. The first stage corresponds to the formation of necks between the silver precursor particles, due to rapid matter redistribution driven by interfacial tensions. This first stage continues until a local equilibrium of the interfacial tensions is established. Then, the "neck growth" stage starts, which corresponds to the interface migration under Laplace pressure gradient due to the surface curvature of the silver grains. This means that, in the transition zone between the first and second regimes, we are at the beginning of the grain size growth stage.

III. CONCLUSION

This study made it possible to map the mechanical and microstructural properties of porous sintered joints that can be obtained by varying 2 parameters of the sintering process: the pressure applied and the rate of temperature rise. The choice of the process parameters levels allowed us to cover all the stages of the silver powder sintering process. The most porous joints remain in the first sintering stage (formation of the necks), while the most dense sintered joints enter the phase of elimination of closed porosity. It was also possible to measure multiple mechanical properties for all configurations of the sintered joint, and a law linking the joint mechanical strength to its flexibility has been established. This makes it possible, for a given application, to find the right compromise between the mechanical strength and the ability of the joint to deform elastically to best cope with the mismatch in thermal expansion in the multi-material system.

The measurements already carried out still need to be supplemented by additional measurements, in particular that of the Young's modulus of the porous sintered joints, using an atomic force microscope (AFM) for example. The thermal and electrical conductivities of these sintered joints also need to be simulated and, if possible, measured.

When all these experimental data are gathered, it will be possible in future works to simulate the behavior of electronic assemblies using these sintered joints subjected to thermal shocks and active cycling. The findings could then be compared with the results for some common solder joints, such as tin-based alloys, and help find the optimal process parameters for the sintered microstructure, i.e. the one that is the most reliable under harsh conditions of electrical and thermo-mechanical stresses.

REFERENCES

[1] M. Abtew, G. Selvaduray, Lead-free solders in microelectronics, Mat. Sci. Eng. R 27 (2000) 95-141.

[2] L. Benabou, V.N. Le, Z. Sun, P. Pougnet, V. Etgens, Effects of voids on thermal-mechanical reliability of lead-free solder joints, Fatigue Design & Material Defects (2014), https://doi.org/10.1051/matecconf/20141204026.

[3] V.N. Le, L. Benabou, Q.B. Tao, V. Etgens, Modeling of intergranular thermal fatigue cracking of a lead-free solder joint in a power electronic module, Int. J. Solids Struct. 106-107 (2017) 1-12.

[4] K. Suganuma, S. Sakamoto, N. Kagami, D. Wakuda, K.S. Kim, M. Nogi, Low-temperature low-pressure die attach with hybrid silver particle paste, Microelectron. Reliab. 52 (2012) 375-380.

[5] S. Fu, Y. Mei, G.Q. Lu, X. Li, G. Chen, X. Chen, Pressureless sintering of nanosilver paste at low temperature to join large area (≥100 mm2) power chips for electronic packaging, Mater. Lett. 128 (2014) 42-45.

[6] W. Guo, Z. Zeng, X. Zhang, P. Peng, S. Tang, Low-temperature sintering bonding using silver nanoparticle paste for electronics packaging, J. Nanomater., 2015. https://doi:10.1155/2015/897142.

[7] K. Kiryukhina, H. Le Trong, P. Tailhades, J. Lacaze, V. Baco, M. Gougeon, F. Courtade, S. Dareys, O. Vendier, L. Raynaud, Silver oxalate-based solders: New materials for high thermal conductivity microjoining, Scripta Mater. 68 (2013) 623-626.

[8] K. Kiryukhina, Pâtes à braser à base d'oxalate d'argent pour applications électroniques fortement dissipatives: de l'intérêt des particules nanométriques issues de la décomposition de l'oxalate d'argent, PhD Dissertation, University of Toulouse, 2014.

[9] K.S. Siow, Are sintered silver joints ready for use as interconnect material in microelectronic packaging, J. Electron. Mater. 43 (2014) 947-961.

[10] Handbook of Surfaces and Interfaces of Materials, H.S. Nalwa (Ed), Academic Press, 2001.

[11] C.A. Schneider, W.S. Rasband, K.W. Eliceiri, NIH Image to ImageJ: 25 years of image analysis, Nat. Methods 9 (2012) 671-675.

Understanding Cu sintering and its role on corrosion behaviour for high-temperature microelectronic application

Juan Ignacio Ahuir-Torres
School of Engineering, Liverpool John
Moores University, Liverpool L3 3AF,
UK
j.i.ahuirtorres@ljmu.ac.uk

Sri Krishna Bhogaraju
Institute of Innovative Mobility,
Technische Hochschule Ingolstadt,
85049 Ingolstadt, Germany
srikrishna.bhogaraju@thi.de

Geoff West
WMG, The University of Warwick,
Coventry CV4 7AL,
UK
g.west@warwick.ac.uk

Gordon Elger
Institute of Innovative Mobility,
Technische Hochschule Ingolstadt,
85049 Ingolstadt, Germany
gordon.elger@thi.de

Hiren R. Kotadia
School of Engineering, Liverpool John
Moores University, Liverpool L3 3AF,
UK
h.r.kotadia@ljmu.ac.uk

Abstract— There has been a significant rise in interest from academic and industry communities on copper nano- and micro-particles-based sinter pastes for harsh environment die attach. However, sintering these particles is a complex process which is affected by many parameters, such as (i) particles size, shape, and distribution, (ii) sintering temperature, pressure, environment, and (iii) organic compounds used for the paste formulation. In literature, various research groups demonstrated that sintered copper layer can achieve good electrical conductivity and high strength. In our previous research [1, 2], we explained that microscale etched flakes are also capable of shear strength >30Mpa while sintering with 10 MPa bonding pressure at 275 °C for just 1 minute with negligible paste cost (in comparison to silver paste). However, very limited knowledge exists on literature about corrosion behaviour in sintered copper interconnects. This, from experience in the industry with silver sintering has proven to be critical to applications. For examples in high power LED headlamp applications, minor exposure to residual Sulphur emerging from vulcanized rubber seals has been observed to cause corrosion in silver sintered interconnects. Therefore, corrosion and its impact on performance and reliability must be considered, especially for safety critical applications. For this reason, we have studied the relationship between copper sintered samples and corresponding corrosion behaviours and compared again to pure copper (conventionally produced). The corrosion resistance of the samples in salt water was assessed via various electrochemical analysis techniques and correlated with sintered microstructure. Sample produced better corrosion resistance when fabricated with higher density (less porosity). However, in case of the microscale copper flakes-based interconnects, a pitting behaviour was observed. The sintered copper paste samples showed to be nobler than commercial copper. In addition, copper oxide and its role in accelerating/preventing oxidation is also studied.

Keywords—Copper sintering, high-temperature electronics, corrosion, porosity, electrochemical analysis.

I. INTRODUCTION

The increasing need for high-temperature electronics (above 175 °C) has driven various advancements in the field of die-attach bonding, ranging from the development of new high-temperature solders to silver (Ag) nanoparticle sintering [1-4]. However, these solutions tend to be relatively expensive compared to the previous solder

solution based on Pb5Sn. As a result, the microelectronic industry is under significant pressure to reduce packaging costs, leading to a growing momentum in the exploration of more affordable alternatives such as copper (Cu) sintering [1, 2].

Cu has emerged as a highly promising option due to its favourable characteristics, such as its high intrinsic electrical and thermal conductivity, which is comparable to Ag but at a substantial cost advantage [5]. The industry has recognised the potential of Cu sinter pastes as a cost-effective solution to meet the requirements of high-temperature electronics while maintaining optimal performance. However, there are significant challenges to overcome [1, 2]. One major challenge is the inherent tendency of Cu to undergo oxidation, which is a spontaneous reaction that becomes more favourable with increasing temperature. Additionally, the higher melting point of Cu (1083 °C) necessitates a higher sintering temperature compared to Ag when working with particles of similar morphologies. Several solutions have been proposed to address these challenges in Cu sintering. These solutions primarily involve the use of a reducing atmosphere (such as H_2, forming gas or formic acid enriched N_2) during the sintering process or the incorporation of reducing binders in pastes [6, 7]. Other approaches include oxidation-reduction bonding processes, the utilisation of Cu core shell particles with outer layers of Ag/Sn, and the application of phosphating techniques to Cu nanoparticles. These strategies aim to mitigate the oxidation issues and optimise the sintering process for Cu, ensuring improved performance and reliability in high-temperature electronics applications. However, some of these solutions also negate the cost advantage of Cu over Ag and pose challenges towards scaling up production. Also, producing Cu nanoparticle is proven expensive and not eco-friendly.

In order to overcome the challenges associated with Cu nanoparticles, our recent development [1, 2] has been made in the form of a copper flakes-based approach. This alternative has demonstrated comprehensive advantages over existing Cu nanoparticle solutions while maintaining a negligible cost. The focus of this research is to investigate the corrosion behaviour of the Cu sintered interconnects. In this paper, a thorough examination of the microstructure has been conducted using advanced characterisation techniques, accompanied by detailed electrochemical analyses. By correlating these results, the significance of porosity in the sintered microstructure is explained. The objective of this study is to provide a better understanding of the role of porosity on the corrosion behaviour and to identify strategies for further improvement in performance.

II. EXPERIMENTAL METHODOLOGY

A. Materials and sample preparation

Copper pastes were formulated by blending specially designed surface-enhanced microscale copper flakes with an organic binder using a planetary rotary mixer. The mixing sequence involved 4 minutes at 1000 rpm, followed by 5 minutes at 500 rpm. Both types of microscale particles had a flake-like morphology. Notably, particles in Paste 1 exhibited slightly greater thickness than those in Paste 2. These pastes demonstrated extended stencil life, no bleed-out during sintering, stability at room temperature, and excellent printability.

The study involved testing two distinct pastes: Paste 1 and Paste 2, both of which employed the same organic binder combination. Notably, Paste 2 contained 17 wt.% less organic content than Paste 1. Additionally, the particles in Paste 1 underwent a surface enhancement process three times less intensive than those in Paste 2.

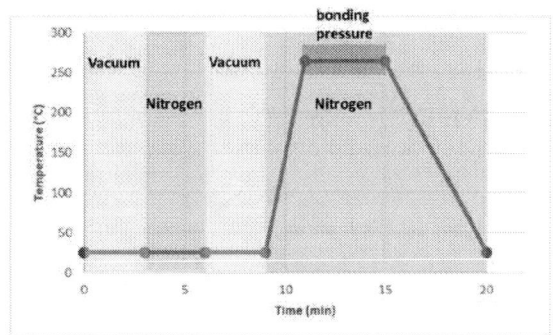

Fig. 1 Sintering profile as performed on the Budatec SP300 sinter press.

Fig. 2 Budatec SP300 sinter press used for copper sintering.

The pastes were manually printed with a 150 μm stencil onto a stainless-steel substrate. Stripes with 5 mm x 50 mm were printed. The oriented traces were pre-dried at 100°C for 5 min in air in a standard convection oven. Sintering was performed using a Budatec SP300 sinter press at 275°C for 5min with 25 MPa bonding pressure under N_2 atmosphere. Two vacuum steps to 10 mbar were executed before introducing N_2 into the sintering chamber. The resulting stripes were measured to have an average thickness of 50 ± 4 μm for Paste 1 and 72 ± 7 μm for Paste 2 sinter pastes. The thermal conductivity of the samples as measured by the LaTIMA equipment from Berliner Nanotest and Design GmbH [3], resulted in a thermal conductivity of 197±26 W/mK and 205±27 W/mK and an electrical conductivity of 20.43 ± 3.24 and 20.45 ± 1.76 MS/m for Paste 1 and Paste 2 respectively. Measurements were performed and averaged across 3 stripes for each paste.

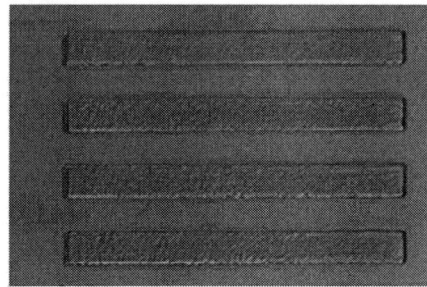

Fig. 3 Optical image of the printed Copper stripe.

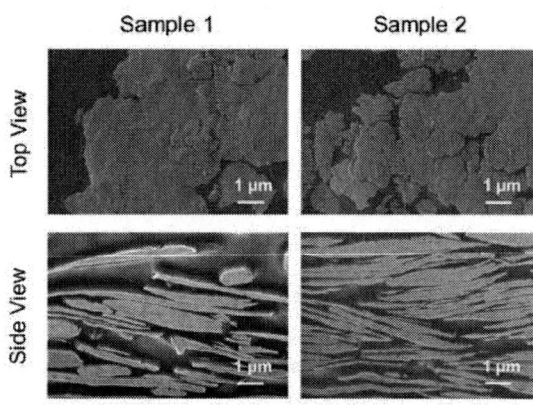

Fig. 4 SEM microstructure of the original surface-enhanced flakes of sample 1 (used for Paste 1) and sample 2 (used for Paste 2) shown from top and side views.

B. Electrochemical analyses

All electrochemical analyses were carried out with a three-electrode cell formed of a reference, counter and working electrode where reference electrode was silver/silver chloride in 3 M KCl (Ag/AgCl 3M KCl), platinum wire of 0.7 mm diameter was counter electrode the sample was the working electrode. Prior the electrochemical testing, the samples were polished at P1200 abrasive grit supplied by Struer. The samples were moreover cleaned with the next subsequent steps; firstly, the samples were cleaned with commercial detergent and fresh water, then soaked with distilled water, sprayed with isopropanol and dried with drier. The samples were let 48 hours before the electrochemical analyses to permit the native passive film generation. RS supplied the isopropanol. The electrochemical testing was conducted using a potentio/galvanostat (Interface1010E). Gamry Framework Version 7.8.4 software was employed to set the potentio/galvanostat parameter for each electrochemical trail. Gamry Echem Analyst software was used to evaluate the electrochemical test outputs. The potentio/galvanostat and software were provided by Gamry Instruments Inc. Two electrochemical tests were performed (i) asymmetric electrochemical noise (AEN) and (ii) potentiodynamic polarisation curve (PPC). The harsh environment was 0.6 M NaCl where the salt was supplied by Merck-Sigma-Aldrich. Total time and acquisition time were 7200 seconds and 0.05 s, respectively for AEN. The potential rate initial and end potential were 0.167 mVs^{-1}, -0.3 V from open circuit potential and 2 V from reference electrode potential, correlatively for PPC. The open circuit potential was at 7200 seconds of immersion in 0.6 M NaCl. Two limit current densities were employed in PPC, being 2.5 10^{-4} Acm^{-2} and 1.0 10^{-2} Acm^{-2} for commercial copper and sintered copper ink, respectively.

C. Microstructural analysis

For microstructure analysis, the samples were prepared using a combination of mechanical grinding, polishing, and, in some cases, ion-beam milling processes. The as-received, sintered stripes were then examined using Scanning Electron Microscope (SEM) coupled with energy-dispersive X-ray spectroscopy (EDS). Additionally, the samples subjected to corrosion analysis were analysed using SEM and FIB to reveal the oxide layer on the sintered samples. The quantitatively analyses of porosity performed using open-source ImageJ software.

III. RESULTS AND DISCUSSION

A. Microstructure analysis of Cu flake and sintered sample

Fig. 4 illustrates the SEM analysis of the as-received surface-enhanced copper flakes. The top view of the two different flakes used in this study shows no noticeable difference. However, the cross-section of the flakes revealed a clear distinction between the two samples, with flakes in sample 1 (used for the Paste 1) being thicker than those in sample 2 (used for the Paste 2). The preparation of both flakes differs significantly, with sample 2 undergoing additional enhancement processes compared to sample 1.

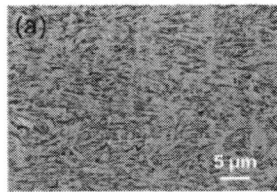 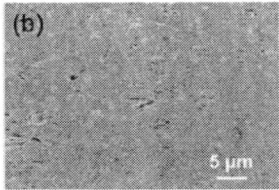

Fig. 5 SEM microstructure of the sintered samples: (a) Sample 1 (using Paste 1) and (b) Sample 2 (using Paste 2), highlighting differences in porosity volume, size, and distribution.

The surface-enhanced flakes were then used for creating the paste in combination with the novel organic binder. Fig. 5 displays cross sections of the sintered samples, revealing a distinctive sintering behaviour, with flakes stacking over each other and exhibiting significantly reduced porosity.

Additionally, a clear contrast in the sintered microstructure was observed between the pastes, as shown in Fig. 4. The results indicate that sample 1 (using paste 1) exhibited a much higher area percentage of porosity (~26%) compared to sample 2 (using paste 2) (~6%).

Fig. 6 SEM cross-section of silicon chip/Cu sintered (Paste 2)/Cu substrate showing a dense and homogeneous interconnect with uniformly distributed porosity.

Paste 2 was also utilised for sintering a silicon chip and Cu substrate, applying 20 MPa bonding pressure for 5 minutes at 275°C. Fig. 6 displays the cross-sectional microstructure of the sintered assembly, which demonstrates a dense and homogeneous interconnect with excellent bonding between the silicon chip and substrate. Porosity analysis of the full cross-section revealed the presence of approximately ~8 vol.% of porosity in the sintered interconnection.

This experimental investigation clearly revealed the influence of flake thickness on sintering behaviour, despite the topographical morphology remaining the same in both samples. The porosity in sample 2 is approximately 5 times lower than in sample 1, attributed to the reduction of flake thickness from ~0.5 μm to ~0.27 μm.

B. Electrochemical analysis of sintered samples

Fig. 7 illustrates AEN results over time. The potential of all samples decreased over time, as depicted in the open circuit potential (OCP) graph (Fig. 7(a)). This, indicate the oxidation of copper. The nobility (potential) of the material typically decreases due to copper oxidation by chloride ions [8]. The potentials varied among the samples, following this order: sintered copper Paste 1 > sintered copper Paste 2 > conventionally manufactured dense copper sheet. The changes in microstructure, such as grain size, impurities, and defects (e.g., porosity), significantly influence the material potentials. The copper with fine microstructure is nobler than copper with coarse microstructure because the grain boundary possesses higher corrosion potential than grain. The surfaces with larger area ratio of the boundary grain thus exhibit nobler for copper. The defects and imperfections are zones with dissimilar surface energy to matrix. This can produce activation or inactivation zones, increasing or reducing the potential of the sample [9, 10]. The sintered copper paste 2 had less porosity and may have a finer microstructure than the sintered copper paste 1. Sintered Paste 2 therefore was nobler than sintered paste 1. In the case of the higher potential of the sintered copper paste samples in comparison with commercial copper, this is owing to their finer microstructures.

In contrast to OCP results, the Zero Resistance Ammeter (ZRA) (Fig. 7(b)) results demonstrate that commercial copper and sintered copper paste-2 exhibit similar behaviours, with constant current density over time. However, in the case of Paste 1, the current density diminishes over time. This reduction can be attributed to the potential decreasing over time, as per Ohm's law [11], suggesting that the corrosion resistance remains constant or experiences a slight reduction over time.

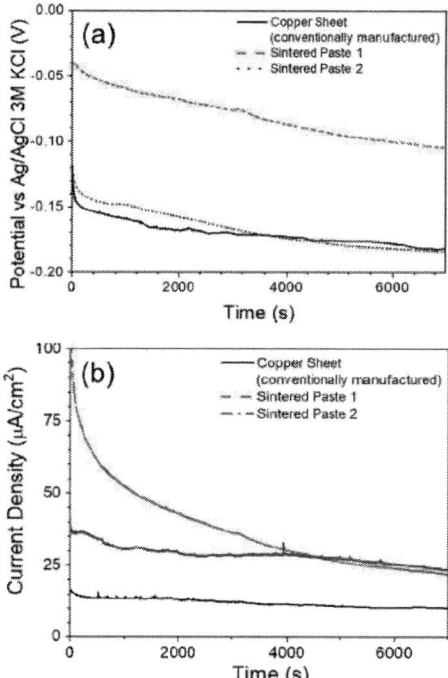

Fig. 7 AEN analysis of the commercial copper sheet, sintered copper Paste 1 and 2: (a) OCP and (b) ZRA evolution over the time for 2 hours in 0.6 M NaCl.

The second analyses preformed on the sample were PPC. The curve behaviours of the samples exhibited striking similarities, as evident from Fig. 8. The cathodic branch displayed a horizontal curve at low potential and a vertical curve at high potential. This behaviour can be attributed to the chemical evolution of hydrogen and water, wherein a portion of the PPC potential is utilized for this evolution [12]. The vertical curve signifies the diffusion control of the cathodic/reduction reactions [8].

On the other hand, the anodic branch appeared as an inclined curve, indicating a mixed control involving activation and passivation/diffusion for the oxidised/anodic reactions [8]. Notably, the PPC did not exhibit a passivation film for copper due to the rapid dissolution of the copper oxide layer caused by chloride ions [13].

The sintered copper Paste 1 exhibited the highest corrosion potential (E_{corr}), while the lowest was observed for the commercial copper. Microstructures with finer boundaries tend to generate nobler materials [9]. In the case of the sintered paste samples, the material's nobility decreases with increasing porosity [10]. These findings were consistent with the results obtained from AEN analysis.

The corrosion current density (I_{corr}) was determined by estimating the intersection of the Tafel lines. The sintered copper ink samples showed similar I_{corr} values among themselves, being higher than that of the commercial dense copper. The microstructural refinement strengthened the cathodic effect of the boundary grain on the grain [4, 14].

The presence of porosity also increased I_{corr} due to the larger contact area between the sample and the aggressive environment [9]. While these values were qualitatively similar to AEN $I_{R.M.S.}$, the quantitative values differed. This discrepancy can be attributed to the asymmetric system, which focuses the electrochemical noise on the cathodic branch [14].

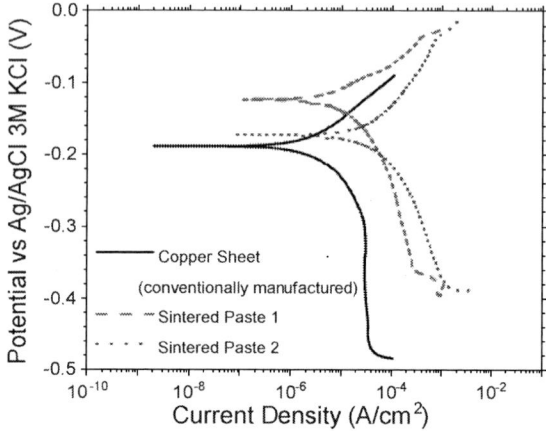

Fig. 8 PPC of the commercial copper, sintered copper Paste 1 and 2 in 0.6 M NaCl at room temperature.

C. Microstructure analysis after corrosion tests

In order to understand the distinct corrosion behaviours of the commercially manufactured dense copper sheet and sinter pastes 1 and 2, SEM analysis was carried out on the surface and cross-section of the samples after the PPC test (Fig. 9).

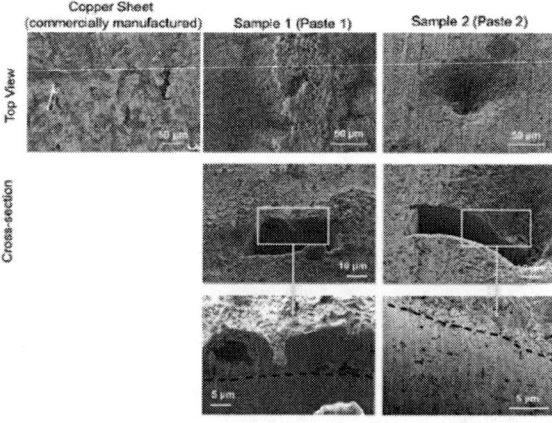

Fig. 9 SEM analysis of the samples after corrosion testing.

In the case of the commercial copper sheet, no crevice or pitting sites were observed. The only noticeable difference was observed at the grain boundaries, likely due to their increased reactivity compared to the bulk material.

Sintered Paste 1 (sample 1), which exhibits approximately 26 area. % porosity, showed pitting and oxide islands at various sites. This porosity is believed to be the cause of the pitting sites, as the native passive film is thinner on the pore, leading to localized corrosion.

On the other hand, highly sintered Paste 2 (sample 2) demonstrated better corrosion resistance, with only a few pitting sites observed. In Paste 2, not only was the porosity reduced from 26% to 6 area. %, but the size, shape, and thickness of the remaining pores were also altered. The channels between the flakes were narrowed due to flake-to-flake diffusion and grain boundary movement during sintering, promoting pore closure and resulting in smaller and fewer pits. This densification process enhanced the resistance of these compacts to pitting corrosion compared to Paste 1.

Moreover, the FIB cross-section offers additional evidence, indicating an unstable oxide layer with porosity and cracks in sample 1, contrasting with sample 2. Notably, the oxide layer is considerably thick, measuring approximately ~5 μm.

IV CONCLUSION

This study focuses on investigating the relationship between copper flakes, sintering, and corrosion in two different samples. The corrosion characteristics of the samples are compared against those of a commercially pure copper sheet.

The thickness of copper flakes significantly influences their sinterability due to the difference in the specific surface area of the flakes, even when the morphology of the flakes remains similar.

The variation in porosity size, shape, and distribution plays a crucial role in determining the corrosion characteristics of the sintered paste. Higher porosity levels resulted in prompting of localized pitting and the formation of a weak native passive film, leading to a higher corrosion rate compared to the densely sintered sample. The sintered copper ink samples are nobler than commercial copper because the fine microstructure.

Further investigations are required to quantify the impact of porosity size and shape on corrosion resistance and to establish a comprehensive understanding of the corrosion mechanisms over time. This knowledge will be essential for optimising sintering processes and enhancing the corrosion resistance of the materials.

ACKNOWLEDGMENT *(Heading 5)*

This work was partially supported by (1) Liverpool John Moore University, Faculty of Engineering and Technology (FET) Pump Prime Awards 2022/23 and (2) Copperfield (Grant number FKZ 13XP5134F) under KMU Innovative from the Federal Ministry of Education & Research, Germany. The authors would like to thank Dr. Ralf Webler from Schlenk SE for support with the SEM imaging.

REFERENCES

[1] S.K. Bhogaraju, H. R. Kotadia, F. Conti, A. Mauser, T. Rubenbauer, R. Bruetting, M. Schneider-Ramelow, G. Elger, Die-Attach Bonding with Etched Micro Brass Metal Pigment Flakes for High-Power Electronics Packaging, ACS Applied Electronic Materials, 3 (2021) 4587-4603.

[2] S.K. Bhogaraju, F. Conti, H.R. Kotadia, S. Keim, U. Tetzlaff, G. Elger, Novel approach to copper sintering using surface enhanced brass micro flakes for microelectronics packaging, Journal of Alloys and Compounds, 844 (2020) 156043.

[3] D. Wakuda, M. Hatamura, K. Suganuma, Novel method for room temperature sintering of Ag nanoparticle paste in air, Chemical Physics Letters, 441 (2007) 305-308.

[4] R. Durairaj, R. Ashayer, H.R. Kotadia, N. Haria, C. Lorenz, O. Mokhtari, S.H. Mannan, Pressure free sintering of silver nanoparticles to silver substrate using weakly binding ligands, in: 2012 12th IEEE International Conference on Nanotechnology (IEEE-NANO), 2012, pp. 1-4.

[5] T.F. Chen, K.S. Siow, Comparing the mechanical and thermal-electrical properties of sintered copper (Cu) and sintered silver (Ag) joints, Journal of Alloys and Compounds, 866 (2021) 158783.

[6] S.A. Paknejad, S.H. Mannan, Review of silver nanoparticle based die attach materials for high power/temperature applications, Microelectronics Reliability, 70 (2017) 1-11.

[7] L. Wang, L. Wang, X. Meng, F.-S. Xiao, New Strategies for the Preparation of Sinter-Resistant Metal-Nanoparticle-Based Catalysts, Advanced Materials, 31 (2019) 1901905.

[8] S. Zor, Sulfathiazole as potential corrosion inhibitor for copper in 0.1 M NaCl, Protection of Metals and Physical Chemistry of Surfaces, 50 (2014) 530-537.

[9] L. Lapeire, E. Martinez Lombardia, I. De Graeve, H. Terryn, K. Verbeken, Influence of grain size on the electrochemical behavior of pure copper, Journal of Materials Science, 52 (2017) 1501-1510.

[10] D. Yang, X. Kan, P. Gao, Y. Zhao, Y. Yin, Z. Zhao, J. Sun, Influence of porosity on mechanical and corrosion properties of SLM 316L stainless steel, Applied Physics A, 128 (2021) 51.

[11] S.J. Kim, J.C. Park, S.K. Jang, Evaluation of Electrochemical Characteristics for Casted AC7AV Aluminum Alloy, Advanced Materials Research, 811 (2013) 54-60.

[12] R.G. Kelly, J.R. Scully, D. Shoesmith, R.G. Buchheit, Electrochemical Techniques in Corrosion Science and Engineering (1st ed.), CRC Press, (2002).

[13] Hamid A. Z. , Gomaa M. H., Hassan H. B., Corrosion Performance of Copper - Diamond Composites in Different Aqueous Solutions, American Journal of Electromagnetics and Applications., 4 (2016) 39-49.

[14] D.-H. Xia, S. Song, Y. Behnamian, W. Hu, Y.F. Cheng, J.-L. Luo, F. Huet, Review—Electrochemical Noise Applied in Corrosion Science: Theoretical and Mathematical Models towards Quantitative Analysis, Journal of The Electrochemical Society, 167 (2020) 081507.

Inspection Techniques Using Scanning Acoustic Microscopy for Silver Sintering Applications in Power Electronic Modules

Heaklig Ayala
School of Engineering
University of Warwick
Coventry, United Kingdom
0000-0001-9828-7289

Jose Ortiz Gonzalez
School of Engineering
University of Warwick
Coventry, United Kingdom
J.A.Ortiz-Gonzalez@warwick.ac.uk

Mohammed-Amer Karout
School of Engineering
University of Warwick
Coventry, United Kingdom
0000-0003-1660-1196

James Cotty
Automotive EV
Custom Interconnect Limited
Andover, United Kingdom
James.Cotty@cil-uk.co.uk

Tim Rumney
Automotive EV
Custom Interconnect Limited
Andover, United Kingdom
Tim.Rumney@cil-uk.co.uk

Philip Mawby
School of Engineering
University of Warwick
Coventry, United Kingdom
P.A.Mawby@warwick.ac.uk

Abstract— **This work presents a practical perspective to Scanning Acoustic Microscopy (SAM) for Ag-sintering applications in power electronic modules and devices. It aims to present typical challenges to consider whilst performing C-SAM for inspecting and analysing microstructures within power electronic devices, with emphasis on silver-sintered interconnects. The study presents a vast number of acoustic images (discrete devices in TO-247 packages, power modules, and also custom samples of substrates and devices), where several defects are identified, including voids, cracks, and delamination. Challenges for Ag-sintered layers inspection related to advanced surface finishes are also revealed and discussed in detail. The suitability of C-SAM for inspecting small and large areas of Ag-sintered interfaces is discussed, and showcased with custom samples, and a commercial power module.**

Keywords— *SAM, C-SAM, Sintering, Ag-Sintering, Power Modules*

I. INTRODUCTION

Ag-sinter technology has proven to be most favourable for power electronic applications requiring high operating temperatures and power cycling, meaning it is a good candidate for power module packages containing wide-bandgap semiconductors [1]. Nevertheless, there are challenges regarding the inspection techniques of Ag-sintered layers, which arise from recent advanced substrate metallisation types, dispensing and curing techniques, and the relatively small thicknesses of Ag-layers compared to the neighbouring structures, just to mention a few. Scanning Acoustic Microscopy (SAM) is an adequate non-destructive inspection technique, that allows to efficiently spot defects such as cracks, voids, and delamination [2]. Therefore, proper inspection by means of SAM can allow to further optimise assembly processes.

The process of Ag-sintering is sensitive to various process parameters such as pressure, temperature, or even different process atmospheres. In order to achieve a reliable attach, the condition of the mating surfaces and the sinter paste also plays an important role [1].

Because the imaging technique is purely based on physical interactions of ultrasound waveforms with mechanical structures, the technique can efficiently find physical defects which would be very hard (sometimes impossible) to identify otherwise. For instance, adhesion problems can occur during sintering due to multiple reasons (e.g., contamination), but they cannot be detected properly using X-Ray. SAM can prove particularly important during process monitoring during specific production stages, because not all the issues during assembly can be identified simply with an end-of-line electrical test [1].

So far, X-Ray is the de facto imaging technology in semiconductor industry. Here, an image contrast is achieved because of the absorption of X-Rays in the sample, meaning this transmission imaging technique can only provide volume information of the defects, and is not as effective for the detection of interface defects (e.g., delamination), or even adhesion problems [1]. Unlike X-Ray, inspection of interfaces is the area where SAM excels in particular compared to other methods, especially in the context of power electronics.

Bringing another technology into the equation, SAM could also be compared to thermography methods. Both SAM and thermography are more sensitive to interface defects compared to X-Ray. However, even though SAM can provide higher resolutions compared to thermography, the sample has to use a water coupling medium [1], which can make it less attractive as it is normally achieved by submerging the sample into the fluid.

A scanning acoustic microscope (SAM) emits an ultrasonic beam onto the specimen, such as the structure of a semiconductor package, and then converts the reflected "echo" intensity at the boundary between different materials into gradient values. The transducer accomplishes the task of sending an ultrasonic signal that hits the sample, which is then received through reflection of the same signal as it hits different interfaces (because of acoustic impedance

---

This work has been supported by the Centre for Doctoral Training (CDT) to Advance the Deployment of Future Mobility Technologies at the University of Warwick, Coventry, UK. The studentship is partially supported by Custom Interconnect Ltd. (Group), Andover, UK.

mismatch). Finally, the signals received are used for imaging and defect detection [3].

The acoustic impedance (Z) can be defined as the resistance a material poses to the propagation of an ultrasound beam as it penetrates through it:

$$Z = \rho \times c \qquad (1)$$

Where: ρ is the material density (in kg/m^3); c is the velocity of sound within the material (in m/s); and Z is related to the acoustic hardness of the material (in Ns/m^3).

Examples of acoustic impedance related properties are provided in TABLE I.

Reflection and scattering happen at the different interfaces of the sample, which depend greatly on the geometry of the defect and the materials involved. The proportion of the reflected waves compared to the transmitted ones is determined from the impedance mismatch of the materials involved. A higher acoustic impedance mismatch at the interface would imply a larger intensity of the reflection [2].

The reflected ultrasound is received using a transducer which then converts it into an electronic signal [4]. The reflection index then given by:

$$R = \frac{(Z_2 - Z_1)^2}{(Z_2 + Z_1)^2} \qquad (2)$$

Similarly, the transmission index can be simplified as:

$$T = 1 - R \qquad (3)$$

The role of the transducer is very important, and as the "lens", its geometry optimises for proper focus on the area of interest, as well as for resolution, and penetration depth [2]. The ultrasonic pulse generated from the transducer is delivered to the sample via a water coupling medium, since air reflects practically all the ultrasound due to its characteristic acoustic impedance.

During the process of sending the ultrasound beam and receiving the reflected echoes, a time duration must be recorded for the series of acoustic reflections. This would be the transit time of the reflections or echoes [2]. The time-of-flight (t) of the pulse reflected from each surface provides the depth information of the sample.

A series of time delays of the pulses from the reflected echoes take place, following interaction with the internals of the sample. The delays relate to the acoustic properties of the material, but mainly to the intrinsic speed of sound of the material and its acoustic impedance [2].

In Fig. 1, one can observe a typical series of echoes against time, and showing relative amplitudes. As it can be noticed from the third echo in the figure, the ultrasonic waves can experience phase inversion, according to the transition type of the acoustic mediums, i.e., from high to low acoustic impedance at the interface, of vice versa. An example interpretation of Fig. 1 could be the following: (1) there are three materials in the structure of the sample, A, B, and C, respectively; (2) there are four interfaces, the first water-to-A interface, the second A-to-B interface, the third B-to-C interface, and the fourth C-to-water interface; (3) the peak of the echoes correspond to the boundaries of the interfaces; (4)

the regions in light red, blue, and orange provide a time difference information that can be related to the thicknesses of each corresponding material; (5) the polarity of the echoes indicate that, for example, material A has a higher acoustic impedance than water (potentially harder), and that material C has a lower acoustic impedance than B (potentially softer).

TABLE I. ACOUSTIC-RELATED PROPERTIES OF DIFFERENT MATERIALS [5]

Material	ρ [kg/m^3]	c [m/s]	Z [$\times$ 10^6 Ns/m^3]
SiC	3,211	13,100	42.1
Sn	7,260	3,320	24.1
Cu	8,960	5,010	44.9
AlN	3,260	10,570	34.46
AlSiC	3,000	8,790	26.4
Al$_2$O$_3$	3,987	9,100	32–41
Tungsten	19,250	5,400	104
Gold	19,300	3,316	64
Silver	10,490	3,650	38.28
Nickel	8,900	6,040	53.76
Zinc	7,140	4,210	30.06
Polymer	–	–	2–7
Air	1.293	343	0
Water	997	1,500	1.5
Glass	2,500	6,000	15

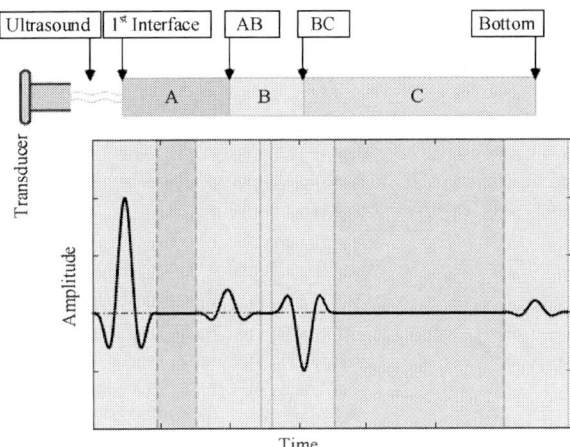

Fig. 1. Example of the A-scan mode of operation, showing ultrasound "echoes" and boundaries

In scanning acoustic microscopy, there are three main modes of operation: A-scan, B-scan, and C-scan. The A-scan provides amplitude, polarity, and time or distance information. In Fig. 1 is shown what an A-scan would represent, and typically, interactive software would display the time-amplitude series by default. On the other hand, the B-scan provides a cross sectional view through the material which area is parallel to the ultrasonic beam. Conversely, the C-scan provides information of an area that is orthogonal to the ultrasonic beam, recording a single image of the selected layer at a specific depth. Fig. 2 shows a conceptual representation of the concepts of B-scan and C-scan.

While operating in the A-scan mode during preparation, the transducer should be moved in the vertical direction in order to moderate the intensity of the echo related to the layer of interest, before committing to running another type of scan. Whilst the A-scan is the most basic mode, all other modes are just a combination of multiple A-scans, which are then further processed to construct a B-scan or C-scan image. The two scan modes that provide the most meaningful information when inspecting Ag-sintered layers are the A-scan and the C-scan.

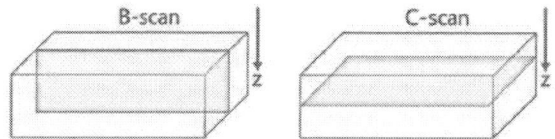

Fig. 2.   B-mode and C-mode planes of operation. Adapted from [2]

As mentioned before, an A-scan shows the amplitude along the time, and the relationship between the position of a particular echo in time (t), the travelling distance of the ultrasound (d), and the intrinsic speed of sound of the material (c), is given by:

$$d = \frac{t \times c}{2} \qquad (4)$$

There is a trade-off which relates to the resolution versus penetration depth of the sample, normally as the transducer frequency is increased. Theoretically, the resolution (w) can be obtained by means of the following expression:

$$w = \frac{c}{2f \times \text{NA}} \qquad (5)$$

Here: c is the speed of sound within the material; f is the frequency of the transducer; NA is the numerical aperture of the transducer. The numerical aperture can then be obtained with the following equation:

$$\text{NA} = \sin \theta \qquad (6)$$

However, it is worth mentioning that although the equations (2) and (3) might suggest the ultrasonic beam is either reflected or transmitted, in reality some ultrasonic energy can either be absorbed by the material, or scattered at the interface. Some materials can be quite absorbent of ultrasonic energy, and because at higher frequencies typically less energy can be carried by the ultrasonic beam, higher frequency transducers are most likely to offer less penetration as the ultrasonic beam can be more easily damped or absorbed by the different materials.

This work aims to demonstrate the challenges that must be taken into account when considering the inspection of power electronic modules and other devices, with Ag-sinter technology in mind.

In Section II, we describe the methods, equipment, and instruments utilised for achieving good quality SAM images, and also a description of the experiments is provided. Section III expands on the results obtained through the experiments, providing extensive interpretation of the findings. Here, recommendation techniques on how to achieve the best results are provided when appropriate, the importance of matching and working with different transducers considering different materials in the structure is discussed (including mould compounds), and also X-Ray imaging of selected parts has been conducted in order to compare and highlight the benefits of SAM. Section IV summarises the findings, and concludes the paper.

II. METHODOLOGY

A Sonoscan Gen 7 C-SAM microscope is utilised to inspect both soldered and Ag-sintered interfaces between substrates to copper plates, and dies to substrates. The various substrates types to be utilised are AlN/bareCu, and Active Metal Brazed (AMB) AMB-Si$_3$N$_4$/Ag-Ni-plated, whilst the dies will be SiC MOSFETs in all cases.

When non-commercial devices are being explored, off-the-shelf Ag-sinter is used, which is adequate for both wet-placing and dry-placing for comparison. Defects will be selectively and purposedly introduced where possible for validation, and large area Ag-sinter attach will be considered. Validation will be conducted using different transducers ranging from 15 to 100 MHz.

TABLE II.   DATA OF THE TRANSDUCERS UTILISED TO PERFORM THE C-MODE SAM SCANS

Frequency	Focal Length	Diameter [in]	F#	ToF [μs]
15 MHz	0.752	0.50	1.5	25.40
50 MHz	1.000	0.25	4.0	33.87
75 MHz	0.750	0.25	5.0	25.40

For the purposes of calibration and identification of different layers, the following steps must be taken:

1. Position

   The transducer must be moved in the XY dimensions in order to be on top of the sample (ideally the centre, initially). During this step, it must be ensured the transducer is operating correctly, with no air bubbles blocking it. Following this, adjustment in the Z direction must be adjusted, until feedback from the TOF is obtained, guaranteeing the transducer is in a proper operating range.

2. Focus on the first surface

   This can either be done automatically by means of software assistance, or manually, by adjusting the Z-height of the transducer against the surface of interest. The goal is to maximise the echo of interest (in this case the first one, as the first surface is being aimed at). This echo should be constantly monitored or displayed in the A-scan mode. Occasionally, the trigger level must be adjusted in order to discard noise or to get good contrast.

3. Scan the surface of interest

This step allows to verify that the settings are correct, by ensuring a balanced image is being obtained. Otherwise, settings might need adjustment. Although this might sound trivial, at times the user does not know the structure of the inspected sample, and he must ensure meaningful results are being obtained from the equipment.

4. Position the "gate"

The gate is the term that is typically used to refer to the range of time of the echoes of interest. As discussed in the introduction, echoes reflected from the internal structure present different times in the time series (or A-scan), hence providing depth and thickness information. This means that by adjusting the "gate" or window of interest we are effectively selecting the interface or interfaces to be imaged. Different software tools often offer a multi-gating option, where several windows can be imaged at a time, hence saving time to the user.

5. Re-focus and adjust gain

By adjusting the Z-height, the transducer penetrates through the structure, effectively focusing more optimally on internal interfaces for which the images will become sharper as it gets closer to the optimal depth of focus. The gain of the transducer can be adjusted as necessary in order to obtain good contrast and balance in the images.

6. Repeat steps 4 – 5 as necessary

When scanning large power modules, a set of fixtures are utilised, which allow for the sample to be scanned from the underside, as a water-jet hits the sample from underneath, so the coupling medium through water is maintained without submerging the part in water.

Apart from SAM, X-Ray imaging is conducted for selected samples. A Dage QUADRA5 X-Ray machine has been used in order to generate images that can be compared to the C-SAM methods utilised.

Since the defect types can be related regardless of the size of the device, or the attaching technology used, we occasionally showcase discrete devices for demonstration purposes, many of which use attaching technologies other than sintering.

In order to demonstrate the capabilities of the C-SAM method, a "story-telling" series of scans are performed, where a 1200 V / 15 A power module with no backplate is scanned to show different layers within the structure.

Following this, a series of C-SAM scans are performed to demonstrate the challenges related to some advanced metallisation types of the substrates. Then, several scans are conducted, this time focused on defect detection, such as voids, cracks and delamination. For this part, both soldered and Ag-sintered devices are utilised, including both discrete devices (TO-247) and custom modules. On the other hand, inspection of large-area Ag-sintered substrates is presented, which also poses its own challenges.

Lastly, a comparison of the performance of different transducers for defect detection is explored. For this part, three 1200 V commercial discrete devices are used (TO-247), all of similar current ratings.

III. EXPERIMENTS & RESULTS DISCUSSION

A. Extraction of individual layer interfaces

As mentioned in the previous section, when proper attach is achieved within the structure, and good co-planarity is maintained, the different layers within a structure can be identified and separated. This is depicted in Fig. 3, where the sample power module has been inspected to show (a) the copper face of the substrate in the underside; (b) the engraved pattern of the copper on top of the ceramic, where the different components are placed and attached; (c) the attach layer of different components; and (d) the dies.

This is something that would have been practically impossible to achieve if , say, a baseplate with pin-fins had been present.

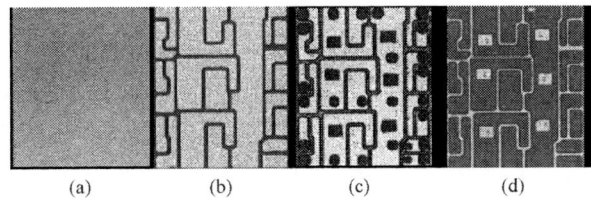

(a)   (b)   (c)   (d)

Fig. 3.   C-SAM of a 1200 V / 15A comercial SiC power module. Substrate: AlN/bareCu. Transducer: 75 MHz | 0.750 in | f5.0. The illustration shows: (a) the copper face of the substrate; (b) the engraved pattern of the; (c) the attach layer of components; and (d) the dies

B. Complex surface finish (metallisation) types

Another aspect that must be taken into consideration, is some of the advanced surface finishes or metallisation types with crystallised microstructures that can be used to plate the copper on substrates, with the objectives of preventing copper oxidisation and improving compatibility with other processes, e.g., Ag-sinter. This, however, can blur the sinter layer being inspected and mislead interpretation. In Fig. 4 (a), an example of this type of metallisation is shown. The same figure also shows a SiC die sintered on a substrate with the same surface finish (b), which has a purposedly introduced defect. Fig. 4 (c) depicts a top C-scan which reveals great details of the physical features, shapes, and surfaces, as the metallisation is not obstructing, which is not the case for (d), where the surface finish blurs the image, preventing good levels of detail to be achieved, such as in the case of Fig. 3.

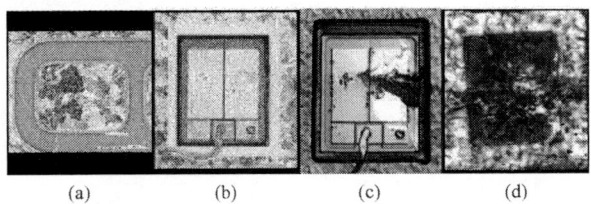

(a)   (b)   (c)   (d)

Fig. 4.   Effects of mixed Ag/Ni/Cu plating on defect detection using C-SAM. Showing: (a) surface finish; (b) die sintered on AMB-Si₃N₄; (c) C-SAM image of the top surface of the die showing a purposedly introduced defect; (d) C-SAM image of the die showing distortion due to surface finish

Obviously, complex surface finishes can pose a major challenge when inspecting Ag-sintered layers and other features, as it makes small defects and details undistinguishable from the blur. Most of the time, the user only has access to the internal structure through the back plate of the module or device (unless, for instance, pin-fins or similar structures are used), hence, this challenge will be present when using similar types of surface finish.

The explanation for this lies in the reflection index of the different elements composing the surface finish metallisation, (e.g., Ag, Cu, Ni, Zn, Au), as reflection index affect the intensity of the echoes at different points, and as a consequence the gradients being imaged vary as well. TABLE III compares the different reflection indexes for the aforementioned materials. The distortion effect caused by the surface finished is magnified significantly when the surface finish is present in both top and bottom sides.

TABLE III. REFLECTION INDEX COMPARISON OF SELECTED MATERIALS, $Z_1 = H_2O$

Material	Z_2 [MNs/m^3]	Reflection [p.u.]	Reflection [p.u., normalised]
Zn	30.1	0.82	0.94
Ag	38.3	0.85	0.98
Cu	44.9	0.87	1.00
Ni	53.8	0.89	1.02
Au	64.0	0.91	1.04

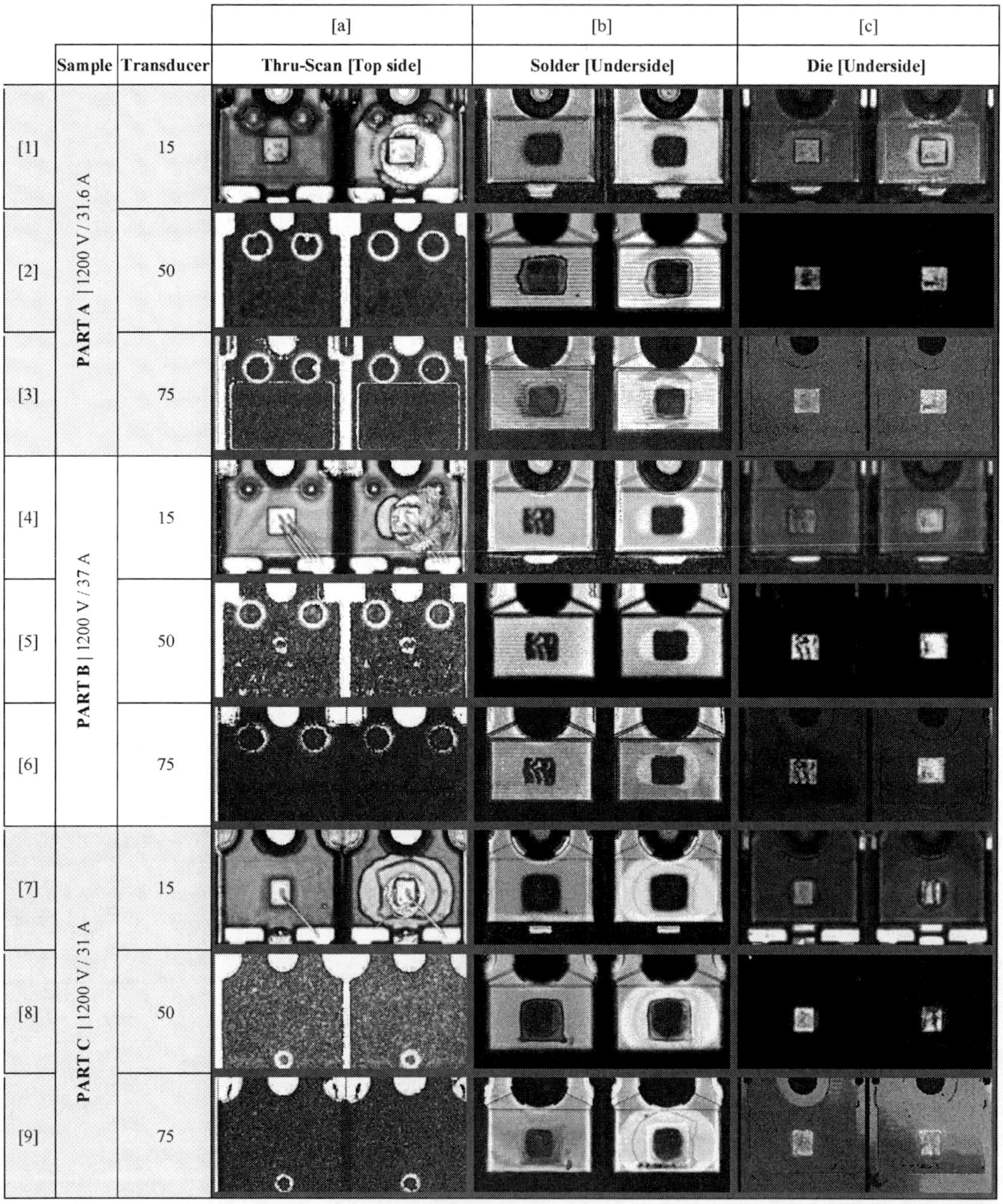

Fig. 5. C-SAM images comparing the effects of different transducers on the image quality, sharpness, and suitability to detect voids, cracks, delamination, and features for three different samples in TO-247 packages, both pristine and failed (short-circuit test). Transducers: 15 MHz | 0.752 in | f1.5; 50 MHz | 1.000 in | f4.0; 75 MHz | 0.750 in | f5.0

C. Defect Detection

As it can be noticed in Fig. 6, a matching transducer must be selected in order to be able to detect defects which are very small. This means, for example, that the small voids shown in Fig. 6 (b) could not have been detected in Fig. 6 (a), which is about 1×1 mm. Furthermore, delamination of the molding compound is shown in (c) for a sample device similar to (a) and (b).

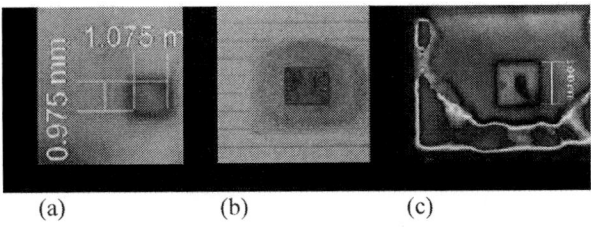

(a)   (b)   (c)

Fig. 6. Examples of poor focus, small voids, and delamination in TO-247 packaged devices. Showing: (a) die; (b) die; (c) delamination effect. Transducer (a), (c) = 15 MHz | 0.752 in | f1.5; (b) = 75 MHz | 0.750 in | f5.0

A broader comparison of different types of defects (including cracks), is presented in Fig. 5. This figure shows two devices with the same part number for every scan. On the left, a pristine device is shown, whilst next to it to the right, a device with the same part number which has failed a short circuit test is also revealed. For example, the image in position [1][a], shows the sample "PART A" scanned with a 15 MHz transducer, and side by side you have a brand new device (left), and a failed device (right). Referring to the same image, the sample on the left reveals minor delamination on the bottom corners of the surface of the die, even though it is new and unstressed. The sample on the right reveals massive changes in the internals of the moulding compound, as a result of the thermal effects generated during short-circuit test. This particular example is sometimes called "popcorn effect" [6], where delamination is initiated and spread very rapidly.

Another example could be the image [5][c] from Fig. 5, which shows a cracked die on the right hand side. The die shown to the left is masked by voiding area in the solder, which can be verified by looking at the solder layer of the same part in positions [4, 5, 6][b]. It is also notoriously evident that the 75 MHz transducer offers much sharper images of the solder layer, whilst the 50 MHz transducer is more suitable for the detection of cracks.

For all three parts, scans through the structure to visualise the internal structures was only possible using the 15 MHz transducer, as the other two transducers being considered were not able to successfully penetrate through the moulding material whilst scanning from the top side.

Similarly, the different images can be observed to compare and interpret other defect types for all three packages. For instance, the image in position [7][a] shows a minor void in the pristine device, and both voids and delamination in the failed device; [8][c] shows a cracked die on the failed device; [6][b] shows significant voiding in the solder layer.

It is worth mentioning that the different scans most be correlated to accurately distinguish in which layer the defects are really found. Also, the performance of the different transducers to detect the different types of defects is not universal, and greatly depend on the materials and geometries involved.

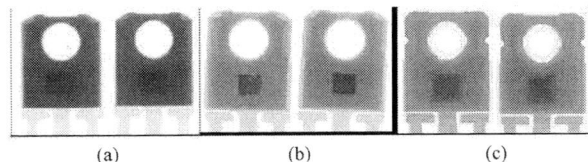

(a)   (b)   (c)

Fig. 7. X-Ray images of sample parts A, B, and C, respectively | 120 kV 10.8W

The Fig. 7 provides X-Ray images of the same parts provided in Fig. 5. Although very shy features can be recognised (such as the voiding area in the solder layer of the "PART B" to the left, in Fig. 5), the level of detail is no were near to the good results that can be achieved using C-SAM. Also, X-Ray is not effective at all at identifying delamination, cracks, or very small voids in the packages.

The reason for this is that the X-Ray is a transmission imaging technique, and the level of energy that is required to penetrate the copper is the same energy that penetrates through everything in the structure, making other materials negligible.

D. Large area sintering

Another important aspect is the inspection of large areas of sintered materials, such as AMB substrates, or very large dies. Fig. 8 (a) depicts what intuitively appears to be a crack, however, it happens to be a "teardrop" resulting from the moisture in a large area wet-placed substrate. Another example of the same situation is shown in (b). Fig. 8 (c) shows non-uniformity in density of the Ag-layer for a dry-placed substrate, demonstrating a process issue related to dispensing. The same issues related complex surface finishes would also apply, should it have been utilised in the substrates.

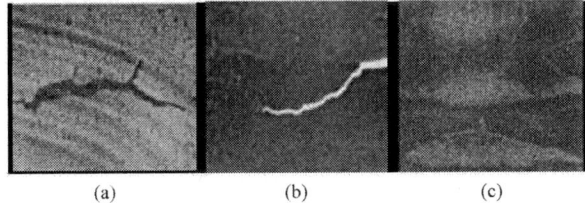

(a)   (b)   (c)

Fig. 8. Defects in large area Ag-sinter layer of substrates: (a), (b) "teardrop"; (c) non-uniform dispensing (manual)

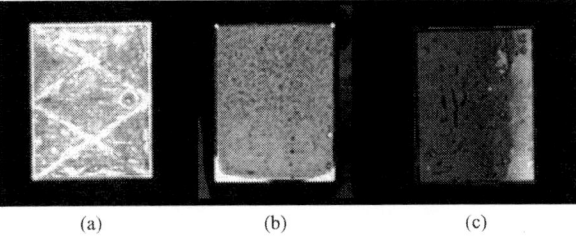

(a)   (b)   (c)

Fig. 9. Issues related to layer identification, dispensing, and lack of co-planarity. Showing: (a) purposely introduced pattern on sinter layer; (b) uncovered areas in the bottom corners; (c) non-uniform sintering, related to tooling issues

Fig. 9 (a) demonstrates an important technique which is possible during the design and development process: to purposedly introduce defects to ensure the right layer is being inspected. While this might sound trivial, it can greatly help set-up and optimise the right settings before massively scanning parts during production. Having done this, Fig. 9 (b) and (c) have been obtained, revealing issues with covering the whole sintering area (dispensing and placing issue), and also the lack of co-planarity, most likely due to problems with the head of the sintering tool.

The final example shows a large area of a commercially available power module. Fig. 10 (a) depicts the solder layer of the substrate to the copper plate. Several voids and poor dispensing in the bottom corners can be seen. Fig. 10 (b) shows the dies on the substrate, which are blurred by the surface finish of the substrate. Also, here some of the voids observed in Fig. 10 (a) can be co-related, as they appear as black dots or patterns. Lastly, Fig. 10 (c) shows an X-Ray image of the same section of the part, for the purposes of comparison.

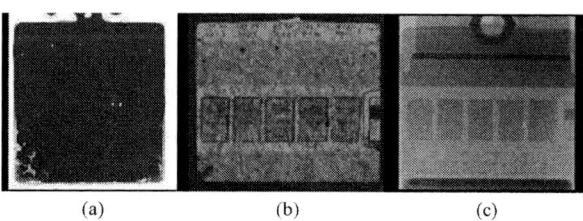

(a) (b) (c)

Fig. 10. C-SAM of an off the shelf SiC power module, 1200 V / 450 A. Showing: (a) solder layer; (b) dies; (c) X-Ray image of the same sample area

IV. CONCLUSION

Several common challenges and defects related to Ag-sintered components have been captured and demonstrated by means of scanning acoustic microscopy (SAM), following a thorough explanation of the background theory behind SAM, which is not always provided from the practical aspects.

The experiments revealed that C-SAM can be far superior compared to X-Ray when aiming at the attach layers, however, its limitations rely on the geometries involved and the ability of matching transducers that can interact well with the materials and the structures.

Also, it has been proven that SAM can be very effective at identifying all three main types of defects that affect performance and degradation in power electronic modules, and hence it is an ideal tool for inspecting Ag-sintered interfaces.

Complex surface finishes have been presented as a major challenge to the inspection of Ag-sintered layers, mainly because they can obscure the ability to distinguish defects, especially voids and cracks.

Finally, this study has demonstrated that SAM can be extremely effective for the inspection of Ag-sintered layers for both small and large area dies or substrates, and that issues related to the sintering process can be identified, such as dispensing issues or lack of co-planarity of the tooling.

ACKNOWLEDGEMENTS

The authors wish to acknowledge the support of Andy Walford, George Rowley, and Lewis Stroud, for their technical assistance and knowledgeable discussions at the premises of Custom Interconnect, Ltd.

REFERENCES

[1] P. Dreher, R. Schmidt, A. Vetter, J. Hepp, A. Karl, and C. J. Brabec, 'Non-destructive imaging of defects in Ag-sinter die attach layers – A comparative study including X-ray, Scanning Acoustic Microscopy and Thermography', *Microelectronics Reliability*, vol. 88–90, no. May, pp. 365–370, 2018, doi: 10.1016/j.microrel.2018.07.121.

[2] H. Yu, 'Scanning acoustic microscopy for material evaluation', *Appl Microsc*, vol. 50, no. 1, 2020, doi: 10.1186/s42649-020-00045-4.

[3] Y. C. Jang, H. E. Kim, A. Schuck, and Y. S. Kim, 'Developed non-destructive verification methods for accelerated temperature cycling of power MOSFETs', *Microelectronics Reliability*, vol. 128, Jan. 2022, doi: 10.1016/j.microrel.2021.114442.

[4] M. Kobayashi, K. Sakai, K. Sumikawa, and O. Kikuchi, 'Signal Processing Method for Scanning-Acoustic-Tomography Defect Detection based on Correlation between Ultrasound Waveforms'.

[5] C. Uhrenfeldt, S. Munk-Nielsen, and S. Bęczkowski, 'Frequency domain scanning acoustic microscopy for power electronics: Physics-based feature identification and selectivity', *Microelectronics Reliability*, vol. 88–90, no. May, pp. 726–732, 2018, doi: 10.1016/j.microrel.2018.07.043.

[6] A. Roy, 'Instrumentation for Studying Real-time Popcorn Effect in Surface Mount Packages during Solder Reflow', *International Journal of Electronics Science and Engineering*, vol. 7, pp. 1126–1130, 2013.

Characterization of a Novel Cost-efficient and Environmentally Friendly Graphene-enhanced Thermal Interface Material

Sihua Guo
SMIT Center, School of Mechatronics
Engineering and Automation
Shanghai University
Shanghai, China
shguo@shu.edu.cn

Kristoffer Harr (Martinsen)
SHT Smart High-Tech AB
Arendals Allé 3, SE-418 79
Gothenburg, Sweden
kristoffer.martinsen@smarthightech.com

Amos Nkansah
SHT Smart High-Tech AB
Arendals Allé 3, SE-418 79
Gothenburg, Sweden
Amos.Nkansah@smarthightech.com

Jiajia Chen
SMIT Center, School of Mechatronics
Engineering and Automation
Shanghai University
Shanghai, China
guyuechuqi@shu.edu.cn

Zhiyang Shen
SMIT Center, School of Mechatronics
Engineering and Automation
Shanghai University
Shanghai, China
szy-gg@shu.edu.cn

Murali Murugesan
SHT Smart High-Tech AB
Arendals Allé 3, SE-418 79
Gothenburg, Sweden
murali.murugesan@smarthightech.com

Hongfeng Zhang
SHT Smart High-Tech AB
Arendals Allé 3, SE-418 79
Gothenburg, Sweden
hongfeng.zhang@smarthightech.com

Lars Almhem
SHT Smart High-Tech AB
Arendals Allé 3, SE-418 79
Gothenburg, Sweden
lars.almhem@smarthightech.com

Arto Ahtonen
SHT Smart High-Tech AB
Arendals Allé 3, SE-418 79
Gothenburg, Sweden
arto.ahtonen@smarthightech.com

Jin Chen
Shanghai Ruixi New Materials High
Tech Co. Ltd.
No 818, Chuhua North Road
Shanghai, China
chenjin@ruixi.name

Johan Liu
SMIT Center, School of Mechatronics
Engineering and Automation, Shanghai
University

&

Department of Microtechnology and
Nanoscience, Chalmers University of
Technology
johan.liu@chalmers.se

Abstract—**With the continuous development of electronic devices, effective heat dissipation has become a major factor affecting service life. Thermal interface materials (TIM) play a key role in controlling heat dissipation of electronic devices and have thus attracted widespread attention. In this study, we used graphene flakes (GF) derived from graphene film that is wasted during the preparation of commercial large-scale graphene-enhanced TIMs as thermally conductive fillers to formulate a new TIM. The thermal conductivity of the developed TIM is 50% higher with GFs than without. Furthermore, the TIM has a tensile strength of 0.46 MPa with an elongation at break of 1225%, a maximum compression strength of 0.64 MPa at 50% compression, and high mechanical cycle stability. This report provides a cost-efficient and environmentally friendly approach to producing high-performance TIMs for electronic cooling applications.**

Keywords—Thermal interface material, Graphene flakes, Cost-efficient, Environmentally friendly

I. INTRODUCTION

The continuous miniaturization and integration of electronic devices lead to a large amount of generated heat. Therefore, fast and effective heat dissipation is a key factor affecting the long-term and reliable operation of electronic equipment [1]. Increasing heat transfer at the interface is an important part of thermal management. The surfaces of rigid objects cannot achieve close contact owing to the tiny concave-convex surface features, in which remaining gaps are filled with air. Therefore, a thermal interface material (TIM) is generally used to fill the gaps between the contact surfaces of two heterogeneous materials and improve heat dissipation [2]. Silicone rubber (SR) is often used as a matrix polymer because of its softness, high elasticity, and thermal stability. However, the thermal conductivity of SR is low, which cannot meet the heat dissipation requirements of modern electronic devices [3]. Traditionally, silicone pads are made of SR loaded with thermally conductive fillers such as metals (including Ag, Cu, Ni, Au), ceramics (including boron nitride (BN), Al_2O_3, aluminum nitride (AlN)), carbon allotropes (including graphite, diamond, carbon nanotubes, graphene), or hybrid fillers.

Recently, fabricating a 3D continuous thermal network in polymers has become a strategy for enhancing heat dissipation in electronic devices [4]. Some researchers have introduced the 3D skeleton into polymers through freeze-drying, high-temperature heat treatment, and impregnation processes. Zhang et al. [5] used graphene aerogel to improve the thermal conductivity of SR. The thermal conductivity of the composite reached up to 1.26 $Wm^{-1}K^{-1}$ at a low filler loading of 0.5 wt%. Xue et al. [6] achieved a thermal conductivity of 0.88 $Wm^{-1}K^{-1}$ in SR modified with BN and reduced graphene sheets. In addition to constructing a skeleton by freeze-drying, Zhao et al. [7] prepared graphene foam on a nickel foam template by chemical vapor deposition and then embedded this foam in carbon fiber/polydimethylsiloxane. The thermal conductivity of the composite reached 0.55 Wm^{-1}K^{-1}. Although these methods have been advantageous for producing interconnected heat transfer networks, several limitations remain. Notably, the cumbersome, expensive, and energy-consuming processes pose a significant challenge for industrial production. It is necessary to develop a cost-effective, simple, green, and scalable strategy to construct high-performance TIMs.

Owing to its beneficial thermal and mechanical properties, graphene has been widely used in the field of heat dissipation of electronics, such as the batteries in smartphones [8]. However, the large-scale use of graphene films generates a large amount of graphene-based waste. Therefore, it is desirable to reduce or reuse this waste. In this study, we formulated a new TIM using graphene flakes (GF) as thermally conductive fillers, which were derived from graphene film waste generated during the preparation of commercial large-scale graphene-enhanced TIM. The thermal conductivity of the new TIM is 1.2 Wm^{-1}K^{-1}, which is 445% and 50% higher than that of the silicone matrix and TIM without GFs, respectively. Additionally, the TIM has a tensile strength of 0.46 MPa with an elongation at break of 1225%, maximum compression strength of 0.64 MPa at 50% compression, and high mechanical cycle stability. This represents a novel approach for producing cost-efficient and environmentally friendly high-performance TIMs for electronic cooling applications.

II. EXPERIMENTS

2.1. Materials

In this work, graphene waste from SHT Smart High-Tech AB, Sweden was used. BN and AlN (10 μm) were obtained from Sigma-Aldrich. SR is formed by mixing a certain proportion of silicone gel A and B, which were purchased from Wacker Chemical. All of the reagents were analytical grade.

2.2. Preparation of graphene flakes

Graphene film waste was processed in a crusher for several seconds to obtain GFs with a relatively uniform texture.

2.3. Preparation of the TIM

First, silicone gel A and B were mixed in a ratio of 1.5:1 and a certain amount of AlN and BN powders were introduced. Then, GFs were added and mixed evenly using a high-speed mixer. The mixture was kept under vacuum for 30 min to remove air bubbles and subsequently poured into a mold. Finally, the sample was placed into an oven at 100 °C for 5 h to obtain the resulting TIM (GT-R/ABG). The fabrication of TIM without GFs (GT-R/AB) was the same, except GFs were not added. The mass fraction of GFs in GT-R/ABG was 1%, 2%, 3%, 4%, and 5%, abbreviated as GT-R/ABG1, GT-R/ABG2, GT-R/ABG3, GT-R/ABG4, and GT-R/ABG5, respectively.

2.4. Characterization

The size distribution of the GFs was characterized using a laser particle size analyzer (Malvern Mastersizer 3000, UK). The microstructure of the GFs was visualized by transmission electron microscope (TEM) (FEI Tecnai F20, USA). The structure of GFs was determined by X-ray diffraction (XRD) (Rigaku Ultima IV, Japan) and Raman spectroscopy (Horiba LabRAM HR Evolution, Japan). The cross-sectional morphology of the TIM was investigated by field emission scanning electron microscopy (FE-SEM) (Helios G4 PFIB). The out-of-plane thermal conductivity of the TIM was determined using a TIM thermal resistance & conductivity measurement apparatus, which is calculated using the following equation:

$$K=BLT/R \qquad (1)$$

Where R is the total thermal resistance, BLT is the bond line thickness of the sample, and K is the thermal conductivity of the sample. The mechanical properties of the TIM were evaluated using a universal testing machine (Instron, USA).

III. RESULTS AND DISCUSSIONS

GFs were obtained by crushing the graphene film waste in a crusher for several seconds. High-resolution transmission electron microscopy (HR-TEM) was used to characterize the microstructure of the resulting GFs. As shown in Fig. 1a, the GFs exhibit a wrinkled tulle shape, and their size is mainly distributed around 35 μm (Fig. 1c). A high-resolution TEM image of the edge morphology reveals that the GFs are composed of six layers of graphene (Fig. 1b). The corresponding XRD and Raman spectra of the graphene film waste and GFs were similar (Fig.1d and e), which indicates that the pulverization process does not change the characteristic structure of the graphene [9].

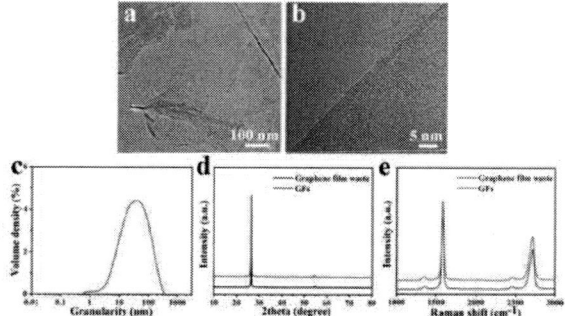

Fig. 1. (a) TEM top view of the surface of GFs. (b) High-resolution TEM image of the edge morphology of GFs. (c) Size distribution diagram of GFs. (d) XRD spectra and (e) Raman spectra of graphene film waste and GFs.

To observe the microstructure of the TIM and the filler distribution, the fractured surface of the TIM was characterized by FE-SEM. The cross-sectional morphology of the silicone matrix is relatively smooth, as a result of brittle fracture (Fig. 2a). However, the introduction of AlN and BN particles changes the morphology of the matrix (Fig. 2b). The corresponding elemental mapping of the image in Fig. 2b is shown in Fig. 2c, confirming that the B, Al, and N are uniformly dispersed in the matrix. Fig. 2(d-h) shows the cross-sectional morphology changes of GT-R/ABG with GF contents ranging from 1 to 5 wt%. The introduction of tulle-like GFs can effectively connect isolated AlN and BN particles. However, GFs begin to aggregate when their content is higher than 2 wt%. At 5 wt%, the degree of aggregation significantly increases, which is confirmed by the corresponding elemental analysis, showing that the C content in red circles is extremely high (Fig. 2i).

310

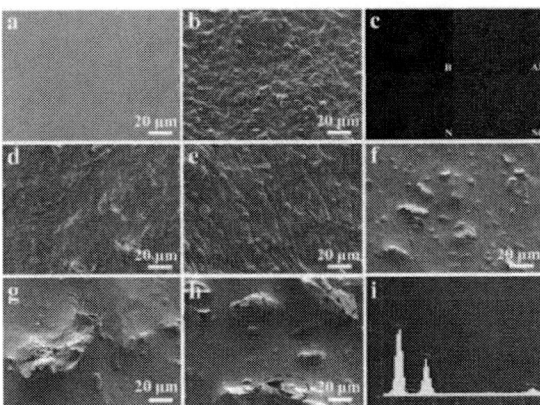

Fig. 2. (a) SEM image of the silicone matrix. (b) SEM image of the cross-section morphology of GT-R/AB. (c) Elemental mappings of (b). (d-h) SEM images of the cross-sectional morphology of GT-R/ABG with different GF contents: (d) 1 wt%, (e) 2 wt%, (f) 3 wt%, (g) 4 wt%, and (h) 5 wt%. (i) Elemental analysis of agglomerations shown in red circles in (h).

In addition to high thermal conductivity, TIMs also need high flexibility and a low elastic modulus to form complete contact between the rough interfaces. GT-R/ABG returns to its original shape after stretching and twisting (Fig. 3a). Fig. 3b shows the stress-strain curves of the TIM. The addition of AlN and BN improves the tensile strength of the silicone matrix. After introducing GFs, the stress applied on GT-R/ABG is uniformly dispersed, which is beneficial to avoid stress concentration. However, when the GF content is excessive, the aggregation of GFs results in a decrease in the tensile strength of GT-R/ABG. Furthermore, the fillers can also increase the compression strength of the silicone matrix. GT-R/AB has an ultra-high compression strength because of the high hardness of BN and AlN particles. However, after introducing GFs, the stress was uniformly dispersed by a small number of GFs, showing a relatively low compressive strength. Higher GF contents caused an increase in the compressive strength of GT-R/ABG (Fig. 3d). Therefore, the GF content has an optimal value, which is 2 wt%, where the GT-R/ABG tensile strength, elongation at break, and compression strength at 50% compression were 0.46 MPa, 1225%, and 0.64 MPa, respectively.

The mechanical stability of GT-R/ABG was evaluated by sequentially performing multiple tensile tests (10 cycles) at 50%, 100%, and 150% maximum strain (Fig. 3c). In the first cycle of every test, the SR molecule chains absorbed on the surfaces of the fillers were stretched, for which the curves showed higher energy loss coefficients. In the second cycle, the molecular chains did not have time to reabsorb on the surface of the filler, resulting in a significant shrinkage of the area enclosed by the curve. Then, subsequent cycles nearly overlap with the second cycle, demonstrating the mechanical stability of the GT-R/ABG [10].

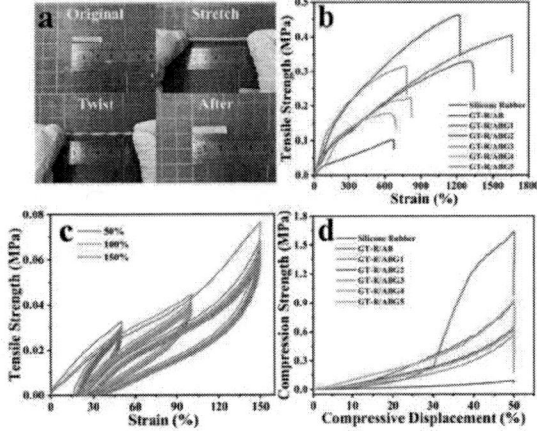

Fig. 3. (a) Optical images of the stretch and twist processes of GT-R/ABG. (b) Stress-strain curves of all samples. (c) Stress-strain curves of GT-R/ABG at the maximum strain of 50%, 100%, and 150%, repeated for 10 cycles. (d) Compression strength of all samples at the maximum compression of 50%.

Fig. 4a showed the thermal conductivity of GT-R/ABG as a function of the GF content. GT-R/ABG2 achieved an excellent thermal conductivity of 1.2 $Wm^{-1}K^{-1}$, which was 445% and 50% higher than that of the pure SR and GT-R/AB, respectively. This TIM outperforms several commercially available thermal pads, as well as some graphene-enhanced TIMs [6,7]. The isolated AlN and BN particles randomly dispersed in GT-R/AB cannot be fully interconnected, and thus, the thermal conductivity of GT-R/AB is limited by the lack of an effective heat conduction network, and severe phonon interface scattering between the fillers and matrix (Fig. 4b). In contrast, the isolated AlN and BN particles can be bridged by GFs, resulting in a more effective heat conduction network and a sharp increase in the thermal conductivity (Fig. 4c). However, excessive GF content leads to its agglomeration, which increases the interfacial thermal resistance between the matrix and fillers, resulting in a decrease in the thermal conductivity of GT-R/ABG.

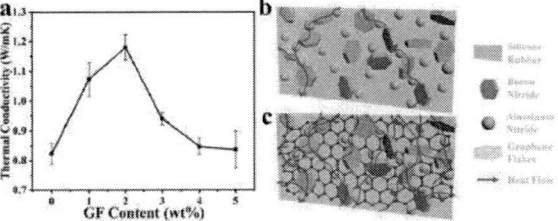

Fig. 4. (a) Variation of thermal conductivity of GT-R/ABG with different GF contents. (b, c) Proposed mechanistic model of the thermal conduction paths for (b) GT-R/AB and (c) GT-R/ABG.

IV. CONCLUSIONS

A new TIM was developed using graphene film waste in a silicone matrix in a simple, cost-effective, and environmentally friendly way, exhibiting beneficial thermal and mechanical properties. The thermal conductivity of TIM reached 1.2 $Wm^{-1}K^{-1}$, which is 445% and 50% higher than that of the silicone matrix and the TIM without GFs, respectively. Furthermore, the TIM displayed a tensile strength of 0.46 MPa, elongation at break of 1225%, a maximum compression strength of 0.64 MPa at 50% compression, and high mechanical stability. This new high-performance TIM has the potential to be applied to the thermal management of modern electronic devices.

ACKNOWLEDGMENT

We acknowledge the financial support from the National Natural Science Foundation of China (No: 51872182), the financial support from the Swedish National Science Foundation with the Contract No: 621-2007-4660, Vinnova SIO Grafen as well as from the Production Area of Advance at Chalmers University of Technology, Sweden, and the financial support from Vinnova SIO Grafen program.

REFERENCES

[1] S. Guo, R. Zheng, J. Jiang, J. Yu, K. Dai, C. Yan. Enhanced thermal conductivity and retained electrical insulation of heat spreader by incorporating alumina-deposited graphene filler in nano-fibrillated cellulose. Compos Part B-Eng. 2019;178:107489.

[2] J. Hansson, T. M. J. Nilsson, L. Ye, Liu J. Novel nanostructured thermal interface materials: a review. International Materials Reviews. 2017;63(1):22-45.

[3] D. Yang, S. Huang, M. Ruan, S. Li, J. Yang, Y. Wu, et al. Mussel Inspired Modification for Aluminum Oxide/Silicone Elastomer Composites with Largely Improved Thermal Conductivity and Low Dielectric Constant. Industrial & Engineering Chemistry Research. 2018;57(9):3255-3262.

[4] M. Loeblein, S.H. Tsang, M. Pawlik, E.J.R. Phua, H. Yong, X.W. Zhang, et al. High-Density 3D-Boron Nitride and 3D-Graphene for High-Performance Nano–Thermal Interface Material. ACS Nano. 2017;11(2):2033-2044.

[5] W. Zhang, Q.-Q. Kong, Z. Tao, J. Wei, L. Xie, X. Cui, et al. 3D Thermally Cross-Linked Graphene Aerogel-Enhanced Silicone Rubber Elastomer as Thermal Interface Material. Advanced Materials Interfaces. 2019;6(12):1900147.

[6] Y. Xue, H. Wang, X. Li, Y. Chen. Synergy boost thermal conductivity through the design of vertically aligned 3D boron nitride and graphene hybrids in silicone rubber under low loading. Materials Letters. 2020;281:128596.

[7] Y-H Zhao, Y-F Zhang, S-L Bai, X-W Yuan. Carbon fibre/graphene foam/polymer composites with enhanced mechanical and thermal properties. Compos Part B-Eng. 2016;94:102-108.

[8] S. Guo, J. Chen, Y. Zhang, J. Liu. Graphene-Based Films: Fabrication, Interfacial Modification, and Applications. Nanomaterials. 2021;11(10):2539.

[9] B. Shen, W. Zhai, W. Zheng. Ultrathin Flexible Graphene Film: An Excellent Thermal Conducting Material with Efficient EMI Shielding. Adv Funct Mater. 2014;24(28):4542-4548.

[10] Q. Hu, X. Bai, C. Zhang, X. Zeng, Z. Huang, J. Li, et al. Oriented BN-Silicone rubber composite thermal interface materials with high out-of-plane thermal conductivity and flexibility. Composites Part A: Applied Science and Manufacturing. 2022;152:106681.

Evolution of Getter Technology in Electronic Hermetic Packaging

Luca Mauri
Getters and Dispenser development laboratory
SAES Getters S.p.A.
Lainate, Italy
luca_mauri@saes-group.com

Giovanni Zafarana
Getters and Dispenser development laboratory
SAES Getters S.p.A.
Lainate, Italy
giovanni_zafarana@saes-group.com

Enea Rizzi
Getters and Dispenser development laboratory
SAES Getters S.p.A.
Lainate, Italy
enea_rizzi@saes-group.com

Alessio Corazza
Getters and Dispenser Business Unit
SAES Getters S.p.A.
Lainate, Italy
alessio_corazza@saes-group.com

Abstract— **Hermetic packaging is a well-established technology used for sealing electronic and optoelectronic devices. A reliable and enabling way to preserve the proper stable atmosphere conditions in hermetic packages is the integration of engineered getter solutions, specifically designed for the characteristics of different packaged systems. Getter solutions may vary from sintered, compressed or sputtered metallic alloys for vacuum sealed packages to dispensable hybrid organic-inorganic materials for gas back filled devices. The paper will present the main features of getter materials suitable for different type of packages and will show their efficiency in absorbing detrimental gases in the devices.**

Keywords— hermetic packaging, getter, gas sorption, gaseous contaminants.

I. INTRODUCTION

Hermetic packaging is a standard technology used for defense, aerospace and telecom applications. The main function of hermetic sealing is to protect the device from external atmosphere and harsh environments, ensuring higher performances and ensuring longer lifetime. The hermetic sealing can be obtained in vacuum or in gas back filling conditions, depending from applications and device features. Nevertheless, in both approaches, hermeticity alone is not always able to guarantee a stable and controlled internal atmosphere needed for proper operation of the devices over their lifetime. Indeed, the gas release inside the package due to internal outgassing or gas generation can significantly change the vacuum level or the back filled gas composition, leading to performance degradation or, in the worst cases, failures [1]. In order to manage such potential issues, a reliable and well proven solution can be the integration of getter materials, that can actively absorb the gaseous contaminants. There are different class of getters that can be integrated: Non Evaporable Getters (NEGs) or Dispensable Getters. NEGs are typically used in devices where vacuum is required. Vacuum level may vary in the range from $1 \cdot 10^{-7}$ mbar to 1.0 mbar. Typical devices that operate under vacuum conditions are X-Ray tubes, vacuum interrupters, cooled or uncooled bolometers, photomultipliers, MEMS (accelerometers, gyroscopes, microbolometers), microwave modules. Getter solutions specifically designed for these class of devices are based on metallic alloys that are milled in powders with different size and the powder can be compressed or sintered to constitute pills, or other structures, of different size: typically few millimeters in diameter and height.

Due to the miniaturization trend and shrinking of dimensions of devices in the electronic industry, new getter solutions have been developed that are based on 2D shapes: sintered powder onto metallic sheets to constitute strips or ribbons with height in the range of hundreds of microns.

MEMS sensors are even more challenging, as miniaturization is in the millimeter or sub-millimeter range, an entirely different type of getter has been designed to be integrated in such small, micron-level dimensions. The getter material is a thin sputtered metallic alloy with thickness in the range of few microns. All these types of getters are able to absorb many different gases like, H_2O, H_2, CO, CO_2, N_2, O_2 and hydrocarbons that are typically left inside the cavity of devices after the hermetic sealing process. The getter material gets rid of these contaminants allowing good vacuum levels inside the package and keeping it through the entire device lifetime, assuring not only the performances of the sensor, but also prolonging the stability and reliability over time [2, 3].

Inside opto and photonic devices, instead, gaseous contaminants present in the filling gas may be responsible of device malfunctioning and performance degradation. The main gaseous species that can be harmful for hermetic sealed devices are moisture, hydrogen and volatile organic compounds (VOCs).

In optical transceiver modules, hydrogen and moisture are the main gas contaminants to be removed. Critical levels of these gases are 1000 ppmv for H_2 and 5000 ppmv for H_2O.

There are several problems induced by the presence of harmful gases. For instance, H2 is responsible of electrical performance degradation, it can diffuse through metal layers of active components, causing shifts in currents and trans-conductance [4, 5], it can reacts with surface oxides inside hermetic packages promoting the formation of moisture [6].

Moisture, instead, can be responsible of electrical shorting or corrosion of solder joints. Photodetectors components may be affected by water from dark-current increase [7].

In laser diodes-based devices, like transmitters, the gases that are considered harmful are water and VOCs. The main issues are related to signal attenuation because of moisture or organic gas condensation.

SAES has been developing for many years engineered sorbing materials for optoelectronic and photonic packaging, combined with special polymeric matrixes and with solventless formulation, easy to be directly dispensed and cured on packaging components or sub-components.

The paper presents the functional performances of two classes of getter materials used in vacuum sealed electronic devices and in gas filled optoelectronic ones. The first class is the so-called "Non Evaporable Getters" family that can interact with many gaseous contaminants, the second one belonging to the ZeDry™ getter family is specifically engineered for sorption of H_2O+H_2, $H_2O+VOCs$ or H_2O only, depending on the specific needs.

II. NON EVAPORABLE GETTERS

Metallic alloys-based getters have been developed in order to interact and absorb many different gas molecules that are generally present inside electronic devices that are hermetically sealed.

The getter surface chemically interact with molecules like O_2, CO_2, CO, N_2, H_2, CH_4 and HCs thus removing them from the inner volume of the package. The interaction is irreversible with all the previous gas species, but hydrogen. Indeed, H_2 gas is the only molecule that diffuses into the bulk of the material forming a solid solution. By heating the getter at elevated temperatures, like 600-900°C, H_2 can be released from the material. Typical operating temperatures of devices are well below such values and therefore the risk of H_2 release is negligible.

One of the main features of getter for electronic devices is the porous structure that allows the gases to diffuse into the inner parts of the getter body, thus optimizing the sorption performances of the material. Metallic alloy powder with selected grain size is compressed and eventually sintered in different size and shapes. Typical products are pills or disks like, 4-10 mm diameter and thickness in the range of 1-6 mm.

These bulky getters are integrated in devices where the available inner space is pretty large, like X-Ray tubes, uncooled bolometers or evacuated tube solar collectors.

In order to absorb gases, the getter material needs to be thermally activated. Heating at high temperature the alloy, the native oxides present on the material surface start diffusing into the bulk, leaving a metallic surface that is reactive and can effectively absorb molecules. Depending on the application and on the selected getter material, the activation temperatures range from 300°C up to 900°C. The activation of getters can take place directly during the vacuum sealing process of the device, if the sealing temperature is compatible and allows a proper activation of the material. In this approach, it is possible to manage both processes simultaneously, thus improving efficiency and reduce the overall manufacturing time. Getter configurations that can be integrated and activated directly during the sealing process are illustrated, as examples, in Fig 1.

Another possible approach to activate the getter is by electrical current. In this alternative solution, the material is not activated during the sealing, but afterwards. A metallic heater is integrated into the getter body in order to be used to heat, by Joule effect, the getter material at the desired temperature (Fig.2).

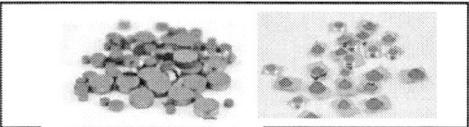

Fig. 1. Getter solutions that can be directly activated during the vacuum sealing process of the device.

Fig. 2. Example of getter solutions that can be activated by electrical current.

Both options allow proper getter functionality and gas absorption capacity. The selection of the proper configuration is mainly related to the specific device architecture and manufacturing process.

MEMS devices require a completely different solutions from the getter point of view because of the miniaturized dimensions. In this case the getter material is integrated at wafer level. It is a thin film deposited by physical vapor deposition directly onto the cap wafer that is used to vacuum seal the opposite wafer where the sensors are manufactured.

The getter thickness is in the micrometer range and the material can be patterned and shaped according to the specific MEMS design (Fig.3).

The getter activation takes place during the hermetic vacuum sealing process of the wafers. The sketch below (fig.4) schematically shows the sealing process of MEMS with the simultaneous getter activation.

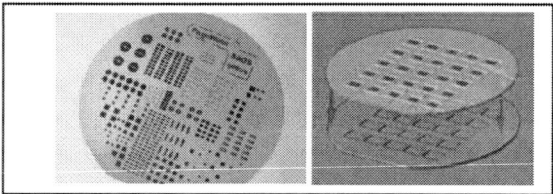

Fig. 3. Thin film getter deposited at wafer level.

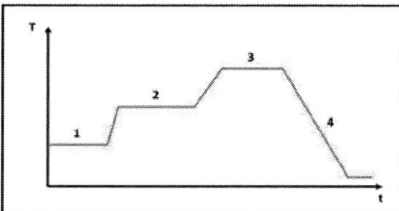

Fig. 4. Vacuum sealing process and getter activation procedure for MEMS: (1)outgassing, (2) getter activation, (3)wafers bonding, (4) cool down.

The first step is to heat the wafers at moderate temperature in order to promote the release of gases from the surfaces, cleaning up as much as possible the wafer before bonding (step 1). The second step is related to the getter activation: the temperature is increased in order to allow the diffusion of native oxides from the surface into the bulk of getter material.

Once the getter is active, the bonding process can take place leading to the hermetic sealing of the parts (sensors wafer and cap wafer).

Regardless the getter material, configuration, size and manufacturing process, the final purpose of the getter integration is the absorption of gases inside a hermetically vacuum sealed device. The following table highlights the residual gases inside an electronic device that integrates or not a getter material (Tab. 1).

This example refers to the residual atmosphere of two identical microbolometers, one without getter and one with getter inside. The getter solution is based on the thin film (about 2 μm thick getter film). The table shows that the getter is able to reduce the total pressure by 3 orders of magnitude or more, absorbing efficiently the active gases (H_2, CO, N_2, CO_2) and significantly reducing at very small levels the CH_4 and HCs partial pressures.

TABLE I.   RESIDUAL GASES INSIDE MICROBOLOMETER, WITH AND WITHOUT GETTER (HCS=HYDROCARBONS, NGS=NOBLE GASES).

Gas	No Getter Residual Pressure (mbar)	With Getter Residual Pressure (mbar)
H_2	$4.9 \cdot 10^{-1}$	-
CO	$5.9 \cdot 10^{-1}$	-
N_2	-	-
CH_4	$2.8 \cdot 10^{-1}$	$3.0 \cdot 10^{-4}$
H_2O	-	-
O_2	-	-
CO_2	7.2	-
*HCs	$2.6 \cdot 10^{-1}$	$2.5 \cdot 10^{-6}$
**NGs	$1.1 \cdot 10^{-4}$	$7.7 \cdot 10^{-4}$
TOTAL	8.8	$1.1 \cdot 10^{-3}$

III.   ZEDRY-FAMILY

A. Products description and main features

Dimensions of optoelectronic devices are typically small and the technological trend is always to get smaller and smaller footprint and volumes.

Consequently, the available space for integration of components and materials is always limited.

In order to accomplish to this trend, the engineered getters materials have been developed in the form of thick film that can be directly applied to the device lid, or eventually to a subcomponent of the package.

Integration can be effectively achieved by dispensing the getter materials on lids and patterning the shape according to specific geometrical requirements (Fig.5).

Fig. 5. Example of metal lids with ZeDry® getters coating.

In table 2 we reported the main characteristics and process information of ZeDry® family products.

After dispensation, a thermal treatment is necessary to cure and consolidate the material. The active getter material that reacts with the gas to be removed (H_2, VOC or H_2O) is dispersed into a polymeric matrix without the need of using solvents.

The solventless formulation ensures that no additional, undesirable gases are generated and released inside the device because of the getter formulation.

TABLE II.   ZEDRY® FAMILY FEATURES

Formulation
* Solventless formulations: no outgassing issues
* Getter coatings can be handled in ambient air
* High thermal resistance: decomposition temperatures > 300 °C
* Compatible with laser/seam welding sealing processes
Process Integration:
* Getter coatings applied on lids or other components
* Thermally cured getter coating with high mechanical stability
* Substrates: kovar lids, ceramic, glass
Getter activation conditions prior sealing
* Vacuum – nitrogen – dry air
* Temperature range: 100 °C – 200 °C
* Time: few hours

The getter composition is compatible with the typical temperatures of the hermetic sealing processes, like laser and seam welding or soldering with eutectic materials, i.e. AuSn preforms. Different substrates, like kovar, plated metals, ceramics and glass can be used: the getter coating has good adhesion and excellent mechanical stability.

To properly work, the getter must be activated by means of a thermal treatment that can be carried out in vacuum, nitrogen or dry air conditions, immediately before sealing the package. Usually, the baking process applied to degas the system components at 105-200°C for a few hours before the final sealing is sufficient to fully activate the getter.

Based on the hermetic package configuration and application, ZeDry® products can be selected in order to interact with the main detrimental gases that are present in the device. Tailored materials were developed for absorption of moisture and hydrogen, or moisture and VOCs, or moisture only if necessary. Depending on the gas species, the interaction can be reversible or irreversible: for H_2 gas the sorption is irreversible, while for VOCs and H_2O the adsorption occurs in a reversible way.

B. ZeDry®-H_2: Getter solution for H_2 and H_2O

ZeDry®/H_2 is a high capacity, solventless, thermally curable, dispensable hydrogen and moisture getter.

It has been designed for use in hermetically sealed optoelectronic devices, microelectronic devices and semi-hermetic packaging of optical and electronic modules. ZeDry®/H_2 works as an irreversible hydrogen getter and as a reversible moisture getter. This product has been engineered also to avoid bleeding tendency on metallic substrates, especially with Au-plated ones.

The specific sorption performances of the ZeDry®/H₂ getter material are reported in Figure 6. The left red graph shows the sorption curve for hydrogen measured at room temperature. Hydrogen uptake signal is followed in a continuous way for 24 hours. Most of the H2 quantity is absorbed after 6 hours of exposure.

The right blue graph highlights the water uptake at room temperature of the ZeDry®/H₂ getter. The kinetics in this case is even faster, since the most uptake of water takes place within 1.5 hours. The nominal hydrogen and moisture capacity are 40 Ncm³/g and 13%wt, respectively.

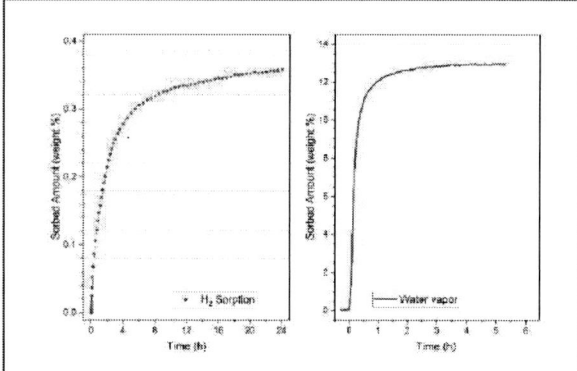

Fig. 6. Hydrogen (left) and water (right) uptake performances of ZeDry/H₂ getter.

C. Getter activation efficiency

In order to assess the efficiency and activation level of the getter material, in terms of water desorption, a series of experiments were performed to investigate different activation conditions and evaluate the performances of the product ZeDry®/H₂.

The water adsorption-desorption capacity and the overall performances were investigated using a digital microbalance (from Hiden) coupled to a temperature controlled vapor analyzer. A proper study of water desorption in isothermal conditions was set up to verify the achievable activation level in terms of recovered water sorption capacity.

TABLE III. RESULTS OF THE ACTIVATION EFFICIENCY-SCREENING PERFORMED ON ZEDRY®-H₂

Low vacuum activation (<1mbar)			High vacuum activation (<1·10⁻⁴mbar)			Dry gas activation		
T(°C)	Efficiency (%)	Time (hr)		Efficiency (%)	Time (hr)		Efficiency (%)	Time (hr)
RT	75.6	27		86	26		48	24
50	90.5	8		93.6	7		70.2	10
100	100	4		100	3		100	8

The data reported in Tab.3 suggest that increasing the temperature is the best and rapid way to efficiently activate the getter material. Concerning the environmental conditions, dry atmospheric gas is the least effective procedure both in terms of time and efficiency. Scroll pump and turbo pump activation conditions improve significantly the efficiency, for each temperature, although at RT the activation time is very long and not exploitable for manufacturing purposes. As reported, green data are the most effective conditions for having a well-activated product, while the yellow ones are much less effective.

D. ZeDry®-H₂ Water Sorption uptake at different working temperature

Some demanding high temperatures (T≈100 °C) operating conditions for opto-electronic devices, suggested to set up specific tests in order to assess the performance of the getter in a pretty warm environment, since the getter could partially release the adsorbed water when heated.

In the below measurements (Fig.7), it is shown the moisture uptake at constant partial pressure of water (15 mbar) when the getter is heated at RT, 50°C, 80°C, 100°C and 120°C. The last one is considered the maximum temperature that opto-electronic devices can reach in operating conditions.

Water uptake at room temperature is 13%wt (corresponding to 100% efficiency). By heating the getter, we observed a weight decrease due to partial water desorption from the material. Despite this, the product still provides more than half of the full efficiency at 120°C (6.79% w/w corresponding to 52.1 % of max performance), guaranteeing good sorption properties even in tough working conditions of the device.

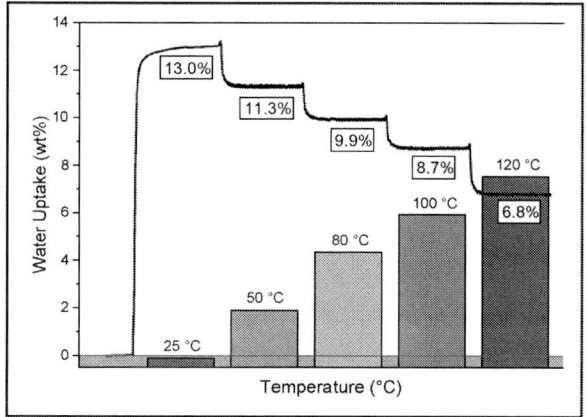

Fig. 7. Zedry®-H2 weight signal variation with exposure to different temperatures, simulating different operating conditions.

E. Water vapor pressure screening

The relative humidity inside devices plays an important role in terms of lifetime and reliability features. We carried out some measurements varying the water partial pressure at constant temperature (RT) in order to get indications about the H₂O uptake by lowering the humidity content and registering the corresponding adsorption signal for Zedry®-H₂.

The range of measured humidity were simulated by varying the operative partial pressure of water inside the microbalance chamber used for the test. In particular 1000, 5000 and 15000 ppmv were used as moisture levels for the test. The material has been activated under high vacuum conditions at 120°C, in order to get rid of all the absorbed moisture, and then the ZeDry®-H₂ lid has been exposed to humid nitrogen gas flow, according to the specific partial pressure of water. The sorption performances are reported in Figure 8.

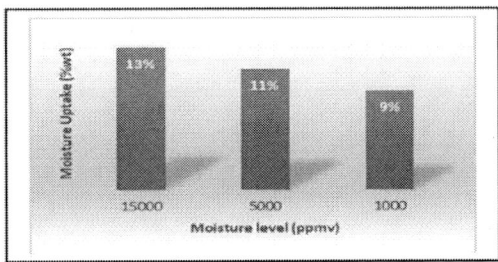

Fig. 8. H₂O uptake of ZeDry®-H₂ with different humidity level

The maximum water performance (13%wt) is reached at 15000 ppmv at room temperature, corresponding to approximately 50% relative humidity. By decreasing the relative partial pressure inside the microbalance chamber, the water absorption decreases to 11%wt and 9%wt, corresponding respectively to 5000 and 1000 ppmv (15% and 3% RH). The ZeDry®-H₂ sorption efficiency is still 70% even at very low moisture levels, demonstrating very good water uptake properties.

F. ZeDry®-VOC: Getter solution for H₂O and VOC

ZeDry®-VOC is a high capacity, solventless, thermally curable, dispensable getter for moisture and volatile organic compounds (VOC).

This product has also low bleed tendency on metallic substrates. It has been designed for use in optoelectronic devices, microelectronic devices and microelectronic devices.

ZeDry®-VOC works as a reversible getter for moisture and VOC (e.g. methyl-ethyl ketone or toluene). In Fig. 9, we reported the results of water and VOC adsorption test in order to highlight the quantity and kinetic of organic molecule uptake.

The sorption curve of Methyl Ethyl Ketone (MEK) is reported as an example of sorption test with VOC and is monitored for prolonged exposure (100 hr), until getter saturation is almost achieved.

For water uptake test ZeDry®-VOC is exposed to 15mbar of constant partial pressure until water saturation is achieved after the activation of the product (120°C for 4hrs).

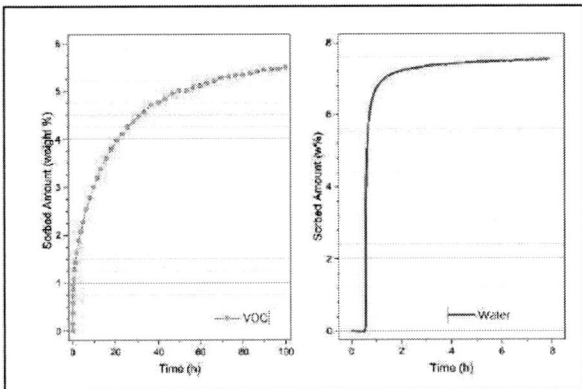

Fig. 9. VOC and Water uptake for ZeDry®/VOC.

The total amount of mix organic adsorption is up to 5.0%wt of cured getter paste recorded in the experimental test and almost 8%wt for water.

G. ZeDry®-VOC: Study of different VOCs adsorption test

The common VOCs can be classified into several groups on the basis of their different properties. Based on the boiling point, the VOCs can be divided into very volatile organic compounds (VVOCs), semivolatile organic compounds (SVOCs) and particulate organic matters (POMs). For the molecular structure, the VOCs include alkanes, alkenes, aromatic hydrocarbons, alcohols, aldehydes, ketones etc. Moreover, the polar and nonpolar VOCs are distinguished according to the degree of molecular polarity.

Taking into account this variety of molecules, we organized a series of experiment in order to highlight the quality of adsorption in terms of different VOC uptake for ZeDry®-VOC. The getter is preliminary activated to cleaning up its surface in order to adsorb the subsequent organic molecules exposed at constant pressure. As reported in Fig. 10 ZeDry®-VOC can face up with different molecules having different chemical features. Even though the kinetic observed was similar for all the compound on the first hour of exposure, on the other hand, the plateau of max adsorption registered was unlike depending on the type of compound.

In fact, for polar compound the maximum uptake recorded was 3.3%wt with a 2.0%wt after 10 hours of VOC exposure, explainable with a poor affinity towards the chemical composition of ZeDry®-VOC.

Linear molecular structure compounds (as ethylene, propane or MEK) resulted in high adsorption value (5.3%wt). Even better results are achieved for medium boiling point molecules showing very high sorption values, 7.4%wt.

This can be ascribed to the small structure of the molecules and easy flowing inside the getter structure, guarantying the best performance in terms of VOC adsorption.

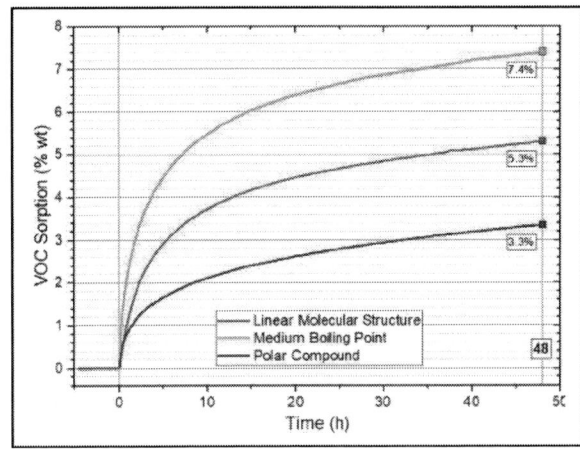

Fig. 10. VOCs adsorption results with different organic molecules families, for Zedry-VOC.

H. ZeDry®-M: High capacity getter solution for H₂O

ZeDry®/M is a high capacity, solventless, thermally curable, dispensable getter for moisture. It was designed to absorb high moisture levels for devices where the only concern is water.

ZeDry/M works as a reversible getter: it can be fully activated with a thermal process at 100°C-150°C, just before the device sealing.

The formulation is optimized for needle dispensing, with blading and die coating as alternate processing techniques.

ZeDry/M is a high capacity getter with a moisture capacity up to 15%wt, as reported in Fig. 11. The plateau is reached after only 1 hour of moisture exposure at 15 mbar, showing an outstanding adsorption kinetic.

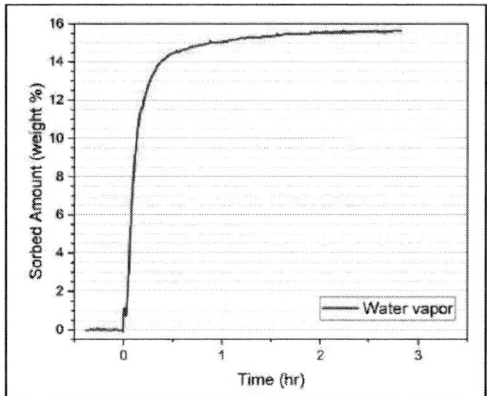

Fig. 11. Water uptake curve for ZeDry/M.

I. ZeDry®-M. Water Sorption uptake at different working temperature

As performed for ZeDry®-H_2, a series of test at different temperatures were carried out, demonstrating the performances of ZeDry®-M at different temperatures. In Fig. 12 is reported the value of water uptake registered with the microbalance, increasing the temperature in order to simulate different operating conditions of devices.

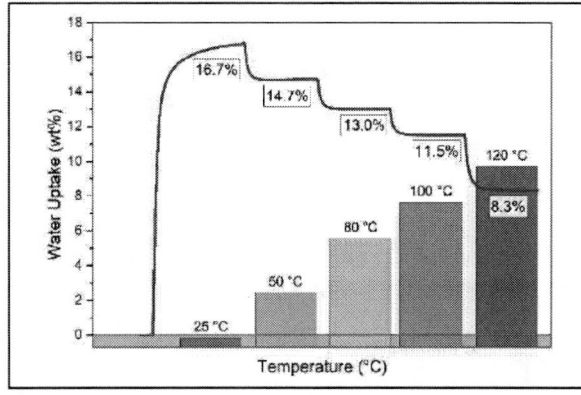

Fig. 12. Weight signal variation with exposure to different temperatures recorded on ZeDry®-M.

In the graph we outline the maximum water uptake at room temperature with an excellent 16.7%wt. As temperature increases, the performance of ZeDry®-M slightly decreases because of moisture partial desorption. At 120°C the water adsorption efficiency is still 50% of the one measured at RT, ensuring proper quality of the device and its reliability at high temperature. ZeDry®-M shows higher sorption performances for every investigated temperature compared to ZeDry®-H_2.

IV. CONCLUSIONS

In this paper, we have presented the evolutions, characteristics and performances of getter products used in hermetically sealed packages for vacuum operated devices and for gas back filled optoelectronic modules.

Hermetic vacuum packaging is achieved by the integration of non evaporable getters. Depending on the application, these materials can be integrated in the form of pills or as thin getter film, deposited by Physical Vapour Deposition, in the case of MEMS sensors. These getters are based on metallic alloy composition and can absorbs many different gases, like H_2, H_2O, CO, CO_2, N_2, Hydrocarbons, and O_2. The main feature of these products is the porous structure that allows the gases to diffuse into the inner parts of the getter body, thus optimizing the sorption performances of the material.

Hermetically sealed optoelectronic devices rely on gas back filling processes. The new products family developed by SAES for these applications, is based on engineered absorber materials with solvent-free formulation: the so called ZeDry® family. The main features are related to the ability to remove significant amounts of water even at elevated temperature conditions or at very low moisture levels; the materials can be easily activated at moderate temperatures in different environments (vacuum or dry gas). Furthermore, the developed products exhibit high sorption performances also for hydrogen and VOCs, ensuring the possibility to remove the detrimental contaminants and guaranteeing the proper operation of the devices along the entire lifetime.

V. REFERENCES

[1] R. K. Lowry, R. C. Kullberg and D. J. Rossiter, "Harsh Environments and Volatiles in Sealed Enclosures," in *Proceedings, Surface Mount Technology Association International Technical Conference*, Orlando, FL, Oct 24-28, 2010.

[2] A. Conte, M. Moraja, G. Longoni, A. Fourrier, "High and stable Q-factor in resonant MEMS with getter film", SPIE Proceedings, Vol. 6111, 2006, DOI: 10.1117/12.647710.

[3] L. Mauri, E. Rizzi, M. Moraja, M. Campaniello, "The discrete vacuum packaging reliability issue in MEMS", 2013 European Microelectronics Packaging Conference (EMPC), 2013, pp. 1-4

[4] M. Albarghouti, N. El Dahdah, G. Perosevic, S. Jain, J.-M. Papillon, and S. Fernandez, "Moisture and Hydrogen Release in Optoelectronic Hermetic Packages", Journal of Microelectronics and Electronic Packaging, 2014, vol. 11, pp. 75-79, DOI: 10.4071/imaps.403

[5] W.O. Camp, R. Lasater, V. Genova, R. Hume, "Hydrogen effects on reliability of GaAs MMICs," Proceedings of 11th GaAs Integrated Circuit Symposium, pp. 203-206, 1989, DOI: 10.1109/GAAS.1989.69326

[6] Ronald L. Pease, Dale G. Platteter, Gary W. Dunham, John E. Seiler, Philippe C. Adell, Hugh J. Barnaby, Jie Chen, "The Effects of Hydrogen in Hermetically Sealed Packages on the Total Dose and Dose Rate Response of Bipolar Linear Circuits", IEEE Transactions on Nuclear Science, Vol. 54, No. 6, pp. 2168-2173, 2007, DOI: 10.1109/TNS.2007.907870

[7] R.R. Blanchard; J.A. del Alamo; A.C. Calveras, "Hydrogen-induced changes in the breakdown voltage of InP HEMTs2", IEEE Transactions on Device and Materials Reliability, Vol. 5, No. 2, June 2005, DOI: 10.1109/TDMR.2005.846825

Advances in Parylene Adhesive Bonding for the Realization of Biocompatible Microsystems

Franz Selbmann
Fraunhofer Institute for Electronic Nano Systems ENAS
Chemnitz, Germany
and
Institute for Electronic and Sensor Materials
Technische Universität Bergakademie Freiberg
Freiberg, Germany
franz.selbmann@enas.fraunhofer.de

Frank Roscher
Fraunhofer Institute for Electronic Nano Systems ENAS
Chemnitz, Germany
frank.roscher@enas.fraunhofer.de

Harald Kuhn
Fraunhofer Institute for Electronic Nano Systems ENAS
Chemnitz, Germany
and
Center for Microtechnologies
Technische Universität Chemnitz
Chemnitz, Germany
harald.kuhn@enas.fraunhofer.de

Maik Wiemer
Fraunhofer Institute for Electronic Nano Systems ENAS
Chemnitz, Germany
maik.wiemer@enas.fraunhofer.de

Yvonne Joseph
Institute for Electronic and Sensor Materials
Technische Universität Bergakademie Freiberg
Freiberg, Germany
yvonne.joseph@esm.tu-freiberg.de

Abstract — **Wafer to wafer and chip to wafer bonding technologies are a key enabling processes for the fabrication of microsystems. The presented paper focuses on the latest research results of Parylene adhesive bonding. Given the established wafer bonding processes based on Parylene C in vacuum, research on how the bonding process alters the Parylene material was performed using x-ray diffraction. Doing so, a significant increase of the crystalline fraction of Parylene was observed. In order to further develop the Parylene bonding beyond its state of the art, new process variants were successfully established with advanced features: Wafer bonding using Parylene C at increased pressure; Wafer bonding using Parylene N and Parylene F, respectively, to address a different bonding temperature range; as well as chip bonding using Parylene C in air.**

Keywords— *Wafer bonding, chip bonding, Parylene, polymer, thermal properties, adhesive bonding, packaging, microsystems, MEMS, biocompatible, microfluidics*

I. INTRODUCTION

Micro-electro-mechanical systems (MEMS) pave the way for current technology trends such as the Internet of Things, Smart Everything or Industry 4.0, which also represents the forecasted growth of the global MEMS market from $ 14.8 billion in 2022 to $ 22.3 billion in 2027. [1] For the fabrication of MEMS as well as of microfluidic systems, reliable and industry compatible bonding processes are crucial. This particularly refers to large and industry standard wafer sizes and short process time.

For the fabrication of advanced microsystems beyond the state of the art, the bonding temperature should be as low as possible to ensure process compatibility with organic materials. At the same time, the usage of biocompatible bonding adhesives is beneficial to enable the fabricated microsystems to be used for medical applications.

Utilizing the outstanding properties of Parylene, within previous studies, Parylene C (Poly[chloro-p-xylylene]) adhesive wafer bonding was established as a bonding process for biocompatible low temperature wafer bonding in industry scale, providing excellent mechanical and thermal stability as well as a good hermeticity of the bonds, and hence, process compatibility with most established microtechnologies. [2,3]

Considering the established Parylene bonding process, the presented work focuses to provide a better understanding how the bonding conditions alter the material properties of Parylene as well as to extend the process flexibility and establish new variants. This includes wafer bonding at increased pressures in order to realize cavities with an adjustable pressure as well as bonding with Parylene N (Poly[p-xylylene]) and Parylene F (Poly[tetrafluoro-p-xylylene]) instead of Parylene C in order to address different bonding temperature ranges, e.g. to enable the bonding of wafer stacks consisting of more than two wafers.

Furthermore, the development of a Parylene chip bonding process is of high interest to enable bonding when the wafers are already singulated, considering also the fact, that there is no single roadmap for the heterogeneous integration of MEMS sensors to follow. [4]

II. EXPERIMENTAL

A. Impact of the bonding conditions on Parylene C

Previous studies have proven that Parylene C does not get chemically altered during the established wafer bonding in vacuum using FTIR analysis. [2] However, other studies have revealed a dependency of the Young's modulus and the hardness on the thermal budget, which is assumed to be caused by recrystallization of the partially crystalline polymer. [5] Hence, free-standing Parylene C membranes of 5 μm thickness were fabricated according to the process and using the equipement given in the literature [2] and stored under the typical conditions of a bonding process, i.e. 30 min in vacuum at 260 °C, 280 °C, 300 °C, and 320 °C, respectively. Changes in the crystallinity of the membranes were traced by X-ray diffraction. The experiments were performed on a Bruker D8 Advance Diffractometer equipped with a sealed X-ray tube with a Cu anode. In order to emphasize the contribution of the thin membrane, the glancing angle XRD (GAXRD) setup was used. Therefore, the device is equipped with a Ni/C multilayer mirror to create a parallel beam. The secondary beam was detected, after passing through an antiscatter and receiving slit, by a scintillation counter. The diffraction experiment was performed in asymmetrical geometry with a fixed angle of incidence of 2° in the range $5° < 2\theta < 25°$ with a step size of 0.02° and a dwell time of 12 s per step.

This work has been supported and partially funded by the European Social Fund and the Fraunhofer Society.

B. Preliminary investigation of the thermal properties of different Parylene types

In order to identify reasonable parameter ranges for the experiments presented in this paper, preliminary investigations of the thermal properties of the different Parylene types were performed. Doing so and considering their different thermal properties [6], Parylene C, Parylene N and Parylene F were investigated by thermogravimetric analysis (TGA) and differential scanning calorimetry (DSC) in oxygen and inert gas (argon) atmosphere. The chemical structures for the three Parylene types are schematically depicted in Fig. 1.

The Parylene samples for the TGA and DSC measurements were received by depositing approximately 10 μm of the respective Parylene type on 150 mm silicon wafers, which were cleaned using standard cleaning. Subsequently, the Parylene was peeled off the wafer using a scalpel. The TGA and DSC measurements, respectively, were performed using an initial mass of 20 mg to 30 mg of each Parylene type. Doing so, the temperature range was set from ambient temperature to 450 °C for Parylene C and from ambient temperature to 600 °C for Parylene N and Parylene F, respectively. The results of the TGA and DSC characterizations are presented in Fig. 2.

C. Wafer bonding in non-vacuum conditions

To enable the sealing of adjustable pressures in cavities, wafer bonding at non-vacuum conditions is required. The TGA and DSC measurements presented in Fig. 2 indicate that all Parylene types decompose more rapidly at oxygen atmosphere in comparison to inertgas. Hence, for Parylene adhesive wafer bonding in non-vacuum conditions, inert nitrogen atmosphere was chosen. Furthermore, to enable a comparison with previous Parylene adhesive wafer bonding results [2], Parylene C was used again. Doing so, 5 x 5 mm² Parylene C bonding frames of 200 μm width and 5 μm height were patterned on 150 mm silicon wafers using the process flow given in the literature. [2] The bonding against plain silicon wafers (Si-P-//-Si) as well as against Parylene C coated silicon wafers (Si-P-//-P-Si) was carried out in an EVG-540 bonder (EV Group Europe & Asia/Pacific GmbH, Austria) at 280 °C and 300 °C, respectively, for 10 min and 1200 mbar nitrogen atmosphere. All other bonding parameters as well as the subsequent routine to measure the tensile and shear strengths, respectively, remained as given in the literature using ten chips per bond. [2]

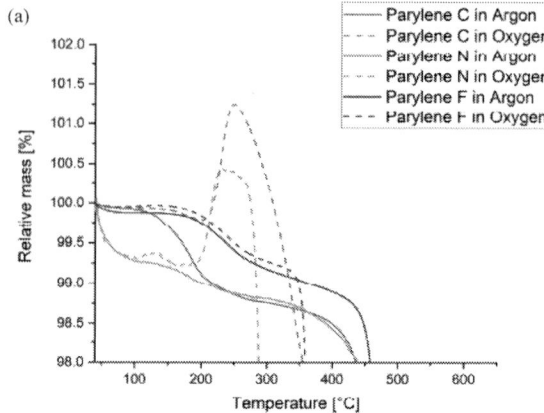

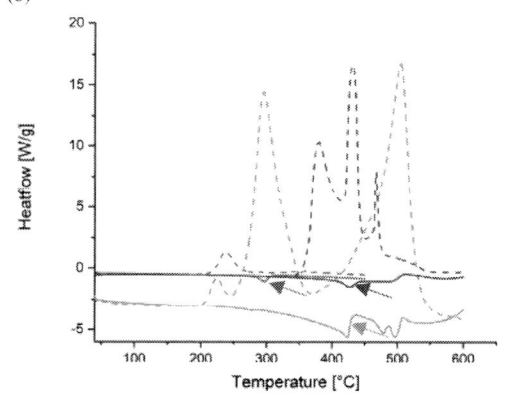

Fig. 2. Results of the TGA (a) and DSC (b) measurements for Parylene C, Parylene N and Parylene F in oxygen and argon. For the DSC results, positive peaks indicate exothermal reactions, whereas negative peaks indicate endothermal reactions. The arrows indicate the melting points of the three different Parylene types in inert gas conditions.

D. Wafer bonding using Parylene N and Parylene F

Even though the lower temperature stability of Parylene N reported in literature [6] suggests lower bonding temperatures, previous bonding experiments using Parylene N at temperatures below the bonding temperatures of Parylene C were not successful. [2] Considering the DSC characterization for Parylene C in inert gas, it can be concluded, that typical bonding temperatures in the range of 280 °C to 300 °C are slightly below and above its melting point at 296 °C, which is indicated by a small endothermal peak in the DSC results. Moreover, the DSC measurement results indicate the melting point for both, Parylene N and Parylene F, at 424 °C. Hence, the optimal bonding temperature can be assumed in this temperature range. However, considering the TGA measurements in parallel, Parylene N shows a higher mass loss, which starts in contrast to the other two Parylene types directly when heating the material and is most significant at its melting temperature. Moreover, Parylene N decomposes completely slightly above its melting point. Hence, for using Parylene N for adhesive wafer bonding, maximum temperatures at 420 °C were defined, which were lowered to 360 °C in steps of 20 K. In contrast to this, for Parylene F, a bonding temperature range between 410 °C and 450 °C slightly below and above its melting point was defined, which is in similarity to Parylene C.

Fig. 1. Chemical structures of Parylene C, Parylene N and Parylene F. [7]

For both Parylene types, several bonding experiments were performed in vacuum in an EVG-540 bonder (EV Group Europe & Asia/Pacific GmbH, Austria). The sample preparation using Parylene bonding frames, all remaining bonding parameters as well as the procedure for the measurement of the bonding strengths were repeated as described earlier and in the literature, respectively, using ten chips per bond. [2]

Additionally to the wafer bonding experiments in vacuum and in similarity to the bonding experiments in non-vacuum conditions described in section C for Parylene C, Parylene F was bonded at 430 °C and at a nitrogen pressure of 1200 mbar.

E. Parylene chip bonding

To enable Parylene adhesive chip bonding at ambient conditions, the Parylene type with the lowest melting and bonding temperature, respectively, as well as the lowest sensitivity to oxygen at elevated temperatures is advantageous. Since Parylene C features the lowest bonding temperature and no mass loss at bonding temperature even in oxygen atmosphere, this Parylene type was used for establishing a chip bonding process. Again, the Parylene bonding frames described above were fabricated as given in literature. [2] Before bonding, the wafers with the Parylene bonding frames as well as plain silicon wafers for the bonding partners were diced into 5 x 5 mm² chips. The bonding process was carried out using a Fineplacer femto 2 (Finetech GmbH & Co. KG, Germany). In similarity to Parylene C wafer bonding, the bonding temperature was chosen to be 280 °C and 300 °C, respectively, whereas the bonding time was varied between 30 s, 1 min and 10 min, and the contact pressure was varied between 6.45 N and 3.23 N. After bonding tensile tests were performed as described in literature again using ten chips per bonding parameter set. [2]

III. RESULTS AND DISCUSSION

A. Impact of the bonding conditions on Parylene C

The obtained results of the XRD measurements are presented in Fig. 3. The position of the Parylene peaks between 13.88° and 14.18° is in good accordance with the values given in literature. [8] For the Parylene, which faced the bonding conditions, a five to six fold higher signal intensity as well as a smaller full width at half maximum is observed compared to the untreated one. This effect is also

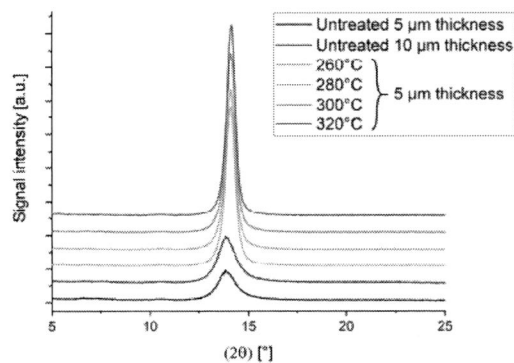

Fig. 3. Results of the XRD measurements of Parylene C annealed at different temperatures in vacuum.

significantly higher than the increase of the signal intensity due to a higher Parylene thickness (1.5 fold when doubling the Parylene thickness). Hence, it can be concluded that the crystallinity of Parylene increases due to the bonding process. According to the literature, it can be assumed that the crystallite size increases. [8] In parallel, the 2θ angle of the peaks slightly increases due to the bonding conditions, which can be correlated to a decrease of the lattice plane spacing. This could possibly be caused by outgassing of volatile components or rearrangement of the structure.

B. Wafer bonding non-vacuum gas conditions

The results for the mechanical strengths of the compounds bonded in inert gas are summarized in Fig. 4 and Table 1. For compounds bonded at 280 °C no difference in the mechanical strengths between bonding in vacuum or nitrogen is observed, whereas for bonding at 300 °C, the compounds bonded in nitrogen show a slightly less mechanical strengths compared to the vacuum bonded ones, but still a higher strengths compared to bonding at 280 °C. Considering that the bonding temperature of 300 °C is above the melting temperature of Parylene C, it can be assumed, that the present nitrogen influences the melting behavior of Parylene C, which leads to decreased bonding strength. This could be also the reason why bonding under nitrogen against Parylene (two bonding partners soften / melt) shows slightly higher mechanical strength compared to bonding against silicon (only one bonding partner softens / melts). Nevertheless, in summary, all bonds in inert gas show good mechanical strengths enabling process compatibility with subsequent wafer processing.

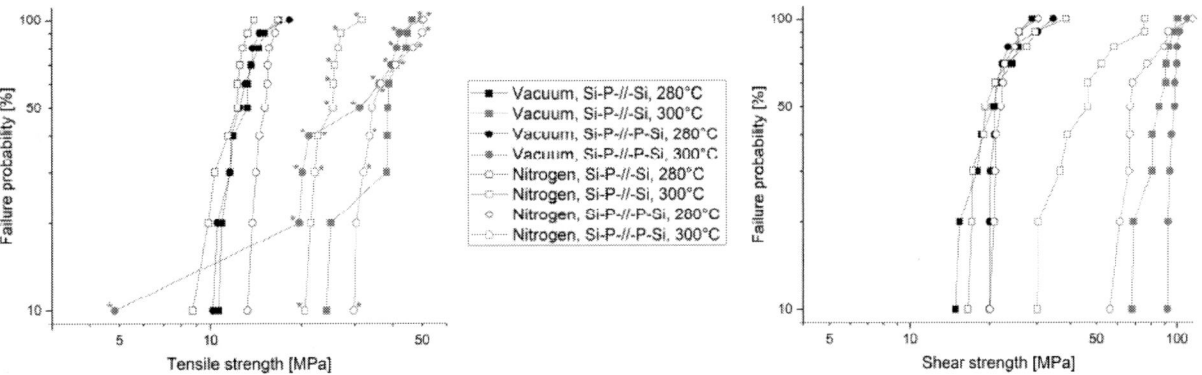

Fig. 4. Tensile and shear strengths of Parylene C based wafer bonding in vacuum and nitrogen, respectively, for bonding against silicon (Si-P-//-Si) and Parylene C (Si-P-//-P-Si). * indicates glue failure between chip and studs.

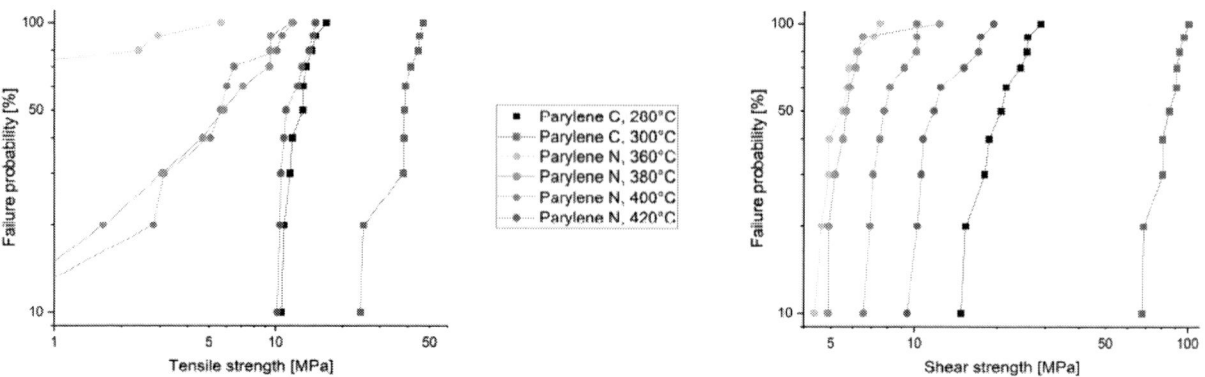

Fig. 5. Tensile and shear strengths of Parylene C and Parylene N based wafer bonding in vacuum. For the values out of the diagram, the pre-force was not reached.

C. Wafer bonding using Parylene N

Fig. 5 and Table 1 depict the mechanical strengths for the wafer bonds with Parylene N. The bonding strength shows a strong dependency on the bonding temperature. Even the compounds bonded at 420 °C, i.e. close to the melting point of Parylene N, show only a comparable tensile strength and still a reduced shear strength compared to Parylene C compounds bonded at 280 °C. Considering the cohesive or cohesively/adhesively mixed fraction mechanism of the Parylene N bonds, its reduced tensile strength of only 45 MPa compared to the one of Parylene C with 69 MPa could be a potential reason. [6] However, considering that the tensile strength of Parylene N is about two third compared to Parylene C and the achieved tensile strength of the Parylene N bonds are only about one third of the ones with Parylene C, the bonding parameters seem to have room for further optimization. Since the bonding temperature is limited according to the TGA and DSC measurements, respectively, the bonding time and the contact pressure seem promising parameters for optimization. Nevertheless, the mechanical strengths of the compounds bonded with Parylene N at 420 °C are sufficient to enable process compatibility with subsequent wafer processing. With the established bonding process, Parylene bonding is possible at two different temperature ranges (420 °C for Parylene N and 300 °C for Parylene C), which can be advantageous for the realization of wafer triple stacks. Furthermore, bonding using Parylene N enables Parylene adhesive bonding without halogen containing materials.

D. Wafer bonding using Parylene F

The results of the mechanical strengths for Parylene F wafer bonds are depicted in Fig. 6 and Table 1. Again, the bonding strength shows a strong dependency on the bonding temperature, whereas the compounds processed at 410 °C and 450 °C show a lower strength compared to the compounds bonded at 430 °C and 440 °C, respectively. Particularly, the bonding strengths for 410 °C and 450 °C are lower than those of the weaker Parylene C compounds bonded at 280 °C. Doing so, the decreasing bonding strength at lower temperature can be caused due to bonding below the melting temperature of Parylene F. A more detailed analysis of the bonding interface after the mechanical characterizations as depicted in Fig. 7 reveals that at 450 °C the Parylene F becomes so a low viscous liquid that it gets squeezed due to the contact pressure in the bonder. Even though the tensile force and shear force for the compounds bonded at 450 °C are similar compared to those compounds bonded at 430 °C, the increased bond interface by a factor of 3.5 to 4 due to squeezing lowers the bonding strengths as per definition. Since the compounds bonded at 440 °C do not feature reduced bonding strengths or squeezing of Parylene frames, respectively, it can be concluded that the temperature, which reduces the viscosity of Parylene F below a sufficient level, is between 440 °C and 450 °C. Considering the cohesively/adhesively mixed fraction mechanism of the compounds bonded at 430 °C, it can be concluded that this temperature is optimal for adhesive bonding using Parylene F.

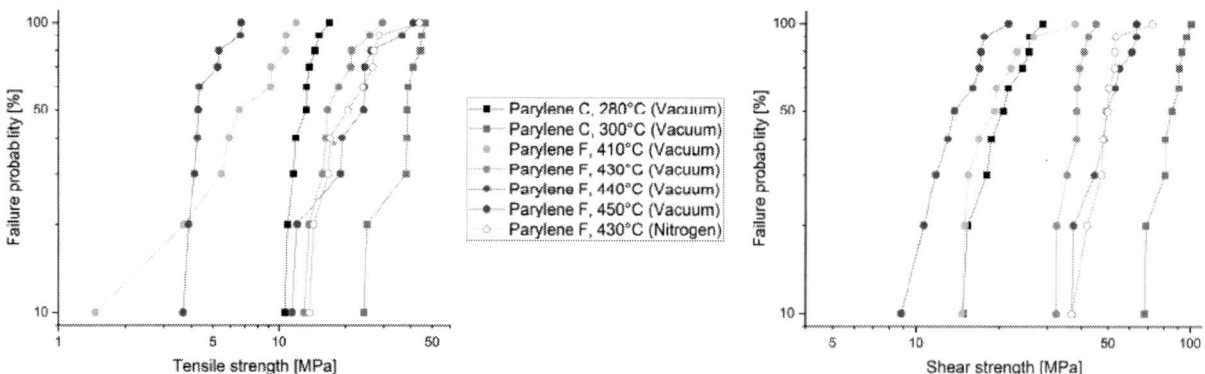

Fig. 6. Tensile and shear strengths of Parylene C and Parylene F based wafer bonding in vacuum and nitrogen. * indicates glue failure between chip and studs.

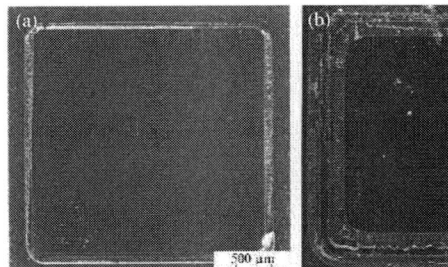

Fig. 7. Parylene F bonding frames after mechanical testing: Compound bonded at 430 °C (a) and 450 °C (b), respectively.

Comparing the results for Parylene F and Parylene C, the bonding strength is in a similar range. Moreover, in similarity to Parylene C, no major difference can be observed between bonding at vacuum conditions or at nitrogen.

In comparison to Parylene N, adhesive bonding using Parylene F has the advantage of higher bonding strengths and less impact of the bonding process on the material itself. Hence, for the realization of triple wafer stacks using two bonding processes at different temperatures, the combination of Parylene F and Parylene C is superior compared to the combination of Parylene N and Parylene C, addressing both the same temperature ranges.

E. Parylene chip bonding

The results for the tensile strengths of the Parylene chip bonds are presented in Fig. 8 and Table 1. Even though the maximum tensile strengths between wafer and chip bonding are comparable for 280 °C and 300 °C, respectively, for the chip bonds, a higher deviation is observed, and hence, the average tensile strengths are reduced. The dependencies of the tensile strengths on the bonding temperature are identical for chip and wafer bonding.

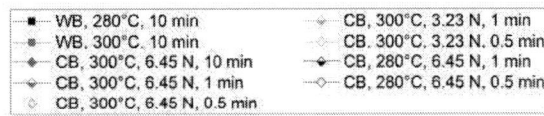

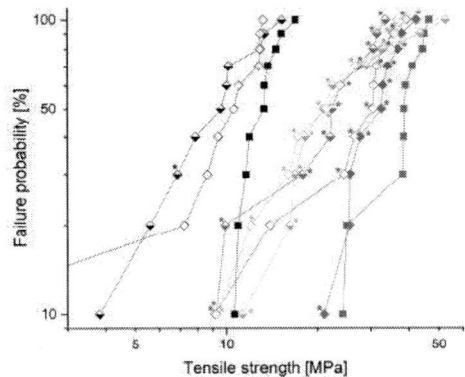

Fig. 8. Tensile strengths of Parylene C based wafer bonding (WB) and chip bonding (CB). * indicates glue failure between chip and studs.

Fig. 8 suggests, that a shortening of bonding time increases the deviation of the tensile strengths, even though it is interesting to note that a bond with only 30 s bonding time is possible. In contrast to this, the tensile strengths are independent on the contact pressure. Considering that the chip bonding is usually one of the last processes of the technology chain, the achieved strengths are sufficient, and hence, a Parylene chip bonding process was successfully established. This process can be advantageous for the bonding of devices such as ultra-thin membranes, which are more sensitive to thermal budgets.

TABLE I.   AVERAGE RESULTS AND STANDARD DEVIATIONS FOR THE MECHANICAL STRENGTHS OF THE DIFFERENT PARYLENE ADHESIVE BONDING VARIATIONS (WAFER BONDING (WB) USING PARYLENE C, PARYLENE N AND PARYLENE F IN VACUUM AND IN NITROGEN, VS. CHIP BONDING (CB) USING PARYLENE C IN AIR).

Bonding parameter							Tensile strength [MPa]	Shear strength [MPa]
Bonding type	Parylene type	Partner material	Atmos-phere	Temper-ature	Yield after dicing	Remarks		
WB	C	Silicon	Vacuum	280 °C	100 %	References [2]	13.2 ± 2.0	21.4 ± 4.7
				300 °C	100 %		38.0 ± 7.6	85.5 ± 11.0
		Parylene		280 °C	100 %		13.0 ± 2.3	23.4 ± 4.9
				300 °C	100 %		30.6 ± 13.8	98.0 ± 5.1
WB	C	Silicon	Nitrogen	280 °C	100 %	-	11.8 ± 1.7	22.9 ± 7.1
				300 °C	100 %		24.9 ± 3.3	49.0 ± 16.6
		Parylene		280 °C	100 %		15.1 ± 1.1	23.1 ± 3.1
				300 °C	100 %		38.5 ± 8.1	76.0 ± 17.9
WB	N	Silicon	Vacuum	360 °C	100 %	-	1.1 ± 2.0	5.7 ± 1.0
				380 °C	100 %		6.2 ± 3.9	6.3 ± 2.2
				400 °C	100 %		6.2 ± 3.8	8.4 ± 1.4
				420 °C	100 %		12.3 ± 1.9	13.4 ± 3.5
WB	F	Silicon	Vacuum	410 °C	100 %	-	7.5 ± 3.4	21.0 ± 7.0
				430 °C	100 %		19.1 ± 5.3	38.2 ± 4.0
				440 °C	100 %		23.8 ± 9.4	51.2 ± 9.8
				450 °C	100 %		4.8 ± 1.1	14.7 ± 3.8
			Nitrogen	430 °C	100 %		23.2 ± 9.0	50.5 ± 9.3
CB	C	Silicon	Air	280 °C	N/A	30 s, 6.45 N	9.4 ± 4.0	N/A
				300 °C			27.4 ± 9.4	N/A

323

IV. CONCLUSIONS

The presented work completes existing studies about Parylene C adhesive bonding and proves, that the crystallinity of Parylene C increases significantly during wafer bonding, even though the chemical composition remains unchanged. [2]

Considering the results of a more detailed characterization of the thermal properties of different Parylene types at oxygen and inert atmosphere, new process variants were established successfully. This includes bonding Parylene adhesives at increased pressures using nitrogen atmosphere, at different temperatures range above 410 °C as well as using Parylene N and Parylene F, respectively, instead of Parylene C. Doing so, Parylene F shows superior bonding strengths in comparison to Parylene N. A fourth process variant established successfully within this study is Parylene chip bonding using Parylene C.

The presented process variants enable the realization of cavities with a defined pressure, e.g. for pressure sensors as well as the realization of wafer compounds consisting of more than two wafers. Furthermore, they allow a reliable bonding after the separation of a wafer into chips.

However, for all four process variants, a more detailed analysis is required on how the bonding conditions impact the Parylene adhesive material.

ACKNOWLEDGMENT

The authors thank R. Kinner and S. Uhlig for bonding the wafers and chips, P. Schwarz for the dicing as well as M. Eichstädt and D. Ullmann for the tensile and shear tests, respectively. Furthermore, the authors thank C. Schimpf at the Institute for Materials Science at TU Bergakademie Freiberg for the XRD measurements and N. Rüffer at the Institute of Chemistry at TU Chemnitz for the TGA and DSC measurements, respectively. Special thank goes to all colleagues at the Center for Microtechnologies at the TU Chemnitz for the wafer processing. Finally, the authors thank all involved colleagues at the Fraunhofer Institute for Electronic Nano Systems, the Center for Microtechnologies at the TU Chemnitz as well as the Institute for Electronic and Sensor Materials at TU Bergakademie Freiberg for their support and fruitful discussions.

REFERENCES

[1] Yole Developpement, "Status of MEMS Industry 2022: Market and Technology Report," 2022.

[2] F. Selbmann, M. Baum, C. Meinecke, M. Wiemer, H. Kuhn, and Y. Joseph, "Low-Temperature Parylene-Based Adhesive Bonding Technology for 150 and 200 mm Wafers for Fully Biocompatible and Highly Reliable Microsystems," ECS J. Solid State Sci. Technol. vol. 10, pp. 074010, 2021.

[3] F. Selbmann, M. Baum, C. Meinecke, M. Wiemer, T. Otto, and Y. Joseph, "Parylene C Based Adhesive Bonding on 6" and 8" Wafer Level For The Realization of Highly Reliable and Fully Biocompatible Microsystems," ECS Trans., vol. 98, 4, pp. 55-66, 2020.

[4] IEEE, "Heterogeneous Integration Roadmap 2021," Chapter 11, https://eps.ieee.org/images/ files/HIR_2021/ ch11_MEMS.pdf, accessed on January 23rd, 2023.

[5] F. Selbmann, C. Scherf, J. Langenickel, F. Roscher, M. Wiemer, H. Kuhn, and Y. Joseph, "Impact of Non-Accelerated Aging on the Properties of Parylene C," Polymers, vol. 14, pp. 5246, 2022.

[6] Plasma Parylene Systems GmbH, Datasheet parylene properties, 2015, downloaded on March 3rd 2015, www.plasmaparylene.de.

[7] F. Selbmann, "Development and Evaluation of a CVD based Parylene deposition process for realization and characterization of thin encapsulation layers of micro systems," Master thesis, Technische Universität Bergakademie Freiberg, Freiberg / Germany, 2015.

[8] J.-M. Hsu, L. Rieth, S. Kammer, M. Orthner, and F. Solzbacher, "Effect of Thermal and Deposition Processes on Surface Morphology, Crystallinity, and Adhesion of Parylene-C," Sens. Mater., vol. 20, 2, pp. 087-102, 2008.

Flip-chip interconnects based on single metal-coated polymer spheres

Van Long Huynh
Department of Microsystems
University of South-Eastern Norway
Horten, Norway
Van.L.Huynh@usn.no

Knut E. Aasmundtveit
Department of Microsystems
University of South-Eastern Norway
Horten, Norway
Knut.Aasmundtveit@usn.no

Hoang-Vu Nguyen
Department of Microsystems
University of South-Eastern Norway
Horten, Norway
Hoang.V.Nguyen@usn.no

Abstract: In microelectronics packaging, minimizing thermo-mechanical stresses is crucial to achieve high reliability. This paper presents a novel approach for flip-chip interconnects, utilizing individual metal-coated polymer spheres at a low bonding temperature. The method involves selectively depositing individual conductive particles onto a polydimethylsiloxane (PDMS) carrier and then transferring the particles to electrical pads on substrate using Ag sintering followed by die placement and introducing underfill material. The deposition process achieved a high yield of 98.8% on the PDMS carrier, while the transferring process resulted in well-defined ink dots with a yield of 96.2%. The sintered Ag formed good bonds between the particles and electrical pads, leading to moderate interconnect resistance (as low as 0.57 Ω). This work has demonstrated the feasibility of interconnects based on a single metal-coated polymer sphere with low thermo-mechanical stress induced, thanks to the low bonding temperature (140 °C) and pressure (0.1 MPa) as well as the mechanical compliance of polymer particles.

Keywords—Flip-chip interconnects, particle deposition, Ag sintering, low bonding temperature

I. INTRODUCTION

In recent years, the miniaturization and increased integration of electronic components have become crucial factors for the advancement of electronics [1–3]. It is because electronic devices continue to shrink in size while simultaneously exhibiting enhanced functionalities, allowing for new applications in demanding environments, such as wearable and implantable devices. Electronic packaging, in particular interconnection technology, has thus been governed by a growing demand for high interconnection density, ultra-fine pitch capability, and high reliability.

Various flip-chip technologies have been developed and employed in the last decades [2–9]. These technologies include solder micro bumps, copper pillars with solder caps, direct metal-metal thermocompression bonding, solid-liquid interdiffusion (SLID) bonding, and electrically conductive adhesives (ECAs). Micro solder bumps enable high-density and fine-pitch interconnections (≤ 130 μm) while possessing self-alignment capabilities between solder bumps and metal pads. Solder reflow occurs at peak temperatures of 230 °C (eutectic bumps) or 260 °C (lead-free bumps) [2–4]. Copper pillars with solder caps utilize a smaller solder volume to prevent bridging between bumps, allowing for ultra-fine pitch capabilities (40 - 130 μm). A typical Cu pillar with a shallow Sn-3.5Ag solder cap is reflowed at 320 °C [5–6]. Direct metal-metal bonding has been explored for extremely fine pitch interconnects by directly joining two pure metal surfaces. For example, Cu-Cu thermocompression bonding enables interconnection pitches below 30 μm but requires a bonding

temperature exceeding 350 °C [8]. SLID bonding has gained attention due to its ability to utilize different substrate interfaces at lower bonding temperatures compared to metal-metal compression bonding, while allowing the use of joint materials in high-temperature environments [7].

When materials with different coefficients of thermal expansion (CTE) are bonded at such elevated temperatures and subsequently cooled down, thermo-mechanical stresses will arise within the package. Metallic joints typically lack mechanical compliance, limiting their capabilities to absorb the stresses generated by the thermal mismatch. Consequently, thermo-mechanical failures, particularly fatigue fractures, can occur in the package. The use of ECAs, including anisotropic conductive adhesives (ACAs) and isotropic conductive adhesives (ICAs), facilitates simplified processing and reduces bonding temperature. Nevertheless, ultra fine-pitch capability has impeded the implementation of ECAs across diverse applications. As a result, there is an increasing demand for interconnection technologies that provide compliance, (ultra) fine-pitch capability, low bonding pressure, and low bonding temperature.

This work presents a novel approach that leverages selective deposition capability of single metal-coated polymer spheres (MPS) to individual interconnects, while utilizing low bonding temperature and low bonding pressure through a sintering method. By employing a low bonding temperature, the thermo-mechanical stress within the package can be significantly reduced. Additionally, the use of MPS enables the absorption of stresses arising from thermal expansion, vibration, or shock during operation, thanks to the elasticity of the polymer core. Previous techniques employed for depositing MPS onto electrical pads have involved various approaches such as applying electric field, magnetic field, dry or wet dispensing, and the utilization of mesh canisters [9–12]. Nonetheless, these methods have encountered limitations in terms of controlling the quantity and placement of the deposited particles. The advancement of the method proposed in this work is presented and discussed.

II. EXPERIMENTAL

A. Overview of assembly process

Fig 1 provides an overview of the assembly process proposed in this work. The process includes four primary steps: i) fabrication of a particle carrier (Fig 1a); ii) deposition of conductive particles onto the fabricated carrier (Fig 1b); iii) transferring and subsequently sintering the deposited particles to electrical pads on substrate using Ag sintering

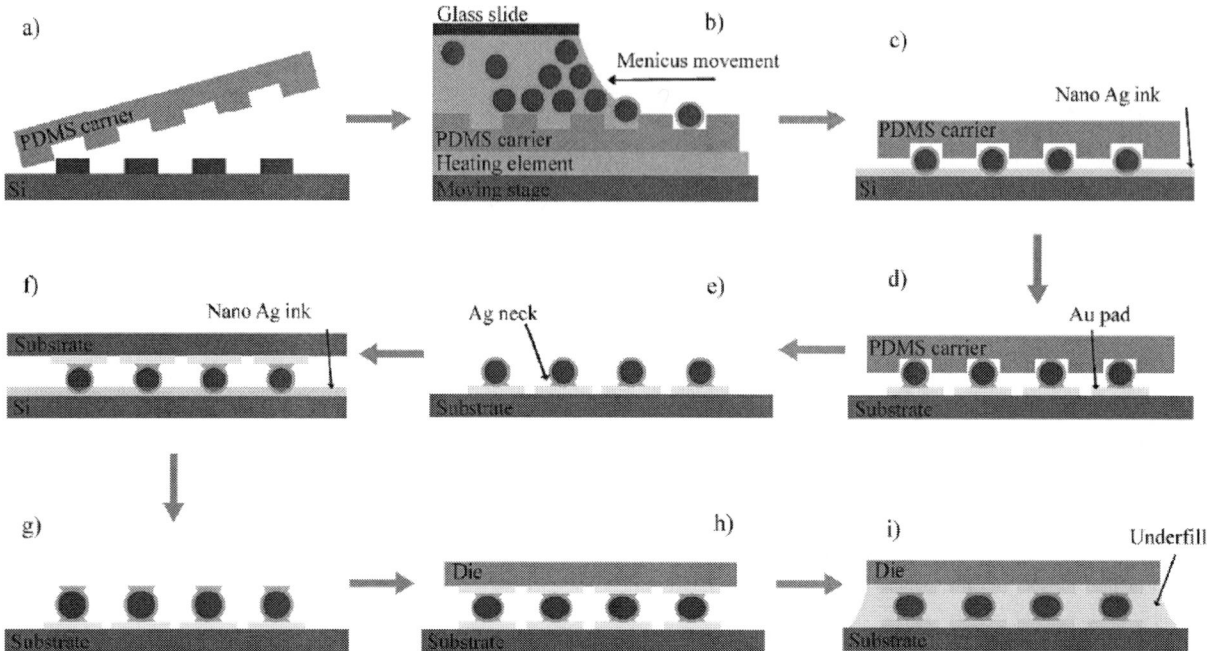

Fig 1. Assembly process for single particle-based interconnects using low-temperature Ag sintering. (a) Fabrication of PDMS carrier with particle traps. (b) Deposition of conductive particle on predefined traps using capillarity-assisted particle assembly. (c), (d), and (e) Deposition of Ag ink and the first Ag joint on substrate using Ag sintering. (f), (g), (h), and (i) Additional ink deposition for the second Ag joint on glass die and the final bonding.

(Fig 1 c, d, and e); iv) die placement and final bonding (Fig 1f, g, h, and i).

B. Design of test samples

The fabrication process of silicon substrates and glass dies followed the layout design shown in Fig 2. The dimensions of the substrates and dies are 7 x 7 mm² and 3 x 3 mm², respectively. The test samples consist of four Kelvin structures located at four corners to monitor interconnect resistance. In addition, three daisy chain structures, each consisting of 18 interconnects, are placed at the central area. The electrical pads and tracks were deposited using sputtered Cr/Au (10 nm/100 nm). The X-Y dimensions of each pad is 150 x 150 μm². The interconnection pitch for the pads is 200 μm for the Kelvin structures, and 200 μm and 175 μm for the daisy chains.

C. Fabrication of PDMS carrier

The PDMS carrier with particle traps was designed to match the position and dimensions of electrical pads on the substrate. First, a Si master mold was fabricated by photopatterning SU-8 photoresist (Kayaku Advanced Materials). PDMS liquid (Sylgard 184 kit Dow Corning) with a 10:1 ratio of prepolymer to cross-linker was then poured onto the SU-8 master mold and cured for 2 hours at 80 °C. The cured PDMS carrier was then peeled off from the mold to allow for particle deposition (Fig 1a).

D. Particle deposition on PDMS carrier

The experimental setup for depositing conductive particles has been described in previous work [14]. In short, a droplet (20 μl) of colloidal Ag-coated polymer spheres (40 μm in diameter, Ag thickness of 120 nm, supplied by Conpart AS, Norway) in 0.01 wt.% polyethylene glycol tert-octylphenyl ether (Triton X-100, VWR Chemicals) was injected between

the moving PDMS carrier and the fixed glass slide (Fig 1b). The velocity of the carrier was controlled by a syringe pump (Chemyx Mirco), and the temperature was maintained at 27 °C using a thermal control module. The carrier with deposited particles was collected when the droplet reached the end of the carrier.

E. Transferring particles to substrate

A thin layer of nano Ag ink (Silverjet DGP 40LT-15C, Advanced nano products) was spin-coated on a Si wafer at 1500 rpm. A semi-automatic flip chip bonder (Finetech, FinePlacer Pico) was used for ink deposition and bonding process. Particles deposited on the carrier were brought into contact with the ink layer and subsequently aligned with the electrical pads on a substrate (Fig 1c, d). A temperature of 140 °C was applied in 2 minutes to form the first Ag sintered necks between the substrate pads and particles (Fig 1e).

F. Die alignment and final bonding

Sintered particles on a substrate were deposited with an additional Ag ink layer on top by dipping in an ink film spin-coated on a Si wafer (Fig 1f, g). A glass die was immediately picked and placed on the prepared substrate, followed by the second sintering of Ag ink at 140 °C for 2 minutes (Fig 1h). A bonding force of 1 N was applied during the second sintering process. Finally, epoxy underfill was introduced between the glass die and Si substrate followed by curing at 140 °C for 10 minutes while maintaining the applied bonding force (Fig 1i). In this work, five samples were bonded using the same bonding profile and characterized.

G. Characterization

The deposition of particles on the PDMS carrier was observed in real-time using an optical microscope (V12, Zeiss). The electrical resistance of individual interconnects

326

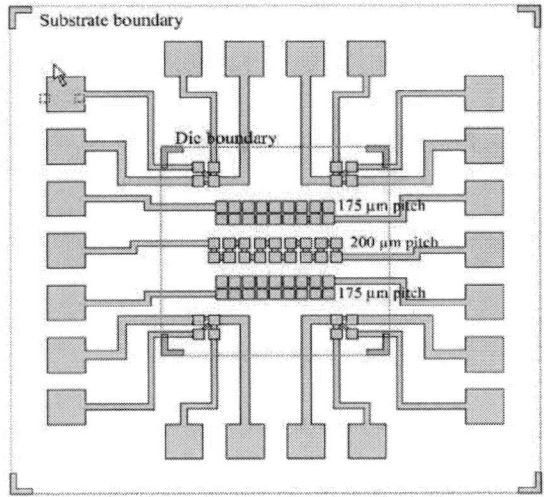

Fig 2. Layout of test sample consisting of Kelvin structures at four corners and three daisy chain structures located at center of the sample.

and daisy chains were measured by means of four-point probe and two-point probe, respectively, using a Keithley 2400 source meter. Cross-sectional microscopy was employed to inspect the deformation of conductive particles between electrical pads on dies and substrates using a scanning electron microscope (SEM SU-3500, Hitachi).

III. RESULTS

A. Fabrication of PDMS carrier and particle deposition

The PDMS carrier with particle traps was successfully fabricated using the replica molding method. The traps had dimensions of 50 µm in diameter and 20 ± 2 µm in depth. After deposition of MPS particles (as sketched in Fig 1b), 692 traps out of 700 were filled with single 40 µm Ag-MPS particles, resulting in a deposition yield of 98.8 %. Fig 3a shows the deposition of 8 particles on predefined traps for the daisy chain structure (175 µm pitch). The position of each deposited particle corresponds to an electrical pad on the substrate.

B. Transferring particles to the substrate using Ag sintering

Fig 3b shows the particles transferred from the PDMS carrier to the electrical pads after sintering at 140 °C, exhibiting the formation of an ink layer between the particle and the electrical pad (Fig 3c). The conductive particles were successfully transferred to the pads while maintaining their arrangement. A shear test demonstrated the creation of ink dots with an ink diameter of 25 ± 5 µm between the particles and electrical pads (Fig 3d). The transfer yield, determined by counting the number of particles successfully transferred to the electrical pads out of the total transferred particles, was found to be 96.2%.

C. Die placement and electrical performance of bonded samples

During the second dip-coating of sintered particles on a substrate to a Ag ink film (Fig 1f), no particles were found detached from the substrate after contacting the wet ink. The results from the electrical characterization of bonded samples are shown in Table I. The resistance of individual interconnects located at the four-point probe structures ranged from 0.57 Ω to 0.74 Ω. The electrical resistance of both daisy

chains with pitches of 200 µm and 175 µm was measured to be 23.8 Ω and 24.3 Ω, respectively. The sintered Ag joint is characterized and shown in Fig 4, with the observation of the Ag neck between the particle and the electrical pad on the substrate.

TABLE I. ELECTRICAL RESISTANCE MEASURED AT THE FOUR-POINT PROBE STRUCTURES AND DAISY CHAINS

Bonding pad structures	Four-point probe	Daisy chains			
		175 µm pitch		200 µm pitch	
		Entire chain	Interconnect *	Entire chain	Interconnect
Resistance [Ω]	0.7 ± 0.1	23.8 ± 1.2	0.8 ± 0.1	24.3 ± 1.4	0.9 ± 0.1

*by using the geometry of the design and the table values for Cr and Au resistivity. Electrical resistance of all metal tracks included in the daisy chains with 175 µm and 200 µm pitch was calculated to be 9.8 Ω and 9.0 Ω, respectively.

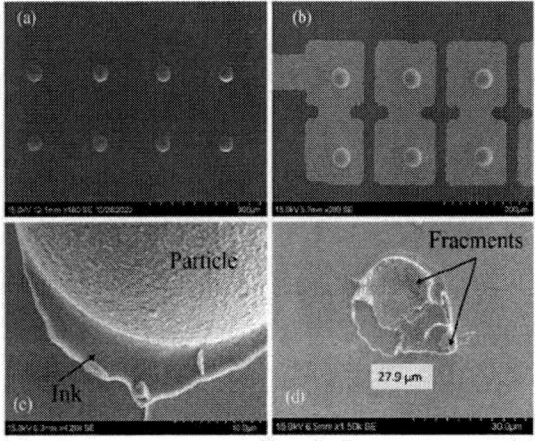

Fig 3. SEM micrographs of (a) particle deposition on PDMS carrier, (b) transferred particles on electrical pads with (c) formation ink dot between particle and pad before and (d) after a shear test.

IV. DISCUSSION

For depositing particles on PDMS carrier, the trap dimensions, particularly the trap depth, are important parameters. On one hand, a shallower depth results in particles being unable to stay in the traps (capillary forces smaller than droplet drag force). On the other hand, increasing the trap depth leads to the formation of particle clusters being deposited in a single trap [14]. Additionally, the trap depth defines the thickness of silver ink to be used in the ink deposition process since the deposited particles on the PDMS carrier come into contact with the ink. The thinner the ink layer, the less ink is deposited on the particles and the faster the evaporation of the solvent in the ink, thus, lowering the transferring yield. In contrast, if the ink is too thick, the surface of the PDMS carrier can be easily contaminated when contacting with thick ink. An optimal trap depth, approximately equal to the particle radius (~20 µm), allows for the deposition of a single particle with a high yield while leaving a reasonable gap (below 20 µm) for Ag ink deposition.

When the particles with deposited ink come into contact with electrical pads, they are deformed, and wet the pad surface, forming ink dots on the pads (Fig 3c). Fragments of

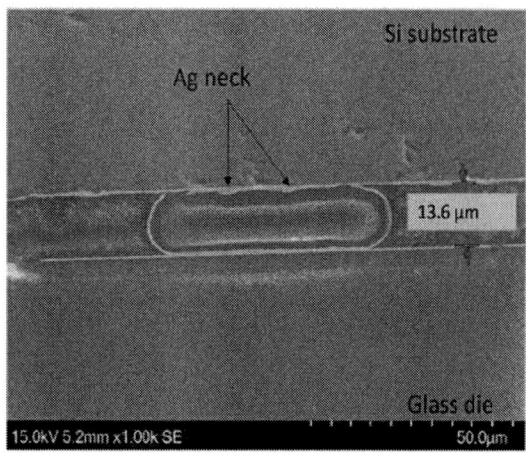

Fig 4. SEM image of cross-section at the four-point probe structure showing particle deformation with the formation of Ag neck

Ag shell on MPS were observed at the bonding interface after the shear test (Fig 3d), indicating that the sintered Ag induces a certain strength between mating particles and pads, although it is challenging to quantify. The sintered Ag plays a crucial role in effectively securing the particles onto the electrical pads, both during the detachment of the PDMS carrier and the subsequent ink deposition process.

The Ag neck was observed between the particle and the electrical pad on a substrate but not between the particle and the electrical pad on a die (Fig 4). This can be explained by the application of 1 N bonding force (corresponding to 14 mN/particle) during the second sintering. The applied force causes the particles to deform and expand, leading to an increased contact area with the electrical pads on the substrate. As a result, an overlap area with the sintered Ag is formed. However, in the case of particles in contact with the electrical pads on the die, the wet ink is drawn up along with the particle deformation. This leads to the formation of a thin sintered layer between the particles and the Au pads on the die.

The estimated electrical resistance of individual interconnects in the daisy chains, with two different pitches of 200 μm and 175 μm, shows similar results. However, these results are slightly higher than the interconnect resistance obtained from the four-point probe measurement (Table I). This difference can be attributed to the contribution of probe contact resistance in the two-point probe measurements as well as uncertainties in the estimation of track resistance. The interconnect resistance achieved as low as 0.57 Ω is significantly higher than other metallic joint technologies (below 50 mΩ), such as micro solder [4], copper pillars with solder cap [16], direct metal-metal [8], and SLID bonding [7] as well as adhesive joints (ICA, ACA) [9], which typically exhibit resistances of 100-200 mΩ. However, while these metallic and adhesive joints require higher bonding temperatures and/or higher pressure over an extended period, the presented approach could be performed at a bonding temperature as low as 140 °C and a bonding pressure as low as 0.1 MPa. To reduce the electrical resistance of interconnects, one can use MPS with a thicker metal coating layer (resistance can be reduced by half if the coating thickness increases threefold, at the same deformation degree) [17]. Achieving finer pitch capability (below 100 μm) is potentially attainable by minimizing the distance between the two closest deposited particles on the PDMS carrier, which can be as close as the particle diameter (40 μm).

V. CONCLUSION

We have demonstrated a new interconnection approach utilizing a single Ag-coated polymer particle for each interconnect. The single-particle interconnects were realized by selectively depositing particles on the PDMS carrier using capillary assembly, and then transferring the particles to mating electrical pads on the substrate and die using thin layers of Ag ink. Bonds between Ag-coated particles and mating pads were achieved by Ag sintering at 140 °C in short time (2 minutes) with low pressure (0.1 MPa). Occupancy rates of 98.8 % for the particle deposition on the PDMS carrier and 96.2 % for the particle transfer to the electrical pads were achieved. The formation of ink dots with well-defined ink dimensions was also accomplished. The resulting interconnect resistance ranged from 0.57 – 0.74 Ω. The measurements of interconnection daisy chains demonstrated reliable sub-ohm resistance across a large number of interconnects. The presented approach enables a low thermo-mechanical stress bonding technique with low bonding temperature and low bonding pressure.

ACKNOWLEDGMENT

The Research Council of Norway is acknowledged for the support to the Norwegian Micro- and Nano-Fabrication Facility (Norfab III, Project number: 295864). We thank Helge Kristiansen (Conpart AS) for providing conductive particles, Daniel Nilsen Wright (SINTEF MiNaLab) for general advice, Thai Anh Tuan Nguyen and Zekija Ramic (USN) for help with the lithography process.

REFERENCES

[1] J. H. Lau, "Recent Advances and Trends in Advanced Packaging," *IEEE Trans. Compon. Packag. Manuf. Technol.*, vol. 12, no. 2, pp. 228–252, Feb. 2022, doi: 10.1109/TCPMT.2022.3144461.

[2] S.-K. Kang, D.-Y. Shih, and W. E. Bernier, "Flip-Chip Interconnections: Past, Present, and Future," in *Advanced Flip Chip Packaging*, H.-M. Tong, Y.-S. Lai, and C. P. Wong, Eds., Boston, MA: Springer US, 2013, pp. 85–154. doi: 10.1007/978-1-4419-5768-9_4.

[3] W. S. Tsai, C. Y. Huang, C. K. Chung, K. H. Yu, and C. F. Lin, "Generational changes of flip chip interconnection technology," in *2017 12th International Microsystems, Packaging, Assembly and Circuits Technology Conference (IMPACT)*, Taipei: IEEE, Oct. 2017, pp. 306–310. doi: 10.1109/IMPACT.2017.8255955.

[4] S. L. Wright *et al.*, "Characterization of Micro-Bump C4 Interconnects for Si-Carrier SOP Applications," in *56th Electronic Components and Technology Conference 2006*, San Diego, CA: IEEE, 2006, pp. 633–640. doi: 10.1109/ECTC.2006.1645716.

[5] J. R. Lee, M. S. A. Aziz, M. H. H. Ishak, and C. Y. Khor, "A review on numerical approach of reflow soldering process for copper pillar technology," *Int. J. Adv. Manuf. Technol.*, vol. 121, no. 7–8, pp. 4325–4353, Aug. 2022, doi: 10.1007/s00170-022-09724-w.

[6] J. Li *et al.*, "Design of Cu nanoaggregates composed of ultra-small Cu nanoparticles for Cu-Cu thermocompression bonding," *J. Alloys Compd.*, vol. 772, pp. 793–800, Jan. 2019, doi: 10.1016/j.jallcom.2018.09.115.

[7] L. Sun, M. Chen, L. Zhang, P. He, and L. Xie, "Recent progress in SLID bonding in novel 3D-IC technologies," *J.*

Alloys Compd., vol. 818, p. 152825, Mar. 2020, doi: 10.1016/j.jallcom.2019.152825.

[8] S. Jangam *et al.*, "Fine-Pitch (≤10 μm) Direct Cu-Cu Interconnects Using In-Situ Formic Acid Vapor Treatment," in *2019 IEEE 69th Electronic Components and Technology Conference (ECTC)*, Las Vegas, NV, USA: IEEE, May 2019, pp. 620–627. doi: 10.1109/ECTC.2019.00099.

[9] R. Aradhana, S. Mohanty, and S. K. Nayak, "A review on epoxy-based electrically conductive adhesives," *Int. J. Adhes. Adhes.*, vol. 99, p. 102596, Jun. 2020, doi: 10.1016/j.ijadhadh.2020.102596.

[10] M. W. Sugden, C. Liu, D. Hutt, D. Whalley, and H. Kristiansen, "Metal-coated mono-sized polymer core particles for fine pitch flip-chip interconnects," in *2012 IEEE 62nd Electronic Components and Technology Conference*, San Diego, CA, USA: IEEE, May 2012, pp. 218–224. doi: 10.1109/ECTC.2012.6248831.

[11] J. Tao, D. Whalley, and C. Liu, "Magnetic deposition of Ni/Au coated polymer core particles for flip-chip interconnection," in *2014 15th International Conference on Electronic Packaging Technology*, Chengdu, China: IEEE, May 2014, pp. 409–413. doi: 10.1109/ICEPT.2014.6922684.

[12] V. C. Duy, H.-V. Nguyen, H. Kristiansen, M. M. V. Taklo, K. E. Aasmundtveit, and N. Hoivik, "Immobilization of metal coated polymer spheres on Indium pads," in *Proceedings of the 5th Electronics System-integration Technology Conference (ESTC)*, Helsinki, Finland: IEEE, Sep. 2014, pp. 1–5. doi: 10.1109/ESTC.2014.6962800.

[13] D. N. Wright, B. D. Belle, and J. S. Graff, "Development of mechanically compliant flip chip interconnect using single metal coated polymer spheres," in *2019 22nd European Microelectronics and Packaging Conference & Exhibition (EMPC)*, Pisa, Italy: IEEE, Sep. 2019, pp. 1–7. doi: 10.23919/EMPC44848.2019.8951849.

[14] V. L. Huynh, K. E. Aasmundtveit, and H.-V. Nguyen, "Selective Deposition of Conductive Particles for Anisotropic Conductive Adhesive Interconnects," in *2022 IEEE 9th Electronics System-Integration Technology Conference (ESTC)*, Sibiu, Romania: IEEE, Sep. 2022, pp. 124–128. doi: 10.1109/ESTC55720.2022.9939469.

[15] J. R. Rumble, Ed., *CRC handbook of chemistry and physics*, 102nd edition 2021-2022. Boca Raton London New York: CRC Press, 2021.

[16] M. Volpert, D. Henry, D. Taneja, T. Chaira, A. Gueugnot, and F. Hodaj, "Cu pillar as interconnect for 10 μm pitch and below: Fabrication issues and assembly results," in *2018 7th Electronic System-Integration Technology Conference (ESTC)*, Dresden: IEEE, Sep. 2018, pp. 1–7. doi: 10.1109/ESTC.2018.8546408.

[17] M. Bazilchuk, S. R. Pettersen, H. Kristiansen, Z. Zhang, and J. He, "Electromechanical characterization of individual micron-sized metal coated polymer particles," *J. Appl. Phys.*, vol. 119, no. 24, p. 245102, Jun. 2016, doi: 10.1063/1.4954218.

20 μm copper micro-bump bonding through a silver metallization for advanced packaging under a low-pressure condition

Zheng Zhang[1], Ming-Chun Hsieh[1], Aiji Suetake[1], Hiroshi Yoshida[1], Rieko Okumuara[1], Noriko Kagami[1], Kazamasa Okamoto[1], Chuantong Chen[1], Kei Hashizume[2], Norihiko Hasegawa[2], Ryuji Yoshida[2], Hidekazu Homma[2], and Katsuaki Suganuma[1]

[1]*Flexible 3D System Integration Lab, SANKEN, Osaka University, 5670047, Japan*
[2]*Okuno Chemical Industries Co. Ltd, 5410045, Japan*
Corresponding author email: zhangzheng@sanken.osaka-u.ac.jp

Abstract—**In this work, we proposed a copper (Cu) micro-bump bonding strategy by using a silver (Ag) metallization layer as an interconnect bridge. Si chips with 20 μm Cu-Ag bumps were bonded under a low pressure (0.4 MPa) and atmospheric conditions at 250 °C. By investigating the surface morphology evolution of the Cu-Ag bump, it is found that severe surface deformation caused by Ag abnormal grain growth and hillock generation plays a critical role in the Cu-Ag bump bonding, which facilitates the approach and bonding of top and bottom Ag surface of the bump structure. Moreover, this bonding strategy is expected to be particularly advantageous for fine-pitch interconnect applications. It not only mitigates the severe bonding condition requirements that are used in thermal compression bonding and hybrid bonding, but also provides higher quality bonding structures for advanced packaging.**

Keywords—*micro-bumps, advanced packaging, low pressure bonding, low temperature bonding*

I. INTRODUCTION

In the last few decades, the rapid development of material science and high precision engineering have enabled us to scale down the transistor size and integrate more transistors in a confined space. However, scaling down the transistor size are becoming more and more difficult due to the limitation of ultraviolet photolithography, making the two-dimensional integrated circuits (ICs) approach the limit of Moore's law. To address such a dilemma, one popular solution is to utilize the vertical space by stacking Si dies, which is also known as advanced packaging. On the one hand, stacking of Si dies can significantly increase the number of transistors that can be accommodated within the limited space. This greatly benefits the continuation of the enhancement in the performance and functionality of the ICs. On the other hand, the wire distance can be reduced using the through Si vias that are buried in the Si die, which enables faster data transmission with less power consumption.

Micro-bumps bonding technology is one of the critical approaches that has been widely applied in advanced packaging, which can provide the vertical interconnect among the stacked dies. The micro-bumps, consisting of a Cu post with a Ag-Sn solder cap, are made by electroplating. During the packaging procedure, the micro-bumps are aligned and in contact with the bonding pad, and the solder caps act as a bridge that provides the connection between micro-bumps and bonding pads after reflowing. Currently, the most advanced bump bonding used for commercial products owns a 40 μm pitch and 20 μm bump size[1]. In order to meet the rapidly increased input and output number in the high-end devices, researchers are engaging in the development of fine pitch bump bonding (less than 40 μm) as well as the availability of them in a realistic application[2, 3].

Although massive efforts have been made on the fine pitch bump bonding, there are many concerns about the reliability of the bump bonding in a longtime service due to brittle intermetallic compounds (IMCs) generation in the bonding structure. The Sn in the solder cap can react with the Cu in the pillar, resulting in brittle Cu-Sn IMCs at the bonding interface. In addition, current density among the bump becomes higher due to the decrease of the bump size, which gives rise to the voids and degradation caused by electrical migration. Therefore, researchers are seeking for alternatives to replace the solder cap, which not only provides a robust bonding but also high reliability.

Advanced packaging using a pure metal passivation layer is a promising alternative to the current solder-capped micro-bump bonding technique. The Cu-to-Cu interconnect can be achieved through the use of a pure metal intermediate layer at high temperatures and pressures, without concerns about the generation of intermetallic compounds (IMCs). Chen et al. have utilized thin layers of Pd[4], Au[5], Cr[6], and Ag[7] to cover the surface of Cu, resulting in high-quality bonding structures under high-vacuum conditions. Chang et al. [8] prepared pure Ag bumps and conducted Ag-Ag direct bonding under atmospheric conditions at high bonding pressures. The bonding exhibited high strength due to the oxidation-reduction properties of Ag. However, requirements such as high bonding strength or vacuum conditions have greatly limited the widespread application of passivation layer bonding.

In this work, we proposed a Cu bump bonding method with the assistance of a Ag passivation layer. Unlike the previously reported works, we achieved the bonding under a low bonding pressure condition of only 0.4 MPa. In addition, we prepared the 20 μm Cu-Ag bumps via sputtering, and demonstrated the feasibility of the Ag layer in the fine pitch bump interconnect.

II. EXPERIMENT

A. Preparation of bonding samples

Two series of bonding samples, sputtered bare Si chip and sputtered Cu-Ag bumps were prepared for the investigation of bonding performance and demonstration of fine pitch bonding.

The bare Si chips with a size of 5×5 mm^2 and 10×10 mm^2 were sputtering with Ti (100 nm)/Cu (500 nm)/Ti (100 nm)/Ag (500 nm) under a sputtering power of 200 W and a chamber pressure of 0.2 Pa, separately. The small Si chip was

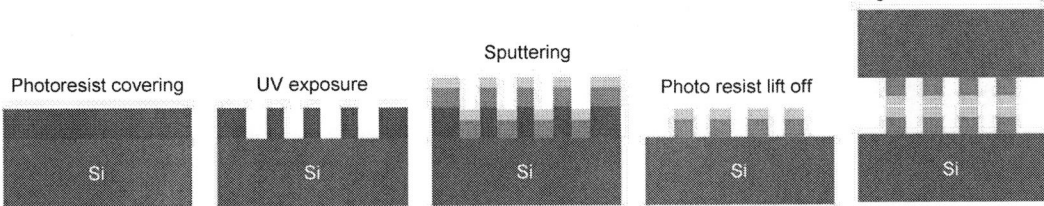

Fig. 1 Schematic drawing of the Cu-Ag bump sample preparation processes.

then bonded to the large Si chip under a low bonding pressure (0.4 MPa) and an atmospheric condition at 250 °C for 30 min. The prepared samples were then carried out the non-destructive and destructive characterizations for investigating the face to face bonding performance.

The process for the preparation of the Cu-Ag bump sample is depicted in Fig.1. At first, bare Si chips (10×10 mm^2) were covered with photoresist and exposed to UV light using a photomask. Once the patterning is complete, the chip is placed in a sputtering machine for metallization with a layer sequence of Ti (50 nm)/Cu (250 nm)/Ti (50 nm)/Ag (250 nm). To enhance the quality of the bumps, the thickness of the metallization layer is reduced, which facilitates the liftoff of the photoresist. At last, the liftoff Si samples are aligned using a flip-chip bonder (T-3000, Tresky) and bonded under low pressure of 0.4 MPa and atmospheric conditions at a temperature of 250 °C for 30 minutes.

B. Characterizations

The bonding quality of the large area bonding sample was investigated via scanning acoustic microscopy (SAM, FineSAT, HITACHI) at first and then cleaved for the cross-sectional structure observation. The cleaved sample was polished by an ion-milling machine (IM4000, HITACHI). Scanning electron microscopy observation was carried out using the SU-8000 (HITACHI high tech.). Surface roughness of the sputtered and annealed Si chip was measured by atomic force microscopic (AFM5000, Hitachi High-tech).

III. RESULTS AND DISCUSSION

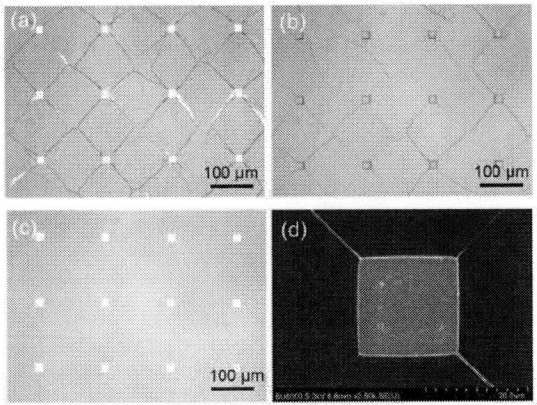

Fig. 2 Images of photoresist after masking (a), after sputtering (b), after liftoff (c), and SEM image of the prepared Cu-Ag bump.

Images in the process of Cu-Ag bump preparation were taken, as shown in Fig. 2. Fig. 2a shows the optical image of the photoresist after masking. Square holes with an edge length of 20 μm and a pitch length of 150 μm were prepared after UV exposure. Then, the Si chips were put into a

sputtering machine for continuous Ti/Cu/Ti/Ag metallization. These square holes were filled with the multi-metallization layers as shown in Fig. 2b. After liftoff, the photoresist with the covered sputtering layer was removed and the metallization in the square holes left on the chip as depicted in Figure C. Figure d shows the SEM observation of the prepared Cu-Ag bump. The bump surface shows a clean surface and there is no photoresist residue left at the edge of the bump, which indicated a total removal of the photoresist.

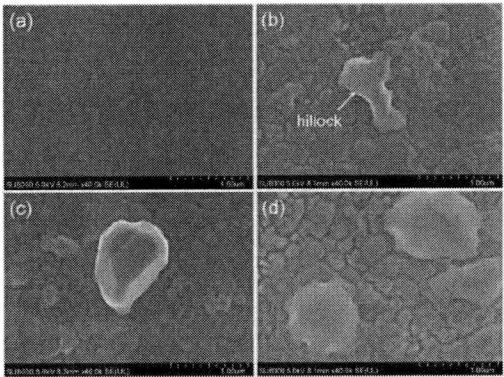

Fig. 3 Surface morphology variation of the Cu-Ag bump. As-sputtering (a), 5 min annealed (b), 15 min annealed (c), and 30 min annealed (d).

Morphologies of the sputtered Ag surface were examined before and after heating, as shown in Fig. 3. The as-sputtered Ag surface (Fig. 3a) appears a smooth surface with fine Ag grains that have a size of around 100 nm. Fig. 3b shows the Ag surface morphology after 5 min annealing at 250 °C. Size of the Ag grains became larger than the as-sputtered grain due to the grain growth at a high temperature. A large Ag grain, approximately 500 nm in size, protrudes from the surface, resulting from abnormal grain growth known as hillock generation. Interestingly, massive nanoparticles are observed on the surface, which can be attributed to the oxidation of Ag grains. Ag trends to absorb the oxygen and generate Ag oxide before 420 K[9]. However, the generated Ag oxide decomposes into pure Ag and oxygen with a further increase of temperature[10]. Thus, this reversible reaction leads to the formation of numerous in-site nanoparticles on the Ag surface. Fig. 3c presents the Ag surface morphology after 15 min annealing. The Ag fine grains merge into together and the hillock becomes larger due to the extended annealing time. With 30 min of annealing, more hillocks appear on the surface due to further abnormal grain growth.

In addition, surface roughness was measured to confirm the variation in Ag surface during the annealing. Fig. 4 presents the AFM images of the initial Ag surface and the surface after 5 min annealing. The as-sputtered Ag surface

331

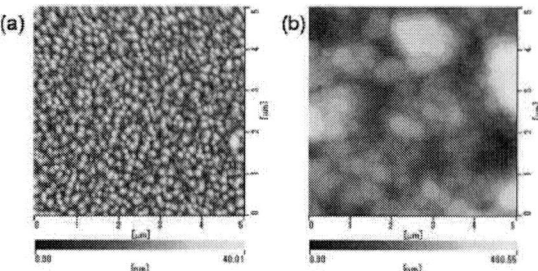

Fig. 4 AFM images of the as-sputtered surface (a) and 5 min annealed surface (b).

shows a uniform morphology with massive fine Ag grains (Fig. 4a). The depth range on the surface is less than 40 nm. However, after 5 min of annealing (Fig. 4b), the surface morphology has completely altered, characterized by the presence of larger Ag grains and hillocks. Simultaneously, the depth range increases to over 460 nm due to severe abnormal grain growth. The surface roughness (arithmetical mean height, Sa) of the as-sputtering Ag surface is 5.12 nm, whereas the roughness significantly increased to 65.84 nm which is almost 13 times higher than the initial state.

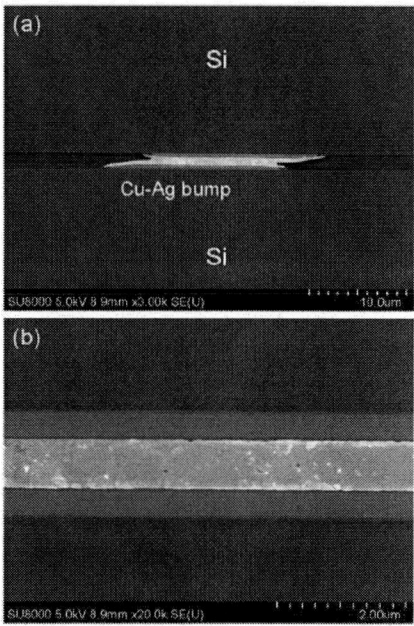

Fig. 5 SEM images of the cross section of the bonded Cu-Ag bump sample.

Cross section of the prepared Cu-Ag bump bonding sample was observed and investigated via SEM. Fig. 5a shows the overall view of the bump structure. The topside Cu-Ag bump intimately bonds to the bottom one through the Ag layer without a gap in between. A slight misalignment of about 4 μm occurred caused by the bonding instrument deviation. A magnified view of the bonding interface is exhibited in Fig. 5b. The multi-layer structure of the Cu-Ag bump can be clearly distinguished. The Ag layer has completely become an entirety and the bonding interface cannot be identified anymore, which is on account of the severe abnormal grain growth and hillock generation. The severe deformation of the

Ag makes a much more easy encounter of the top and bottom surface, leading to the bonding of the Cu-Ag bumps. This uniform bonding structure suggests the Ag layer can provide a good bonding quality for the chip to chip interconnect.

IV. CONCLUSIONS AND OUTLOOK

In this work, we realized a 20 μm Cu bump bonding by applying a Ag intermediate layer as a bonding bridge. The 20 μm bump structure, consisting of Ti/Cu/Ti/Ag was prepared by continuous sputtering. The bonding of the prepared Cu-Ag bump sample was accomplished under low pressure (0.4 MPa) and atmospheric condition at 250 °C by using a flip-chip bonder. By investigating the Ag surface morphology variation, severe deformation due to the abnormal grain growth and hillock generation occurred after the annealing process. The AFM results suggest that surface roughness changed from 5.12 nm to 65.84 nm after the annealing. The deformed surfaces allow the contact of the top and bottom, thus creating a robust and uniform bonding interface. This bonding strategy shows great potential for achieving the mild bonding condition for the advanced packaging, which is expected to cut down manufacturing costs and improve the throughput of the packaging.

ACKNOWLEDGMENT

This paper is based on results obtained from a project from "Research and Development Project of the Enhanced Infrastructures for Post-5G Information and Communication Systems" JPNP20017, subsidized by the New Energy and Industrial Technology Development Organization (NEDO).

REFERENCES

[1] M. Datta "Manufacturing processes for fabrication of flip-chip micro-bumps used in microelectronic packaging: An overview". *Journal of Micromanufacturing*, 2020, 3(1): 69-83.

[2] M. Mariappan *et al*., "Tight-Pitched 10 μm-Width Solder Joints for c-2-c and c-2-w 3D-Integration in NCF Environment," *2022 IEEE 72nd Electronic Components and Technology Conference (ECTC)*, San Diego, CA, USA, 2022, pp. 1138-1143.

[3] M. H. Chan *et al*., "Comparison of 3D Packages with 20μm bump pitch using reflow soldering and thermal compression bonding," *2021 IEEE 71st Electronic Components and Technology Conference (ECTC)*, San Diego, CA, USA, 2021, pp. 481-486.

[4] Y. -C. Tsai, H. -W. Hu and K. -N. Chen, "Low Temperature Copper-Copper Bonding of Non-Planarized Copper Pillar With Passivation," in *IEEE Electron Device Letters*, 2020,vol. 41, no. 8, pp. 1229-1232.

[5] Z. -J. Hong *et al*., "Room Temperature Cu-Cu Direct Bonding Using Wetting/Passivation Scheme for 3D Integration and Packaging," *2022 IEEE Symposium on VLSI Technology and Circuits (VLSI Technology and Circuits)*, Honolulu, HI, USA, 2022, pp. 387-388.

[6] D. Liu, P. -C. Chen and K. -N. Chen, "A Novel Low-Temperature Cu-Cu Direct Bonding with Cr Wetting Layer and Au Passivation Layer," *2020 IEEE 70th Electronic Components and Technology Conference (ECTC)*, Orlando, FL, USA, 2020, pp. 1322-1327.

[7] T. -C. Chou *et al*., "Electrical and Reliability Investigation of Cu-to-Cu Bonding With Silver Passivation Layer in 3-D Integration," in *IEEE Transactions on Components, Packaging and Manufacturing Technology*, vol. 11, no. 1, pp. 36-42, Jan. 2021.

[8] Chang, Leh-Ping, et al. "Low temperature Ag-Ag direct bonding under air atmosphere." *Journal of Alloys and Compounds* 862, 2021pp. 158587.

[9] M. L. Zheludkevich, et al. "Oxidation of silver by atomic oxygen." *Oxidation of metals* 61.1-2, 2004, pp. 39-48.

[10] J.Assal, B. Hallstedt, L. J. Gauckler. "Thermodynamic assessment of the silver–oxygen system". *Journal of the American Ceramic Society*, 1997, 80(12). pp. 3054-3060.

Practical study to demonstrate an increase in the reliability of flip chip connections by adding nanoparticles to solder

David M. Harvey,
General Engineering Research Institute
Liverpool John Moores University
Liverpool, UK
d.m.harvey@ljmu.ac.uk

Teresa Partida Manzanera
University of Liverpool
Liverpool, UK
T.Partida-Manzanera@liverpool.ac.uk

Kangkana Baishya
Assam Engineering College
Guwahati, India
kbaishya010@gmail.com

Guangming Zhang
General Engineering Research Institute
Liverpool John Moores University
Liverpool., UK
g.zhang@ljmu.ac.uk

Mohd Arif Anuar Mohd Salleh Arif
UniMAP
Perlis, Malaysia
arifanuar@unimap.edu.my

Y. C. Chan
Liverpool John Moores University
Liverpool, UK
eeycchan@gmail.com

Nduka Ekere
Liverpool John Moores University
Liverpool., UK
n.n.ekere@ljmu.ac.uk

Derek Braden
Aptiv
Coventry, UK
derek.braden@apiv.com

Abstract— **The aims of the research are to improve the reliability and lifetimes of solder joints to contribute to the green economy, by reducing waste and extending recycling times. The objectives of the work are; to improve the materials properties of various solders used for attaching die/flip chip packages to PCBs by adding nanoparticles to the mix, to fabricate industrial test samples using flip chips soldered to PCBs, to test the reliability of the solder through accelerated testing, to examine failure modes and lifetime extensions.**

Keywords—nanoparticles in solder, solder joint reliability

I. INTRODUCTION

For over 50 years, SnPb solder has been used for electronic packaging due to its good solderability, good reliability and low cost[1][2][3][4]. However, lead/Pb is toxic for humans and harmful for the environment. The European Union's (EU) Restriction of Hazardous Substances (RoHS)[5] and Waste of Electrical and Electronic equipment (WEEE)[6] Directives mandated the elimination of SnPb solder from consumer electronics sold in EU market in 2006, and later from other products worldwide. Since then, Sn3.0Ag0.5Cu (SAC305) has become a popular lead-free solder alternative in the packaging industry for attaching electronic devices to printed circuit boards (PCBs)[7]. The main advantage of SAC305 is its lower melting point with respect most Pb-free solder alloys [8]. However, the high cost of Ag has driven the industry to look at less expensive alternatives such as low-silver alloys with Ag content <1.0 w% or Ag free alternatives such as SnCu-based solders[9]. Currently, SnCu solders with dopants such as Ni[10], Ge[11][12], Co[13] and Bi[14] or nanoparticles (NPs) such as Al2O3[15], TiO2[16], SiO2[17], or ZnO[18] to improve their performance, are the least expensive and most promising lead-free alternative[9]. Currently, there is a need to evaluate their reliability outside consumer applications, for example automotive applications where electronic components are often exposed to harsher environments[19]. This work will examine factors for improving solder reliability in the round, by considering electrical, mechanical and materials factors required, and which current research has made available for connections between components and circuit on a PCB.

Firstly a solder joint must satisfy the electrical connection required between circuits components. This must be reliable over a product's lifetime, and have the lowest possible electrical resistance. Secondly, the mechanical joint must be strong enough to keep all components attached during any conditions a product may face during its lifetime. Thirdly, the solder material must be designed to be solderable enabling easy manufacture of solder connections, and remain sufficiently stable to maintain the electrical and mechanical connection throughout a product's lifetime. With the thousands of solder joints in modern products it is worth reminding all users that if only one joint fails a product could fail, so reliability of every solder connection is imperative.

Research work conducted at LJMU and elsewhere during the last twenty years has found that not only the individual solder joints, but other factors can influence the reliability and degradation of electrical connection over time. It has been shown that in normal operation and in accelerated testing solder joints may crack and become open circuit rendering a circuit not fit for use. This can have catastrophic results such as plane crashes, car accidents and major disasters. It is fairly well known that materials sets and factors in the manufacturing such as choice of PCB surface finish, UBM on chip connections, as well as the solder chosen will influence reliability. It is less well known that the position of chips on a PCB will effect solder joint lifetimes, so PCB floorplan/layout and solder joint positions on a chip relative to the PCB edges can accelerate failure. Double-sided PCBs are interesting as failure modes based on chip position on a PCB can be different for thermal cycling and vibration tests. This makes the optimum choices for each chip's position difficult and application dependent, to be considered by the circuit designers and reliability engineers.

There now follows a short review of the research into the addition of nanoparticles into lead-free solder, some highlights demonstrating their value to the integrity of the solder joint, practical test results from the team, future research work to refine the solder design, manufacture and test, and brief conclusions.

II. SHORT REVIEW OF NANOPARTICLE RESEARCH FOR SOLDER JOINTS

New solders have been developed by the academic partners using a variety of nanoparticles (NPs) introduced into various base solders. These include SAC 305, Sn0.7Cu and low temperature solders such as Sn58Bi. Solder balls were manufactured to connect industrial grade flip chips to FR4 PCBs with devices having 109 and 12 balls. Once connections made the reliability of interconnection was testing through thermal cycling and through vibration. Aggressive test profiles were used to accelerated the tests and suitable for application in the automotive sector. Solder joints were scanned using ultrasound for cracks at appropriate intervals as the test progressed to non-destructively monitor progression of any failures from manufacture to end-of-life.

Previous work has introduced nanoparticles of silver or tungsten to Sn58Bi and will be briefly reported here.

Sn58Bi is a contender when low temperature soldering is required for use where low temperature packaging, heat sensitive devices and low CTE mismatch is necessary. Studies have been concluded to show that by the addition of nanoparticles to Sn58Bi solder the mechanical strength can be enhanced [28]. For 1% weight of silver/Ag nanoparticles added to Sn58Bi it can be seen that the nanostructure in Figure 1 for 76nm Ag particles is more refined than for the standard solder and for 31nm or 133nm Ag particles, and has the best mechanical properties.

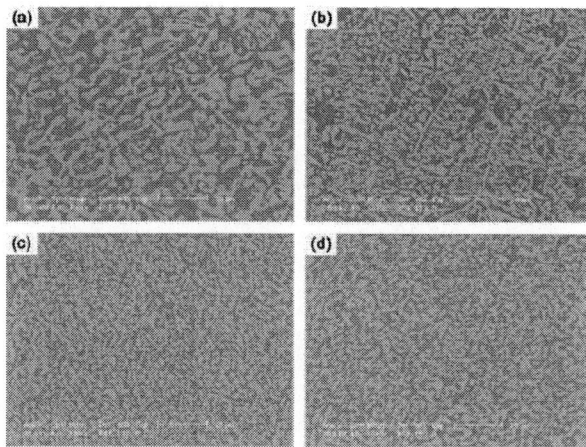

Fig.1. Microstructures of solders: (a) Sn58Bi, (b) 31 nm Ag particle doped, (c) 76 nm Ag particle doped and (d) 133 nm Ag particle doped. [20]

Fig. 1 illustrates how nanoparticles can refine microstructure in Sn58Bi solder when an optimum nanoparticle size found,

and Fig. 2 how silver nanoparticles have increased micro hardness and shear strength in Sn58Bi solder [20].

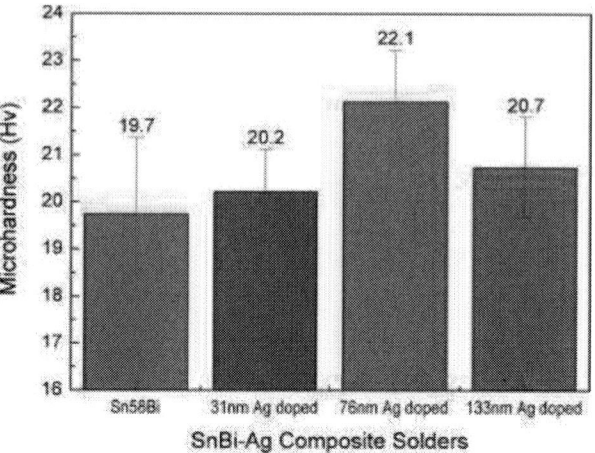

Fig. 2. Vickers microhardness of Ag-free and Ag-containing composite solders.[20]

Another interesting study illustrated that the addition of 1% weight of Tungsten/W to Sn57.6Bi0.4Ag improved the properties [22] Fig. 3. The improved micro structure contributes to enhanced physical properties. The microhardness increased by 6.2% and the shear strength of the solder was enhanced by 8.2%.

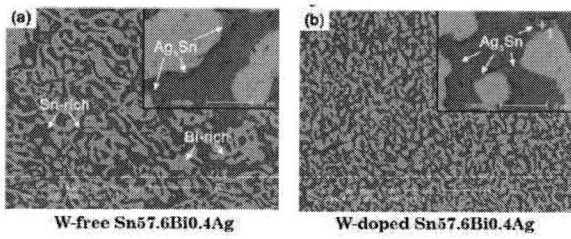

Fig. 3. Microstructure of (a) Sn57.6Bi0.4Ag solder and (b) Sn57.6Bi0.4Ag-W nanocomposite solder. [22]

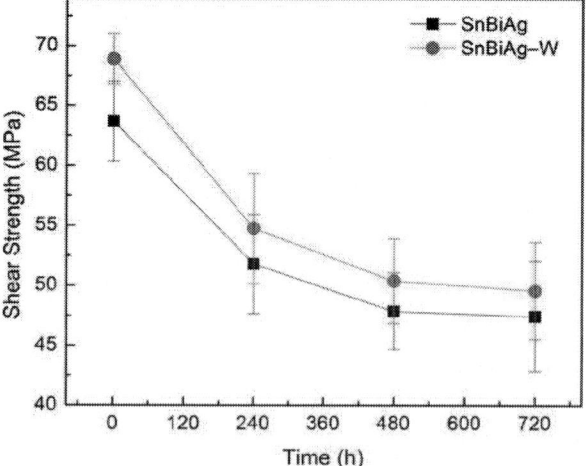

Fig. 4. Shear strength v aging time for SnBiAg and SnBiAg with W nanoparticles solder balls [22]

Fig. 4 shows for 720 hours aging at 100 degrees C shear strength performance was improved throughout the test by the addition of W nanoparticles.

In an attempt to create a greener solder the popular SAC305 solder alloy was reinforced with kaolin geopolymer ceramic particles [24]. The particles were at the micro scale rather than nano scale but the results were significant. SAC305 solder powder with mean particle size of 31.557 microns was mixed with small amounts of kaolin geopolymer ceramic powder with a mean particle size of 18.41 microns. Various weight percentage (wt.%) of kaolin geopolymer ceramics (0, 0.5, 1.0, 1.5 and 2.0 wt.%) were used to form the solder samples.

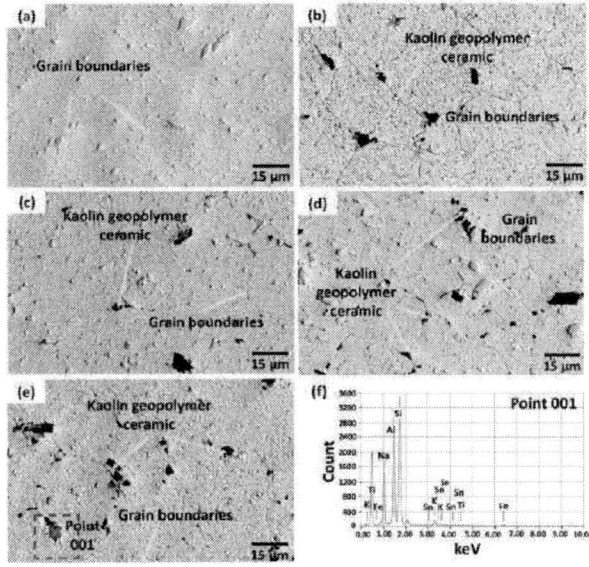

Fig. 5. SEM micrograph of sintered SAC305/kaolin geopolymer ceramic samples showing distribution of kaolin geopolymer ceramic at (a) 0 wt.% kaolin geopolymer ceramic, (b) 0.5 wt.% kaolin geopolymer ceramic, (c) 1.0 wt.% kaolin geopolymer ceramic, (d) 1.5 wt.% kaolin geopolymer ceramic, (e) 2.0 wt.% kaolin geopolymer ceramic and (f) EDX point analysis at Point 001.

The average shear strength of kaolin geopolymer ceramic enhanced SAC305 is given in Fig. 6 (a). All additions of the geopolymer from 0.5 weight % to 2 weight % enhanced the strength with a peak of 13.01 MPa when 1% weight of geopolymer added. Fig. 6 shows surface features for all the solders tested after fracture. More dimples were observed in the samples containing kaolin geopolymer ceramic which correlates with better plastic properties for ductile materials. However, as the addition of kaolin geopolymer ceramic beyond 1.0 wt.%, large dimples were observed at the fracture surface which may explain the drop in the shear strength of the samples at 1.5 wt.% and 2.0 wt.% .

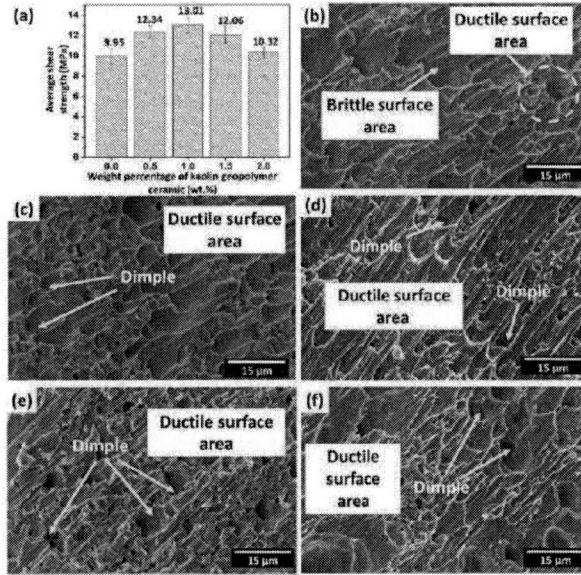

Fig. 6. (a) Average shear strength of SAC305 lead free solder with different wt.% kaolin geopolymer ceramic, SEM micrograph of fracture surfaces of pure SAC305 and kaolin geopolymer ceramic reinforced SAC305 composite solders at different weight percentage of kaolin geopolymer ceramic; (b) 0 wt. % kaolin geopolymer ceramic, (c) 0.5 wt.% kaolin geopolymer ceramic, (d) 1.0 wt.% kaolin geopolymer ceramic, (e) 1.5 wt.% kaolin geopolymer ceramic and (f) 2.0 wt.% kaolin geopolymer ceramic.

III. PRACTICAL RESULTS DEMONSTRATING VALUE OF NANOPARTICLES TO IMPROVE SOLDER MECHANICS

A recent review of lead-free solders that have incorporated micro or nano particles was completed in 2020/2021 [21]. Work to date has demonstrated the potential to improve some properties of solder joints by:

1. Incorporation of a small percentage of nanoparticles to a solder material. Typically up to 1% by weight, but smaller fractions have been reported with positive results.
2. Addition of a very small percentage of nanoparticles has a negligible influence on the melting points of each solder alloy.
3. The formation and growth during a solder joints lifetime of inter-metallic compounds (IMCs) is limited by suitable nanoparticles.
4. Optimum size of nanoparticles has to be found for each additive, as particle agglomeration and adsorption need to be considered for each case.
5. Mechanical strength of solder can be increased, reported at up to 18.9% [20].
6. Solder with nanoparticles fatigue life extended by up to 41.9% [21, 25]
7. Electrical conductivity will be changed but only marginally.
8. Electro-migration is hampered by the new structures that contain nanoparticles, increasing reliability
9. Thermo-migration is also reduced, improving longevity.

IV. ONGOING RESEARCH WORK

This leads to research to evaluate the impact of the addition of the geopolymer ceramic NPs on the reliability of Sn0.7Cu solder. For this two custom designed flip-chip packages were fabricated using Sn0.7Cu and Sn0.7Cu+nanoparticle solder joints, respectively. Test fixtures were made with 9 silicon flip-chips with area ~9 x 4 mm and thickness 525 μm mounted on a ~ 10 cm x 10 cm area and 1.6 mm thick FR4 PCB, using twelve 600 μm diameter solder balls to solder each chip onto the PCB. Accelerated thermal cycling (ATC) tests were set according to the JEDEC standard 22-A104D test condition B with a maximum temperature of 125 °C, a minimum temperature of - 55 °C, and a dwell time of 15 min at each temperature. The total cycle time was 71 min, with average ramp up rate of 9.7 °C/min and ramp down rate of 8.8 °C/min. At this time reliability has proved to be 100% as after 500 hours of thermal cycling no errors have occurred. Further reliability results will be presented as soon as any failures occur.

V. CONCLUSIONS

The addition of nanoparticles to solder definitely has benefits. By the addition of < 1% weight of a suitable nanoparticle many physical and engineering properties can be enhanced. Although research work is still expanding and solder mixtures are constantly being developed to meet various applications, so far no perfect solution has been disseminated or knowingly taken up by manufacturing industries. It is amazing that such a small percentage of added nanoparticles can increase the mechanical strength on solder bonds, slow down degrading of IMC layers and slow electromigration and thermo-migration. The joint working between materials scientists and engineers will continue to increase our understanding of how and why solder joints fail and how we can extend solder joint lifetimes, improve reliability and hence make electronics greener as less failure means less recycling.

ACKNOWLEDGEMENT

This research was partially supported by the British Council Newton-Ungku Omar Fund Institutional Links [Green Electronics Project ID: 332397914], and the EU H2020-MSCA-RISE -2019 [Reactive Too Project: 871163].

REFERENCES

[1] A. E. Geckinli and C. R. Barrett, "Superplastic deformation of the Pb-Sn eutectic," *J. Mater. Sci.*, 1976, doi: 10.1007/BF00540932.

[2] M. M. I. Ahmed and T. G. Langdon, "Ductility of the superplastic Pb-Sn eutectic at room temperature," *J. Mater. Sci. Lett.*, 1983, doi: 10.1007/BF00725431.

[3] R. H. & W. J. P. H. Jiang, "High-strain fatigue of Pb-Sn eutectic solder alloy," *J. Mater. Sci.*, vol. 31, pp. 6455–6461, 1996.

[4] F. A. Stam and E. Davitt, "Effects of thermomechanical cycling on lead and lead-free (SnPb and SnAgCu) surface mount solder joints," *Microelectron. Reliab.*, 2001, doi: 10.1016/S0026-2714(01)00087-7.

[5] European Parliament and the Council of the European Union, "Directive 2002/95/EC: Restriction of the use of Certain Hazardous Substances in Electrical and Electronic Equipment (RoHS)," *Off. J. Eur. Union*, 2003.

[6] European Parliament and European Council, *Directive 2002/96/EC: Waste electrical and electronic equipment (WEEE)*. 2003.

[7] IPC, "Round Robin Testing and analysis of lead free Solder pastes with alloys of Tin, Silver and Copper," *Ipc Solder Products Value Council*, 2013.http://www.ipc.org/3.0_Industry/3.5_Councils_Associations/3.5.1_Industry_Assoc/spvc_bro/SVPC_Final_ExecSumm.pdf.

[8] H. R. Kotadia, P. D. Howes, and S. H. Mannan, "A review: On the development of low melting temperature Pb-free solders," *Microelectronics Reliability* 2014, doi: 10.1016/j.microrel.2014.02.025.

[9] S. Cheng, C. M. Huang, and M. Pecht, "A review of lead-free solders for electronics applications," *Microelectronics Reliability.* 2017, doi: 10.1016/j.microrel.2017.06.016.

[10] A. Gyenes, A. Simon, P. Lanszki, and Z. Gacsi, "Effects of nickel on the microstructure and the mechanical properties of Sn-0.7Cu lead-free solders," *Arch. Metall. Mater.*, 2015, doi: 10.1515/amm-2015-0151.

[11] N. Superior, "SN100 high reliability." [Online]. Available: http://nihonsuperior.co.jp/english/wp-content/themes/nihonsuperior/pdf/pdoduct/download/catalog/preform/catalog_sn100c_preform_eng.pdf.

[12] M. Hasnine and N. Vahora, "Microstructural and mechanical behavior of SnCu–Ge solder alloy subjected to high-temperature storage," *J. Mater. Sci. Mater. Electron.*, 2018, doi: 10.1007/s10854-018-8908-4.

[13] C. Andersson, P. Sun, and J. Liu, "Tensile properties and microstructural characterization of Sn-0.7Cu-0.4Co bulk solder alloy for electronics applications," *Journal of Alloys and Compounds.* 2008, doi: 10.1016/j.jallcom.2007.03.028.

[14] M. I. I. Ramli, M. A. A. Mohd Salleh, H. Yasuda, J. Chaiprapa, and K. Nogita, "The effect of Bi on the microstructure, electrical, wettability and mechanical properties of Sn-0.7Cu-0.05Ni alloys for high strength soldering," *Mater. Des.*, 2020, doi: 10.1016/j.matdes.2019.108281.

[15] X. L. Zhong and M. Gupta, "Development of lead-free Sn-0.7Cu/Al2O3 nanocomposite solders with superior strength," *J. Phys. D. Appl. Phys.*, 2008, doi: 10.1088/0022-3727/41/9/095403.

[16] L. C. Tsao, C. H. Huang, C. H. Chung, and R. S. Chen, "Influence of TiO 2 nanoparticles addition on the microstructural and mechanical properties of Sn0.7Cu nano-composite solder," *Mater. Sci. Eng. A*, 2012, doi: 10.1016/j.msea.2012.03.025.

[17] Z. Fathian, A. Maleki, and B. Niroumand, "Synthesis and characterization of ceramic nanoparticles reinforced lead-free solder," *Ceram. Int.*, 2017, doi: 10.1016/j.ceramint.2017.01.067.

[18] A. F. Abd El-Rehim, H. Y. Zahran, and A. M. Yassin, "Microstructure evolution and tensile creep behavior of Sn–0.7Cu lead-free solder reinforced with ZnO nanoparticles," *J. Mater. Sci. Mater. Electron.*, 2019, doi: 10.1007/s10854-018-0492-0.

[19] K. Choi, D. Y. Yu, S. Ahn, K. H. Kim, J. H. Bang, and Y. H. Ko, "Joint reliability of various Pb-free solders under harsh vibration conditions for automotive electronics," *Microelectron. Reliab.*, 2018, doi: 10.1016/j.microrel.2018.05.006.

[20] Yi, L.; Chan, Y.C. Effect of silver (Ag) nanoparticle size on the microstructure and mechanical properties of Sn58Bi-Ag composite solders. J. Alloy. Compd. 2015, 645, 566–576.

[21] Mu-lan Li, Liang Zhang, Nan Jiang, Lei Zhang, Su-juan Zhong, Materials modification of the lead-free solders incorporated with micro/nano-sized particles: A review,Materials & Design, Volume 197, 2021, 109224, doi.org/10.1016/j.matdes.2020.109224.

[22] Yi Li, Kaiming Luo, Adeline B.Y. Lim, Zhong Chen, Fengshun Wu, Y.C. Chan, Improving the mechanical performance of Sn57.6Bi0.4Ag solder joints on Au/Ni/Cu pads during aging and electromigration through the addition of tungsten (W) nanoparticle reinforcement, Materials Science and Engineering: A, Volume 669,2016, pp 291-303,doi.org/10.1016/j.msea.2016.05.092.

[23] Zhang, P.; Xue, S.; Wang, J.; Xue, P.; Zhong, S.; Long, W. Effect of Nanoparticles Addition on the Microstructure and Properties of Lead-Free Solders: A Review. Appl. Sci. 2019, 9, 2044. https://doi.org/10.3390/app9102044.

[24] N.S. Mohamad Zaimi, M.A.A. Mohd Salleh, M.M.A.B. Abdullah, R. Ahmad, M. Mostapha, S. Yoriya, J. Chaiprapa, G. Zhang, D.M. Harvey, Effect of kaolin geopolymer ceramic addition on the properties of Sn-3.0Ag-0.5Cu solder joint, Materials Today Communications, Volume 25, 2020,doi.org/10.1016/j.mtcomm.2020.101469.

[25] Zhang, L., Han, Jg., Guo, Yh. et al. Creep behavior of SnAgCu solders containing nano-Al particles. J Mater Sci: Mater Electron 26, 3615–3620 (2015). doi.org/10.1007/s10854-015-2876-8.

Development and characterizations of fine pitch flip-chip interconnection using silver sintering

Julie Gougeon
ICMCB UMR 5026
Univ. Bordeaux CNRS, Bordeaux INP
33600 Pessac France
Univ. Grenoble Alpes, CEA, Leti
38000, Grenoble, France
julie.gougeon@cea.fr

Céline Feautrier
Univ. Grenoble Alpes, CEA, Leti
38000, Grenoble, France
celine.feautrier@cea.fr

Jean-Charles Souriau
Univ. Grenoble Alpes, CEA, Leti
38000, Grenoble, France
jean-charles.souriau@cea.fr

Rémi Franiatte
Univ. Grenoble Alpes, CEA, Leti
38000, Grenoble, France
remi.franiatte@cea.fr

Edouard Deschaseaux
Univ. Grenoble Alpes, CEA, Leti
38000, Grenoble, France
edouard.deschaseaux@cea.fr

Laurent Mendizabal
Univ. Grenoble Alpes, CEA, Leti
38000, Grenoble, France
laurent.mendizabal@cea.fr

Mona Tréguer-Delapierre
ICMCB UMR 5026
Univ. Bordeaux CNRS, Bordeaux INP
33600 Pessac France
mona.treguer@icmcb.cnrs.fr

Abstract — **Flip-chip interconnects made of silver are promising candidates to overcome the intrinsic limits of solder-based interconnects and match the demand for increased current densities of high-performance microprocessors. Dip-based interconnects have been demonstrated to be a promising approach to form electrical interconnects by sintering paste between copper pillars and pads. However, the quality of the process is limited by residual porosity and poor performances of the sintered joint formed between the pillar and the pad during sintering if a pressure > 50 MPa is not applied in order to decrease the final porosity. In this study, development has been focused on varying key dipping process parameters allowing a pressureless sintering process. Dip-transfer process was optimized on test vehicle and has shown electrical continuity over 700 interconnections with diameter down to 50 μm. We demonstrate high reliability of the process with microstructural observations, tomography X and thermal cycle up to 200 cycles without breakdown.**

Keywords — packaging, flip-chip, silver sintering, fine pitch, low temperature

I. INTRODUCTION

The constant growth of the electronic market and the increase in processor performance, thanks to advanced nodes, require the development of new packaging interconnect solutions (process and material) to preserve electronics device functionality. [1] Depending on the application, interconnects need to transmit signals at high bandwidths (e.g. for artificial intelligence) or high currents (e.g. for power devices). Furthermore, in SiP or 3D packaging, interconnections must be able to withstand multiple annealing during the integration process. This leads to a current interest for flip-chip interconnect, showing pitches below 200 μm. [2]

Some strategies have been developed to reduce interconnects' pitch in flip-chip assembly, such as micro-tubes [3] or hybrid bonding [4]. Currently, the most extensive strategy used to reduce the size and pitch of interconnects is the copper (Cu) pillar technology, capped on the tip by solder bump (micro-bump). [5-8] The bonding is formed by reflow, with temperature up to 250°C, in order to melt the solder alloy capping the Cu-pillars. This range of temperature can generate thermomechanical stress in the previous material interconnect used, for instance, in 3D stacking of components. In addition, the presence of solder joint in the interconnect leads to the formation of intermetallic compounds (IMCs) that are brittle, and to the formation of Kirkendall voids that reduce interconnects' reliability during operating time. [9-10]

Silver sintering process is currently a promising alternative to overcome intrinsic limits of solder bumps as die-attach technique in terms of thermal stability, while benefiting of better electrical and thermal performances. Compared to solder alloys, silver sintering allows to process at lower temperature and to stack components with temperature-resistant interconnects. Hence, recent developments of low-temperature pressureless sintering (LTPS) process have extended the application area of this process to mechanically sensible assemblies, as well as high-temperature power devices. [11-13]

Presently, the sintering process interconnect is mostly used to bond the back of chip onto substrate. In order to adapt this process to flip-chip interconnects such as nanofoam on top of Cu-pillars. [14-15] However, the elaboration of the nanofoam would increase cost, time and complexity of fabrication, whereas a significant bonding pressure (100 MPa) has to be applied to form interconnects. Also, electrically bond has not yet be demonstrated. Another strategy recently developed, is the photolithography process, associated to screen-printing, in order to form precise pattern made of sintering paste. [16-17] Drying of the deposited paste is necessary in order to strip off the resin from the substrate without damaging paste patterns. LTPS process needs highly reactive metallic content, such as nanoparticles. Whereas pre-drying step allows fixing deposited paste, huge loss of reactivity happens and makes it difficult to ensure LTPS process.

Dip-transfer of sinter paste between Cu pillars and pads is a promising alternative to form flip-chip assembly using soft conditions process in order to reduce pads' size below 100 µm and avoiding waste of sintering paste as in photolithography. Currently, dip based interconnect formation have been studied using copper paste for all-copper interconnects by sintering between 200 °C and 250 °C in a batch oven under formic atmosphere. [18] Whereas sintering without pressure is considered, the best performances are obtained with a bonding pressure up to 50 MPa. The consequence of this bonding pressure is thinner sintered joints (1-3 µm) that would hardly compensate the inhomogeneity of the pillar height. A work of Zurcher *et al.* showed a resistivity almost 5 times of the copper resistivity without using pressure bonding. [19] Besides, ageing study has not yet be carried out.

This work presents an advance dip-based silver sintering approach for 50 µm interconnect. We investigate the robustness of the dip transfer process with variation of critical parameter process, aiming for the thickest silver interconnect in order to compensate Cu pillars height inhomogeneity while processing in soft conditions. Then, we report satisfying repeatability of the process with a significant number of electrical continuity obtained over 700 interconnects, using as optimized dip-transfer process. Finally, thermal cycling study displayed interconnects reliability in soft conditions.

II. MATERIALS & METHODS

A. Ag commercial paste

A commercial silver paste was used as referenced throughout this study (CT2700R7M, KYOCERA, Tokyo, Japan), made of micro triangular platelets, showing nanometer thickness, as presented in Fig.1.

Fig. 1. SEM picture of KYOCERA paste before sintering.

B. Interconnect test vehicule

A specific test silicon chip was designed and fabricated for this study. It contains Cu pillars with a diameter of 50 µm, a pitch of 150 µm and a height of 75 µm. A 100 nm Au finish was deposited on the Cu pillars to prevent oxidation. (Fig.2.a,b) The 11.9 x 11.9 mm² sized chips were placed to a 22.1 x 14 mm² sized silicon substrates presenting Au-pads of 50 µm diameter. The global pattern test shows three Kelvin patterns (K) and five daisy chain patterns (DC) including 709 interconnects for 4-probes measurements, as presented Fig.2.c Aluminium-based routing has been passivated with 500 nm SiO₂ layer to ensure the precise measurement of interconnect in case of paste overflow. In addition, four lines of interconnect have been designed with Cu pillar with same diameter, height and pitch, and complete electric track along DCs pattern in order to detect the formation of bridges between interconnect after bonding pressure by an eventual spread of the silver paste. Further tests were performed with

Cu pillars of smaller diameter and pitch of 35 µm and 105 µm respectively, with the same pattern test.

C. Ag interconnect formation

The process to form the silver interconnect is based on a dip-transfer method which is described Fig. 3. The chips containing the Cu pillars and the substrate were handled and aligned with a Datacon 2200 Evo (Besi, Netherlands). The commercial paste is applied in the device's cavity depth of 50 µm using its fixed gap applicator. The pillars were dipped into the silver paste film using the built-in dipping function of the datacon. Then, the pillars were withdrawn, aligned, and placed onto the substrate in order to transfer the silver paste collected. The flip-chip interconnect assembly was next transferred into an oven (UF55plus, Memmert) and heated for 1h under air at 250°C.

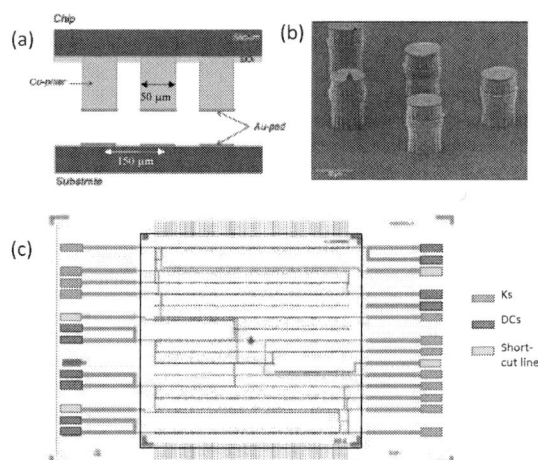

Fig. 2. (a) Scheme of test silicon chip containing Cu-pillar of 50 µm with 150 µm pitch ; (b) SEM images of Cu-pillar with Au-pad ; (c) scheme of the pattern test with 3 Ks, 5 DCs and 4 'short-cut lines'.

D. Interconnect characterization

After thermal treatment, tomography X analyses were carried out on the assembled parts in order to be assured that each interconnect was well formed (Nanotom S180, General Electric Research, US). Manual probes were used to carry out electric measurements onto the measurement pads of Kelvin and daisy chain pattern (PM8, SUSS Micro Tec, Germany). A Keithley 2400 multimeter was used to carry out 4-wire resistance.

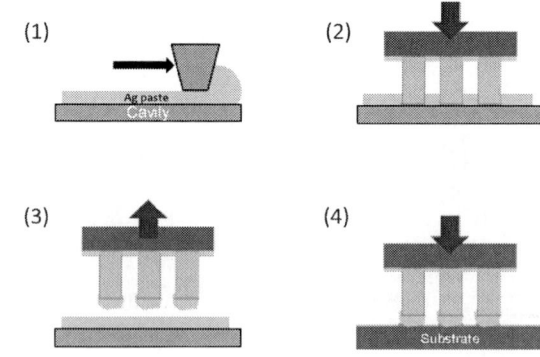

Fig. 3. Steps of the dip-transfer process in order to obtain silver sintered flip-chip assembly.

III. RESULTS & DISCUSSION

A. Optimization of dip-transfer process

Control over the collection of Ag paste on the pillar tips during dip-transfer is necessary to form silver interconnects with high yield, enough thickness to compensate height Cu-pillar mismatch (± 2 µm) and to avoid the paste spreading between interconnects.

The main parameters of the dip-transfer process were evaluated, such as the dipping time in the Ag-paste film, withdrawal velocity from the paste and bonding force, using the interconnect test vehicles described above. Other parameters were kept constant. Three chips were tested for each parameter. We found that the velocity withdrawal do not affect the morphology and thickness of the paste collected on the Cu-pillar tips (from 0.2 to 4 mm/s). This phenomenon was already observed in [19]. The increase of the bonding force do not favor the homogeneous formation of interconnect (1N, 3N and 5 N), hence, obtaining electric continuity overall the daisy chain is difficult (respectively 100 %, 50% and 0% viable). The highest bonding force used during the report of the chip favors a drag force between the tool of the datacon's head and the back of the chip that stays stuck. In hindsight, reducing the approach velocity of the datacon's head (50ms) at the bonding step, leading to a 'stamp effect', has shown increase in the interconnect formation yield.

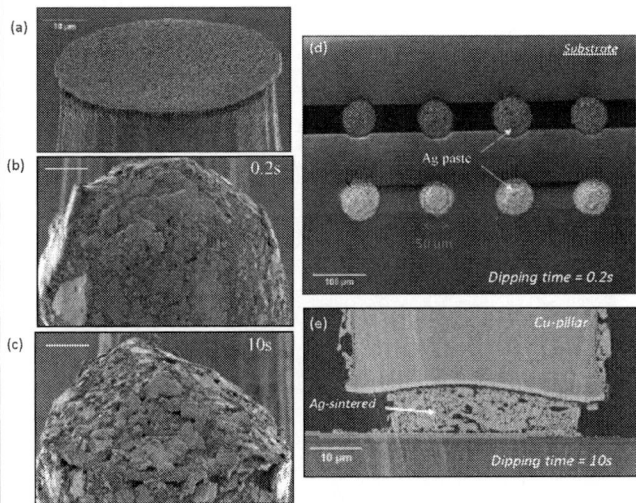

Fig. 4. (a)(b)(c) SEM observations of Cu-pillar tip before dipping, after 0.2 and 10s dipping ; (d) SEM observation of the 'wetting diameter' of the silver paste using 0.2s dipping time ; (e) Cross-section of silver sintered interconnect using 10s dipping time, showing 8-10 µm of thickness.

The dipping time has been shown to be the parameter with the most influence over the dipping process to obtain a precise coating of the Cu-pillar tips. (Fig.4.a, b, c) Lower dipping time (0.2s) favors inhomogeneous coating of the pillars. While the coating tends to be thicker and rounder, we estimated the 'wetting area' of the paste deposits after assembly, as they are squeezed between Cu-pillar and Au-pads on the substrate. The 'wetting diameters' were greater than the pillars' diameter and rather inhomogeneous (65 ± 5 µm vs. 50 µm), but no bridges were observed. (Fig.4.d.) Substantial increase of the dipping time (10s) reduces the coating to a reasonable amount of paste at the top of the Cu-pillars. Cross-sections show most interconnect with diameters below 35 µm, reducing paste

spreading risks, with a satisfying thickness of 8-10 µm. (Fig.4.e) Finally, the dip-transfer process as optimized was proved robust with precise coating on the Cu-pillar tips in a single dip.

B. Electrical performances of the silver sintered interconnects

Silver interconnects were formed using the optimized parameters of the dip-transfer process. The final silver sintered bond were about 8-10 µm, with a diameter from 25 µm to 35 µm, sandwiched between Cu-pillar tips and the Au-pads of the substrate. No signal were obtained overall the batch of 4 'short-cut lines' on each sample, meaning that no bridges were ever formed due to paste spreading after assembly.

The overall electrical continuity of a batch of 8 samples was satisfying, with 5 out of 5 DCs (709 interconnects) were functional. Tomography X characterizations were carried out upon the assemblies showing defective DCs but also upon perfectly functional DCs. Electric continuity rely on each interconnect of the pattern : one faulty interconnect is enough to discriminate hundreds of them. We observed that each time a DC was defective, a Cu-pillar was missing. (Fig. 5.) The loss of Cu-pillar could be explain either by defect of fabrication process (missing aperture in the masks), or relative trauma when wafer were handled before experiments. This is not uncommon in industrial process scale to get a certain amount of partially damaged components. In our case, it would be easily spotted through optic microscopy observations.

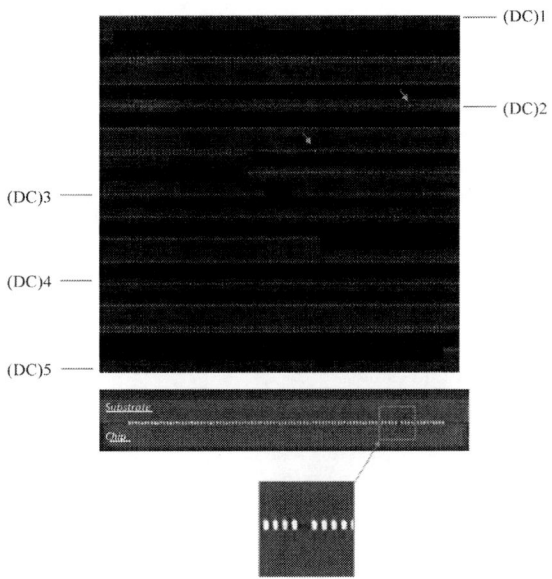

Fig. 5. Reconstruction after tomography X analyses of functional DCs (green) and dysfunctional DCs (red), showing two missing pillars correlating with only 3 functional DCs.

The average value measured for the 24 Ks patterns was 33.14 mOhms, with best value at 6.5 mOhms. The resistivity of one Cu-pillar obtained by electrolytic growth was estimated at 2.95 µOhm.cm. Hence, it is possible to estimate the global resistivity of the silver sintered layer and the two imperfect interfaces with Au-pads of the substrate and the Cu-pillar. The value of the resistivity estimated is around 56.7 µOhms.cm at best, i.e. 35 times the resistivity value of the Ag bulk.

Compared to the literature on this research topic for pressureless sintered interconnect, it is 100 times better.

C. Thermal cycling of the silver interconnects

Thermal cycling from -50°C to 150°C was carried out upon a batch of four samples presenting each 5 DCs. Over 200 cycles were achieved, leading to the failure of only one DC of one sample. After each 24 cycles, electric measurement of each DC showed slight decrease of the overall resistance value. (Fig.6.) This is one of the great advantages of sintered interconnects which benefit from the first heating to tend to a better densification throughout atomic diffusion between highly reactive particles and interdiffusion with metal finishes.

IV. CONCLUSION

In this study, by varying key dipping process parameters, a reliable dip-transfer method has been developed, leading to the formation of relatively thick silver interconnect (8-10 µm) allowing to compensate eventual mismatch height between Cu-pillars. Dip-transfer process as optimized has shown electrical continuity over 700 interconnections with diameter down to 50 µm, 150 µm pitch, suitable to endure over 200 cycles from -50°C up to 150°C with slight increase of the electrical performances during the thermal cycling. We have demonstrated with tomography X analyses that dysfunctional DCs comes from previously missing Cu pillar.

V. OUTLOOK

First tests have been done on interconnects of 35 µm with great success, showing similar issue with previously missing Cu pillars. Further mechanical characterizations before and after thermal cycling in order to estimate the reliability of the silver interconnect as obtained. In order to soften the sintering conditions with temperature at 185 °C, and to favor well-densified bond, a nanopaste made of silver nanocubes will be elaborated with, this time, the rheological properties needed for dip-transfer process. [20]

ACKNOWLEDGMENT

The authors thank Catherine Brunet-Manquat and Lucile Magnier for their skills and expertise that they put at the service of this study.

REFERENCES

[1] Sohel Murshed, S.M.; Nieto de Castro, C.A. "A critical review of traditional and emerging techniques and fluids for electronics cooling." Renew. Sustain. Energy Rev., vol. 78, pp 821–833, 2017.

[2] R. Ghaffarian, "Microelectronics packaging technology roadmaps, assembly reliability, and prognostics," Facta Univ.-Ser., Electron. Energetics, vol. 29, no. 4, pp. 543–611, 2016.

[3] B. G. de Brugière et al., "A 10µm pitch interconnection technology using micro tube insertion into Al-Cu for 3D applications," 2011 IEEE 61st Electronic Components and Technology Conference (ECTC), Lake Buena Vista, FL, USA, 2011, pp. 1400-1406.

[4] S. Moreau et al., "Recent Advances on Electromigration in Cu/SiO2 to Cu/SiO2 Hybrid Bonds for 3D Integrated Circuits," 2023 IEEE International Reliability Physics Symposium (IRPS), Monterey, CA, USA, 2023, pp. 1-7.

[5] A. Roshanghias and A. D. Rodrigues, "Low-Temperature fine-pitch flip-chip bonding by using snap cure adhesives and Au stud bumps," 2019 22nd European Microelectronics and Packaging Conference & Exhibition (EMPC), Pisa, Italy, 2019, pp. 1-4.

[6] P. A. Totta, "History of flip chip and area array technology," in Area Array Interconnection Handbook. New York, NY, USA: Springer, 2001, pp. 1–35.

[7] M. Gerber et al., "Next generation fine pitch Cu Pillar technology—Enabling next generation silicon nodes," in Proc. IEEE 61st Electron. Compon. Technol. Conf., May/Jun. 2011, pp. 612–618.

[8] A. Garnier et al., "Electrical Performance of High Density 10 µm Diameter 20 µm Pitch Cu-Pillar with Chip to Wafer Assembly," 2017 IEEE 67th Electronic Components and Technology Conference (ECTC), Orlando, FL, USA, 2017, pp. 999-1007

[9] Xiong, My., Zhang, L. Interface reaction and intermetallic compound growth behavior of Sn-Ag-Cu lead-free solder joints on different substrates in electronic packaging. J Mater Sci 54, 1741–1768 (2019).

[10] Y.-S. Lai, Y.-T. Chiu, and J. Chen, "Electromigration reliability and morphologies of Cu pillar flip-chip solder joints with Cu substrate pad metallization," J. Electron. Mater., vol. 37, no. 10, pp. 1624–1630, 2008.

[11] Yan J. A "Review of Sintering-Bonding Technology Using Ag Nanomaterials for Electronic Packaging." Nanomaterials (Basel). 2021 Apr 6;11(4):927.

[12] H. Yan, P. Liang, Y. Mei and Z. Feng, "Brief review of silver sinter-bonding processing for packaging high-temperature power devices," in Chinese Journal of Electrical Engineering, vol. 6, no. 3, pp. 25-34, Sept. 2020.

[13] H.-M. Tong, Y. S. Lai, and C. P. Wong, Advanced Flip Chip Packaging. New York, NY, USA: Springer, 2013.

[14] K. Mohan et al., "Demonstration of Patternable All-Cu Compliant Interconnections with Enhanced Manufacturability in Chip-to-Substrate Applications," 2018 IEEE 68th Electronic Components and Technology Conference (ECTC), San Diego, CA, USA, 2018, pp. 301-307.

[15] R. A. Sosa, K. Mohan, A. Antoniou, V. Smet, D. Thienpont and Y. Tan, "Low-temperature all-Cu interconnections formed by pressure-less sintering of Cu-pillars with nanoporous-Cu caps," 2021 IEEE 71st Electronic Components and Technology Conference (ECTC), San Diego, CA, USA, 2021, pp. 390-394.

[16] B. Zhang, Y. C. P. Carisey, A. Damian, R. H. Poelma, G. Q. Zhang and H. W. van Zeijl, "3D interconnect technology based on low temperature copper nanoparticle sintering," 2016 17th International Conference on Electronic Packaging Technology (ICEPT), Wuhan, China, 2016, pp. 1163-1167.

[17] X. Ji, H. Van Zeijl, J. Romijn, H. J. Van Ginkel, X. Liu and G. Zhang, "Low Temperature Sapphire to Silicon Flip Chip Interconnects by Copper Nanoparticle Sintering," 2022 IEEE 9th Electronics System-Integration Technology Conference (ESTC), Sibiu, Romania, 2022, pp. 368-372.

[18] L. Del Carro, J. Zürcher, U. Drechsler, I. E. Clark, G. Ramos and T. Brunschwiler, "Low-Temperature Dip-Based All-Copper Interconnects Formed by Pressure-Assisted Sintering of Copper Nanoparticles," in IEEE Transactions on Components, Packaging and Manufacturing Technology, vol. 9, no. 8, pp. 1613-1622, Aug. 2019.

[19] J. Zürcher et al., "Nanoparticle assembly and sintering towards all-copper flip chip interconnects," 2015 IEEE 65th Electronic Components and Technology Conference (ECTC), San Diego, CA, USA, 2015, pp. 1115-1121.

[20] M. Bronchy et al., "Low-temperature silver sintering by colloidal approach," 2020 IEEE 8th Electronics System-Integration Technology Conference (ESTC), Tønsberg, Norway, 2020, pp. 1-5, doi: 10.1109/ESTC48849.2020.9229830.

Analysis of the Impact of Environmental Conditions on the Reliability in 5G PCB Assemblies

Hans Walter
Dept.: ERE
Fraunhofer IZM, Berlin
Berlin, Germany
hans.walter@izm.fraunhofer.de

Olaf Wittler
Dept.: ERE
Fraunhofer IZM, Berlin
Berlin, Germany
olaf.wittler@izm.fraunhofer.de

Marius van Dijk
Dept.: ERE
Fraunhofer IZM, Berlin
Berlin, Germany
marius.van.dijk@izm.fraunhofer.de

Saskia Huber
Dept.: ERE
Fraunhofer IZM, Berlin
Berlin, Germany
saskia.huber@izm.fraunhofer.de

Julia-Marie-Köszegi
Dept.: R3S
Fraunhofer IZM, Berlin
Berlin, Germany
julia-marie.koeszegi@izm.fraunhofer.de

Martin Schneider-Ramelow
Res. Cent. of Microperipheric
Technologies, TU Berlin
Berlin, Germany
martin.schneider-ramelow@tu-berlin.de

Abstract

The designs of integrated systems are increasingly compact while the requirements of application, especially in the automotive sector, remain demanding. One main challenge is posed by the environmental conditions. In the present study, their impact on mmWave 5G PCB assemblies is evaluated with respect to both mechanical and high-frequency properties. To that end, material samples were stored at three temperatures for various durations in order to accelerate the ageing effects. Afterwards, the samples were tested by use of mechanical, dielectric and high-frequency test methods. Thermo-oxidative ageing processes were observed and lead to significant changes in mechanical and dielectric properties, with simultaneous changes of sample color. The combined approach of mechanical and high-frequency analysis allows for three innovative advantages: First, a correlation of the degradation in performance with other parameters, e.g. sample change of color, which is much easier to detect. Second, the definition of a Safe Area of Operation and a methodology to estimate the behaviour of a mmWave 5G PCB assembly over its lifetime. Finally, the study opens the pathway to a design for robustness of mmWave communication system in automotive applications.

Keywords—PCB, RF, Robustness, 5G, Warpage, Dielectric, Mechanical, Ageing effects

I. INTRODUCTION

The current developments in automotive systems show a strong trend towards increased vehicle-to-vehicle (V2V) and vehicle-to-everything (V2X) communication. The communication systems play a vital role in improving road safety, and lead to optimized traffic flow. V2V communication enables vehicles to exchange information with other (nearby) vehicles in real-time or vehicles connecting to cloud-based platforms to access a wide range of services and applications. The progress is fostered by the promising capabilities of mmWave systems. The mobility of future decades builds heavily on interconnected participants. V2V/V2X communication represents a basic requirement for advanced automotive driver assistances systems, autonomous driving and smart transport systems. As a consequence, Printed Circuit Boards (PCB), as a substrate material for such advanced systems, must provide stable mechanical, electrical and dielectric properties for high and long-term temperature

and humidity loadings. Therefore, in the design of such systems different questions arise. First it is of interest, if the PCB material is suitable for the intended use case? To answer this question, it needs to be identified what the lifetime limit of the material is with respect to the intended application conditions. In order to identify the lifetime limit, we need to estimate, what functional properties limit the lifetime and which degradation mechanisms are associated.

Material used in PCBs are laminates based on glass fibres embedded in a polymer matrix of either epoxy or polyimide. Since the materials are subjected to the harsh environmental conditions, a detailed investigation of possible changes in the properties due to the conditions is needed. For example, the exposure to elevated temperatures leads to a degradation in the materials microscopic structure and in turn decreases the mechanical integrity. Finally, the thermomechanical reliability of the laminate is compromised. In addition, a degradation of radio frequency (RF) performance is to be expected. All these changes are referred to as ageing.

It is known, that exposure to high temperature and oxygen loading initiates various chemical and physical processes in the PCB, referred as thermo-oxidative ageing. This process starts with the absorption of oxygen in the polymer, followed by a diffusion process. The oxygen then leads to a reaction, which can cause for example chain scission. The reaction may furthermore lead to a change in colour, as well as alterations in mechanical and dielectric properties. In total, optical properties and thermo-mechanical material parameters (glass transition temperature (T_g), coefficient of thermal expansion (CTE), modulus of elasticity as well as dielectric parameters (permittivity ε_r and losses $\tan\delta$) react sensitively to these ageing mechanisms and can be used for evaluating the ageing resistance [1, 2]. In operation, integrated systems degrade slowly with time. Accelerated ageing tests and lifetime predictions are of crucial importance for the reliability of PCB assemblies, as their long-term performance often requires a lifetime of over 15 years and cannot be tested in laboratories for their actual lifespan.

The controlled accelerated ageing tests furthermore allow for an analysis of the underlying mechanisms. In additions to the direct physical and chemical effects, the temperature loading plays a crucial role in all chemical reactions. An increase of temperature leads to an acceleration of the chemical ageing process. The reaction kinetics can be described by the well-known Arrhenius Equation [3, 4]. This relation is typically modified and used to estimate the time to failure. Because in our case a failure criterion is not defined, we determine the time to degradation (t_d). The equation is being used as follows:

$$t_d = K_0 e^{-\frac{\Delta H}{RT}} \qquad [1]$$

In this equation, the time to degradation, t_d, is described as a function of the temperature T in Kelvin, the relative gas constant R, the activation energy ΔH and a material constant K_0. In order to estimate the constants ΔH and K_0, material samples were aged at different elevated temperatures for specific time intervals, followed by an extensive characterisation, as listed below. A threshold value was extracted, at which significant ageing did occur in order to estimate t_d for each material property. With this results the time until degradation of the material is going to be expected, can be estimated for different application temperatures by simple extrapolation. Therefore, a Safe Area of Operation (time, temperature) can be defined, in which no significantly measurable degradation effects may impact the material behaviour.

II. Experimental and Results

A. Analysed Materials

For the investigations, an RF compatible PCB laminate with low Dk glass fabric and low dielectric losses was chosen. Depending on the respective materials characterisation either an application-oriented layer structure with a total thickness ~900 μm or a single layer of the same material with a thickness of ~300 μm was utilised. The embedded glass fabric provides a high dimensional stability for the PCB and adapts the thermal expansion behaviour to copper. Furthermore, the PCB material is characterized by a relatively high glass transition range (~180°C). The structure of the used multi-layer laminates is shown in **Error! Reference source not found.**. The datasheet values from the supplier are summarized in Table 1.

Fig. 1: Structure of used multilayer laminates (total thickness ~900 μm).

TABLE 1: PCB MATERIAL DATA FROM SUPPLIER DATASHEET.

T_g [°C]	CTE ($<T_g$) [ppm/K]	ε_r at 10 GHz	tanδ at 10 GHz
>185	14...16	3.6	0.004

All samples were stored in temperature chambers under steady-state conditions of 125°C, 150°C and 175°C for defined times of 20h, 100h, 500h, 1000h, 2000h. Visual inspections of the stored sample was conducted, indicating a significant change in color, see Table 2, which is a result of the reaction between the polymer and the oxygen from the environment. As expected, the changes of color on the multilayer PCB materials (substrate surface) exhibited different rates of discoloration, depending on the exposure temperature and time. At 125°C, darkening was observed only

after 1000h of storage. In contrast, samples stored at 175°C already exhibited a significant darkening after 100h. Samples stored at 150°C began to darken on the surface after 500h.

TABLE 2: EFFECT OF TEMPERATURE STORAGE ON PCB COLOR.

Time [h] / Temperature	0	20	100	500	1000	2000
125°C						
150°C						
175°C						

The time for a significant change of color on the sample surface from yellow to brown, dependent on storage temperature, is defined as the time to degradation, t_{d_color} and is summarized in Table 3.

TABLE 3: TIME FOR DEGRADATION OF THE SURFACE COLOR.

	Exposed temperature		
	125°C	150°C	175 °C
t_{d_color}	>1000h	>100h	>20h

B. Test methods

In the study, various test methods were employed in order to identify and correlate the material parameters that exhibit sensitivity to the oxidation effects. The choice of test methods was guided by the common parameters of interest for reliability analysis.

TABLE 4: SELECTED PROPERTIES AND RELEVANT TEST METHODS.

Properties	Test method
Coefficient of Thermal Expansion (CTE)	TMA
Glass Transition temperature (T_g)	TMA/DMA
Temperature induced warpage behaviour	Warpage analyzer
Permitivity ε_r, dielectric loss factor tanδ (non-metallised)	Split-Cylinder Resonator
Permitivity ε_r, dielectric loss factor tanδ (with Cu layer)	Planar resonators
RF performance	Single element patch antenna

The thermo-mechanical analysis (TMA) method, the dynamical mechanical analysis (DMA) method and the warpage analysis method of the aged samples were used to determine the thermo-mechanical material parameters depending on temperatures and storage time. On the other hand, the dielectric properties of aged samples, the permittivity ε_r and the losses tanδ were characterized by means of dielectric analysis. The permittivity describes the amount to which a material can be polarized by an electric field and the losses describe the dissipated energy which occurs during the process of changing the polarization of the materials in an alternating field. For any RF circuit, the permittivity is a crucial parameter since it is one factor that defines the impedance of transmission lines and other interconnects. A change in ε_r will lead to a change in impendence of a line and hence a mismatch in the signal path, resulting in additional reflections. Furthermore, the performance (e.g. resonance frequency) of components will be affected by a change in ε_r.

The value of tanδ determines the percentage of RF power which can be transmitted or radiated in a system with respect to the signal source. An increase in tanδ will lower the available power for the intended operation. A system which uses a lossy dielectric material will require more active components for amplification in order to achieve sufficient functionality. In order to limit the number of active components, RF systems usually employ low loss dielectric substrates and as short as possible transmission lines between signal source and e.g. antenna.

C. Thermo-Mechanical Parameters

a) Thermomechanical Analysis (TMA) for determination of CTE and Tg

The glass transition range, indicated by the glass transition temperature (T_g) marks the division between rubbery and glassy behaviour for polymer materials. At T_g a change of slope of specific volumes respectively length versus temperature exists.

It can be exemplarily shown that the storage temperature of 150°C up to 100h does not significantly change the linear thermal expansion coefficient (below T_g) and the T_g. Only for longer exposure times, the expansion coefficient decreases and the glass transition temperature increases noticeably. These changes in thermal properties can likely be attributed to the influence of the thermo-oxidative degradation process in the polymer matrix. It was assumed that with longer storage times, a decrease in adhesion between the polymer matrix and glass fabric takes place. Since the glass fabric has a much higher stiffness and lower expansion behaviour, this reduction in adhesion leads to a significant shift of the glass transition temperature towards higher temperatures. Further investigations (e.g. DSC) are necessary to confirm this. These changes in thermal expansion behaviour occur earlier (accelerated) at storage temperatures of 175°C and later at 125°C [5].

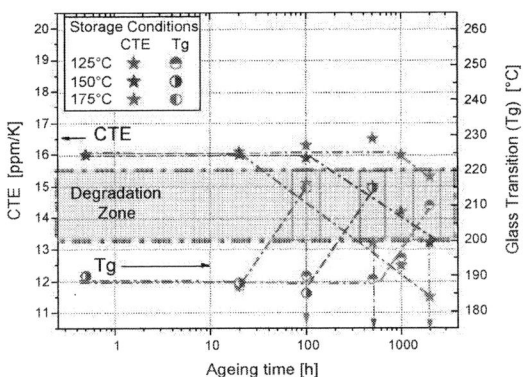

Fig. 2: Influence of storage conditions on thermomechanical properties (TMA).

b) Determination of time to degradation (t_{d_CTE} / t_{d_Tg}).

The time it takes to reach a significant change of glass transition (T_g) to values increased by +15 K (T_g) and of CTE to a lower value by -1 ppm/K are extracted from Figure 2 and listed in Table 5 as a function of storage temperature.

TABLE 5: TIME TO DEGRADATION EXTRACTED FROM TMA DATA.

	Exposed temperature		
	125°C	150°C	175°C
t_{d_CTE}	>1000h	>100h	>20h
t_{d_Tg}	>500h	>100h	>20h

c) Dynamical Mechanical Analysis (DMA) for determination of modulus and T_g

The results of the DMA in Figure 3 show the storage modulus (E') and the glass transition temperature (T_g). The maximum of the mechanical loss function (E'/E'') is defined as the T_g, which for the initial samples is at 215°C. It should be noted that each measurement method provides different characteristic T_g results, which is why it is essential to specify the measurement conditions and evaluation procedures in order to make meaningful comparisons. It can be observed that the temperature storage significantly effects the thermo-mechanical properties. At a storage temperature of 150°C there is a slightly increase of E' for a storage time up to 500h, whereas there is a significantly increase of T_g for exposure time up 100h. For the other storage conditions were observed a moderate changing of modulus but a significantly shift of T_g under 125°C and 175°C storage temperature. This behaviour is characterised by a further decrease of molecular chain mobility and an increasing dominance of glass fibre due to diminishing adhesion behaviour matrix and glass fibre, resulting in sligthly increasing stiffness

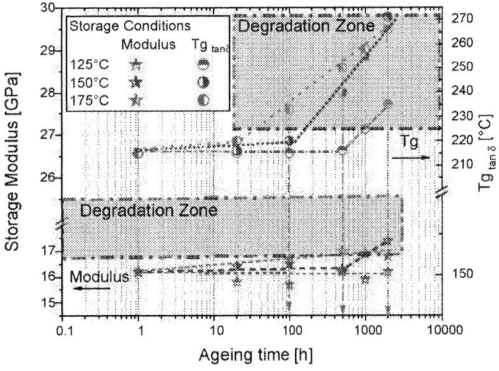

Fig. 3: Influence of storage conditions on thermo-mechanical properties (DMA) of core material.

Similar to the results from the TMA analysis, a shift in glass transition temperature towards higher values is also observed. Depending on the storage temperature, these changes in thermo-mechanical behaviour are achieved at different rates.

d) Determination of time for degradation ($t_{d\_E'}$ / t_{d_Tg})

The time it takes to reach a significant change of T_g to higher value of +15 K depending on storage temperature and for significant change of storage modulus (below T_g) to higher values (+/- 1 GPa) is referred as the time to degradation.

TABLE 6: TIME TO DEGRADATION EXTRACTED FROM DMA DATA.

	Exposed temperature		
	125°C	150°C	175°C
$t_{d\_E'}$	n.d.	>500h	>100h
t_{d_Tg}	>1000h	>100h	>20h

e) Warpage Analysis

The warpage measurements were performed on copper coated PCB samples which were stored in the same manner as described above. Warpage occurs due to the intrinsic stresses and the CTE mismatch of the different thin metal films and the substrate by variation of temperature. As thermo-oxidation effects lead to a lower CTE, the mismatch with the CTE of copper increases over ageing time. The initial warpage at room temperature is subtracted in order to visualize and compare the relative warpage changes occurring during temperature cycling.

In Figure 4, the relative warpage at the respective measurement temperatures for different ageing temperatures and times is shown. It can be observed that the samples stored at 125°C for 2000h show a relative warpage of approx. 250 µm. This value was defined as a maximum (uncritical) deflection. Measurement results at other storage temperatures indicate that max. deflection of 250 µm at higher storage temperatures already occurs after 100h at 175°C and 500h for 150°C.

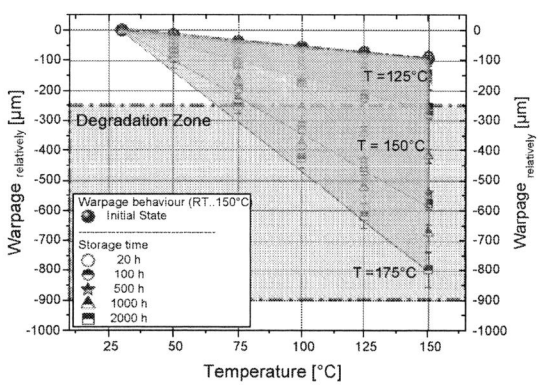

Fig. 4: Influence of storage conditions on warpage behaviour on copper-coated core material during temperature change between RT and 150°C for different storage temperatures (125°C, 150°C, 175°C) and times

f) Determination of time for degradation ($t_{d\_warpage}$).

The storage time it takes to reach a deflection of 250 µm depends on storage temperature and is referred to as the time to degradation $t_{d\_warpage}$.

TABLE 7: TIME TO DEGRADATION EXTRACTED FROM WARPAGE ANALYIS.

	Exposed temperature		
	125°C	150°C	175 °C
$t_{d\_warpage}$	>2000 h	>500 h	>100 h

Overall, it is evident that temperature storage significantly influences the mechanical property profile. A notable increase in modulus values and a significant shift or decrease in the mechanical loss factor are observed, but only after varying storage durations. This is attributed to a further decrease in

molecular chain mobility and an increasing dominance of the glass fabric due to diminishing adhesion between the matrix and glass fabric, leading to higher stiffness. Similar to the results from the TMA analysis, a shift in glass transition temperatures towards higher values was also observed.

Depending on the storage temperatures, these changes in thermo-mechanical properties are achieved at different rates. It can be assumed that this ageing process is not yet complete.

D. High-frequency Parameters

a) Dielectric Analysis

The set of samples for the RF analysis provides a number of physical parameters which were evaluated and which allow for an in-depth discussion on the critical performance. First, the changes of ε_r and tanδ were measured without any metalized layer (pure dielectric analysis). Second, planar resonators were structed on metallised substrates. The ring and fork resonators provide information on the interaction of dielectric and metal layer when under accelerated ageing and the effective values of ε_r and tanδ were extracted.

RF Analysis of Pure Dielectric Analysis: The analysis of the pure dielectric is commonly performed with an external test fixture. In our case a split-cylinder resonator was employed. The detailed description of the characterization methods is out of the scope of the present paper and presented in [5-7]. The data from the pure dielectric characterisation is displayed in Figure 5. A significant increase of ε_r and tanδ with the different ageing time is observed. Furthermore, the higher the temperature, the higher the increase. The samples for 175°C and 1000h and 2000h however deviated from the trend. Further tests resulted in non-repeatable results for these two samples. A very high susceptibility to humidity was observed which was most likely caused by a proceeding decomposition of the material under test.

RF Analysis of Metallised Dielectric Analysis: The pure dielectric analysis provides the basis for the performance estimation of any interconnect or component. However, the layout of the individual structure and the metalized layer will impact the performance as well. In the present study, the effect resulting from the accelerated ageing is also changed depending on e.g. the percentage of the surface which is covered by the copper layer.

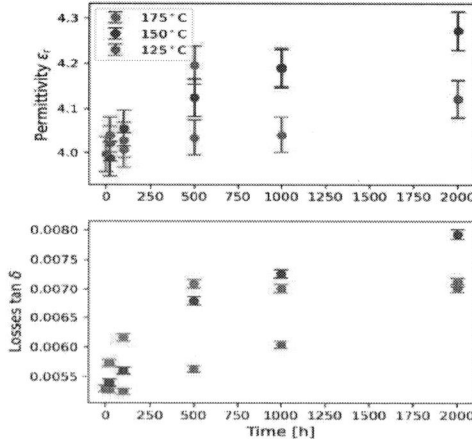

Fig. 5: Results of the characterization of the pure dielectric material by use of a split-cylinder resonator.

In order to quantify the impact of the accelerated ageing on actual RF components, two types of resonators were designed: a ring and a fork resonator. The underlying principle, the design and the evaluation of such resonators is detailed [5]. Figure 6 displays selected examples of the manufactured and aged structures.

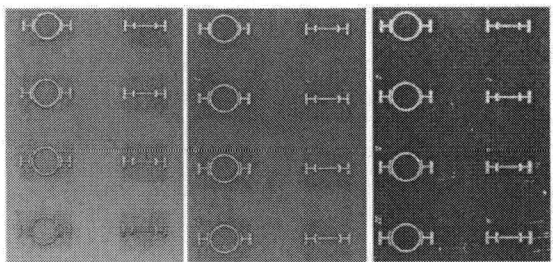

Fig. 6: Ring (left) and fork (right) resonators for the metalized dielectric analysis. Samples were stored at 175°C for 20h (left), 100h (centre) and 500h (right).

Based on the S-parameters of each resonator, the effective values for ε_r and tanδ were extracted. It is important to stress that the effective values include any changes to the materials due to the manufacturing process and any changes in the conductor performance, for example increased losses due to the surface finish or due to the oxidation of the copper during the ageing process.

Figure 7 displays the extracted effective dielectric properties from the ring and fork resonators. The initial losses are already higher when compared with the pure dielectric characterisation while the effective permittivity is initial lower; both for the reasons discussed above. Furthermore, the copper structures are more fragile than the non-metalized samples hence the increased variation within the dataset (e.g. effective tanδ extracted from a fork resonator after 20h at 125°C). Despite all these effects, the impact of ageing is still pronounced. We refrain from further detailed evaluation of the data due to the limited number of ageing temperatures.

b) Determination of time for degradation (t_{d_losses}).

Since the consequences of the changes in ε_r strongly depend on the design and functionality of any component which is build from the material, it is not possible to define a time for degradation based on the permittivity itself. However, the losses which are measured by tanδ can be evaluated in a general manner. We propose that an increase of the losses by 25% can most likely not be tolerated by any RF circuit. This criterion was used to extract the time to degradation (t_{d_losses}) for each storage temperature from the combined data of the pure and the metallised dielectric analysis. The values are listed in Table 8.

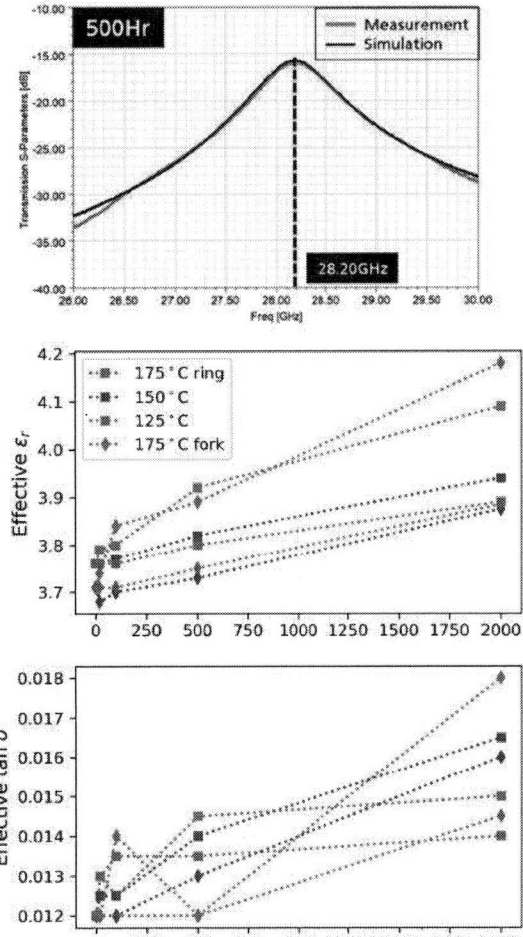

Fig. 7: Top: Example for the extraction of the effective ε and tanδ values from the S-parameter measurement via full wave simulation (fork resonator, 500h at 150°C).
In the two Graphs below: Results of the ring (squares) and fork (diamonds) resonator measurements. The lines are added to guide the eye.

TABLE 8: TIME FOR DEGRADATION EXTRACTED FROM DIELECTRIC CHARACTERISATION

	Exposed temperature		
	125°C	150°C	175°C
t_{d_losses}	>2000h	>500h	>100h

c) Antenna Analysis

The final step of the RF analysis is the measurement of antennas. To that end, single element patch antennas were designed, manufactured and aged together with the samples for the dielectric characterization. Figure 8 displays the initial state of the antenna together with the aged state for 2000h at 150°C. It is clearly visible how the main resonance is shifted to lower frequencies as a result of the accelerated ageing. Furthermore, the magnitude is significantly decreased which is most likely a consequence of the elevated losses in the dielectric substrate.

d) Determination of time to degradation ($t_{d\_antenna}$).

The extent to which the change in the dielectric properties influences a given interconnect or component depends strongly of the respective design. It is not possible to define a global threshold for the acceptable drift in dielectric properties. However, for components with specific applications, a threshold can be argued for. In the case of the current antennas for mmWave 5G communication systems, the antenna was designed for 29.5 GHz.

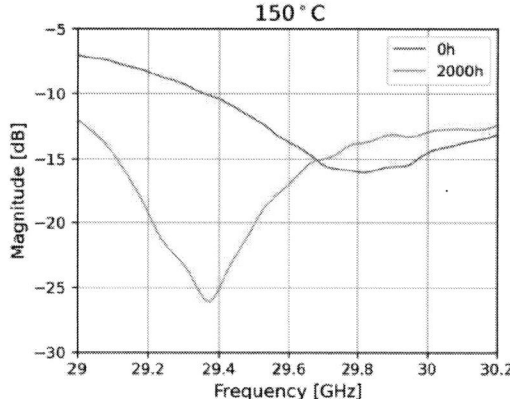

Fig. 8: S-Parameter of the single element patch antenna in the initial state and after 2000h at 150°C.

The threshold was set to the time when no proper resonance, and hence no usable radiation within the targeted frequency band was present in the S-parameters. The extracted $t_{d\_antenna}$ values are listed in Table 9.

TABLE 9: TIME FOR DEGRADATION EXTRACTED FROM ANTENNA MEASUREMENTS.

	Exposed temperature		
	125°C	150°C	175 °C
$t_{d\_antenna}$	>2000h	>500h	>100h

III. DISCUSSION

The storage of samples under different temperatures leads to a significant change of color depending on the storage temperature and duration. This color change is caused by thermo-oxidative degradation processes in the polymer matrix. Selected thermo-mechanical and dielectrical parameters were used to evaluate the ageing process, which provided evidence of ageing-related material changes. In combination, these ageing-related changes in material properties of the studied substrate material, also lead to a significant change in the warpage behaviour of the metal coated laminates. In particular, the increased temperature-dependent warpage behaviour correlates with the change in CTE to lower values and glass transition temperature to higher values.

Table 10 lists the extracted threshold values (t_d) for selected mechanical and dielectrical properties of different ageing temperatures. For all material properties, a significant change could be observed over the duration of the accelerated ageing process.

In Figure 9, the results from Table 10 are summarized. It can be observed that significant changes in mechanical and dielectric properties occur in relation to the storage temperatures, similar to the time to degradation of color changes, a property much easier to characterize. The graph shows further that the temperature loading conditions play an important role in the chemical reaction behaviour of the investigated laminate.

TABLE 10: TIME TO DEGRADATION OF RELEVANT PARAMETERS FOR DIFFERENT TEMPERATURES.

	Exposed temperature		
	125°C	150°C	175°C
t_{d_color} (color change)	>1000h	>100h	>20h
$t_{d\_E'}$ (modulus (E'))	n.d.	>500h	>100h
t_{d_Tg} (T_g DMA)	>1000h	>100h	>20h
t_{d_CTE} (CTE)	>1000h	>100h	>20h
t_{d_Tg} (T_g TMA)	>500h	>100h	>20h
$t_{d\_warpage}$ (warpage)	>2000h	>500h	>100h
t_{d_losses} (dielectric losses)	>2000h	>500h	>100h
$t_{d\_antenna}$ (antenna shift)	>2000h	>500h	>100h

Storage at high temperatures accelerated the chemical ageing process. In contrast, the chemical processes slowed down for lower temperatures. This reaction kinetic can be described by the Arrhenius approach. By applying Equation 1, the activation energy ΔH was calculated for diffferent elevated temperatue and time intervals on the example on Tg degradation and color change.

ΔH = 13,983 kJ/mol → 1,2 eV (Tg_DMA/Color change)

The activation energy, represented by ΔH, is the minimum energy required for a chemical reaction to occur. It is the energy barrier that must be overcome for a significant degradation of the material.

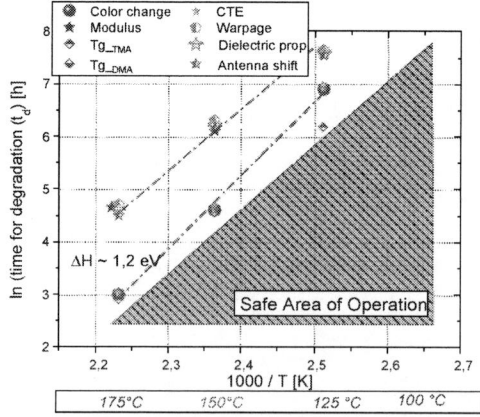

Fig. 9: Arrhenius plot of time to degradation (t_d) vs. inverse temperature.

Below the determined critical times and durations, a zone is defined, where no degradation can be observed and which can be referred to as Safe Area of Operation. Based on extrapolation of the Arrhenius equation, there will be no significant degradation effects expected in this zone which may impact the material properties of the used laminates. By application of the calculated activations energies, it is furthermore possible to predict the lifetime of the PCB material with accelerated lifetime tests. Thus reliability can be assured for lower temperatures than the test temperatures and much longer lifetimes than testable in the lab.

IV. CONCLUSIONS

In this study the thermo-oxidative ageing processes of a laminate material was investigated. Thermo-oxidative ageing processes were observed under defined laboratory conditions. Significant changes in mechanical and dielectric properties were observed, with simultaneous changes of sample color. Using of Arrhenius approach, it become possible to calculate activation energy ΔH for selected mechanical and dielectric parameters. For this purpose, times to degradation were determined depending on storage conditions (temperature, time) for the significant changes of properties and color. Outside of the so-called degradations zone a Safe Area of Operation (SAoO) were defined, where no measurable degradation effects may impact the material behaviour. By means of both, SAoO- concept and activation energy, critical times to degradation can be derived for other, e.g. lower temperatures. With these data, estimates can be made for long-term stable assemblies, devices using experimentally validated data from laboratory and applied to field applications.

V. ACKNOWLEDGEMENTS

The authors would like to thank all involved colleagues for their contributions to the presented work, especially Robert Stöcker, Sutharsan Balasubramaniam, Abhijeet Kanitkar and Thi Huyen Le for the RF testing.

Thank you for the financial support provided by the Fraunhofer Society: "Leistungszentrum Digitale Vernetzung".

REFERENCES

[1] M. van Dijk et.al.:, Modification of Prony Series Coefficients to Account for Thermo-Oxidative Ageing Effects within Numerical Simulations In 2023 24rd International Conference on Thermal, Mechanical and Multi-Physics Simulation and Experiments in Microelectronics and Microsystems (EuroSimE), 2023.

[2] M. van Dijk et.al.:, Numerical simulation of transient thermomechanical ageing effects. In 2022 23rd International Conference on Thermal, Mechanical and Multi-Physics Simulation and Experiments in Microelectronics and Microsystems (EuroSimE), 2022.

[3] S.Arrhenius. Über die dissociationswärme und den einfluss der temperatur auf den dissociationsgrad der elektrolyte. *Z. Phys. Chem.* 4U, 96–116 (1889).

[4] R. Martin,"Ageing of composites",Woodhead Publishing in materials, (2008)

[5] H.Walter et al.: "Bewertung des Werkstoffverhaltens von Leiterplattenmaterialien unter Anwendung von beschleunigten Alterungstests"; EBL 2022 (German)

[6] A. Kanitkar, et al., "A Comparative Analysis of Two Dielectric Extraction Methods of a PCB Material for D-Band Applications," 9th Electronics System-Integration Technology Conference (ESTC), IEEE, 2022.

[7] I. Ndip, et al., "A comparative analysis of 5G mmWave antenna arrays on different substrate technolgies," 22nd International Microwave and Radar Conference (MIKON), IEEE, 2018.

[8] T. Braun, et al., "Fan-out wafer level packaging for 5G and mm-Wave applications," International Conference on Electronics Packaging and iMAPS All Asia Conference (ICEP-IAAC), IEEE, 2018.

Evaluation of the environmental impact within semiconductor packaging materials

Andrew Bainbridge, Kathleen Grant, Lewis Clark, Jeff Kettle

James Watt School of Engineering
University of Glasgow
Glasgow, Scotland
Contact email: jeff.kettle@glasgow.ac.uk

Abstract—**This paper reports the sustainability of materials and components required for PCB manufacturing, namely the substrate material, die-attach method and IC chip design. Electronics manufacturing processes are generally decided by the technical and economic requirements, and not sustainability, resulting in poor practices and unintended outcomes. Alternative methods could be used and this paper reviews some of the drivers and opportunities for sustainable PCB manufacturing. A lifecycle assessment (LCA) was performed on simple Printed Circuit Boards (PCB) and a comparison of the manufacturing approach using FR-4, polyimide and paper-based substrates is made. The impacts of different processing steps were considered, all scored using the ReCiPe Endpoint. The results found solder material has a significant influence, with lower-temperature solders reducing the overall environmental impact. However, it has been found that the component that has the biggest environmental impact overall is the IC. We discuss how die size, technology node and packaging formats effects this rating, and also show how the move towards alternative materials can reduce the environmental impact of electronic products.**

Index Terms—**LCA, degradable electronics, additive printing, PCB**

I. INTRODUCTION

Electronic waste (e-waste) is quickly becoming a major, global problem. In 2020, 53.6 million tonnes of e-waste was generated, and only less than 20 % of that was collected for recycling. [1] The amount of e-waste generated is only growing [2]. On top of the end-of-life waste, there is a significant amount of greenhouse gas emissions produced during the production of electronic components - throughout the supply chain.

Current technologies used in printed circuit board assemblies (PCBA) have evolved over more than 50 years to maximise the performance of the electronics, with little consideration for the sustainability and environmental impact of the product. [3] Conventional PCB substrates utilise composite materials that cannot easily be recycled, as well as a number of critical rare Earth elements that have underdeveloped recycling processes. [4]

This paper evaluates the environmental impacts of each part of the PCBA separately. The widely used FR4 substrate is compared to alternative flexible substrates, which have clearer recycling pathways. Secondly, the current most popular, high-energy-use solder material is discussed and compared to silver

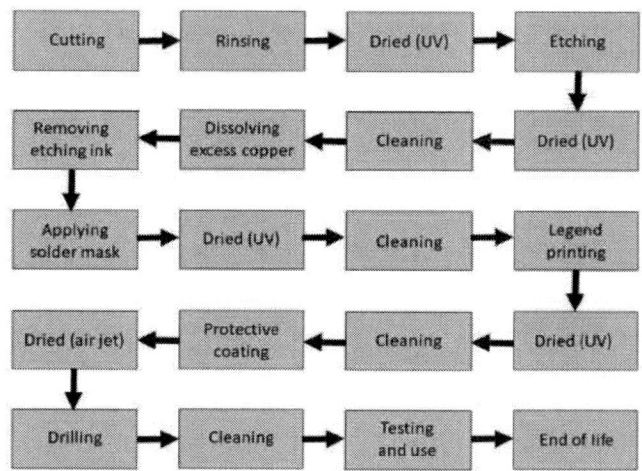

Fig. 1: Fabrication process flow of a standard PCB board. Also, the flow used for the substrate comparison LCA calculations in this work.

conductive adhesives with lower energy processing requirements. The final part of the PCBA is the integrated circuit. Although this is a small percentage of the PCBA by weight, it is a large part of the required input energy and raw materials.

II. METHODS

Our model is based on the work of Grant *et al.* [5]. Using this, an initial model was created in a commercial software packaged "GaBi" from Sphera, Germany. The results were scored using the ReCiPe Endpoint. The initial material inputs into the PCB manufacturing process include an epoxy based polymer (representing FR-4), copper, glass fibre, and additional resin to model the four key layers of a printed circuit board. ICs data was sourced from the GaBi database. Solder matrials were gernated form in-house or from the GaBi database.

III. SUBSTRATE

The most-used, conventional manufacturing approach to make a PCB is with FR4, which uses a combination of glass fibre, epoxy resin and copper. The standard manufacturing flow for an FR4-based PCB is shown in figure 1. This manufacturing flow starts with an FR4 board and the other input materials and transforms it into a finished PCB board.

There are alternative materials that could replace FR4 in the future and are currently commercially available. Polyethylene terephthalate (PET) is one material that has been reported to be enzymatically degradable. [6], [7] Other alternatives include fluorinated ethylene propylene (FEP), Poly ether ether ketone (PEEK), flexible glass (CMG) and paper. A key parameter for the usability of these materials within the manufacturing processes is the glass transition temperature of the material. For PET the value is quite low, less than $70\,^{\circ}$C, which might cause the material to be incompatible with many of the common manufacturing processes that require heating, e.g. soldering. For FEP, PEEK, CMG and paper the glass transition temperature is significantly higher, at more than $200\,^{\circ}$C.

To evaluate the environmental impact of some of the alternative substrates, a life cycle assessment has been carried out on the conventional PCB, as well as PET and paper alternatives. The standard manufacturing flow was used to compare each of the substrate materials. Modelling was carried out using GaBi software from Sphera and utilised the inbuilt database, as well as custom-built processes. For the FR4 PCB, the processes were designed based on values from Ozkan et al. [8] For the PET and paper substrates input materials from the GaBi database were used, 'EU-28: Polyethylene terephalate fibres' for PET, and 'EU-28: Corrugated board' for paper.

The substrate materials are compared using the ReCiPe life cycle impact assessment method. Figure 2a shows the global warming potential (GWP) of the three materials and indicates a clear decrease in GWP could be made by swapping FR4 for paper within the PCB. There is a smaller decrease in GWP to be gained from swapping from FR4 to PET, but this benefit is offset by a much higher human toxicity cancer potential for the PET material - figure 2b. The source of this high toxocity cis primarily contributed by the production of the raw materials of terephthalic acid and ethylene glycol.

IV. Solder and Die Attach

The material used to attach components to the PCB is another area where a move to less energy-intensive materials reduces the environmental impact of the final, populated PCB. In the past, lead-based solders were most commonly used within the electronics industry, but currently, lead-free solders are dominant after the harmful effects of lead were realised. The most commonly used solder alloy is tin-silver-copper (SnAgCu or SAC) which is considered to have lower human toxicity than lead-based solders.

From the perspective of the global warming potential of the materials, a significant parameter of the solder alloy is the required reflow temperature. As an example, the melting point of lead-based SnPbAg solder is $179\,^{\circ}$C, whereas for the lead-free SnAgCu the melting point is much higher at $217\,^{\circ}$C. So, the energy input required for the lead-free solder is consequentially higher. This energy input for the soldering

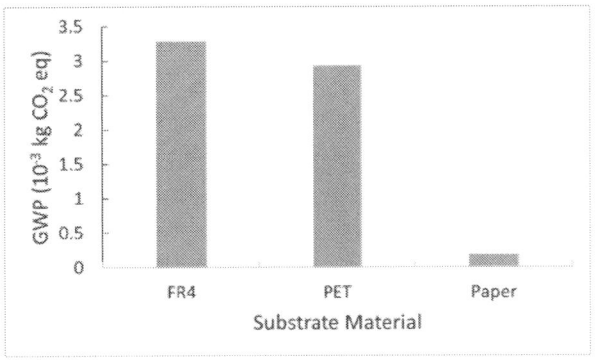

(a)

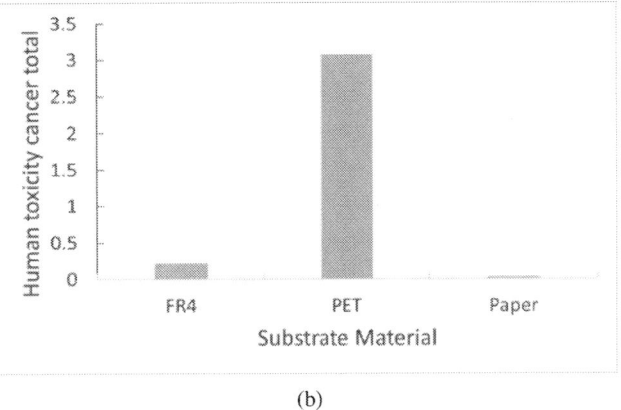

(b)

Fig. 2: a) Global warming potential of the FR4, PET and paper within a $25\,\mathrm{cm}^2$ PCB. Results using the ReCiPe method. b) Human toxicity cancer total of the same materials.

process is usually significantly higher than the embodied energy from the production of the solder.

Figure 3 shows the global warming potential for each solder material. Both production and soldering process are taken into account for soldering a 100-pin IC to a PCB. The GWP due to solder production is dependent on the constituent components of each solder paste and the processes used in production. The GWP due to the energy consumption is majority dependent on the temperature required to melt the solder; a standard solder oven requires about $8\,\mathrm{W}/^{\circ}$C.

To reduce the overall GWP of the die-attach process it is important to reduce the required reflow temperature of the adhesive material. This will result in a direct reduction in the process energy consumption and could be a significant reduction in the overall energy consumption as well. One option that has been explored is silver conductive adhesive. This adhesive is epoxy based, with silver nanoparticles as the conductive material. The curing temperature is $140\,^{\circ}$C, much lower than the reflow temperature of lead and tin-based solder pastes. Figure 4 compares the energy consumption of the silver conductive adhesive to that of SAC solder paste - the current most used die attach material.

The embodied energy cost of the silver ICA production is actually slightly higher than for the SAC solder, however,

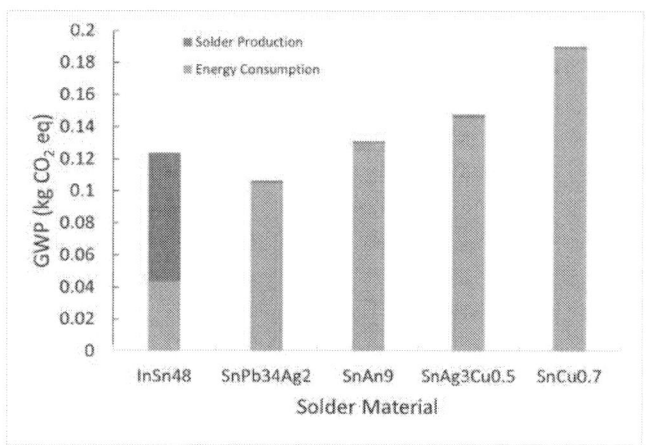

Fig. 3: Global warming potential for the production of solder and energy consumption for soldering. All measurements are for soldering an IC with 100 pins.

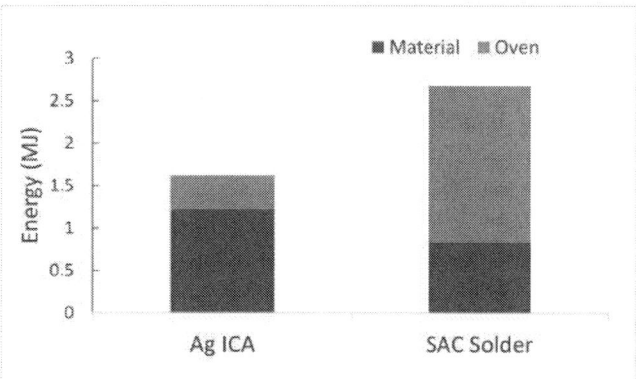

Fig. 4: Energy requirements for production and process use of both standard SAC solder and alternative silver conductive adhesive. Data from A. S. G. Andrae *et al.* [9]

the overall energy required is significantly lower. The curing energy required is only 22 % of the solder process energy. The overall energy saving is 40 % for this material combination, although a higher energy saving could be achieved by reducing the embedded energy of the silver ICA. This could be done through improved production processes or changes in the composition of the adhesive to minimise energy use during production. There are many ways of fabricating the required silver nanoparticles, which could be investigated from an energy usage viewpoint.

V. IC LAYOUT

The final part of a populated PCB is the integrated circuits that are attached to the PCB. The layouts that underpin the IC design have been decreasing in size as fabrication technology has evolved. However, a smaller scale doesn't equate to fewer resources used in the manufacturing process. In fact, by chip area, the GWP increases as node size decreases. This is shown to be the trend for all the ICs in the GaBi database, see figure

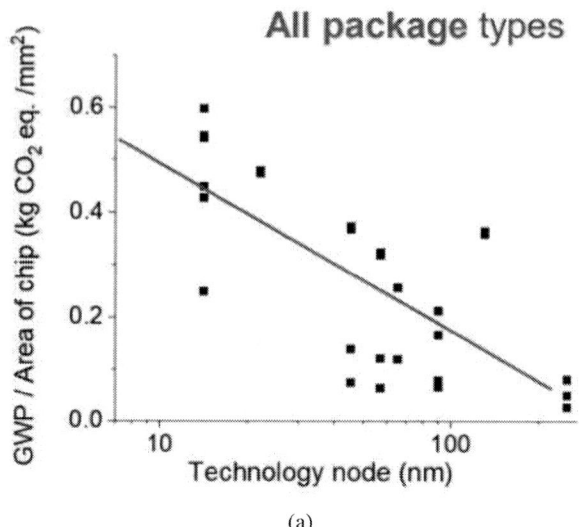

(a)

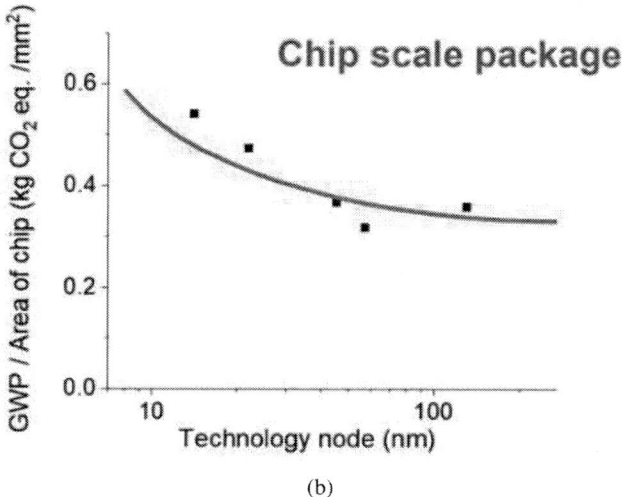

(b)

Fig. 5: Comparisons of the global warming potential by area of IC chip for all ICs in the GaBi database. a) chip-scale packages. b) all types of packages.

5. The results shown here are plotted by area, but this is affected if a different parameter is used. For example, if GWP per megabyte on the IC is considered the results are more favourable to the lower technology node lengths. It is worth noting that, in general, as node size has decreased, the size of the dies has increased. So, GWP per die shows a similar trend to the GWP per area shown in the figure.

One limitation of the results discussed here is the lack of data from current and recent generations of ICs. The fabrication process data is considered proprietary and of high value by chip manufacturers. Therefore, the available doesn't include the recent technology node sizes and the smallest size shown in the results is 14 nm. The current lowest technology node size is 3 nm. Clearly further research is needed in this area.

VI. CONCLUSION

The increasingly urgent need to reduce emission and electronic waste requires all aspects of PCBAs to be optimised for sustainability. To this end, alternative materials for each of the major components have been evaluated. Life cycle assessments of substrate materials have shown that a paper-based substrate can reduce the global warming potential by 94 % compared to the standard FR4 material. Similarly, the total energy consumption for the die attach process can be almost halved through the use of a silver conductive adhesive instead of the most popular SAC solder paste in use today. Finally, the current trend of IC designs has been shown to be increasing global warming potential with each technology advancement. Since the IC is the most significant part of the PCBA from a global warming perspective, alternative designs to combat this trend is an important goal.

ACKNOWLEDGMENT

We would like to acknowledge the support of the EPSRC grant 'GEOPIC' (EP/W019248/1)

REFERENCES

[1] Forti, V., Baldé, C., Kuehr, R., and Bel, G., "The global e-waste monitor 2020: Quantities, flows and the circular economy potential," *United Nations University (UNU)/United Nations Institute for Training and Research (UNITAR) – co-hosted SCYCLE Programme, International Telecommunication Union (ITU) International Solid Waste Association (ISWA)* (2020).

[2] Nogueira, G. L., Kumar, D., Zhang, S., Alves, N., and Kettle, J., "Zero waste and biodegradable zinc oxide thin-film transistors for uv sensors and logic circuits," *IEEE Transactions on Electron Devices* **70**(4), 1702–1709 (2023).

[3] Chakraborty, M., Kettle, J., and Dahiya, R., "Electronic waste reduction through devices and printed circuit boards designed for circularity," *IEEE Journal on Flexible Electronics* **1**(1), 4–23 (2022).

[4] Jowitt, S. M., Werner, T. T., Weng, Z., and Mudd, G. M., "Recycling of the rare earth elements," *Current Opinion in Green and Sustainable Chemistry* **13**, 1–7 (2018).

[5] Grant, K., Zhang, S., and Kettle, J., "Improving the sustainability of printed circuit boards through additive printing," *2023 IEEE Conference on Technologies for Sustainability (SusTech)* , 86–90 (2023).

[6] Kawai, F., Kawabata, T., and Oda, M., "Current knowledge on enzymatic pet degradation and its possible application to waste stream management and other fields," *Applied Microbiology and Biotechnology* **103**, 4253–4268 (Jun 2019).

[7] Taniguchi, I., Yoshida, S., Hiraga, K., Miyamoto, K., Kimura, Y., and Oda, K., "Biodegradation of pet: Current status and application aspects," *ACS Catalysis* **9**(5), 4089–4105 (2019).

[8] Ozkan, E., Elginoz, N., and Germirli Babuna, F., "Life cycle assessment of a printed circuit board manufacturing plant in turkey," *Environmental Science and Pollution Research* **25**, 26801–26808 (Sep 2018).

[9] Andrae, A. S. G., Itsubo, N., Yamaguchi, H., and Inaba, A., "Screening life cycle assessment of silver-based conductive adhesive vs. lead-based solder and plating materials," *Materials Transactions* **48**(8), 2212–2218 (2007).

Measurement and simulation of mechanical strength of Back-End-Of-Line layer in advanced CMOS dies

Bart Vandevelde
Imec
Leuven, Belgium
bart.vandevelde@imec.be

Riet Labie
Imec
Leuven, Belgium
riet.labie@imec.be

Kris Vanstreels
·Imec
Leuven, Belgium
kris.vanstreels@imec.be

Kevin Cox
Tektronix Component Solutions
Beaverton, US
kevin.cox@tektronix.com

Jason Krantz
Tektronix Component Solutions
Beaverton, US
jason.krantz@tektronix.com

Mario Gonzalez
Imec
Leuven, Belgium
mario.gonzalez@imec.be

Reza Moloudi
Imec
Leuven, Belgium
reza.moloudi@imec.be

Matt Borden
Tektronix Component Solutions
Beaverton, US
matt.borden@tektronix.com

Abstract— The back-end-of-line (BEOL) layers of ICs are subjected to considerably high mechanical forces during processing and when operating in harsh conditions. Without proper design and flip chip process control, the forces induced from flip chip bumps may exceed the ultimate strength of the BEOL resulting in critical failures. It is therefore of high interest to have a measurement method to quantify the strength of the BEOL structure of a functional IC. Ultimate fracture stress values are extracted from these measurements which can be used in thermo-mechanical simulations of flip chip assemblies for survivability evaluations.

Keywords—BEOL, thermo-mechanical stress, flip chip assemblies,

I. THERMO-MECHANICALLY INDUCED STRESS FAILURES IN BACK END OF LINE

Over the last few years, Tektronix Component Solutions has been working to advance package reliability through selecting optimal material sets, using best-in-class simulation methods, and refining processing techniques. In this research, Tektronix partnered with IMEC to establish die strength characterization methods and to evaluate the risk of failure of an advanced node CMOS die in a flip chip package.

Due to temperature changes during processing and in operation, forces and bending moments are induced on the solder joints of flip chip assemblies (Fig. 1). The origin of these forces and moments are the thermal expansion mismatch between the silicon chip (~ 3 ppm/°C) and substrate (~ 14 ppm/C°).

In reference [1], an analytical model is available which calculates the forces and bending moments acting on solder joints and at the interfaces with chip and substrate. These forces cause the bending of the structure.

The analytical model is based on linear elastic equations, and obviously and luckily, solder joints will deform plastically and by creep which reduces the forces acting on the structure. However, the forces and bending moments are still significant.

In a historical case (not part of the work reported hereafter), fractures in the BEOL were found after flip chip assembly processing. As shown in the scanning acoustic microscopy (SAM) picture in Fig. 2, a lot of so-called "white bumps" [2] are seen which show fracture in the BEOL of silicon die. The cross-section confirms the fracture in the BEOL below the solder bump. On other cross-sections, underfill material was seen in the fracture which indicates that the fractures occurred during the solder flip chip reflow, and before underfilling.

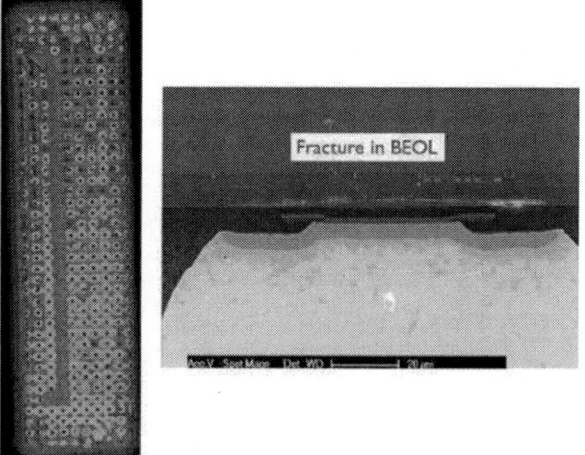

Fig. 2. Left: Scanning Acoustic Measurement (SAM) showing white bumps after flip chip assembly: Right: Cross-section showing the fracture inthe BEOL.

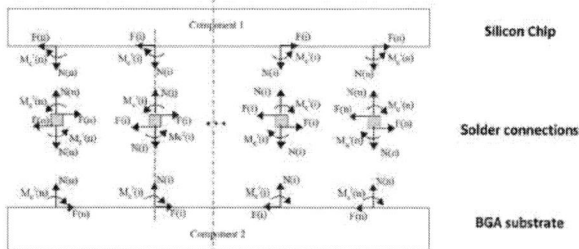

Fig. 1. Schematic drawing of forces and bending moments acting in a flip chip assembly during temperature variations

The fractures could be explained very well through calculating the forces and bending moments induced after cooling down from solder reflow temperature down to room temperature. Fig. 3 shows the bending moments and forces for three flip chip assemblies. It was seen that the bending moments were largest when there was a large pitch between the solder joints. This was more important than the size of the chip. This was also confirmed in the processing tests.

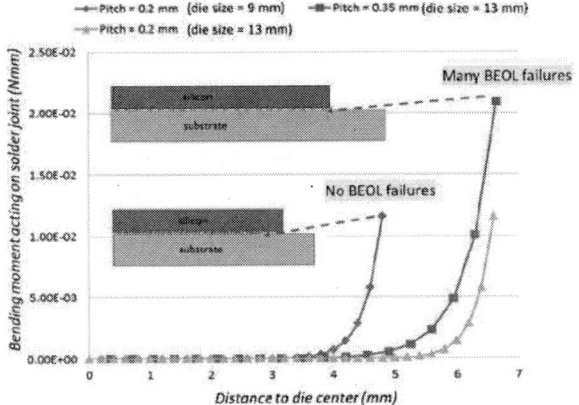

Fig. 3. Calculated bending moments in solder joints with varyaing die size and solder bump pitch

Underfills obviously reduce the forces acting on the solder joints and underlying BEOL layers. But there is still the solder reflow phase where underfill is not yet applied and therefore may be a critical phase in the assembly process. But also, the underfill must be well-chosen. Underfills with large CTE (coefficient of thermal expansion) mismatch with the solder bump material can cause high vertical forces which can also result in BEOL fractures.

A method to avoid these potential BEOL fractures during flip chip processing and thermal cycling qualification testing is through finite element modelling (FEM) of the thermo-mechanically induced stresses in the flip chip structure [3] [4] [5] . The outputs of the simulation study are stresses and deformations acting in the BEOL during the solder reflow process and during operational or qualification temperature cycling. To know if these stresses are causing fractures, measurements are needed to quantify the strength of BEOL layers. In this paper, a method is described and applied to a flip chip assembly structure.

II. FOUR POINT BENDING METHOD TO CHARACTERISE THE STRENGTHS OF BEOL

The method of measuring the BEOL strength utilized a four-point bending mechanism to generate delamination in the BEOL layers (Fig. 4) [6]. To create this mode of failure, two silicon strips were adhered to each other, and a notch was cut to act as the crack initiator during loading. The intent is that the crack should first grow downward through the silicon until it meets the BEOL layers where it then propagates laterally to create BEOL delamination. This is a highly dynamic process, and the crack propagates towards the weakest interface. It is no guarantee that the crack propagates through the BEOL layer. Therefore, inspection of the fractured samples after the bending test was needed.

The outcome of the four-point bending experiment is the force displacement curve (Fig. 5). There are different stages

in this curve which refer to the fracturing at the notch, the fracturing towards the BEOL layer and the delamination of the BEOL itself. In the graph, the plateau refers to delamination growth of the BEOL; this plateau force can be transferred into an energy release rate value.

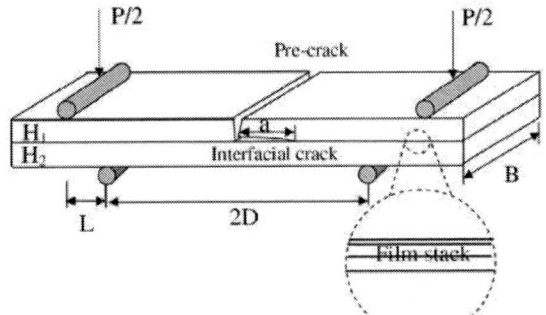

Fig. 4. 4-point bending setup on a stack of two dies

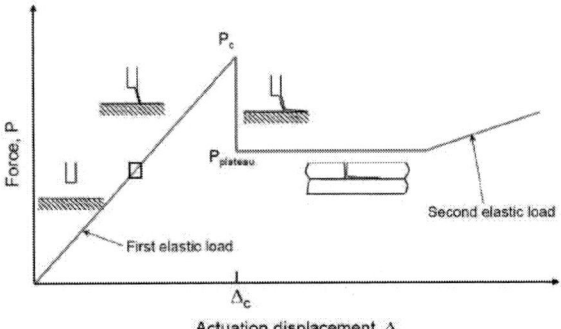

Fig. 5. A typical force versus displacement curve measured during a bending test

The critical strain energy release rate, G_c, can be calculated from the steady state fracture plateau load P as follows:

$$G_c = \frac{21}{16} \frac{(1 - v_{substrate}^2)P^2L^2}{E_{substrate}B^2H^3}$$

where L is the distance between an inner and an outer load pin (i.e. the moment arm), and 2D is the distance between the two inner load pins. The two substrates have thickness H_1 and H_2, width B, Young's modulus E, and Poisson's ratio v; H = $(H_1 + H_2)/2$.

The energy release rate is an appropriate value for the average strength of a specific BEOL structure. It allows to benchmark different BEOL technologies (oxide, low-k, ultra-lowK, airgap etc), different suppliers, and different number of stack-up layers.It is however less obvious to use this value in thermo-mechanical simulation studies. There are basically two methods in FEM which can handle these energy release rate values: cohesive zone material elements (CZM) and virtual crack closure technique (VCCT).

Cohesive zone material elements (CZM) are special interface elements that are inserted in the model. The idea behind cohesive elements is that fracturing is a gradual phenomenon in which separation takes place across an extended crack 'tip', or cohesive zone, and is resisted by cohesive tractions. Consequently, cohesive elements do not represent any physical material, but describe the cohesive forces which occur when material elements (such as grains) are being pulled apart [6].

The VCCT technique is based on the principle that the work required to extend a crack by an infinitesimal distance is equal to the work required to close the crack to its original length [7]. Fracture growth is then simulated by comparing energy release rate G to critical energy Gc, set as a criterion. An advantage of this technique is that it calculates the different energy mode components at the crack tip, and that mode separation can be included in the failure criterion.

Implementation of both techniques in a complex 3D structure with 2000 solder bumps is basically impossible. For VCCT, you need to implement a crack which is not obvious in 3D. The cohesive element also substantially increases the model size.

At the end, the modelling is meant as a first order sanity check to assess the risk for BEOL fracture for a specific flip chip assembly. In this work, it was chosen to extract a critical stress value from the four point bending experiment.

III. EXTRACTING ULTIMATE STRESS VALUES FOR BEOL

Four-point bending experiments were performed on a Tektronix-designed advanced node CMOS that goes into a Tektronix product. The IC was 10x8 mm in size with 180-micron bump pitch and included low-k and ultra-low-k layers. Around 50% of the samples showed a plateau force which indicated a delamination in the BEOL structure. One curve representative for all measurements is shown in Fig. 6. In that case, a plateau force was measured at 10.5N. Due to complex multilayer structure of the BEOL, the force was not constant and slightly varied during the continued displacement as shown in the figure (between ~400 to 500 µm displacement).

For the samples that showed a plateau force, the values were found to be within 10 and 15N for this specific bending setup. SEM (Scanning Acoustic Microscopy) analyses confirmed that the crack propagated from the notch, through the glue, and into the thickness of the BEOL (Fig. 7). Delamination had occurred between the BEOL layers: specifically, between silicon oxide passivation and low-k dielectric passivation. (Fig. 8)

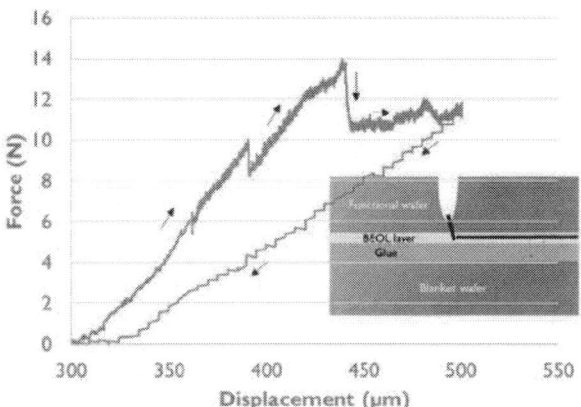

Fig. 6. Representative force displacement curve with a plateau force around 10.5N

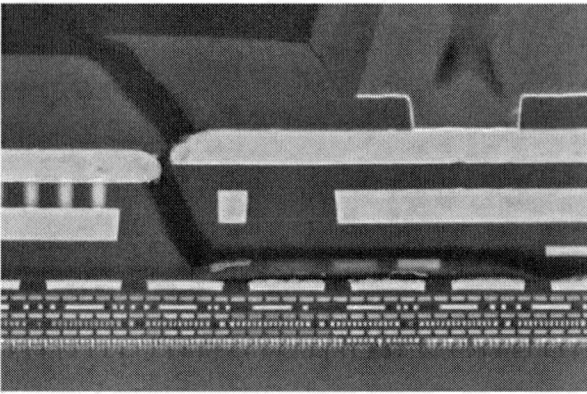

Fig. 7. Picture shows how fracture enters through the bulk chip through the BEOL to further propagate in the weakest area of the BEOL

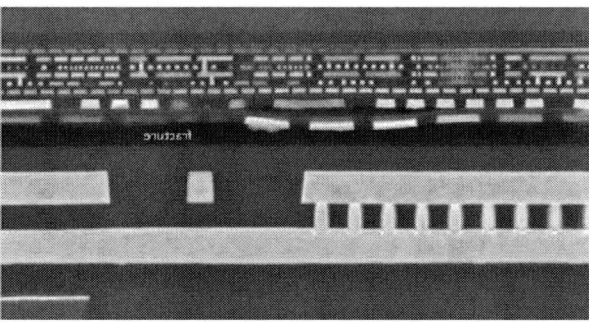

Fig. 8. Fracture zone in the BEOL after four point bending delamination test

Having verified that the test method correctly induced the delamination failure mode, a 2D plain strain FEM simulation of the four-point bending test was created to extract critical stress values (Fig. 9). It simulated the structure which had already cracked at the notch and has about 1 mm delamination in the BEOL.

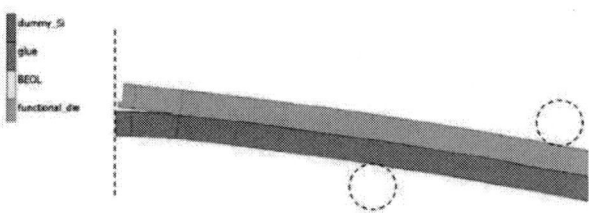

Fig. 9. 2D plain strain FEM of the 4 point bending experiment

With this approach, there was the issue of the stress singularity at the crack tip which made the stress value dependent on the mesh density. Therefore, a pragmatic approach was followed by having a similar mesh density in this 2D model as was used in a complex 3D model for flip chip assembly.

Fig. 10 shows the out-of-plane (or vertical) stress distribution at crack tip for a 10N bending. The stress was also linear to the applied force.

This critical stress is also just an average value for the BEOL. Some zones in BEOL will be weaker, other zones will be stronger. Typically, in the region of (solder) bumps, BEOL are made stronger with higher copper density.

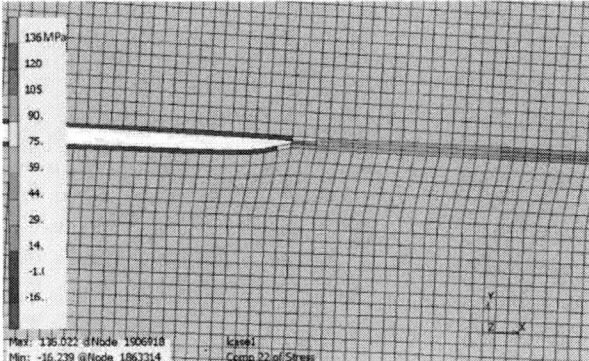

Fig. 10. 2D plain strain FEM of the 4 point bending experiment

IV. APPLICATION OF MEASURED ULTIMATE STRESS VALUES TO A FLIP CHIP ASSEMBLY

A detailed simulation model was made for a standard solder flip chip assembly package with about 2000 solder bumps (Fig. 11). A visco-plastic model was applied to the solder elements while the underfill and BGA substrate had visco-elastic models. Additional details about the flip-chip simulation model are described in [10].

The aim was to calculate the stress evolution in the BEOL over the complete process of the flip chip BGA and also during temperature cycling (Fig. 12). This meant that in the first phases of the modelling, the underfill and stiffener ring were deactivated.

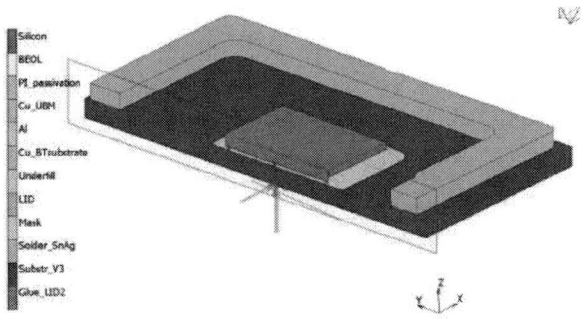

Fig. 11. Full 3D FEM of a flip chip assembly

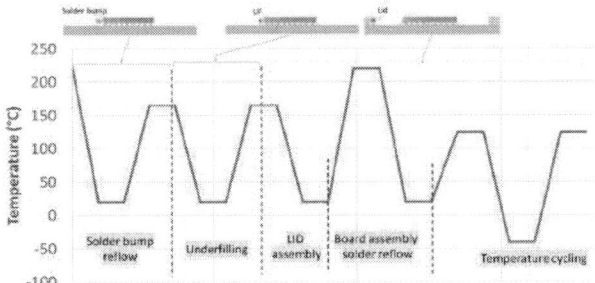

Fig. 12. Simulation process including all temperature variations

The highest vertical (peel) stress was seen in the BEOL during the cooling down from solder reflow to room temperature. Due to the large CTE mismatch between the large die and the substrate. As shown in Fig. 13, the highest value occurred near the die corner and was substantially below the ultimate stress value measured in the 4pt bending experiments.

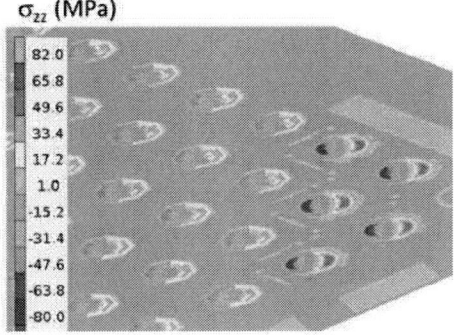

Fig. 13. Map of the vertical (out-of-plane) stress in the BEOL in the corner area during the cooling down from solder reflow to room temperature (first step in the processing)

The stresses during underfilling, lid assembly and temperature cycling were lower, but this was dependent on the chosen underfill and stiffener ring material. With worse choices of these materials, the BEOL stress may be even higher than in the first phase of the process and could cause BEOL damage.

V. CONCLUSIONS

Four-point bending experiments provided quantitative data on BEOL strengths. It allows for searching for the weakest interface in BEOL and has the capability to benchmark different technologies and suppliers of CMOS wafers.. These data can be used to perform sanity checks in thermo-mechanical simulations whether BEOL fractures are possible during packaging processes and during qualification and operational cycles.

ACKNOWLEDGMENT

The authors would like to thank Myriam Vandepeer for the sample preparation and bending experiments and Jan d'Haen from the IMOMEC group within imec for the SEM study on the tested samples.

REFERENCES

[1] Vandevelde, B., Beyne, E., Vandepitte , D., and Baelmans, M., Analytical Thermo-Mechanical Model for Non-Underfilled Area Array Flip Chip Assemblies, ASME. J. Electron. Packag. September 2004; 126(3): 351–358. https://doi.org/10.1115/1.1772416.

[2] J. Kawahara et al., A simple model-base prediction method for delamination failures in Low-k/cu interconnects with flip chip packages, 2013 IEEE International Interconnect Technology Conference - IITC, Kyoto, Japan, 2013, pp. 1-3, doi: 10.1109/IITC.2013.6615560.

[3] M. Gonzalez et al., "Chip package interaction (CPI): Thermo mechanical challenges in 3D technologies," 2012 IEEE 14th Electronics Packaging Technology Conference (EPTC), Singapore, 2012, pp. 547-551, doi: 10.1109/EPTC.2012.6507142.

[4] Lei Fu, Milind Bhagavat, Ivor Barber, Bumping Process Impact on the Chip Package Interaction (CPI) Reliability, IMAPS 2018 - 51st

International Symposium on Microelectronics - Pasadena, CA USA - Oct. 8-11, 2018

[5] Bart Vandevelde, A. Ivankovic, B. Debecker, M. Lofrano, K. Vanstreels, W. Guo, V. Cherman, M. Gonzalez, G. Van der Plas, I. De Wolf, E. Beyne, Z. Tokei, Chip-Package Interaction in 3D stacked IC packages using Finite Element Modelling, Microelectronics Reliability, Volume 54, Issues 6–7, 2014, Pages 1200-1205, ISSN 0026-2714, https://doi.org/10.1016/j.microrel.2014.02.026.

[6] Ortiz M. and Pandol A. "Finite-Deformation Irreversible Cohesive Elements for ThreeDimensional Crack-Propagation Analysis", Int. J. Num. Meth. Eng., Vol. 44, pp. 1267-1282, 1999

[7] Krueger R., "The Virtual Crack Closure Technique: History, Approach and Applications", Appl. Mech. Rev., Vol. 57(2), pp. 109-143, 2004

[8] B. Debecker, K. Vanstreels, M. Gonzalez and B. Vandevelde, "Delamination in BEOL: Analysis of interface failure by combined experimental & modeling approaches," 2013 IEEE International Reliability Physics Symposium (IRPS), Monterey, CA, USA, 2013.

[9] Luka Kljucar, Mario Gonzalez, Kris Vanstreels, Andrej Ivankovic, Michael Hecker, Ingrid De Wolf, Effect of 4-point bending test procedure on crack propagation in thin film stacks, Elsevier, Microelectronic Engineering, Volume 137, 2 April 2015, Pages 59-63.

[10] Kevin Cox, Jason Krantz, Matt Borden, Steven Tonthat, "Validating Flip Chip package models through experimental deflection measurements", IMAPS 19th International Conference & Exhibition on Device Packaging, Fountain Hills, AZ USA, Mar 13-16, 2023.

Application of machine learning methods for process optimization in electronic packaging processes

Corinna Niegisch
Corporate Sector Research and
Advance Engineering
Robert Bosch GmbH
Renningen, Germany
Corinna.Niegisch@de.bosch.com

Sabine T. Haag
Corporate Sector Research and
Advance Engineering
Robert Bosch GmbH
Renningen, Germany
Sabine.Haag@de.bosch.com

Tanja Braun
Fraunhofer IZM, Departement System
Integration and Interconnection
Technologies
Berlin, Germany
Tanja.Braun@izm.fraunhofer.de

Ole Hølck
Fraunhofer IZM, Departement System
Integration and Interconnection
Technologies
Berlin, Germany
Ole.Hoelck@izm.fraunhofer.de

Martin Schneider-Ramelow
Technical University Berlin
Faculty IV Electrical Engineering and
Computer Science
Berlin, Germany
Martin.Schneider-Ramelow@tu-berlin.de

Abstract´ Epoxy resins are commonly used as encapsulation materials in electronic packaging processes. Fluctuations in the materials lead to both changes in processability and to varying quality. Ideally, these variations should be identified, and measures taken as quickly as possible to reduce scrap parts and thus costs. A promising optimization approach for encapsulation processes are machine learning models, which have already shown good results in quality predictions, especially for injection molding. Subsequent quality measurements are avoided with good quality prediction models. With this type of models, not only predictions can be made, but also optimal parameter combinations can be found. In this paper, models for predicting quality criteria warpage and residual enthalpy of the epoxy molding compound were set up, trained and validated. Time series of in-situ sensors were used, from which relevant features were extracted and which, together with machine parameters, provide a dataset for prediction. The most promising prediction models are random forest regression and gradient boosting regression. They predict warpage with an accuracy of 90 % to 91 % and the residual enthalpy with an accuracy of 95 %. Subsequently, optimization models of the machine parameters were set up. All relevant target variables were considered in a cost function, through the minimization of which an optimal parameter set was found. The gradient boosted tree and Bayesian optimization were determined to be the most promising models, as they lead to the lowest values of the respective cost function.

Keywords´ machine learning methods, warpage prediction, enthalpy prediction, process optimization, molding processes, epoxy molding compounds, electronic packaging

I. INTRODUCTION AND MOTIVATION

Encapsulation is an essential process step in electronic packaging technology. It involves encapsulating electronics to protect the electronic components from environmental influences, insulate electrical contacts and provide structural support [1]. As encapsulation materials mainly epoxy molding compounds are used in a molding process such as transfer or compression molding [2]. A major problem in this process is batch variation or drifts in material properties, which affects both processability and quality [3]. The quality also depends on the specific process parameters that are selected depending on the material properties of the epoxy resin. Quality criteria for electronic encapsulation are usually warpage,

delamination, wire sweep, voids, incomplete filling or curing state of the epoxy molding compound [4].

In recent years, extensive research has been done to achieve a deeper comprehension of these type of processes. For this purpose, various sensors were installed in molding tools to obtain process information during the encapsulation. Classical sensors are temperature and pressure sensors [5], dielectric sensors (DEA) [3][5], ultrasonic sensors [6] or Fourier-transform infrared sensors [7]. For process optimization, mainly simulation models or statistical experimental designs are performed, or adjustments are made purely on empirical data [8]. However, another promising approach is offered by data-driven methods such as machine learning models [9]. They can be used to find underlying structures in data that are not yet perceivable to humans [10].

Machine learning models have already been applied to various plastics processing applications. The most common use case is the injection molding process, where there is already diverse research work on quality prediction or process optimization with machine learning models [11][12]. Tree-based algorithms, regression-based algorithms, or neural networks are applied to perform classifications or regressions [13][14]. An interesting approach is described in [15], where the use of sensor data for prediction models is outlined. The authors identified the cavity pressure as essential for quality, which is why sensor data provide important information. They extracted key features of the recorded time series as additional information for quality predictions. For machine learning models in thermoset processes there is still very few research work. However, there are initial approaches for transfer molding. By applying classification methods, it is possible to predict the filling behavior or voids during transfer molding [9]. Another approach is described in [8] where the authors performed process optimization for transfer molding with neural networks.

In the following, the use of machine learning models for molding processes is presented. First, methods for the prediction of selected quality criteria such as warpage and the residual enthalpy of the molding material are shown. After that, different approaches for the optimization of the process parameters of molding processes are presented.

II. METHODOLOGY

The main objective of this work is the implementation and validation of quality predictive supervised learning models and the optimization of process parameters for molding processes. Therefore, information about data acquisition and the acquired dataset are presented. In addition, the pre-processing of the data and the implemented machine learning models including their evaluation are shown.

A. Data acquisition

In this study, the so-called local pressure molding was used as encapsulation process. This is a combination of conventional transfer molding and compression molding [7][16]. The experiments were carried out with a press from the company Lauffer GmbH & Co. KG (UVKU 125-SO). In this process, a rectangular area with dimensions of 65x35x1 mm is pressed onto a substrate material (200 µm FR4 R1566, Panasonic Corporation, Kadoma, Japan) in a heated tool. The molding material is a highly filled epoxy molding compound with a filler content of approximately 90 %, which is available in a pre-pressed sheet geometry (Shin-Etsu Chemical Co., Ltd., Tokyo, Japan). A release film (200 µm single-side matte LM-ETFE film, AGC Chemicals Europe, Ltd., Amsterdam, Netherlands) was used for the experiments. The pressing process is achieved by a piston that compresses the material. During the process, the movement for filling is position-controlled first, which changes to a pressure-controlled profile as soon as the cavity is filled. During the molding process, there are various machine parameters that have an influence on the processability and thus also on the quality of the encapsulation. These are the mold temperature, the cavity pressure, the piston feed speed at which the material is pressed into the cavity, and the cycle time.

To acquire data during the molding process, two temperature sensors T_1 and T_2 (type 6195B, Kistler IInstruments AG, Winterthur, Switzerland), three pressure sensors p_1, p_2 and p_3 (type 6167A, Kistler IInstruments AG, Winterthur, Switzerland) and two dielectric sensors DEA_1 and DEA_2 (4 mm monotrode-sensor 4/3RC, Netzsch-group GmbH & Co. KG, Selb, Germany) were integrated into the molding tool. The sensors are in direct contact with the epoxy molding compound and therefore provide information of the molding material during the process. In addition, the pressure p_4 and the piston position s_1 were also read out directly from the machine. A trigger signal from the machine ensures that the start of the measurements is automated, and the measurements can be correlated with the elapsed process time.

After the encapsulation process, the quality criteria were evaluated, which are the warpage of the pressed parts on the one hand and the residual enthalpy of the epoxy resin on the other. The warpage was evaluated with a 3D profilometer (VR-5000, Keyence AG, Osaka, Japan), obtaining height data. The residual enthalpy of the molding compound was determined by differential scanning calorimetry (DSC 204 F1 Pheonix, Netzsch-Gerätebau GmbH, Selb, Germany). The residual enthalpy can then be correlated with the curing stage of the molding material.

B. Dataset

In order to obtain an extensive dataset with a large range of different warpage and enthalpy values, the machine parameters were varied. The parameter limits were determined experimentally resulting in the following limits:

- ¿ Temperature between 150 éC and 175 éC,
- ¿ Pressure between 25 bar and 100 bar,
- ¿ Piston feed rate between 0.2 mm/s and 1.0 mm/s and
- ¿ Cycle time between 180 s and 900 s.

No statistical experimental design was performed purposefully, as these are often not ideal for machine learning models. Instead, each parameter was varied individually while the others remained constant. A total of 23 parameter combinations were performed with 5 repetitions each. The standard parameter combination at 170 éC, 50 bar, 0.2 mm/s and 300 s was performed more often with 20 replicate trials, resulting in a total dataset size of 135 data points. Warpage and enthalpy were measured and evaluated for each part produced.

C. Data pre-processing

First, features of the time series measured were extracted. Common statistical parameters such as minimum (min), maximum (max) or mean value were calculated as well as additional key features which are all listed in TABLE I. It results in 29 features extracted from the time series in addition to 4 machine parameters.

TABLE I.

FEATURES EXTRACTED FROM THE TIME SERIES

Sensor data	Feature extraction	
	Sensor	Extracted feature
Sensors installed in the tool	T_1, T_2	min, max, mean
	p_1, p_2, p_3	min, max, mean
	DEA_1, DEA_2	min, permittivity at change point (cp), time stamp at change point (ts_cp)
Sensors from machine	p_4	min, max, mean
	s_1	min, max, mean, slope, time stamp at switching point (ts_sp)

After feature extraction, the dataset was split into 60 % training data and 40 % test data. The dataset was grouped so that each experimental combination performed is present at least once in the test data. The subsequent machine learning models are trained using only the training data and finally tested with the test data.

D. Machine learning models

For the prediction of warpage and the residual enthalpy of the molding material different machine learning models were trained. Existing models from the scikit-learn library were applied. Random forest regression, support vector regression, gradient boosting regression and k-nearest neighbors were trained and tested with the previously presented dataset. No hyperparameter optimization was performed as the models were applied with their standard hyperparameters.

The best quality prediction model in each case is required for the following optimization models. The models random search (RS), gradient boosted tree (GBT) and Bayesian optimization (BO) from the scikit-learn library were applied and trained for optimization. The optimization targets were defined as the quality criteria warpage and residual enthalpy. The machine parameters were selected as the parameters to be optimized.

The optimization models are designed to maximize or minimize one target variable. If several variables are to be taken into consideration, they must be combined in a cost function. The optimization models then optimize the cost function by minimization. The definition of a cost function is essential, as it significantly influences the success of the optimization models substantially [17]. All target variables were thus scaled between zero and one. In addition, the scaled quantities can subsequently be weighted depending on the application.

In this study, two different cost functions were implemented and tested. First, the target variables warpage and residual enthalpy were used. The warpage w_{sc} was scaled between 0 µm and 1300 µm since a warpage of 0 µm is targeted and 1300 µm is the maximum occurring warpage. For the enthalpy, a value of 0 J/g, i.e., complete curing, is targeted. The lowest residual enthalpy possible after molding is 8 J/g, so the enthalpy h_{sc} was scaled to these limits. The weights of both quantities w_w and w_h were set to 1 since both target quantities are considered equally important. This results in the cost function described by (1).

$$cost_1 = w_w \Theta w_{sc} + w_h \Theta h_{sc} \qquad (1)$$

The second cost function also includes the cycle time t_{sc} in addition to the two target variables. This is limited by the curing of the material and therefore scaled between 60 s and 300 s. The weighting of the cycle time w_t was also set to the value 1, resulting in the cost function shown in formula (2).

$$cost_1 = w_w \Theta w_{sc} + w_h \Theta h_{sc} + w_t \Theta t_{sc} \qquad (2)$$

E. Model evaluation

To compare the prediction models, the accuracy of each model was determined by calculating the R^2 value, and the mean absolute error (MAE). In addition, the feature importance was calculated to identify the most relevant features for prediction. Then, the plot of predicted over real values was considered in more detail for the model with the highest accuracy.

For the optimization models, the predicted process parameters and the value of the respective cost function were compared with each other. In addition, the training history is plotted and evaluated for all optimization models presented.

III. RESULTS AND DISCUSSION

Warpage and enthalpy prediction models were trained and tested with the acquired dataset. Different models were evaluated and compared with each other. Subsequently, the most promising predictive models were used for optimization models, which were tested and compared with two different cost functions.

A. Warpage prediction model

The first prediction model is implemented for the quality criterion warpage. The accuracy results and the calculated MAE are shown in TABLE II. The prediction was first performed only with the machine parameters (MP) and then including the extracted features from the sensor timeseries (ts-features).

The warpage predictions with random forest regression and gradient boosting regression show an accuracy of 86 % and an MAE of 36 µm when the prediction was made with machine parameters alone. With a prediction value of 87 %, the model k-nearest neighbors is almost equally as accurate but has a larger MAE. The support vector regression however predicts warpage with an accuracy of 48 %. This is the lowest prediction value, and it causes an MAE of 88 µm. Since the measurement device for warpage measurements exhibits an error up to 10 µm, the prediction with an error of 36 µm to 46 µm is concluded to be satisfying.

TABLE II.
COMPARISON OF DIFFERENT WARPAGE PREDICTION MODELS

Model	Evaluation test data (MP)		Evaluation test data (MP and ts-features)	
	Accuracy (R^2)	MAE	Accuracy (R^2)	MAE
Random forest regression	86 %	36 µm	90 %	45 µm
Support vector regression	48 %	88 µm	72 %	67 µm
Gradient Boosting regression	86 %	36 µm	91 %	38 µm
K-nearest neighbors	87 %	46 µm	86 %	53 µm

Warpage is a very sensitive parameter which, despite supposedly constant machine parameters, shows fluctuations in the warpage measurements. Process deviations for example in the temperature are very common. Despite the same temperature set in the machine, the parts are produced with different temperatures which is why in-situ measurements are necessary and provide important information. Therefore, warpage prediction was performed with machine parameters and extracted features from sensor data. Whereas the prediction accuracy with the model k-nearest neighbors is almost the same for both datasets, all the other models show improvements in their prediction accuracy. Random forest regression and gradient boosting regression show an accuracy of 90 % respectively 91 %. The accuracy with support vector regression is even improved by 24 % to an accuracy of 72 %. The improvement in the prediction accuracies is reasoned to be due to the additional information provided by the extracted features. There is however a risk of overfitting the models since the dataset has only 135 samples. Overfitting can be avoided by adding more samples to the dataset or by hyperparameter optimization of the prediction models (e.g., early stopping).

To identify the most relevant features for the prediction of warpage the feature importances are plotted in Fig. 1. The most important feature for random forest regression, gradient boosting regression and k-nearest neighbors is by far the cycle time. For random forest regression besides the cycle time, DEA_2_min is considered as an important feature whereas the importance is factor 7 lower compared to the cycle time. The gradient boosting regression model considers T_2_max as the second important feature and k-nearest neighbor DEA_ts_cp with a factor of 4 and 10 less importance compared to cycle time, respectively. In the dataset it is visible, that long cycle times lead to low warpage values which is why the cycle time is considered as the most important feature.

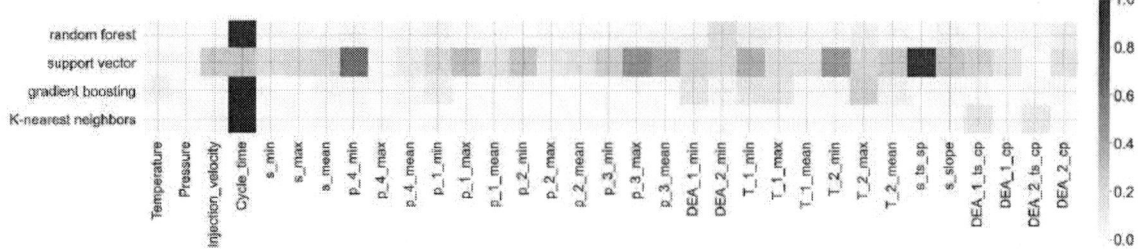

Fig. 1 Feature importance of warpage prediction models

The feature importance of the support vector regression model is different compared to the other models. Multiple features are considered as important whereas the features s_ts_sp and p_4_min are of utmost importance. Apparently, the support vector regression model does not correctly recognize the significant correlations in the dataset, which is why the prediction values for this model are lower than for the other models tested.

Features extracted from time series in addition to the machine parameters are helpful to identify outliers since warpage prediction accuracy scores are higher when features of time series are included. However, Fig. 1 shows that the cycle time is the most important feature for warpage prediction. This is why the prediction accuracies are hardly lower when only machine parameters are used for prediction.

In Fig. 2 the predicted against the real warpage values of the test data for random forest regression is shown since this model exhibits a high accuracy at 90 %. It is clear that the predicted warpage values are close to the real warpage values, indicating an accurate prediction model. The model could be improved by extending the dataset since there are only a few datapoints above 900 ı m.

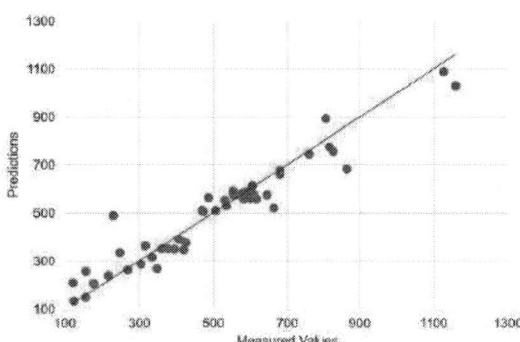

Fig. 2. Comparison of predicted and real warpage values for random forest regression

B. Enthalpy prediction model

The second prediction model is implemented for the quality criterion enthalpy. TABLE III. shows the accuracy results and the calculated MAE for the prediction with machine parameters and after sensor timeseries are included. The enthalpy prediction with random forest regression and gradient boosting regression displays an accuracy of 95 % when the prediction is made with machine parameters. The accuracy values do not change when the features are applied

for prediction. This leads to an MAE of 0.15 J/g. The model k-nearest neighbors predicts the residual enthalpy with an accuracy of 85 % when machine parameters are applied, and 96 % when the features from the sensor timeseries are included for prediction. The model support vector regression is not able to predict any reasonable prediction values. It is concluded that this model fails for enthalpy prediction.

TABLE III.
COMPARISON OF DIFFERENT ENTHALPY PREDICTION MODELS

Model	Evaluation test data (MP)		Evaluation test data (MP and ts-features)	
	Accuracy (R^2)	MAE	Accuracy (R^2)	MAE
Random forest regression	95 %	0.15 J/g	95 %	0.15 J/g
Support vector regression	-	-	-	-
Gradient Boosting regression	95 %	0.13 J/g	95 %	0.12 J/g
K-nearest neighbors	85 %	0.24 J/g	96 %	0.12 J/g

The residual enthalpy is determined based on DSC measurements of the produced parts. On the material side, the minimum cycle time is limited to 180 s, which corresponds to a maximum residual enthalpy of 8 J/g. Even with a slight increase in the cycle time, the residual enthalpy drops to values close to 0 J/g, which corresponds to complete curing. Therefore, the dataset contains mainly enthalpy values close to 0 J/g. Since the measurement device for enthalpy measurements exhibits an error up to 2 J/g, the prediction with an error below 1 J/g is concluded to be satisfying. It is concluded that prediction with accuracies of more than 85 % are very accurate.

Fig. 3. shows the importance of the individual features for the prediction of the residual enthalpy. It is clear, that the most important feature is the cycle time, when random forest regression, gradient boosting regression or k-nearest neighbors is applied. For k-nearest neighbors the feature DEA_2_ts_cp is considered as relevant with the importance being by factor 20 lower compared to the cycle time. This is the reason why the prediction accuracy improves by 11 % when the features are applied for prediction. The model support vector regression considers multiple features as important but since the prediction for residual enthalpy fails it is not considered further.

Another approach which is not presented in this work, is the prediction of the residual enthalpy with spectral data. Regression models can be trained on the change of infrared spectra, since the spectra change with increasing curing of the

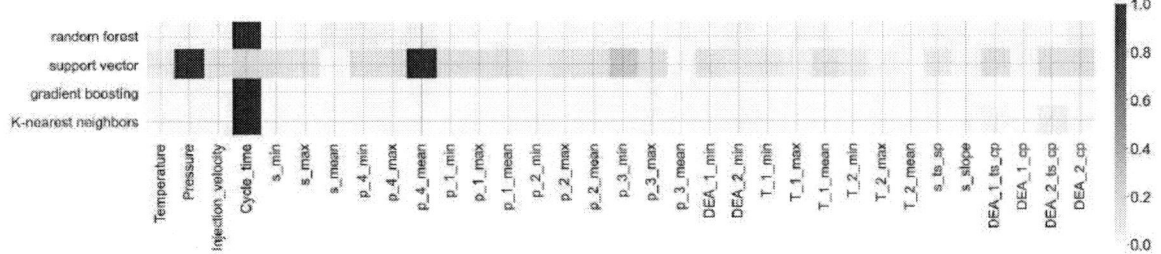

Fig. 3 Feature importance of enthalpy prediction models

material. The changes in spectra can be correlated with the residual enthalpy. This approach has been proven several times in literature [18] and could lead to better prediction results.

C. Optimization approach

First, cost function (1) is minimized with the presented optimization models. For these models a prediction model for each quality criteria needs to be defined. The prediction models are trained on the presented dataset and then applied for the parameter combinations, that the optimization models suggest during optimization. The random forest regression model was chosen as a prediction model for warpage and enthalpy since it performed well for both quality criteria. The prediction models were trained on machine parameters only.

TABLE IV displays the optimized results for random search, Bayesian optimization and gradient boosted trees. The three optimization models predict similar optimized process parameters. Cycle time and temperature are the two decisive factors regarding warpage and enthalpy whereas the cavity pressure and injection velocity have no noticeable influence on the two quality criteria. All optimization models predict long cycle times which is clear since the cycle time is not considered in cost function (1) and long cycle times lead to low warpage values. The model gradient boosted trees reaches the lowest value of the cost function. The optimized process parameters lead to a predicted warpage value of 121 ı m and a predicted enthalpy value of 0.0 J/g which are reasonable predictions.

TABLE IV.
OPTIMIZED PROCESS PARAMETERS WITH COSTFUNCTION (1)

Model	Predicted process parameters				Optimized value cost function
	T	p	t	v	
RS	159éC	54 bar	838 s	0.2 mm/s	0.0944
BO	158 éC	100 bar	832 s	0.2 mm/s	0.0938
GBT	159 éC	64 bar	824 s	0.3 mm/s	0.0635

To compare the models with each other, a convergence plot of the three models was created in Fig. 4 by visualizing the value of the cost function against the number of iterations. It is clear, that all models achieve an optimization of the cost function after a few iteration steps. The lowest value is achieved with gradient boosted trees after 28 iteration steps. The success of these optimization models depends on several factors. One factor, is the setup of the cost function, for which there are unlimited possibilities. Additional process knowledge is required, such as for the scaling of the variables

to be optimized. Furthermore, the weights can be varied, which also influence the result of the optimization.

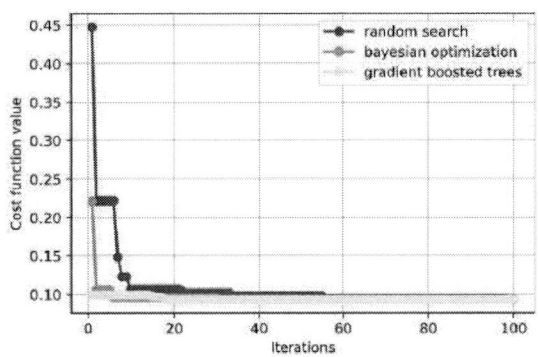

Fig. 4. Development of cost function 1 against iteration steps

For consideration of the cycle time, the optimization models were applied to cost function (2). The optimized process parameters are presented in TABLE V. The models behave similarly to cost function (1) with the Bayesian optimization achieving the lowest value of the cost function. Bayesian optimization predicts a cycle time of 396 s compared to 832 s when cost function (1) was considered. This is a reduction of 52.4 % in cycle time. The optimized process parameters lead to a predicted warpage of 180 ı m and a predicted residual enthalpy of 0.0 J/g. The predicted warpage increases compared to cost function (1) because cycle time was added in cost function (2).

TABLE V.
OPTIMIZED PROCESS PARAMETERS WITH COSTFUNCTION (2)

Model	Predicted process parameters				Optimized value cost function
	T	p	t	v	
RS	160 éC	86 bar	404s	0.2 mm/s	0.4496
BO	162 éC	39 bar	396 s	1.0 mm/s	0.4412
GBT	158 éC	29 bar	421 s	0.2 mm/s	0.4537

Fig. 5 shows the convergence plot of all models for the optimization of cost function (2). All models achieve minimization in the cost function after a few iteration steps whereas the decrease in cost function (2) is visible after more iteration steps compared to cost function (1). The lowest value in cost function is achieved by Bayesian optimization after 25 iteration steps indicating the most promising optimization method for cost function (2).

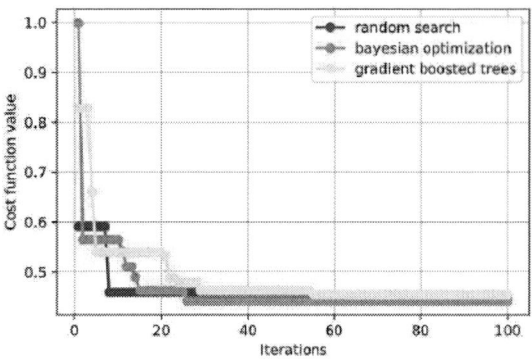

Fig. 5. Development of cost function 2 against iteration steps

IV. CONCLUSION AND OUTLOOK

The models presented in this work allow accurate quality predictions for warpage and residual enthalpy. Several prediction models were tested and compared, with gradient boosting regression and random forest regression achieving promising predictions for both quality criteria. It is possible to test other models such as neural networks in order to make better predictions. Also, it is possible to expand the dataset for better prediction, as many machine learning models tend to perform better on large datasets. The prediction of warpage with the use of machine parameters and sensor data appears to be useful and promising.

In addition, different optimization models were implemented and validated to find the optimized process parameters. All relevant criteria were considered in a cost function. The best optimization model is found in the gradient boosted trees model for cost function (1), and in Bayesian optimization for cost function (2). These models achieve the lowest value for each cost function and predict reasonable process parameters. A disadvantage of the presented models is that they are necessarily not able to predict values outside their data range. For future optimizations it is also possible to use another kind of models like reinforcement learning models. These models do not need a large dataset at the beginning, but learn continuously. This could be a completely new approach for optimization of molding processes.

ACKNOWLEDGMENT

The authors would like to thank Ms. Yasmina Kit for her assistance in performing the measurements for this publication.

REFERENCES

[1] H. Sasajima, I. Watanabe, M. Takamoto, K. Dakede, S. Itoh, Y. Nishitani, J. Tabei, T. Mori, ‗New Development of Epoxy Molding Compound for Encapsulating Semiconductor Chips,‗ Springer, 2009.

[2] S. Komori, Y. Sakamoto, `Development Trend of Epoxy Molding Compound for Encapsulation Semiconductor Chips, ‗ Springer, 2009.

[3] B. Kaya, J. Kaiser, K. Becker, T. Braun, K. Lang, ` Evaluation of the Dielectric Cure Monitoring of Epoxy Molding Compound in Transfer Molding Process for Electronic Packages,‗ In IEEE Proceedings of the European Microelectronics Packaging Conference (EMPC), Friedrichshafen, Germany, 14‾16 September 2015.

[4] B. Kaya, `Concept Development and Implementation of Online Monitoring Methods in the Transfer Molding Process for Electronic Packaging,‗ Dissertation, Technische Universität Berlin, Berlin, 2018.

[5] G. Tuncol, M. Danisman, A. Kaynar, and E. M. Sozer, `Constraints on monitoring resin flow in the resin transfer molding (RTM) process by using thermocouple sensors, ‗ Compos. Part Appl. Sci. Manuf., vol. 38, no.5, pp. 1363-1386, May 2007.

[6] E. Schmachtenberg, J. Schulte zur Heide, and J. TØpker, `Application of ultrasonics for the process control of Resin Transfer Moulding (RTM),‗ Polym. Test., vol. 24, no. 3, pp. 330-338, May 2005.

[7] C. Niegisch, S. Haag, T. Braun, O. HØlck and M. Schneider-Ramelow, `In-situ infrared spectroscopy for chemical analysis in electronic packaging processes,‗ In IEEE Proceedings of the 9th Electronics Systems-Integration Technology Conference (ESTC), Sibiu, Romania, 13-16 September 2022.

[8] K.W. Tong, C.K. Kwong, K.M. Yu, `Process optimisation of transfer moulding for electronic packages using artificial neural networks and multiobjective optimisation techniques,‗ The International Journal of Advanced Manufacturing Technology, vol. 24, pp. 675‾685, 1 November 2004.

[9] J. Mendikute et al., `Impregnation quality diagnosis in Resin Transfer Moulding by machine learning,‗ Composites Part B: Engineering, vol. 221, 15 September 2021.

[10] D. Weichert et al., `A review of machine learning for the optimization of production processes,‗ The International Journal of Advanced Manufacturing Technology, vol. 104, pp. 1889‾1902, 20 June 2019.

[11] S. Selvaraj, A. Raj, R. Mahadevan, U. Chardha, V. Paramasivam, `A Review on Machine Learning Models in Injection Molding Machines,‗ Advances in Materials and Science Engineering, vol. 2022, 5 January 2022.

[12] S. Hwang, J. Kim, `Injection mold design of reverse engineering using injection molding analysis and machine learning,‗ Journal of Mechanical Science and Technology, vol 33, issue 8, pp. 3803‾3812, 14 May 2019.

[13] H. Jung, J. Jeon, D. Choi, J-Y. Park, `Application of Machine Learning Techniques in Injections Molding Quality Prediction: Implications on Sustainable Manufactring Industry,‗ Sustainability, vol. 13, issue 8, 7 April 2021.

[14] P. Nagorny et. al, `Quality Prediction in Injection Molding,‗ In IEEE Proceedings of the International Conference on Computational Intelligence and Virtual Environments for Measurement Systems and Applications (CIVEMSA), Annecy, France, June 2017.

[15] R. P®izs, D. TØr®k, T. Ageyeva, J. Kov®cs, `Machine Learning in Injection Molding: An Industry 4.0 Method of Quality Prediction,‗ Sensors, vol. 22, issue 7, 1 April 2022.

[16] U. Schaaf, A. Kugler, ‗Vorrichtung und Verfahren zur Herstellung von mit einer Gieómasse zumindest bereichsweise aberdeckten Bauelementen,` (DE 10 2017 216 711 A1 2019.03.21). Deutsches Patent- und Markenamt, 2019.

[17] A. Akbari, M. Awais, M. Bashar, J. Kittler, `How Does Loss Function Affect Generalization Performance of Deep Learning? Application to Human Age Estimation,‗ In PMLR Proceedings of the 38th International Conference on Machine Learning, vol. 139, pp. 141‾151, 2021.

[18] M. Salzmann, Y. Blóól, A. Todorovic, R. Schledjewski, `Usage of Near-Infrared Spectroscopy for Inline Monitoring the Degree of Curing in RTM Processes,‗ Polymers, vol. 13, Issue 18, 3145, pp. 1‾13, 17. September 2021.

Partial Discharge Characterization of Ceramic Power Electronics Circuit Carriers Assisted by Machine Learning

Johannes Drechsel
Systems Integration and Electronic Packaging
Fraunhofer Institute for Ceramic Technologies and Systems IKTS
Dresden, Germany
johannes.drechsel@ikts.fraunhofer.de

Henry Barth
Systems Integration and Electronic Packaging
Fraunhofer Institute for Ceramic Technologies and Systems IKTS
Dresden, Germany
henry.barth@ikts.fraunhofer.de

Lars Rebenklau
Systems Integration and Electronic Packaging
Fraunhofer Institute for Ceramic Technologies and Systems IKTS
Dresden, Germany
lars.rebenklau@ikts.fraunhofer.de

Abstract— **This paper presents an approach for transferring knowledge about partial discharges in polymer insulators to ceramic insulators with the aid of machine learning. It is shown how various machine-learnable features can be generated from partial discharge measurement data and processed in varying artificial neural networks for classification. It is found that polymer-based partial discharges can be classified using this method. In addition, the Long Short-Term Memory based artificial neural network enables partial discharge cause finding and thus fault detection in ceramic power electronics substrates.**

Keywords—Ceramics, Partial Discharges, Machine Learning, High Voltage, High Frequency

I. INTRODUCTION

Electronics are used in almost all areas of life, whereby the electronic assemblies are usually mounted on rigid carriers. The common green fiberglass composite boards, for example in PCs or smartphones, are well known in this context. This material, called FR-4, is used extensively in the consumer sector. Here, the requirements on thermomechanical, chemical, and electrical stability are comparatively low while cost pressure is high. The situation is different for high-voltage switchgear or power electronics assemblies, widely used in the e-mobility sector. Here, ceramic carrier boards are used due to the increased requirements. These include higher operating voltage, circuit waste heat, and harsh environmental conditions in an often limited package size. Despite the excellent material properties of ceramics, certain high-voltage systems are based on polymer insulators. Operation of these systems has shown repeatedly that the occurrence of so-called partial discharges (PD) can impair their insulation properties. Over time, this phenomenon can lead to an electrical breakdown and thus total failure of the system. Therefore, close monitoring is necessary, which places high voltage and frequency requirements on the measurement technology due to the PD-inherent occurrence characteristics. Furthermore, vast amounts of data are recorded during PD measurements, which must be processed, visualized, and interpreted. To address these matters, this work evaluates the following points:

- Knowledge transfer from polymer-based insulation systems to ceramics

- Use of machine learning in ceramics-based PD classification

II. PARTIAL DISCHARGES

A. Occurence

The phenomenon of failing insulation is commonly known in the form of lightning during a thunderstorm. Simply put, the trigger for such a breakdown is the potential difference between the high voltage side and ground, which allows a starting electron to build up enough kinetic energy to carry other electrons along in an avalanche effect. This leads to the formation of a conduction channel in the previously poorly conducting medium, which transports electrical charge. So, concluding, the electric field strength must exceed the breakdown limit of the insulation material for a discharge to occur. Scaled down, this means that discharges can also occur in partial regions of the insulator as long as the maximum local intrinsic field strength is exceeded and a start electron is available. The discharges occurring in a part of the total insulator are consequently called partial discharges. Despite the local confinement, these, as well as complete breakdowns, can damage the insulator by means of heat emission, vibrations, photons or chemically.

B. Model

For a better understanding of this behavior in solid insulators such as ceramics, the capacitive equivalent circuit for partial discharges by Philippoff and Gemant (Fig. 1) is used.

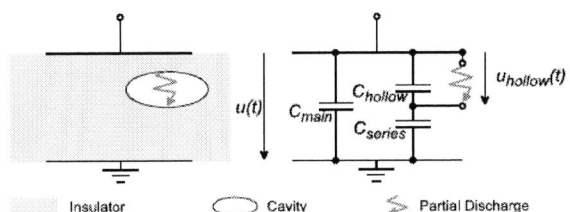

Fig. 1. Capacitive PD model view [1].

In this model conception, it is assumed that both the ideal insulation path (C_{main}) and possible imperfections (C_{hollow}) as well as the surrounding, undisturbed insulation material (C_{series}) are described as capacitors. The PDs that occur when an alternating voltage of sufficient amplitude is applied are represented by the spark gap. Inherent in both the model and the real test specimen is that the transported charge (q_{real}) cannot be measured directly but can be described in relation to the charge measured at the substrate (q_{app}) by (1) [2].

$$q_{real} = q_{app} * \frac{C_{hollow} + C_{series}}{C_{series}} \qquad (1)$$

C. Measurement

The measurement setup shown schematically in Fig. 2 is suitable for inducing and measuring partial discharges.

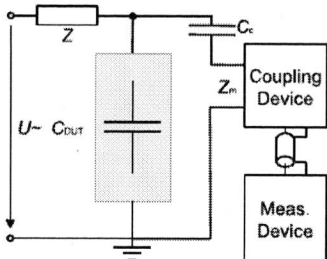

Fig. 2. PD measurement setup schematic conforming to IEC 60270 [3].

In order to be able to determine the apparently transported charge q_{app}, the measuring device must be calibrated to the entire setup including the test specimen. This is done using pulses of known amplitude applied directly to the device under test without applying the test voltage. Just like the PDs induced in the test setup when a high voltage is applied, these pulses manifest themselves as high-frequency current spikes in the supply line. To determine this charge quantitatively, the high-frequency high-voltage PD signal is decoupled via coupling capacitors (C_c) as well as filters and calculated from voltage peaks across the measuring impedance Z_m according to (2).

$$q_{app} = \int_{t_1}^{t_2} i(t)\,dt = \frac{1}{z_m}\int_{t_1}^{t_2} u(t)\,dt \qquad (2)$$

A real-time measurement device that implements both peak detection and integration is shown schematically in Fig. 3.

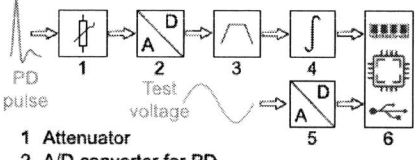

1 Attenuator
2 A/D converter for PD
3 Digital bandpass   5 A/D converter for AC
4 Numerical integrator   6 Processing unit

Fig. 3. Real-time PD measuring device schematic conforming to [3].

In addition, such a device detects the alternating test voltage and sets the time of occurrence of each PD in phase relation to it. Thus, the measuring device provides the following data triplet for each PD:

- Apparent charge
- Time of occurence
- Phase

D. Data Evaluation

In order to gain insights into the partial discharge origins from the generated data (up to several million events per second), two visualization methods have been established. One is an amplitude-phase correlation called phase resolved partial discharge pattern (PRPD, Fig. 4), the other is the correlation of successive pulses, called pulse sequence analysis (PSA, Fig. 5).

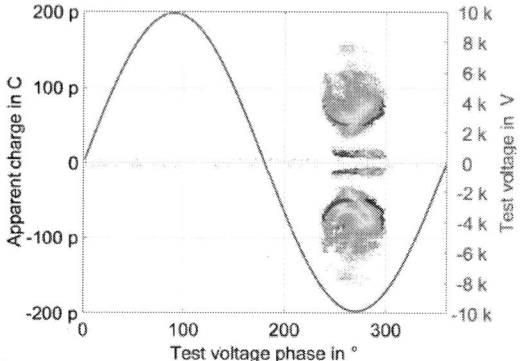

Fig. 4. Exemplary PRPD.

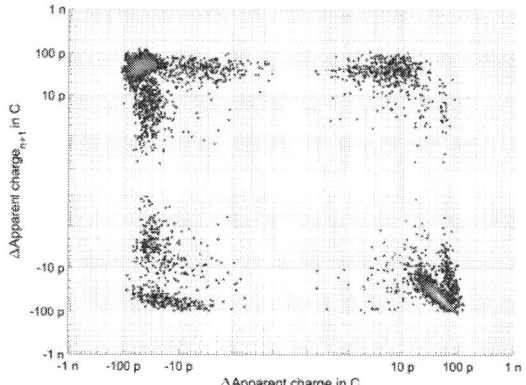

Fig. 5. Exemplary PSA.

Established PD literature states that the patterns shown exhibit significant differences depending on the PD origin (e.g., point-plate structure, cavity discharge, surface-gliding discharge) [4, 5]. Since classifying millions of pulse shapes by hand is impractical, some form of machine pattern recognition must be employed.

III. APPROACH

A. Test Specimens

To enable knowledge transfer from PDs in gaseous and polymer insulators to PDs in ceramic circuit carrier boards, the test specimens shown in Tab. I are used.

TABLE I. TEST SEPCIMENS

No.	Descrip-tion	Schematic	Expected PD types
1	Point-plate in air	HV ↓ GND	Corona discharges
2	Cavity in FR-4	HV FR-4 Air GND	Cavity discharges, external corona discharges
3	Inclusion in FR-4	HV Polymer GND	Cavity discharges, boundary layer discharges, external corona discharges
4	Cavity in ceramic	HV Al₂O₃ Air GND	Cavity discharges, , external corona discharges
5	Point-point in ceramic	HV GND	Cavity discharges, boundary layer discharges, external corona discharges

The use of comparatively easy-to-manufacture polymer substrates with defined inclusions enables precise knowledge about the main PD origins in each measurement. This allows for their labeling for the subsequent machine learning. The need for manual labeling arises from the use of supervised learning approaches in Section III D. These approaches require predefined classes for each sample in order to learn the dependency between label and input data in the training process.

B. Partial Discharge Source Differentiation

In real applications, depending on the type and number of superimposed PD origins, differentiation between origins is difficult. This makes a meaningful PSA impossible, since actually independent PDs are set in relation to each other due to their temporal position. Consequently, an in-depth understanding of possible PD sources remains difficult. Therefore, the individual pulse shapes are used for clustering in literature [6, 7]. Fig. 6 shows data from air and polymer measurements in two possible planes. The purpose here is to show the usual overlapping within the planes, the Time/Bandwidth map is explained in more detail in the following section.

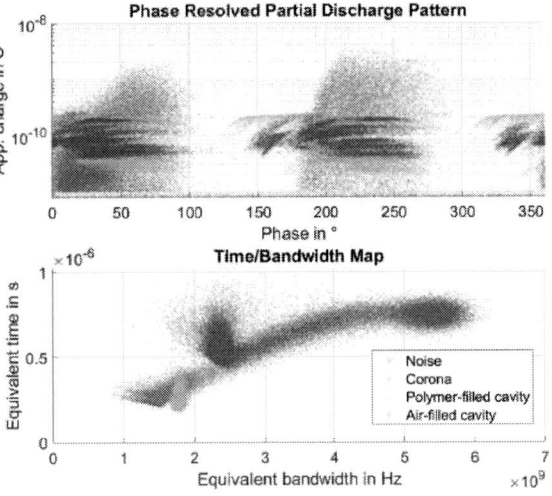

Fig. 6. Data from 4 different sources as PRPD and T/W map.

C. Feature Extraction

Since the standardized measuring device used only supplies data triplets, a transient recorder sampling at 100 MSamples/s is triggered for 10 s as soon as a PD level of 25 pC is reached. As the trigger time is influenced by operating system and communication delays, recorded data are synchronized according to Fig. 7.

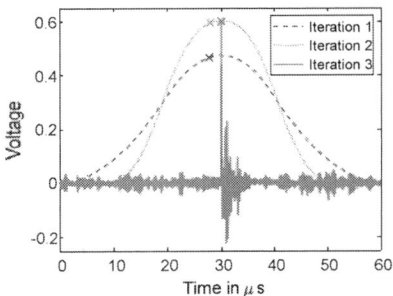

Fig. 7. Synchronization algorithm.

The synchronization algorithm shifts the most prominent time-charge samples in three steps:

- Shift over minimum-maximum-filtered raw data between -0.01 s … 0.65 s in 400 ns steps

- Shift over minimum-maximum-filtered raw data between -0.065 s … 0.065 s in 200 ns steps

- Shift over raw data between -0.013 s … 0.013 s in 10 ns steps

The time-charge samples are shifted by a time determined from the maximum of the summed raw data values at each shift step following each iteration. This approach achieves high precision with comparatively low computation time due to the multi-step method.

As the overlap in the two usual charge-phase and time-bandwidth planes is still significant, additional features have to be calculated. From the extracted single pulses, the features shown in Tab. II are calculated with $s(t)$ being the signal in time domain, $\tilde{S}(f)$ being the signal in frequency domain and $A(i)$ being the signal sampled at N discrete times.

TABLE II. CALCULATED FEATURES

Marker	Name	Equation
	Normalized signal (helper) [6]	$\tilde{s}(t) = \dfrac{s(t)}{\sqrt{\int_0^T s(t)^2 dt}}$
	Gravity center (helper) [6]	$t_0 = \int_0^T t\tilde{s}(t)^2 dt$
✖	Equivalent time [6]	$\sigma_T = \sqrt{\int_0^T (t - t_0)^2 \tilde{s}(t)^2 dt}$
▦	Equivalent bandwidth [6]	$\sigma_F = \sqrt{\int_0^\infty f^2 \left|\tilde{S}(f)\right|^2 df}$
	Mean (helper)	$\mu = \dfrac{1}{N}\sum_{i=1}^N A_i$
	Standard deviation (helper)	$S = \sqrt{\dfrac{1}{N-1}\sum_{i=1}^N |A_i - \mu|^2}$
◆	Equivalent wave	$\sigma_W = \dfrac{\sigma_F}{S}$
●	Mean steepness	$\mu_\Delta = \dfrac{1}{N}\sum_{i=1}^N |\Delta A_i|$
△	Overshoot	$\widehat{D} = \dfrac{|A_{min}|}{|A_{max}|}$ for $|A_{min}| < |A_{max}|$ $\widehat{D} = \dfrac{|A_{max}|}{|A_{min}|}$ for $|A_{min}| > |A_{max}|$
▷	Equivalent pulselength	$\sigma_{\widehat{T}} = \dfrac{\sum_{i=1}^N |A_i| - A_{min}}{N}$

To visualize these features, Fig. 8 shows three differently shaped pulses as well as their normalized features.

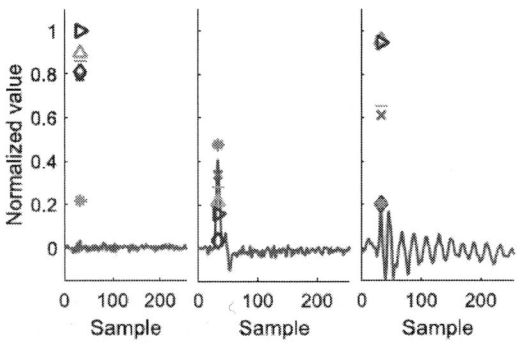

Fig. 8. Exemplary pulse shapes and their normalized features.

It is worth noting that the significance of the extracted features depends on the signal-to-noise ratio and the length of the time series evaluated per pulse [7].

To illustrate this, Fig. 9 and Fig. 10 show Time/Bandwidth maps of 4 different measurements calculated from time series lengths of 2.56 µs and 10.24 µs, respectively.

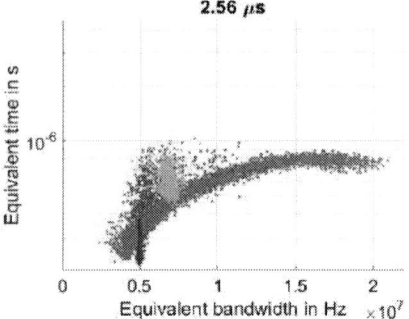

Fig. 9. Time/Bandwidth map using 2.56 µs extraction length.

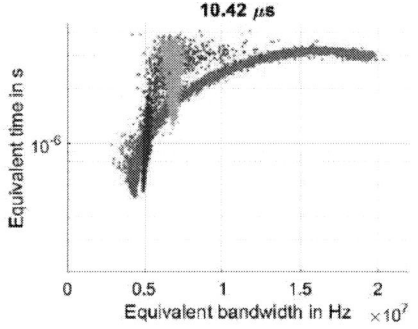

Fig. 10. Time/Bandwidth map using 10.42 µs extraction length.

It is shown that the selection of an appropriate evaluation length reduces the dispersion of the data within each cluster and thus enables a better separation between the clusters.

To enable evaluation of different machine learning techniques, the time series of each pulse is extracted (Fig. 11).

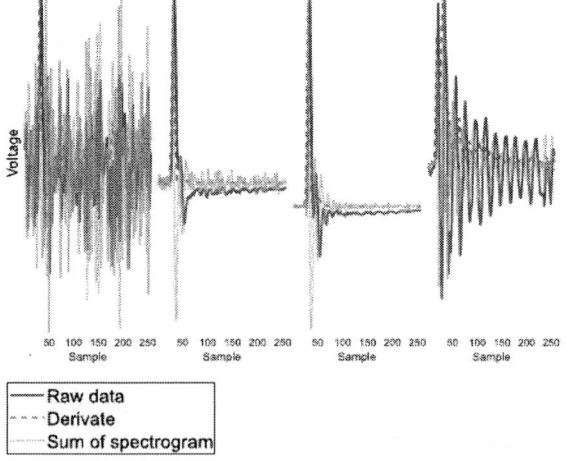

Fig. 11. Extracted time series of noise, corona discharge, filled cavity discharge and polymer-filled discharge (left to right).

Finally, a 192x192 pixel image consisting of the spectrogram superimposed with the pulse shape is generated for each discharge (Fig. 12).

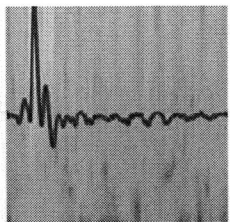

Fig. 12. Image input generated from spectrogram and pulse shape.

D. Machine Learning

The machine learning implementation must accomplish two tasks:

- Learn classification from air- and polymer-based measurements
- Classify samples acquired using ceramic substrates

For this purpose, the polymer- and air-based measurement data are manually labeled as four classes shown in Tab. III.

TABLE III.　LABEL OVERVIEW

Name	Origin	Pulse Shape
Noise	Test setup	
Corona in air	Test setup	
External corona	Substrate metallization	
Cavity	Gas-filled cavities inside substrate	

No clear class could be established for gliding boundary layer discharges. This could be due to their non-occurrence or lack of differentiability from external corona discharges. The following three approaches are trained using these data.

1) Multilayer Perceptron (MLP)

In this network of fully connected layers, only the features are processed at each iteration, which makes it resource-efficient and thus fast (Fig. 13).

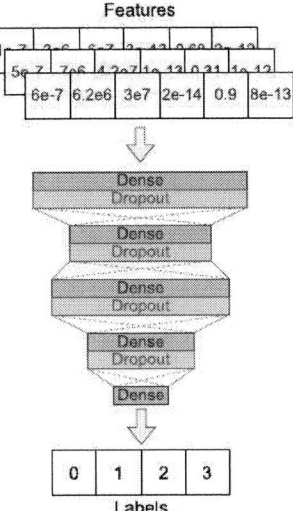

Fig. 13. Fully connected Multilayer Perceptron schematic.

2) Long Short-Term Memory Network

This type of network consists of a block of Long Short-Term Memory (LSTM) cells which enable learning of long-term dependencies as well as a stack of fully connected layers (Fig. 14).

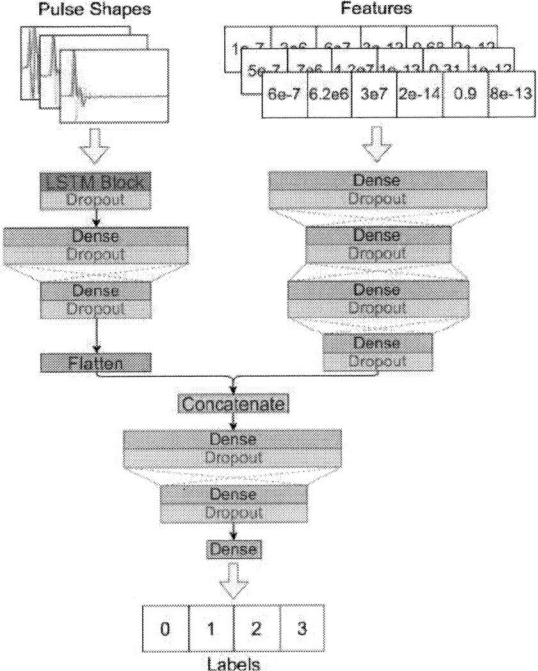

Fig. 14. MLP-induced LSTM schematic.

This model includes more hidden nodes as well as a more complicated structure and thus requires comparatively higher processing time.

3) Image Recognition Network

This approach includes a simplified version of the image recognition network proposed in [8], combined with a fully connected multilayer perceptron (Fig. 15).

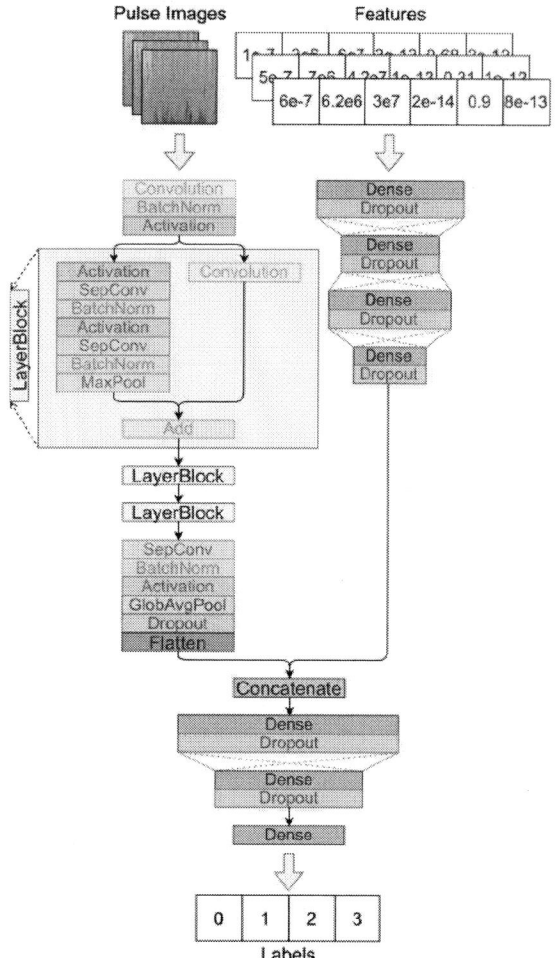

Fig. 15. Xception-based, MLP-induced image recognition schematic.

This network has the highest memory and computation requirements. Furthermore, image generation duration exceeds that of the pulse shape export roughly 20-fold.

IV. RESULTS

A. Training

Fig. 16 and Fig. 17 show the training histories of the different networks.

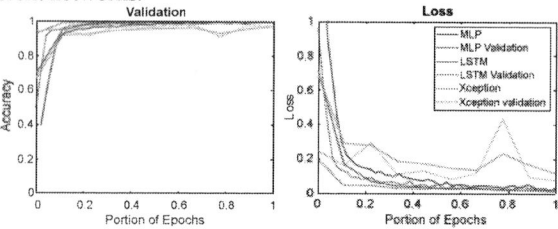

Fig. 16. Training histories.

The graphs show valid learning curves and acceptable accuracy for all networks. Especially the validation accuracy shows that all the approaches can classify data acquired from air- and polymer-based specimens. To evaluate their applicability to ceramics, various parameters are taken into consideration:

- In-class similarity

- Inter-class overlap

- Concordance of predictions using pulse shapes

- Concordance of predicted regarding expected PD sources

- Differentiability of classes compared to training data

B. Classification of Ceramic Measurement Data

To compare the networks regarding the first four parameters, Fig. 17 - 20 show PRPDs, T/W maps, and pulse shapes.

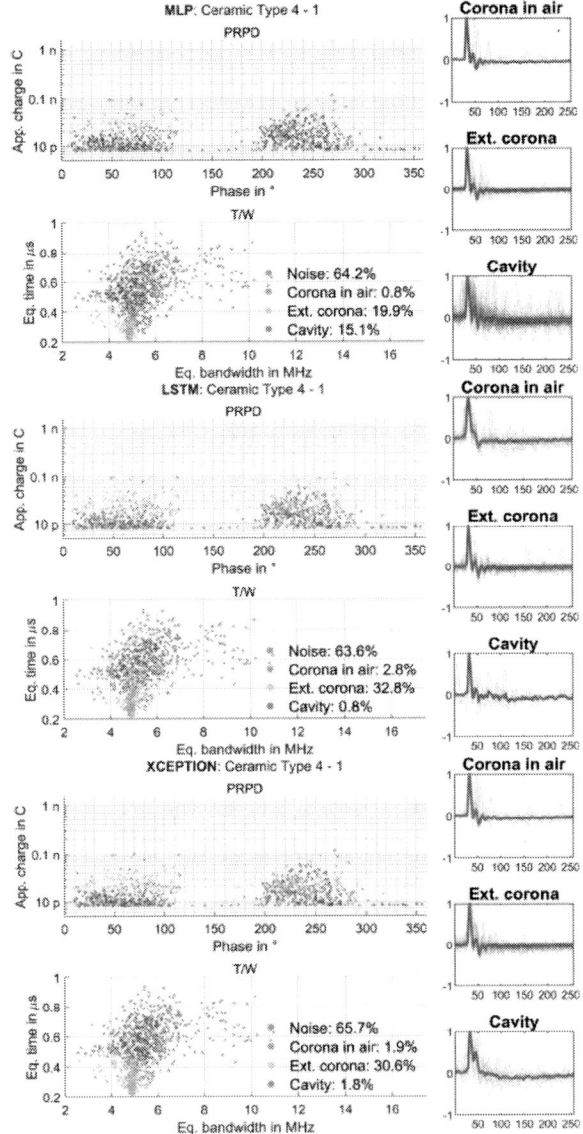

Fig. 17. PRPDs and T/W maps of Ceramic Type 4 - 1 measurements, labeled by the three different networks.

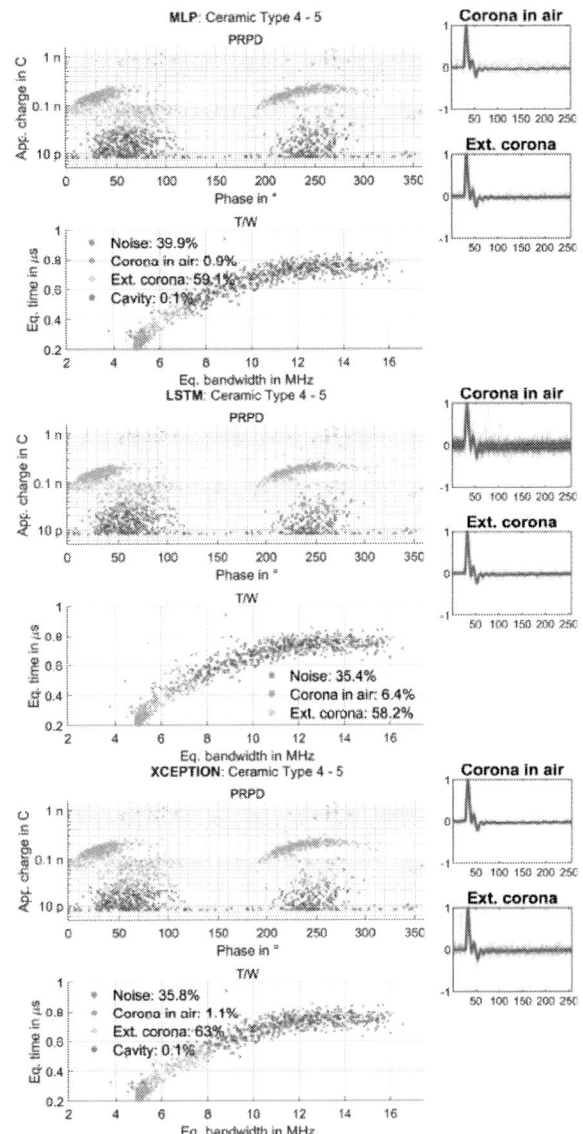

Fig. 18. PRPDs and T/W maps of Ceramic Type 4 - 5 measurements, labeled by the three different networks.

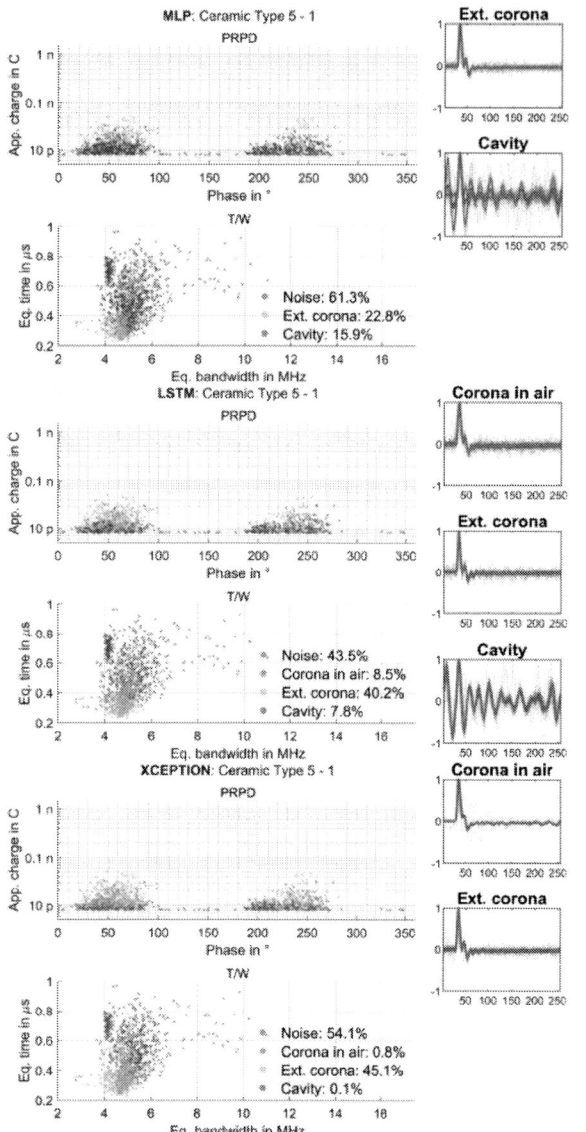

Fig. 19. PRPDs and T/W maps of Ceramic Type 5 - 1 measurements, labeled by the three different networks.

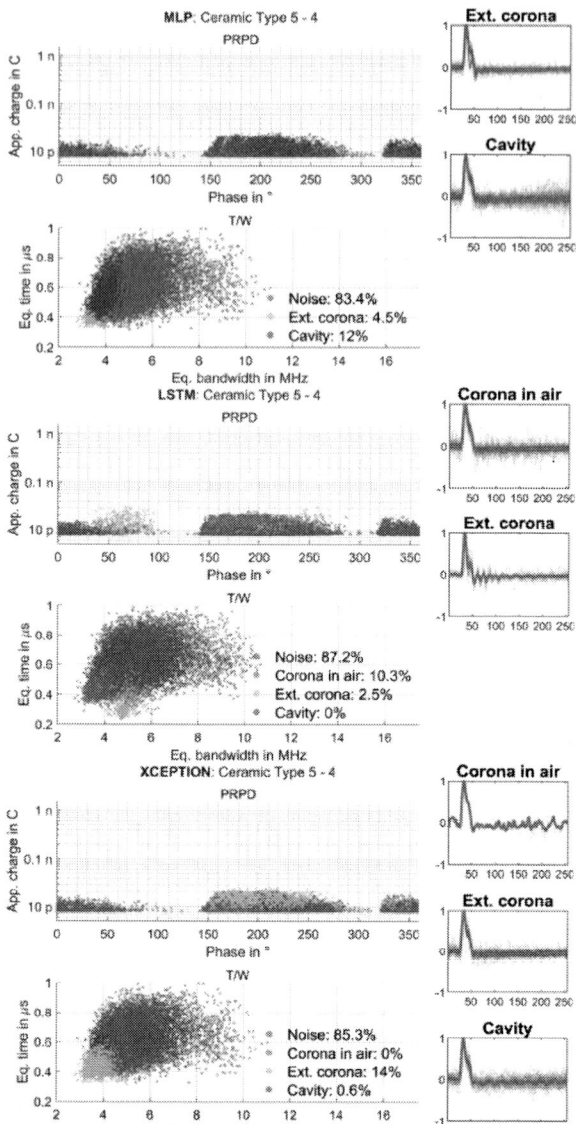

Fig. 20. PRPDs and T/W maps of Ceramic Type 5 - 4 measurements, labeled by the three different networks.

This overview demonstrates that considering solely features is not sufficient. The MLP figures show this by strongly scattered clusters as well as by pulses clearly recognizable as misassigned based on their pulse shapes (e.g. Fig. 17). The two pulse shapes-considering networks show a higher susceptibility to misclassification of individual pulses as seen by sub-percentage sample counts. Since the Xception-based network misclassifies some visible clusters as noise (e.g. Fig. 19) whilst also requiring significantly more computing power, the LSTM-based network seems to be most suitable regarding the first four parameters.

C. Evaluation of Group Separation Based on Pulse Sequence Analysis

To examine the LSTM's differentiability of the detected classes compared to training data, apparent charge PSA is performed on the data, shown in Fig. 21 to Fig. 23.

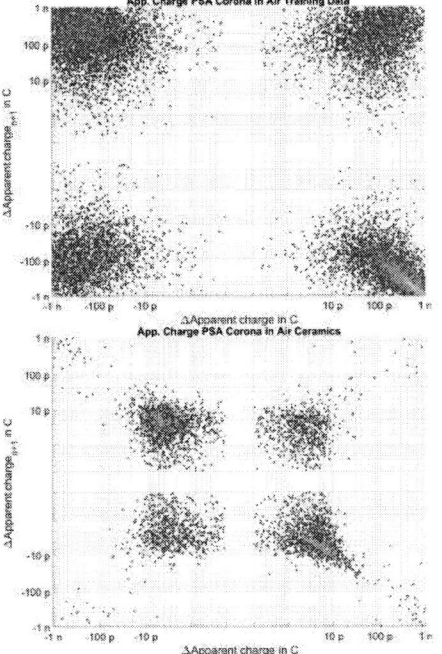

Fig. 21. Apparent charge PSA of accumulated auto-labeled corona in air data acquired from training data (top) and ceramics (bottom).

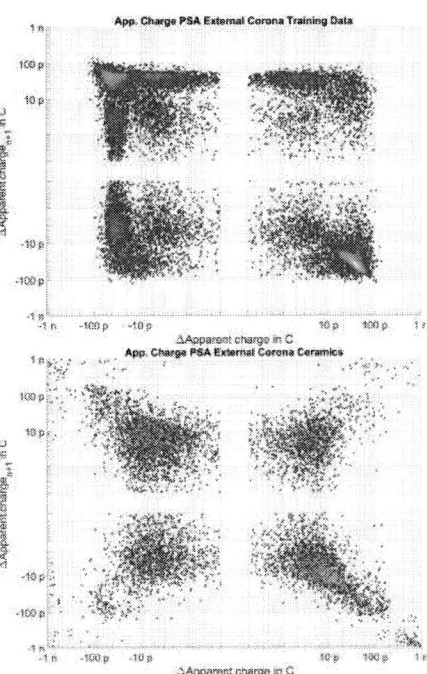

Fig. 22. Apparent charge PSA of accumulated auto-labeled external corona data acquired from training data (top) and ceramics (bottom).

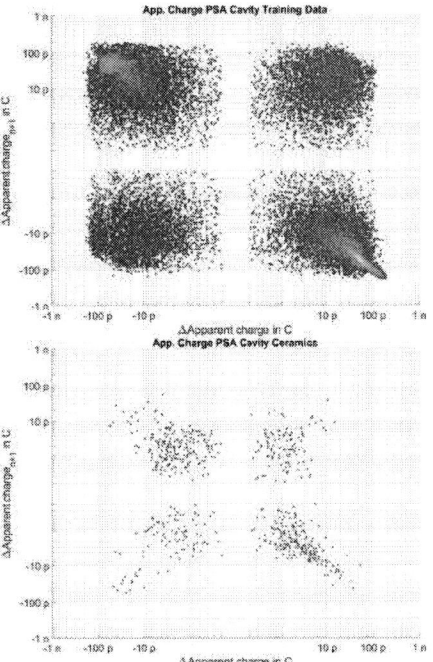

Fig. 23. Apparent charge PSA of accumulated auto-labeled cavity data acquired from training data (top) and ceramics (bottom).

Although the shapes of the LSTM-based corona and cavity clusters are roughly similar to those of the training data, significant differences are apparent. Visually, this can be explained by the differences in the number of data points. For the density-independent differences, two explanations are suggested:

- Misclassification

- Material-dependent partial discharge repetition behavior of identical discharge sources

V. DISCUSSION AND FUTURE WORK

Since it is evident in the LSTM-classified pulse shapes that at least corona and cavity discharges are correctly detected, fundamental misclassification can be ruled out. It is conceivable that, due to the lack of manual differentiability between external corona and internal glide discharges (if present), this differentiation is also impossible for the network. This can be prevented in the future by fabricating and measuring polymer-based substrates with assured glide discharge occurrence. To be able to investigate the differences in PD behavior between polymers and ceramics given the same discharge types in more detail, defined ceramics measurement samples are needed. However, with this constraint, several key points were demonstrated:

- The developed synchronization solution enables targeted raw data extraction using a variety of measurement devices not designed for interoperation

- The calculated features are suitable for a deeper PD measurement data analysis

- The demonstrated approach allows for knowledge transfer from air- and polymer-based measurements to ceramics

- The developed LSTM-based artificial neural network is able to distinguish PD shapes and thereby detect different PD sources with moderate computational resource requirements

This allows the demonstrated toolchain to be trained and used on other test beds or for different substrates such as complete power electronics assemblies. Consequently, new possibilities in defect detection or early failure detection that were previously impossible due to the demonstrated superimpositions and abundance of data are opened up. Work is also underway to develop an FPGA-based measuring device that performs both feature extraction and LSTM classification directly at the point of measurement, allowing each charge-time sample to be immediately labeled for advanced defect detection.

ACKNOWLEDGMENT

Research was done within the project „PROGNET - IDA / Datengenerierung und Messtechnik" funded by the German Federal Ministry for Economic Affairs and Climate Action (grant number 16KN054235).

REFERENCES

[1] A. Gemant, W. Philippoff, „Die Funkenstrecke mit Vorkondensator.", „Z. techn. Phys.", Berlin, Vol. 13, 1932, pp. 425-430.

[2] W. Hauschild, E. Lemke, "High-Voltage Test and Measuring Techniques", Springer Vieweg Berlin Heidelberg, Vol. 1, 2014, pp. 157-162.

[3] International Electrotechnical Commission, IEC 60270, "High-voltage Test Techniques: Partial Discharge Measurements", International Electrotechnical Commission, Vol. 3, 2014.

[4] F. H. Kreuger, E. Gulski, A. Krivda, "Classification of Partial Discharges", IEEE Trans. on EI, 28, (1993), 917-31.

[5] R. Patsch, D. Benzerouk, "Characterization of Partial Discharge Processes- What Parameters work best?", ICSD'04, Toulouse, France, (2004), 636-9.

[6] A. Cavallini, G. C. Montanari, D. Fabiani, L. Testa, "Advanced technique for partial discharge detection and analysis in power cables", Int. Conf. on Condition Monitoring & Diagnostic Engineering Management of Power Station / Substation Equipment, 2009.

[7] A. R. Mor, L. C. Heredita, F. A. Muñoz, "Effect of acquisition parameters on equivalent time and equivalent bandwidth algorithms for partial discharge clustering", International Journal of Electrical Power & Energy Systems, Volume 88, 2017, pp. 141-149.

[8] F. Chollet, "Xception: Deep Learning with Depthwise Separable Convolutions," in 2017 IEEE Conference on Computer Vision and Pattern Recognition (CVPR), Honolulu, HI, USA, 2017 pp. 1800-1807

Impact of Pad Layouts and Solder Volume on Self-Alignment of Micro Solar Cells

Elisa Kaiser
Fraunhofer Institute
for Solar Energy Systems ISE
Freiburg, Germany
elisa.kaiser@ise.fraunhofer.de

Maike Wiesenfarth
Fraunhofer Institute
for Solar Energy Systems ISE
Freiburg, Germany
maike.wiesenfarth@ise.fraunhofer.de

Victor Vareilles
Université Grenoble Alpes, CEA LITEN
Campus INES,
Le Bourget du Lac, France
victor.vareilles@cea.fr

Henning Helmers
Fraunhofer Institute
for Solar Energy Systems ISE
Freiburg, Germany
henning.helmers@ise.fraunhofer.de

Abstract—In micro-concentrating photovoltaics (micro-CPV), tiny solar cells (<1 mm²) are assembled on a circuit board on glass. To mount thousands of dies per square meter, a high throughput process is required which still yields alignment accuracy in the order of 10 μm. We developed a micro-CPV module technology where micro cells are accurately mounted directly on a circuit board on glass using a high throughput pick and place process combined with self-alignment based on the restoring force due to surface-tension of the liquid solder. In this work, we study how the self-alignment accuracy is influenced by the pad layout, solder volume and initial position of the die. We show that self-alignment significantly benefits from an induced motion, whether induced by initial displacement of the die or by flowing of molten solder along tracks. Furthermore, we found that lower solder volumes lead to higher alignment accuracies for dies on rectangular pads, whereas the dependence on solder volume is lower for pads with connected tracks where the solder flows out. Overall, alignment accuracies below 15 μm are demonstrated despite initial displacements up to 150 μm. Thus, self-alignment using surface tension of the liquid solder is suitable for micro-CPV.

Keywords—Micro-concentrating photovoltaics (CPV), solar cell assembly, soldering, self-alignment, restoring force

I. INTRODUCTION

Concentrating photovoltaics (CPV) enable higher solar conversion efficiencies than conventional flat-plate solar panels due to the application of high-efficiency multijunction solar cells and improved efficiency under concentration [1]. At the same time, the amount of energy-intensive semiconductor material is reduced by the concentration factor of the concentrating optics used. In this process, the direct sun light is collected with lenses and focused on the solar cells. One major trend in CPV is the miniaturization of all components (micro-CPV), with typical solar cell dimensions well below 1 mm²[2, 3].

Fig. 1 (a) shows our micro-CPV module concept developed at Fraunhofer ISE [4, 5]. Sun light is concentrated on the solar cell by a factor of 1000 using a silicone-on-glass lens. Each cell is equipped with a spherical lens as secondary optical element. The solar cells are mounted directly on a circuit board on glass (chip-on-board approach). Given the small size of the components and the high concentration ratio, even small displacements can cause significant optical losses as the focused light

misses the active cell area. Hence, precise alignment is crucial. In this work, we investigate the alignment accuracy of the solar cells on the circuit board on glass experimentally.

The circuit board serves for electrical interconnection and heat dissipation. The full area rear side contact of the cells is connected directly, and the front side contact is wire bonded to the circuit. The edges of the pad cannot be covered with solder resist. Misalignment of components or tracking uncertainties shift the focus point away from the cells center. Often a solder stop mask is used to define the die position. In this case, however, misalignment could burn conventional solder resist, which thus should be avoided. As shown in Fig. 1 (b) and (c) the direct focus on a glass plate with solder resist already yields to a visible burn spot after 30 minutes of sunlight. Due to the avoidance of a structured solder resist the accurate geometry of the receiving pads themselves must be used to define the solar cell position. For electrical interconnection of the rear side, the pads have partly connected tracks and partly gaps to isolate the front side from the rear. Further design considerations and a description of the manufacturing processes of the module concept are discussed in [4, 5].

In many sectors like flip-chip bonding, the standard for chip assembly is a pick and place process. Manufacturers of high throughput machines claim impressive assembly rates of more than 100'000 dies per hour with accuracies between 22 and 40 μm [6, 7]. The affection of the pick-and-place accuracy is determined with Raytrace3D, a raytracing tool developed at Fraunhofer ISE [8]. A displacement of the solar cells of 40 μm lead to an optical loss of 0.6%abs. Thereby, for the solar cell displacement an optical loss of 0.1%abs is acceptable, which means an accuracy of ±16.6 μm [4].

For precise alignment, thus, we apply a combination of standard pick and place with surface tension driven self-alignment. First, solder paste is deposited on the receiving pads. Then, the solar cells are placed with low precision on the receiving pads (see Fig. 2, a). Next, during reflow, the solder paste melts and wets the surface of the pad and the rear side of the cell (see Fig. 2, b). The liquid solder develops a meniscus whose force induced by the surface tension, pulls the cell to the pad (see Fig. 2, c).

This work was funded by the German Federal Ministry for Economic Affairs and Climate Action (BMWK) under the "micro-CPV" project (#03EE1046A) and by the Institut Carnot Energies du Futur in the framework of the PVμCo project.

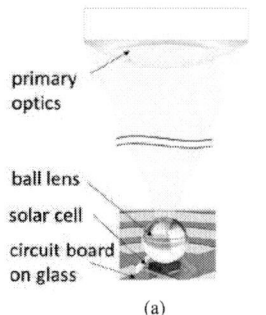
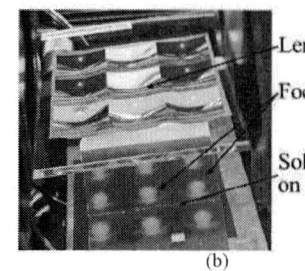
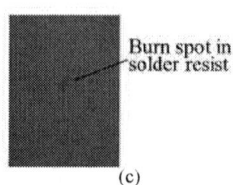

Fig. 1. (a) Sketch of a cell-lens unit of the micro-CPV module, developed at Fraunhofer ISE. The sun is focused with a silicone-on-glass primary lens onto a micro solar cell. The cell is assembled on a circuit board on glass and equipped with a secondary spherical lens. (b) Outdoor test: A 3x3 lens array focuses the light onto a glass plate which is covered with a green solder resist. (c) Microscope picture of the base plate after 30 min sun exposure. The burn spot is visible in the solder resist.

The self-alignment process is influenced by the solder volume, surface tension and viscosity, the wetting properties of the pad and the cell surface, the shape of die and pad and the initial displacement [9–12]. This work focuses on the influence of solder volume, pad layout and initial displacement. Note that a more detailed study is in preparation [13].

A force-based model can be used to analytically describe the self-alignment of a die [9, 12]. Fig. 3 shows a simplified sketch of the investigated case with a die (S) on a liquid depot (L) with straight flanks. The contact area between the liquid solder and the die is the rear side of the die. For self-alignment, the lateral forces are relevant, namely the force induced by the surface tension of the liquid solder F_{st} and the dynamic friction force F_v. The surface tension force acts on the boundary areas between the gaseous- liquid- solid (G-L-S) phases and depends on their surface tensions γ_i. It can be described by Eq. (1). The die and the pad have the same shape with the length l and the width w. The die is displaced by x from the pad and has the solder height h. The solder flank between the die and the pad has the angle α and it is assumed that the surfaces are completely wetted. Therefore, gradients of the surface tension between the liquid-solid phase are neglectable and the surface tension force can be described by Eq. (2) [9]

$$\vec{F}_{st} = \sum_i^{LS,LG,SG} \int \gamma_i \, d\vec{l} \qquad (1)$$

$$\vec{F}_{st,x} \approx 2\gamma_{LG} w \cos(\alpha) \approx 2\gamma_{LG} w \frac{x}{\sqrt{x^2 + h^2}} \qquad (2)$$

| (a) Die placed inaccurately on solder paste | (b) Melting process: Liquid solder wets pad and die surface | (c) Self-aligned die to pad |

Fig. 2. Mechanism of self-alignment using surface-tension of the liquid solder. Pictures of a high-speed camera of a micro solar cell (chip size 885x685 µm²) placed on a solder depot on a freestanding rectangular pad. Solder was melted by a heating plate.

The dynamic friction force between the liquid and the solid phases acts against self-alignment. The friction mainly originates from the viscosity η_L of the liquid solder. Simplified, the velocity v of the liquid solder is set equal to the die velocity and the area corresponds to the die rear side surface A_{LS} (3) [12].

$$\vec{F}_{v,x} \approx \eta_L A_{LS} \frac{v}{h} \qquad (3)$$

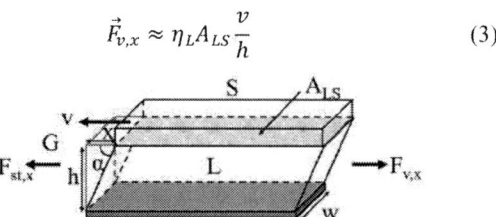

Fig. 3. Simplified sketch, sizes not to scale. A die (S) is displaced by the distance x on a liquid depot (L) with the height h. The contact area of die rear side and the solder is named A_{LS} and the angle between the gaseous (G) and liquid (L) phase is α. The die moves with the velocity v.

II. MATERIALS AND METHODS

Micro solar cell dies with an edge length of 885x685 µm² and a height of approximately 190 µm are placed on four different solder volumes and on two different pad layouts.

The pads are manufactured with a subtractive printed circuit board (PCB) process on a 3 mm glass substrate and consist of 30 µm copper and a chemical nickel/ gold finish. Fig. 4 (b) and (c) show pictures of the samples. One is a rectangular pad with an edge length by 27 µm smaller than the cell, named in the following R-27. The other pad is also rectangular but has additional connected tracks. Here, the edge length is smaller by 15 µm, named I-15. Before soldering, the pads are pre-treated with acetone and then ethanol.

Dies were mounted using stencil printing and die assembly at CEA LETI. SnAg3.0Cu0.5 solder paste with a flux content of 11.5±1.0% and particle size between 20 and 45 µm is applied. The stencil has a thickness of 200 µm and four different sizes for the openings between 615x454 µm² and 765x565 µm² (see TABLE I). The dies are assembled with a pick and place machine with an accuracy <5 µm. The dies are initially placed either in center position or intentionally displaced by 50 µm, 100 µm and 150 µm. Afterwards the solder is melted in a reflow oven, where the samples are kept 2.5 min above liquidus temperature of 217 °C. The resulting solder layer thickness for the rectangular pad R_{-27} is between 28.1 and 50.9 μm and for layout I_{15} between 5.3 and 15.4 μm. Based on the solder layer thickness on the pad R_{-27} the solder volumes were calculated and are between 16.0 and 28.9 nl. Note that these volumes are calculated from the solid state after soldering, and the liquid solder volumes are larger. For statistics, 20 samples are produced for each solder volume and initial displacement variation.

The alignment accuracy of the dies to the pads is measured with a coordinate measurement device that has an accuracy below 2 μm. First, the size and x-y-positions of the pads with respect to fiducials are measured. After soldering, the size and x-y-positions of the dies are determined in the same manner. From this data the offset is determined as the difference between cell center and pad center.

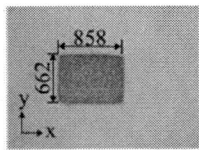 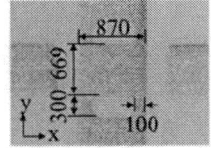

    (a) Rectangular pad   (b) Interconnected
         R_{-27} rectangular pad I_{-15}

Fig. 4.   Microscope pictures of the different pad layouts. All sizes in μm.

TABLE I.   STENCIL DESIGN

	Opening			
	A	*B*	*C*	*D*
Opening size [μm]	615x454	665x491	715x528	765x565
Measured solder thickness R_{-27} [μm]	28.1±3.2	35.3±4.5	40.1±2.6	50.9±5.4
Measured solder thickness I_{-15} [μm]	5.3±3.2	7.1±4.8	8.0±5.0	15.4±4.3
Volume based on measured solder thickness R_{-27} [nl]	16.0	20.1	22.8	28.9

III. RESULTS

The purpose of this study is to determine the influence of the pad layout, different solder volumes and initial displacements on the self-alignment accuracy of dies.

First, dies are initially displaced with four different solder volumes on the rectangular pad R_{-27}. Fig. 5 (a) shows the resulting offsets as a function of the initial displacement. The boxes represent the 25-75 percentiles, and the lines represent the entire observed range from minimum to maximum. Self-alignment is considered to be successful, when the cell is fully placed above the slightly smaller pad. Note that this includes left edge alignment, center alignment, and right edge alignment. Dashed lines mark this range, spun across $\pm\Delta l_R/2$. For pad R_{27}, $\pm\Delta l_R/2$ is ± 13.5 μm. As a first observation, for the lowest solder volume of 16.0 nl (top), most of the dies are within this range. The median offset is close to the zero line up to an initial displacement of 100 μm. With an initial displacement of 150 μm, the median offset jumps to the edge of the self-alignment range $\pm\Delta l_R/2$ and scattering is reduced.

In Fig. 5 (a), from the top to the bottom graph, the solder volume is increased from 16.0 to 28.9 nl, i.e. the solder layer thicknesses are increased from 28.1 μm to 50.9 μm. For the medium solder volumes (blue and red data) a similar behavior as for the lowest solder volume is observed. In contrast, for the largest solder volume, the median offset shows no clear trend and with increasing solder volume, also the scattering increases. Increasing the solder thickness from 28.1 to 50.9 μm results in a relative increased scattering of 59% for ideal initial placement and of 60% for an initial displacement of 150 μm. Overall, for all solder volumes we observe decreased scattering with increased initial displacement. For a solder thickness of 28.1 μm the scattering reduces from ideal initial placement to 150 μm initial displacement by 46%, for 35.3 μm by 41%, for 40.1 μm by 38% and for 51 μm by 46%. Note that for all solder volumes, the highest measured offset is below 35 μm despite initial displacements up to 150 μm.

Fig. 5 (b) shows the behavior of dies on rectangular pads I_{-15}, again as a function of initial displacements up to 150 μm with increasing solder volume from top to bottom. Fig. 5 (c) shows a microscope picture of the interconnected after solder melting. As can be seen in the micrograph, after melting the solder leaves the pad and flows a certain distance along the connected tracks. Hence, a significant fraction of the solder is not underneath the chip anymore and the solder layer thickness is reduced to 5.3 to 15.4 μm. For pad I_{-15}, the self-alignment range $\pm\Delta l_I/2$ is given by ± 7.5 μm. Most of the data and all medians are within this range. With the exception of ideal initial placement, all offsets are below 15 μm. In detail, 56% of the offsets are within ± 5 μm and 87% within ± 10 μm.

A comparison between the rectangular pad R_{-27} (Fig. 5, a) and the pad with connected tracks I_{-15} (Fig. 5, b) shows, that the scattering is reduced over all variations with pad I_{-15}. For example, for a solder volume of 20.1 nl, the scattering for ideal initial placement is reduced by 52% and for 150 μm initial displacement by 36%.

Besides the lateral offsets, out-of-plane rotations are visible dependent on the solder volumes, as sketched in the top of Fig. 5 (d). For the rectangular pad R_{-27} out-of-plane rotations in the horizontal axis were measured up to 2.6° for the lowest solder volume and up to 6.4° for the largest volume. The bottom of Fig. 5 (d) shows an x-ray image of a die placed on pad R_{-27}. The out-of-plane rotation can be recognized by the change in contrast from gray to black due to increasing solder thickness. In contrast, for the pads I_{-15} with connected tracks, no out-of-plane rotation was observed for the lowest solder volume. For the largest volume maximal outliers were 2.6°.

IV. DISCUSSION

Our study showed that for dies placed on rectangular pads R_{-27}, the cell to pad offsets are below 35 μm despite initial displacements up to 150 μm. A comparison of different solder volumes with equal initial displacement showed, that the scattering of the offsets increases with increased solder volume.

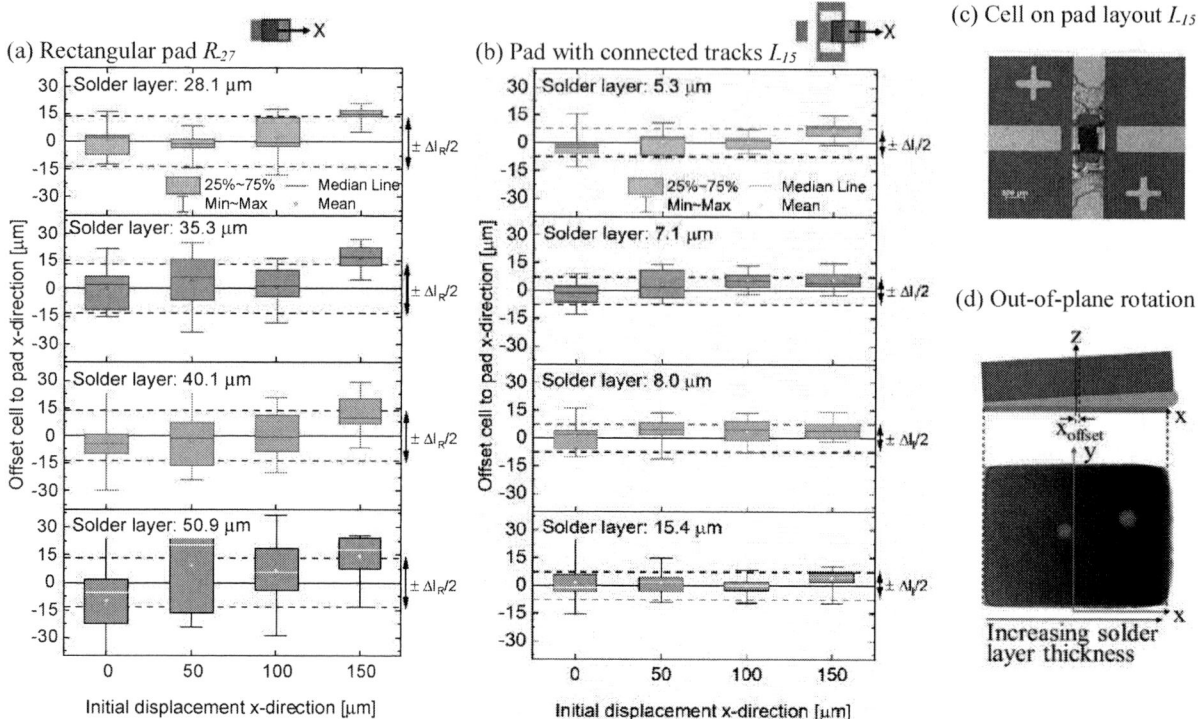

Fig. 5. (a), (b) Offset of the cell to pad dependent on the initial displacement in x-direction. From top to bottom: Solder volume is increased. The dashed line indicates the range of plus/minus half of cell to pad length difference. (a) For the rectangular pad R_{27} the solder thickness is between 28.1 μm (top) and 50.9 μm (bottom). (b) For the rectangular pad L_{15} the solder thickness is between 5.3 μm (top) and 15.4 μm (bottom). (c) Microscope image of a cell assembled on pad layout L_{15} after the melting process. (d) X-ray image of a die mounted on a rectangular pad R_{27}. Die is aligned on the left edge of the pad and is out-of-plane rotated in the horizontal axis. Above, an exaggerated sketch (side view) visualizes the out-of-plane rotation.

First, we conclude, that self-alignment using surface-tension of molten solder in principle works for all dies on rectangular pads with solder volumes between 16.0 and 28.9 nl. Yet, the observed scattering differs notably with the solder volume. This dependency can be explained by the restoring force. According to Eq. (2), the restoring force generally decreases with higher solder layer thickness h ($\sim x/\sqrt{x^2 + h^2}$). Thus, the restoring force is influenced stronger by the thickness at small displacements ($x<<h \rightarrow \sim x/h$) than at larger displacements ($x>>h \rightarrow \sim 1/x$). For the investigated cases, the solder thickness h in the range of 28 to 56 μm and the initial displacements x up to 150 μm differ by only half an order of magnitude at maximum. Still, the relation between restoring force and solder volume is valid and explains, why the scattering increases with increased solder volume and especially for small offsets. However, the dynamic friction force, which acts again the restoring force, decreases with a thicker solder layer ($\sim \Delta v/h$) (3). Note, that the dynamic friction force is several orders of magnitude lower and thus, the surface tension force is dominant.

For all solder volumes, the median offsets are close to zero for low initial displacements, i.e. die center is to pad center alignment, whereas for larger initial displacement the median shifts towards the edge of the self-alignment range, which means edge-alignment. In general, the motion of the dies stops, when the restoring force vanishes or becomes overruled by the friction force. For the investigated samples it differs dependent on the initial displacement. Higher initial displacements tend to align on the edges and lower initial displacements to align center-to-center.

Furthermore, our study demonstrates for dies placed on pads L_{15} with connected tracks, that all die to pad offsets are below 15 μm despite initial displacements up to 150 μm. Thus, the layout with connected tracks apparently improved the self-alignment.

It is important to emphasize that for pad L_{15} we observe no clear dependence on initial displacement nor solder volume. Here, independent of the initial displacement, the liquid solder flows underneath the dies along the tracks. As a result of that, the solid solder thicknesses after assembly are reduced from 28-51 μm to 5-15 μm. The previously observed increased scattering with increasing solder volume vanishes. Overall, we conclude that induced motion, whether induced by initial misalignment or by flowing of the molten solder underneath the chip, is beneficial for the self-alignment process.

Finally, our observation of larger out-of-plane rotations for the rectangular pad layout R_{27} compared with the layout L_{15} with connected tracks is in line with the findings of Berthier et al. [14]. There, rectangular chips were soldered onto rectangular pads of the same size. During self-alignment, the out-of-plane rotation was not restored. An adjustment of the pads with additional bands decreased the out-of-plane rotation. However, in our case an additional effect has to be considered, which is the thinner solder layer for L_{15} than for R_{27}. Hence, lower out-of-plane rotations are possible.

V. CONCLUSION

Self-alignment of a die using surface tension of the molten solder was experimentally investigated regarding two pad layouts, four solder volumes, and initial displacements up to 150 μm. We find initial displacement of dies on rectangular pads to be beneficial for the self-alignment accuracy. The motion of the die due to surface tension induced restoring force supports overcoming opposing friction force and benefits the alignment accuracy. Furthermore, a lower solder volume is observed to be beneficial and reduces the scattering by approximately 60%. For pads with connected tracks, flowing solder positively affects the self-alignment, by helping to overcome static friction. For all investigated variations of solder volume and initial displacements, alignment accuracies below 15 μm were reached, 56% below 5 μm and 87% below 10 μm. It is remarked that the solder flow also reduces the resulting solder layer thickness to below 15.4 μm. Such thin solder layers can suffer from thermal cycling loads and long-term stability needs to be investigated in the future. Finally, ray tracing analysis of our micro-CPV module suggests that an alignment accuracy of 15 μm results in optical losses of less than 0.1%$_{abs}$.

ACKNOWLEDGMENT

The authors sincerely thank Philippe Voarino and Romain Cariou for hosting Elisa Kaiser at CEA INES and for enabling the die assembly at CEA LETI. We thank Alexander Dilger at Fraunhofer ISE for pick-and-place support, and all colleagues of the "III-V Photovoltaics and Concentrator Technology" department for continuous support. We thank EVYTRA GmbH (formerly FELA GmbH) for providing circuit boards on glass, and AZUR SPACE Solar Power GmbH for providing micro solar cells. The authors are responsible for the content of this work.

REFERENCES

[1] M. Wiesenfarth, I. Anton, and A. W. Bett, "Challenges in the design of concentrator photovoltaic (CPV) modules to achieve highest efficiencies," *Applied Physics Reviews*, vol. 5, no. 4, p. 41601, 2018.

[2] C. Domínguez, N. Jost, S. Askins, M. Victoria, and I. Antón, "A review of the promises and challenges of micro-concentrator photovoltaics," in Ottawa, Canada, 2017, p. 80003.

[3] S. Paap, V. Gupta, A. Tauke-Pedretti, P. Resnick, C. Sanchez, G. Nielson, J. L. Cruz-Campa, B. Jared, J. Nelson, M. Okandan, and W. Sweatt, "Cost analysis of flat-plate concentrators employing microscale photovoltaic cells for high energy per unit area applications," in *2014 IEEE 40th Photovoltaic Specialist Conference (PVSC)*, 2014, pp. 2926–2929.

[4] E. Kaiser, P. Schöttl, M. Wiesenfarth, P. Nitz, and H. Helmers, "Effects of Manufacturing Tolerances on Micro-CPV Module Performance," in *18TH INTERNATIONAL CONFERENCE ON CONCENTRATOR PHOTOVOLTAIC SYSTEMS (CPV-18)*, to be published.

[5] M. Wiesenfarth, D. Iankov, J. F. Martínez, P. Nitz, M. Steiner, F. Dimroth, and H. Helmers, "Technical boundaries of micro-CPV module components: How small is enough?," in *17TH INTERNATIONAL CONFERENCE ON CONCENTRATOR PHOTOVOLTAIC SYSTEMS (CPV-17)*, Freiburg, Germany / Online, 2022, p. 30008.

[6] JUKI Smart Solutions, "RX-8," [Online] Available: https://www.juki-smt.com/produkte/placement/placement-rx-8/. Accessed on: May 28 2023.

[7] ASMPT GmbH & Co. KG, "SPILACE X S: Maximale Kapazität für die Integrated Smart Factory," Feb. 2023. [Online] Available: https://smt.asmpt.com/de/produkte/placement-solutions/placement-heads/siplace-speedstar/. Accessed on: May 28 2023.

[8] P. Schöttl, G. Bern, P. Nitz, F. Torres, and L. Graf, "Raytrace3D by Fraunhofer ISE: Accurate and efficient ray tracing for concentrator optics," Fraunhofer Institut für Solare Energiesysteme ISE, 2022. [Online] Available: https://www.ise.fraunhofer.de/content/dam/ise/de/downloads/pdf/raytrace3d.pdf. Accessed on: May 20 2022.

[9] M. Mastrangeli, Q. Zhou, V. Sariola, and P. Lambert, "Surface tension-driven self-alignment," (eng), *Soft matter*, vol. 13, no. 2, pp. 304–327, 2017.

[10] H.-P. Park, G. Seo, S. Kim, and Y.-H. Kim, "Effects of Solder Volume and Reflow Conditions on Self-Alignment Accuracy for Fan-Out Package Applications," (En;en), *JEM*, vol. 47, no. 1, pp. 133–141, https://link.springer.com/article/10.1007/s11664-017-5883-0, 2018.

[11] S. Härter, "Qualifizierung des Montageprozesses hochminiaturisierter elektronischer Bauelemente," FAU University Press, 2020.

[12] O. Krammer and Z. Illyefalvi-Vitéz, "Investigating the self-alignment of chip components during reflow soldering," *Per. Pol. Elec. Eng.*, vol. 52, no. 1-2, p. 67, 2008.

[13] E. Kaiser, M. Wiesenfarth, V. Vareilles, M. Schneider-Ramelow, S. W. Glunz and H. Helmers, "Forced Motion Activated Self-Alignment of Micro-CPV Solar Cells," *IEEE J. Photovoltaics*, to be published.

[14] J. Berthier, K. A. Brakke, S. Mermoz, C. Frétigny, and L. Di Cioccio, "Stabilization of the tilt motion during capillary self-alignment of rectangular chips," *Sens. Actuators, A Phys.*, vol. 234, pp. 180–187, 2015.

Novel low temperature and low pressure sintering of ADAS radar sensor antenna stack

Sri Krishna Bhogaraju[1*], Nihesh Mohan[1], Fabian Steinberger[1], Hueseyin Erdogan[2], Philipp Hadrava[2], Gordon Elger[1]

[1*]Institute of Innovative Mobility, Technische Hochschule Ingolstadt, Esplanade 10, 85049, Ingolstadt, Germany

[2] Conti Temic microelectronics GmbH, Ringlerstraße 17, 85057, Ingolstadt

*Email: srikrishna.bhogaraju@thi.de; Tel: +49-841-9348-6449

Abstract— **A low temperature and low pressure sintering of antenna stacks with the promising prospect of achieving a single step bonding process of multi-layer wave guide antennas for automotive radar sensors is demonstrated. Silver nanoparticles (Ag-NP) paste with the ability to sinter at 200°C in air allows for easy integration into standard surface mount device reflow process. The paste drying and sintering process of the Ag-NP paste is first optimized using dummy test chips to reduce the typical voiding and channeling in the interconnect when sintered without pressure. In especially, a slow ramp rate with an integrated pre-drying step is applied after placement of the test chips into the wet paste. Afterwards, the process is transferred to the antenna. However, pre-existing bending in the large antenna layers challenges the sintering process without pressure, e.g. larger amounts of paste need to be applied for compensation of the bending. Therefore, the use of a low bonding force in observed to improve the quality of the sintered interfaces. Near bulk like Ag sintered interconnects were realized.**

Keywords — sintering, ADAS, nano Ag sintering, low pressure, low temperature, antenna stack, radar

I. INTRODUCTION

Autonomous driving is a megatrend of today and advanced driver assistance system (ADAS) sensors play a critical role enabling this megatrend, being the fastest growing segment in automotive today. Cameras, radio detection and ranging (RADAR) and light detection and ranging (LiDAR) are the sensors enabling the environmental perception of future autonomous cars. Radar is considered indispensable since it can work under any weather conditions due to its unique ability to penetrate smoke, fog, dust, rain and/or snow. It also has a strong anti-interference characteristic and high accuracy of speed and distance measurement [1].

In the standard automotive ADAS applications, radars can be characterized into three main categories: (1) short range radar (SRR), medium range radar (MRR) and long-range radar (LRR). SRR is typically used in applications such as parking assist and has a detection range of ~ 30m and a wide field of view (FOV). MRR has a similar FOV to SRR, but a higher detection range of upto 130m and are typically used for applications such as blind spot detection, lane keeping assist and lane changing alert. LRR are used in applications with a range > 180m but have a comparatively small FOV of ± 20°. These are typically used in applications such as adaptive cruise control, collision avoidance systems and automated emergency braking [1].

Considering the stringent safety and reliability requirements, it is essential to design and develop robust sensors while at the same time ensuring low cost, excellent process traceability and improved reliability and performance. In addition, reduction of complex assembly processes is also desired. Therefore, novel materials and efficient and effective processes need to be designed and developed.

A key aspect of the radar architecture is the radio frequency antenna. For automotive radar sensors microstrip patch antennas were dominating, due to their compact size, low profile, and directional radiation pattern. Enabled by improvements of accuracy of injection molding, plastic waveguide antennas were introduced [2] and getting more and more popular. After molding, the individual plastic layers of the antenna are metallized and soldered on top of each other. The fully assembled antennas are tested and finally attached to the printed circuit bord (PCB). The antenna considered in the paper, consists of a multi-layer wave guide antenna. Due to the plastic material (ABS) of the antenna, it is challenging to realize an appropriate solder hierarchy, i.e., solder the single layers of the antennas on to each other using a solder with higher melting point than standard SAC solder used for board level attachment. Therefore, the antenna stack layers are first soldered using a low temperature solder and then the stack is soldered on to the populated PCB to undergo a second reflow process where the surface mount devices (SMDs) are soldered on to the PCB. The second reflow process can cause remelting the low temperature solder that bonds the antenna stacks together and risks reduction of antenna performance due to reposition or misalignment of the stacks.

Low temperature sintering is, therefore, an attractive alternative. The present paper deals with the development of a novel low temperature & low-pressure sintering process for injection molded antenna stacks. A sintering process using a low temperature and in air sinterable Ag sinter paste is investigated, providing the possibility of easy integration into a standard SMD soldering line. The goal is to realize a pressureless sinter process wherein the individual antenna stack layers are sintered under pressureless conditions in air and at temperature substantially below the peak soldering temperature of 255°C.

II. MATERIALS AND METHODS

A. Sinter pastes and device under test

Two different commercially available Ag sinter pastes were considered for the study. A pure nanoparticle (NP) based Ag paste and a nanoparticle and salt (NP+S) based Ag paste were initially evaluated to ascertain their performance. Both pastes are specified for pressureless sintering. They can be stencil printed or dispensed. While the pure NP based Ag paste was capable of sintering at temperature as low as 175°C under air, the NP+S paste has a minimum sintering temperature of 250°C.

The antenna stacks have an overall surface area of 1000mm² and Ag coating on the ABS plastic as shown in Fig 1. The pastes is dispensed as 1mm wide track with a measured thickness of approximately 80 μm onto the bottom stack layer and the top layer is assembled onto it manually using the alignment holes as provided in the design.

For the development of the sinter process, first, small 1mmx1mm test chips with a 120 μm thickness were used. Afterwards the process was applied for the assembly of the antenna stacks.

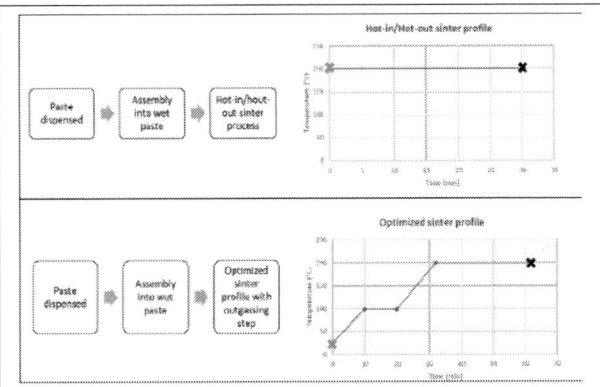

Fig. 1 Sinter process: The green marks indicate the point of placement of the chips and the black marks indicate when the assembly is removed from the oven after sintering. For NP the sinter temperature is 200C and for NP+S 250C.

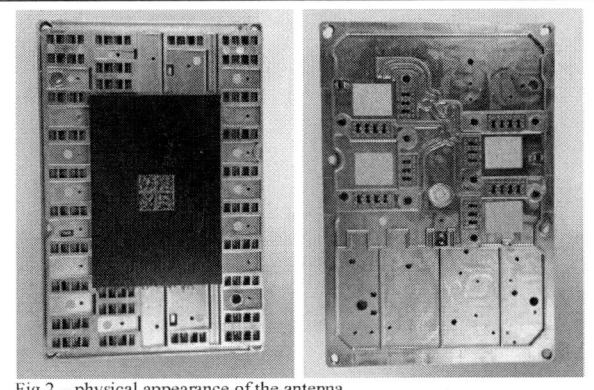

Fig 2 – physical appearance of the antenna.

B. Sinter pastes and device under test

The sinter process is depicted in Fig. 2. For sintering without pressure, the chip is placed into the wet sinter paste. The hot-in/hot out process describes the process in which the device is placed into the hot oven. Therefore, the heating up to sintering temperature is fast but depends on the thermal mass of the product.

C. Sample preparation & characterization techniques

For the initial process investigation 1mm² test chips with a thickness of 120 μm and an Au metallization surface finish were used as standard test vehicles. Aluminum insulated metal substrates, also with Au surface finish were used. The commercial Ag pastes was carried out by stencil printing using a PBT-Uniprint-PMGo3v semi-automatic printer and a 75 μm stencil. The sintering was performed without applying pressure. in air in using a Memmert convection oven with 100% fan speed.

For the assembly of the antenna stacks the pastes were dispensed using a Musashi dispenser with a dispensing pressure of 20kPa and nozzle diameter of 0.15mm. Considering the limitations of the temperature as discussed above, solely the NP Ag sinter paste was considered for the sintering of the antenna stacks. Sintering was performed without and with small pressure in the Memmet convection oven. To apply pressure, a tape was attached tightly at four positions around the stack to equalize bending and hold the antenna stacks in place.

Optical microscopy imaging was performed on a Keyence Digital Microscope. Profilometry is performed using a Nanofocus profilometer. Scanning Electron Microscopy (SEM) analysis was performed using a ZEISS EVO MA15 microscope.

III. RESULTS AND DISCUSSIONS

A. Development pressureless sinter process

As mentioned earlier, the initial tests to understand the voiding behavior of the pastes was carried out with 1mm² test chips (120 μm thickness). The test chips were placed into the wet paste and the hot-in/hot-out process was applied. As can be seen in Fig. 2, the hot-in/hot-out sintering process resulted in a high rate of defects in both the NP and the NP+S pastes. However, the defects manifested differently between the pastes. While in case of the NP paste a high degree of voiding and channeling defects were seen. This is also evident from the SEM imaging of the cross-section of the sintered interconnects as shown by the red arrows in Fig. 3. The channels are observed to follow the entire thickness of the interconnect i.e., from the substrate to the chip interface. The hypothesis for explanation of this phenomenon is that during too fast evaporation of organics sinter particle lose contact and the material shrinkage leads to the formation of these channels. One possible process step to reduce these effects is by introducing a pre-drying and/or a slow ramp step [3].[4] which is discussed later. It is important to note that without a pre-drying step or a slow ramp rate to sintering temperature and by following a hot-in/hot-out process, even within a small chip area of 1mm², large number of defects are observed. The waveguide sealing ring, has a similar width of 1 mm. Therefore, the voiding and channeling is expected to be similar.

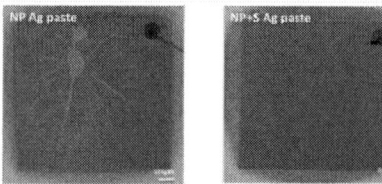

Fig. 6 – X-ray analysis of the sintered interconnects after introducing the two-step sintering process after placing the chip in the wet-paste.

As can be observed from Fig. 5, the introduction of the two-step sintering process drastically reduced the defects in the NP+S paste and considerably reduced the voiding in the NP paste. It must be noted that unlike a pressure-sintering process where the paste is printed, pre-dried and the chip is placed in the pre-dried paste, in this pressureless sinter process, the chip was placed in the wet paste at room temperature with only 0.1N placement force, but instead of a hot-in/hot-out process as earlier followed, the two-step sintering process was introduced to allow for the efficient outgassing of the binder. This is similar to a pre-drying step in a pressure sinter process, but here with the test chip already placed in the wet paste. This step ensures a slow and improved degassing of the organics in the paste. Then, the sintering temperature of 200°C for NP and 250°C for NP+S is reached at a ramp rate of ~8°C per min and an isothermal sintering step at 200°C for 30min is applied. Finally, the samples are removed from the hot-oven.

Ag NP paste

Hot-in/hot-out profile

Ag NP+S paste

Hot-in/hot-out profile

Fig. 3 – X-Ray analysis of the NP and NP+S paste after hot-in/hot-out sintering process following die-placement in wet paste with 0.1N placement force.

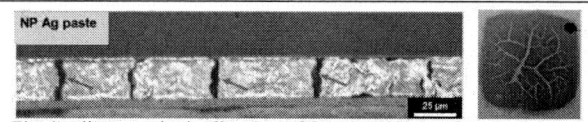

Fig. 4 – discrepancies in alignment of the top and bottom part due to pre-existing bending in the test material as shown by the red arrows.

The inhomogeneities in case of the NP+S paste were large voids that appear to be highly localized as shown by the red arrows in Fig. 4, indicative of the inhomogeneous dispersion of the Ag salt in the paste and the localized decomposition of the same. Therefore, contrary to the NP paste where a very dense interconnect is observed between the vertical channels, in case of the NP+S paste the sintered microstructure shows poor sintering between the particles. This has also been reported to lead to poor performance under thermal shock conditions [5].

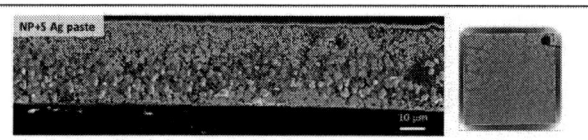

Fig. 5 – Localized sintering and high voiding observed in the NP+S paste following the hot-in/hot-out sintering process, consistent with features observed under X-ray analysis.

To improv the sintering results, a two-step sintering process was evaluated by introducing an intermediate step at 100°C and a slow ramp to the sintering temperature. The NP and NP+S pastes was sintered at 200°C and 250°C in air respectively.

B. Bonding of antenna stacks

Following the preliminary analysis with the two Ag sinter pastes, the NP Ag paste was chosen for further analysis despite the risk of the development of vertical channels as observed in the preliminary study with the 1mm² test chips. This is because, the paste is capable of sintering in air at 200°C. Further, although in this study, the NP Ag paste was dispensed, a dipping/stamping process can also be realized with this paste. However, that would limit the bond line thickness (BLT) and is part of a larger ongoing study to understand the impact of the BLT on the performance of the antenna stack and the reliability.

The NP+S paste required a minimum sintering temperature of 250°C which is critical for the antenna stack material which is desired to withstand a maximum of 250°C and only for short durations (typically durations as defined for peak soldering temperatures in standard SMD solder profiles). Further, the paste required a minimum sintering time of 30 min and therefore under the present design restrictions, such a process is not desirable.

The NP Ag paste is composed of NPs of ~80nm and requires deep freeze (-40°C) storage. Therefore, it is allowed to thaw at room temperature for 1 hour before dispensing. The paste is dispensed by a Musashi dispenser with parameters as described earlier by inputting a CAD file to the printer which replicates the wave guide antenna structure on the stack.

Fig. 7 – Optical image of the assembled antenna stack with a standard soldering process.

Fig. 8 – optical image of the cross-section of the soldered antenna stack, showing the multi-layer structure.

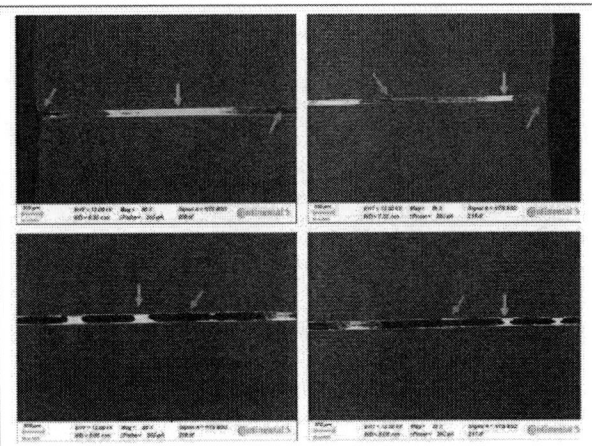

Fig. 9 – SEM analysis of the cross-sections showing sporadically well sintered areas (green arrows) among large voids (red arrows) after pressureless sintering for 30min in air at 200°C.

The profilometer analysis of the dispensed trace showed a thickness of ~80µm after dispensing. Warpage is a well-known and commonly observed phenomenon in microelectronics packaging for PCB substrates. In the sinter process no liquid phase forms. Therefore, contact between the sinter particles and the metallized antenna stacks is more critical compared to soldering. Unlike under pressure sintering, where bending of the chip or the substrate can be compensated to a certain extent by the application of the external pressure, within a pressureless sintering process, any pre-existing bending can compromise the contact between the bonding interfaces and thereby lead to poor sinter contacts. The antenna stacks used in this study showed pre-existing bending between the stacks.

A pressureless sintering step was followed after printing the NP Ag paste onto the individual layers of the antenna stack and assembling the stacks over each other manually. Therefore, the squeeze out of the paste during assembly could not be avoided. A slow ramp two step sintering process with the first step at 100°C reached at 7°C/min, followed by sintering at 200°C in air for 30min with a ramp rate of 8°C/min from the intermediate step of 100°C to the isothermal sintering step of 200°C.

The SEM analysis of the sintered interconnect (Fig. 8), cross-sectioned along the yellow dotted line as shown in Fig. 6, however shows poor sintering along both the edges and the presence of large voids between the stacks.

Therefore, to further improve the process, a small force of 8N is applied across the entire antenna stack, by placing a bare Cu block on the antenna stack. The SEM analysis of the sintered interconnect shows a more homogeneous and well sintered interconnect compared to the pressureless sintered sample. As can be observed from Fig. 11, well sintered interconnects are realized also along the edges in comparison to Fig. 10.

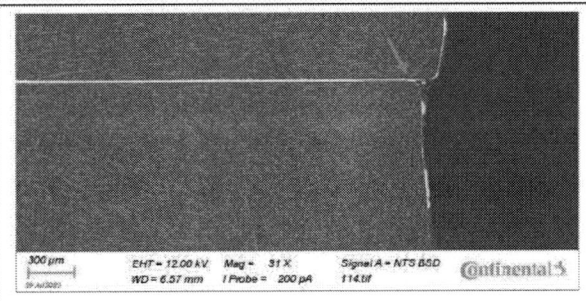

Fig. 10 – SEM analysis of the cross-sections showing well sintered areas interconnect after sintering for 30min in air at 200°C while applying a force of 8N over the entire area of the antenna stack.

Further analysis, as depicted in Fig. 12, that bond line thickness (BLT) across the sample is not uniform. Also, the presence of the defects similar to those observed earlier for the test chips, i.e. vertical cracks that exist along the entire thickness of the interconnect, between the interfaces.

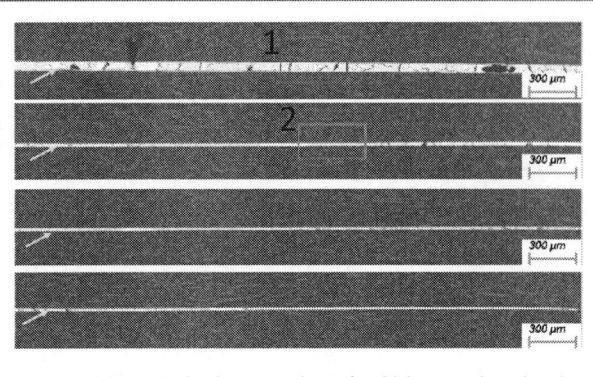

Fig. 11 – SEM analysis of cross-sections of multiple areas along the edge shown in Fig.6, showing sintered microstructure after sintering for 30min in air at 200°C while applying a force of 8N over the entire area of the antenna stack.

As can be seen in Fig. *13*, the defects are cracks. Similar observations were made in the preliminary studies with the test chips (Fig. 3). However, near bulk like interconnects are observed around these defects and in other areas (Fig. 11 image 2), indicating the potential of the NP Ag paste of realizing bulk Ag interconnects even under low pressure bonding conditions. It must be noted such bulk like sintered areas were also observed under pressureless conditions as seen in Fig. 8, but were sporadic in nature owing to the pre-existing warpage in the antenna stack layers. Therefore, if this issue can be addressed, the NP Ag paste has the potential to realize good sintered microstructure even under pressureless conditions. While it is desirable to eliminate the vertical cracks in the sintered microstructure, its impact on the reliability is unknown at the moment and needs further detailed analysis. In general, due to the shrinkage during sintering, it might not be completely possible to eliminate these vertical cracks. Therefore, as in case of soldering where in a standard SMD soldering profile no vacuum steps are included and instead an acceptable void rate is determined for specific solder materials and applications, the acceptable defect rate for sintered interconnects must also be defined. Thereby, a balance between process times and acceptable defects can be determined.

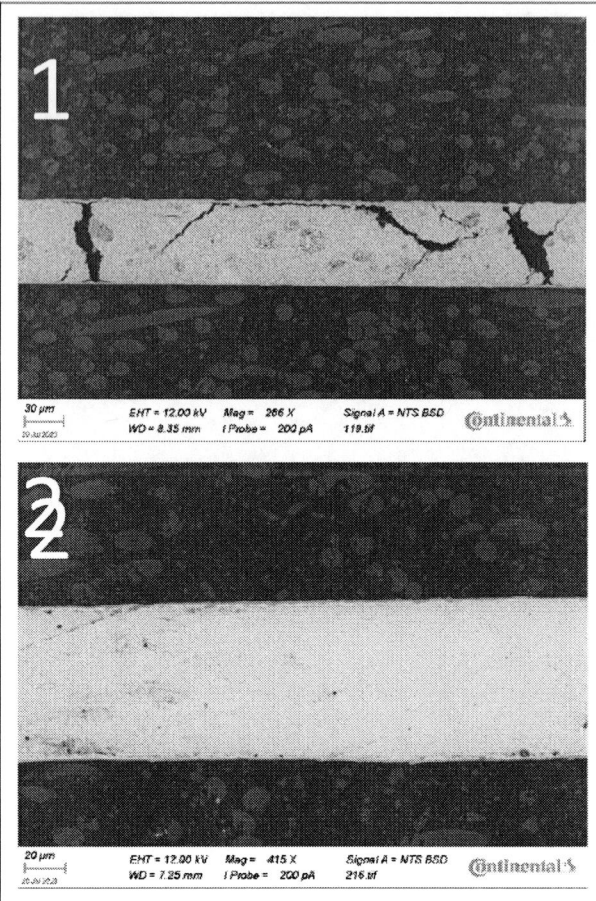

Fig. 12 – Magnified SEM images of the areas 1 & 2 shown in Fig.10.

IV. CONCLUSIONS

Low pressure and eventually pressureless sintering with NP Ag sinter paste offers a promising alternative to soldering of antenna stacks in ADAS Radar sensors, eliminating the risk of remelting of the solder.

The possibility to sinter in air and at a temperature of 200°C allows the sintering of the antenna stacks without damaging the plastic material by high process temperature. Furthermore, there is no risk of remelting the solder in the followed SMD reflow process to attach the antenna to the PCB.

Therefore, low pressure, and low temperature sintering in air with NP Ag paste is shown to be a promising approach to bonding antenna stacks together and eventually to the PCB in a single step without affecting the standard soldered SMD components. Near bulk like sintered microstructure can be realized and a potential towards miniaturization can be evaluated.

For a required high yield process, like in case of soldering, the acceptable defect rate for sintered interconnects must be determined, specifically the impact of vertical cracks as observed in this case with the NP Ag paste, on the reliability must be ascertained.

In addition, potentially, the sinter process could be integrated into the existing SMD soldering process: The antenna layer are not bonded in a separated process but stacked onto the PCB. Within the SMD process also the sintering would be performed. However, to implement such a process the sinter time would need to be reduced. Furthermore, warpage in the antenna stacks is a challenge to realize well sintered interconnects and must be addressed.

ACKNOWLEDGMENT

The authors would like to thank Ms. Elena Scheiermann from Conti temic microelectronics GmbH, Ingolstadt for the support with the SEM imaging. This project was carried out under the funding through "Additive Fertigung zur Miniaturisierung hochperformanter und robuster Radarsensoren – ADDIRA", under the grant number DIE-2107-005//DIE0159/01.

REFERENCES

[1] Y. Li and H. Shi, Advanced Driver Assistance Systems and Autonomous Vehicles From Fundamentals to Applications. 2022.

[2] U. Huegel, A. Garcia-Tejero, R. Glogowski, E. Willmann, M. Pieper and F. Merli, *"3D Waveguide Metallized Plastic Antennas Aim to Revolutionize Automotive Radar"*, Microwave Journal, September 2022, https://www.microwavejournal.com/

[3] A. Hanss, M. Schmid, S. K. Bhogaraju, F. Conti, and G. Elger, "Reliability of Sintered and Soldered High Power Chip Size Packages and Flip Chip LEDs," in 2018 IEEE 68th Electronic Components and Technology Conference (ECTC), May 2018, pp. 2080–2088. doi: 10.1109/ECTC.2018.00312.

[4] A. Hanss, M. Schmid, S. K. Bhogaraju, F. Conti, and G. Elger, "Process Development and Reliability of Sintered High Power Chip Size Packages and Flip Chip LEDs," 2018, pp. 479–484.

[5] S. K. Bhogaraju, M. Schmid, H. R. Kotadia, F. Conti, and G. Elger, "Highly reliable die-attach bonding with etched brass flakes," in 2021 23rd European Microelectronics and Packaging Conference & Exhibition (EMPC), Sep. 2021, pp. 1–6. doi: 10.23919/EMPC53418.2021.9584967.

Characterizations for eWLB (Embedded Wafer Level Ball Grid Array) Antenna in Molded Package Integrations in 77GHz Automotive Applications

Ming-Che Hsieh
Field Applications Engineering
JCET Group Co., Ltd.
Singapore
mc.hsieh@jcetglobal.com

Frank Zhang
Chief Executive Officer
Andar Technologies Co., Ltd.
Hangzhou, China
frankz@andartechs.com

Forest Zhu, Kang Liu
Research & Design Dept.
Andar Technologies Co., Ltd.
Hangzhou, China

Linda Chua, Yaojian Lin
R&D Engineering
JCET Group Co., Ltd.
Singapore

Joel Damalerio, Hin Hwa Goh, Kai
Chong Chan
Process Engineering
JCET Group Co., Ltd.
Singapore

Zihao Chen
College of Electronics and Information
Engineering
Harbin Institute of Technology
Shenzhen, China

Abstract— The worldwide IC market has had a significant growth rate in recent years, driving diversified package solutions in the segments of communication, computing, consumer, automotive, industrial, etc. Among these market segments, the automotive market is forecast to have the strongest CAGR (Compound Annual Growth Rate) of any of the end-use segments in the next 5 years, which results in more IC packages and components being designed into automotive vehicle systems such as infotainment, ADAS (Advance Driving Assistance System), instrumentation, body systems, battery control systems and so on. In these applications, the 77GHz short-range radar sensor application that can enable higher EIRP (Effective Isotropic Radiated Power) and adaptive cruise control performance has become more attractive to be commercialized. Due to the benefits of shorter interconnection length, lower conductive loss with smooth Cu surface as well as a lower dielectric constant and dielectric loss of materials in a fan-out wafer level package (FOWLP) technology, the FOWLP is proven as a suitable packaging solution to achieve the high speed and frequency requirements in 5G and mmWave applications. The eWLB Embedded Wafer Level Ball Grid Array) is a versatile FOWLP technology, providing a robust package platform in a space-efficient design with the very dense interconnection and routing of multiple dies as well as a smaller footprint and lower package profile providing a reliable package solution. Meanwhile, since the utilization of an antenna that is integrated into the packages has been studied as a low-cost solution because the packages can be tested in normal FR4 printed circuit board (PCB) instead of the RF function integrated PCB, the eWLB integrated with molded antenna chip packages (called eWLB antenna in molded package) technology is introduced in this paper. The characterizations for assembly process challenges are evaluated and illustrate the process optimizations of equipment handling, antenna molded chip package fabrication, reconstitution process control and warpage characterization of the molded wafer. In addition, the radio frequency (RF) test validation is performed to verify the performance of scattering parameter (S-parameter), radiation pattern, impedance and detection range of the antenna. In order to prove the reliable quality and yield of the evaluated eWLB antenna in molded packages, package-level reliability tests are studied. Through these results, it will show that this eWLB antenna in molded package integration is a robust and cost-effective solution for 77GHz automotive applications.

Keywords—Fan-out wafer level package, FOWLP, eWLB, antenna in package, AiP, 77GHz, automotive.

I. INTRODUCTION

The semiconductor industry is growing faster than ever, driving the need for various advanced packaging solutions to achieve higher performance requirements in multiple market segments, such as communication, computing, consumer and automotive applications. Among these semiconductor markets, the automotive applications have been predicted to have tremendous growth with a 2022-2027 CAGR of around 7.4% according to an industry report [1]. In order to achieve the requirements of high speed, high frequency, fine routing pitch and efficiency enhancement, the FOWLP technology has been studied as the proper packaging solution not only in 5G and mmWave communication applications but also in 77GHz (and above) automotive radar applications [2-3]. The eWLB (Embedded Wafer Level Ball Grid Array) technology is one of the commercialized FOWLP technologies that utilizes the simplest process flow with chip first and face down architecture as well as lower dielectric constant and dielectric loss re-passivation materials. In addition, the smooth electroplated Cu surface with lower conductive loss and shorter interconnection length is also adopted to reduce the parasitic effect and insertion loss at mmWave high frequency in the eWLB technology [4-6].

For the purpose of satisfying the demand for a higher performance and cost-effective packaging solution for 77GHz automotive radar applications, the FOWLP technology with integrated antenna has been widely discussed. Because the use of antenna integrated with a FOWLP will result in many complicated processes and higher assembly cost, the related function tests can be performed with the normal FR4 PCB instead of the RF function integrated PCB. The PCB cost will be reduced more than several dozens of times. In this paper, the technology of eWLB integrated antenna in a molded package and the corresponding assembly characterizations as well as electrical test verifications are studied. The antenna in molded chips are designed with the backside metal antenna patterning process not having any connection to the eWLB redistribution layer (RDL) and are fabricated internally. The CMOS die is integrated with several antenna molded chips in the eWLB to form a single channel (one transmitter and one receiver, 1TX1R) and dual channel (two transmitters and two receivers, 2TX2R) packages. In order to validate the robust design and the reliable packages of this evaluated eWLB integrated antenna in the molded package, the characterizations for optimizations of equipment handling, fabrications of antenna molded chip packages, the process flows of reconstitution and warpage control as well as the package reliability tests are presented. In addition, the electrical test validation includes the evaluation of S-parameter, radiation pattern, and impedance and detection range of the antenna verification. These results are proven to meet 77GHz automotive requirements. With these results, it is believed that this present eWLB antenna in molded package integration will be a proper package solution to meet high frequency, and high reliability requirements in 77GHz automotive application and an overall cost-effective solution to minimize the assembly and test board cost.

II. PROCESS FLOWS FOR eWLB ANTENNA IN A MOLDED PACKAGE

With the benefit of eWLB technology that can integrate multi-chips, enable high speed and high-frequency requirements as well as achieve reliable performance in 5G and mmWave applications, the innovation of this study is to design antenna molded chips that can integrate into a eWLB package with the current compatible integration process flow.

A. Antenna molded chip fabrication

The antenna molded chip is designed with the backside metal antenna patterning process on the molded wafer and will be viewed as the other die that integrates with the CMOS die using a pick and place process that places them onto a carrier simultaneously. This will be followed with the compression molding process in the standard eWLB reconstitution process. These antenna molded chips are fabricated internally and there is no signal routing connection between antenna molded chips and the eWLB redistribution layer (RDL). The uses of the antenna molded chips are not only compatible to the existing eWLB process flows, but also reduce the antenna fabrication cost. The optical and cross-sectional image of an antenna molded chip is illustrate in Fig. 1.

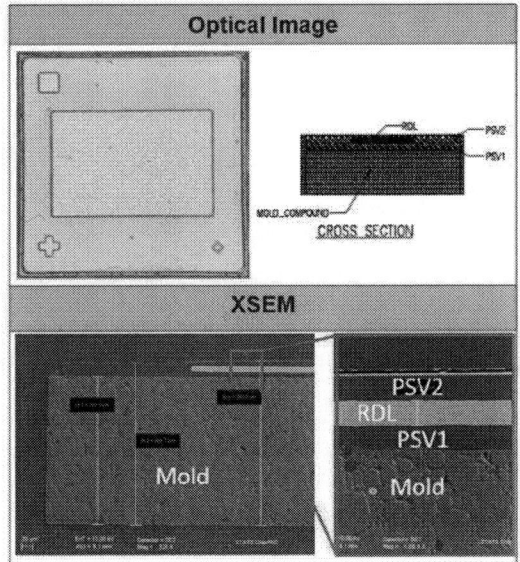

Fig. 1. The optical and cross-sectional image of an antenna molded chip.

B. Process Flow for eWLB Antenna in a Molded Package

Two devices with the utilizations of eWLB antenna in molded package integrations are studied. One is AiMP-1TX1R, the single channel (one transmitter and one receiver) design with a pair of antenna molded chips and the other is AiMP-2TX2R, the dual channel (two transmitters and two receivers) design with two pairs of antenna molded chips. The evaluated devices' specifications and final package images are illustrated in Fig. 2 and 3, respectively. The assembly processes of these evaluated eWLB antenna in molded package integrations are illustrated in Fig. 4 and described below.

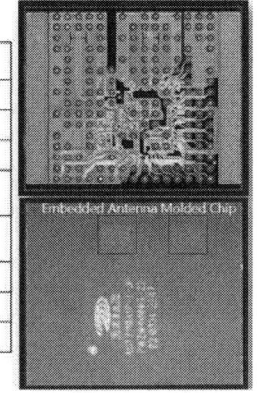

Items	Specifications
Package size	10x8.5mm
Die size	~4x4mm
Die thickness	250um
Antenna molded chips size	1.75x1.75mm
Antenna molded chips #	2
Ball pitch	0.5mm
Ball diameter	0.3mm
Package thickness	<0.6mm

Fig. 2. AiMP-1TX1R specifications and package image.

Items	Specifications
Package size	10x8.5mm
Die size	~4x4mm
Die thickness	250um
Antenna molded chips size	1.75x1.75mm
Antenna molded chips #	4
Ball pitch	0.5mm
Ball diameter	0.3mm
Package thickness	<0.6mm

Fig. 3. AiMP-2TX2R specifications and package image.

(1) Pre-assembly processes: includes incoming wafer thinning with backside grinding as well as wafer singulation by laser grooving and mechanical sawing;

(2) Reconstitution process: starts from the processes of laminating an adhesive foil onto a carrier then placing active die face down and antenna molded chips with top side patch antenna with the pick and place tool. After the process of die pick and place, the compression molding process is implemented and then followed by a de-bonding process to remove the carrier and foil, resulting in a reconstituted wafer with the molding compound surrounding all antenna molded chips and active dies surfaces.

(3) WLP process on a reconstitution wafer: includes dielectric layer (PSV1 & PSV2) coating, exposure, development, curing and dielectric layer opening etching; seed layer sputtering, RDL and UBM (under ball metallization) plating and etching; solder ball mounting and package singulation.

(4) Backend process: includes final test, laser marking, EVI/AOI inspection, packing, etc.

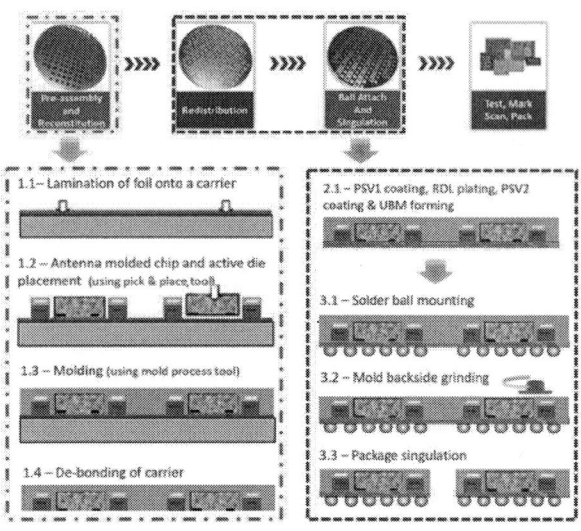

Fig. 4. Process flow of eWLB antenna in molded package integration.

III. PACKAGE ASSEMBLY CHARACTERIZATION FOR EWLB ANTENNA IN MOLDED PACKAGE

As the pre-assembly process of the incoming wafer is needed before the eWLB reconstitution process, the characterization data of the monitored items in pre-assembly (includes wafer grinding and dicing) is listed in Table I. Table II and III shows the process monitored items in the reconstitution processes with the integrations of antenna molded chips and an active die, which include PSV1 and PSV2 lithography, RDL and UBM plating as well as the related warpage control data, respectively. The warpage specification requirements is controlled within -1.5mm to 2.0mm in each process in eWLB reconstitution processes. Through these results, it is not only observed that the maximum warpage after annealing, PSV1, RDL, PSV2 and UBM plating processes are all less than 2 mm and meet the in-line specification of well warpage control but also indicates both pre-assembly and reconstitution molded wafer processes are consistently under control.

In addition, the ball shear test is also performed to verify the strength of solder ball after the solder ball mount (ball drop) process. The ball shear results for both eWLB antenna in molded package integrations, AiMP-1TX1R and AiMP-2TX2R are illustrated in Table IV, which shows that all the examined solder balls exceeded the spec. of 260gf with good strength. After the solder ball mounting process, the mold backside grinding process is utilized to achieve the targeted mold thickness with the specification requirement of 0.355 ±0.025mm after backside grinding. With this target mold thickness, it can be ensured that the overall maximum package thickness (includes solder ball stand-off height) will be less than 0.6mm. After the package singulation process, the final eWLB packages can be obtained. These characterization data are illustrated in Table V and VI, respectively, which showed all are passed specification requirements and the cross-sectional images of the final AiMP-2TX2R package is showed in Fig.5.

TABLE I.   PROCESS CONTROL ITEMS IN PRE-ASSEMBLY PROCESS

Process	Monitoring Items	Criteria	Average	Stdev
Si Backgrind	Wafer Thickness	250 ± 10um	249.26	0.664
Si Dicing	Kerf Width	20 ± 12um	21.5 (X) 21.5 (Y)	0.689 (X) 0.599 (Y)
	Kerf Shift	± 5um	0.1 (X) -0.1 (Y)	0.551 (X) 0.721 (Y)

TABLE II.   PROCESS CONTROL ITEMS IN RECONSTITUTION PROCESS

Process	Monitoring Items	Criteria	Average	Stdev
PSV1	Thickness	12.3 ± 2um	11.82	0.561
	Opening (CD)	22.5um ± 3um	23.84	0.562
RDL1	Thickness	8 ± 1.6um	8.17	0.111
	Opening (CD)	20um± 3um	21.07	0.531
PSV2	Thickness	13.0 ± 2um	12.90	0.670
	Opening (CD)	250um ± 6um	249.51	0.849
UBM	Thickness	12.0 ± 2.6um	11.60	0.182
	Opening (CD)	290um ± 6um	288.65	0.706

TABLE III.   MAX. WARPAGE DATA FOR RECONSTITUTION MOLDED WAFER

Process	Criteria	Max. Warpage (mm)
Molding		1.88
PSV1		-0.37
RDL1	-1.5mm ~ 2um	0.53
PSV2		0.94
UBM		1.16

TABLE IV.   BALL SHEAR TEST AFTER BALL DROP PROCESS

Device	Criteria	Min. (gf)	Max. (gf)	Result	Image
AiMP-1TX1R	Min. 260gf	592.7	660.9	Pass	
AiMP-2TX2R	Min. 260gf	600.2	633.7	Pass	

TABLE V.   MOLD BACKSIDE GRINDING RESULT

Device	Item	Mold thickness (um)
AiMP-1TX1R	Before Backside Grinding	~610
	After Backside Grinding	355.1
AiMP-2TX2R	Before Backside Grinding	~610
	After Backside Grinding	357.8

TABLE VI.   PACKAGE SINGULATION RESULT

Device	Monitoring Items	Criteria	Average	Stdev
AiMP-1TX1R	X-Dim.	10.0 ± 0.05mm	9.999	0.003
	Y-Dim.	8.5 ± 0.05mm	8.502	0.002
AiMP-2TX2R	X-Dim.	10.0 ± 0.05mm	9.997	0.003
	Y-Dim.	8.5 ± 0.05mm	8.500	0.002

IV. ELECTRICAL PERFORMANCE CHARCTERIZATIONS

To fully characterize the designed AiMP-1TX1R and AiMP-2TX2R samples, some test structures are also fabricated. For testing purposes, the S-parameter performance of the designed testing structure is characterized first. The measurement process was conducted using on-wafer G-S-G probing up to 100 GHz by means of a vector network analyzer

N5290A from Keysight and a 200-μm pitch Air Coplanar Probe (ACP) probe with 1-mm connectors from FormFactor. The on-wafer calibration was made by using a conventional short-load-open-thru (SLOT) approach to move the reference planes from the connectors of the equipment to the tips of the RF probes.

The die microphotograph, testing setup, and comparison results between the simulation and measured data are given in Fig. 6. As it can be seen from Fig. 6(c), a good agreement between simulated and measured results is obtained. More than 15-dB return loss is achieved across a bandwidth of 76-84 GHz. It indicates a good impedance matching for the designed antenna as well. The impact due to back-side metal, and BGA balls on impedance variations of the designed testing structure is marginal.

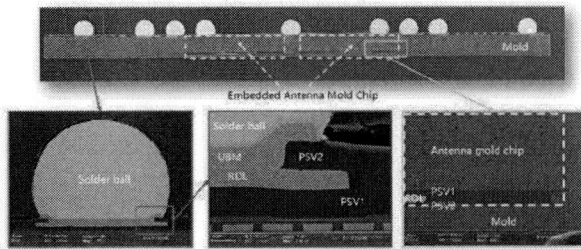

Fig. 5. Cross-sectional images for AiMP-2TX2R.

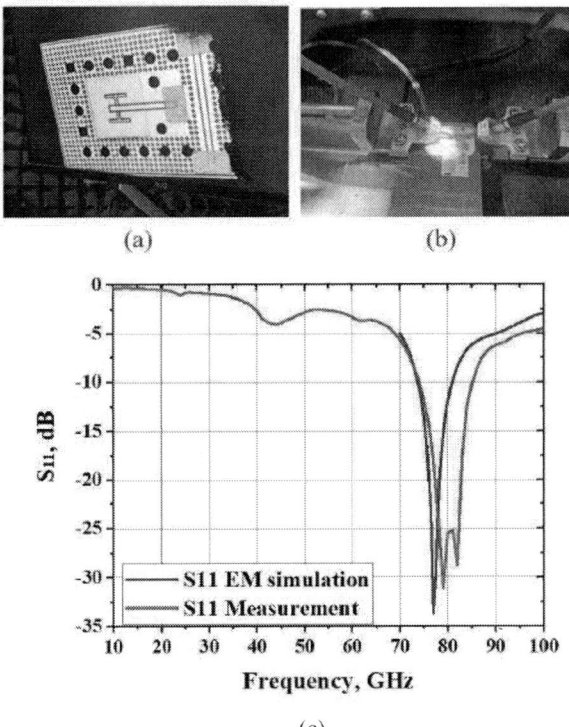

Fig. 6. Measurement results and set up for testing structures using on-wafer probing. (a) die microphotograph, (b) set up on probe station, (c) comparison results between simulation and testing.

To further verify the performance of the designed AiMP, the prototype is also tested at the PCB level, which is shown in Fig. 7(a). In Fig. 7(b), the comparison results between simulation and measurement results are illustrated. A good

agreement between them is also obtained. It once again indicates that the designed AiMP is fully functional as expected.

(a)

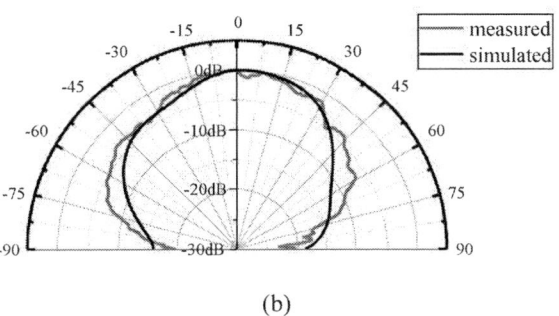

(b)

Fig. 7. Measurement results and setup for testing the packaged antenna. (a) circuit board with the packaged chipset; (b) radiation pattern at 79 GHz.

V. RELIABILITY TEST DATA

For the purpose of demonstrating the examined reliable eWLB antenna in molded package integrations, long term reliability tests such as pre-condition of moisture sensitivity level (MSL3) with 30°C/60%RH and 192 hours condition, unbiased highly accelerated stress test (uHAST, test condition at 130°C/85%RH), pressure cooker test (PCT, test condition at 121°C/100%RH) and thermal cycling test condition B and C (TCB, -55°C to 125°C and TCC, -65°C to 150°C) with pre-condition of MSL3 as well as high temperature storage test (HTST, test condition at 150°C) were performed. The long-term package reliability results of all examined legs passed uHAST 168 hours, PCT 168 hours, TCB and TCC 1000 cycles as well as HTST 1000 hours through the T-SAM inspection. The test condition, result and sample size are listed in Table VII. Fig. 8 illustrates the T-SAM result for final reliability reading points of AiMP-1TX1R and it clearly shows that there is no abnormality and delamination observed through these images. Bases on these reliability test results, it shows not only the examined eWLB antenna in molded package integrations can achieve the robust antenna design and compatible fan-out wafer level package assembly process flows but can also meet package level reliability criterion. Therefore, it is believed that with the technology of eWLB antenna in mold package integrations established in this paper will provide a suitable packaging solution for when highly integrated, high speed, high frequency and cost-effective purposes are required in the semiconductor industry.

TABLE VII.   LONG-TERM RELIABILITY TEST RESULT

READOUT POINTS	TEST CONDITIONS	SAMPLE SIZE	PRECON DITION (Y / N)	AiMP-1TX1R	AiMP-2TX2R
MSL3	30degC / 60% RH / 192hrs	44 units	Y	0/44	0/44
uHAST 168 hrs	130degC / 85% RH	11 units	Y	0/11	0/11
PCT 168 hrs	121degC / 100% RH	11 units	Y	0/11	0/11
TC(B) 1000x	-55degC / 125degC	11 units	Y	0/11	0/11
TC(C) 1000x	-65degC / 150degC	11 units	Y	0/11	0/11
HTS 1000 hrs	150degC	22 units	N	0/22	0/22

Remarks: No delamination found on all readout points for both AiMP-2TX2R & AiMP-1TX1R packages.

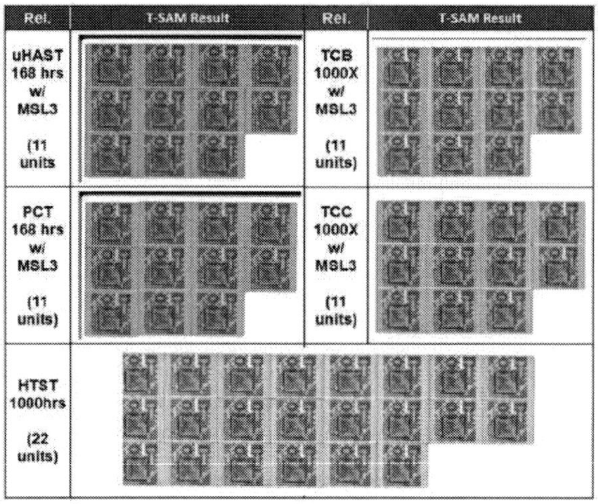

Fig. 8.   T-SAM result for AiMP-1TX1R.

VI. CONCLUSIONS

The technology of eWLB antenna in molded package integrations was introduced to deliver a robust and highly integrated performance packaging solution with high speed, high frequency, reliable as well as cost-efficient requirements in 77GHz automotive applications;

(1) The eWLB antenna in molded package (eWLB AiMP) adopts the pre-fabricated antenna in molded chips to integrate with the eWLB package. As the antenna in molded chips are fabricated internally, the compatible integration process flow has been observed.

(2) Two eWLB antenna in molded package integrations are studied for process, mechanical, electrical and reliability characterizations. The examined eWLB AiMP integrations passed the process specification requirements, S-parameter, impedance, radiation pattern performance verification as well as long term reliability tests. The results show the good assembly quality and electrical performance of this robust and cost-efficient eWLB AiMP integration solution.

Bases on the investigated results above, it is believed that this eWLB AiMP integration technology can be an overall low-cost solution to minimize both assembly and test board cost as well as a proper package solution to meet high frequency and reliability requirements.

ACKNOWLEDGMENT

The authors would like to express their sincere appreciation to Andar Technology Co. Ltd. and the Process Engineering team in JCET Group Co. Ltd. in Singapore for their helpful advice and support of this work.

REFERENCES

[1]   The Semiconductor and Packaging Report, Primask Partners LLC, Fourth Quarter 2022.

[2]   M. Brunnbauer, E. Fürgut, G. Beer, T. Meyer, H. Hedler, J. Belonio, E. Nomura, K. Kiuchi and K. Kobayashi, "An Embedded Device Technology Based on a Molded Reconfigured Wafer," 56th Electronic Components and Technology Conference (ECTC 2006), 2006.

[3]   G. Haubner, W. Hartner, S. Pahlke and M. Niessner, "77 GHz automotive RADAR in eWLB package: From consumer to automotive packaging," Microelectronics Reliability, vo. 68, pp. 699-704, Sep. 2016.

[4]   S.W. Yoon, Roger Emigh, Kai Liu, SJ Lee, Ray Coronado and Flynn Carson. "Thermal and Electrical Characterization of eWLB (embedded Wafer Level BGA)," Electronic Components and Technology Conference (ECTC), 2010.

[5]   S.W. Yoon, "Advanced eWLB/FO-WLP (embedded Wafer Level Ball Grid Array/FanOut-Wafer Level Package) for High Frequency Applications," Semicon Korea, 2017.

[6]   D. Wang, F. An, S.W. Yoon, "Advanced eWLB (embedded Wafer Level Ball Grid Array) Solutions for mmWave Applications," International Wafer Level Packaging Conference (IWLPC), 2018.

A Dual-Band Dual-Polarized 2x2 Antenna array with beamforming for 5G AiP and mmWave Applications

Sheng-Chi Hsieh
Cetral Reserch and Development
ASE Gruop
Kaohsiung, Taiwan
Rick_Hsieh@aseglobal.com

Hong-Sheng Huang
Cetral Reserch and Development
ASE Gruop
Kaohsiung, Taiwan
SamHS_Huang@aseglobal.com

Wen-Chun Hsiao
Cetral Reserch and Development
ASE Gruop
Kaohsiung, Taiwan
Stanley_Hsiao@aseglobal.com

Cheng-Yu Ho
Cetral Reserch and Development
ASE Gruop
Kaohsiung, Taiwan
Derk_Ho@aseglobal.com

Chen-Chao Wang
Cetral Reserch and Development
ASE Gruop
Kaohsiung, Taiwan
Alexcc_Wang@aseglobal.com

Abstract—**In this study, we present a dual-band (28/39GHz) 2 by 2 antenna array design on a 10(4+2+4) multi-layer organic substrate with broad bandwidth and higher isolation onto a compact AiP module. In addition, the H-type slot antenna structure can improve interference and isolation between 28 and 39GHz bands. The measured result shows the isolation can larger than 15dB between low and high band. For antenna measurement, the spherical of probing chamber is utilized to validate 28/39GHz antenna pattern and performance with a package level passive testing. Our AiP measured result shows the return loss is better than 10 dB in 23.5-30.5 GHz range, with ~7 GHz bandwidth and provides a high-gain per element (above ~6 dBi) radiation pattern for 28GHz applications. For 39GHz band, the antenna has 7 GHz bandwidth and provides a 5 dBi gain between 38-45 GHz. Comparison of S-parameter between simulated and measured results, there is a good correlation to obtain the quality of manufacturing. Furthermore, a 28GHz beamforming array is designed and demonstrated. The beamfoming array consists of a four-channel transceiver with 2 x 2 antenna array. The beamformer IC is implemented by TSMC 90-nm CMOS technology and the flip-chip bonded on BT substrate. For beamforming test, an integrated socket and load board are designed for support the mmWave module and then connection between DUT and test system. The load board substrate includes the IF, LO, and DC distribution network for respectively. There are 4-channel transmit beamforming array front-end module for testing. Finally, the 3D beam steering of 2 by 2 antenna array is measured with multi-states at 28 GHz, with a maximum realized gain of 11.8 dBi achieved in the main beam (θ=0°). It shows that beamforming can be generated at a specific beam direction by controlling the phase of the signal in each antenna element.**

Keywords—Antenna in package(AiP), Dual-band Antenna, mmWave Antenna, Broad band antenna

I. INTRODUCTION

The new radio (NR) frequency bands of 5G communication are distributed in two defined frequency ranges (FR), which are FR1: 450 MHz to 6 GHz and FR2: 24.25 GHz to 52.6 GHz [1]. There are three dimensions to improve the performances, which are massive IOT, low latency, and the enhanced mobile broadband (eMBB), for the usage of massive connectivity, ultra-high reliable and low latency, and capacity enhancement, respectively. As shown in Fig. 1, the physical wavelengths for mmWave range are about one to ten millimeters so that the antenna size in the

millimeter-wave band is obviously small to design into package. The antenna in package (AiP) technology is applied to satisfy electronic devices of the fifth generation(5G) communication nowadays in the millimeter-wave range. Recently, mmWave frequency range is widely discussed and expected to provide higher capability of handling larger data volume and higher speed for the 5G and beyond 5G(B5G) wireless communications. The technology trend shows a strong demand for a larger bandwidth (BW), multi-band, low cost materials into antenna in package technology (AiP) [2-3]. Compared with glass substrate [4], low-temperature co-fired ceramic (LTCC) [5], on-chip [6] or organic substrate [7] fabrication technologies, the BT-based organic substrate is a relatively cost-effective solution. Although some mmW AiP studies are demonstrated on the BT-based substrates applications. However, the multi-frequency bands are a key issue when a lot of the antennas designed onto a one package. In this work, a compact size of dual-band 2 by 2 antenna array on 10(4+2+4) multi-layer organic substrate that achieves broad bandwidth and higher isolation on a 13 mm x 13 mm AiP module.

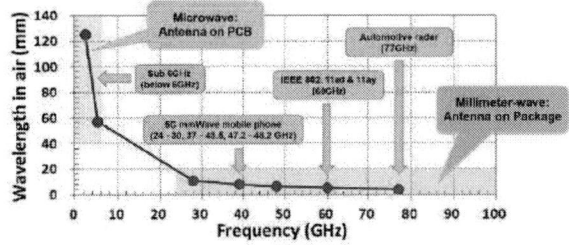

Fig. 1. Comparison of wavelength in air at microwave band and millimeter-wave band.

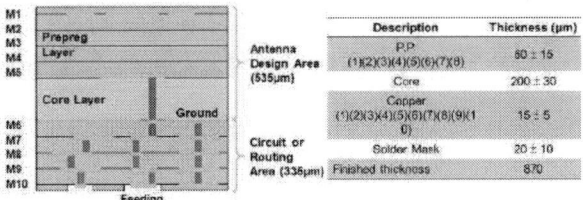

Fig. 2. The cross-section view of the AiP stack-up structure.

(1), then the relative permittivity (ε_r) of the material is converted by the following equation [8] for $W/h \geq 1$:

$$\varepsilon_{\mathit{eff}} = \frac{\varepsilon_r + 1}{2} + \frac{\varepsilon_r - 1}{2}\left(1 + \frac{12h}{W}\right)^{-\frac{1}{2}} \qquad (2)$$

where W is the width of the microstrip line and h is the thickness of the substrate.

As shown in Fig. 3, the designed T-resonator and microstrip transmission lines are used to extract the material characteristic of 4+2+4 multi-layer substrate. The extracted material characteristic of 4+2+4 multi-layer substrate are shown in Fig. 4.

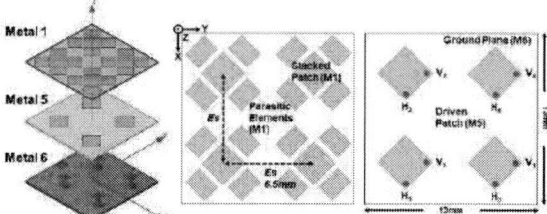

Fig. 5. Geometry of the 2 by 2 antenna

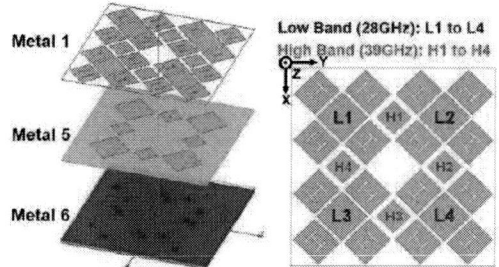

Fig. 6. Geometry of the dual-band antenna

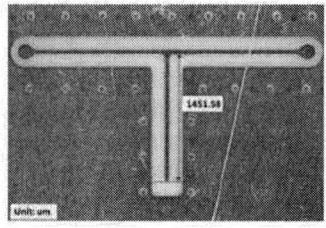

(a) T-resonator

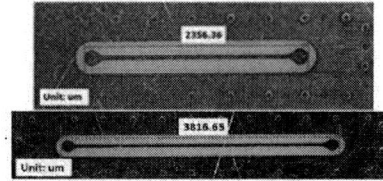

(b) Microstrip transmission line

Fig. 3. The test kits are implemented on 4+2+4 multi-layer substrate. (a) T-resonator, (b) Microstrip transmission line TL1 and TL2

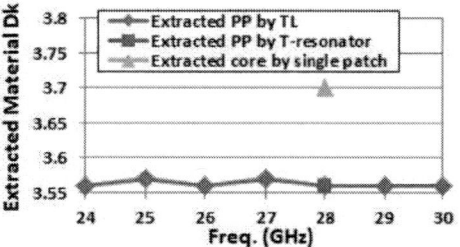

Fig. 4. The extracted material characteristic of 4+2+4 multi-layer substrate by designed test kits.

II. ANTENNA DESIGN

A. Extracted Material Property

The stack up of 10 layers (4+2+4) with low Dk and Df of core/pregrep(p.p) is shown in Fig. 2. The BT and core material are GHPL970LF and CCL-HL972LF respectively. Two 50-Ω microstrip lines with different trace length are employed to extract the material property. After applying multimode TRL calibration, the probe pad and the transition taper effect are eliminated and resulting in a simple transmission line model with a 15 mm length difference between the long and the short lines. The relationship between the effective relative permittivity ($\varepsilon_{\mathit{eff}}$) and the phase constant (β) of a microstrip line is shown as below:

$$\frac{c}{\sqrt{\varepsilon_{\mathit{eff}}}} = \frac{\omega}{\beta} \qquad (1)$$

where c is the speed of light in vacuum and ω is the angular frequency. $\varepsilon_{\mathit{eff}}$ is obtained from the measured phase of S_{21} by

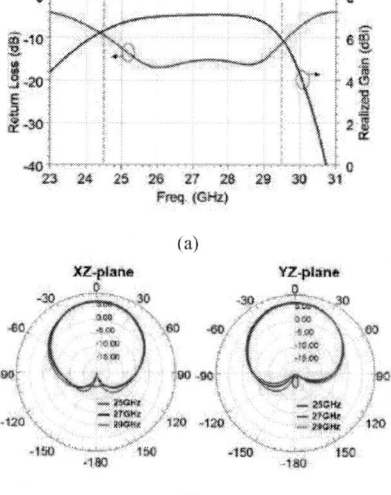

(a)

(b)

Fig. 7. The simulated result for low band antenna (a) return loss and realized gain (b)2D pattern for XZ-plane and YZ-plane at 25GHz, 27GHz, and 29GHz.

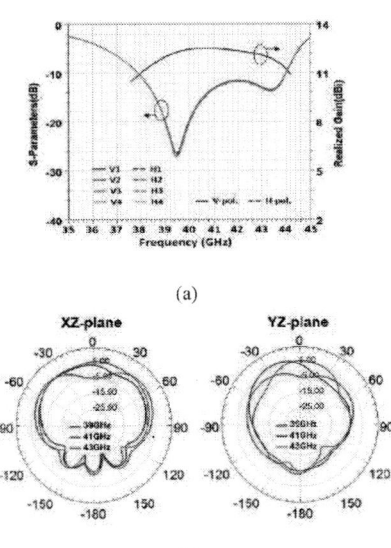

(a)

(b)

Fig. 8. The simulated result for high band antenna (a) return loss and realized gain (b)2D pattern for XZ-plane and YZ-plane at 39GHz, 41GHz, and 43GHz.

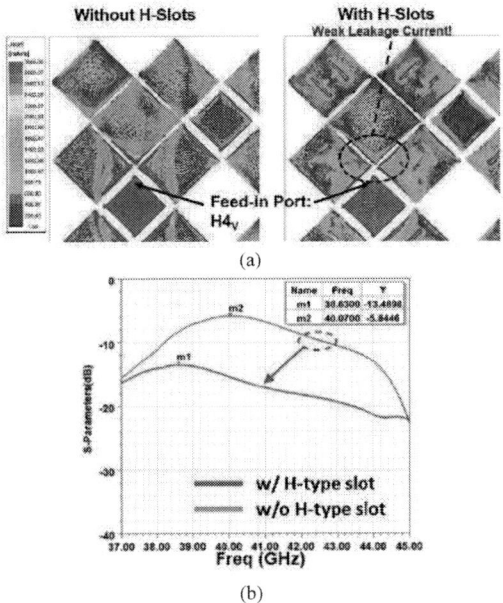

(a)

(b)

Fig. 9. The EM simulation of dual band antenna (a) current distributions (b) The isloation with and withoiut H-slot

B. Dual-Band Antenna Design

The Dual-Band antenna design is based on a stacking patch antenna with parasitic structure. In order to improve the performance of patch antenna, the stacking patch antenna with parasitic elements has advantages of wider bandwidth and higher antenna gain. For transmitting performances, this features improve the beam scanning range and array gain, which will enhance the transmission quality while beamforming. On the antenna design, the driven patch is printed on the Copper 5 layer and coupled to the stacked patch which is printed on the Copper 1 layer. The parasitic patches located close to the staked patch on the Copper 1 layer as well. The ground plane is printed on Copper 6. Fig. 5 shows the proposed 2 by 2 antenna with multi-parasitic elements, where

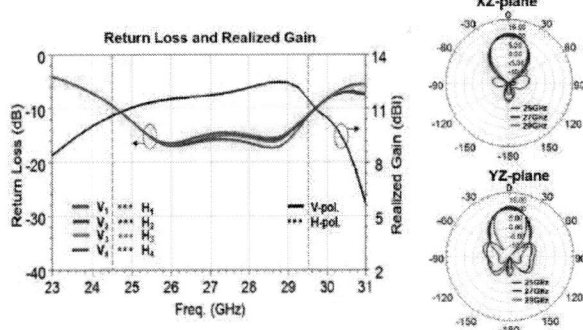

Fig. 10. The simulated 2D radiation pattern for 28GHz 2x2 antenna array

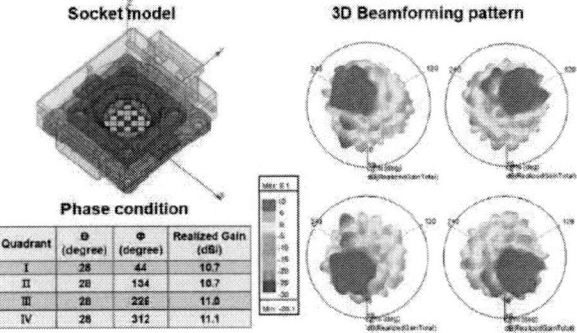

Fig. 11. The simulated 3D beam steering pattern with test socket model

the multi-parasitic elements are located around the stacked patch to provide multi-modes responses. The dimension of the driven patch and staked patch with parasitic elements for maximum gain is given by half of the wavelength. The antenna can be extended to a 2 by 2 array design as adjusting the spacing between array elements. In order to obtain the best directivity and linearity, the linear arrays are designed to have half-wave spacings to give identical results. The spacing between the elements is optimized to 6.5mm. The 2x2 antenna array consists of four antenna elements with two feeding ports placed orthogonally in each antenna element for dual-polarization. Its dimension is 13 x 13 x 0.87 mm^3. For 39 GHz antenna design, the construction of antenna is a stacked patch antenna as well. It is embedded and placed between 28GHz elements with the same ground plane to combine as a dual band antenna with different input feeding. Fig. 6 presents the geometry of the proposed dual band 28/39 GHz antenna. Fig. 7 and Fig. 8 show the simulated return loss and 2D radiation pattern for low band and high band antenna. In this work, we add a H-slot to a patch antenna which enables it to adjust current distributions and improve isolation between two bands of antennas. Fig. 9(a) shows the surface current distributions of the dual band antenna with H-slot. The isolation is effectively improved from 5dB to 15dB as shown in Fig. 9(b).

C. Beamforming Simulation with Test Socket

In order to drive 2x2 antenna array for beamforming generation at the desired XZ and YZ directions, each antenna of array must have an appropriate phase difference with respect to its adjacent antenna where θ and Φ denote the relative phase difference between the antenna located in the XZ and YZ directions. For the beamforming performance, the simulated 2D radiation patterns of the V-pol are presented in the XZ- and YZ- planes at 25GHz, 27GHz, and 29GHz,

391

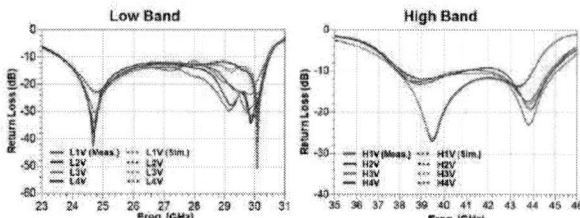

Fig. 12. Simulated and measured return loss

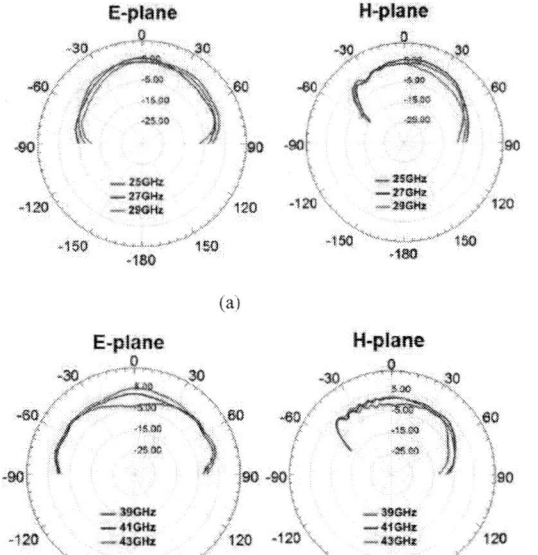

(a)

(b)

Fig. 13. The measured peak gain and 2D radiation pattern(a) Low band (b) High band

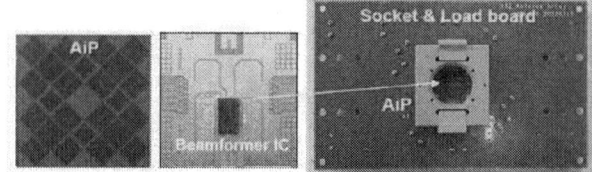

Fig. 14. The photographs of the fabricated AiP and socket

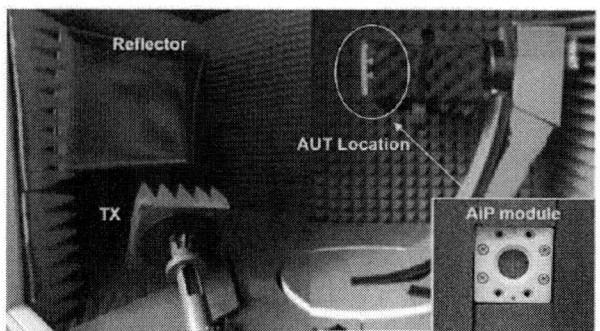

Fig. 15. mmWave AiP measurment setup in CATR chamber

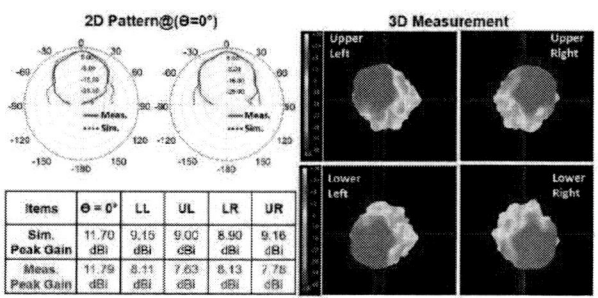

Fig. 16. The measured 2D & 3D beam steering pattern in CATR chamber

respectively, as shown in Fig. 10. It is essential to obtain the 3D beam steering pattern to ensure the radiation performance with phased antenna array that can be integrated with a beamformer circuit for active test in the chamber. In Fig. 11, the 3D beam steering is simulated in all four quadrants at 28GHz. The simulated model is not only 2x2 antenna array but also include the test socket for performance evaluation with a real test in the chamber. The simulation shows the maximum realized gain of 11.1dBi is achieved in the beam direction (θ=28°, Φ=312°) of quadrant IV, and shows the beamforming can work properly by controlling the phase of signal in each antenna element.

III. MEASUREMENT RESULT

In this section, the beamformer IC and setup are firstly introduced. And then the simulated and measured results for the return loss, bandwidth, peak gain and the radiation patterns are presented and compared. The AUT is measured in an far field and a spherical of probing chamber that the AUT is placed on a low permittivity foam and measured by probing. Fig. 12 shows the measured and simulated return loss of the each elements for 2 by 2 antenna array. The results show our AiP design has better than 10 dB return loss in 23.5-30.5 GHz range, with ~7 GHz bandwidth for 28GHz band antenna. For 39GHz band antenna, the antenna has 7GHz bandwidth from 38-45 GHz. The measurement result demonstrates a good

agreement with simulation. Fig. 13 shows the measured gain pattern and peak gain for low band and high band antenna. The measured peak gain is about 6 dBi and 5dBi for low band and high band antennas, respectively.

For beamforming test, the beamformer IC can be flip-chip bonded on the BT substrate, combined with the antenna array. The beamformer IC is implemented by TSMC 90-nm CMOS technology. It is composed of one-way 4x4 Butler matrix, four LNA (low noise Amplifier), two VGA (Variable Gain path Selector), one Coupler and one SPDT Switch. Since the whole beamformer contains 4x4 Butler matrices, the circuit size of the single Butler matrix must be considerable minimized. The Chip size is 2.97x1.75mm². Fig. 14 shows the photographs of the fabricated AiP with a beamformer IC and test socket on load-board for beamforming test. The simulated progressive phase differences between the adjacent port are 45°±5°. The overall integrated 2x2 AiP is measured by CATR chamber (Compact Antenna Test Range), as shown in Fig. 15. The AiP module can be mount in the test socket. The absorbers are placed around the module to avoid the holder scattering. All the setups and acquired data are integrated on a measurement panel programming by NI LabVIEW. The calibration phases for each channels needed before beamforming measurement. Then, the circuit calibration and the realized gain calibration have to correct. Once the calibration is done, the beam performance are obtained. The measurement of beam steering

392

in all four quadrants at 28GHz, the maximum realized gain of 11.8 dBi is achieved in the main beam (θ=0°). The 3D beam steering is measured in all four quadrants at 28GHz as shown in Fig. 16.

IV. CONCLUSIONS

This paper demonstrates a 2x2 dual-band antenna array using 4+2+4 multi-layer buildup substrates for 5G mmWave applications. As a result, our stacked patch antenna with parasitic elements achieves broadband operation. The proposed 2 by 2 dual-band antenna exhibits a 7 GHz bandwidth with a 6 dBi peak gain per element within the 23.5-30.5 GHz for low band. For 39GHz band, the antenna has 7 GHz bandwidth and provides a 5 dBi gain per element from 38-45 GHz. In addition, we design the H-type slot structures onto antenna that it can improve interference and isolation between 28 and 39GHz bands. The measured result shows the isolation can larger than 15dB between low and high band. We also measured beam steering by controlling the phase of each element, which ensures that the phased array is working properly. The 3D beam steering of 2 by 2 antenna array is presented with multi-states at 28 GHz, with a maximum realized gain of 11.8 dBi achieved in the main beam (θ=30°).

ACKNOWLEDGMENT

The authors would like to acknowledge National Chung Cheng University, Lab543, Professor Chang for technical support to overcome a lot of testing issues, and successfully realize the proposed beamforming test.

REFERENCES

[1] 3GPP, "User Equipment (UE) radio transmission and reception; Part 1: Range 1 Standalone (Release 15)." Tech. rep vol. 3GPP TS 38.101-1 V1.0.0: NR, Dec. 2017.

[2] Y. Zhang and J. Mao, "An Overview of the Development of Antenna-in-Package Technology for Highly Integrated Wireless Devices," in *Proceedings of the IEEE*, vol. 107, no. 11, pp. 2265-2280, Nov. 2019.

[3] A. Fischer, Z. Tong, A. Hamidipour, L. Maurer and A. Atelzer, "77-GHz Multi-Channel Radar Transceiver With Antenna in Package," IEEE *Trans Antennas Propag.*, vol. 62, no. 3, pp. 1386 – 1394, March 2014.

[4] K. -Q. Huang and M. Swaminathan, "Antennas in Glass Interposer for sub-THz Applications," *in Proc. IEEE 71st Electron. Compon. Technol.Conf. (ECTC)*, pp. 1150-1155, 2021.

[5] R. Kulke et al., "24 GHz radar sensor integrates patch antenna and frontend module in single multilayer LTCC substrate," in Proc. *Eur. Microelectronics and Packaging Conf.*, Jun. 2005, pp. 239–242.

[6] X. Gu, D. Liu, and B. Sadhu, "Packaging and antenna integration for silicon-based millimeter-wave phased arrays: 5g and beyond," *IEEE Journal of Microwaves*, vol. 1, no. 1, pp. 123-134, 2021.

[7] S. -C. Hsieh, F. -C. Chu, C. -Y. Ho, W. -Y. Chen and C. -C. Wang, "mmWave AiP Measurement Turnkey Solution in Millimeter-Wave Wireless Communication Applications," *2020 IEEE 70th Electronic Components and Technology Conference (ECTC)*, pp. 114-119, 2020.

[8] I. J. Bahl and R. Garg, "A designer's guide to stripline circuits," Microwaves, pp. 90–96, Jan. 1978.

Analysis and Characterization of Castellated Holes as RF Interconnects for Modular Millimeter-Wave Devices

Paul Perlwitz
Research Center of Microperipheric
Technologies
Technische Universität Berlin
Berlin, Germany
paul.perlwitz@tu-berlin.de

Tschoban, Christian
RF & Smart Sensor Systems
Fraunhofer Institute for Reliability and
Microintegration IZM
Berlin, Germany
christian.tschoban@izm.fraunhofer.de

Ndip, Ivan
RF & Smart Sensor Systems
Fraunhofer Institute for Reliability and
Microintegration IZM
Berlin, Germany
ivan.ndip@izm.fraunhofer.de

Pötter, Harald
RF & Smart Sensor Systems
Fraunhofer Institute for Reliability and
Microintegration IZM
Berlin, Germany
harald.poetter@izm.fraunhofer.de

Schneider-Ramelow, Martin
Research Center of Microperipheric
Technologies
Technische Universität Berlin
Berlin, Germany
martin.schneider-ramelow@tu-
berlin.de

Abstract´ This paper presents an innovative approach to the use of Castellated Holes or Plated Half Holes (PHH) as transmission elements for high frequency interconnections between printed circuit boards in a frequency range of up to 40 GHz. The basic design was examined using commonly used design parameters for PCB modules with Castellated Holes which were then investigated and subsequently optimized to improve their RF performance, with a particular focus on design considerations according to today's industrial manufacturing capabilities. Therefor the FEM-based full wave simulation technique was used to compare and optimize the design parameters in the aim to improve the capability of PHHs as low-loss, low-disturbant RF interconnection alternative in modular PCB designs. Based on these results test samples were developed, which were subsequently measured and thus verified the findings of the simulation. To provide a general comparability, this study focused on the S-Parameter to quantify the performance in terms of transmission losses and reflection. With an optimized design feasible results were achieved with a maximum transmission loss of less than 2 dB at the maximum of the desired frequency range and about 1dB at 20 GHz over the entire signal path. These insights make PHHs suitable for seamless integration into modular designs in RF and millimeter-wave applications.

Keywords´ castellated holes, PHH, millimeter wave application, board to board soldering, RF

I. INTRODUCTION

In modern PCB design, the approach of modular design has emerged as a standard process due its the benefits of improving efficiency, flexibility and scalability throughout the product development lifecycle. By breaking down complex systems into individual subgroups with specific functions, developer and manufacturers are able to optimize and improve design, prototyping and production processes. The modular approach plays also a significant role in RF design. Sensitive circuits and components can be better shielded in contained modules, expensive RF substrates and special design rules do not have to be applied to the entire system, thus reducing the complexity and cost of the entire system. In addition, a modular design offers further key advantages: It allows the parallel development and testing of individual functional groups and also simplifies the

prototyping, debugging and validation processes. It also increases the reliablity, as many new developments often make use of existing and well-known subgroups and modules. This considerably reduces the development effort and fasten design processes. It also streamlines the production process and reduces the need for complex multi-layer PCBs, thereby reducing costs, simplifying assembly and increasing overall efficiency.

For connecting separate modules with each other or with the main system, several options exist: the widespread method of using connectors (with cables or directly to the board) or board-to-board soldering as a common technique among developers and manufacturers to integrate smaller PCBs into larger ones. Soldered connections serve hereby as an interface for establishing a fixed mechanical connection between the modules and the main system as well as providing the electrical connections for power supply or signal transmission. This can be achieved by using Ball Grid Arrays (BGA) or - the most common method - by using castellated holes, also known as PHHs. The latter approach is one important factor in establishing modular design processes, enabling scalability and facilitating customization of products, as it uses standard PCB manufacturing and assembly processes to providing an effective electronic packaging solution.

A. Comparison of RF interconnection approaches

Modular design is also well established in RF applications [4]. To establish a RF interconnection between two boards, specific RF connectors are used as the primary method. Despite their explicit function, their usage can result in signal degradation due to impedance mismatch and insertion loss, particularly at high frequencies or have to be optimized with great effort [1][2]. They also require more board space, leading to size constraints, and are susceptible to mechanical issues. In addition, the assembly process of connectors can be very complicated. BGA-based methods bring challenges, such as soldering complexity, limited accessibility after soldering, and potential thermal issues. Design complexity increases, especially for higher density BGAs, requiring careful unbundling and consideration of signal integrity [3].

PHHs are a good alternative for connecting two boards. These are inexpensive to manufacture and do not increase the complexity of the assembly. Despite their common usage as module interconnections, PHH are not yet quantified as RF transmission elements. To use the full capabilities of the modular approach, investigations in this field of view are needed. So the primary objective of this study is to comprehensively evaluate their applicability and effectiveness in terms of reflection parameters and transmission losses in this domain.

B. Via interconnects

Numerous studies have been published, investigating the profound impact of geometry variations in via interconnects within the RF domain [5][6][7]. These studies show relationships between PTH interconnect geometries - such as via diameter, length, and spacing - and their consequential effects on signal integrity, impedance matching, and overall RF circuit performance. Findings from these studies were used to understand and examine the influences of design parameters and the importance of well-defined return current paths.

II. Modeling and Simulation

In order to examine the usage of PHHs as RF interconnects, an investigation on which design parameters are typical for modules with castellated holes was conducted and used as a baseline for a virtual model. The primary goal is to explore the performance of the baseline model in the targeted frequency range, represented by S-Parameter with S11 for the reflection and S21 for the transmission. As secondary object an optimization of various design parameters for the baseline model is targeted. Based on this, the optimized virtual model should be transferred back to reality in which an optimized PHH design is to be created, manufactured and measured.

A. Simulation model

For the investigation of the high-frequency behavior of PHH, full wave simulations based on the finite element method (FEM) were performed in this work. This method provides the possibility to detect and observe local effects such as near-field coupling and impedance matching, which are critical in high frequency applications. In particular, the simulation software Ansys HFSS V 2023.1 was used for this purpose. As simulation results, the S-Parameters were analyzed and evaluated. As the solution setup, the Driven Modal solution type was employed. The simulation was carried out at a frequency of 40 GHz, employing a maximum of 15 passes and a targeted Delta S of 0.002.

In order to create a realistic scenario, a model consisting of two boards was designed, shown in Fig.1. The upper board features the PHH while the lower one provides the landing pads where the vias are soldered. To excite the signal properly, a short 50 'n impedance-matched grounded coplanar microstrip line was introduced for providing a connection between the PHH and the waveports as well as serving as contact pads for the later measurements with GSG probes. While this arrangement induced losses to the microstrip line, effective matching ensured that these losses remained within a few hundredths˘ decibel up to 40 GHz. In the subsequent measurement phase, these lines served the purpose of contacting the sample with the GSG probes.

As a typical RF design setup, a six-layered stack up was used while the upper substrate layer serves as the RF layer. Megtron 7 was therefore chosen as high frequency laminate, a commonly used material in RF applications due to its constant dielectric constant of 3.31 from 14 to 59 GHz [1] and exceptionally low loss tangent of 0.0025 at 14 GHz respectively 0.0028 at 36 GHz. As substrates for the layers below standard FR4 laminates were used. The RF substrate layer has a thickness of 300 ı m while the FR4 layers have a thickness of 200 ı m each. As a standard configuration for modules with castellated holes, a via diameter of 1 mm and a

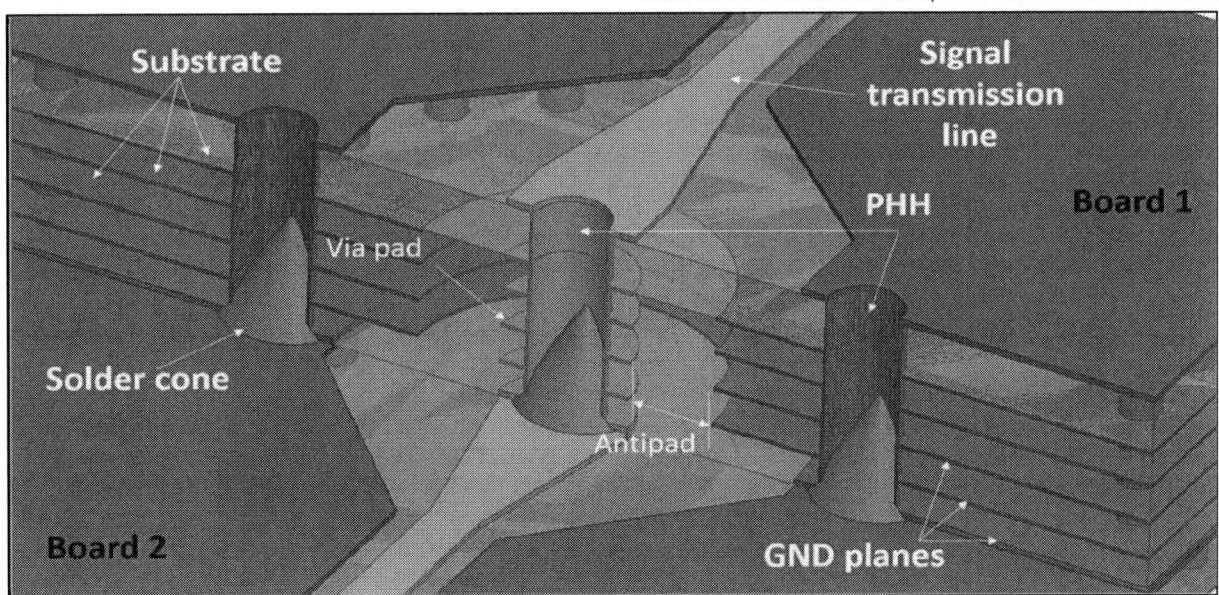

Fig. 1: Simulation model of a 6 layered and a 2 layered PCB with PHH, solder cones, and transmission line

---

[1] Panasonic Industry, Megtron7 Laminate R-5785(N) Datasheet, No. 22040127, pp. 2-3, Apr. 2022

pitch of 2.54 mm were used. The via pads have a restring of 125 ı m and the antipads a gap of 150 ı m. Additional shielding vias with a diameter of 200 ı m were distributed strategically within the PCB.

A separation of 10 ꞊m was established between the two boards models in vertical direction to represent the separation of the boards after the soldering process. To provide a realistic electrical connection between those two boards, solder cones (Solder Lead-Free Sn-3.5Ag with 8.12x10⁶ siemens/m) were introduced as a simple approximation to the later experimental situation.

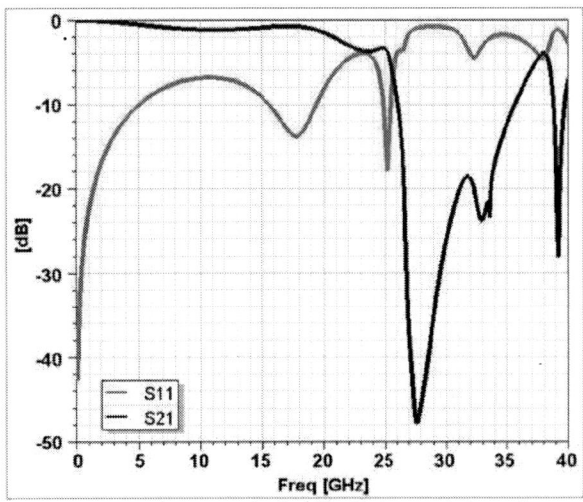
Fig. 2: Simulation results for standard PHH design values

The simulation shows a decent result for S21 (transmission) in a frequency range up to 25 GHz with less than -2 dB and -1 dB up to 20 GHz (Fig. 2). However, the S11 parameter shows rather unsuitable values with crossing the crucial -10 dB limit already at 5 GHz. For frequencies higher than 25 GHz this PHH configuration is completely unusable due to the high transmission losses and reflection, represented by S21 and S11 respectively. In [6] it was described, that these drops result from resonances occurring due to uncontrolled return currents.

B. Optimization by design parameter variations

To achieve better transmission performance by adjusting various design parameters, the specifications of PCB manufacturers for a conventional production process were considered to reach a realistic set-up and provide a practical relevance. In order to avoid reaching a multitude of solutions by examining a seemingly random variety of parameters and their combinations, a more methodical approach was chosen in which certain solutions were already determined from previous findings, so that the number of parameters to be varied is limited to a few.

In order to reach a S21 of better than -3 dB and a S11 of less than -10 dB over the whole frequency range, first measures were taken to ensure that the return current path is as short and homogeneous as possible, based on the results of previous studies [4][7]. For this purpose, the smallest possible via pitch for the PHH of 1.5 mm according to standard manufacturing guidelines was selected.

In order to reduce impedance mismatches and current path discontinuities in the transition between PHH and transmission line, the smallest possible via diameter, which is 400 ı m was chosen. By setting the viapad width of 125 ı m restring to a uniform and smallest possible value across all layers, several aspects are taken into account: Firstly, like mentioned above, the transition to the microstrip line is more advantageous since the latter is narrower than the viapad itself. On the other hand, it results in the via having a symmetrical structure, which leads to fewer discontinuities and achieves greater stability in the event of later production.

The remaining parameter that needs to be considered in this approach is the antipad diameter. An antipad stands for the distance between the pad and the pad surrounding ground plane in an inner layer. In detail the hereinafter mentioned diameter dimensions the outer diameter in the ground plane, while the inner dimeter is given by the via pad (see fig. 1). There are four inner layers whereby it is assumed that the two middle ones are also to be treated uniformly, which was checked with a short validation. This leaves the antipads on layer 2, layer 3 and layer 5, whose variation in diameters should be investigated. In the following, the antipad is described by its radius, starting from the center of the vias they surround.

The simulation result for the antipad radius combinations optimized by themselves and together are given as S21 in Table I; some specific are plotted in Fig. 3.

TABLE I.   S21 FOR SPECIFIC ANTIPAD RADII

Antipadradius [ı m]			S21 [dB]	
Layer 2	Layer 3 & 4	Layer 5	20 GHz	40 GHz
675	750	900	-0.38	-1.52
675	675	675	-0.46	-1.21
675	550	550	-0.64	-1.15
675	425	425	-1.25	-3.05
550	675	675	-0.4	-1.78
550	550	550	-0.576	-1.36
425	425	425	-1.06	-3.72

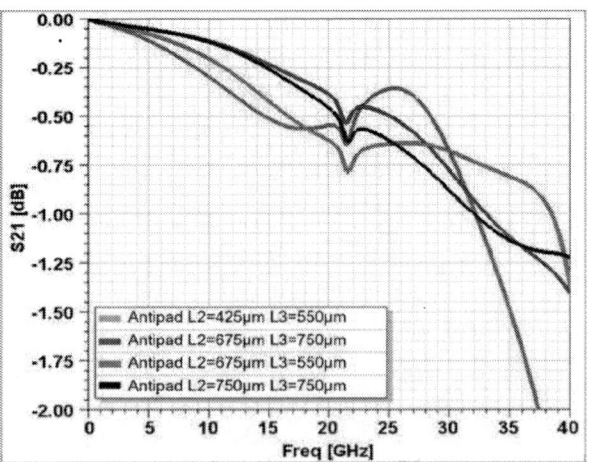
Fig. 3: Variation of antipad sizes to optimize the transmission with antipad diameter in layer 5 of 675ı m

Due to the considerations mentioned above and the simulation assisted optimization of the antipads a remarkable improvement of the transmission performance represented by S21 could be achieved. With optimized parameters the limit value of at least -3 dB can be maintained over the entire frequency range. The best values are ranging at -0.5 dB at 20 GHz and -1.2 GHz at the upper frequency range.

The sharp drop of S21 at around 21.3 GHz is an indicator that at some point in the system starts to resonate. It reveals one crucial point in this construction: The area in between both boards. Their only junction are the three soldering areas where the PHHs from the upper board are connected to the pads of the lower board. The two outer PHHs are connected to GND and providing the return current path. The area between the two boards acts like parallel plate capacitors, while the solder connection represents an inductance. At resonance frequency, return current is floating via displacement current through this gap and incite standing waves at resonance frequency. Depending on the height, surface condition and the lateral expansion, the impedance can be changed und thereby shifting the resonance frequency.

Parallel plate modes need to be avoided or reduced at the target frequency since they can cause a strong radiation of the PCB which would lead to serious EMI problems and a strong degradation of the transmission signal at this specific frequency as it is been shown in Fig. 4.

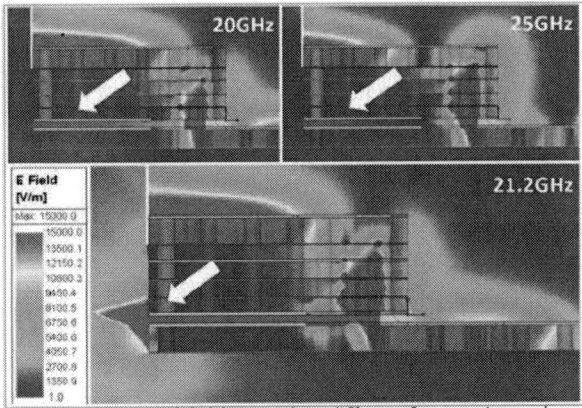

Fig. 5: Electrical field strength at different frequencies and at resonance frequency of 21.2 GHz

The parallel plate modes can be influenced by adding additional GND connections between both boards near the signal path to reduce the return current discontinuities. For PHH technolog this is only possible by adding additional connections in this area at the lower board. Due to its lack of feasibility with a focus on keeping the design as simple as possible, this approach is just mentioned here but not pursued further. Changing the size of the parallel plate section is another possibility to influence the resonances. By changing their overlapping areas, we can can reduce the strength of the signal degradation or shifting the resonance frequencies into more uncritical areas (Fig. 5), which becomes relevant in development practice when just specific frequencies are used. In practice a decrease of the overlapping area can be reached by a cavity underneath the RF board.

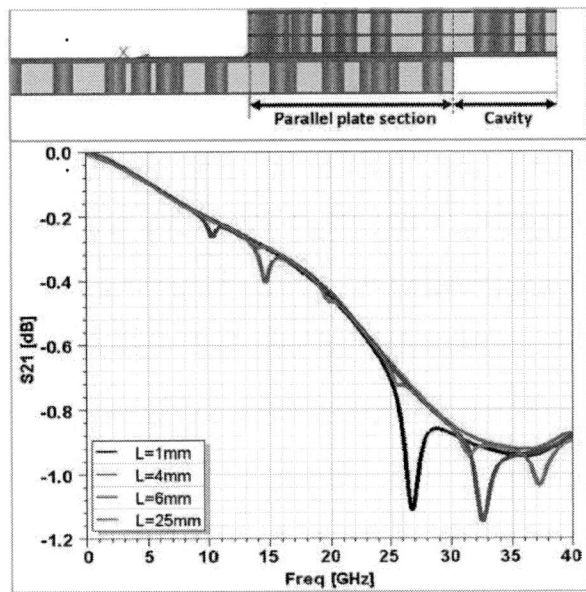

Fig. 4: Influence on S21 of varying the length L of the parallel plate section

C. Consideration of unintendet parameter variations

Fig.6 shows a typical via interconnect with surrounding copper and substrate layers. The drilling process and subsequent plating leads typically to non-exact via diameters. The offset of via to the previously manufactured pads in the individual layers can be up to 50 ı m. Additional smearing effects and copper erosion caused by the milling process in order to create the PHH are difficult to represent in a simplified model and can therefore not be considered in the simulations. Additional factors such as the etching factor and the inherent tolerances of the lithography process can

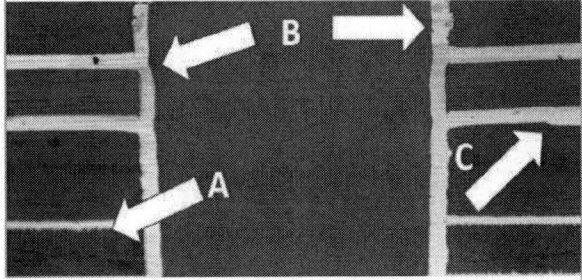

Fig. 6: Microsection of a via interconnect with various manufacturing effects. A: Increased roughness in a single layer B: Discontinuities in via cylinder C: Discontinuities in copper layer

contribute to the degradation of the originally designed structures and the resulting PCB samples. These cumulative effects have the potential to collectively influence the final performance of PCB samples beyond the confines of idealized simulations.

Surface roughness emerged as a relevant factor [8], initially ignored in the model for simplicity. Subsequent investigations revealed a certain impact, shown in a comparative simulation of a 6.8 mm long coplanar microstrip line - representing the overall transmission line length in the PHH model - simulated with different surface roughness models: Huray Surface Roughness Model (Nodule

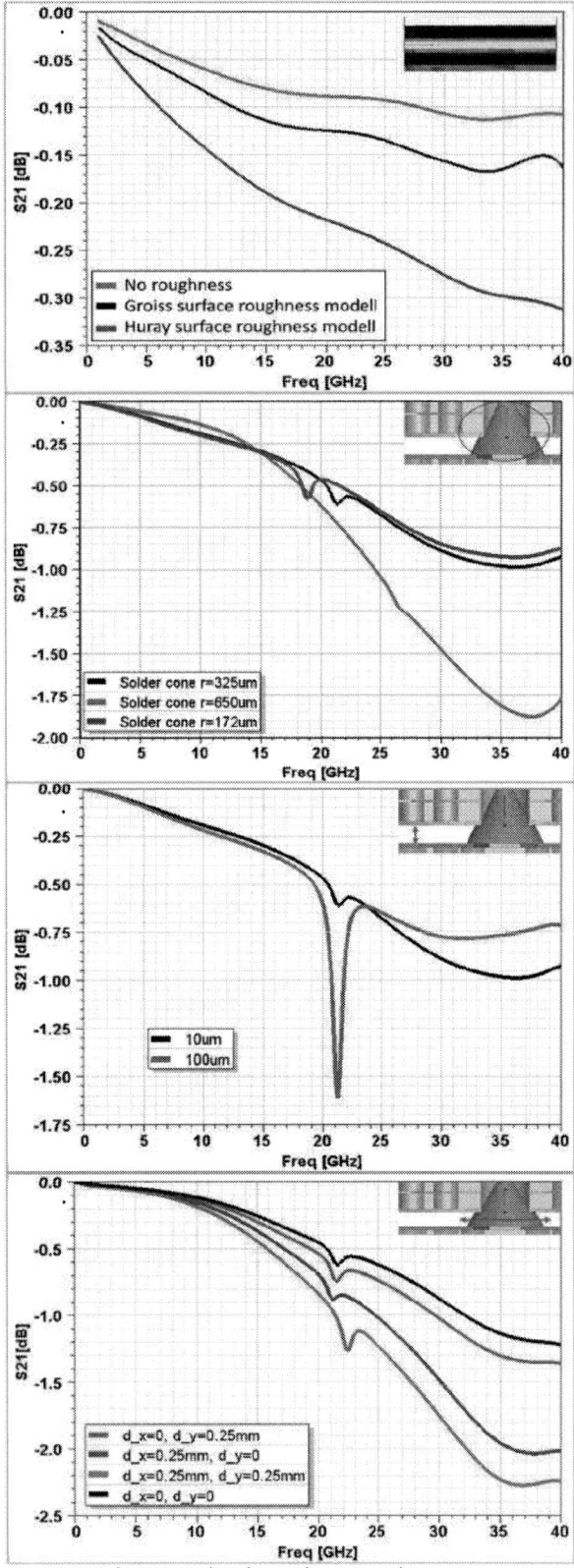

Fig. 7: Simulation results of manufacturing tolerance parameters. a) surface roughness of a microstrip line b) solder mass c) airgap d) misalignment between boards

Radius=2 ι m, Hail-Huray-Surface-Ratio =2.9) and the Groiss Surface Roughness Model (Surface roughness=2 ι m). Clearly, roughness has a impact on transmission losses of around 0.1 to 0.2 dB in the simulated frequency range, confirming its importance (Fig. 7a).

The assembly process will lead in the reality to problems with accurate component alignment, solder distribution and overall solder quality. To quantify the influences of the solder mass distribution, a simulation was done with different sized solder cones. The influence of this can be shown by varying the diameter of solder cone. In the frequency range considered here, a doubling of the losses can be observed over most of the range, with an absolute increase in transmission loss of about 1 dB at the upper end of the frequency range for a bigger solder mass distribution. Decreasing the solder mass size has es smaller effect on the transmission characteristics though we observed a small performance increase in the lower half of the frequency range and a better one in the upper half (Fig. 7b).

We have identified the gap between the two PCBs as another possible cause of a change in transmission behavior, which can occur if the PCBs are not placed exactly on top of each other, for example due to warping, coatings on the PCBs in other areas or an inaccurate soldering process. Therefore, a simulation was conducted to investigate the influence of changing the airgap (Fig. 7c). In the lower frequency range, minimal differences are observed. However, in the vicinity of the resonance frequency described earlier, a noticeable deterioration of the transmission quality occurs.

At the higher end of the considered frequency range, the changed parallel plate capacitance resulting from the larger gap actually improves the transmission behavior by up to 0.25 dB in comparison to the smaller gap.

Further investigations were carried out to evaluate the influence of misalignment of the two boards, a typical source of error caused by manufacturing tolerances from manual assembly. A simulation setup was created in which the lower PCB was either shifted laterally (y-direction), forward or backward (x-direction), or a combination of both. The displacement value was assumed to be 250 ι m. The simulation results indicate a clear influence on the transmission behavior. A shift in the y-direction has a stronger influence than x-direction, since discontinuities in signal propagation direction having a stronger effect. A slight shift in the resonant frequencies can also be seen (Fig. 7d), caused by the changed current path and impedance in the area of the return vias and the transition between both boards. In the upper range of the frequencies considered here, there is a significant degradation of the transmission performance of up to 1.1 dB additional losses for a simultaneous misalignment in both directions.

These investigations show that in some scenario's significant deviations from the idealized simulation model are to be expected in a realistic setup. The large number of parameters and possible sources of error and their mutual amplification and compensation do not allow an exact simulation of the test samples and mean that only an expected range of transmission losses can be given. It could be shown that this influence is clearly noticeable, but through a proper optimization of the PHH model based on the design parameters considered above, compliance with the limits of -3dB for S21 is achievable despite negative influences that cannot be influenced by the design.

III. Measurment and Discussion

In order to prove the applicability of the PHH we have also conducted measurements in the considered frequency range. Therefor a board set up was created, which included a selection of PCBs with different structures to cover the range of variation that includes several variations of the pad/antipad ratios as well as variations in line width and different solder mask configurations (Fig. 8).

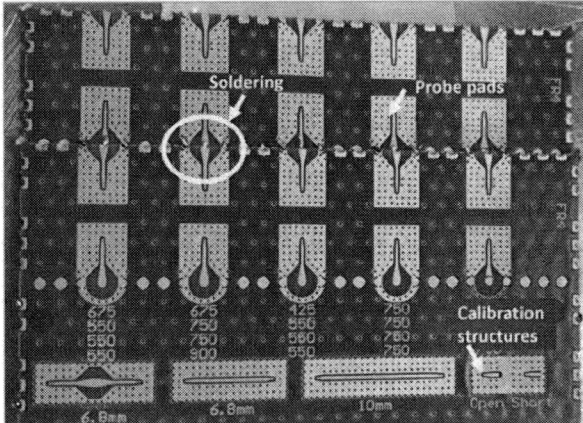

Fig. 9: Soldered test samples

It consists of the layer stack up described in the previous chapters with Megtron7 on top and 4 FR4 substrate layers. In addition, structures were applied for calibration and comparative measurements. They were manufactured in a standard process, also to ensure their practicality for possible future applications. The samples were subsequently soldered manually with a standard soldering station.

High-resolution S-Parameter measurements were carried out using the Keysight PNA-X Series Network Analyzer instrument with the Millimeter Test Set and the LRRM Calibration to test the actual S parameters of the PHH, utilizing 250ı m GSG probes. The frequency range considered was 1 to 40 GHz, with a monitored IF Bandwidth 30 Hz (Fig. 9).

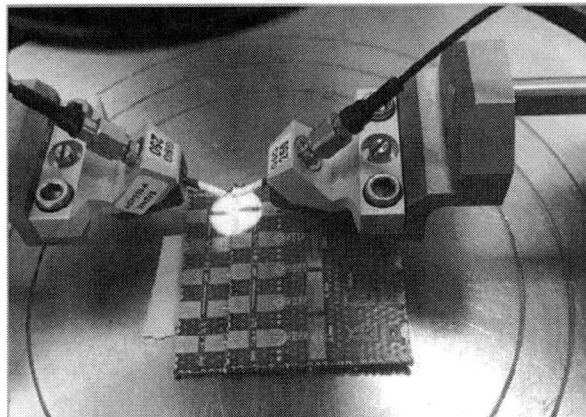

Fig. 10: Measurement of test sample on probe station with 250um GSG probes

To quantify the quality of the PCB and the test setup we measured the 6.8 mm microstrip line on the test sample, which corresponds to the total length of the test set-up of the test structures excluding the PHH. For the Megtron 7 we are ranging at 0.1 to 1.4 dB transmission losses (S21) and a maximum of -10 dB reflection (S11) at 33.4 GHz (Fig. 10).

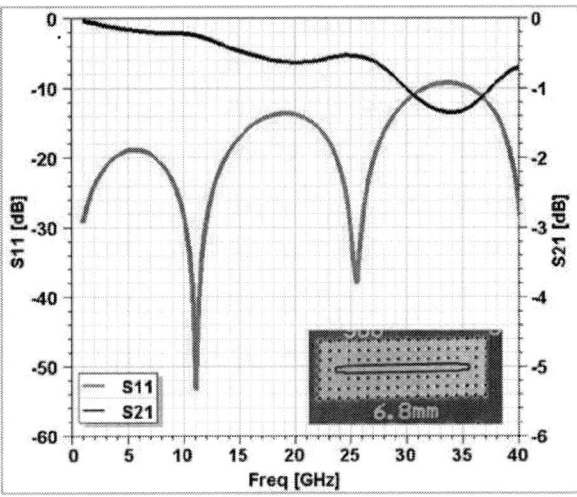

Fig. 8: S11 and S21 parameters of the transmission line

Subsequently the PHH structures were measured. Here we distinct between optimized und unoptimized structures to reach a comparability of the various design parameter and their improvements, based on the findings from the simulation analysis. Optimized means, according to simulation results, a improved viapad-antipad ratio to provide a proper distribution of the electrical field and currents in the PCB. Be varying the radii of the antipads in different layers, the improvement in the transmission results could be increased, so different variations of combinations of antipads were used, following referred as `antipads equal_ and `antipads vary_ configuration.

The measurement of the PHH structures themselves gives a good approximation to the performance in the observed frequency range according to the simulation results, so the S21 ranges about -1 dB at 20 GHz and -1.5 to -2 GHz at the upper side of the range for the optimized structure and is therefore significantly better than the target limit of -3dB as it is been shown in Fig. 11. The reflection, represented by the S11, ranks below the limit value for the optimized structure with equal antipads, in this case 625ı m. Comparing this with the measurement on the pure microstrip line of the same length, it can be shown that the signal degradation due to the PHH can be significantly reduced through an appropriate design of it. The conducted measurements show clearly, which improvement can be done by optimizing the design parameter of PHH, even more if the single antipads were optimized. However, even the unoptimized structures in terms of antipad radii are showing acceptable values for a frequency range up to 30 GHz.

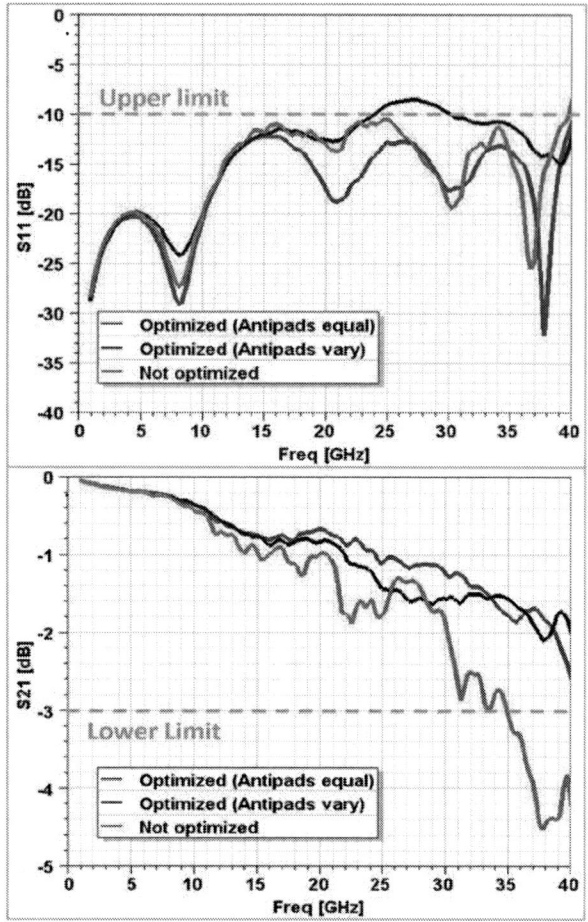

Fig. 11: Measurement of a) S11 (reflection) and b) S21 (transmission) of the optimized and not optimized PHH test samples

IV. Conclusion

In this study it could be demonstrated that the integration of Castellated Holes as RF interconnections for board-to-board transmission element up to 40 GHz is a viable option. The simulations on an idealized model show a good performance with losses of around 1.5 dB at 40 GHz which is comparable to other connection methods like BGA board-to-board soldering or the usage of connectors. Furthermore, measurements confirm that a S21 of around -1 dB at 20 GHz respectively 2 dB at 40 GHz are feasible with the

optimizations described here. The insertion, represented by S11 could be significantly reduced to values below -10dB compared to not optimized structures. With this results a good alternative for board-to-board connections could be evaluated, which provide a powerful feature for modular design of RF applications in terms of cost efficiency and design effort. Further design improvements to increase the efficiency were suggested but need further research as well as examinations in order to widening the frequency range.

References

[1] R. Ben, S. Hu, X. Li and Z. Fu, "Signal intergrity analysis for SMA via on the PCB," 2017 IEEE 9th International Conference on Communication Software and Networks (ICCSN), Guangzhou, China, 2017, pp. 865-869, doi: 10.1109/ICCSN.2017.8230235.

[2] S. Park et al., "Signal integrity analysis of high-speed board-to-board floating connectors for automotive systems," 2017 IEEE 26th Conference on Electrical Performance of Electronic Packaging and Systems (EPEPS), San Jose, CA, USA, 2017, pp. 1-3, doi: 10.1109/EPEPS.2017.8329718.

[3] S. Dilek et al., "Influence of Ball Size and Geometry on the Reliability and RF Performance of mmWave System-in-Package: A Simulation Approach," 2021 22nd International Conference on Thermal, Mechanical and Multi-Physics Simulation and Experiments in Microelectronics and Microsystems (EuroSimE), St. Julian, Malta, 2021, pp. 1-4, doi: 10.1109/EuroSimE52062.2021.9410849.

[4] C. Tschoban et al, "Development of a 60GHz MIMO Radar Packaging Concept," 2021 23rd European Microelectronics and Packaging Conference & Exhibition (EMPC), Gothenburg, Sweden, 2021, pp. 1-5, doi: 10.23919/EMPC53418.2021.9584998.

[5] M. Cracraft, S. Connor and B. Archambeault, "Unintended effects of asymmetric return vias and via array design for reduced mode conversion," 2014 IEEE International Symposium on Electromagnetic Compatibility (EMC), Raleigh, NC, USA, 2014, pp. 250-255, doi: 10.1109/ISEMC.2014.6898979.

[6] I. Ndip et al., "Modeling, Quantification, and Reduction of the Impact of Uncontrolled Return Currents of Vias Transiting Multilayered Packages and Boards," in IEEE Transactions on Electromagnetic Compatibility, vol. 52, no. 2, pp. 421-435, May 2010, doi: 10.1109/TEMC.2010.2049069.

[7] P. Perlwitz, et al., `Analytical modeling and measurement of vias in PCB-Technology up to 20 GHz‚, in Smart Systems Integration 2018. International Conference and Exhibition on Integration Issues of Miniaturized Systems, April 2018

[8] B. Curran, I. Ndip, S. Guttowski and H. Reichl, "On the quantification and improvement of the models for surface roughness," 2009 IEEE Workshop on Signal Propagation on Interconnects, Strasbourg, France, 2009, pp. 1-4, doi: 10.1109/SPI.2009.5089851.

High-Q Ku-band Microstrip Spiral Resonator in Fan-out Wafer-Level Packaging (FoWLP) Technology for VCO Applications

Mykola Chernobryvko
Fraunhofer Institute for Reliability and Microintegration
Berlin, Germany
mykola.chernobryvko@izm.fraunhofer.de

Michael P. Kaiser
Fraunhofer Institute for Reliability and Microintegration
Berlin, Germany
michael.kaiser@izm.fraunhofer.de

Kavin Senthil Murugesan
Fraunhofer Institute for Reliability and Microintegration
Berlin, Germany
kavin.senthil.murugesan@izm.fraunhofer.de

Dan Kuylenstierna
Chalmers University of Technology,
Gothenburg, Sweden
dan.kuylenstierna@chalmers.se

Julia-Marie Köszegi
Fraunhofer Institute for Reliability and Microintegration
Berlin, Germany
julia-marie.koeszegi@izm.fraunhofer.de

Robert Gernhardt,
Fraunhofer Institute for Reliability and Microintegration
Berlin, Germany
robert.gernhardt@izm.fraunhofer.de

Tanja Braun
Fraunhofer Institute for Reliability and Microintegration
Berlin, Germany
tanja.braun@izm.fraunhofer.de

Ivan Ndip
Fraunhofer Institute for Reliability and Microintegration
Berlin, Germany
ivan.ndip@izm.fraunhofer.de

Martin Schneider-Ramelow
Fraunhofer Institute for Reliability and Microintegration
Berlin, Germany
martin.schneider-ramelow@izm.fraunhofer.de

Abstract— In this paper, a planar spiral resonator operating at 13 GHz is investigated. The resonator was fabricated using fan-out wafer level packaging (FoWLP) technology. The field analysis highlights the importance of via fence. Two thick film resistors suppress an undesired resonance in proximity of targeted one. The measurements demonstrate a very good correlation in comparison with full-wave simulations. The measured Q-factor defined based on analysis of feedback oscillatory system is about 48. The proposed configuration is suitable for voltage-controlled oscillators (VCOs) with a resonator realized in package.

Keywords—resonator, Q-factor, Ku-band, fan-out wafer level packaging, VCO

I. INTRODUCTION

Modern trends in miniaturisation and demands for higher resolution have prompted radar technologies to explore and utilise W-band frequencies, i.e. 75-110 GHz, to exploit the higher available bandwidth [1], [2]. Moreover, due to low atmospheric losses W-band frequencies provide better Signal-to-Noise ratio (SNR) than the previously used frequency bands at around 24 GHz and 60 GHz. Consequently, current research activities are concentrated on the development of components for novel radar systems operating at the W-band [3].

A crucial part of radar systems is the frequency synthesis with low phase noise, e.g. by applying Phase-Locked-Loops (PLLs), where voltage-controlled oscillators (VCOs) are required in order to relate the phase of the output signal to a reference input signal [4]. Therefore, improving the phase noise performance of the VCO is beneficial for the purity of the frequency synthesiser's output signal. A VCO can be implemented on chip-level in combination with a package-integrated resonator. Due to the advantageous packaging material properties, a high Q-factor can be achieved providing better phase noise performance in contrast to resonators realized on-chip. To overcome insufficient gain of active devices at high frequencies, the frequency synthesis can occur at lower frequencies and the output signal is then transferred

to the actual carrier through frequency multipliers. As shown in [5] a 77 GHz carrier can be generated at Ku-band, i.e. at about 13 GHz, followed by a x6 frequency multiplication. Thus, the requirements on the VCO design become more cost-efficient and flexible as well as its external resonator is more robust towards fabrication tolerances providing wider tuning ranges.

Various resonator configurations being suitable for package integration were investigated in literature. Substrate integrated waveguide (SIW) resonator [6] has the potential to be a good candidate due to its excellent power handling characteristics. However, such resonator requires manufacturing of via rows with high precision in the alignment. Oscillators based on dielectric resonator (DROs) are known as preferable for signal synthesis [7]. Though, such configurations are complex and expensive due to challenges in realization of low-loss dielectrics with high dielectric permittivity. Alternatively, the planar spiral microstrip resonator described in [8] is simple in realisation. Nevertheless, a thorough analysis of this configuration is essential in which Q-factor is defined based on the phase slope of the transfer function. This is necessary for developing of VCO with low phase noise performance.

In order to realise a VCO with an off-chip resonator a suitable packaging technology has to be chosen. Recently, Fan-out Wafer-Level Packaging (FoWLP) has emerged as a key technology for the integration of passive high frequency components [9], [10] offering a great opportunity for planar resonator realisation. Such structures can benefit from the advantageous shorter RF connection path to IC avoiding wire bonds or bumps, what is crucial in passive structures integration. Some research activities have already demonstrated the importance of passive components integration into a FoWLP showing an improved Q-factor [11], [12].

In this contribution, a FoWLP-based resonator is investigated operating at 13 GHz. The feeding microstrip line (MSL) and electro-magnetically coupled to it spiral form a planar resonator in the redistribution layer (RDL). The resonator's reference layer is located on a metallised epoxy molding compound (EMC) on top of the RDL. To suppress an undesired resonance of the planar resonator, two thick film resistors were embedded into the EMC. The measurements of the fabricated sample are in good correlation with modelling showing high Q-factor.

The paper, first, presents the definition of Q-factor and the FoWLP-based planar spiral resonator configuration with detailed explanations of the stack-up in Section II. Moreover, the importance of field analysis is discussed. In Section III, results of the full-wave simulation and the measurement of the fabricated structure are highlighted. Section IV finalises the paper with conclusions of the investigation.

II. RESONATOR CONFIGURATION AND MODELING

A. Definition of Q-Factor

The investigated Ku-band resonator is developed to serve as an off-chip LC tank for a VCO. Therefore, a proper definition of the term Q-factor is essential. The package-integrated single port spiral resonator can be considered as a feedback system. For the feedback oscillatory system, the open loop Q-factor is defined based on the phase of the transfer function

$$Q = (\omega_0/2)(d\phi(H)/d\omega), \qquad (1)$$

with ω_0 defining the resonance frequency of LC tank and $d\phi(H)/d\omega$ is the phase slope of the resonator transfer function [13], [14]. Taking into account that the designed resonator is a single port structure, Z_{11} of impedance parameters is used as a transfer function, which leads to the final definition of the Q-factor to be

$$Q = (\omega_0/2)(d\phi(Z_{11}(\omega))/d\omega). \qquad (2)$$

It is worth to mention that the reflection S_{11} of S-parameters cannot serve as a transfer function due to the dependency on the port impedance.

B. Stack-Up and Materials

The proposed planar microstrip spiral resonator was developed for the integration in the RDL of a FoWLP-based platform [15]. A stack-up of the applied package configuration is illustrated in Fig. 1. The RDL consists of three polyimide (PI) layers and two copper layers. The thickness of each PI layer is 8 μm. The copper layers are denoted as Cu Layer 2 and Cu Layer 3 are 3 μm and 5 μm, respectively. On top of the RDL, an EMC of 300 μm

thickness is applied being metallised with another Cu layer of 3 μm thickness (Cu Layer 1). The PI material was characterized using the procedure described in [16] with corresponding dielectric permittivity $\varepsilon_{r,PI} = 3.5$ and dielectric loss $\tan\delta_{PI} = 0.008$ at 12 GHz. Similarly, the EMC properties were extracted at 12 GHz yielding the permittivity $\varepsilon_{r,EMC} = 3.66$ and the dielectric loss $\tan\delta_{EMC} = 0.004$.

The Cu Layer 1 acts as the reference for the resonator in the RDL. The MSL and the electromagnetically coupled spiral were realized on Cu Layer 3. For the vertical interconnections between Cu Layer 3 and Cu Layer 2 standard single via transitions were implemented, whereas the interconnections between Cu Layer 2 and Cu Layer 1 are challenging, since via rows through the PI Layer 1 must be properly aligned with the through mold via (TMV). In this case, the TMVs have not been realised by drilling and metallizing the openings, but have thus been manufactured separately as pick-and-place components. Further details of this procedure are described in [17].

C. Modelling and Simulations

The design of a planar microstrip spiral resonator is shown in Fig. 2. Fig. 2 (a) illustrates a top view of the simulation model. As can be seen, the feeding of the resonator starts with a coplanar waveguide (CPW) section to ensure a proper connection to the pads of an integrated circuit (IC), which are typically in Ground-Signal-Ground (GSG) configuration. Another reason for the CPW feeding is the need to measure the planar resonator in stand-alone test structures without any VCO in a wafer probing station. The measurements were performed using RF probes whose tips are also available in GSG configuration. The CPW is followed by an MSL section. Both transmission line sections designed to be 50 Ω were matched using tapering. The resonator itself consists of a non-uniform, spirally bent transmission line electro-magnetically coupled to the MSL. The gap between MSL and resonator is 20 μm. The resonator produces series and parallel resonances in proximity to each other. One of them should be damped to ensure that a VCO employing such resonator generates oscillations without spurious components. In order to supress spurious resonance, two 80 Ω surface mount device (SMD) resistors are placed in parallel at the end of the MSL and tied to ground. The side view of the modelled configuration including one of the SMD resistors is shown in Fig. 2 (b).

The electromagnetic simulations were carried out with Ansys HFSS, a 3D Finite Element Method (FEM) full wave solver. The modelling includes also optimisation of the resonator. Particularly, the position of the spiral along the MSL feeding line and spiral shape were tuned to the best Q-factor. Moreover, the field analysis is essential for such a configuration, since the resonator is not shielded. Therefore, the design should ensure that the structure operates without significant leakage, which potentially influences the operation of IC or further system components. In Fig. 2 (c), the magnetic field at the resonance frequency of 12.6 GHz is visualised showing a high magnitude at the spiral cross section, indicating a proper functioning of the modelled resonator. Additionally, the H-field is shown in the plane crossing the protective via fence to analyse its effectiveness. As can be seen, the field is only strong at the region of the CPW feeding line demonstrating that the via fence blocks properly undesired penetration of electromagnetic fields. The via fence

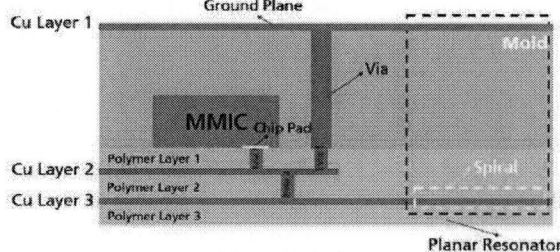

Fig. 1: Stack-up for a planar spiral resonator in FoWLP technology.

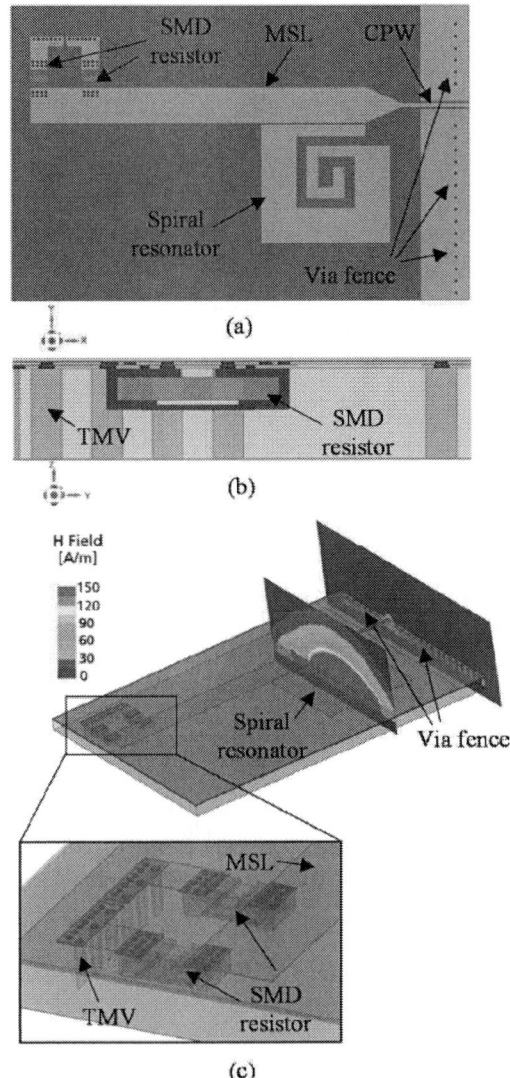

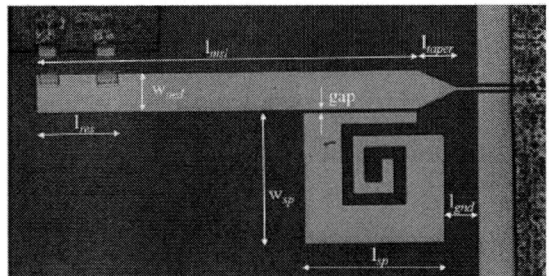

Fig. 3. Top view on fabricated Ku-band planar spiral resonator in FoWLP technology.

room temperature. The VNA was calibrated with the short-open-load-through (SOLT) method by means of a standard calibration substrate. This ensures the reference impedance of 50 Ω and the reference planes to be near the tip of the probe.

The VNA measures S-parameters that can be converted to Z-parameters to apply (2) for Q-factor extraction. The simulated and measured curves are shown in Fig. 4. Note that the model was re-simulated using the actual measured values from the fabricated sample. Fig. 4 (a) displays the reflection S-parameters. A resonance at 12.9 GHz can be identified in the simulated results, whereas the measured structure reaches its resonant behaviour at approximately 13.1 GHz, hence, a shift of about 200 MHz can be identified. The measured return losses are below -10 dB in the frequency ranges 10.6-11.9 GHz and 13.6-13.8 GHz while simulated return losses are lower than -10 dB in the bands of 10-12 GHz and 13.2-15 GHz. Lower values of return losses outside the resonance frequency result in better purity of oscillations generated by VCO. Furthermore, the small offset can be seen even more clearly in Fig. 4 (b), where the comparison between simulated and measured Q-factor of the planar spiral resonator is shown. The spurious resonance at approximately 13.55 GHz could not be suppressed perfectly. Nevertheless, a measured Q-factor of 48 is achieved in this configuration, which is approximately two times higher than on-chip resonators demonstrate [18].

The deviation between simulation and measurement is associated with the modelling of the SMD resistors. During the design phase, they were assumed to be ideal components. However, real resistors include some parasitics, e.g. due to

Fig. 2. Ku-band planar spiral resonator configuration operating at 12.6 GHz. (a) Top view. (b) Cross sectional view including SMD resistor. (c) Isometric view with cross sections representing H field and details for embedded SMD resistors.

was introduced on design stage in the region where leakage of fields at resonance frequency was noticed.

III. RESULTS OF SIMULATION AND MEASUREMENT

Fig. 3 shows a top view on the fabricated planar spiral resonator including the embedded SMD resistors. A comparison between the design and the actual manufactured geometrical values are provided in TABLE I. In general, the fabricated structure shows a good correlation to the design values with a maximum deviation of approximately 4%. However, the spacing between the MSL and the spiral (gap) being the smallest value in TABLE I is doubled due to over-etching.

The RF measurements were performed using the Keysight Vector Network Analyser (VNA) E8361A, a wafer prober system and a 250 μm pitch GSG probe. The measurements were carried out from 10 GHz to 15 GHz at

TABLE I.   COMPARISON OF MEASURED GEOMETRICAL PARAMETERS WITH RESPECT TO NOMINAL DESIGN OF THE SPIRAL RESONATOR

	Design value, (mm)	Measured value, (mm)	Var. w.r.t. design value (%)
l_{msl}	5.6	5.58	-3.5
w_{msl}	0.6	0.585	-2.5
l_{taper}	0.6	0.59	-1.7
l_{res}	1.15	1.165	1.3
gap	0.02	0.04	100
l_{sp}	2.14	2.12	-0.9
w_{sp}	2	1.98	-1
l_{gnd}	0.49	0.51	-4.1

403

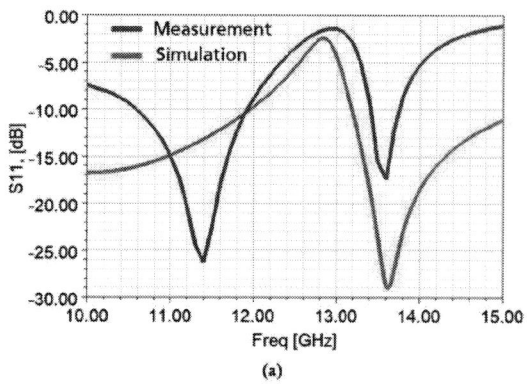

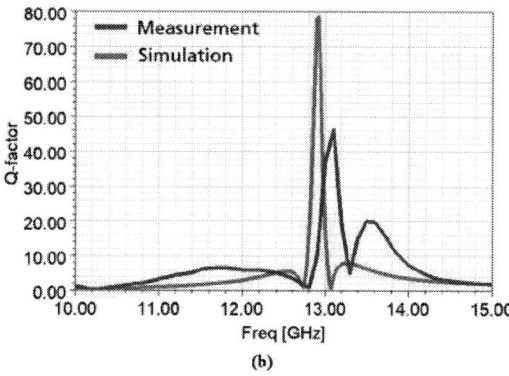

Fig. 4. Comparison of simulated and measured results for the planar spiral resonator in FoWLP. (a) Reflection S-parameters. (b) Q-factor.

their packaging, which can be modelled as series inductance and parallel capacitance to the resistor. Therefore, the SMD resistors must also be thoroughly investigated to model more accurate the planar resonator. In addition, since Q-factor is dependent on how fast the phase of Z_{11} is changing (5), any inaccuracy in modelling might significantly influence the phase slope close to a resonant frequency. Consequently, measured Q-factor values are different from the simulation results.

IV. CONCLUSION

In the present work, a planar spiral resonator was successfully designed, fabricated and characterised in FoWLP. The measurements show a good correlation to the full-wave simulations. A measured Q-factor of 48 shows a significant improvement compared to on-chip solutions. This reveals a great potential of FoWLP for the implementation of VCOs with package-integrated resonators. Starting point for future investigations is an improved modelling of the SMD resistors, which considers their parasitic properties or even approaches making the discrete components obsolete. In addition, a sensitivity analysis should be performed for the expected process and manufacturing tolerances to determine limits within which the planar spiral resonator demonstrates stable performance.

ACKNOWLEDGMENT

This work was financed by the CleanSky 2 program within the European Union's Horizon 2020 research and innovation framework, grant agreement no. 821270 "GaN mm-wave Radar Components Embedded - GRACE".

REFERENCES

[1] D. A. Murrell, S. A. Lane, N. P. Tarasenko and C. Christodoulou, "A review of spaced based RF propagation experiments and examination of a new interest in W/V band (40–110 GHz) studies," in 2016 IEEE International Symposium on Antennas and Propagation (APSURSI), 2016.

[2] I. Kallfass, et. al., "Towards the Exploratory In-Orbit Verification of an E/W-Band Satellite Communication Link," in 2021 IEEE MTT-S International Wireless Symposium (IWS), 2021.

[3] T. N. Thi Do, Y. Yan and D. Kuylenstierna, "A low phase noise W-band MMIC GaN HEMT oscillator," in 2020 IEEE Asia-Pacific Microwave Conference (APMC), 2020.

[4] C. Zech, et. al., "A compact W-band LFMCW radar module with high accuracy and integrated signal processing," in 2015 European Microwave Conference (EuMC), 2015.

[5] J.-H. Song, C. Cui, S.-K. Kim, B.-S. Kim and S. Nam, "A Low-Phase-Noise 77-GHz FMCW Radar Transmitter With a 12.8-GHz PLL and a x6 Frequency Multiplier," IEEE Microwave and Wireless Components Letters, vol. 26, no. 7, p. 540–542, 2016.

[6] Z. Chen, W. Hong, J. Chen and J. Zhou, "Design of High-Q Tunable SIW Resonator and Its Application to Low Phase Noise VCO," IEEE Microwave and Wireless Components Letters, vol. 23, no. 1, p. 43–45, 2013.

[7] S. Huang, Z. Shi, X. Zhang, X. Yu and P. Chen, "DRO Design for Millimeter Wave Terminal Communication System," in 2022 International Conference on Microwave and Millimeter Wave Technology (ICMMT), 2022.

[8] M. S. Abouyoussef, A. M. El-Tager and H. El-Ghitani, "Quad spiral microstrip resonator with high quality factor," in 2018 35th National Radio Science Conference (NRSC), 2018.

[9] T. Braun, K.-F. Becker, M. Wohrmann, M. Topper, L. Bottcher, R. Aschenbrenner and K.-D. Lang, "Trends in Fan-out wafer and panel level packaging," in 2017 International Conference on Electronics Packaging (ICEP), 2017.

[10] T. Braun, et. al. "Fan-out wafer level packaging for 5G and mm-Wave applications," in 2018 International Conference on Electronics Packaging and iMAPS All Asia Conference (ICEP-IAAC), 2018.

[11] K. S. Murugesan, et. al., "High Quality Integrated Inductor in Fan-out Wafer-Level Packaging Technology for mm-Wave Applications," in 2020 50th European Microwave Conference (EuMC), 2021.

[12] M. Wojnowski, V. Issakov, G. Knoblinger, K. Pressel, G. Sommer and R. Weigel, "High-Q embedded inductors in fan-out eWLB for 6 GHz CMOS VCO," in 2011 IEEE 61st Electronic Components and Technology Conference (ECTC), 2011.

[13] T. Ohira, "What in the World Is Q? [[Distinguished Microwave Lecture}," IEEE Microwave Magazine, vol. 17, no. 6, p. 42-49, 2016.

[14] B. Razavi, "A study of phase noise in CMOS oscillators," IEEE Journal of Solid-State Circuits, vol. 31, no. 3, p. 331–343, 1996.

[15] R. Gernhardt, et. al., "An Overview About the Excimer Laser Ablation of Different Polymers and Their Application for Wafer and Panel Level Packaging," in 2019 International Wafer Level Packaging Conference (IWLPC), 2019.

[16] A. Kanitkar, et. al., "Fork-Coupled Resonators for Characterization of Mold Material for 5G Applications," in 2020 23rd International Microwave and Radar Conference (MIKON), 2020.

[17] T. Braun, et. al., "3D stacking approaches for mold embedded packages," in 18th European Microelectronics & Packaging Conference (EMPC), 2021.

[18] I. Mansour et. al., "70 % Improvement in Q-Factor of Spiral Inductor and its Application in Switched K-Band VCO Using 0.18 µM CMOS Technology" in 2018 Asia-Pacific Microwave Conference (APMC), 2018.

Thermal-Mechanical Analysis of a Power Module with Parametric Model Order Reduction

Sheikh Hassan[1,*], Pushparajah Rajaguru[1], Stoyan Stoyanov[1], Christopher Bailey[2]

[1]School of Computing and Mathematical Sciences, University of Greenwich, London, United Kingdom.
[2]School of Electrical, Computer and Energy Engineering, Arizona State University, Arizona, United States.
*Corresponding Author: s.r.hassan@greenwich.ac.uk.

Abstract—This paper presents parametric model order reduction (pMOR) by the Lagrange approach of matrix interpolation for the thermal-mechanical and reliability study of a power electronics module (PEM) with nonlinear behaviours. Most previous research in model order reduction (MOR) studies reports thermal-mechanical simulations using a sequentially coupled method. In this research, a direct-coupled thermal-mechanical analysis, which simultaneously solves the thermal and structural governing equations, has been used to obtain thermal and deformation results. Furthermore, for pMOR, the linear approach of matrix interpolation is limited to linear changes between sampled-parametric points. Hence, a new way of interpolating system matrices using the Lagrange interpolation method has been adopted to implement the matrix interpolation efficiently. The parametric reduced-order model (pROM) solution by the Lagrange approach of matrix interpolation agrees well with the full-order model (FOM) and takes similar computational time as the linear (bi-linear) approach of matrix interpolation. pROM simulations offer up to 85.5% reduction in computational time.

Index Terms—Finite Element Method, Thermal-Mechanical Analysis, Power Electronics Module, Reliability Assessment, Parametric Model Order Reduction.

I. INTRODUCTION

New advanced technologies in industries, e.g., space & defence, energy, renewable energies and transportation, introduce complex and expensive projects to engineers and scientists. Reliability analysis is one of the most crucial factors of these technologies in keeping them safe and operational. Luckily, mathematical models can simulate the physical behaviours of domains and provide in-depth data for analyses and designs. Engineering sciences and their derived PDEs (partial differential equations) are capable of successfully describing physical behaviours of systems. One of the most widely used computational methods to solve PDEs is FEM/FEA (finite element method/analysis). For FEM computation, PDEs are discretised to algebraic equations via approximate unknown variables, which have complex and notably high dimensional systems of differential equations [1]. FEM provides excellent prediction, but in terms of simulating large-scale models or design points explorations, the computational time requirement is challenging, which is why model order reduction (MOR) is vital. In present literature, Krylov subspace-based MOR methods have been commonly used as they are "semi-automatic" compared to the Modal truncation method, a classic reduced order modelling method [2, 3].

Thermal behaviours of electric systems can cause performance degradation and reliability issues. Thermoelastic damping is one of the critical causes of these systems' component deterioration. Therefore, reliability assessment based on thermal-mechanical analysis is a crucial subject of study. Previously, Eblen [4] carried out a coupled thermomechanical analysis-based reliability study of electronics components with the help of FEM and CARES, a computing tool for reliability assessments. Thermomechanical analysis of an electronic package was presented by Codecasa et al. [5], focusing on reducing computing time with TRIC, a projection-based solver. Coupled thermomechanical analysis has been extended to electrical-thermal-mechanical analysis to study an electrical contact site by Shen and Ke [6].

Thermal analysis with MOR techniques is also helpful in reducing computing time for reliability assessments. Krylov subspace-based MOR techniques have been utilised for thermal analysis of an electric converter assembly by Liu et al. [7]. Thermal boundary condition independency of Krylov subspace-based reduced order modelling approach has been examined for thermal and coupled thermal analysis by Rogié et al. [8] and Codecasa et al. [9, 10].

Reduced order modelling for coupled problems is highly desirable due to the complexity of systems. Choi et al. [11] used Krylov subspace-based MOR technique to achieve a coupled thermal-mechanical ROM (reduced order model) of a micro-resonator. Rajaguru et al. [12] examined electrical loading in a PEM structure using ROMs built with Krylov subspace-based MOR techniques. A coupled electrical-thermal-mechanical model of a MEMS microgripper was studied by Binion and Chen [13], exploring several MOR techniques.

Parametric modelling is crucial for design point explorations and optimisations, and the parametric reduced order model (pROM) presents an appropriate solution to overcome computational time requirement issues that arise during parametric modelling. Bissuel et al. [14] surveyed numerous thermal models of electronic boards to optimise the device by building pROM based on the modal approach of MOR. Baur et al. [15] demonstrated parameterisation of the physical properties of a micro-thruster through various pMOR (parametric model order reduction) approaches in its frequency response investigation. Feng et al. [16] have developed electrical and coupled electrothermal pROMs to enhance the geometric

405

parameters of a nanoelectronics structure. A superposition principle-based pMOR approach has been used by ter Maten et al. [17] to change several variables of coupled electromagnetic-thermal models of electronic devices. Bouhedma et al. [18] built a Krylov subspace-based pROM for a piezoelectric energy harvester to modify the physical dimensions of the model. Schütz et al. [19] looked at several ROMs of a micro-actuator to transform its magnetostatic variables. An optimised model of a miniaturised thermoelectric generator has been created by Yuan et al. [20] exercising multiple pMOR methodologies.

Considering nonlinear behaviours in a model is a central part of an analysis as it provides in-depth data, and building a nonlinear ROM is advantageous as nonlinear computation requires relatively more time. Scognamillo et al. [21] exercised a MOR code, FANTASTIC, to evaluate the nonlinear thermal activities of a coupled electrothermal PEM model. A multiphysical ROM of a piezoelectric actuator considering nonlinear inputs was achieved, through Krylov subspace-based MOR method, by Schütz et al. [22]. In a recent investigation [23], the temperature-dependent coefficient of thermal expansion (CTE) of the wire material of a PEM has been parametrised in direct coupled thermal-mechanical parametric reduced order modelling.

The sequential coupling method has been widely exercised in most afore-mentioned studies for thermal-mechanical ROMs. Direct coupled thermal-mechanical analyses offer more sensible and precise insights into systems and should be considered in reduced-order modelling. The direct coupling method, which concurrently solves thermal and structural equations, has been utilised in this work for the thermomechanical ROM. Temperature-dependent material properties have been varied here for design point exploration, as prior pMOR studies only focused on changing constant model parameters. Nonlinear plasticity behaviour is also evaluated for the wire material of the PEM structure. The matrix interpolation method has been expanded to a new way of interpolating matrices, based on Lagrange interpolation, to create the pROM with a Krylov subspace-based pMOR approach for the current analysis.

II. Problem Formulation

A. Parametric Full Order Model (pFOM)

The state-space representation of the parametric full-order model (pFOM) is stated as the followings [20, 24]:

$$
\begin{aligned}
\boldsymbol{E}(\boldsymbol{p}_i)\dot{\boldsymbol{x}}(\boldsymbol{p}_i, t) &= \boldsymbol{A}(\boldsymbol{p}_i)\boldsymbol{x}(\boldsymbol{p}_i, t) + \boldsymbol{B}(\boldsymbol{p}_i)\boldsymbol{u}(\boldsymbol{p}_i, t) \\
\boldsymbol{y}(\boldsymbol{p}_i, t) &= \boldsymbol{C}(\boldsymbol{p}_i)\boldsymbol{x}(\boldsymbol{p}_i, t)
\end{aligned}
\tag{1}
$$

$\boldsymbol{E}(\boldsymbol{p}_i), \boldsymbol{A}(\boldsymbol{p}_i) \in \mathbb{R}^{N \times N}$ are parametric point ($\boldsymbol{p}_i$) dependent system matrices, with $\boldsymbol{B}(\boldsymbol{p}_i) \in \mathbb{R}^{N \times M}$ and $\boldsymbol{C}(\boldsymbol{p}_i) \in \mathbb{R}^{P \times N}$ representing input and output matrices. $\boldsymbol{u}(\boldsymbol{p}_i, t) \in \mathbb{R}^M$ and $\boldsymbol{y}(\boldsymbol{p}_i, t) \in \mathbb{P}$ are parametric point ($\boldsymbol{p}_i$) and time (t) dependent inputs and outputs of the model together with $\boldsymbol{x}(\boldsymbol{p}_i, t) \in \mathbb{N}$, which defines the states. $\boldsymbol{p}_i$ is the vector of model parametric points for design point explorations with $i = 0, 1, \ldots k$, and k identifies the total parametric points.

B. Projection-based Model Order Reduction (MOR)

The state-space representation of the full-order model (FOM), which is the non-parametric-point-dependent form of the model shown in (1), is stated as the followings [20, 24]:

$$
\begin{aligned}
\boldsymbol{E}\dot{\boldsymbol{x}}(t) &= \boldsymbol{A}\boldsymbol{x}(t) + \boldsymbol{B}\boldsymbol{u}(t) \\
\boldsymbol{y}(t) &= \boldsymbol{C}\boldsymbol{x}(t)
\end{aligned}
\tag{2}
$$

$\boldsymbol{E}, \boldsymbol{A} \in \mathbb{R}^{N \times N}$ are system matrices, with $\boldsymbol{B} \in \mathbb{R}^{N \times M}$ and $\boldsymbol{C} \in \mathbb{R}^{P \times N}$ representing input and output matrices, they are parametric point ($\boldsymbol{p}_i$) independent. $\boldsymbol{u}(\boldsymbol{p}_i, t) \in \mathbb{R}^M$ and $\boldsymbol{y}(\boldsymbol{p}_i, t) \in \mathbb{P}$ are only time (t) dependent inputs and outputs of the model together with $\boldsymbol{x} \in \mathbb{R}^N$, which defines the states. The order of the model, $N \in \mathbb{N}$, is significantly high.

The reduced order model (ROM) of the system expressed in (2) can be specified as follows [20, 24]:

$$
\begin{aligned}
\boldsymbol{E}_r\dot{\boldsymbol{x}}_r(t) &= \boldsymbol{A}_r\boldsymbol{x}_r(t) + \boldsymbol{B}_r\boldsymbol{u}_r(t) \\
\boldsymbol{y}_r(t) &= \boldsymbol{C}_r\boldsymbol{x}_r(t)
\end{aligned}
\tag{3}
$$

The matrices in the reduced model, established in (3), are obtained by the following operations: $\boldsymbol{E}_r = \boldsymbol{V}^T\boldsymbol{E}\boldsymbol{V}$, $\boldsymbol{A}_r = \boldsymbol{V}^T\boldsymbol{A}\boldsymbol{V}$, $\boldsymbol{B}_r = \boldsymbol{V}^T\boldsymbol{B}$ and $\boldsymbol{C}_r = \boldsymbol{C}\boldsymbol{V}$, utilising PRIMA [25, 26], a Krylov subspace-based MOR procedure. $\boldsymbol{E}_r, \boldsymbol{A}_r \in \mathbb{R}^{q \times q}$, $\boldsymbol{B} \in \mathbb{R}^{q \times m}$, and $\boldsymbol{C} \in \mathbb{R}^{p \times q}$ have incredibly lower dimensions, $q \ll N$, as they are transformed via the projection matrix $\boldsymbol{V} \in \mathbb{R}^{N \times q}$. The transfer function of the FOM in (2) is used to obtain the projection matrix ($\boldsymbol{V}$), and then the transfer function of the ROM in (3) is determined. The full-order and reduced-order models' transfer functions are expressed as the followings [25, 26]:

$$
\boldsymbol{Y}(s) = \boldsymbol{C}(s\boldsymbol{E} - \boldsymbol{A})^{-1}\boldsymbol{B}
\tag{4}
$$

$$
\boldsymbol{Y}_r(s) = \boldsymbol{C}_r(s\boldsymbol{E}_r - \boldsymbol{A}_r)^{-1}\boldsymbol{B}_r
\tag{5}
$$

C. Interpolation of Sparse Matrices

Linear matrix interpolation of the system's sparse matrices in (2), described by $\boldsymbol{X} = \boldsymbol{E}, \boldsymbol{A}, \boldsymbol{B}$, can be utilised to construct a state-space system in (1) as the following [24, 27]:

$$
\boldsymbol{X}(\boldsymbol{p}_i) = \boldsymbol{X}(\boldsymbol{p}_0) + \omega(\boldsymbol{p}_i)\left[\boldsymbol{X}(\boldsymbol{p}_k) - \boldsymbol{X}(\boldsymbol{p}_0)\right]
\tag{6}
$$

The values of weighting functions, $\omega(\boldsymbol{p}_i)$, are determined through the linear interpolation method. The bi-linear (or multilinear) approach, capable of sampling more than two parametric points, has been exercised previously for a study to implement matrix interpolation [23]. For this study, $i = 0, 1, \ldots, k$, where $k = 6$. The implemented bi-linear matrix interpolation method can be expressed as the followings [23]:

$$
\begin{aligned}
\boldsymbol{X}(\boldsymbol{p}_{i=0,1,2,3}) &= \boldsymbol{X}(\boldsymbol{p}_0) + \omega(\boldsymbol{p}_i)\left[\boldsymbol{X}(\boldsymbol{p}_{\frac{k}{2}}) - \boldsymbol{X}(\boldsymbol{p}_0)\right] \\
\boldsymbol{X}(\boldsymbol{p}_{i=4,5,6}) &= \boldsymbol{X}(\boldsymbol{p}_{\frac{k}{2}}) + \omega(\boldsymbol{p}_i)\left[\boldsymbol{X}(\boldsymbol{p}_k) - \boldsymbol{X}(\boldsymbol{p}_{\frac{k}{2}})\right]
\end{aligned}
\tag{7}
$$

A new way of matrix interpolation, based on the Lagrange method, has been applied then as the following:

$$
\boldsymbol{X}(\boldsymbol{p}_i) = \omega_0(\boldsymbol{p}_i)\boldsymbol{X}(\boldsymbol{p}_0) + \omega_{\frac{k}{2}}(\boldsymbol{p}_i)\boldsymbol{X}(\boldsymbol{p}_{\frac{k}{2}}) + \omega_k\boldsymbol{X}(\boldsymbol{p}_k)
\tag{8}
$$

D. Parametric Reduced Order Model (pROM)

The interpolated matrices illustrated in (7) and (8) and the MOR method outlined in (3) have been utilised to create pROMs. The pROM can be written as the followings [20]:

$$E_r(p_i)\dot{x}_r(p_i,t) = A_r(p_i)x_r(p_i,t) + B_r(p_i)u_r(p_i,t)$$
$$y_r(p_i,t) = C_r(p_i)x_r(p_i,t)$$
$$(9)$$

The pROM in (9) has been solved by the generalized trapezoidal rule (GTR) [28, 29].

E. Parametric Thermal-Mechanical Model

The coupled thermal-mechanical model considered for this analysis is a second-order system; this is attained using FEM discretisation. The parametric-point-dependent discretised thermal-mechanical model is stated in the following forms [29, 30]:

$$M(p_i)\ddot{z}(p_i,t) + D(p_i)\dot{z}(p_i,t) + K(p_i)z(p_i,t)$$
$$= G(p_i)u(p_i,t) \quad (10)$$
$$y(p_i,t) = L(p_i)z(p_i,t)$$

$M(p_i), D(p_i), K(p_i) \in \mathbb{R}^{n \times n}$ signify mass, damping and stiffness matrices, where $2n = N$, with $G(p_i) \in \mathbb{R}^{n \times M}$ and $L(p_i) \in \mathbb{R}^{P \times n}$ representing input and output matrices. $u(p_i,t) \in \mathbb{R}^M$ and $y(p_i,t) \in \mathbb{R}^P$ define the inputs and outputs of the model, with $z(p_i,t) \in \mathbb{R}^n$ depicting states. In general, matrices are described as the followings [29]:

$$M = \begin{bmatrix} M_s & 0 \\ 0 & 0 \end{bmatrix}, D = \begin{bmatrix} D_s & 0 \\ D^{tu} & D^t \end{bmatrix}, K = \begin{bmatrix} K_s & K^{ut} \\ 0 & K^t \end{bmatrix},$$
$$G = \begin{bmatrix} F \\ Q \end{bmatrix}, \ddot{z} = \begin{bmatrix} \ddot{z}_{ut} \\ \ddot{T} \end{bmatrix}, \dot{z} = \begin{bmatrix} \dot{z}_{ut} \\ \dot{T} \end{bmatrix}, z = \begin{bmatrix} z_{ut} \\ T \end{bmatrix}$$
$$(11)$$

with,
$$K^t = K^{tb} + K^{tc}, F = F^{nd} + F^{pr} + F^{ac},$$
$$Q = Q^{nd} + Q^g + Q^c$$
$$(12)$$

M_s denotes structural mass matrix. D_s, D^{tu} and D^t correspond to structural and thermoelastic damping and thermal-specific heat matrices, respectively. K_s, K^{tu} and K^t stand for structural and thermoelastic stiffness and thermal conductivity matrices, accordingly, while K^{tb} and K^{tc} indicate thermal conductivity matrices of material and convection surfaces. F signifies structural (mechanical) load vectors, with F^{nd}, F^{pr} and F^{ac} describing nodal force and pressure load vectors and force vectors caused by acceleration effects correspondingly. Q represents the thermal load vector, whilst Q^{nd}, Q^c and Q^g define nodal heat flow rate and convection surface vectors and heat generation rate vector without Joule heating, in that order. z_{ut} and T symbolise potential displacement and thermal vectors.

The state space model, presented in (1), can be formed by transforming the thermal-mechanical model, explained in (10), as the followings [30]:

$$\begin{bmatrix} F & 0 \\ 0 & M \end{bmatrix}\begin{bmatrix} \dot{z} \\ \ddot{z} \end{bmatrix} = \begin{bmatrix} 0 & F \\ -K & -D \end{bmatrix}\begin{bmatrix} z \\ \dot{z} \end{bmatrix} + \begin{bmatrix} 0 \\ G \end{bmatrix}u$$
$$y = \begin{bmatrix} L & 0 \end{bmatrix}\begin{bmatrix} z \\ \dot{z} \end{bmatrix}$$
$$(13)$$

with,
$$E = \begin{bmatrix} F & 0 \\ 0 & M \end{bmatrix}, A = \begin{bmatrix} 0 & F \\ -K & -D \end{bmatrix}, B = \begin{bmatrix} 0 \\ G \end{bmatrix},$$
$$C = \begin{bmatrix} L & 0 \end{bmatrix}, \dot{x} = \begin{bmatrix} \dot{z} \\ \ddot{z} \end{bmatrix}, x = \begin{bmatrix} z \\ \dot{z} \end{bmatrix}$$
$$(14)$$

F has to be non-singular here. $F = I_n$ is assumed, for the presented system, with I_n as an $n \times n$ identity matrix.

III. POWER ELECTRONICS MODULE (PEM)

For this analysis, we focused on a *2D-plane* model of a PEM to explore its thermal-mechanical behaviours for a set of parameteric points. This PEM has *SiC* as semiconductors. The physical dimensions of the PEM structure and boundary conditions set for the direct-coupled analysis are described in Fig. 1. For the model built in ANSYS, *Al* (alloy) has been assumed as the wire material, and the properties of this material will be parametrised in this pMOR study.

A. FEM Model

In the FEM model, a direct coupled transient thermal-mechanical analysis has been exercised for the present investigation. The *SiC* bodies of the model act as thermal sources and have an isothermal boundary condition with a maximum value of $T_{SiC} = 200°C$ (labelled A in Fig. 1). The bottom surface of the baseplate, made of *Cu* material, has a convection boundary condition with a maximum convection coefficient of $h = 5W/mm^2 \cdot °C$ and a maximum ambient temperature of $T_C = 50°C$ (labelled C in Fig. 1) and reflects assumed ambient temperatures. The left and right corner vertices of the baseplate (*Cu*) are fixed with no anticipated deformation (labelled B in Fig. 1). The analysis has 11 loading steps as shown in Fig 2, and the heat-generating body and the convection surface have differing temperature and coefficient values throughout the simulation. Material properties data are retrieved from ANSYS booklet [31]. The wire material, *Al* (alloy), has nonlinear plasticity behaviour, which follows the nonlinear power hardening law. The power hardening law, based on Gurson's Model, is stated in the following [29, 32]:

$$\frac{\sigma_Y}{\sigma_0} = \left(\frac{\sigma_Y}{\sigma_0} + \frac{3G}{\sigma_0}\bar{\varepsilon}^p\right)^{N_S}$$
$$(15)$$

Here, σ_Y and σ_0 are current and initial yield strengths, and G is the shear modulus. $\bar{\varepsilon}^p$ represents the microscopic equivaent plastic strain, and N_S is the stress ratio. Initial yield stress ($280MPa$) and exponent (0.134) values for nonlinear plasticity modelling are assumed based on the approaches described in [29, 33].

B. Parametric Points

The thermal expansion coefficient, CTE_{Al}, and Young's Modulus, E_{Al}, of the *Al* (alloy) are temperature dependent. These properties of the material add further nonlinear behaviours to the body. CTE_{Al} and E_{Al} have been parametrised, for the current study, for design point exploration. Uniformly spaced parametric points have been preferred here to implement pMOR with matrix interpolation, as presented in Fig. 3. The parametric points have evenly spaced CTE_{Al} and E_{Al}

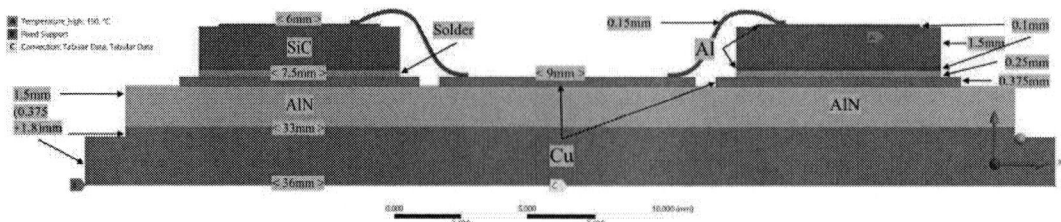

Fig. 1: *2D*-plane of the PEM structure and boundary conditions.

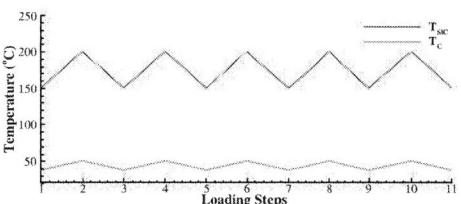

Fig. 2: Heat-generating body (T_{SiC}) and Ambient (T_C) temperatures during loading steps.

values corresponding to their temperature values. The ranges of CTE_{Al} and E_{Al} values considered for the parametric study are explained in Fig. 4 and Table I.

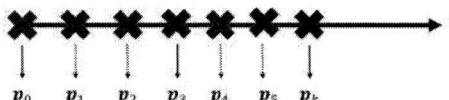

Fig. 3: Uniform parametric points.

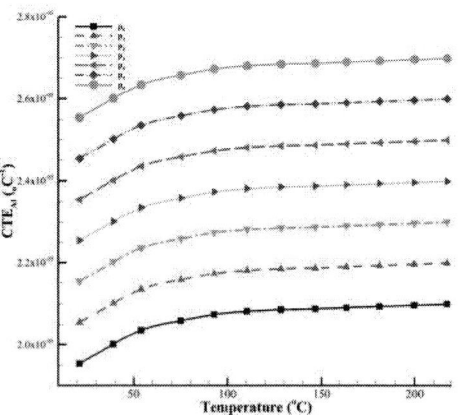

Fig. 4: Parametric points for the temperature dependent coefficient of thermal expansion of *Al* (alloy), CTE_{Al}.

C. Reduced Model

The total degrees of freedom (DOFs) of the state-space full-order model (FOM) is $N = 2n = 27,094$, with $N \times N$ system matrices. On the other hand, the reduced order model (ROM) has a total DOFs of $q = 8$, with $q \times q$ system matrices. This results in ROM demanding substantially less time for computation against its FOM. The overall simulation time

TABLE I: Parametric points for the temperature dependent Young's Modulus of *Al* (alloy), E_{Al}.

Parametric Points	Young's Modulus (E_{Al}) in GPa		
	$19.85\,^{\circ}C$	$116.3\,^{\circ}C$	$212.95\,^{\circ}C$
p_0	71.06	70.12	69.18.
p_1	71.05	70.11	69.17
p_2	71.04	70.10	69.16
p_3	71.03	70.09	69.15
p_4	71.02	70.08	69.14
p_5	71.01	70.07	69.13
p_6	71.00	70.06	69.12

in pROM, including order reductions of system matrices, is $608s$, which provides the solution for all seven parametric points. The pFOM-ANSYS solution, in comparison, would require around $4200s$ on the same computer. By means of the pMOR approach, a decline of 85.5% in computational time requirements has been accomplished.

Fig. 5 exhibits a flow chart summarising the stages of the pMOR study. The ANSYS Workbench/Mechanical has been used to create the PEM model, FEM discretisation and obtain the system matrices for three sampled-parametric points (p_0, p_3 and p_6). System matrices are extracted as sparse matrices due to their compactness. The matrices have been then imported into MATLAB to develop the pROM with the pMOR method.

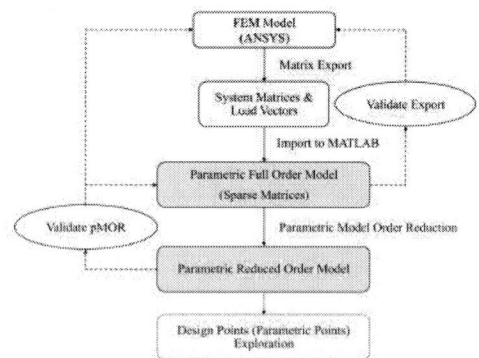

Fig. 5: The organizational process to build the pROM.

D. Results and Discussions

The pROM outcomes must align with the FOM-ANSYS solution to verify that the valid pROM has been built. The model's DOFs are temperature and directional deformations, in this case, directional deformations in the x and y-axis. Fig. 6 illustrates the temperature and total deformation distributions for the left section of the PEM structure, assessing the results from FOM-ANSYS and pROM with Lagrange interpolation. In addition to the MOR approach, the means of matrix interpolation must be confirmed, so the interpolated point's (p_2) solution from the pROM has been evaluated in this figure. The pROM solution demonstrates excellent agreement compared to the FOM-ANSYS solution.

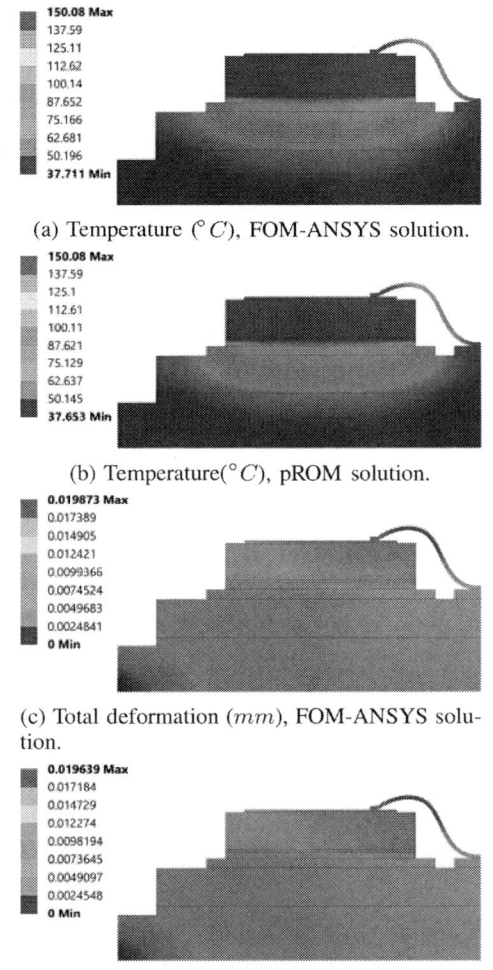

(a) Temperature ($^{\circ}C$), FOM-ANSYS solution.

(b) Temperature($^{\circ}C$), pROM solution.

(c) Total deformation (mm), FOM-ANSYS solution.

(d) Total deformation (mm), pROM solution.

Fig. 6: Temperature ($^{\circ}C$) and Total Deformation (mm) distribution from FOM-ANSYS and pROM (Lagrange) solutions in the left part of the PEM structure for the interpolated-parametric point p_2.

The wire bond site experiences one of the highest temperatures, seen in Fig. 6a and 6b. The temperature here goes up to that of the heat-generating body, the SiC-based semiconductor. The *Min* temperature values for the PEM structure differ only by 0.15% between the FOM-ANSYS and pROM solutions, and there is no difference in the *Max* temperature. This wire bond site will be the focus of the presenting result analysis.

Figs. 6c and 6d indicate that the wire and wire bond site of the PEM structure undergo a significant amount of deformation. The FOM-ANSYS and pROM solutions vary by about 1.8% in terms of the total deformation peak value. The wire body was expected to show the peak deformation in the PEM model.

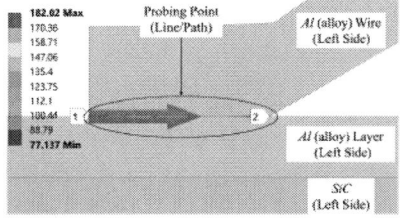

(a) Equivalent (von-Mises) stress (MPa), obtained by the FOM-ANSYS solution, along the probing point (line/path) in the PEM for interpolated-parametric point p_2.

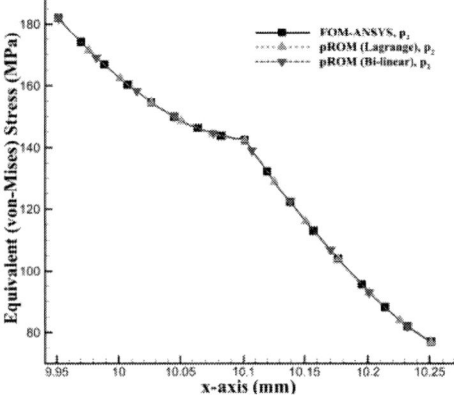

(b) Equivalent (von-Mises) stress, FOM-ANSYS vs pROM solution, for the interpolated-parametric point p_2.

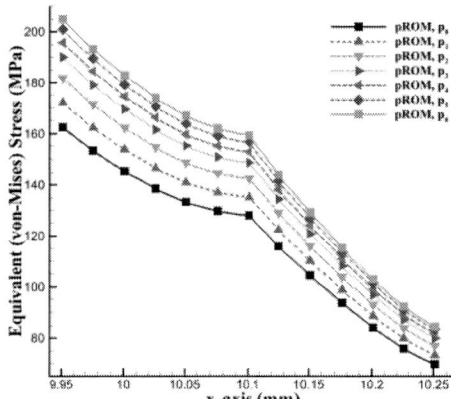

(c) Equivalent (von-Mises) stress, obtained by the pROM (Lagrange) solution, for all the parametric points.

Fig. 7: Maximum Equivalent (von-Mises) stress over time along a probing point in wire bond site (in wire body) of the PEM structure.

The maximum equivalent (von-Mises) stresses over time in the wire bond site (in wire body) are presented in Fig. 7. The depicted results are obtained from a probing point, which is a line/path along the wire bond shown in Fig. 7a, considering the significance of the site [34]. FOM-ANSYS and pROM (Lagrange and Bi-linear) stress results are displayed in Fig. 7b for an interpolated-parametric point p_2. Stress results obtained from the pROMs are in excellent agreement with the FOM-ANSYS result, with only a 0.1% average difference between the solutions. Thus, the pMOR approach is a suitable approximate modelling approach for reliability analysis-based design points exploration. Fig. 7c evaluates stress results obtained by the pROM (Lagrange) for all parametric points. *Max* stresses along the probing point extent from $181MPa$ to $206MPa$ for the studied parametric points, which is $< 280MPa$, the yield strength of the material. Non-zero equivalent plastic strain values are seen in this site at and after p_3 and reach a maximum of $4.4 \times 10^{-4} mm/mm$ at p_6 for current loading.

IV. CONCLUSION

The pMOR method has been exercised in this work to explore different parametric points of a directly coupled thermal-mechanical PEM model, considering nonlinear plasticity behaviours in the wire material. The temperature-dependent coefficient of thermal expansion and Young's modulus of the *Al* (alloy), the wire material, have been parametrised for the pROM. The MOR approach PRIMA has been utilised here to reduce computational time requirements. A new matrix interpolation technique, based on the Lagrange interpolation, which provides a better process of matrix interpolation for multiple sampled-parametric points compared to the linear/multilinear approach, has been offered here to build the pROM. The pROM requires 85.5% less computing time with only a 0.1% disparity vs FOM-ANSYS in stress results. The pMOR method can help reduce the time requirements of reliability analysis-based design explorations for large-scale models. Future studies will focus on applying this method to models with rate-dependent material nonlinearities.

REFERENCES

[1] G. Dhatt, E. Lefrançois and G. Touzot, *Finite element method*, John Wiley & Sons, 2012.

[2] E. Davison, "A method for simplifying linear dynamic systems," IEEE Transactions on automatic control, vol. 11, no. 1, pp. 93-101, 1966.

[3] R. W. Freund, "Model reduction methods based on Krylov subspaces," Acta Numerica, vol. 12, pp. 267-319, 2003.

[4] M. Eblen, "Applied FEM techniques in ceramic feedthru package design," in 54th Electronic Components and Technology Conference (IEEE Cat. No. 04CH37546), 2004.

[5] L. Codecasa, *et al.*, "Towards the Extension of TRIC for Thermo-Mechanical Analysis," in 27th International Workshop on Thermal Investigations of ICs and Systems (THERMINIC), 2021.

[6] F. Shen and L. Ke, "Numerical study of coupled electrical-thermal-mechanical-wear behavior in electrical contacts.," Metals, vol. 11, no. 6, pp. 955, 2021.

[7] W. Liu, H. Torsten and J. Drobnik, "Effective Thermal Simulation of Power Electronics in Hybrid and Electric Vehicles," World Electric Vehicle Journal, vol. 5, no. 2, pp. 574-580, 2012.

[8] B. Rogié, *et al.*, "Multi-port dynamic compact thermal models of dual-chip package using model order reduction and metaheuristic optimization," Microelectronics Reliability, vol. 87, pp. 222-231, 2018.

[9] L. Codecasa, *et al.*, "Versatile MOR-based boundary condition independent compact thermal models with multiple heat sources," Microelectronics Reliability, vol. 87, pp. 194-205, 2018.

[10] L. Codecasa, V. d'Alessandro and D. D'Amore, "Altering MOR-based BCI CTMs into Delphi-like BCI CTMs," In 26th International Workshop on Thermal Investigations of ICs and Systems (THERMINIC), 2020.

[11] J. Choi, M. Cho and J. Rhim, "Efficient prediction of the quality factors of micromechanical resonators," Journal of Sound and Vibration, vol. 329, no. 1, pp. 84-95, 2010.

[12] P. Rajaguru, M. Bella and C. Bailey, "Applying Model Order Reduction to the Reliability Prediction of Power Electronic Module Wirebond Structure," in 27th International Workshop on Thermal Investigations of ICs and Systems (THERMINIC), 2021.

[13] D. Binion and X. Chen, "Coupled electrothermal-mechanical analysis for MEMS via model order reduction". Finite elements in analysis and design, vol. 46, no. 12, pp.1068-1076, 2010.

[14] V. Bissuel, *et al.*, "Multi-port Dynamic Compact Thermal Models of BGA package using Model Order Reduction and Metaheuristic Optimization.," in 18th IEEE Intersociety Conference on Thermal and Thermomechanical Phenomena in Electronic Systems (ITherm), 2019.

[15] U. Baur, *et al.*, "Comparison of methods for parametric model order reduction of instationary problems," Max Planck Institute for Dynamics of Complex Technical Systems, 2015.

[16] L. Feng, *et al.*, "Parametric modeling and model order reduction for (electro-) thermal analysis of nanoelectronic structures," Journal of Mathematics in Industry, vol. 6, no. 1, pp. 1-16, 2016.

[17] E. ter Maten, *et al.*, "Nanoelectronic COupled problems solutions-nanoCOPS: modelling, multirate, model order reduction, uncertainty quantification, fast fault simulation.," Journal of Mathematics in Industry, vol. 7, no. 1, pp. 1-19, 2016.

[18] S. Bouhedma, *et al.*, "System-level model and simulation of a frequency-tunable vibration energy harvester," Micromachines, vol. 11, no. 1, pp. 91, 2020.

[19] A. Schütz, *et al.*, "Parametric system-level models for position-control of novel electromagnetic free flight microactuator," Microelectronics Reliability, pp. 114062, 2021.

[20] C. Yuan, D. Hohlfeld and T. Bechtold, "Design optimization of a miniaturized thermoelectric generator via parametric model order reduction.," Microelectronics Reliability, vol. 119, pp. 114075, 2021.

[21] C. Scognamillo, *et al.*, "Compact Modeling of a 3.3 kV SiC MOSFET Power Module for Detailed Circuit-Level Electrothermal Simulations Including Parasitics," Energies, vol. 14, no. 15, pp. 4683, 2021.

[22] A. Schütz, S. Maeter and T. Bechtold, "System-Level Modelling and Simulation of a Multiphysical Kick and Catch Actuator System," Actuators, vol. 10, no. 11, pp. 279, 2021.

[23] S. Hassan, *et al.*, "Parametrising Temperature Dependent Properties in Thermal-Mechanical Analysis of Power Electronics Modules using Parametric Model Order Reduction", in 46th International Spring Seminar on Electronics Technology (ISSE), 2023.

[24] H. Panzer, *et al.*, "Parametric model order reduction by matrix interpolation," Automatisierungstechnik, vol. 58, no. 8, pp. 475-484, 2010.

[25] A. Odabasioglu, M. Celik and L. Pileggi, "PRIMA: Passive reduced-order interconnect macromodeling algorithm.", IEEE Transactions on computer-aided design of integrated circuits and systems, vol. 17, no. 8, pp. 645-654, 1998.

[26] S. Race, *et al.*, "Circuit-based electrothermal modeling of SiC power modules with nonlinear thermal models," IEEE Transactions on Power Electronics, vol. 37, no. 7, pp. 7965-7976, 2022.

[27] M. Geuss, H. Panzer and B. Lohmann, "On parametric model order reduction by matrix interpolation," in European Control Conference (ECC), 2013.

[28] T.J. Hughes, *The finite element method: linear static and dynamic finite element analysis*, Courier Corporation, 2012

[29] ANSYS, *Mechanical APDL 2023 R1 - Theory Reference*, ANSYS, 2023.

[30] B. Lohmann and B. Salimbahrami, "Reduction of second order systems using second order Krylov subspaces," in IFAC Proceedings Volumes, 2005.

[31] M. Ashby, *Material property data for engineering materials*, ANSYS, 2021

[32] A. L. Gurson, "Continuum theory of ductile rupture by void nucleation and growth: Part I—Yield criteria and flow rules for porous ductile media", Brown Univ., Providence, RI (USA). Div. of Engineering, 1977.

[33] T. Wierzbicki, *Structural Mechanics (Wierzbicki)*, Massachusetts Institute of Technology via MIT OpenCourseWare, 2023. [Online]. Available: https://eng.libretexts.org

[34] K. Nwanoro, *et al.*, "Advantages of the extended finite element method for the analysis of crack propagation in power modules," Power Electronic Devices and Components, vol. 4, pp. 100027, 2023.

Thickness effect of copper clips on power module packaging design

Haiyong Wan
School of Engineering
University of Warwick
Coventry, UK
Haiyong.Wan@warwick.ac.uk

Nikolaos Iosifidis
School of Engineering
University of Warwick
Coventry, UK
Nikos.Iosifidis@warwick.ac.uk

Xu Zhang
School of Engineering
University of Warwick
Coventry, UK
Xu.Zhang.1@warwick.ac.uk

Rui Rong
MacMic Science & Technology Co., Ltd
Changzhou, China
rrong@macmicst.com

Marina Antoniou
School of Engineering
University of Warwick
Coventry, UK
Marina.Antoniou@warwick.ac.uk

Philip Mawby
School of Engineering
University of Warwick
Coventry, UK
P.A.Mawby@warwick.ac.uk

Abstract—The superior switching performance of SiC MOS-FETs compared to Si devices is limited by packaging-related issues. High electromagnetic parasitics which lead to high voltage overshoots, and higher power losses are the main challenges of high frequency operation. Hence, packaging design modification and optimisation are necessary to accommodate faster switching. Copper clips replacing the bonding wires show promising results in reducing stray inductances, while enhancing the current conduction and heat dissipation since the clip are in contact with a larger area of the semiconductor chip. Silver sintering instead of soldering further improves the electrothermal behaviour. This paper investigates numerically various thicknesses of the copper clips and their effect on double-sided silver sintering, half-bridge 1200V/400A SiC MOSFET power module. Moreover, the silver layer thickness is varied to reduce the maximum stresses on the chip - silver interface.

Index Terms—SiC MOSFETs, copper clips, parasitic inductance, thermal stress

I. Introduction

The decarbonisation of the transport section with the widespread adoption of electric vehicles (EVs) has led to more stringent requirements for power electronic converter technologies. Power electronic converters are crucial to the EV operation and are used as rectifiers for battery charging and as inverters for motor drivers.

At system level, the power electronics voltage rating and the power requirements are increasing as the battery DC level is increasing towards 800V, or even higher in the future. This trend aims to achieve higher system efficiency, higher mile autonomy, at a smaller system footprint. Further, to achieve charging duration of several minutes (<30mins), DC fast chargers are pushed towards hundreds of kWs. At a power electronic module level, to supply the higher load demands, power modules tend to use higher switching frequencies which are currently in the range of tens of kHz, with a potential to reach hundreds of kHz in the future.

Currently, most of the EVs use Si-based technologies such as Si insulated gate bipolar transistor (IGBT). However, con-

ventional Si IGBT-based technologies have material limitations at higher operating frequencies and higher operating temperatures. There is an ongoing trend of EV companies to replace Si IGBT power modules with advanced SiC MOSFETs modules. SiC has 10 times higher dielectric breakdown voltage and 3 times higher thermal conductivity than Si. Further, SiC's bandgap is 3.3eV for 4-H SiC structures, which enables higher temperature operation. Moreover, SiC devices have low junction capacitance and gate charge, which allows for higher switching frequency. As unipolar devices, MOSFETs enable even faster switching frequency, thereby reducing the size of the whole system and increasing power density [1].

However, higher switching frequencies lead to higher di/dt, which combined with high parasitic inductances cause higher voltage overshoots, increased heat power losses, and increased electromagnetic interference (EMI) [2]. Even though the switching frequency is an operating parameter that can be changed, the parasitic inductances are inherent in electrical circuits and are subject to the power module design. Further, due to the miniaturisation of the SiC MOSFET chips, thermal issues rise due to the higher power density and subsequent high and localised heat fluxes. Therefore, the commercial packaging designs need to consider the implications that rise at higher switching frequencies from both the electromagnetic and the thermal perspective.

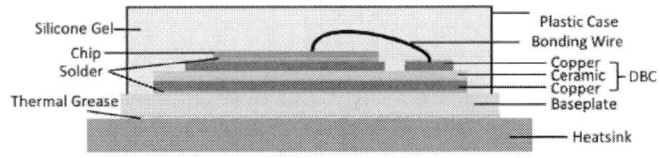

Fig. 1. Traditional power module packaging structure.

The traditional structure of a power module consists of 7 layers, Fig. 1. The semiconductor chips are soldered on top of Direct-Bonded-Copper (DBC) substrates. The top side copper of the DBC is used for the electrical connections that carry the current, the DBC ceramic is used for electrical insulation between the current conducting parts and the rest of the package. The bottom DBC copper is soldered to a metallic baseplate to provide mechanical support and enable heat dissipation and connection with external cooling solutions such as water heatsinks or air fans. Aluminium wires are used to link the power terminals with the chips. A plastic case surrounds the package, and is filled with a dielectric material such as silicone gel or epoxy to provide electrical insulation and protection against harmful environment such as humidity and dust.

Wire bonding is the most widely adopted method for electrical interconnections due to its low-cost, flexibility, and established assembly infrastructure. However, the bonding wires are usually thin aluminium wires and cause a long current path to provide enough height and space for the connection, which results in very high parasitic inductance. Further, due to the small diameter of the bonding wires, many wire connections are required to achieve higher current levels, lower on-state resistance and increased power density; resulting in productivity, material cost and reliability issues. Large number of wires cause accumulation of current and heat on the contact points. The higher temperature on the contact points could lead to wire bond degradation or even lift-off, which would result in uneven distribution of the current through the remaining wires. Moreover, as the total area of the chip is shrinking with SiC technology, and for increased current rating modules with multiple wires bonded on the SiC MOSFET's source, the reliability of the wires decreases at high switching frequencies due to the increased heat power losses [3].

Practical realisation of high-temperature operation of power modules is also restricted by the rest of the packaging materials. The interface materials are usually tin-lead or led-free solder alloys, which have relatively low melting points. Considering the higher heat fluxes and elevated temperature of a package at higher frequency operation, these interface materials might degrade over time, thus decreasing the lifetime of the device. The thermal stresses that are generated during operation in the multiple layers of the power modules vary according to their material properties. The mismatch between the coefficient of thermal expansion of adjacent layers causes increased stresses and increased degradation during temperature cycling. These stresses are exacerbated during high frequency operation due to the faster power cycling. Often power module failure mechanisms associated with temperature cycling include wire bond lift-off, solder degradation and cracking, semiconductor die cracking [4].

To this end, advanced packaging designs with lower parasitic inductance and capable of continuous operation at higher temperatures should be explored to enable higher switching frequency operation of SiC MOSFET modules to meet the EV application power demands. Innovative interconnection methods such as wire bondless structures using directly de-

posited copper or copper clip connection, flexible PCB connection, novel 3D layouts, and component integration have been proposed in literature to reduce the parasitic inductances [5]. To enhance the thermal performance of package, new interconnection methods are used to reduce thermal resistance, such as welding or sintering multiple copper pillars to the mold metal, using a lead frame to increase the contact area. To achieve high temperature working capability, new materials and innovative methods to replace traditional soldering have been proposed. For instance, gold alloy solders such as eutectic AuSn have been used in small mold conditions [6]. Silver sintering has been proposed due to its high melting point (961°C). It also has high thermal and electrical conductivity, high temperature reliability, and the sintering process takes places in relatively low temperatures.

In this paper, copper clips are adopted as interconnections replacing the bonding wires. Double-sided silver sintering is proposed as bonding method for die and copper clip attachment. Copper clips offer many advantages compared with bonding wires. Copper clips enable larger contact area for current conduction, which results in lower parasitic inductances, lower contact resistance, lower thermal resistance, and more uniform heat dissipation. However, introducing a copper layer on top of the SiC chip would result in thermal stresses during operation [7]. This study analyses the electromagnetic, thermal, and thermomechanical effect for various thicknesses of copper clips to optimise the performance of the SiC MOSFET power module. Therefore, the copper clip geometric and thickness need to be carefully selected so that the thermal stresses generated during operation do not outweight the electrical and thermal advantages of copper clips. The design under investigation, Fig. 2, is a 1200V/400A MOSFET half-bridge introduced in [8].

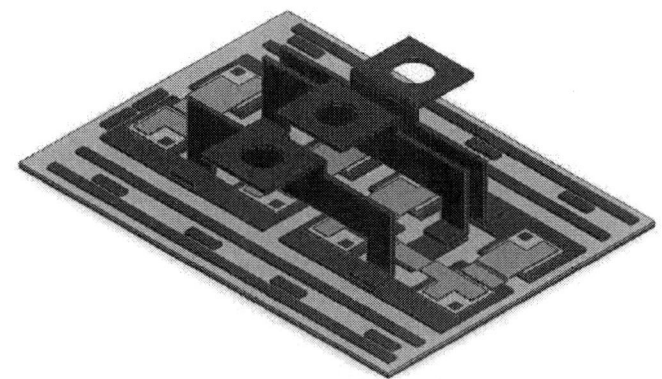

Fig. 2. Half-bridge structure under investigation.

II. ELECTROMAGNETIC PERFORMANCE ANALYSIS

This section analyses the electromagnetic performance of the power module for thicknesses of copper clips ranging from 0.1mm to 1mm. ANSYS Q3D is used to extract the parasitic inductance at various frequencies. The DC links are set to the

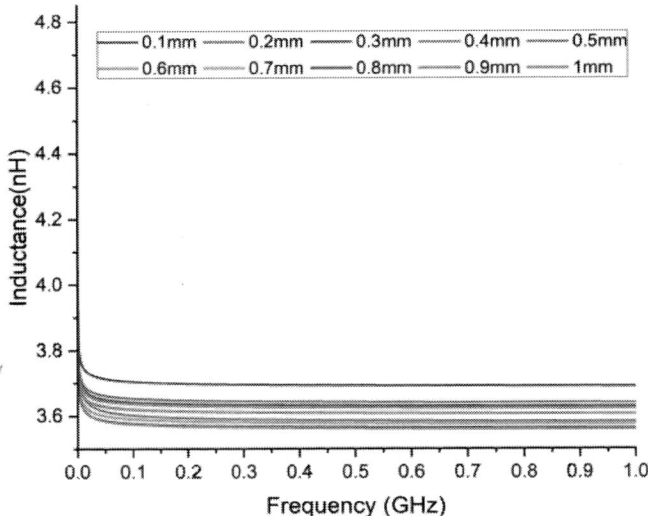

Fig. 3. Total inductance with frequency for various copper clip thicknesses.

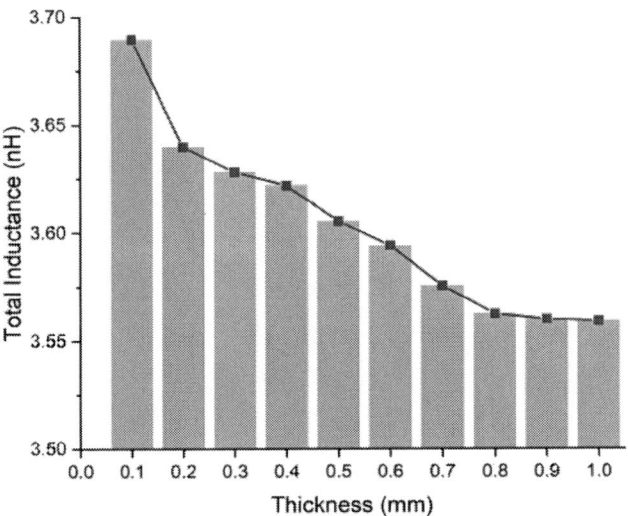

Fig. 4. Total inductance for various copper clip thicknesses at 1GHz.

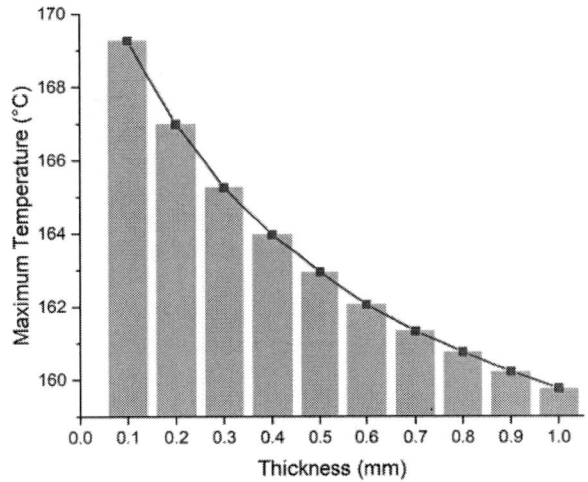

Fig. 5. Maximum junction temperature for various copper clip thicknesses.

electric source and sink, respectively, and the total parasitic inductance is calculated.

The inductance decreases with frequency according to the skin effect [?] Fig. 3, with maximum inductance at DC point. The turn-on and turn-off process time of SiC MOSFETs are usually in the nanosecond or GHz scale. Fig. 4 suggests that at 1GHz, increasing the copper clip thickness from 0.1mm to 1mm, the total inductance decreases slightly, with a total of 0.13nH difference or 3.6% reduction.

III. THERMAL PERFORMANCE ANALYSIS

Replacing the bonding wires with copper clips leads to better current and heat flow paths. The copper clips are in direct contact with a larger area of the SiC MOSFET's source, thus the current and the heat are distributed more evenly.

Moreover, the localised current and heat accumulation that is often present on the bonding wire - SiC source interface is avoided. To further increase the electrothermal performance, the bond between SiC die and copper clip is realised by silver sintering. Despite the thermal enhancement, the addition of copper clips on the top side of the SiC die induces stresses on the sintered layer and on the SiC die. Therefore, a sensitivity analysis of the copper clip thickness on the thermal and structural performance of the power module is required to find the optimal thickness of copper clips.

A. Junction temperature

To examine the thermal performance of the copper clips for various thicknesses, the SiC MOSFETs are used as heat sources. Since each SiC MOSFET is rated at 100A with a $15m\Omega$ on-state resistance, a total of 150W is dissipated for short-circuit conditions. The copper clip thickness is varied from 0.1mm to 1mm, and the silver sintering layer is set to 25μm. Fig. 5 shows the maximum SiC junction temperature for various copper clip thicknesses at steady state. As the copper clip thickness is increasing, the MOSFET chip's temperature is decreasing; from 169°C to 159°C for copper clip thicknesses of 0.1mm to 1mm, respectively. Fig. 6 indicates a uniform distribution of temperatures in the semiconductor area.

B. Thermomechanical Analysis

To examine the structural integrity of the structure, a thermomechanical analysis is performed for the range of copper clip thicknesses. The maximum stresses are located on the edge of the chip - silver sintering layer interface. Fig. 7 shows the Von Mises stress distribution for 0.5mm copper clips and 25μm silver layer. Moreover, as the copper clip thickness increases, the maximum thermal stresses increase. For 0.1mm thick copper clips, a maximum stress of 469MPa was calculated, whereas the stresses reach 923MPa for 1mm

413

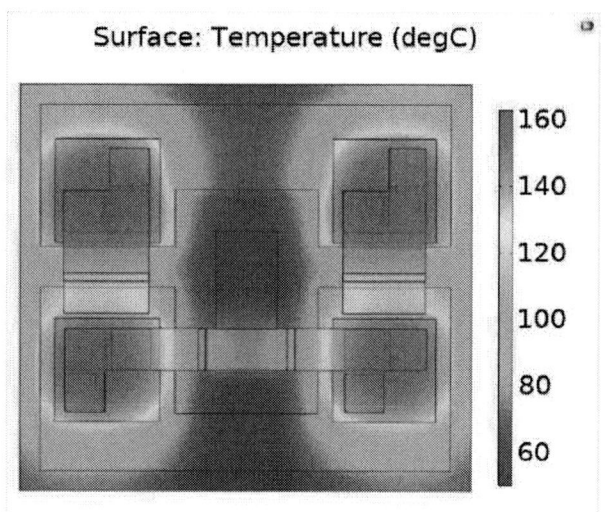

Fig. 6. Temperature distribution of a 0.5mm copper clip design at steady state.

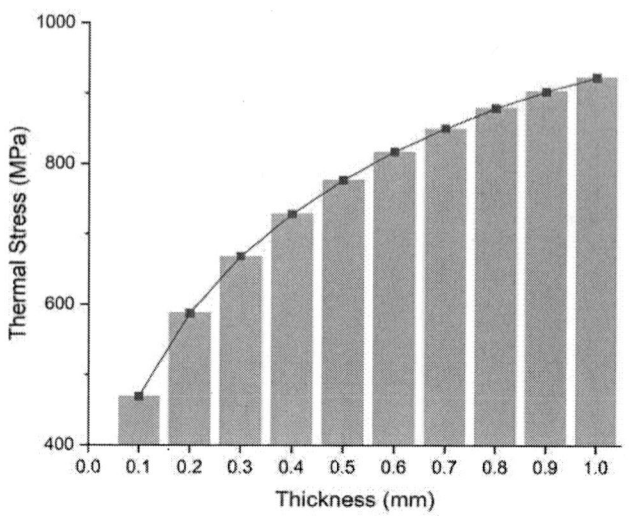

Fig. 8. Maximum thermal stress for various copper clip thicknesses.

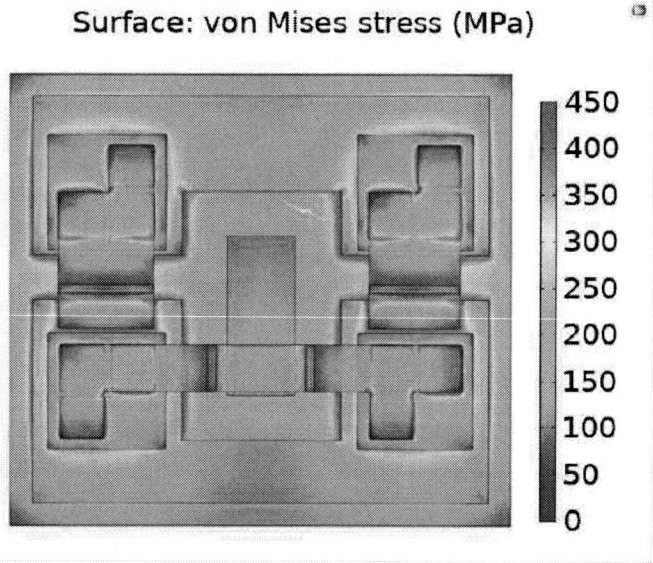

Fig. 7. Maximum thermal stress distribution of a 0.5mm copper clip design at 163°C.

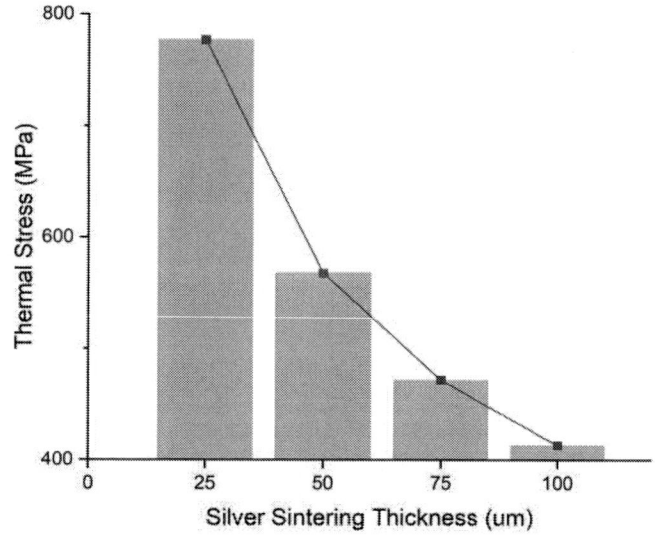

Fig. 9. Maximum thermal stress for various silver thicknesses.

thick copper clips, Fig. 8. Therefore, there is an apparent trade-off between the thermal benefits of using copper clips against the induced thermal stresses. For a 10°C chip temperature reduction, the maximum thermal stresses on the chip - silver interface are doubled.

To further optimise the structure and reduce the thermal stresses, the silver layer thickness could be modified. Fig. 8 demonstrates the maximum thermal stresses of the package for a design with 0.5mm thick copper clips, and silver layers of $25\mu m$, $50\mu m$, $75\mu m$, $100\mu m$. Lower thermal stresses are found for thicker silver layers, with a nearly 50% reduction in the von Mises stresses from 777MPa to 410MPa in the

0.5mm copper clip design. The chip temperature is not affected significantly by introducing thicker silver layers, and remains between 155°C to 155.5°C for the 0.5mm thick copper clip design; hence, suggesting that thicker silver layers could act as stress dampeners whilst maintaining the same thermal performance. The simulation results are summarised in a colourmap which demonstrates the chip temperature and the maximum Von Mises stresses for various copper clip and silver thickness combinations, Fig. 10. Finally, Fig. 11 indicates that the maximum displacement occurs on the copper clip interconnection, where the copper clips are bent. A maximum displacement of $9\mu m$ is found, which is less than 2% compared to the 0.5mm thickness of the clip.

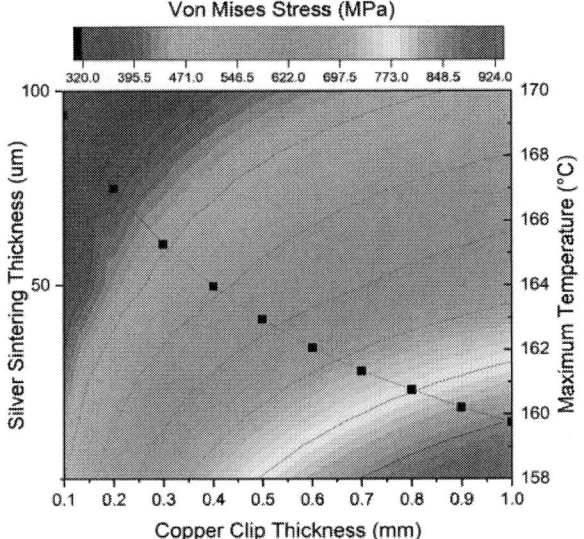

Fig. 10. Thermomechanical performance of the power module for various copper clip and silver thicknesses.

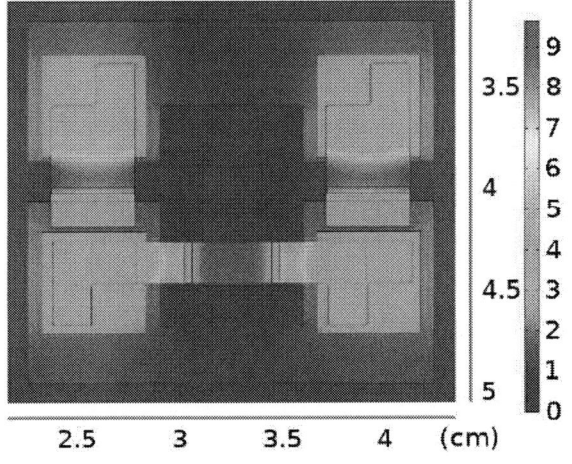

Fig. 11. Total displacement of a 0.5mm copper clip design at 163°C.

IV. CONCLUSION

Copper clips replace traditional bonding wires to reduce the parasitic inductances and enhance the current and heat conduction. The copper clip bonds are realised by double-sided silver sintering, which further improves the electrothermal behaviour of the power module. This paper investigates the effect of different thicknesses of copper clips and silver layers on the electromagnetic, thermal, and thermomechanical behaviour of a 1200V/400A SiC MOSFET half-bridge power module with four chips in parallel. Results suggest that increasing the thickness of the copper clips results in a minor improvement in parasitic inductances, from 3.69nH to 3.56nH at 1GHz for copper clip thicknesses from 0.1mm to 1mm,

respectively. For the same copper clip thickness range, the maximum junction temperature of the SiC MOSFET reduces for thicker copper clips, from 169°C to 159°C. However, as the copper clips become thicker, the thermal stresses are increasing significantly, with Von Mises stresses reaching a maximum of 920MPa on the edge of the chip - silver interface for 1mm thick copper clips. To further optimise the package, the silver layer is increased from an initial thickness of 25μm to 100μm, which resulted in a 50% reduction of stresses, from 777MPa to 410MPa for a 0.5mm copper clip design. The maximum displacement is negligible compared with the copper clip dimensions. The results indicate a trade-off between thermal and thermomechanical behaviour. However, the design should consider the assembly process and the implications that would rise in a practical assembly process of the power module, i.e., the fabrication of very thick preformed copper clips, and the porosity in thick silver sintering layers.

ACKNOWLEDGMENT

This research is supported by the Centre for Doctoral Training (CDT): to Advance the Deployment of Future Mobility Technologies at the University of Warwick, UK. The studentship is partially supported by MacMic Science & Technology Co., Ltd, China.

REFERENCES

[1] Renz, A. B., et al. "The improvement of Mo/4H-SiC Schottky diodes via a P2O5 surface passivation treatment." Journal of Applied Physics 127.2 (2020).

[2] H. Lee, V. Smet and R. Tummala, "A Review of SiC Power Module Packaging Technologies: Challenges, Advances, and Emerging Issues," in IEEE Journal of Emerging and Selected Topics in Power Electronics, vol. 8, no. 1, pp. 239-255, March 2020, doi: 10.1109/JESTPE.2019.2951801.

[3] P. D. Reigosa, H. Wang, Y. Yang and F. Blaabjerg, "Prediction of Bond Wire Fatigue of IGBTs in a PV Inverter Under a Long-Term Operation," in IEEE Transactions on Power Electronics, vol. 31, no. 10, pp. 7171-7182, Oct. 2016, doi: 10.1109/TPEL.2015.2509643.

[4] B. Hu et al., "Failure and Reliability Analysis of a SiC Power Module Based on Stress Comparison to a Si Device," in IEEE Transactions on Device and Materials Reliability, vol. 17, no. 4, pp. 727-737, Dec. 2017, doi: 10.1109/TDMR.2017.2766692.

[5] C. Chen, Z. Huang, L. Chen, Y. Tan, Y. Kang and F. Luo, "Flexible PCB-Based 3-D Integrated SiC Half-Bridge Power Module With Three-Sided Cooling Using Ultralow Inductive Hybrid Packaging Structure," in IEEE Transactions on Power Electronics, vol. 34, no. 6, pp. 5579-5593, June 2019, doi: 10.1109/TPEL.2018.2866404.

[6] J. Cui, R. W. Johnson and M. C. Hamilton, "Reliability of AuGe Die Attach on DBC Substrates With Different Ni Surface Finishes," in IEEE Transactions on Components, Packaging and Manufacturing Technology, vol. 7, no. 10, pp. 1598-1607, Oct. 2017, doi: 10.1109/TCPMT.2017.2733383.

[7] R. Clemente, E. N. Tolentino and M. A. Azman, "Reliability considerations of sintered silver paste on clip semiconductor packages," 2016 IEEE 37th International Electronics Manufacturing Technology (IEMT) & 18th Electronics Materials and Packaging (EMAP) Conference, Georgetown, Malaysia, 2016, pp. 1-6, doi: 10.1109/IEMT.2016.7761965.

[8] H. Wan, M. Antoniou, N. Iosifidis, R. Rong and P. Mawby, "A symmetric low-inductance high-power density SiC power module for EV applications," 2023 35th International Symposium on Power Semiconductor Devices and ICs (ISPSD), Hong Kong, 2023, pp. 298-301, doi: 10.1109/ISPSD57135.2023.10147454.

24th European Microelectronics Packaging Conference

High Frequency Thin Film Magnetics-on-Silicon with Improved Inductance and Resistance

Martin Sittner, Mahmoud Shousha, Martin Haug
MagI³C PU, Würth Elektronik eiSos Group, Munich, Germany
martin.sittner@we-online.de, mahmoud.shousha@we-online.de, and martin.haug@we-online.de

1. Background

In the electronics industry there is a significant trend towards miniaturization of electronic devices. At the same time, system functionality should continue to be guaranteed or improved. To support the increasing level of integration, magnetics-on-silicon is introduced [1]. The main goal of this paper is to show improvement compared to previous works [1] by increasing the magnetic core and copper thicknesses.

2. Method

The approach is to use magnetics on silicon technology based on a thin-film fabrication process known from CMOS technology. The fabrication is done by a basic lithography process combining plasma deposition and electroplating processes. To improve the performance of the magnetics the existing 15µm copper layers as well as the 4.5µm magnetic core are increased to 25µm copper and 6µm core. By using a special CZT magnetic core material the BH loop is optimized to make it compatible with high-frequency applications. With the narrow BH loop, the core AC losses can be kept low at high frequencies enabling this technology to operate in applications with hundreds of megahertz.

3. Results

The increased thicknesses result in improved inductance. Comparing the same design with both stacks, the final test on the wafer level shows a maximum improvement in inductance of 15% for the biased test and single frequency (f=30MHz; DC bias = 0 – 340mA) improvement up to 10% for an unbiased frequency sweep up to 300MHz.

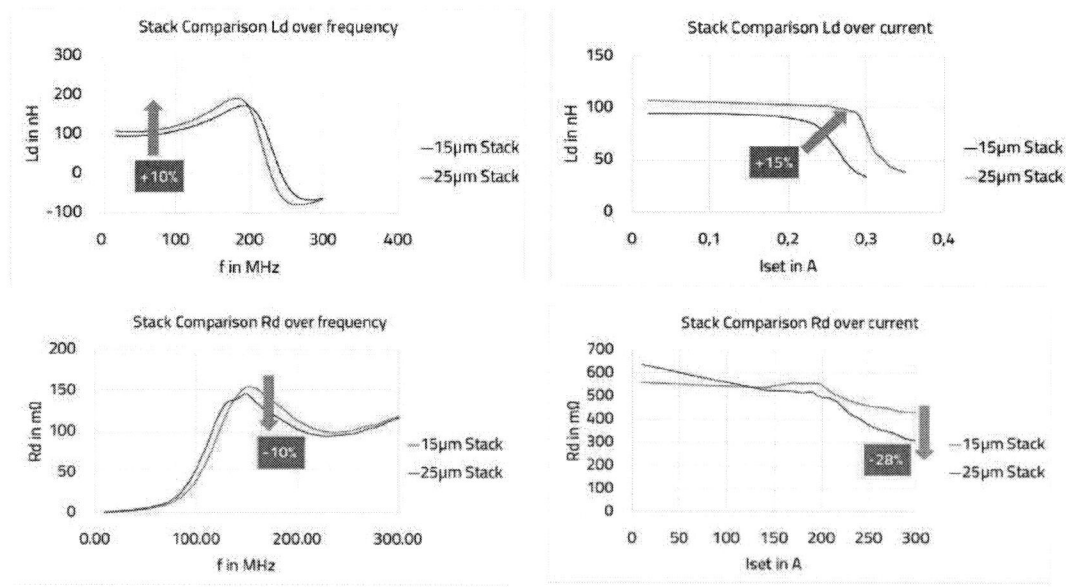

Figure 1: Comparison 15µm stack vs. 25µm stack: Unbiased frequency sweep (left), current sweep (right).

The overall resistance is decreased with high frequency (up to -10% at f > 150MHz) as well as for biased situation. At 300mA the resistance is decreased by 28%. Major differences can be found looking at the RDC. For the shown test structure measurements the RDC value of the 15µm stack sample is 310mΩ whereas the 25µm stack achieved an RDC value of 203mΩ. This RDC reduction of

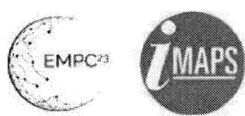

24th European Microelectronics Packaging Conference

34.5% matches to the 40% Cu thickness difference of both stacks. Figure 1 shows the stack comparison.

The flexible design and packaging approaches allows for usage in power applications such as common DC/DC converters or gate drive supplies as well as signal applications like high speed signal lines (i.e. USB 3.1). Contrary to the broad range of possible applications, technology limitations can be seen as listed in table 1:

Table1: Technology limitations.

Parameter	Microinductor	Microtransformer
Inductance	nH range (no economic use case with L in µH range)	
Operation current	< 3A (due to RDC)	
Frequency	Below 1GHz (high power losses in silicon substrate above 1GHz)	

Frequency limitation for the current technology comes from the use of undoped silicon as the substrate. In the GHz range, the power loss within the substrate material needs to be taken into account. Several challenges for changing to different substrate material become apparant. One possible option is to use

Figure 2: 20MHz prototype using a thin film magnetic-on-silicon device

glass as a substrate material since the resistivity is higher than undoped silicon. The downside of using glass is the higher complexity in production processing such as optical issues during packaging. Additionally, wafers with glass substrate are commonly 7" wafer whereas the production process for the introduced technology is based on a 12" wafer. Another option is to change to doped silicon with high resistivity. The disadvantage of the specially high resistive material is the higher production cost leading to an uneconomical wafer cost and therefore to a non-competitive pricing of the final component. The technology's major advantage comes to play operating in high frequency applications. The described core material is able to efficiently operate in the range of up to several hundreds of megahertz. This enables the component values and sizes in power applications to be significantly miniaturized while preserving signal integrity in signal applications since the core material isn't running

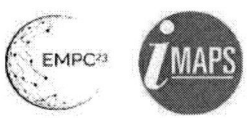

24th European Microelectronics Packaging Conference

in saturation. A prototype was built and tested to prove the effectiveness of the presented thin-film silicon-based magnetic technology by implementing the magnetic device in a high switching frequency non-isolated DC/DC converter meant to be used as an auxiliary supply for power SIP and power SOC applications. The prototype consists of the SC220 synchronous step-down regulator IC which implements an internal 20MHz oscillator block, external input and output capacitors from the Würth Elektronik portfolio (2x 1µF 885012206002 for both input and output capacitors), the resistive feedback divider (set to 2.5V) as well as the silicon-based magnetic device. The converter can operate from 2.7V – 5.5V with an adjustable output voltage from 1V – 5V. Maximum load current is 400mA which fits for auxiliary supply applications. Figure 2 shows the build prototype PCB. The regulator IC requires a minimum external inductance of 220nH. Table 2 shows the magnetic device specifications:

Table2: Microtransformer specifications.

Parameters	Values
L	360nH
Rdc	1.65 Ω
Q-factor	21.4 at 20MHz
Isat	450 mA
Chip Size	3.2 x 2.5 (corresponding to 1210 SMD package size)

At room temperature the converter can achieve an efficiency of up to 82.3% converting 3.3V to 2.5V.

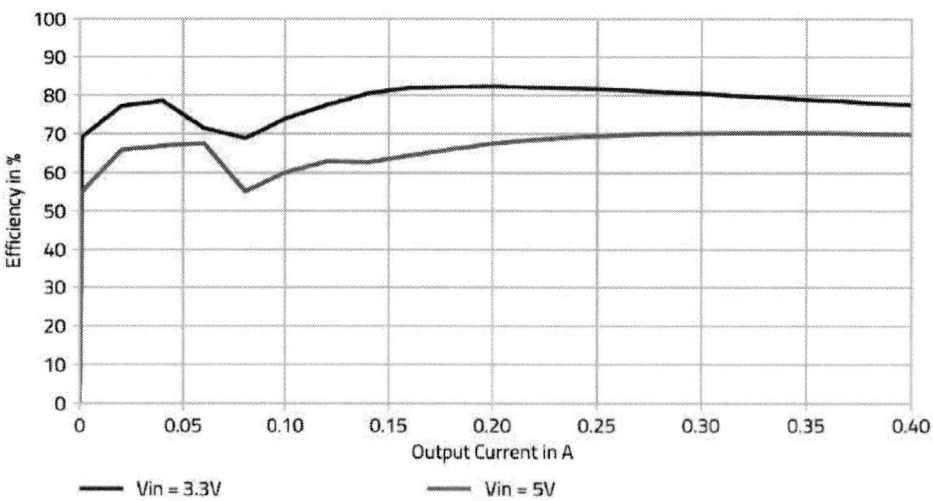

Figure 3: Efficiency at T_A = 25°C

While converting 5V to 2.5V a peak efficiency of 70% is achieved. These results can be seen in figure 3 The thin film magnetic technology is already qualified according to AEC-Q 200 grade 1 which ensures operation for a wide temperature range from -40°C to 125°C. Since the IC is only specified

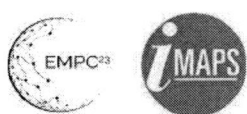

24th European Microelectronics Packaging Conference

from -40°C to 85°C, the efficiency is again measured at these borders. The attained

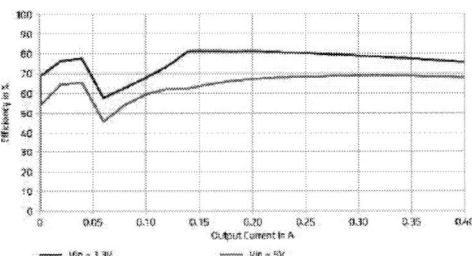

Figure 4a: Efficiency at T_A = 85°C

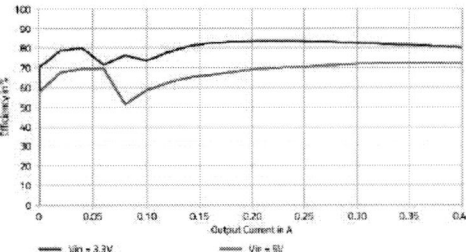

Figure 4b: Efficiency at T_A = -40°C

efficiency can be seen in Figures 4a and 4b. The lack of significant deviation in efficiency shows the temperature robustness of the magnetic component. To validate these results, the shown prototype in Figure 2 was adjusted by replacing the microinductor with a power molded chip inductor from Würth Elektronik WE-PMCI series. The inductor is also packaged in a 1210 SMD package. Figure 5 shows the measured efficiency at room temperature.

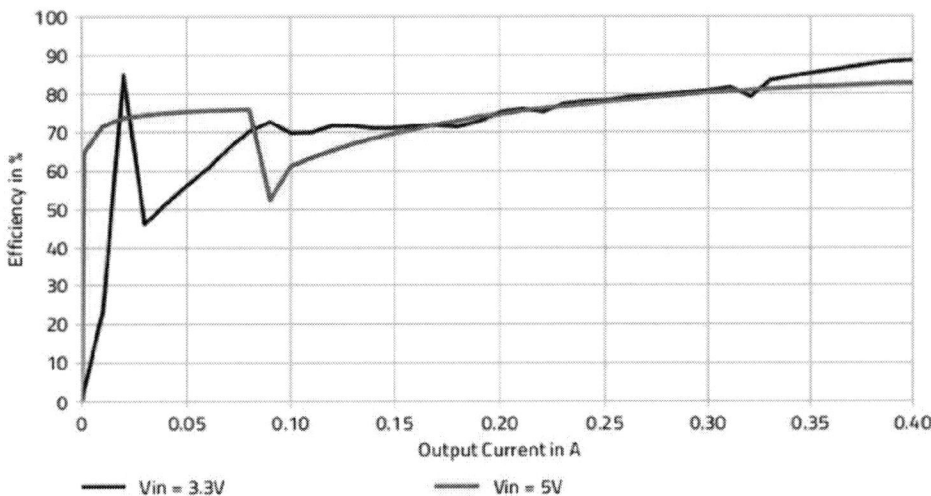

Figure 5: Efficiency at T_A = 25°C

It can be noted from Figure 5 that despite the fact that these inductors have higher inductance and better DCR, the efficiencies of both prototypes are very similar with even higher overall efficiency of the prototype with the thin-film magnetic working with 3.3Vin. For the discussed targe application space is the biggest constrain. Therefore, a magnetic component with little space consumption by keeping decent efficiency to avoid the need for external heat sinking as well as additional cooling cost. The implemented thin-film magnetic chip size reflects the standard SMD 1210 package size but with a reduced height of 200µm height. In comparison, an inductor with a 1210 package typically has a height of 1mm. Therefore, with the shown thin-film magnetic device a chip volume reduction of 80% can be achieved.

4. Discussion

As the packaging approach for this magnetics type is highly flexible, there are myriad possibilities for integration such as WLCSP for PCB soldering and bare dies for wire bonding or embedding.

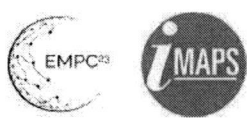

24th European Microelectronics Packaging Conference

Experimental experience shows that as of today WLCSP integration with a given solder ball diameter of 230μm is more suitable for the prototyping phase since handling within production processes turns out to be difficult resulting in a low assembly yield. Therefore, for mass volume production, bonding or embedding is recommended.

5. Conclusion

This paper shows improvement to the already achieved results. Improving inductance and resistance is a key enabler for the adoption of this technology within a wider range of applications. Further, a fully functioning DC/DC converter prototype operating with very high frequency (20MHz) is described and measured proving the technology's high frequency capability. It's proven that the magnetic component is able to handle a broad temperature range. The converter can operate with efficiency over 82% range with a magnetic component volume reduction of 80%.

6. References

[1] S. L. Selvaraj *et al.*, "On-Chip Thin Film Inductor for High Frequency DC-DC Power Conversion Applications," *2020 IEEE Applied Power Electronics Conference and Exposition (APEC)*, 2020, pp. 176-180, doi: 10.1109/APEC39645.2020.9124544.

Enhanced reliability for power modules via a new Ag/Si sinter joining strategy

1st Yang Liu
Flexible 3D System Integration Laboratory
Osaka University
Osaka, Japan
liu-yang@sanken.osaka-u.ac.jp

2nd Chuantong Chen
Flexible 3D System Integration Laboratory
Osaka University
Osaka, Japan
https://orcid.org/0000-0002-2365-4387

3rd Koji S. Nakayama
Flexible 3D System Integration Laboratory
Osaka University
Osaka, Japan
https://orcid.org/0000-0002-9165-388.5

4th Minoru Ueshima
Daicel Corporation
Osaka, Japan
mn_ueshima@jp.daicel.com

5th Takeshi Sakamoto
Daicel Corporation
Osaka, Japan
tk2_sakamoto@jp.daicel.com

6th Naoe Takuya
Joining and Welding Research Institute
Osaka University
Osaka, Japan
naoe@jwri.osaka-u.ac.jp

7th Hiroshi Nishikawa
Joining and Welding Research Institute
Osaka University
Osaka, Japan
https://orcid.org/0000-0002-4363-9520

8rd Katsuaki Suganuma
Flexible 3D System Integration Laboratory
Osaka University
Osaka, Japan
https://orcid.org/0000-0003-2137-9905

Abstract—In this work, with an appropriate adding amount of 20 vol% Si, the novel Ag-Si sintering demonstrated enhanced reliability during thermal shock cycling, specifically including the improved bonding strength stability as well as good microstructure maintenance. The high reliability can be attributed to the reduced CTE value of composited Ag-Si sintering inhibition of thermomechanical stress generation, allowing for the ultra-stable joining structure without coarsening, delamination and deformation. Additionally, the well-adhered interface between sintered Ag and Si additive was also verified by examining the microstructure, ensuring the reliable bonding characteristics of the Ag/Si sintered joint.

Keywords—die-attach, Ag sintering, CTE mismatch, thermo-mechanical stress

I. INTRODUCTION

Owing to the high conversion efficiency and switching frequency, the WBG materials (SiC, GaN, Ga_2O_3) permit power modules to be smaller, faster and more efficient than the counterpart Si-based devices by supplying higher power in a reduced semiconductor area. The merits of WBG semiconductors also create a harsh operation environment at temperature as high as 600 °C [1-2]. A major requirement for further development of WBG devices is to make packaging materials compatible with the harsh working environment. Recently, Ag particles sintering is a novel bonding technique that provides a service temperature up to the melting point of Ag (960°C) in theoretical and high electrical ($63*10^6$ S/m) and thermal (429 W/mK) conductivities, because of which, it is highly advantageous than the currently solders. However, a major challenge lies in the mismatch of coefficient of thermal expansion (CTE) between sintered Ag (18.9×10^{-6}/K) and SiC chip (4.1×10^{-6}/K)/ DBC substrate (3.2×10^{-6}/K) [3-5]. Such CTE mismatch would induce strong thermo-mechanical stress as the module is operated under temperature swings from repetitive switch on and off, which would subsequently be relieved by defection growth and material bending and the following joining failure.

Herein, a novel composite paste is proposed wherein silicon microparticles were added into Ag sintering matrix because of such superior properties as low density and CTE value. Si has intrinsic low CTE value of 2.6×10^{-6}/K quite close to those of SiC chip and DBC substrate. Thereby, an Ag/Si composite paste is expected to possess a more compatible CTE value with module components and thus mitigated thermo-mechanical properties as compared to pure sintered Ag.

II. METHOD

A. Fabrication of composited Ag/Si sinter paste

The micro-Ag flakes and bulk Si particles were adopted as the raw materials to prepare the composite joining materials. To systematically analyze the effect of Si additives on the bonding properties, the Ag/Si pastes with different Si volume ratios were prepared, that is, Ag100%-Si0%, Ag92%-Si8%and Ag80%-Si20% (hereafter denoted as, Ag100%, Ag-Si8% and Ag-Si20%). These particles were initially mixed using magnetic stirring and ultrasonication in ethanol. The mixtures were then dispersed in CELTOL-IA (CxHyOz, x>10, Diacel Corporation) to obtain three types of sinter pastes of Ag100%, Ag-Si8% and Ag-Si20%, respectively. The SiC/DBC die-attached structure was then established using the as-prepared pastes through a pressure-less sintering process at 250 °C.

B. Thermal shock test and evaluations

A harsh thermal shock test was conducted at the temperature range from -50 °C to 250 °C for 500 cycles. the variation in bonding strength was measured by shear tester (DAGE 4000 bonds tester, UK) at 50 μm/s. The microstructure evolution during thermal cycling was then visualized using SEM (SU-8020, Hitachi). The interface between sintered Ag and Si particles within the sintered joining layer was examined using TEM (TEM, JEM-ARM200F, JEOL).

III. RESULTS AND DISCUSSIONS

In a composite sinter joining system, the interface between sintered silver (Ag) and silicon (Si) additive is a critical factor that determines the bonding quality. If the interface is weak, it can easily lead to crack initiation and degradation on bonding strength. Therefore, ensuring a robust Ag/Si bonding requires special attention to the interface between sintered Ag and Si. Thereby, the Ag/Si interface was pre-checked, and a thermal reliability evaluation was subsequently conducted. This evaluation involved assessing the die shear strength and observing the evolution of microstructure.

A. Analysis of Ag/Si interface of composite sinter network

Fig.1a shows the cross-section of initially formed Ag-Si20% sinter structure. It can be observed the uniform porous Ag structure was integrated with some Si bulk particle. The local enlarged view at the interface between sintered Ag and Si was presented in Fig. 1b. Si additive was found to tightly adhered to sintered Ag structure without apparent voids or gap detected. A further TEM observation was conducted by thinning the sintered Ag-Si20% joint structure through Focus ion beam milling. It is clearly seen the well-connected interface formed between the porous Ag and Si as shown in Fig. 1c. was indeed connected at certain area. The high-resolution TEM in Fig. 1d shows that no gap can be found even in the nanoscale observation. This result demonstrate an integrated Ag-Si sintering structure was obtained which is promising to provide reliable bonding quality.

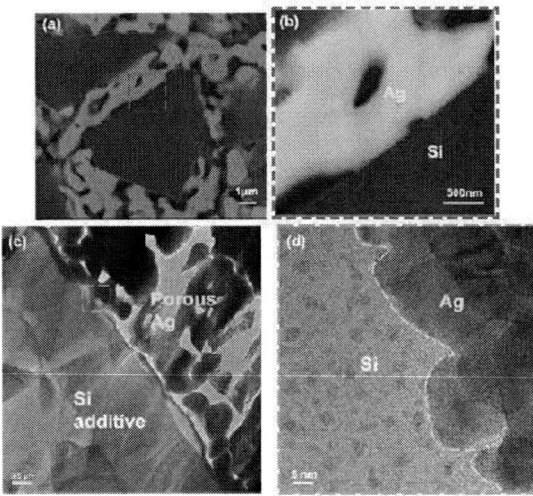

Fig. 1. (a) The cross-sectional images of Ag-Si20% sintering structure at the initial state. (b) The local enlarged view at the interface between Ag and Si (c), (d) TEM observation at the Ag/Si interface and its local enlarged view.

B. Evaluation on thermal properties

Considering the critical role of die-attach materials on thermal management, the change in thermal conductivity of the composited sintering structure were investigated as well. Fig. 2 shows that the thermal conductivity decreased gradually with the increase in Si amount due to the inherent low thermal conductivity of Si of 148 W/K·m. Although the thermal conductivity of Ag-Si20% dropped significantly to approximately 80 W/K·m that is almost half that of Ag100% sintered structure. This thermal property is still much better than the poor thermal conductivity of current Sn-based solders below 50 W/K·m. On the other hand, adding silicon at a volume ratio of 20% can significantly enhance the cost-effectiveness of Ag sinter joining. With the merits of both reduced cost and acceptable thermal conductivity, it is believed that the Ag/Si composited sintering holds promise as a reliable die-attachment used for high-temperature application.

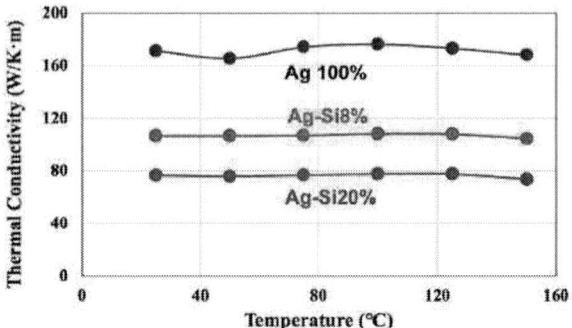

Fig.2. Thermal conductivity of Ag100%, Ag-Si8%, Ag-Si20% sintered structure with temperature.

C. Evaluation on bonding properties

Fig. 3a shows the die-attached module structure joined by three types of sinter pastes, Ag100%, Ag-Si8% and Ag-Si20%. The as-bonded module structures were then subjected to a harsh thermal shock cycling. Fig. 3b presents the corresponding die shear results. It can be seen that the initial bonding performance decreased with the increase in the amount of Si additives, where the Ag100% exhibited the highest shear strength of 60.1 MPa and Ag-Si20% exhibited the lowest bonding strength of 33.9 MPa. This is attributed to the inevitable decrease due to the integration of a heterogenous element into Ag homogeneous system. Of noted, different variation trend of shear strength occurred between joining structures during thermal cycling. With increasing number of thermal cycles, A similar rapid decline trend was found in the Ag100% and Ag-Si8% joint structures. However, a more gently decline trend appears in the Ag-Si20% joint. The specific difference was presented via retained ratio in bonding strength after 500 thermal cycles as shown in Fig. 3c. The strength retention rate of Ag-Si20% joint is 71.9%, which is obviously superior to the other joint structures. This result indicate that the addition of 8 vol% Si hardly affect the changes of bonding strength of sintered Ag, while the addition of 20 vol% Si can slow down the decline trend in shear strength. The detailed mechanism should be investigated through further microstructural analysis.

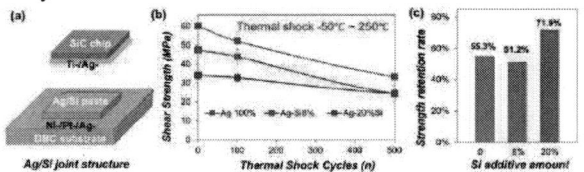

Fig.3. (a) The schematic for die-attachment via Ag-Si sinter joining. (b) Die-shear strength of joints by three types of Ag sintered pastes during thermal cycles. (c) Retention rate of shear strength after 500 thermal cycles.

D. Anlysis on microstruture evolution

The microstructural changes of the overall cross-section of three die-attached structures during thermal shock cycling are displayed in Fig. 4. All initial structures presented a porous network typical of press-less Ag sintering as shown in Fig. 4a. Meanwhile, some randomly distributed Si particles can be clearly seen in the Ag-Si8% and Ag-Si20% joining layers as shown in Fig. 4c and 4e. After 500 thermal cycles, distinct differences appeared in the defect growth among the three joining layers. Figure 4b shows that many cracks grow in the vertical direction with their size even spanning the whole joining layer. Additionally, the long cracks at horizontal direction that easily cause interface delamination were also found. The severe degradation of the microstructure of Ag-Si8% joint was similar to that of Ag100% joint, with both large vertical and horizontal cracks occurred, as shown in Fig. 4d. However, Fig. 4f shows that these serious degradation was significantly mitigated in the Ag-Si20% joint where only a few short vertical cracks was detected. This result indicates that 8 vol% Si particles are not sufficient to inhibit the deterioration of the microstructure, while 20 vol% Si has a significant effect on the maintenance of the microstructure. The improvement in microstructure deterioration of Ag-Si20% joint indicated the alleviated thermo-mechanical stress within the module structure. The severe microstructure evolution in Ag100% and Ag-Si8% joints was consistent well with their rapid decrease in shear strength.

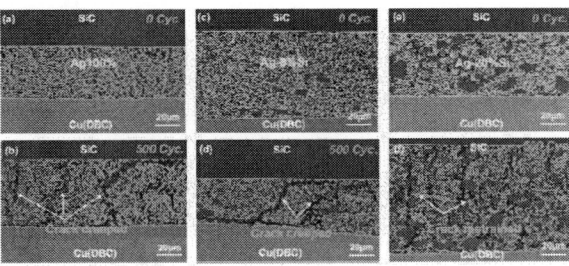

Fig. 4 The cross-sectional images of three die-attached structures before and after 500 thermal shock cycles. (a, b) Ag100% joint; (c, d) Ag-Si8% joint (e, f) Ag-Si20% joint.

The enlarged microstructure views at bonding interface were provided for a better understanding of the effect of Si additive on microstructure evolution. Since Ag100% and Ag-Si8% showed similar changes after thermal cycling, Ag100% was used as a representative for comparison with Ag-Si20% joint. Fig. 5a and b show the bonding structure at the interfaces of Ag100% sinter layer with SiC chip and DBC substrate after 500 thermal cycles, respectively. It can be clearly seen that large cracks of micron size appear at both the upper and lower interfaces, which should be the primary cause of bonding degradation and fracture of the Ag100% joint structure. In contrast, in the Ag-Si 20% joint, the upper bonding interface was well preserved and neither cracks nor delamination were observed. The crack size at the bottom interface is also reduced as displayed in Fig. 5d, further confirming the weakened structural damage caused by thermo-mechanical stress in the Ag-Si20% joint.

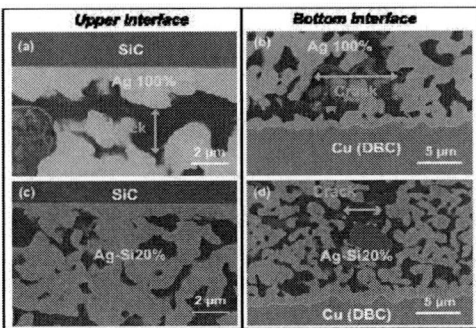

Fig. 5 The cross-sectional images at bonding interfaces with chip and substrate after 500 thermal cycles. (a, b) Ag 100% joint; (c, d) Ag-Si20% joint.

IV. SUMMARY

In summary, the Ag100% and Ag-Si8% joint showed decreased bonding strength accompanied by degraded microstructure, while a significant improvement in reliability was achieved for the Ag-Si20% joint structure during the harsh thermal cycling. Based on a comprehensive investigation of strength evaluation and microstructural stability, it can be concluded that, an appropriate amount of Si additive play a critical role in alleviating the thermal stress derived from CTE mismatch. It is believed that the Ag/Si composite sintering with well-connected interfaces is a promising die-attach solution superior to sintered Ag for future high-power electronics applications.

ACKNOWLEDGMENT

This work was supported in part by the New Energy and Industrial Technology Development Organization NEDO under Grant Number JPNP20004, and JSPS KAKENHI Grant Number JP22K04243. The author acknowledges both of financial support and material supply from Daicel Corporation.

REFERENCES

[1] X. Long, Y. Guo, Y. Su, K.S. Siow, C. Chen, Constitutive, creep, and fatigue behavior of sintered Ag for finite element simulation of mechanical reliability: a critical review, Journal of Materials Science: Materials in Electronics (2022) 1-17.

[2] C. Chen, K. Suganuma, Large-scale ceramic–metal joining by nano-grained Ag paste sintering in low-temperature pressure-less conditions, Scripta Materialia 195 (2021) 113747.

[3] K. Siow, Y. Lin, Identifying the development state of sintered silver (Ag) as a bonding material in the microelectronic packaging via a patent landscape study, Journal of Electronic Packaging 138(2) (2016).

[4] C. Chen, K. Suganuma, T. Iwashige, K. Sugiura, K. Tsuruta, High-temperature reliability of sintered microporous Ag on electroplated Ag, Au, and sputtered Ag metallization substrates, Journal of Materials Science: Materials in Electronics 29(3) (2018) 1785-1797.

[5] T. Fan, H. Zhang, P. Shang, C. Li, C. Chen, J. Wang, Z. Liu, K. Suganuma, Effect of electroplated Au layer on bonding performance of Ag pastes, Journal of Alloys and compounds 731 (2018) 1280-1287.

Authors' Index of all technical papers presented at the 24th European Microelectronics and Packaging Conference (EMPC2023)

Name	Paper-IDs	Session #
Aasmundtveit, Knut Eilif	168, 191, 197, 129	S 2A:4, P:10, P:12, S 8A:1
Abdi, Salim	107	S 6A:1
Abu-Hamdeh, Ghassan	185	S 3B:2
Adolfo, Annbel	103	S 2B:3
Ahtonen, Arto	111	S 7B:1
Ahuir-Torres, Juan Ignacio	150, 152	S 5A:2, S 7A:2
Ali, Ishpa	130	S 4B:1
Almhem, Lars	111	S 7B:1
Altuntaş, Fırat	149	P:7
Andena, Luca	193	P:6
Andreassen, Erik	197	P:12
Antoniou, Marina	166	S 10B:2
Araki, Hitoshi	121	S 3A:2
Araki, Noritoshi	175	S 2A:2
Ayala, Heaklig	173	S 7A:3
Bailey, Christopher	180	S 10B:1
Bailey, Richard	143	S 4B:2
Bainbridge, Andrew	195	S 8B:2
Baishya, Kangkana	162	S 8A:3
Barth, Henry	188	S 9A:3
Bazilchuk, Molly	168	S 2A:4
Bechtold, Franz	183	S 1A:3
Behrmann, Ole	199	S 2B:4
Belponer, Federico	140	S 5A:1
Benabou, Lahouari	131	S 7A:1
Benešová, Andrea	113	P:2
Bhogaraju, Sri Krishna	140, 150, 152, 141	S 5A:1, S 5A:2, S 7A:2, S 9B:2
Binner, Ralph	175	S 2A:2
Bock, Karlheinz	189, 198, 146, 174	S 1B:3, P:14, S 6A:3, S 6A:4
Bock, Volker	118	S 5B:1
Bogoni, Antonella	112	S 2B:1
Bolk, Jeroen	107	S 6A:1
Booker, Paul	178	S 2A:1
Borden, Matt	185, 181	S 3B:2, S 8B:4
Böttcher, Lars	109	S 1B:1
Braden, Derek	162	S 8A:3
Braun, Silvia	156	P:5
Braun, Tanja	171, 184	S 9A:1, S 10A:4
Breuer, Dirk	198	P:14
Brookes, Toby G.	178	S 2A:1
Burt, David	178	S 2A:1
Carminati, Roberto	112	S 2B:1
Catalano, Guendalina	154	S 4A:2

Authors' Index of all technical papers presented at the
24th European Microelectronics and Packaging Conference (EMPC2023)

Name	Paper-IDs	Session #
Chan, Kai Chong	102	S 10A:1
Chan, Y. C.	162	S 8A:3
Chee, Vincent	175	S 2A:2
Chen, Chuantong	194, 161, 187	P:15, S 8A:2, S 10B:4
Chen, Jiajia	111	S 7B:1
Chen, Jin	111	S 7B:1
Chen, Ming-Hung	205	S 4B:4
Chen, Zihao	102	S 10A:1
Chernobryvko, Mykola	184	S 10A:4
Chiesa, Marco	112	S 2B:1
Choi, Gwang-Mun	108, 125	P:3, S 6B:2
Choi, Halim	136	S 6B:1
Choi, Kwang-Seong	108, 125	P:3, S 6B:2
Cholaj, Aneta	160	P:16
Chua, Linda	102	S 10A:1
Cirulis, Imants	156	P:5
Clark, Lewis	195	S 8B:2
Corazza, Alessio	135	S 7B:2
Cotty, James	173	S 7A:3
Cox, Kevin	185, 181	S 3B:2, S 8B:4
Cyriax, Andrea	164	S 2B:2
Damalerio, Joel	102	S 10A:1
De Girolamo, Julia	122	S 4A:1
De Wit, Ruud	153	S 5A:3
Del Sarto, Marco	103, 112	S 2B:3, S 2B:1
Deschaseaux, Edouard	182	S 8A:4
Despax-Ferreres, Auriane	122	S 4A:1
Do, Nu Bich Duyen	197	P:12
Donaldson, Nick	143	S 4B:2
Drechsel, Johannes	188	S 9A:3
Du, Yuekang	158	P:8
Ekere, Nduka	162	S 8A:3
Elger, Gordon	140, 150, 152, 141	S 5A:1, S 5A:2, S 7A:2, S 9B:2
Ellinger, Frank	189	S 1B:3
Elliott, Andrew	170	S 5B:3
Elrifai, Emad	104	S 1A:1
Eom, Yong-Sung	108, 125	P:3, S 6B:2
Erdogan, Hüseyin	141	S 9B:2
Eto, Motoki	175	S 2A:2
Feautrier, Céline	182	S 8A:4
Fedeli, Patrick	103	S 2B:3
Feldmann, Jochen	175	S 2A:2
Fischer, Michael	164	S 2B:2

Authors' Index of all technical papers presented at the
24th European Microelectronics and Packaging Conference (EMPC2023)

Name	Paper-IDs	Session #
Franiatte, Rémi	182	S 8A:4
Franke, Jörg	115	S 3A:1
French, Marcus J.	178	S 2A:1
Friedrich, Aline	128	P:11
Galinski, Stefan	160	P:16
Gausden, Johannes	143	S 4B:2
Gernhardt, Robert	184	S 10A:4
Ghorbanian, Navid	178	S 2A:1
Glinski, Krzysztof	160	P:16
Goh, Hin Hwa	102	S 10A:1
Gojdka, Björn	199	S 2B:4
Goldberg, Adrian	183	S 1A:3
Gonzalez, Mario	181	S 8B:4
Gougeon, Julie	182	S 8A:4
Grant, Kathleen	195	S 8B:2
Greci, Alessio	140	S 5A:1
Gritti, Alex	103	S 2B:3
Guina, Mircea	123	S 6A:2
Guo, Sihua	111	S 7B:1
Guo, Yusong	142	S 4B:3
Gutzeit, Nam	164	S 2B:2
Haag, Sabine	171	S 9A:1
Hadrava, Philipp	141	S 9B:2
Hagen, Gunter	183	S 1A:3
Haibara, Teruo	175	S 2A:2
Hamacek, Ales	138	P:1
Hangen, Ude	120	S 2A:3
Harr (Martinsen), Kristoffer	111	S 7B:1
Harvey, David Mark	162	S 8A:3
Hasegawa, Norihiko	161	S 8A:2
Hashizume, Kei	161	S 8A:2
Hassan, Sheikh Rokibul	180	S 10B:1
Häußler, Felix	115	S 3A:1
Hecht, Christoph	115	S 3A:1
Helmers, Henning	147	S 9B:1
Herrmann, Jeffrey W.	155	S 6B:3
Hibino, Chika	121	S 3A:2
Hintz, Michael	164	S 2B:2
Hirman, Martin	113	P:2
Hlina, Jiri	138	P:1
Ho, Cheng-Yu	159	S 10A:2
Hoff, Lars	139	P:9
Hölck, Ole	171	S 9A:1

Authors' Index of all technical papers presented at the
24th European Microelectronics and Packaging Conference (EMPC2023)

Name	Paper-IDs	Session #
Hollingham, James	178	S 2A:1
Homma, Hidekazu	161	S 8A:2
Hoyer, Christian	189	S 1B:3
Hsiao, Wen-Chun	159	S 10A:2
Hsieh, Ming-Che	102	S 10A:1
Hsieh, Ming-Chun	161	S 8A:2
Hsieh, Sheng-Chi	159	S 10A:2
Huang, Chengliang	170	S 5B:3
Huang, Chien-Ming	155	S 6B:3
Huang, Hong-Sheng	159	S 10A:2
Huber, Saskia	177	S 8B:1
Huynh, Van Long	129	S 8A:1
Illés, Balázs	136	S 6B:1
Imenes, Kristin	197	P:12
Iosifidis, Nikolaos	166	S 10B:2
Ishii, Masami	148	S 3B:1
Jan Geluk, Erik	107	S 6A:1
Jang, Ki-Seok	108, 125	P:3, S 6B:2
Jiang, Dai	130	S 4B:1
Jiao, Yuqing	107	S 6A:1
Johnson, Greg M.	170	S 5B:3
Joo, Ji-Ho	108, 125	P:3, S 6B:2
Joseph, Yvonne	196	S 7B:3
Jousseaume, Vincent	122	S 4A:1
Kaczynski, Jaroslaw	120	S 2A:3
Kagami, Noriko	161	S 8A:2
Kaiser, Elisa	147	S 9B:1
Kaiser, Michael P.	177, 184	S 8B:1, S 10A:4
Kalkan, Olcay	149	P:7
Kaneki, Takayuki	121	S 3A:2
Karout, Mohamed-Amer	173	S 7A:3
Kettle, Jeff	195	S 8B:2
Kleinholz, Cathleen	164	S 2B:2
Klej, Tomasz	160	P:16
Klengel, Robert	175, 186	S 2A:2, P:13
Klengel, Sandy	175, 186	S 2A:2, P:13
Kobayashi, Koji	194	P:15
Koç, Emrah	149	P:7
Koscielski, Marek	160	P:16
Köszegi, Julia-Marie	177, 184	S 8B:1, S 10A:4
Kotadia, Hiren	150, 152	S 5A:2, S 7A:2
Krantz, Jason	181	S 8B:4
Krengel, Markus	128	P:11

Authors' Index of all technical papers presented at the
24th European Microelectronics and Packaging Conference (EMPC2023)

Name	Paper-IDs	Session #
Krieger, Uwe	183	S 1A:3
Kristiansen, Helge	168	S 2A:4
Kuechenmeister, Frank	198	P:14
Kuhn, Harald	156, 196	P:5, S 7B:3
Kuramochi, Satoru	117	P:17
Kurth, Steffen	156	P:5
Kuylenstierna, Dan	184	S 10A:4
Labie, Riet	181	S 8B:4
Lancashire, Henry	130	S 4B:1
Le, Duy Hoang	139	P:9
Lee, Chan-Mi	108, 125	P:3, S 6B:2
Lee, S. W. Ricky	142	S 4B:3
Leone, Federico	148	S 3B:1
Lepukhov, Evgenii	123	S 6A:2
Li, Jianfeng	158	P:8
Lin, Yaojian	102	S 10A:1
Lipp, John D.	178	S 2A:1
Lisec, Thomas	199	S 2B:4
Litschke, Oliver	128	P:11
Liu, Feixiang	158	P:8
Liu, Johan	111	S 7B:1
Liu, Kang	102	S 10A:1
Liu, Liangjie	158	P:8
Liu, Ran	194	P:15
Liu, Yang	187	S 10B:4
Lo, Jeffery C. C.	142	S 4B:3
Löher, Thomas	109	S 1B:1
Lohse, Sascha	200	S 4A:4
Lorenz, Lukas	118	S 5B:1
Ludewig, Thomas	118	S 5B:1
Machani, Kashi Vishwanath	198	P:14
Maggi, Luca	103, 112	S 2B:3, S 2B:1
Maierna, Amedeo	112	S 2B:1
Manh, Tung	139	P:9
Manhica, Birgit	116	S 1B:2
Manzanera, Teresa Partida	162	S 8A:3
Mariani, Stefano	193	P:6
Mauri, Luca	135	S 7B:2
Mawby, Philip	173, 166	S 7A:3, S 10B:2
Meier, Karsten	189, 198	S 1B:3, P:14
Mellina Gottardo, Alessandro	154	S 4A:2
Mendicino, Gianluca	112	S 2B:1
Mendizabal, Laurent	182	S 8A:4

Authors' Index of all technical papers presented at the
24th European Microelectronics and Packaging Conference (EMPC2023)

Name	Paper-IDs	Session #
Menjo, Toshiaki	119	S 5B:2
Mistry, Akash Sunilkumar	146	S 6A:3
Mizuno, Jun	190	P:4
Mohan, Nihesh	150, 141	S 5A:2, S 9B:2
Mohd Salleh Arif, Mohd Arif Anuar	162	S 8A:3
Moloudi, Reza	181	S 8B:4
Moon, Seok-Hwan	108, 125	P:3, S 6B:2
Moon, Yoon-Hwan	125	S 6B:2
Morath, Helmuth P. E.	189	S 1B:3
Müller, Jens	172, 164	S 1A:2, S 2B:2
Murugesan, Kavin Senthil	184	S 10A:4
Murugesan, Murali	111	S 7B:1
Nakayama, Koji S.	187	S 10B:4
Naoe, Takuya	187	S 10B:4
Napolitano, Teresa	103	S 2B:3
Navrátil, Jiří	113	P:2
Ndip, Ivan	165, 184	S 10A:3, S 10A:4
Nghiem, Giang	168	S 2A:4
Nguyen, Hoang-Vu	191, 197, 129	P:10, P:12, S 8A:1
Niederhoffer, Thomas	130	S 4B:1
Niegisch, Corinna	171	S 9A:1
Nieweglowski, Krzysztof	189, 174, 146	S 1B:3, S 6A:4, S 6A:3
Nishikawa, Hiroshi	190, 187	P:4, S 10B:4
Nkansah, Amos	111	S 7B:1
O'Neill, Antony	143	S 4B:2
Oh, Jin-Heuk	108	P:3
Oh, Jin-Hyuk	125	S 6B:2
Ojanen, Samu-Pekka	123	S 6A:2
Okamoto, Kazamasa	161	S 8A:2
Oklej, Jan	160	P:16
Okumuara, Rieko	161	S 8A:2
Ollmann, Henrik	118	S 5B:1
Ortiz-Gonzalez, Jose	173	S 7A:3
Ortlepp, Thomas	164	S 2B:2
Osanai, Hideyo	194	P:15
Ostaszewski, Dariusz	160	P:16
Ostmann, Andreas	109	S 1B:1
Öznazlı, Nihan	149	P:7
Pang, Yong	158	P:8
Pantou, Remi	156	P:5
Park, Byungho	190	P:4
Passagrilli, Carlo	193	P:6
Perez Henriquez, Carlos	130	S 4B:1

Authors' Index of all technical papers presented at the
24th European Microelectronics and Packaging Conference (EMPC2023)

Name	Paper-IDs	Session #
Perlwitz, Paul	165	S 10A:3
Peters, Henk	175	S 2A:2
Pötter, Harald	165	S 10A:3
Radestock, Moritz	156	P:5
Rajaguru, Pushparajah	180	S 10B:1
Ratti, Andrea	103	S 2B:3
Razkordanisharahi, Amirabbas	118	S 5B:1
Rebenklau, Lars	188	S 9A:3
Reboun, Jan	138	P:1
Reinhardt, Kathrin	116	S 1B:2
Ringelstetter, Severin	133	S 3B:3
Rizzi, Enea	135	S 7B:2
Rohlfs, Patrick	198	P:14
Rong, Rui	166	S 10B:2
Roscher, Frank	196	S 7B:3
Roshanghias, Ali	120	S 2A:3
Rotta, Davide	112	S 2B:1
Rovitto, Marco	193	P:6
Rummel, Andreas	170	S 5B:3
Rumney, Tim	173	S 7A:3
Saito, Mikiko	190	P:4
Sakamoto, Takeshi	187	S 10B:4
Sandmann, Gunther	198	P:14
Sattari, Romina	202	P:18
Schadow, Eric	115	S 3A:1
Scheer, Achim	175	S 2A:2
Schirmer, Patrick	133	S 3B:3
Schletz, Andreas	116	S 1B:2
Schneider-Ramelow, Martin	177, 171, 165, 184	S 8B:1, S 9A:1, S 10A:3, S 10A:4
Schneider, Andreas	178	S 2A:1
Schroeter, Annett	183	S 1A:3
Schulz, Alexander	172	S 1A:2
Scott, Travis	200	S 4A:4
Seckel, Manuel	109	S 1B:1
Seki, Hidetoshi	148	S 3B:1
Selbmann, Franz	156, 196	P:5, S 7B:3
Serrano, Aina	112	S 2B:1
Shah Idil, Ahmad	130, 143	S 4B:1, S 4B:2
Shaw, Mark Andrew	112, 103	S 2B:1, S 2B:3
Shen, Zhiyang	111	S 7B:1
Shoji, Yu	121	S 3A:2
Simon, Olivier	131	S 7A:1
Simon, Winfried	128	P:11

Authors' Index of all technical papers presented at the
24th European Microelectronics and Packaging Conference (EMPC2023)

Name	Paper-IDs	Session #
Simoncini, Daniele	103	S 2B:3
Simonovsky, Marek	138	P:1
Singh, Prashant Kumar	198	P:14
Sittner, Martin	106	S 10B:3
Skwarek, Agata	136	S 6B:1
Smits, Edsger	153	S 5A:3
Souriau, Jean-Charles	122, 182	S 4A:1, S 8A:4
Sprenger, Mario	115	S 3A:1
Stegmann, Heiko	170	S 5B:3
Steinberger, Fabian	141	S 9B:2
Steiner, František	113	P:2
Stephan, Tino	186	P:13
Steplewski, Wojciech	160	P:16
Stoll, Thomas	115	S 3A:1
Stoyanov, Stoyan	180	S 10B:1
Suetake, Aiji	161	S 8A:2
Suganuma, Katsuaki	194, 161, 187	P:15, S 8A:2, S 10B:4
Swiecinski, Kai	118	S 5B:1
Tai, Takahiro	117	P:17
Takano, Ken	119	S 5B:2
Takyu, Shinya	119	S 5B:2
Talledo, Jefferson	103	S 2B:3
Tanaka, Masaya	117	P:17
Tiquet, Pascal	122	S 4A:1
Tréguer-Delapierre, Mona	182	S 8A:4
Tschoban, Christian	165	S 10A:3
Ueshima, Minoru	187	S 10B:4
Ugolini, Francesco	140	S 5A:1
Uhlig, Peter	128	P:11
Uusitalo, Topi	123	S 6A:2
van Dijk, Marius	177	S 8B:1
van Zeijl, Henk	202	P:18
Vandevelde, Bart	181	S 8B:4
Vanstreels, Kris	181	S 8B:4
Vareilles, Victor	147	S 9B:1
Varghese, Jobin	116	S 1B:2
Viheriälä, Jukka	123	S 6A:2
Villa, Claudio Maria	193	P:6
Villa, Riccardo	154	S 4A:2
Virtanen, Heikki	123	S 6A:2
Viswanathan, Vignesh	170	S 5B:3
Vivet, Laurent	131	S 7A:1
Viviani, Fulvio	148	S 3B:1

Authors' Index of all technical papers presented at the 24th European Microelectronics and Packaging Conference (EMPC2023)

Name	Paper-IDs	Session #
Vlasov, Aleksandr	123	S 6A:2
Vogel, Klaus	156	P:5
Wager, Jens	189	S 1B:3
Walter, Hans	177	S 8B:1
Wan, Haiyong	166	S 10B:2
Wang, Chen-Chao	159	S 10A:2
Wang, Xingzhi	158	P:8
West, Geoff	152	S 7A:2
Weyers, David	174	S 6A:4
Wieland, Marcel	198	P:14
Wiemer, Maik	196	S 7B:3
Wiese, Steffen	172	S 1A:2
Wiesenfarth, Maike	147	S 9B:1
Williams, Kevin	107	S 6A:1
Wiss, Erik	172	S 1A:2
Wittler, Olaf	177	S 8B:1
Wunderle, Bernhard	156	P:5
Xu, Qianwen	142	S 4B:3
Xue, Fei	130	S 4B:1
Yamada, Tadatomo	119	S 5B:2
Yamada, Takashi	175	S 2A:2
Yin, Ran	189	S 1B:3
Yoshida, Hiroshi	161	S 8A:2
Yoshida, Ryuji	161	S 8A:2
Yuile, Adam	172	S 1A:2
Zafarana, Giovanni	135	S 7B:2
Zalaffi, Samuele	193	P:6
Zhang, Frank	102	S 10A:1
Zhang, Guangming	162	S 8A:3
Zhang, Guoqi	202	P:18
Zhang, Hongfeng	111	S 7B:1
Zhang, Xu	166	S 10B:2
Zhang, Zheng	194, 161	P:15, S 8A:2
Zhu, Forest	102	S 10A:1
Ziesche, Steffen	116	S 1B:2
Zozulia, Aleksandr	107	S 6A:1
Zschenderlein, Uwe	156	P:5